# THE DICTIONARY OF
# Physical Geography

## Third Edition

Edited by

David S.G. Thomas
and Andrew Goudie

International Advisory Panel

David Dunkerley (Australia)
Michael Meadows (South Africa)
Robert Balling (USA)
Douglas Sherman (USA)
Athol Abrahams (USA)

**Blackwell** Publishing

350 Main Street, Malden, MA 02148-5018, USA
108 Cowley Road, Oxford OX4 1JF, UK
550 Swanston Street, Carlton South, Melbourne, Victoria 3053, Australia
Kurfürstendamm 57, 10707 Berlin, Germany

First edition published 1985 by Blackwell Publishing Ltd
Second edition, revised and updated, published 1994
Third edition published 2000
Reprinted 2002, 2003

*Library of Congress Cataloging-in-Publication Data*

The dictionary of physical geography / edited by David S. G. Thomas and
   Andrew Goudie. — 3rd ed.
      p.  cm.
   Includes bibliographical references.
   ISBN 0-631-20472-5 (hb : alk. paper) — ISBN 0-631-20473-3 (pb. : alk. paper)
   1. Physical geography—Dictionaries.  I. Thomas, David S. G.  II. Goudie, Andrew.

GB10 .D53 2000
910'.02'03— dc21                                                    99-049818

A catalogue record for this title is available from the British Library.

Set in 9 on 10pt MPlantin
by Kolam Information Services Pvt Ltd., Pondicherry, India.
Printed and bound in the United Kingdom
by T. J. International Ltd., Padstow, Cornwall

For further information on
Blackwell Publishing, visit our website:
http://www.blackwellpublishing.com

# Contents

# Preface to the Third Edition

This edition of this dictionary represents substantial evolution from the second edition. Following consultation with the international advisory panel, whose composition reflects many key areas of physical geography including biogeography, climatology, environmental change and key areas of geomorphology, 200 entries from the second edition have been removed, and replaced with 450 new entries. These were chosen from an original list of possible new entries over twice this length, with the final selection representing changes within the discipline, an increased international flavour, and the need to maintain the final volume at a certain length. The total list of contributors is increased by 34, with new experts drawn in to add their knowledge to the volume.

Managing a volume of this size is a complex task, the size of which I did not quite realize when approached by John Davey, formerly of Blackwell Publishers, and Andrew Goudie. Completion of the task has been made much simpler due to the help of the advisory panel, the goodwill of contributors, and especially to the assistance of Jill Landeryou and Sarah Falkus at the publishers, and, in the final stages, the considerable help given in the preparation of the final manuscript by Lucy Heath. All are thanked enormously.

DSGT

# Preface to the First Edition

The preparation of a dictionary of this complexity has involved many people, and all deserve thanks for the efficiency with which they have prepared their material on time and in the format required. We have been fortunate in having as a model our companion volume, *The Dictionary of Human Geography*, which was so expertly edited by R.J. Johnston and his team. I would like to express particular thanks to Janet Godden for having taken over so much of the organizational burden, and to Andrew Watson for being willing to prepare many of the short entries.

ASG

# Preface to the Second Edition

In this second edition we have taken the opportunity to update many of the entries and their illustrations, and have added a substantial number of new entries. These new entries include some that should doubtless have been in the first edition, but most are entries that relate to new developments that have taken place in the discipline, especially with respect to increasing concerns over major environmental issues. We have also made substantial additions to the list of acronyms and abbreviations, and have updated many of the references and guides to further reading.

ASG

# Introduction

This dictionary provides definitions of terms and explanations of key ideas, concepts, and issues in physical geography. It draws upon the wealth of knowledge of over 90 contributors and is aimed for the use of professionals, students, teachers and researchers in geography and allied environmental and life sciences.

Entries are organized alphabetically, but to aid further understanding, they are, where appropriate, cross-referenced to other relevant entries, which are shown in small capitals in the text. An index allows the identification of other entries in which a term is referred to, allowing a wider sense of its usage to be gained.

Many entries are referenced and/or accompanied by suggestions for further reading. Together, this allows source material, examples of usage and extended explanations to be explored.

# Acknowledgements

The editors and publishers wish to thank the following for permission to use copyright material.

Addison Wesley Longman Limited for the figure in **groundwater** from Jones, *Global hydrology: processes, resources and environmental management*, 1997; reprinted by permission of Pearson Education Limited

Blackwell Publishers Ltd for the figure in **floristic realms** from Goudie, *The nature of the environment*, 1993

Blackwell Science Ltd for the figure in **carrying capacity** from Begon, Harper and Townsend, *Ecology, individuals, population and communities*, 1986

David Bridgland for the figure in **river terrace** from Bridgland, Allen and Haggart (eds), *The Quarternary of the lower reaches of the Thames: field guide*, Quaternary Research Association, 1995

Butterworth Heinemann Publishers, a division of Reed Educational & Professional Publishing Ltd, for the figures in **water mass** from Open University Oceanography Course Team, *Ocean Circulation*, 1989

Cambridge University Press for the table in **geological time-scale** from Harland, Armstrong, Cole, Craig, Smith and Smith, *A geological time scale*, 1989

Chapman & Hall for the table in **Beaufort scale** from Oliver and Fairbridge, *The encyclopedia of climatology*, 1987; and figures in **global ocean circulation** and **water mass** from Tolmazin, *Elements of dynamic oceanography*, 1985

David Evans for the figure in **crag and tail forms** from Benn and Evans, *Glaciers and glaciation*, 1998

Longman Group Ltd for figures and tables in **plate tectonics, rift valley, stress** and **karren** from Summerfield, *Global geomorphology*, 1991; and for the table in **glacio-fluvial** from Price, *Glacial and fluviolglacial landforms*, 1973, Oliver & Boyd

Macmillan for the figure in **association, plant** from Whittaker, *Communities and ecosystems*, second edition, 1975

Routledge for the figure in **tropical rain forest** from Parks, *Tropical rainforests*, 1992

The Geological Society for the table and figure in **weathering profile** from Fookes, *Tropical residual soils*, 1997

Every effort has been made to trace copyright holders. The publishers apologize for any errors or omissions in the above list and will be pleased to make the necessary arrangement at the first opportunity.

# Contributors

AD    Angus Duncan
University of Luton

AHP    Allen H Perry
University of Wales, Swansea

AH-S    A Henderson-Sellers
MacQuarie University

ALH    A Louise Heathwaite
University of Sheffield

AMG    Angela M Gurnell
University of Birmingham

AP    Adrian Parker
University of Oxford

ARH    Alan R Hill
York University, Ontario

ASG    Andrew S Goudie
University of Oxford

AW    Andrew Watson
formerly University of Oxford

AWE    Andrew W Ellis
Arizona State University

BAK    Barbara A Kennedy
University of Oxford

BGT    Bruce G Thom
University of Sydney

BJS    B J Smith
Queen's University of Belfast

BWA    B W Atkinson
Queen Mary & Westfield
College, London

CAMK    Cuchlain A M King
formerly of University of
Nottingham

CDC    Chris D Clark
University of Sheffield

CTA    Clive T Agnew
University of Manchester

DB    Denys Brunsden
formerly of King's College
London

DES    David Sugden
University of Edinburgh

DEW    D E Walling
University of Exeter

DW    David Watts
University of Hull

DGT    David G Tout
University of
Manchester

DH    David Higgitt
University of Durham

DJN    David J Nash
University of Brighton

DJS    Douglas Sherman
University of Southern
California

DLD    David L Dunkerley
Monash University

DSGT    David S G Thomas
University of Sheffield

DTP    David T Pugh
IOS Deacon Laboratory,
Godalming

ECB    Eric C Barrett
University of Bristol

GCN    Gerald C Nanson
University of Wollongong

GFSW    Giles F S Wiggs
University of Sheffield

HAV    Heather Viles
University of Oxford

HMF    Hugh M French
University of Ottawa

HvL    Harry van Loon
National Center for
Atmospheric Research,
Boulder, Colo

IAC    Ian A Campbell
University of Alberta

IB    Ian Burton
Lasalle Academy, Ottawa

IE    Ian Evans
University of Durham

IGS    I G Simmons
University of Durham

JAD    John A Dearing
University of Liverpool

JAM    John A Matthews
University of Wales, Swansea

JAS    John Shaffer
Arizona State University

JEA    J E Allen
Queen Mary & Westfield
College, London

JEL    Julie E Laity
California State University

JET    John E Thornes
University of Birmingham

JGL    John G Lockwood
formerly of University of Leeds

JL    John Lewin
University College of Wales,
Aberystwyth

JLB    Jim Best
University of Leeds

JM    Judith Maizels
University of Aberdeen

JO    J Orford
Queen's University of Belfast

JSAG    John S A Green
University of East Anglia

KB    Katherine Brown
University of Tasmania

KEB    Keith Barber
University of Southampton

KEI    Keith Idso
Arizona State University

KJG    Kenneth J Gregory
Goldsmiths College, London

KJW    Keith J Weston
University of Edinburgh

KS    Keith Smith
University of Stirling

KSR    Keith S Richards
University of Cambridge

LN    Lynn Newman
Arizona State University

| | | | |
|---|---|---|---|
| MAS | M A Summerfield<br>University of Edinburgh | PSh | Paul A Shaw<br>University of Luton |
| MDB | Mark D Bateman<br>University of Sheffield | PWW | Paul W Williams<br>University of Auckland |
| MEM | Michael E Meadows<br>University of Cape Town | RCB | Robert C Balling Jr<br>Arizona State University |
| MFT | Michael F Thomas<br>University of Stirling | RGB | Roger G Barry<br>University of Colorado, Boulder |
| MHU | Michael H Unsworth<br>Institute of Terrestrial Ecology,<br>Midlothian | RHS | Rodney H Squires<br>University of Minnesota,<br>Minneapolis |
| MJK | Mike J Kirkby<br>University of Leeds | RH-Y | Roy Haines-Young<br>University of Nottingham |
| MJS | M J Selby<br>University of Waikato | RID | Ronald I Dorn<br>Arizona State University |
| MLH | Mark L Hidebrandt<br>University of Arizona | RLJ | Robert L Jones<br>Coventry University |
| NJM | Nick J Middleton<br>University of Oxford | RR | Ross Reynolds<br>University of Reading |
| NJS | Nancy J Selover<br>Arizona State University | RSC | Randell S Cerveny<br>Arizona State University |
| PAB | Peter Bull<br>University of Oxford | RSW | Richard S Washington<br>University of Oxford |
| PAF | Peter A Furley<br>University of Edinburgh | SAC | Stanley A Changnon<br>Illinois Department of Energy,<br>Champaign Ill |
| PAS | Philip A Stott<br>School of Oriental and African<br>Studies | SLO | Sarah L O'Hara<br>University of Nottingham |
| PHA | Patrick H Armstrong<br>University of Western Australia | SMP | Susan M Parker<br>London |
| PJC | Paul J Curran<br>University of Southampton | SMW | Stephen M Wise<br>University of Sheffield |
| PS | Peter Smithson<br>University of Sheffield | SNL | Stuart N Lane<br>University of Leeds |

| | | | |
|---|---|---|---|
| SS | Stephen Stokes<br>University of Oxford | WDS | William D Sellers<br>University of Arizona,<br>Tucson |
| TRO | T R Oke<br>University of British<br>Columbia, Vancouver | WLG | William L Graf<br>Arizona State University |
| WBW | W Brian Whalley<br>Queen's University of<br>Belfast | | |

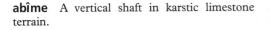

**abîme** A vertical shaft in karstic limestone terrain.

**abiotic** The abiotic components of an ECOSYSTEM are those which are not living. These include mineral soil particles, water, atmospheric gases and inorganic salts; sometimes simple organic substances that have resulted from excretion or decomposition may be included. The term abiotic is also used for physical and chemical influences upon organisms, for example humidity, temperature, pH and salinity. An abiotic environment is one which is devoid of life.                                    PHA

**ablation** The process by which snow or ice is lost from a GLACIER, floating ice or snow. Examples are melting and run-off, calving of icebergs, evaporation, sublimation and removal of snow by wind. Melting followed by refreezing at another part of a glacier is not regarded as ablation because the glacier does not lose mass. Melting is the most important process in temperate and subpolar regions and accounts for seasonal and diurnal meltwater floods. Most such ablation occurs at the glacier surface, and at the snouts of glaciers in many mid-latitude areas it lowers the ice surface by the order of 10 m each year. A small amount of melting occurs within and beneath glaciers whose ice is at the pressure melting point. In the Antarctic the most important ablation process is the calving of ice shelves, though considerable losses may also occur through bottom melting of ice shelves and the removal of snow by offshore katabatic winds.                              DES

**Reading**
Paterson, W.S.B. 1981: *The physics of glaciers*. 2nd edn. Oxford: Pergamon.

**abrasion** The process of wearing down or wearing away by friction as by windborne sand or material frozen into glacial ice.

**absolute age** The age of an event or rock, mineral or fossil, measured in years.

**absolute humidity** See HUMIDITY.

**abundance** The total number of individuals of a particular species present in an area. Various methods are used to measure the abundance of organisms but in view of the time and effort involved it is usually impractical to count all individuals within an area. Instead, population size is often estimated by collecting data from small plots (quadrats) selected by a random sampling procedure. Population size is influenced by a complex array of factors which include, for example, the physical environment, weather conditions, available resources (food, nesting sites, etc.), competition both within and between species, and predation.    ARH

**Reading**
Mueller-Dombois, D. and Ellenberg, H. 1974: *Aims and methods of vegetation ecology*. New York and London: Wiley. Chapter 6, pp. 67–92. · Watts, D. 1971: *Principles of biogeography*. New York and London: McGraw-Hill. Chapter 5, pp. 197–241.

**abyss** *a*. A deep part of the ocean, especially one more than about 3000 m below sea level. *b*. A ravine or deep gorge.

**abyssopelagic zone** The deep water zone of oceans and lakes, usually lying below 4000 m depth. This open water zone, associated in the oceans with the so-called abyssal plains, is characterized by low temperatures, high water densities and the complete absence of light, conditions which severely limit its range of life forms. More than half of the earth's oceans lie below 4000 m and yet the zone remains largely unexplored. Spatial variations in temperature, salinity and density of ocean water in this zone play an important role in the generation of large scale deep water currents.    MEM

**Reading**
Knauss, J.A. 1997: *Introduction to physical oceanography*. 2nd edn. New Jersey: Prentice-Hall.

**accessory mineral** The mineral components of a rock which do not occur in sufficient quantities to merit their inclusion in the definition or classification of the rock, that is, not an essential mineral.

**accordant junctions, law of** The law which states that tributary rivers join main rivers at the same level, that is, there is usually no sudden drop (Playfair's law).

**accordant summits** The phenomenon of hill crests and mountain peaks in a region being within a similar plane, horizontal or inclined, attesting that they are remnants of a former plain or plateau.

**accretion** *a.* The gradual increase in the area of land as a result of sedimentation.
*b.* The process by which inorganic objects increase in size through the attachment of additional material to their surface as with the growth of hailstones.

**accumulated temperature** Normally the total number of days (or hours) since a given date, during which the mean temperature has been above or below a given threshold. The threshold value for agriculture is usually 6 °C and accumulated mean temperatures above this value can be correlated with the growth of vegetation. For heating purposes the threshold is usually 15.5 °C and accumulated mean temperatures below this value can be correlated with energy use. Generally accumulated temperature is used in agriculture and DEGREE DAYS are used in energy management.       JET

**acid deposition** Rainfall is naturally acid (pH 5.65) owing to the dissolution of carbon dioxide in water:

$$CO_2 + H_2O(H_2CO_3)$$

where:

| | |
|---|---|
| $CO_2$ | carbon dioxide |
| $H_2O$ | water |
| $H_2CO_3$ | carbonic acid. |

Therefore, the functional definition of acid precipitation is a pH $< 5.65$. Acidity is generated by the presence of hydrogen (H+) ions and is measured in units of pH on a logarithmic scale. A one-unit difference in solution pH is equivalent to a 10 times difference in the concentration of H+.

Acidifying substances may be deposited from the atmosphere by two main pathways: (1) wet deposition of material entrained in rain, snow and fog-water – often referred to as ACID PRECIPITATION; (2) dry deposition of particulate aerosols (any solid particulate matter transported through the atmosphere), and including uptake of certain gases by vegetation, soil and water surfaces. Atmospheric inputs occur mainly from point sources, and are derived from human activities such as fossil fuel combustion and intensive livestock holdings, and natural sources such as volcanic emissions. The balance between wet and dry deposition pathways varies geographically and according to the dominant wind direction. In general, wet deposition is more important in upland areas with higher rainfall amounts. In north and west UK, for example, inputs of H+ in precipitation may exceed 1 kg ha$^{-1}$a$^{-1}$. Dry deposition of $SO_2$ may exceed 2.4 kg ha$^{-1}$a$^{-1}$ in industrial areas of the midlands and northern England and the major urban areas of the Scottish central lowlands (Fowler *et al.* 1985). In Europe, the relative contributions from wet and dry deposited sulphur change with distance from the coast, with wet deposition becoming relatively less important with distance south from the North Sea.

*Wet deposition*: may occur via two processes, depending on where atmosphere scavenging occurs. Rainout describes acid inputs, which originate within the cloud system; washout describes the removal of solutes by falling precipitation. Mist and fog are generally more acidic than rain, thus at high altitudes, acid deposition is enhanced by the contribution of mist, fog and cloud water.

Most of the excess acidity in precipitation is generated from sulphuric and nitric acids. The presence of excess sulphur and nitrogen in the atmosphere is largely derived from anthropogenic sources from the oxidation of $SO_2$ (sulphur dioxide) and $NO_x$ (nitrous oxide [NO] + nitrogen dioxide [$NO_2$]). Key sources of $SO_2$ include fossil fuel burning and metal smelting. Nitrogen oxides are generated during combustion by oxidation of atmospheric nitrogen; the main anthropogenic source is vehicle emissions. Whilst sources of $SO_2$ have been controlled and atmospheric concentrations are decreasing, atmospheric $NO_x$ continues to rise.

*Dry deposition*: occurs in the interval between precipitation events. The physics of the mechanisms involved in dry deposition are complex and include gravitational settling and filtering of particulate aerosols together with the direct uptake of gases such as $SO_2$ and $NO_x$ onto vegetation, water and/or soil surfaces. Factors such as surface wetness, and vegetation size, growth patterns and surface roughness are important controls on the magnitude of dry deposition. Acidity derived from dry deposition is often generated by secondary chemical reactions. Thus dry deposited $SO_2$ is oxidized to the anion sulphite ($SO_3^{2-}$) and rapidly oxidized to sulphate ($SO_4^{2-}$), electrochemical balancing releases an equivalent number of hydrogen ions which in turn generates acidity. Similarly, dry deposited $NO_x$ may be oxidized to nitrate ($NO_3^-$) which again generates hydrogen ions. The atmospheric sources are not necessarily acidic in themselves: thus acidity may be generated at the ground surface where gaseous ammonia or the cation ammonium ($NH_4^+$) is deposited and oxidized to nitrate, which releases

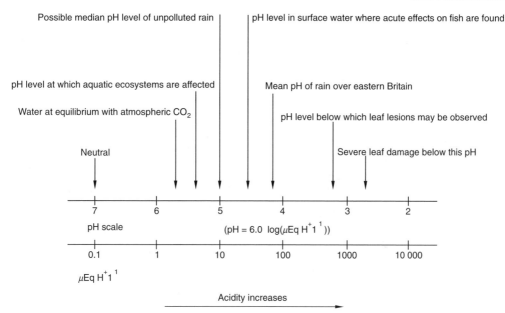

Possible median pH level of unpolluted rain | pH level in surface water where acute effects on fish are found

pH level at which aquatic ecosystems are affected | Mean pH of rain over eastern Britain

Water at equilibrium with atmospheric $CO_2$ | pH level below which leaf lesions may be observed

Neutral | Severe leaf damage below this pH

pH scale    $(pH = 6.0 \; log(\mu Eq \; H^+ l^{-1}))$

| 7 | 6 | 5 | 4 | 3 | 2 |

| 0.1 | 1 | 10 | 100 | 1000 | 10 000 |

$\mu Eq \; H^+ l^{-1}$

Acidity increases

Scale of pH and H+ concentration; acidity (H+) increases, while pH decreases on a 'log scale', i.e. 10:1. *Source*: G. Howells 1990: *Acid rain and acid waters*. Chichester: Ellis Horwood.

hydrogen ions. Quantitative estimation of dry fluxes is difficult, thus this pathway of acidification is less understood than that of wet deposition. Problems in measurement include wide spatial and temporal variations in atmospheric gases and particulates, and difficulties in estimating rates of deposition to natural ecosystems with inherently complex structures. Dry deposition is more affected by distance from emission source and tends to dominate close to source.

Acid deposition is not the only source of acidification of terrestrial and aquatic environments. Other sources include afforestation and forest clearance, livestock grazing, cultivation techniques on agricultural land and the use of fertilizers.                                                    ALH

**Reading and References**

Howells, G. 1990: *Acid rain and acid waters*. Chichester: Ellis Horwood Limited. · Fowler, D., Cape, J.N. and Leith, I.D. 1985: Acid inputs from the atmosphere in the United Kingdom. *Soil use and management* 1, pp. 70–2.

**acid precipitation**   Rain and snow with a pH of less than 5.65. The latter is the hydrogen ion concentration of natural precipitation subject to normal concentrations and pressures of atmospheric carbon dioxide. As the pH scale is logarithmic, a one-point change on it represents a tenfold increase or decrease in acidity (Kemp 1990). The slight natural acidity of precipitation is largely due to weak carbonic acid formed by dissolved atmospheric carbon dioxide, and to

sulphur compounds from volcanic eruptions which are converted to sulphuric acid in the atmosphere. The chemical analysis and dating of fossil ice has revealed that some two centuries ago, precipitation possessed a pH that was generally in excess of 5. Since that time, industrial-urban development, particularly in northern hemisphere mid-latitudes, has resulted in the release of increasing quantities of sulphur and nitrogen oxides into the atmosphere. These emissions are caused by fossil fuel burning and sulphide ore smelting, the oxides being transformed into sulphuric and nitric acids in the atmosphere. These relatively strong acids undergo ionic separation in weakly acidic natural precipitation, with the dissociated hydrogen ions causing its pH to fall below 5.6. (Likens *et al.* 1979). *Sensu stricto*, acid precipitation is thus wet deposition. However, a related process, dry deposition, whereby oxides of sulphur and nitrogen fall out from the atmosphere either as dry gases or adsorbed on other AEROSOLS such as soot, is also operational. These particles become acidic when they join with moisture; fog or surface water, for example (Park 1987; Kemp 1990).

Atmospheric circulation patterns mean that pollutants (POLLUTION) can travel substantial distances before being deposited as acid precipitation. As the loci of acidic pollution are within the westerly wind-belt, their discharges are usually routed eastward. The rate and distance of movement are associated with the

3

height of pollutant emission (Kemp 1990). Tall stacks enhance long-distance transfer (Elsom 1987), with upper westerly winds or jet streams more effective than boundary-layer circulation, both in this respect and in increasing the residence time of pollutants in the atmosphere (Kemp 1990).

The phenomenon of acid precipitation was first recognized in England during the mid-nineteenth century, but has been studied in detail only during the past three decades (Park 1987). Its role is still imperfectly understood. Complex interactions between environmental factors (geology, hydrology and land use, for example) mean that acid precipitation is probably one of numerous components operative in a particular locality (Kemp 1990). Aquatic ecosystems appear to respond more rapidly to acidification than terrestrial ones. Acid water is thought to diminish biological productivity in lakes and rivers developed on siliceous substrates, with the reproductive capacity of fish being impaired (Pearce 1982). The interception of acid precipitation by trees is considered to increase the chance of tissue death, nutrient leaching and chlorophyll degradation in leaves (Shriner and Johnston 1985). Increased acidity of soil water seems to check bacterial activity, results in the replacement of nutrient cations by hydrogen ions and the displacement of the former in solution, and stimulates the mobilization of toxic heavy metals such as aluminium and lead (Kemp 1990). The liberation of heavy metal cations from soils and sites of toxic waste disposal can contaminate drinking water, while supply pipes of the latter may be leached of copper and lead by acidic water (Elsom 1987). Building stone containing calcium and magnesium carbonates could be subject to the reaction of these with the sulphuric component of acid precipitation and to the production of soluble sulphates. In urban areas, dry deposition is frequently of most importance, with chemical reactions initiated by the addition of moisture. In towns and cities, sulphuric acid in the atmosphere, inhaled during episodes of smog, can lead to respiratory difficulties (Kemp 1990).

In a global climatic context, nitrogen oxides are part of the process whereby tropospheric ozone (a greenhouse gas) is produced. Thus more or less of these oxides could contribute to higher or lower ozone concentrations and increase or reduce climatic warming (Martin 1989). Between 1970 and 1984 there was a 40 per cent decrease in the emission of $SO_2$ in Britain (Caulfield and Pearce 1984). Chemical and biological evidence of a slight reduction in the acidity of water in Galloway since 1980 may reflect this trend (Battarbee *et al.* 1988). RLJ

**Reading and References**
Battarbee, R.W., Flower, R.J., Stevenson, A.C., Jones, V.J., Harriman, R. and Appleby, P.G. 1988: Diatom and chemical evidence for reversibility of acidification of Scottish lochs. *Nature* 332, pp. 530–2. · Caulfield, C. and Pearce, F. 1984: Ministers reject clean-up of acid rain. *New Scientist* 104, pp. 1433–6. · Elsom, D. 1987: *Atmospheric pollution: causes, effects and control policies*. Oxford and New York: Basil Blackwell. · Kemp, D.D. 1990: *Global environmental issues: a climatological approach*. London and New York: Routledge. · Likens, G.E., Wright, R.F., Galloway, J.N. and Butler, T.J. 1979: Acid rain. *Scientific American* 241, pp. 39–47. · Martin, H.C. 1989: The linkages between climate change and acid rain. In J.C. White ed., *Global climate change linkages: acid rain, air quality and stratospheric ozone*. New York: Elsevier. Pp. 59–66. · Park, C.C. 1987: *Acid rain: rhetoric and reality*. London: Methuen. · Pearce, F. 1982: The menace of acid rain. *New Scientist* 95, pp. 419–23. · Shriner, D.S. and Johnston, J.W. 1985: Acid rain interactions with leaf surfaces: a review. In D.D. Adams and W.P. Page eds, *Acid deposition: environmental, economic and policy issues*. New York and London: Plenum Press. Pp. 241–53. · Wellburn, A. 1988: *Air pollution and acid rain: the biological impact*. Harlow: Longman Scientific and Technical.

**acid rocks** Commonly used term for igneous rocks which contain more than 66 per cent silica, free or combined, or any igneous rock composed predominantly of highly siliceous minerals (see BASIC ROCKS).

## acid susceptibility/acid neutralizing capacity

Generally, ACIDIFICATION decreases the acid neutralizing capacity of a system, and surface waters that are susceptible to acidification have a small alkalinity or acid neutralizing capacity. Usually, hydrogen (H+) is absorbed by the freshwater system until a buffering threshold is exceeded, after which there is a rapid decrease in pH until another buffering system operates. Inflowing H+ ions react with bicarbonate ($HCO_3-$) ions present in lake waters to form carbonic acid $H_2CO_3$; this weak acid does not appreciably alter the pH of the lake waters. A continued deposition of acid in the form of hydrogen ions will eventually override the buffering capacity of the lake waters and a shift in pH will occur.

The acid neutralizing capacity, sometimes referred to as ANC, is the difference between strong base anions and strong acid cations. It may be represented as alkalinity which is the obverse of acidity:

$$\text{Alkalinity} = [HCO_3-] + 2[CO_3-] = [A-] \\ + [OH-] - [H+]$$

where:

| | |
|---|---|
| $HCO_3-$ | bicarbonate |
| $CO_3-$ | carbonate |
| $A-$ | contribution from weak acids |
| $OH-$ | hydroxyl |
| $H+$ | hydrogen |

In the pH range 6 to 8, bicarbonate alkalinity is the critical buffering system that may be depleted by acid deposition. Here the alkalinity equation is simplified to:

$$\text{Alkalinity} = [HCO_3-] + [A-]$$

In this pH range, bicarbonate alkalinity ($HCO_3-$) reacts with $H+$ to form $H_2O$ and $CO_2$, thus neutralizing the added hydrogen ions. Usually there is little or no change in pH unless the supply of alkalinity is exhausted. The concentration of bicarbonate in surface waters is influenced by catchment geology and soils, especially contact with mineral carbonates (e.g. calcite [$CaCO_3$] and dolomite [$Ca, MgCO_3$]) which generates a large acid neutralizing capacity because alkalinity is produced by the dissolution of these minerals. Surface waters in CATCHMENTS with large terrestrial sources of calcium and magnesium carbonate are unlikely to acidify even in regions with high rates of acid deposition.

Apart from carbonate and bicarbonate, the major ions in acid waters are chloride, sulphate and nitrate; less important ions include silicate, fluoride, aluminium and organic anions. For acid waters below pH 5.0, the alkalinity is dictated by the balance between a few major ions:

$$\text{Alkalinity} = \{Ca_2 + +Mg_2+\}$$
$$-\{SO_{42-} + NO_{3-}\}$$

The influence of chloride is usually discounted because it is assumed to be largely derived from sea salts entrained in rain where the presence of sodium counteracts the activity of chloride. Sulphate is primarily derived from atmospheric deposition whilst nitrate is derived both from human-derived nitrogen gases and nitrate and ammonium in deposition, and biological activity.

Where catchment bedrock and soils consist primarily of oligotrophic minerals such as granite, gneiss and quartzite that do not contain appreciable quantities of carbonate minerals, the acid neutralizing capacity of the catchment is small. In these systems, the *in situ* alkalinity generated by the biological reduction of sulphate and nitrate, and the exchange of $H+$ for $Ca_2+$ in sediments is important but provides only a small amount of alkalinity; thus the acid neutralizing capacity remains small. Here, atmospheric deposition of acidifying substances generates acidity relatively easily. Acidification is often exaggerated because catchments susceptible to acidification often have a small sulphate adsorption capacity that is easily overwhelmed by sulphate inputs from atmospheric sources. When this occurs, elevated sulphate plus hydrogen and aluminium inputs to surface waters may result in conditions that are toxic for aquatic biota.

The acidification of surface waters broadly takes place in three stages.

At stage 1, the natural buffering capacity provides resistance to pH change. By stage 2, all buffering capacity has been utilized; this results in large seasonal variations in the pH of freshwaters, particularly lakes. By stage 3, stabilization to a new, lower pH occurs resulting in loss of species diversity and for lakes, few fish species. In the UK, the effect of acid deposition on the acid susceptibility of freshwaters is largely indirect because most UK freshwater is alkaline and well-buffered. For example, for groundwaters, the most serious consequence of a significant fall in groundwater pH is an increase in metal solubility in potable waters and associated risk of corrosion of the water supply network (Cresser and Edwards 1987). In upland areas, acid flushes may be associated with high concentrations of sulphate and nitrate in stream waters – especially during snowmelt.      ALH

**Reading and Reference**
Mason, B.J. 1990: *The surface waters acidification programme*. Cambridge: Cambridge University Press. · Cresser, M. and Edwards, A. 1987: *Acidification of freshwaters*. Cambridge: Cambridge University Press. Pp. 136.

**acidity profile**   The acid concentration in ice core layers as a function of depth as determined from electrical measurements. The magnitudes of some volcanic eruptions in the northern hemisphere have been estimated from the acidity of annual layers in ice cores taken in Greenland. This methodology is sometimes referred to as 'acidity signal' or 'acidity record'.      ASG

**aclinic line**   The magnetic equator, an irregularly curved line near the equator along which the compass needle does not dip from the horizontal.

**actinometer**   An instrument used for measuring the chemical and heating influences of the sun's radiation.

**active layer**   The top layer of ground above the permafrost table which thaws each summer and refreezes each autumn. In temperature terms, it is the layer which fluctuates above and below 0 °C during the year. In permafrost areas seasonally frozen and thawed ground can be equated with the active layer. Other synonyms include 'depth of thaw', 'depth to permafrost', and 'annually thawed layer'. These terms are acceptable in areas where the active layer extends downwards to the permafrost table, but they are misleading where the active layer is separated from the permafrost by a layer of ground which remains in an unfrozen state throughout the year. The thickness of the active layer varies from as little as 15–30 cm in high latitudes to over 1.5 m in subarctic continental regions. Thickness depends on many factors, including the degree and orientation of the slope, vegetation, drainage, snow cover, soil and rock type, and ground moisture conditions.

Processes operating in the active layer include FROST CREEP and FROST HEAVE or cryoturbation, the lateral and vertical displacement of soil which accompanies seasonal and/or diurnal freezing and thawing. During thaw, water movement through the active layer assists various mass wasting processes, especially gelifluction. Most patterned ground phenomena form in the active layer.                                HMF

**Reading**
Brown, R.J.E. and Kupsch, W.O. 1974: *Permafrost terminology*, Publication 14274. Ottawa: National Research Council of Canada. · French, H.M. 1976: *The periglacial environment*. London and New York: Longman. · —1988: Active layer processes. In M.J. Clark ed., *Advances in periglacial geomorphology*. Chichester: Wiley. Pp. 151–79.

**active margin**  A continental margin which coincides approximately with a plate boundary. It has much more tectonic and igneous activity than a PASSIVE MARGIN, which lies within a plate and on the edge of a spreading ocean basin such as the Atlantic, Indian, Arctic or Antarctic Oceans. Active margins rim most of the Pacific, where the volcanic activity gave rise to the term 'Ring of Fire': they show EARTHQUAKE activity at various depths in BENIOFF ZONES or at shallow depths along transform faults. Plate motions are either convergent, with SUBDUCTION at an orogen (the Andes and Central America) or at an island-arc (the Kurile Islands, Japan and the Philippines), or oblique, with a transform fault (the San Andreas in California or the Alpine in New Zealand).   IE

**Reading**
Summerfield, M.A. 1991: *Global geomorphology*. Harlow: Longman. Chapter 3.

**activity (ratio)**  An empirical relationship of a soil, defined by Skempton as the plasticity index divided by percentage weight less than 2 $\mu$m in size.

As the plasticity index often varies according to cations in the clay mineral structure, the activity is a useful measure of the swelling potential of a soil (its ability to take up moisture). Skempton suggested three main classes of activity: active, normal, and inactive. Most British soils tend to be in the normal to inactive ranges. Active soils tend to have high CATION EXCHANGE capacities. There are sometimes problems with using the activity values,

especially if particles aggregate, so the AGGREGATION RATIO has also been used to express percentage of clay mineral relationships.    WBW

**Reading**
Bell, F.G. 1992: *Engineering properties of soils and rocks*. 3rd edn. Oxford: Butterworth Heinemann. · Skempton, A.W. 1953: The colloidal activity of clays. *Proceedings of the Third International Conference of Soil Mechanics*. Zurich.

**actualism**  See UNIFORMITARIANISM.

**adaptive radiation**  The evolutionary diversification of a group of organisms in response to the ecological pressures of different habitats. When new groups of organisms evolve or occupy a newly accessible environment they will tend to fill, over time, all the available niches; they will thus 'radiate' along evolutionary lines in a genetical response to the stimulus of environmental diversity. The sum of the various lines leading away from the ancestral stock comprises an adaptive radiation. Among mammals, for example, there have arisen specialist grazers and carnivores, burrowers, fliers and aquatic species, with different species filling the equivalent NICHE in various BIOMES, e.g. the bison in North America and the kangaroo in Australia and many others. Adaptive radiation is an important process in the diversification of island floras and faunas (see ISLAND BIOGEOGRAPHY).                                PAS

**Reading**
Stebbins, G.L. 1977: *Processes of organic evolution*, 3rd edn. Englewood Cliffs, NJ: Prentice-Hall.

**adhesion ripple**  An irregular sand ridge transverse to wind direction, formed when dry sand is blown across a smooth moist surface. It may be 30–40 cm long and a few centimetres high. The crest is symmetrical and migrates upwind. The stoss (windward) side is steeper than the lee side.                                ASG

**adhesion warts**  Small-scale aeolian BEDFORMS resulting from SALTATION over a wet sand surface. The moisture captures the falling grains, and capillary action wets that deposit. With unidirectional winds, adhesion ripples may form, and migrate upwind. With polymodal winds, irregularly shaped and oriented adhesion warts result.                                DJS

**Reading**
Reineck, H.-E., and Singh, I.B. 1986: *Depositional sedimentary environments*. 2nd edition. Berlin: Springer-Verlag.

**adiabatic**  An adiabatic process is a thermodynamic change of state of a system in which there is no transfer of heat or mass across the boundaries of the system. In the atmosphere the most commonly related variables in the adiabatic process are temperature and pressure. If a mass of air experiences lower pressure than in its initial condition it will expand and do mechanical work on the surrounding air. The energy required to do this work is taken from the heat energy of the air mass and consequently the temperature of the air falls. Conversely, when pressure increases, work is done on the mass of air and the temperature rises. A diabatic process is a thermodynamic change of state of a system in which there is transfer of heat across the boundaries of the system.                    BWA

**adobe**   Sun-dried bricks, the clay soil from which they are made or the buildings made from them. The term is probably Spanish-Moorish (Spanish, *adobar*, to plaster, but it may have an older, Arabic root). Adobe was initially used in arid and semi-arid parts of Africa but is now especially prominent in the American south-west, into which it was introduced by colonial settlers. The bricks are made from clods of clay soil to which is added water and a small quantity of straw to prevent shrinkage on drying. The material is then trampled with bare feet and moulded by hand into bricks. Correctly constructed, adobe buildings may survive several centuries of weathering. In the semi-arid south-west and Mississippi valley of USA, the term corresponds to LOESS of Europe and Asia.                    MEM

**adret**   The side of a hill or valley that receives the most sunlight (see UBAC), and which may therefore in high-altitude areas have the most intensive land use and settlement.

**adsorption**   The process in which substances, often ions, leave a liquid and accumulate upon the surface of a solid. In the soil, adsorption occurs when ions and other charged entities move from the soil water and attach themselves close to the surfaces of CLAY particles. This takes place because the clay surface normally carries an excess negative charge arising from ionic substitution within the lattice structure, and because of unsatisfied valence at crystal edges. Adsorbed materials are most commonly cations like $Ca^{2+}$ and $Mg^{2+}$, but may also include oxyanions like $NO_3^-$ as well as organic compounds. Adsorbed materials may be difficult to differentiate from precipitates, so that in some cases materials held on a surface may be referred to in a more all-encompassing way as *sorbed*.
                    DLD

**Reading**
Sposito, G. 1984: *The surface chemistry of soils*. Oxford: Oxford University Press.

**advection**   The movement of a property in the fluid natural environment (air and water) due solely to the velocity field of the fluid. Thus heat is transferred through the atmosphere by winds. (We should note that heat is also transferred by radiation – a non-advective process.) Advection may be resolved into two components: horizontal and vertical. In meteorology, advection refers frequently only to the horizontal motion whereas vertical advection, particularly on the scale of the individual cloud, is often called CONVECTION.                    BWA

**advection fog**   See FOG.

**adventitious**   The term applied to the roots and buds of plants that grow from unusual portions of the plant, e.g. roots growing from tree trunks and branches.

**aeolation**   It has been suggested that in the late nineteenth and early twentieth century there was a tendency to identify AEOLIAN processes as the dominant geomorphic mechanism that shaped the world's DESERTS. This view is exemplified by the work of C.R. Keyes (1908, 1909) whose explanations of the shaping of the American southwest were in turn influenced by scientific descriptions of wind eroded forms in hyper-arid areas of Asia and southern Africa. While Keyes may well have over-emphasized the power of aeolian erosion, the past importance attached to the aeolation paradigm may itself have been over-stressed in recent years.

If the power of the wind to shape dry landscapes was portrayed excessively by early geomorphologists, it was soon succeeded by views emphasizing the role of runoff as an erosive agent in environments with sparse or partial vegetation covers. This view might itself have been over-played, especially since much of the work leading to this paradigm shift was conducted in the south-west USA, itself 'not very typical of the world's dry lands' (Peel 1966). Today, aeolation and runoff erosion are not viewed as competing paradigms in the explanation of the formation of dryland landscapes. The expansion of process based studies at the expense of simple description, the increase in data sources and the availability of high resolution aerial imagery, all allow the relative importance of different processes in space and time to be considered, rather than a unitary explanation of dryland evolution to be sought.                    DSGT

**Reading and References**
Breed, C.S., McCauley, J.F., Whitney, M.I., Tchakarian, V.P. and Laity, J.E. 1997: Wind erosion in drylands. In D.S.G. Thomas ed, *Arid zone geomorphology: process, form and change in drylands*. Chichester: John Wiley. Pp. 437–64. · Keyes, C.R. 1908: Rock-floor of intermont plains of the arid region. *Bulletin of the Geological Society of America* 19, 63–92. · — 1909: Erosional origin of the Great Basin Range. *Journal of geology* 44, pp. 201–13. · Peel, R.A. 1966: The landscape of aridity. *Transactions, Institute of British Geography* 38, 1–23. · Thomas, D.S.G. 1997: Arid zone geomorphology: perspectives and challenges. In D.S.G. Thomas ed., *Arid zone geomorphology: process, form and change in drylands*. Chichester: John Wiley. Pp. 691–6.

**aeolian** (**eolian** in the USA)  Of the wind; hence aeolian processes and aeolian sediments and landforms. Derived from Aeolus, Greek god of the winds. DSGT

**aeolianite**  Cemented dune sand, calcium carbonate being the most frequent cement. The degree of cementation is very variable, the end product being a hardened dune rock with total occlusion of pore space. Aeolianite of Quaternary age is generally found in coastal areas within 40° of the equator, especially those that experience at least one dry season. The balance between leaching and lime production is the prime control of this overall distribution. Most examples contain between 30 and 60 per cent calcium carbonate, although not all of this may occur as cement. According to Yaalon (1967) a minimum of 8 per cent calcium carbonate is required for cementation of dune sands under semi-arid conditions. Sources of calcium carbonate include biogenic skeletal fragments, dust, spray, and groundwater. ASG

**Reading and Reference**
Gardner, R.A.M. 1983: Aeolianite. In A.S. Goudie and K. Pye eds, *Chemical sediments and geomorphology*. London: Academic Press. · Yaalon, D. 1967: Factors affecting the lithification of aeolianite and interpretation of its environmental significance in the coastline plain of Israel. *Journal of sedimentary petrology* 37, pp. 1189–99.

**aeration zone**  In the context of the HYDROLOGICAL CYCLE, the zone between the soil moisture zone and the capillary zone immediately above the water table. In this zone vadose water is moving downwards under the influence of gravity and the zone may vary in thickness from 0 to several hundred metres in arid regions. KJG

**aerial camera**  A camera designed specially to hold aerial film for the taking of AERIAL PHOTOGRAPHY. The five main types of aerial camera are the mapping camera, reconnaissance camera, strip camera, panoramic camera and multiband camera.

The majority of aerial photographs are taken with high quality *mapping cameras*. These are relatively simple in design and comprise a low distortion lens and a very large film magazine. The *reconnaissance camera* is cheaper than the mapping camera both to buy and to operate. Their disadvantages are their relatively large levels of geometric distortion and their unsuitability for colour aerial film.

The *strip camera* focuses light onto an adjustable slit, under which the film moves at a speed that is proportional to the ground speed of the aircraft. The resulting long strips of photography are not usually appropriate for use in physical geography. The lens in the *panoramic camera* oscillates, scanning from horizon to horizon while focusing the light onto a cylinder, upon which a photographic film is held. This results in distorted photographs which are rarely used by physical geographers.

The *multiband camera* makes use of the fact that objects on the earth's surface vary in the way in which they reflect ELECTROMAGNETIC RADIATION. To use this phenomenon to differentiate between objects, a scene is photographed through different filters using either a multi-lens camera or a multi-camera array. A multi-lens camera uses one film which is exposed by radiation passing through several filters and lenses whereas a multi-camera array comprises a number of small format cameras, each camera having its own film and filter. PJC

**Reading**
Paine, D.P. 1981: *Aerial photography and image interpretation for resource management*. New York: Wiley. · Wolf, P.F. 1974: *Elements of photogrammetry*. New York: McGraw-Hill.

**aerial photography**  is taken using aerial film in an AERIAL CAMERA that is usually mounted in an aircraft. It is the most widely used type of REMOTE SENSING. The characteristics of aerial photography that make it so popular are:

1 *Availability*: aerial photographs are readily available at a range of scales for much of the world.
2 *Economy*: aerial photographs are cheaper than field surveys and are often cheaper and more accurate than maps for many countries of the world.
3 *Synoptic viewpoint*: aerial photographs make possible the detection of both small features and spatial relationships that would not be evident on the ground.
4 *Time freezing ability*: an aerial photograph is a record of the earth's surface at a particular moment and can therefore be used as an historical record.

5 *Spectral and spatial resolution*: aerial photographs are sensitive to ELECTROMAGNETIC RADIATION in wavelengths and for areas that are outside the spectral sensitivity range of the human eye.

6 *Three-dimensional perspective*: a stereoscopic view of the earth's surface can be created and measured both horizontally and vertically.

The angle from which the aerial photography is taken determines whether it is vertical, high oblique or low oblique. Vertical aerial photography results when the camera axis is pointing vertically downwards and oblique aerial photography results when the camera axis is pointing obliquely downwards. Low oblique aerial photography incorporates the horizon into the photograph, while high oblique aerial photography does not. Vertical aerial photographs are the most widely used type as they have an approximately constant scale over the whole photograph and can be used for mapping and measurement. Oblique aerial photographs have their advantages, as they cover many times the area of a vertical aerial photograph taken from the same height using the same focal length lens and in addition present a view that is more natural to the interpreter.

Aerial photographs are taken at a wide range of scales. A small-scale aerial photograph at 1–50,000 will provide a synoptic, low spatial resolution overview of a large area, while a large-scale aerial photograph at 1–2000 will provide a detailed and high spatial resolution view of a small area.

Once obtained, aerial photographs are interpreted for the identification of objects and assessment of their significance. During this process the interpreters usually undertake several tasks of detection, recognition and identification, analysis, deduction, classification, idealization and accuracy determination. Detection involves selectively picking out visible objects. Recognition and identification involve naming the objects or areas, and analysis involves trying to detect their spatial order. Deduction involves the principle of convergence of evidence in order to predict the occurrence of certain relationships on the aerial photographs. Classification is used to arrange the objects and elements identified into an orderly system before the photographic interpretation is idealized using lines which are drawn to summarize the spatial distribution of objects or areas. The final stage is accuracy determination in which random points are visited in the field to confirm or correct the interpretation.

Recognition and identification of objects or areas comprise the most important link in this chain of events. An interpreter uses seven characteristics of the aerial photography to help with this stage: tone, texture, pattern, place, shape, shadow and size. Tone is the single most important characteristic of the aerial photograph as it represents a record of the electromagnetic radiation that has been reflected from the earth's surface onto the aerial film. Texture is the frequency of tonal changes which arise within an aerial photograph when several features are viewed together. Pattern is the spatial arrangement of objects on the aerial photograph. Place is a statement of an object's position on the aerial photograph in relation to others in its vicinity. Shape is a qualitative statement of the general form, configuration or outline of an object on an aerial photograph. Shadows of objects on an aerial photograph are used to help in identifying them, e.g. by enhancing geological boundaries. Size of an object is a function of the scale of the aerial photograph. The sizes of objects can be estimated by comparing them with objects for which the size is known. PJC

**Reading**

Lo, C.P. 1976: *Geographical applications of aerial photography*. Newton Abbot: David & Charles; New York: Crane, Russak. · Ritchie, W., Wood, M., Wright, R. and Tait, D. 1988: *Surveying and mapping for field scientists*. Harlow: Longman Scientific and Technical.

**aerobic** Describes either conditions of the environment, or of a metabolic process, in which oxygen is freely available. Most frequently it is the term applied to a form of RESPIRATION in which the gaseous or dissolved form of oxygen is the principal agent of oxidation, as opposed to the ANAEROBIC form in which other substances participate in the energy release. MEM

**aerobiology** The study of the recognition, conveyance and behaviour of passive airborne organic particles (Gregory 1973). These particles, or aerosols, may be viable or non-viable and occur both in and out of doors. They are dispersed in air and move, dependent upon atmospheric properties and conditions. Both living and non-living organisms are considered in an aerobiological context (Edmonds 1979). (See also POLLEN ANALYSIS). RLJ

**Reading and References**

Edmonds, R.L. ed. 1979: *Aerobiology: the ecological systems approach*. Stroudsburg, Pa.: Dowden, Hutchinson & Ross. · Gregory, P.H. 1973: *The microbiology of the atmosphere*. 2nd edn. Aylesbury: Leonard Hill. · Knox, R.B. 1979: *Pollen and allergy*. London: Edward Arnold.

**aerodynamic ripples** are a type of aeolian bedform that are thought to differ in formation from other ripple types. Normal or CURRENT RIPPLES form at right angles to the direction of fluid flow, whether the fluid is water or air, and their formation is likely to relate to the impact of saltating particles – hence the term impact ripple also being prevalent in the aeolian literature. By way of contrast, Wilson (1972) believed that aerodynamic ripples (wavelength c.0.015–0.25 m, as opposed to a range of 0.05–2.0 m for normal ripples) could form either transverse or parallel to the wind direction, as a result of wave-like instability in the turbulent elements of secondary flow close to the ground surface. As Cooke *et al.* (1993) note, there is little if any empirical support for this theory.                                    DSGT

**References**
Cooke, R.U., Warren, A. and Goudie, A.S. 1993: *Desert geomorphology.* London: UCL Press. · Wilson, I.G. 1972: Aeolian bedforms: their development and origins. *Sedimentology* 19, pp. 173–210.

**aerography** Describes studies of the geographical ranges of species, genera, families, etc.

**aerology** The study of the atmosphere, especially the study of the atmosphere above the surface layers.

**aeronomy** The branch of atmospheric physics which is concerned with those regions, generally above 50 km, where ionization and dissociation are fundamental properties.

**Reading**
Brasseur, G. and Solomon, S. 1986: *Aeronomy of the middle atmosphere: chemistry and physics of the stratosphere and mesosphere.* Dordrecht: Reidel.

**aerosol** An intimate mixture of two substances, one of which is in the liquid or solid state dispersed uniformly within a gas. The term is normally used to describe smoke, condensation nuclei, freezing nuclei or fog contained within the atmosphere, or other pollutants such as droplets containing sulphur dioxide or nitrogen dioxide. Aerosols tend to obscure visibility by scattering light. They tend to vary in size between about a milli-micron ($10^{-9}$ m) and one micron ($10^{-6}$ m). Clouds are not normally considered to be aerosols because the droplets are too large and tend to fall due to gravity. Aerosols can remain in the atmosphere for long periods, collisions with air molecules keeping them aloft.                                    JET

**aestivation** The dormancy of certain animals during the summer season, the dry season or prolonged droughts. It is an important means of adaptation for desert animals, and can be contrasted with hibernation – a state of dormancy in the winter months.

**affluent** A stream or river flowing into another, a tributary.

**aftershocks** A series of small earthquakes following a major tremor and originating at or near its focus. Aftershocks generally decrease in frequency over time but may occur over a period of several days or months.

**Agenda 21** The declaration, signed by world leaders at the 1992 UN Conference on Environment and Development (UNCED) which took place in Rio de Janeiro, Brazil, and aimed at achieving global SUSTAINABLE DEVELOPMENT. It is described as an environmental action programme and includes proposals for, among others, monitoring and reducing chemical waste, the treatment and elimination of radioactive waste, the protection of forests and other natural vegetation, the development of sustainable farming and measures to combat soil degradation. Moreover, the Agenda lists programmes in the fields of oceans, water and coastal management, combating poverty, health care and price and trade policies linked to environmental objectives. Agenda 21 is meant to provide a basis for global co-operation and is an attempt to penetrate much deeper than the traditional development aid provided by developed nations countries to poorer ones. It represents co-operation based on common interests, mutual needs and shared responsibilities. Agenda 21 is an action programme that is intended to be implemented by governments, UN agencies, local and regional administrators, organizations in the community and the public at large. National governments are required to promote the dialogue among these players but local government, too, is expected to play a major role in providing information, education and mobilizing the general public to achieve sustainable development. One of the major objectives of Agenda 21 is that every local government should draw up its own 'Local Agenda 21' in close consultation with its citizens. The key principle is that of sustainable development.                 MEM

**Reading**
Lafferty, W. 1998: *From Earth Summit to Local Agenda 21.* Carbondale: Earthscan. · Sitarz, D. ed. 1993: *Agenda 21: the Earth Summit strategy to save our planet.* Carbondale: Earthscan.

**ageostrophic** A type of atmospheric motion in the troposphere in which the horizontal pressure gradient is not balanced with the deviating force (CORIOLIS FORCE) owing to the wind velocity. It is associated with vertical motion and hence the formation of cloud and weather.

**agglomerate** A rock composed of angular fragments of lava, generally more than 20 mm in diameter, which have been fused by heat.

**aggradation** The vertical growth of the land surface by sediment accumulation. Aggradation may occur across unconstrained land surfaces (e.g. in SAND SHEETS that result from aeolian accumulation) or in constrained settings (e.g. in a river channel). Any deposition process can result in aggradation which can occur at a range of spatial scales and over varying time-scales.

**aggregation ratio** of a soil is the ratio of the percentage weight of clay minerals, determined by mineralogical analysis, to the percentage weight of clay particles, determined by sedimentation methods. The ratio is meant to account for problems with the ACTIVITY of a soil in which clay mineral particles aggregate and act as clays, but have a size corresponding to a value in the silt size range. WBW

**aggressivity** In the context of limestone solution, the propensity of water to dissolve calcium carbonate. When water comes in contact with air it dissolves an amount of carbon dioxide into the water. The resultant carbonic acid can dissolve calcium carbonate (via the theoretical compound $CaHCO_3$) until the aggressiveness of the water diminishes (when the $CO_2$ is used up). It is then said to be saturated with respect to calcium carbonate. When two saturated solutions of calcium carbonate are mixed it is possible for the resulting water to be aggressive. This phenomenon, first identified by Bögli (1971, English version) is called mixing corrosion or, in German, *Mischungskorrosion*. PAB

**Reference**
Bögli, A. 1971: Corrosion by mixing of karst water. *Transactions of the Cave Research Group of Great Britain* 13, pp. 109–14.

**agonic line** A shifting, irregular imaginary line running through the earth's north and south magnetic poles along which the compass needle points to true north, hence the line of no magnetic variation.

**agricultural drought** is a particular consideration or calculation of DROUGHT conditions that assesses the effect of moisture deficits upon agricultural productivity. Various definitions exist, for example it may simply be considered as a period of dry weather of sufficient duration to cause at least partial crop failure. Unlike meteorological drought which specifically reflects precipitation deficits, agricultural drought focuses upon soil moisture deficits during the growing season (Wigley and Atkinson 1977), so that precipitation and potential evaporation are assessed in conjunction with soil moisture levels. If the moisture needs of particular crops are known, their respective drought tolerances and thresholds may be calculated (Agnew 1982). It is possible to assess the impact of meteorological drought upon actual crop production levels. The timing of crop planting and harvesting, and the occurrence of rain during the year, are also important considerations, perhaps more so than absolute rainfall deficits. In areas vulnerable to drought, changing the crops that are sown may reduce the susceptibility of production to meteorological drought (e.g. Barrow 1987). DSGT

**References**
Agnew, C. 1982: Water availability and the development of rainfed agriculture. In S.W. Niger, *Transactions, Institute of British Geographers* NS7, pp. 419–57. · Barrow, C. 1987: *Water resources and agricultural development in the tropics*. London: Longman. · Wigley, T.M.L. and Atkinson, T.C. 1977: Dry years in south-east England since 1698. *Nature* 265, pp. 431–4.

**agroclimatology** The study of the interaction between climatological and hydrological factors and agriculture, including animal husbandry and forestry. Its aim is to apply climatological information for the purpose of improving farming practices and increasing agricultural productivity in quantity and in quality. Agroclimatology and AGROMETEOROLOGY share nearly the same aims, scope and methodology. However, in their application the latter tends to emphasize weather forecasting in dealing with daily problems, whereas the former is more concerned with the use of mean data as a guide to long-range planning. ASG

**Reading**
Chang, J.H. 1968: *Climate and agriculture*. Chicago: Aldine.

**agroforestry** Any system where trees are deliberately left, planted or encouraged on land where crops are grown or animals grazed. It includes practices as diverse as slash-and-burn agriculture, the growth of shade trees and the use of living fences either to contain or to exclude animals. Deep-rooted trees tap nutrient sources that are out of reach of most crops; these nutrients become readily available when the leaves fall. Leguminous trees improve soil fertility directly through nitrogen fixation. Tree roots help to bind the soil and increase aeration. Mixtures of trees and crops provide more complete ground cover which helps to prevent soil erosion and weed invasion, while making full and productive use of available solar radiation. Leaf litter from the trees adds organic matter to the soil and acts as a mulch. Tree cover helps to regulate temperatures, reducing extremes. Farming communities benefit from a regular supply of wood and other tree products. Multi-purpose trees can provide fodder for livestock, edible fruits and nuts, fuel, timber, supports for climbing vegetables and medicinal products. ASG

**agrometeorology** The science concerned with the application of meteorology to the measurement and analysis of the physical environment in agricultural systems. The influence of the weather on agriculture can be on a wide range of scales in space and time, and this is reflected in the scope of agrometeorology. At the smallest scale the subject involves the study of microscale processes taking place within the layers of air adjacent to leaves of crops, soil surfaces or animals' coats (see MICROCLIMATE; MICROMETEOROLOGY). These processes determine rates of exchange of energy, and mass between the surface and the surrounding air. Such exchange rates are the essential link between the biological response and the physical environment. For example, the capture of radiant energy and its use to convert carbon dioxide and water into carbohydrates are essential elements in crop growth. Agrometeorologists have studied how the structure of leaf canopies affects the capture of light and how measurements of the atmospheric carbon dioxide concentration may be used to determine rates of crop growth.

On a broader scale agrometeorologists attempt to use standard weather records to analyse and predict responses of plants and animals. An area of particular interest concerns the estimation of water use by crops as a basis for planning irrigation requirements. Methods based on empirical correlations with windspeed, sunshine, temperature and humidity have been superseded by methods with a sounder physical basis, often using new measuring techniques

(see EVAPOTRANSPIRATION). Other examples of this scale of agrometeorology include procedures for forecasting the occurrence of damaging frosts, or plant or animal diseases. The agrometeorologist is also commonly interested in the soil environment because of the large influence which the weather can have on soil temperature and on the availability of water and nutrients to plant roots. MHU

**Reading**
Smith, L.P. 1975: *Methods in agricultural meteorology.* Amsterdam: Elsevier.

**aiguille** A sharply pointed rock outcrop or mountain peak. Often applied to pinnacles which are the products of frost action.

**air mass** A body of air which is quasi-homogeneous in terms of TEMPERATURE and HUMIDITY characteristics in the horizontal plane and has similar LAPSE RATE features. An ideal air mass is a barotropic fluid in which isobaric and isosteric (constant specific volume) surfaces do not intersect. Air mass classification (see diagram) is based on the nature of the source area (e.g. polar, tropical) and the characteristics of the surface during the outward trajectory (e.g. maritime, continental). Thermodynamic and dynamic factors will modify the properties of air masses in their transit from source areas. AHP

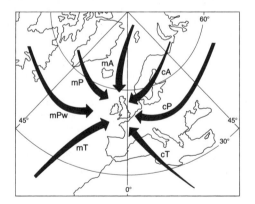

**Air mass** classification and properties with examples of source regions.

Air mass classification

| | *Tropical* | *Polar* | *Arctic/Antarctic* |
|---|---|---|---|
| *Maritime* | Maritime tropical (mT) warm and very moist; near Azores in N. Atlantic | Maritime polar (mP) cool and fairly moist; Atlantic south of Greenland | Arctic or Antarctic (A)    (AA) very cold and dry; |
| *Continental* | Continental tropical (cT) hot and dry; Sahara desert | Continental polar (Cp) cold and dry; Siberia in winter | frozen Arctic Ocean central Antarctica |

**Reading**
Balasco, J.E. 1952: *Characteristics of air masses over the British Isles*. Meteorological Office, Geophysical Memoir 11. London: HMSO. · Harvey, J.G. 1976: *Atmosphere and ocean*. Sussex: Artemis. · Miller, A.A. 1953: Air mass climatology. *Geography* 38, pp. 55–67.

**air parcel** An imaginary body of air to which may be assigned any or all of the dynamic and thermodynamic properties of the atmosphere. Investigation of the STABILITY of the atmosphere is made most simply by the 'parcel method' in which it is hypothesized that a test parcel of air moves vertically with respect to its environment as represented by an ascent curve on a TEPHI-GRAM. AHP

**Reading**
Barry, R.G. and Chorley, R.J. 1992: *Atmosphere, weather and climate*. 6th edn. London: Routledge.

**air pollution** If people had never evolved on this planet the composition of the earth's ATMO-SPHERE would be different from what it is today. Air pollution can therefore be defined as the presence in the atmosphere of natural or human caused contaminants in a given quantity, and for a given time, that are damaging to human, plant or animal life; or to property; or sensually interfere with the comfortable enjoy-ment of life. Common air pollutants include smoke, sulphur dioxide, carbon monoxide, car-bon dioxide, nitric oxide, nitrogen dioxide, ozone and lead. JET

**Reading**
Lyons, T.J. and Scott, W.D. 1990: *Principles of air pollution meteorology*. London: Belhaven. · Seinfeld, J.H. 1986: *Atmospheric chemistry and physics of air pollution*. New York and Chichester: Wiley.

**aklé** A network of sand dunes, comprising overlapping TRANSVERSE DUNE ridges that totally enclose the interdune areas. Sometimes described as producing a fishscale pattern. See DUNE NETWORK DSGT

**alas** A steep-sided, flat-floored depression, sometimes containing a lake, found in areas where local melting of permafrost has taken place. It is one manifestation of thermokarst.

**albedo** A measure of the reflectivity of a body or surface derived from the Latin *albus* white. The albedo is defined as the total RADIATION reflected by the body divided by the total incid-ent radiation. Numerical values are expressed between the ranges of either 0–1 or 0–100 per cent. It is therefore wavelength integrated across the full solar spectrum, while the term reflectivity is generally associated with a single wavelength or narrow waveband, i.e. a spectral reflectivity. The term albedo was already in common use by astronomers at the beginning of the twentieth century (e.g. Russell 1916) when it referred to the whole-planet value. The earth's albedo is close to 0.3, contrasting strongly with its highly reflective neighbour, Venus, which has an albedo of 0.7. The advent of orbiting satellites permitted measurement of reflected radiation at the top of the atmosphere above specific geographical locations. Maps of albedo are now produced illustrating the varia-tion of reflectivity over the globe. This top-of-the-atmosphere albedo is termed the *planetary* or *system albedo*. The satellite-sensed radiation is composed of the reflected beams from the sur-face, the atmosphere and clouds. Strictly, surface albedos can be measured only with albedo-meters mounted close to the surface but they may also be calculated from clear-sky albedos measured by satellites. Surface albedos range in value: oceans: 0.07, dense forests: 0.10, grass and farmlands: 0.16–0.20, bright deserts: 0.25–0.40, and highly reflective ice: 0.40–0.60

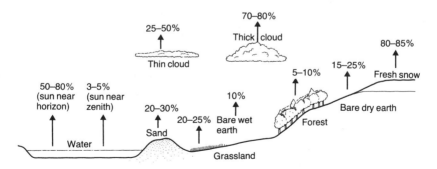

**The albedo of the earth's surface**. The fraction of the total radiation from the sun that is reflected by a surface is called its albedo. The albedo for the earth as a whole, called the planetary albedo, is about 53 per cent. The albedo varies for different surface types. Note also that the angle at which the sun's rays strike a water surface greatly affects the albedo value.
*Source*: A.S. Goudie 1984: *The nature of the environment*. Oxford and New York: Basil Blackwell. Figure 2.2.

Albedos for selected objects

| Water surfaces | | % |
|---|---|---|
| Winter: | 0° latitude | 6 |
| | 30° latitude | 9 |
| | 60° latitude | 21 |
| Summer: | 0° latitude | 6 |
| | 30° latitude | 6 |
| | 60° latitude | 7 |
| Bare areas and soils | | |
| Snow, fresh-fallen | | 75–95 |
| Snow, several days old | | 40–70 |
| Ice, sea | | 30–40 |
| Sand dune, dry | | 35–45 |
| Sand dune, wet | | 20–30 |
| Soil, dark | | 5–15 |
| Soil, moist grey | | 10–20 |
| Soil, dry clay or grey | | 20–35 |
| Soil, dry light sand | | 25–45 |
| Concrete, dry | | 17–27 |
| Road, black top | | 5–10 |
| Natural surfaces | | |
| Desert | | 25–30 |
| Savannah, dry season | | 25–30 |
| Savannah, wet season | | 15–20 |
| Chaparral | | 15–20 |
| Meadows, green | | 10–20 |
| Forest, deciduous | | 10–20 |
| Forest, coniferous | | 5–15 |
| Tundra | | 15–20 |
| Crops | | 15–25 |
| Cloud overcast | | |
| Cumuliform | | 70–90 |
| Stratus (500–1000 ft thick) | | 59–84 |
| Altostratus | | 39–59 |
| Cirrostratus | | 44–50 |
| Planets | | |
| Earth | | 34–42 |
| Jupiter | | 73 |
| Mars | | 16 |
| Mercury | | 5.6 |
| Moon | | 6.7 |
| Neptune | | 84 |
| Pluto | | 14 |
| Saturn | | 76 |
| Uranus | | 93 |
| Venus | | 76 |
| Human skin | | |
| Blond | | 43–45 |
| Brunette | | 35 |
| Dark | | 16–22 |

*Source*: Sellers, W.D. 1965: *Physical climatology*. Chicago: University of Chicago Press. P. 21.

and snow surfaces: 0.50–1.0. CLOUDS generally have high albedos though they too exhibit a considerable range from *cumuliform* clouds (0.80) to some *cirriform* clouds which barely reflect solar radiation making their detection from space very difficult. AH-S

**Reference**
Russell, H.N. 1916: On the albedo of the planets and their satellites. *Astrophysical Journal* 43.3, pp. 173–96.

**alcove** A steep-sided, arcuate cavity on the flank of a rock outcrop, typically produced by water erosion processes such as SOLUTION and spring SAPPING or seepage. DJN

**alcrete** Aluminium-rich duricrusts, often in the form of indurated bauxites. Generally the products of the accumulation of aluminium sesquioxides within the zone of weathering.

**alfisol** Relatively young, acid soils characterized by a clay-enriched B horizon, commonly occurring beneath deciduous forest in humid, subhumid, temperate and subtropical climates. A soil order of the US SEVENTH APPROXIMATION.

**algae** A large group of photosynthesizing organisms, many of which are unicellular and microscopic, although some are multicellular and conspicuous such as the seaweeds. They are most frequently aquatic, common in either marine or freshwater habitats; some are adapted to lower levels of moisture and may be found, for example, as the green coating of trees or in the soil. They are subdivided into several distinctive phyla on the basis of structure, pigmentation and characteristics of the cell wall. The major groupings are the red algae (e.g. the coralline species associated with coral reefs), brown algae (mainly marine, for example all the common seaweeds including kelp which may exceed 60 m in length), diatoms and green algae (frequently occurring as the PLANKTON of freshwater habitats). MEM

**Reading**
Sze, P. 1997: *The biology of the algae*. New York: McGraw-Hill.

**algal bloom** A spontaneous proliferation of microscopic algae in water bodies as a result of changes in water temperature or chemistry. Algal blooms may be characteristic of lakes where eutrophication has been caused by the addition of pollutants.

**alidade** *a*. The sighting device, index and reading device of a surveying instrument.
*b*. A straight-edged rule with a sighting device mounted parallel to the ruler used to plot the direction of objects.

**aliens** Organisms either deliberately or accidentally introduced by people into regions outside their natural distribution range. The term is commonly used to describe those exotic species able to propagate themselves in their new habitats without human intervention, so that they are invasive and may become serious pests. Alien species during the twentieth century, in parallel with the increased scale of human impact, have become major factors in land transformation and degradation, have disrupted ecosystem functioning and threaten biological diversity. Organisms transported to remote regions by people often arrive without the significant natural enemies that control population numbers within their natural ranges. In certain cases, these organisms can become successful invaders of natural and semi-natural habitats in their new localities and, as so-called 'environmental weeds' become hazardous to the invaded communities. Mediterranean-type ecosystems appear to have been especially susceptible to such invasions. For example, the species-rich FYNBOS heathlands of the south-western Cape of South Africa has been severely impacted by invasive alien trees, among the most significant of which are several Australian varieties of the genus *Acacia*. Islands are also strongly impacted by aliens; the introduction of the European rabbit to Australia has had serious ecological consequences and the unique flora and flora of the Galápagos is threatened by the introduction of mammals such as domestic cat and dog. More recently, attempts to control the populations of alien species have been intensified and biological control techniques are now frequently utilized. MEM

**Reading**
Cronk, Q.C.B. and Fuller, J.L. 1995: *Plant invaders: the threat to natural ecosystems*. London: Chapman and Hall. · Drake, J.A., Mooney, H.A., di Castri, F., Groves, R.H., Kruger, F.J., Rejmánek, M. and Williamson, M. 1989: *Biological invasions: a global perspective*. Chichester: Wiley.

**alimentation** The accumulation in quantity of ice, through snowfall or avalanching, in a firn field contributing to a glacier.

**allelopathy** The production of chemicals by plants in order to inhibit or depress the growth of competing plants. The phenomenon is probably far more widespread than generally thought and is known to occur in plant communities as diverse as desert shrubs, tropical and temperate forests and heathlands. Examples include the checking of spruce growth on heather moorland by the production of a chemical by the heather roots which inhibits the growth of the mycorrhizal fungi essential for good growth of the trees; the inhibition of herbaceous species by the shrubs of the Californian chaparral, and the suppression of their own seedlings (autotoxicity) by a number of forest trees, such as black walnut (*Juglans nigra*) and the silky oak (*Grevillea robusta*). Several plants also use chemical defences against herbivores, e.g. oak leaves with a high tannin content are less palatable to certain defoliating caterpillars. KEB

**Reading**
Ashton, D.H. and Willis, E.J. 1982: Antagonisms in the regeneration of Eucalyptus regnans in the mature forest. In E.I. Newman ed., *The plant community as a working mechanism*. British Ecological Society: special publication no. 1. Oxford: Blackwell Scientific. Pp. 113–28. · Krebs, C.J. 1985: *Ecology: the experimental analysis of distribution and abundance*. 3rd edn. New York: Harper & Row.

**Allen's rule** The rule which states that the relative size of the limbs and other appendages of warm-blooded animals tends to decrease away from the equator. This correlates with the increased need to conserve body heat.

**Allerød** The name given to an INTERSTADIAL of the *Late Glacial* of the last glaciation of the Pleistocene in Europe. The classic threefold division into two cold zones (I and III) separated by a milder interstadial (zone II) emanates from a type section at Allerød, north of Copenhagen, where an organic lake mud was exposed between an upper and lower clay, both of which contained pollen of *Dryas octopetula*, a plant tolerant of severely cold climates. The lake muds contained a cool temperature flora, and the milder stage which they represented was called the Allerød Interstadial. The interstadial itself and the following Younger Dryas temperature reversal are sometimes called the Allerød Oscillation. The classic date for the interstadial is 11,350–12,000 BP, but its exact date and status are in dispute. ASG

**Reading**
Lowe, J.J. and Gray, M.J. 1980: The stratigraphic subdivision of the Lateglacial of NW Europe: a discussion. In J.J. Lowe, M.J. Gray and J.E. Robinson eds, *Studies in the Late glacial of north-west Europe*. Oxford: Pergamon. Pp. 157–75. · Mercer, J.H. 1969: The Allerød oscillation: a European climatic anomaly. *Arctic and Alpine Research* 1, pp. 227–34.

**allochthonous** Refers to the material forming rocks which have been transported to the site of deposition, whereas an autochthonous sediment is one in which the main constituents have been formed *in situ* (e.g. evaporites, coal etc.).

**allogenic stream** A stream which derives its discharge from outside the local area. The term is particularly used where local conditions do not generate much streamflow, for example in arid areas or ones with permeable rocks. Here streamflow may be derived from distant parts of the topographic catchment where precipitation and run-off are effective. Appearances can sometimes be misleading because streamflow may be augmented by contributions from local groundwater that are not readily appreciated. JL

**allogenic succession** This process is caused by an external environmental factor rather than by the organisms themselves. Instances are the change in vegetation induced by the inflow and accumulation of sediment in a pond (a geomorphological process), or a change in regional climate. (See also AUTOGENIC SUCCESSION; CLISERE.) JAM

**allometric growth** A biological concept which derives from 'the study of proportional changes correlated with variation in size of either the total organism or the part under consideration. The variates may be morphological, physiological or chemical' (Gould 1966, p. 629). Allometric growth therefore defines a condition in which a change in size of the whole is accompanied by scale-related changes in the proportions of aspects of the object under study. In terms of physical geography, investigation of such scale-related changes has generally concentrated on morphological variables so that, as Church and Mark (1980) point out in their major review of the concept, one is dealing with scale distortions of geometric relationships (compare D'Arcy Thompson 1961). If no such distortions occur, *isometric* growth has taken place. To give two examples: on the one hand, it is widely observed that the gradient of the principal stream channel in a drainage basin is reduced at an ever-decreasing rate as the drainage area enlarges; on the other hand, the relationship between channel width and the wavelength of meanders appears to be roughly constant, regardless of actual channel dimensions.

Following Church and Mark (1980), we can define some basic concepts. First, allometry refers to a proportional relationship of the form: $A_1/A_2 = b$. If the resulting ratio is constant for all values of $A_2$, isometry exists. If $A_1$ increases at a faster rate than $A_2$, there is positive allometric growth; if the reverse, negative. It is often the case that $b$ represents the exponent in the general form of the power equation $y = ax^b$ (where $a$ is a constant dependent upon the units of measurement): if $y$ and $x$ have the same scale dimensions, $b$ will be 1.0. If the relationship is positively allometric $b$ will be $> 1.0$, and negatively allometric if $< 1.0$. If $y$ and $x$ have different scale dimensions, then the value of $b$ indicative of isometry will vary accordingly. (For example, if $y$ is a length, $L^1$ and $x$ is an area, $L^2$, a value of $b$ of 0.5 would indicate isometry.) A clear departure from isometry is also indicated if the relationship between $x$ and $y$, plotted as a power function, is curved rather than linear.

Church and Mark further indicate (1980, p. 345) that *dynamic* and *static* allometry should be distinguished. In the former case one is dealing with the changing proportions of an individual landform over time; in the latter data are taken from a number of individuals of different sizes at one moment. Strictly, a study of static allometry should include only individuals of equivalent age. In practice, this requirement may be difficult to meet and has frequently been ignored. It should be clear, however, that the interpretation of the results of studies which aim to investigate static allometry can be very readily complicated by extraneous sources of inter-individual variation, particularly those due to differences in materials and detailed history.

The concept of allometric growth was explicitly introduced into geomorphological literature by Woldenberg (1966) in an investigation of HORTON'S LAWS of drainage basin composition. The widest application, however, has been suggested by Bull (1975), who suggested that *all* proportional relationships of the form $y = ax^b$ are essentially allometric. The preceding definitions should have made it clear that this is far from necessarily the case and Bull's suggestion seems likely simply to obscure the true nature of the underlying concept.

Given the persistent concern of geomorphologists with the size and shape of landforms the idea of allometry is of obvious potential interest. Which relationships are scale-dependent? Which are isometric? Why? All these would seem to be valid and valuable questions to ask. Moreover, as landforms *do* alter in size over time, the adoption of this particular concept from biology appears to be quite permissible, especially since it involves very little modification of the underlying biological ideas. Nevertheless, not all geomorphological use seems to have been governed by a clear grasp of the basic principles and, in particular, workers have sometimes failed to appreciate the need for equations to be properly balanced in dimensional terms.

Church and Mark (1980), in their painstaking discussion of cases of allometric and isometric relationships in geomorphology, conclude with a very important theoretical proposition: that isometry indicates relationships which are, physically, completely defined. Allometric situations suggest the intrusion of as yet unidentified, scale-dependent controls. In brief, they consider that a reduction of allometric equations to properly dimensioned, isometric forms should produce valuable insights into the manner in which 'growth' influences 'form'. BAK

**References**
Bull, W.B. 1975: Allometric change of landforms. *Geological Society of America Bulletin* 86, pp. 1489–98. · Church, M.A. and Mark, D.M. 1980: On size and scale in geomorphology. *Progress in physical geography* 4, pp. 342–90. ·

D'Arcy Thompson, A.W. 1961: *On growth and form*, ed. J.T. Bonner. Cambridge: Cambridge University Press. · Gould, S.J. 1966: Allometry and size in ontogeny and phylogeny. *Biological review* 41, pp. 587–640. · Woldenberg, M.J. 1966: Horton's laws justified in terms of allometric growth and steady state in open systems. *Geological Society of America bulletin* 77, pp. 431–4.

**allopatric**  Descriptive of the condition whereby two closely related species (a 'species pair'), or sub-species occur in separate geographical areas, for example, *Plantago ovata* (Canary Islands to India across North Africa) and *Plantago insularis* (southwestern USA). It is also used to describe the process of speciation when populations of the same species become isolated from each other through the establishment of a geographical 'barrier', for example as a result of climatic or geological change. The resulting genetic isolation allows the separate populations to evolve independently so that, even upon removal of the geographical barrier, the populations are sufficiently distinctive as to be classified as separate species.        MEM

**allophane**  An amorphous hydrated aluminosilicate gel. The chemical composition is highly variable. The name is applied to any amorphous substance in clays.

**alluvial channel**  A river channel that is cut in ALLUVIUM. This applies to most larger natural rivers; ones that are entirely developed in bedrock may be found in high-relief areas, but even there incising streams transporting the material they have eroded may have a discontinuous veneer of alluvial material. More generally, even those channels that have bedrock on the floors of their deepest scour pools have banks in alluvial materials which have been deposited during floods or during the lateral movement of such rivers.

Non-alluvial channels may be different in form and development from alluvial rivers. Bedrock channels may be constrained by rock outcrop, while meltwater streams flowing on glacier ice create channels by removing ice rather than interacting with the bed materials. This is an essential characteristic of alluvial channels: they are self-formed in their own transportable sediments and can adjust their morphology according to discharges and the sediment sizes and loads present.

Alluvial channels can be characterized in terms of their cross-section, planform and long profile properties. These are interrelated. For example, a BRAIDED RIVER planform pattern is usually associated with a shallow cross-section and with relatively steep gradients. But these form elements are also dependent on river discharge and stream power and on sediment properties. Alluvial channel systems are therefore complex ones to study. In the short term stable or equilibrium forms may be developed and these may alter when controlling conditions vary. Unfortunately, equilibrium states are not easy to define, nor are changes precisely predictable. The study of alluvial channels is a large area of scientific enquiry. (See also CHANNEL CLASSIFICATION.)        JL

**Reading**
Richards, K. 1982: *Rivers*. London and New York: Methuen. · Schumm, S.A. 1977: *The fluvial system*. New York: Wiley.

**alluvial fans**  See FANS.

**alluvial fill**  Sedimentary material deposited by water flowing in stream channels. During the seventeenth century the term included all water-laid deposits (including marine sediments), but in 1830 Lyell restricted its use to materials deposited by rivers (Stamp 1961). Particle sizes range from fine clays deposited by overbank waters to boulders deposited in the channel bed by large floods. Materials may be massively bedded if deposited by a single event, or they may occur in a variety of bed forms related to flow variation or location related to the active channel.

**Reading and References**
Happ, S.C. 1971: Genetic classification of valley sediment deposits. *American Society of Civil Engineers: journal of the hydraulics division* 97, pp. 43–53. · Schumm, S.A. 1977: *The fluvial system*. New York: Wiley. · Stamp, L.D. 1961: *A glossary of geographical terms*. London: Longman.

**alluvial terrace**  See RIVER TERRACE.

**alluvium**  Material deposited by running water. The term is not usually applied to lake or marine sediments and may be restricted to unlithified, size-sorted fine sediments (silt and clay). Fine material of marine and fluvial origin may not in practice be easy to distinguish; on geological maps this is often not attempted and both are classed together. Coarser sediment may not by convention be included, but there is no good reason for this if no particular grain size is intended. The term can be prefixed by 'fine-grained' or 'coarse-grained'. Studies of alluvial channels do not imply any particular grain sizes.

Distinguishing characteristics of alluvium are its stratification, sorting and structure. Coarser sediment deposited on the channel bed or in bars is overlain by finer materials deposited from suspension, either in channel slackwater areas or as an OVERBANK DEPOSIT following floods. In detail, sedimentary structures may be complex and dependent on the type of river activity. Large-scale contrasts are often drawn, for example between the deposits of braided and meandering rivers, while different scales and types of CURRENT BEDDING may be present. The size of sediments involved may depend on that supplied to streams from slopes or BANK EROSION and on the distance from such sources that a particular reach may be, because rivers sort and modify alluvial materials as they are transported. Thus fine-grained alluvium can be dominant in the lower courses of present rivers; coarser alluvium can be found close to the supply points for such material (e.g. in alluvial cones in high-relief semi-arid areas or at glacier margins) and in earlier Pleistocene deposits in mid-latitudes. Here the slope- and glacier-derived coarser materials produced under former cold-climate conditions contrast with the finer alluvium coming from more recent slope inputs.                                                JL

**Reading**
Allen, J.R.L. 1970: *Physical processes of sedimentation*. London: Allen & Unwin. · Reading, H.G. ed. 1986: *Sedimentary environments and facies*. 2nd edn. Oxford: Blackwell Scientific.

**alp**  *a.* A shoulder high on the side of a glacial trough.
*b.* (In Switzerland) a summer pasture below the snow level.

**alpha diversity**  The richness of species *within* habitats, i.e. the number of species located within relatively small, environmentally coherent, geographical areas, as opposed to BETA DIVERSITY which measures species turnover *between* habitats and gamma diversity which measures turnover between different geographical areas. Alpha diversity is effectively a measure of the numbers of different types of organisms living in a particular locality under similar circumstances. For example, the total number of plant species found in a small (say, 100 m$^2$) sample plot of Mediterranean shrubland in southern France represents its alpha diversity. If the sample plot size is very small (1 m$^2$), then the measure of richness may be referred to as point diversity. Because of the positive relationship between species richness and area, greater sample plot sizes are likely to contain higher species numbers. Consequently, scale is an integral component of this kind of diversity measure. There is, however, a maximum richness value related to the total number of species available in the geographical area concerned (the species pool), which varies principally with latitude (the tropics have generally greater spe-

cies complements), taxonomic group under consideration (some groups are richer at the poles) and habitat circumstances (some habitats are especially harsh). Alpha diversity is spatially highly variable, but is especially elevated in the neotropical rain forests approaching up to 400 different species of tree alone in a sample plot of 100 m × 100 m.                                        MEM

**Reading**
Groombridge, B. 1992: *Global biodiversity: status of the earth's living resources*. London: Chapman and Hall.

**alpine**  The zone of a mountain above the tree line and below the level of permanent snow.

**alpine orogeny**  The period of mountain-building during the Tertiary era, ending during the Miocene, that produced the Alpine-Himalayan belt.

**altimetric frequency curve**  A frequency curve constructed by dividing an area into squares, determining the maximum altitude in each square and plotting the frequency of these determinations. A useful technique for rapid determination of generalized altitudes in an area, and much used by denudation chronologists for identifying erosion surface remnants.

**altiplanation**  A form of solifluction, i.e. earth movement in cold regions, that produces terraces and flat summits that consist of accumulations of loose rock. An alternative term is cryoplanation.

**altithermal**  During the Holocene there was a phase, of varying date, when conditions were warmer (perhaps by 1–3 °C) than at present. In the Camp Century Ice Core (Greenland) a warm phase lasted from 4100 to 8000 BP, whereas in the Dome Ice Core (Antarctica) the warmest phase was between 11,000 and 8000 BP (Dansgaard *et al.* 1970; Lorius *et al.* 1979). Rainfall conditions may also have changed, aridity having triggered off both renewed sand dune activity and a decline in human population levels in areas such as the High Plains from Texas to Nebraska, USA.                        ASG

**References**
Dansgaard, W., Johnsen, S.J. and Clausen, H.B. 1970: Ice cores and paleoclimatology. In I.U. Ollson ed., *Radiocarbon variation and absolute chronology*. New York: Wiley. Pp. 337–51. · Lorius, C., Merlivat, L., Jouzel, J. and Pourclet, M. 1979: A 30,000-year isotope climatic record from Antarctic ice. *Nature* 280, pp. 644–8.

**altocumulus**  See CLOUDS.

**altostratus**  See CLOUDS.

**alveolar** Or honeycomb weathering features take the form of small hollows in rock surfaces that may occur as individual features but more commonly are found in clusters. They may be related to TAFONI and indeed the two forms can often be seen on the same rock surfaces, suggesting that alveoles (*c.*1–50 cm in diameter) may evolve into larger cavernous features in some circumstances (Mellor *et al.* 1997). There is no clear morphometric boundary between the two and the terms have sometimes been used in an interrelated manner. There may be a tendency for alveoles to form on vertical or near-vertical surfaces (distinguishing them from GNAMMAS, which occur in more horizontal surfaces). Although alveoles have often been cited as a particular feature of sandstones and of drier and coastal environments, they can be found in many different lithologies and in many environments. There is a tendency towards explaining their development in terms of SALT WEATHERING and granular disintegration (Mustoe 1982), which may favour their occurrence in drylands and coastal situations.                    DSGT

**References**
Mellor, A., Short, J. and Kirkby, S.J. 1997: Tafoni in the El Chorro area, Andalucia, Southern Spain. *Earth surface processes and landforms* 22, pp. 817–33. · Mustoe, G.E. 1982: The origin of honeycomb weathering. *Geological Society of America, Bulletin*, 93, pp. 108–15.

**ambient** Preceding or surrounding a phenomenon, e.g. ambient temperature refers to the temperature of the surrounding atmosphere, water or soil.

**amensalism** A kind of interspecific interaction in which there is a negative influence of one species (the inhibitor) but no reciprocal negative impact upon the other (the amensal). This kind of interaction is illustrated by the African proverb: 'When elephants fight, the grass suffers.' In essence, one organism harms the other as a byproduct of its activities. Such interactions are common, indeed in the case of humans many of our interactions with other organisms could be described as amensal, for example the negative impact of acid rain on coniferous forests.                    MEM

**amino acid racemization** A method in GEO-CHRONOLOGY. This dating technique is based on the fact that the amino acid building blocks which make up proteins in skeletal remains of animals undergo time-dependent chemical reactions. Amino acid racemization dating is a relative-age method involving measurement of the extent to which certain types of amino acids within protein residues have transformed from one of two chemically identical forms (isomers) to the other. Materials which may preserve such protein residues within sediment bodies include bones and other body components (e.g. mummies), mollusc shells and eggshells. At formation, only *L*-form amino acids are present. Over time, and in part controlled by temperature and other factors (hydrolysis, pH), some of the *L*-form acids are converted to a *D*-form until an equilibrium is reached. *Racemization* and *epimerization* reactions differ in that racemization involves only amino acids with a single chiral carbon atom, whereas epimerization involves amino acids with two chiral carbon atoms (Aitken 1990). The use of a range of amino acids which undergo racemization or epimerization over a range of timescales makes it possible to apply the method over timescales ranging from a few years to hundreds of thousands. The degree of change in amino acid composition, however, depends on factors other than time (e.g., temperature) which may result in substantial errors in amino acid dates.                    SS

**Reference**
Aitken, M.J. 1990: *Science-based dating in archaeology.* London: Longman.

**amphidromic point** The node around which KELVIN WAVES rotate. The rotation is caused by the CORIOLIS FORCE acting on the flow induced by a standing tidal wave in large basins. The amplitude of the wave is nil at the amphidromic point, and at a maximum at the boundaries of the tidal basin. Rotation of the TIDES is counter-clockwise in the northern hemisphere, and reversed in the southern hemisphere. *Cotidal* lines connect points experiencing simultaneous high tides, and these radiate from the amphidromic point. *Corange* lines connect points of equal tidal range, and these are concentric about the amphidromic point. *Amphidromic* points are approximately constant in location, but may show seasonal or other periodic migrations.                    DJS

**Reading**
Cartwright, D.E. 1999: *Tides: a scientific history.* Cambridge: Cambridge University Press. · Thurman, H.V. 1991: *Introductory oceanography*, 6th edn. New York: Macmillan Publishing Company.

**AMS radiocarbon dating** See CARBON DATING.

**anabatic flows** Upslope winds usually produced by local heating of the ground during the day. The most common type is the VALLEY WIND. Anabatic flows develop best on east or west facing slopes on days with clear skies. The air along the slope is heated by contact with the warm surface much more rapidly than air at the same elevation away from the slope. The resulting temperature difference sets up a thermal circulation with the air ascending along the slope and descending over the adjoining plain or valley. Under ideal conditions anabatic winds can reach speeds of $10-15 \, \text{m s}^{-1}$ and can be a factor in the spreading of forest fires in dry weather. If the air is moist the anabatic flow may produce anabatic clouds above the crest of the slope. WDS

**Reading**
Atkinson, B.W. 1981: *Meso-scale atmospheric circulations.* New York and London: Academic Press. · Geiger, R. 1965: *The climate near the ground.* Cambridge, Mass.: Harvard University Press. · Oke, T.R. 1987: *Boundary layer climates.* 2nd edn. London: Routledge.

**anabranching** A channel planform that resembles braiding but where the individual channels (anabranches) are separated by islands that divide flow at bankfull rather than by bars in a braided channel that are overtopped below bankfull (Nanson and Knighton 1996). Brice (see CHANNEL CLASSIFICATION) describes anabranching as having islands of more than three times the width of the river at average discharge. Individual anabranches can be meandering, braided or straight. *Anastomosing* is a term commonly used by sedimentologists to define a group of fine grained, low energy anabranching rivers with a distinctive alluvial architecture sometimes associated with coal and hydrocarbon preservation. GCN

**Reference**
Nanson, G.C. and Knighton, A.D. 1996: Anabranching rivers: their cause, character and classification. *Earth surface processes and landforms* 21, pp. 217–39.

**anaclinal** Refers to a feature, especially a river or valley, which is transverse to strike and against the dip of strata.

**anaerobic** Term used to describe conditions in which oxygen is absent or, in the case of metabolic processes, respiration in the absence of oxygen (see AEROBIC). Most organisms obtain their energy through aerobic respiration involving the breakdown of sugars in the presence of oxygen. Some organisms, however, for example certain types of bacteria or mould, do so in the absence of oxygen. Such organisms are referred to as obligate anaerobes if they live permanently in oxygen-deficient conditions and facultative anaerobes if they respire aerobically when oxygen is present and resort to anaerobic respiration if oxygen is scarce or absent. Also known as fermentation, the process generates carbon dioxide and either organic acids

(e.g. lactic acid in the case of anaerobic respiration in vertebrate skeletal muscles) or alcohols (e.g. ethanol production from yeast). Fermentation of yeast is perhaps the most familiar form of anaerobic respiration, since it is responsible for the production of alcohol for human consumption in the form of beer or wine. Biochemically, the forms of respiration may be compared as follows, note the variation in energy production (identified here in terms of kilojoules), which indicates the relative efficiency of aerobic respiration:

*Anaerobic respiration with ethanol formation (alcoholic fermentation)*

$$C_6H_{12}O_6 \rightarrow 2CH_3CH_2OH + 210kJ$$
ethanol

*Anaerobic respiration with lactic acid formation*

$$C_6H_{12}O_6 \rightarrow 2CH_3CH(OH)COOH + 150kJ$$
lactic acid

*Aerobic respiration*

$$C_6H_{12}O_6 + O_2 \rightarrow 6H_2O + 6CO_2 + 2880kJ$$

In still other organisms, for example bacteria of the genus *Sulfobolus*, energy is obtained anaerobically through the oxidation of inorganic substances, for example hydrogen sulphide. MEM

**ana-front** A front which has ascending air at one side, particularly one experiencing the unusual phenomenon of rising cold air.

**analemna** A scale drawn on a globe to show the daily declination of the sun, enabling the determination of those parallels where the sun is directly overhead at any specific time of year.

**anamolistic cycle** The tidal cycle, normally taken as lasting 27.5 days, which is related to the varying distance between the earth and the moon.

**anaseism** The vertical component of the waves moving upwards from the focus of an earthquake.

**anastomosing** A category of fine grained, low energy ANABRANCHING rivers. They have a distinctive alluvial architecture consisting of arenaceous channels and argillaceous overbank deposits and are sometimes associated with coal and liquid hydrocarbon preservation in the stratigraphic record. They should not be confused with BRAIDED RIVERS. GCN

**anastomosis** Braiding of rivers, i.e. the tendency of some streams to divide and reunite producing a complex pattern of channels. (See BRAIDED RIVER.)

**andosols** Dark soils developed on volcanic rock and ash.

**andromy** The migration of some fish species from salt to fresh water for breeding.

**anemograph** A self-recording instrument for measuring the speed and sometimes the direction of the wind.

**anemometer** A mechanical or electronic device for measuring wind speed. The simplest of these devices is the cup-anemometer which consists of a rotating set of cups (or vanes) measuring the 'run of wind' on a continuous counter. This is the distance the wind travels per unit time and is commonly expressed as km/day. More sophisticated cup-anemometers use a reed switch which completes an electrical circuit on each rotation, providing an electrical pulse which can be monitored by a DATA LOGGER. Cup-anemometers suffer from inertia and friction, both of which limit their accuracy in low velocity winds or TURBULENT FLOW, particularly when measurements are required over short timespans such as 10 s or less (Kaganov and Yaglom 1976). Wholly accurate measurements are inherently problematic to obtain with all mechanical anemometers as it is always necessary to place the sensor in the wind, hence disrupting it.

Hot-wire anemometers determine high-frequency velocity fluctuations from the resistance of a thin wire placed in the flow, the electroconductivity of which is controlled by its temperature (Castro 1986). The wire is usually made of platinum and has a diameter of about 5 microns, the active portion of the wire being of the order of 1 mm. In practice, the voltage required to keep the wire at a constant temperature is measured and this is calibrated against the velocity of flow. Such hot-wires are very delicate and have been restricted to use in wind tunnels, although armoured versions can now be used in the field. Laser Doppler anemometry measures the velocity of airflow by sensing the change in frequency of light scattered back from tiny particles introduced to the flow. The great advantage of this technique is that it is non-intrusive and enables high frequency measurements (> 100 Hz), but it is very expensive and currently restricted to wind tunnel use.                                        GFSW

**Reading and References**
Castro, I.P. 1986. The measurement of Reynold's stresses. *Encyclopaedia of fluid mechanics*. Houston: Gulf. · Clifford, N.J. and French, J.R. 1993. Monitoring and modelling turbulent flow: historical and contemporary perspectives. In N.J. Clifford, J.R. French and J. Hardisty eds, *Turbulence: perspectives on flow in sediment transport*. Chichester: Wiley. Pp. 1–34. · Kaganov, E.I. and Yaglom, A.M. 1976. Errors in windspeed measurement by rotation anemometers. *Boundary-layer meteorology* 10, pp. 229–44.

**angiosperms** Flowering plants, a subdivision of seed-producing plants (*Spermatophyta*), whose main characteristic is the presence of the flower. Angiosperms are distinguished from the 'naked-seed' plants (gymnosperms) by the fact that, following fertilization, the seeds are developed in a protected ovary. Flowers are essentially modified shoots comprising four series arranged as whorls. From the outside of the flower inwards, these series are (1) sepals, (2) petals, (3) stamens (the pollen-producing male component) and (4) carpels (female structures from which the seeds develop). Flowering plants represent the dominant group of plants today and there are approximately 250,000 described species including all the commonly occurring herbs, shrubs and trees. Of the two groups of seed plants, angiosperms appear much later in the fossil record, although they have diversified to occupy virtually every ecological NICHE and, moreover, are of major economic importance in that they provide much of human food supply as wheat, rice, maize etc. Gymnosperms, of which there are approximately 700 species, have a more ancient evolutionary history and have been outcompeted in most contemporary environments by their more successful flowering relatives, although they remain prominent in certain environments (e.g. the boreal forests of North America and Eurasia). Angiosperms are divided into two classes based on the number of leaves in the embryo, one in the case of the *Monocotyledoneae* (e.g. lily) and two in the *Dicotyledoneae* (e.g. oak).                                        MEM

**Reading**
Rudall, P. 1993: *Anatomy of flowering plants: an introduction to structure and development*. Cambridge: Cambridge University Press.

**angle of dilation** (θ) The angle by which the grains of a granular material are displaced and reorientated on a shearing surface (SHEAR STRENGTH) cutting through the mass of particles. The reorientation movement is a response to the interlocking of particles which provide the frictional resistance or shear strength. The angle (θ) is related to the ANGLE OF INTERNAL SHEARING

RESISTANCE ($\phi$) and the static ANGLE OF PLANE SLIDING FRICTION ($\phi_{us}$) by $\theta = \phi - \phi_{us}$. WBW

**Reading**
Statham, I. 1977: *Earth surface sediment transport*. Oxford: Clarendon Press.

### angle of initial yield ($\varphi_i$)

The angle of a slope of granular material at which movement, often as 'avalanching', is seen to start. The angle depends upon the type of material, particularly its packing and bulk density, which together create interlocking particles. It has a higher value than the ANGLE OF RESIDUAL SHEAR ($\varphi_r$). WBW

### angle of internal shearing resistance ($\phi$)

The angle, usually measured in a TRIAXIAL or SHEAR BOX apparatus to give the friction angle $\varphi$, for granular materials, of the MOHR-COULOMB equation. It is not a constant for any material but depends upon the VOID RATIO or POROSITY, as well as other frictional properties which relate to the interlocking of particles. WBW

**Reading**
Statham, I. 1977: *Earth surface sediment transport*. Oxford: Clarendon Press.

### angle of plane sliding friction ($\phi_{us}$)

The angle at which non-cohesive (granular) particles just begin to slide down a surface. Strictly, this is the static angle; if it is the angle at which particles just stop moving it is the dynamic angle. It can apply to individual granular particles, a mass of such particles, or a slab of rock. In the latter case it is related to joint friction. For any of these, the static angle $\phi_{us}$ and the dynamic angle $\phi_{ud}$ are approximately constant but $\phi_{us}$ is greater than $\phi_{ud}$. WBW

**Reading**
Statham, I. 1977: *Earth surface sediment transport*. Oxford: Clarendon Press.

### angle of repose

The angle at which granular material comes to rest (also called the angle of rest); it approximates to the angle of scree slopes. Strictly, natural slopes for one material may have a variable angle of repose according to whether the material has just come to rest or is about to move, hence it is also related to the ANGLE OF INITIAL YIELD or the ANGLE OF RESIDUAL SHEAR. WBW

### angle of residual shear ($\phi_r$)

The angle at which granular material comes to rest after movement. The angle is less than the ANGLE OF INITIAL YIELD ($\phi_i$) and is comparable to the ANGLE OF INTERNAL SHEARING RESISTANCE ($\phi$)

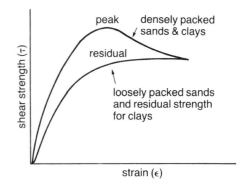

for cohesionless material in its most loosely packed state; thus the difference between $\phi_r$ and $\phi_i$ represents a loss of strength of the material to define a residual shear strength. WBW

**Reading**
Statham, I. 1977: *Earth surface sediment transport*. Oxford: Clarendon Press.

### angular momentum

A rather abstract quantity of great generality, essentially the product of the mass of a particle, the distance to an axis, and tangential components of velocity. It can be shown that forces directed towards (or away from) the axis cannot change the angular momentum. Many natural systems, including hurricanes and tornadoes, are dominated by forces acting towards a centre of low pressure. The (large) tangential components of velocity $V$ at distance $r$ are constrained by conservation of angular momentum to have $V$ nearly proportional to $1/r$. This gives such systems a characteristic shape, flow field, and qualities of persistence. Angular momentum on the global scale is similarly constrained. (See also MOMENTUM BUDGET.) JSAG

### angular unconformity

A stratigraphic unconformity represented by younger strata overlying older strata which dip at a different angle, usually a steeper one.

### anisotrophy

The condition of a mineral or geological stratum having different optical or physical properties in different directions.

### annual series

A term used in flood frequency analysis for the series of discharges obtained by selecting the maximum instantaneous discharge from each year of the period of record. The annual series is therefore equal to the number of years of hydrological record analysed. (See also FLOOD FREQUENCY.) KJG

**annular drainage** A circular or ring-like drainage pattern produced when streams and rivers drain a dissected dome or basin.

**antecedent drainage** A drainage system which has maintained its general direction across an area of localized uplift.

**antecedent moisture** The soil moisture condition in an area before a rainfall, which moderates the area's run-off response to the rainfall. Antecedent moisture condition is usually expressed as an index and may be estimated by weighting past rainfall events to derive an antecedent precipitation index (API). The API is a weighted sum of preceding rainfall within given time units and if a daily time base is used the API is often calculated as:

$$API_t = API_t^{-1} \cdot k + P_t$$

where $API_t$ is the antecedent precipitation index for day $t$, $P_t$ is the precipitation on day $t$, and $k$ is a decay factor ($k < 1.0$ and usually $0.85 < k < 0.98$).

Alternatively, a water budget based upon the preceding rainfall and evapotranspiration rates can be used to provide an estimate of soil moisture storage (see WATER BALANCE) or base flow in the area may be used as an index of soil moisture or antecedent moisture condition. AMG

Reading
Chow, V.T. 1964: *Handbook of applied hydrology.* New York: McGraw-Hill.

**antecedent precipitation index** An index of moisture conditions in a catchment area used to assess the amount of effective rainfall that will form direct surface run-off. If there has been no rain for several weeks less rainfall will get into the streams than if the ground surface is already saturated. The index is calculated on a daily basis and assumes that soil moisture declines exponentially when there is no rainfall. Thus we have:

$$API_t = k \cdot API_{t-1}$$

where $API_t$ is the index $t$ days after the starting point. The value of $k$ will depend upon the potential loss of moisture so has a seasonal variation between 0.85 and 0.98. An allowance is made for any precipitation input during the period. PS

**anteconsequent stream** A stream which flows consequent on an early uplift but antecedent to later stages of the same tectonic uplift.

Reading
Shelley, D. 1989: Anteconsequent drainage: an unusual example formed during constructive volcanism. *Geomorphology* 2, pp. 363–7.

**anthropochore** Plant introduced to an area by humans (see also ALIENS). This may take the form of an intentional introduction, for example the introduction of various types of European and North American trees to the southwestern Cape of South Africa by colonial settlers. These trees were planted to augment dwindling supplies of locally available wood for fuel and shelter. In other situations, introduction may be entirely an accidental by-product of human activity; seeds of exotic plant species may be transported to a novel locality in, for example, aircraft tyre treads or in containers onboard ship. In some instances, plants introduced intentionally may disperse from their new localities and into places where they become ecologically problematic. For example, the ornamental garden azalea, *Rhododendron ponticum*, originally an Asian species, has become a significant invasive pest in the woodlands of Britain. MEM

**anthropogene** A primarily Russian term for the period during which man has been an inhabitant of the earth (i.e. the past two to three million years).

**anthropogeomorphology** The study of the role of humans as a geomorphological agent. There are very few spheres of human activity which do not create landforms (table 1). There are those landforms produced by direct anthropogenic processes. These are relatively obvious in form and origin and are frequently created deliberately and knowingly (table 2). Landforms produced by indirect anthropogenic processes are often less easy to recognize, not least because they do not so much involve the operation of a new process or processes as the acceleration of natural processes. They are the result of environmental changes brought about inadvertently by human technology. ASG

Reading
Brown, E.H. 1970: Man shapes the earth. *Geographical journal* 136, pp. 74–85. · Goudie, A.S. 2000: *The human impact.* 5th edn. Oxford: Basil Blackwell; Cambridge, Mass.: MIT Press. · Haigh, M.J. 1978: Evolution of slopes on artificial landforms, Blaenavon, UK. *University of Chicago, Department of Geography research paper* 183. · Jennings, J.N. 1966: Man as a geological agent. *Australian journal of science* 28, pp. 150–6. · Sherlock, R.L. 1922: *Man as a geological agent.* London: Witherby.

**Table 1** Some anthropogenic landforms

| Feature | Cause |
|---------|-------|
| Pits and ponds | Mining, marling |
| Broads | Peat extraction |
| Spoil heaps | Mining |
| Terracing, lynchets | Agriculture |
| Ridge and furrow | Agriculture |
| Cuttings | Transport |
| Embankments | Transport; river and coast management |
| Dikes | River and coast management |
| Mounds | Defence, memorials |
| Craters | War; *qanat* construction |
| City mounds (*tells*) | Human occupation |
| Canals | Transport, irrigation |
| Reservoirs | Water management |
| Subsidence depressions | Mineral and water extraction |
| Moats | Defence |

**Table 2** Classification of anthropogenic landforming processes

*1 Direct anthropogenic processes*
  1.1 Constructional
     tipping: loose, compacted, molten
     graded: moulded, ploughed, terraced
  1.2 Excavational
     digging, cutting, mining, blasting of cohesive or non-cohesive materials
     cratered
     trampled, churned
  1.3 Hydrological interference
     flooding, damming, canal construction dredging, channel modification
     draining
     coastal protection

*2 Indirect anthropogenic processes*
  2.1 Acceleration of erosion and sedimentation
     agricultural activity and clearances of vegetation
     engineering, especially road construction and urbanization
     incidental modifications of hydrological regime
  2.2 Subsidence: collapse, settling
     mining
     hydraulic
     thermokarst
  2.3 Slope failure: landslide, flow, accelerated creep
     loading
     undercutting
     shaking
     lubrication
  2.4 Earthquake generation
     loading (reservoirs)
     lubrication (fault plane)

*Source*: Haigh 1978.

**antibiosis** A specific form of antagonism which involves the formation by one organism of a substance which is harmful to another organism.

**anticentre** The point opposite the epicentre, above the focus, of an earthquake.

**anticline** A type of geological FOLD that is convex-upwards in shape (i.e. the strata form an arch) which develops as a result of laterally applied compression resulting from tectonic activity.     DJN

**anti-cyclone** An extensive region of relatively high atmospheric pressure, typically a few thousand kilometres across, in which the low level winds spiral out clock-wise in the northern hemisphere and counter-clockwise in the southern. Anti-cyclones are common features of surface weather maps and are generally associated with calm, dry weather.

They originate either from strong radiative cooling at the earth's surface or from extensive subsidence through the depth of the troposphere. The first kind, known as cold anti-cyclones or highs, form across the wintertime continents and are shallow features produced by cold, dense air which is confined to the lower troposphere. The mobile ridges which occur in the polar air between frontal systems are also cold highs. The second kind, called warm anti-cyclones, are semipermanent features of the subtropical regions of the world. Here the descending branch of the HADLEY CELL ensures the persistence of a large downward flow of mass to supply the outflowing surface winds. The compression of the subsiding air leads to a deep, anomalously warm troposphere within which the anti-cyclonic circulation persists with height. Fine summers in Britain are normally associated with an unusual north-eastward excursion of the warm Azores anti-cyclone.

Although the subsidence and dry air in the highs tend to dampen convective activity, extensive and sometimes persistent low layer cloud can occur in some regions.     RR

**Reading**
Palmer, E. and Newton, C.W. 1969: *Atmospheric circulation systems*. New York and London: Academic Press.

**antidune** A ripple on the bed of a stream or river similar in form to a sand dune but which migrates against the direction of flow (i.e. upstream).

**antiforms** Upfolds of strata in the earth's crust; *synforms* are downfolds. In both cases the precise stratigraphic relationships of the rocks are not known, whereas in the case of anticlines and synclines they would be.

**antipleion**  An area or a specific meteorological station where the mean annual temperature is lower than the average for the region.

**antipodal bulge**  The tidal effect occurring at the point on the earth's surface opposite that where the pull of the moon's gravity is strongest. Hence it is the tidal effect at the point where lunar attraction is weakest.

**antipodes**  Any two points on the earth's surface which are directly opposite each other so that a straight line joining them passes through the centre of the earth.

**antitrades**  A deep layer of westerly winds in the TROPOSPHERE above the surface TRADE WINDS. In a simple way they represent the upper limits of the HADLEY CELL within which occurs the poleward transfer of heat, and momentum and water vapour.                                BWA

**antitriptic wind**  See WIND.

**aphanitic**  Microcrystalline and cryptocrystalline rock textures. Pertaining to a texture of which the crystalline components are not visible with the naked eye.

**aphelion**  The point of the orbit of a planet, or other solar satellite, which is farthest from the sun.

**aphotic zone**  The portion of lakes, seas and oceans at a depth to which sunlight does not penetrate.

**aphytic zone**  The portion of the floor of lakes, seas and oceans which, owing to their depth, are not colonized by plants.

**apogee**  The point of the orbit of the moon or a planet that is farthest from the earth.

**aposematic coloration**  The conspicuous and distinctive markings on a plant or animal which communicate that it is poisonous or distasteful to potential predators.

**applied geomorphology**  The application of geomorphology to the solution of miscellaneous problems, especially to the development of resources and the diminution of hazards (Hails 1977). The great American geomorphologists of the second half of the nineteenth century –

Powell, Dutton, McGee and Gilbert – were employed by the US Government to undertake surveys to enable the development of the West, and R.E. Horton, one of the founders of modern quantitative geomorphology, was active in the soil conservation movement generated by the 'Dust Bowl' conditions of the 1930s. It is notable that all these workers made fundamental contributions to theory, and thus to 'pure' geomorphology, even though much of their work was concerned with the solution of immediate environmental problems. More recently there has been much interest in the USA in engineering geomorphology and environmental impact assessment (Coates 1976): in the UK in the geomorphological problems of arid environments (Cooke *et al.* 1983), in Australia in the preparation of inventories of environmental conditions in different parts of that island, in Canada in the development of permafrost areas (Williams 1979), and among French geomorphologists working for Office de la Recherche Scientifique et Technique Outre-Mer (ORSTOM) and other bodies, etc. In the past two decades the role of the geomorphologist has developed, partly because with increasing population presures and technological developments the impact of man on geomorphological processes has increased (Goudie 1993) and partly because the increasing competence of geomorphologists in the study of materials and processes proved to be more valuable than denudation chronology to the solution of environmental problems.

The role of the geomorphologist in environmental management (Cooke and Doornkamp 1990) can be subdivided into six main categories. First of these is the mapping of geomorphological phenomena. Landforms, especially depositional ones, may be important resources of useful materials for construction, while maps of slope angle categories may help in the planning of land use, and maps of hazardous ground may facilitate the optimal location of engineering structures. Secondly, because landforms are relatively easily recognized on air photographs and remote sensing imagery, they can be used as the basis for mapping other aspects of the environment, the distribution of which is related to their position on different landforms. An important example of this is the use of landform mapping to provide the basis of a soil map (Gerrard 1992; see also CATENA). The third category is the recognition and measurement of the speed at which geomorphological change is taking place. Such changes may be hazardous to man. By using sequential maps and air photographs, or archival information, and by monitoring processes with appropriate instrumentation (Goudie 1990), areas at potential risk can be identified, and predictions can be made as to the amount and direction of change. For example, by calculating rates of soil erosion in different parts of a river catchment an estimate can be made of the likely life of a dam before it is silted up, and measures can be taken to reduce the rates of erosion in the areas where the erosion is highest. Indeed, the fourth category of applied geomorphology is to assess the causes of the observed changes and hazards, for without a knowledge of cause, attempts at amelioration may have limited success. Fifthly, having decided on the speed, location and causes of change, appropriate solutions can be made by employing engineering and other means. Sixthly, because such means may themselves create a series of sequential changes in geomorphological systems, the applied geomorphologist may make certain recommendations as to the likely consequences of building, for example, a groyne to reduce coastal erosion. Examples of engineering solutions having unforeseen environmental consequences, sometimes to the extent that the original problem is heightened and intensified rather than reduced, are all too common, especially in many coastal situations (Bird 1979).                                   ASG

**Reading and References**
Bird, E.C.F. 1979: Coastal processes. In K.J. Gregory and D.E. Walling eds, *Man and environmental processes*. Folkestone: Dawson, Pp. 81–101. · Coates, D.R. ed. 1976: *Geomorphology and engineering*. Stroudsburg, Pa.: Dowden, Hutchinson & Ross. · Cooke, R.U., Brunsden, D., Doornkamp, J.C. and Jones, D.K.C. 1983: *Urban geomorphology in drylands*. Oxford: Oxford University Press. · Cooke, R.U. and Doornkamp, J.C. 1990: *Geomorphology in environmental management*. 2nd edn. Oxford: Oxford University Press. · Gerrard, J. 1992: *Soil geomorphology*. London: Chapman and Hall. · Goudie, A.S. ed. 1990. *Geomorphological techniques*. 2nd edn. London: Unwin Hyman. · —1993: *The human impact*. 4th edn. Oxford: Basil Blackwell; Cambridge, Mass.: MIT Press. · Hails, J.R. ed. 1977: *Applied geomorphology*. Amsterdam: Elsevier. · Verstappen, H.Th. 1983: *Applied geomorphology*. Amsterdam: Elsevier. · Williams, P.J. 1979: *Pipelines and permafrost; physical geography and development in the circumpolar North*. London: Longman.

**applied meteorology** The use of archived and real-time atmospheric data to solve practical problems in a wide range of economic, social and environmental fields. The need to analyse and apply atmospheric information for such purposes arises because weather and climate impinge directly on vital human concerns such as agriculture, water resources, energy, health and transport. Specific operational problems are usually raised by clients or managers involved with weather-sensitive activities but a

Applied meteorology: sectors and activities where climate has significant social, economic and environmental significance

| Primary sectors | General activities | Specific activities |
|---|---|---|
| Food | Agriculture | Land use, crop scheduling and operations, hazard control, productivity, livestock and irrigation, pests and diseases, soil tractionability |
| | Fisheries | Management, operations, yield |
| Water | Water disasters | Flood/droughts/pollution abatement |
| | Water resources | Engineering design, supply, operations |
| Health and community | Human biometeorology | Health, disease, morbidity and mortality |
| | Human comfort | Settlement design, heating and ventilation, clothing, acclimatization |
| | Air pollution | Potential, dispersion, control |
| | Tourism and recreation | Sites, facilities, equipment, marketing, sports activities |
| Energy | Fossil fuels | Distribution, utilization, conservation |
| | Renewable resources | Solar/wind/water power development |
| Industry and trade | Building and construction | Sites, design, performance, operations, safety |
| | Communications | Engineering design, construction |
| | Forestry | Regeneration, productivity, biological hazards, fire |
| | Transportation | Air, water and land facilities, scheduling, operations, safety |
| | Commerce | Plant operations, product design, storage of materials, sales planning, absenteeism, accidents |
| | Services | Finance, law, insurance, sales |

*Source*: Thomas, M.K. 1981: The nature and scope of climate applications. Canadian Climate Centre (unpublished).

continuing dialogue between meteorologist and customer may well be necessary to ensure the optimum application of atmospheric knowledge in any particular field or industry. For example, atmospheric data can be usefully applied throughout the construction industry from the initial planning of location, through the design of buildings and the on-site construction phase to the control of energy and other running costs when the building is complete.

The comprehensive scope of applied meteorology is illustrated in the table, which lists the activities which can benefit from the application of atmospheric knowledge. Such applications cover all time-scales of atmospheric behaviour.

KS

**aquaculture**  The commercial cultivation of plants and animals in aquatic environments, both freshwater and marine (mariculture). Fish aquaculture includes the production of food crops and ornamental (i.e. aquarium) fish. Plant aquaculture includes hydroponic farming as well as open environment farming (e.g. seaweed production). Fish farming refers to closed system, and pen or cage production. Fish ranching refers to the cultivation of migratory species that are seeded, released, and harvested upon their return to the 'ranch'. Aquaculture is often touted as a solution to feeding the world's hungry populations.

DJS

**Reading**
Stickney, R.R. 1996: *Aquaculture in the United States*. New York: Wiley.

**aquatic macrophyte**  Freshwater or marine plant defined on the basis of its large size (i.e. nominally large enough to be seen without the aid of a microscope). In marine conditions, in which case they are known as seaweeds, such plants may be extremely large, for example kelp, commonly occurring brown ALGAE of coastal waters, which may exceed 60 m in length. Seaweeds are most abundant in the inter-tidal and sub-tidal zone where environmental conditions are such that specialized adaptations are required for survival. For example, twice a day the intertidal seaweeds are exposed to the atmosphere and must cope with a desiccating atmosphere and direct solar radiation, whereas during twice-daily tidal inundation periods they must endure submersion by seawater and the physical impact of waves in the surf zone. In order to deal with the rigours of survival under such circumstances, seaweeds have evolved unique metabolic and anatomical characteristics. For example, the body, or thallus, of the organism consists of a root-like holdfast and a stem-like stipe supporting leaf-like blades used in photosynthesis. Cell walls are composed of cellulose and gel-forming (e.g. agar) substances accounting for the slimy and rubbery feel of many seaweeds, a feature which may cushion the thalli against wave impact and also help them resist drying during low tide. Especially in Asia, seaweeds, which are rich in nutrients such as iodine, are used as food. While the macrophytes of the oceans are usually algae, aquatic plants in fresh water, for example the cosmopolitan duckweeds (family *Lemnaceae*), may belong to higher evolutionary groups. Under conditions of nutrient enrichment of lakes due to inwashed agricultural fertilizer, such plants may rapidly increase in population and contribute to the problem of EUTROPHICATION.

MEM

**Reading**
Sundaralingam, V.S. 1990: *Marine algae: morphology, reproduction and biology*. Port Jervis, New York: Lubrecht and Cramer.

**aquiclude**  See AQUIFUGE; GROUNDWATER.

**aquifer**  Refers either to a permeable or porous subsurface rock that holds water, or to the body of water itself. This water is termed GROUNDWATER and is important both as a resource for human use and as a component of the hydrological cycle. A confined aquifer occurs between two impermeable rocks (termed aquicludes), while the upper limit of an unconfined aquifer is marked by the water table.

DSGT

**aquifuge**  An impermeable rock incapable of absorbing or transmitting significant amounts of water. Unfissured, unweathered granite is an example. Certain rocks such as clay and mudstone are very porous and absorb water, but when saturated are unable to transmit it in significant amounts under natural conditions. Such formations are known as *aquicludes*. The term *aquitard* is also sometimes used to describe the hydrological characteristics of the less permeable bed in a strategic sequence, that may be capable of transmitting some water, but not in economically significant quantities. PWW

**Reading**
Freeze, R.A. and Cherry, J.A. 1979: *Groundwater*. Englewood Cliffs, NJ: Prentice-Hall. · Ward, R.C. 1975: *Principles of hydrology*. 2nd edn. Maidenhead, Berks.: McGraw-Hill.

**aquitard**  See AQUIFUGE; GROUNDWATER.

**arboreal**  Pertaining to trees.

**arches, natural**  A bridge or arch of rock joining two rock outcrops which has been produced by natural processes of weathering and erosion.

**archipelago**  A sea or lake containing numerous islands or a chain or cluster of islands.

**arctic**  Variously defined depending on one's interest. The popular definition is to include all areas north of the Arctic Circle (Lat. $66\frac{1}{2}°$ N) which is the latitude at which the sun does not rise in mid-winter or set in mid-summer. A more useful natural definition includes land areas north of the tree line and oceans normally affected by Arctic water masses.       DES

**Reading**
Sugden, D.E. 1982: *Arctic and Antarctic*. Oxford: Basil Blackwell; Totowa, NJ: Barnes & Noble.

**arctic-alpine flora**  A group of plants displaying a disjunct geographical distribution that embraces both the lowland regions of the arctic and the high-altitude mountain areas of the temperate and even the tropical zones. The main mountain systems involved are the Rockies, the Alps, the Himalayas and high intertropical mountains, such as those of East Africa. Classic examples of the flora include *Anemone alpina* (Alps/Arctic), *Polygonum viviparum* (Alps, Altai, Himalayas/Arctic), *Ranunculus pygmaeus* (Alps, Rockies/Arctic), *Salix herbacea* (Alps, Urals, Rockies/Arctic) and *Saxifraga oppositifolia* (Alps/Arctic).       PAS

**Reading**
Löve, A. and Löve, D. 1974: Origin and evolution of the arctic and alpine floras. In J.D. Ives and R.G. Barry eds, *Arctic and alpine environments*. London: Methuen. Pp. 571–603.

**arctic haze**  A reddish-brown atmospheric haze, which is often observed in the Arctic, especially in the winter and spring, when atmospheric conditions are calm. It consists primarily of atmospheric pollutants derived from industrial sources in Europe and northern Asia. These include sooty and acidic particles.       ASG

**arctic smoke**  See FROST SMOKE.

**areic**  Without streams or rivers.

**arena**  A shallow, broadly circular basin hemmed in by a rim of higher land.

**arenaceous**  Pertaining to, containing or composed of sand. Applied to sedimentary rocks composed of cemented sand, usually quartz sand.

**areography**  The study of the geographical ranges of plant and animal taxa. It focuses on the form and size of taxonomic ranges, and differs in emphasis from BIOGEOGRAPHY (concerned with the delimitation of floral and faunal sets and the origins of their constituent elements) and ecogeography (concerned with the reasons for the form and size of taxonomic ranges).       ASG

**Reading**
Rapoport, E.H. 1982: *Areography: geographical strategies of species*. Oxford: Pergamon.

**arête**  A fretted, steep-sided rock ridge separating valley or cirque glaciers. The basic form is the result of undercutting or BASAL SAPPING by glaciers which evacuate any rock debris and thus maintain steep rock slopes. Arêtes are common whenever mountains rise above glaciers, for example in mountain chains and as nunataks protruding above ice sheets.       DES

**argillaceous**  Pertaining to, containing or composed of clay. Applied to rocks which contain clay-sized material and clay minerals.

**aridisols**  The soils of dry climates, which are grouped as one of the eleven soil orders of the US system of soil taxonomy. Aridisols show limited differentiation into HORIZONS, owing to lack of water to break down and translocate materials, and of organic matter, which normally darkens the upper parts of other soils. Aridisols may have subsurface layers rich in salts, calcium carbonate, gypsum (calcium sulphate) or other materials that are normally removed by water from the soils of wetter regions. Part of the mineral fraction of many aridisols is composed of materials derived from elsewhere and delivered by long-distance wind transportation.       DLD

**Reading**
Skujins J. ed. 1991: *Semiarid lands and deserts. soil resource and reclamation*. New York: Marcel Dekker.

**arkose**  A sandstone containing more than 25 per cent feldspar. Any feldspar-rich sandstone.

**armoured mud balls**  (also called clay balls, pudding balls, mud pebbles and mud balls). Roughly spherical lumps of cohesive sediment, which generally have diameters of a few centimetres, though much larger examples have been reported. Many examples are lumps of clay or cohesive mud that have been gouged from stream beds or banks by vigorous currents. They often occur in badlands and along ephemeral streams, but can also be found on beaches, in tidal channels, etc.       ASG

**armouring** A term used of heterogeneous river bed material when coarse grains are concentrated sufficiently at the bed surface to stabilize the bed and inhibit transportation of underlying finer material. An armour layer is coarser and better sorted than the substrate, and is typically only 1–2 grains thick. Genetically, a distinction exists between armoured and paved beds. An armoured bed is mobile during floods, but the coarser veneer reforms on or after the falling limb of a flood capable of disrupting and transporting the armouring grains as finer grains are winnowed and the protective layer is recreated. The substrate remains protected until the next event capable of entraining the armour grains. A paved surface, however, is more stable and is markedly coarser than the substrate. Whereas an armour layer is characteristic of an equilibrium channel in heterogeneous sediment, a paved layer often occurs in a channel experiencing degradation, and arises because of scour and removal of a substantial thickness of bed material such that the coarsest component is left as a lag deposit. The immobility of a paved surface is often indicated by discoloration and staining of the component grains; the occasional transport of armour grains keeps them clean.                                    KSR

**Reading**
Gomez, B. 1984: Typology of segregated (armoured paved) surfaces: some comments. *Earth surface processes and landforms* 9, pp. 19–24.

**arroyo** A trench with a roughly rectangular cross-section excavated in valley-bottom alluvium with a through-flowing stream channel on the floor of the trench (Graf 1983). Although the term gully is used for similar features, a gully is V-shaped in cross-section instead of rectangular and is excavated in colluvium instead of alluvial fill. The term arroyo in the sense of a stream bed has been used in Spanish since at least the year 775, but its modern use in English physical geography dates from the exploration and survey of the American West in the 1860s. Dodge (1902) first used the term in geomorphological research, indentifying arroyos as the product of stream-channel entrenchment resulting from changes in climate and land use that caused increased run-off.                                    WLG

**Reading and References**
Cooke, R.U. and Reeves, R.W. 1976: *Arroyos and environmental change in the American South-West*. Oxford: Clarendon Press. · Dodge, R.E. 1902: Arroyo formation. *Science* 15, p. 746. · Graf, W.L. 1983: The arroyo problem – paleohydrology and paleohydraulics in the short term. In K.J. Gregory ed., *Background to palaeohydrology*. Chichester: Wiley.

**artesian** A term referring to water existing under hydrostatic pressure in a confined aquifer. The water level in a borehole penetrating an artesian aquifer will usually rise well above the upper boundary of the water-bearing rocks and may even flow out of the borehole at the surface, in which case it is known as a flowing artesian well.

Water moves in artesian aquifers as a result of differences in fluid potential (see EQUIPOTENTIALS), water moving towards areas of lower HYDRAULIC HEAD. Natural outflow points are artesian springs, where water boils up under pressure. As a consequence, the surface of an artesian spring is usually domed upwards. Artesian waters are often highly mineralized as a result of a long residence time underground. Hence artesian springs may sometimes build mounds of chemical precipitates deposited from emerging GROUNDWATER, especially if they are located in a tropical arid environment where evaporation is great.                                    PWW

**artificial recharge** See RECHARGE.

**aspect** The orientation of the face of a SLOPE. Aspect is an important control on slope MICROCLIMATE. In the northern hemisphere, for example, south facing slopes receive more solar radiation than north facing slopes. Therefore, south facing slopes tend to be hotter and drier, support different vegetation assemblages, and develop different soils. This is one cause of valley asymmetry. Aspect may also be a control on precipitation. In the mid-latitudes of the northern hemisphere, west facing slopes tend to receive more precipitation than east facing slopes, because the latter are in a RAIN SHADOW.                                    DJS

**Reading**
Birkeland, P.W. 1999: *Soils and geomorphology*. 3rd edn. New York: Oxford University Press.

**association, plant** Basic unit of classification of plant COMMUNITY in which dominant or typically co-occurring species are used as the defining characteristic. For example, the temperate deciduous forests of large areas of the Appalachians of eastern North America are characterized by an association, in this instance co-dominance, of oak and chestnut. The approach has been especially widely applied in Europe, where the BRAUN–BLANQUET SCALE system has been used to classify common plant associations at a range of spatial scales, although in North America the term is more generally applied to large scale vegetation formations regarded as CLIMAX VEGETATION.

In essence, the concept of a plant association derives from the ecologist's desire to account for and, in so doing, classify the relationships between different plant species, i.e. PHYTOSOCIOLOGY, and to explain their spatial distribution within the environment. It is, of course, impossible to represent accurately the complex totality of plant distributions, so there exists the need to simplify the situation by identifying particularly conspicuous and commonly recurring species assemblages. Fundamentally, this is a floristic approach to community classification which attempts to compensate for the fact that, while the full species complement of communities better expresses their relationships to each other, some species provide a more sensitive expression of these relationships, i.e. they typify a particular set of habitat conditions. The approach, therefore, seeks to identify patterns and emphasizes certain diagnostic species as indicators of given environments or relationships. Two kinds of diagnostic species may be identified within an association, namely: *character* species, which are centred upon, or have their dominant areas of distribution within, a particular association, as against their absence from other associations; *differential* species are present in one association but absent from most or all other associations (although they need not be centred upon the association which they define). This is illustrated in the figure, which shows how nine different plant species form associa-

tions along an environmental gradient, in this instance of moisture. Species 1 and 3 are character species for association B and, indeed, have their populations centred within that association. Species 4, on the other hand, while having its distribution centred in association A, is a differential species for sub-association Ba, and distinguishes the 'moist' group of species within association B as a whole. In this way, classification is a form of gradient analysis, since the pattern of plant associations is seen to mirror that of gradients in environmental characteristics.

When more strictly applied (following a Braun–Blanquet cover survey, for example) the approach is also a hierarchical one that parallels that used for a taxonomic classification of individual organisms. In order to arrange distribution and plant cover data in the form of a hierarchy, vegetation samples (recorded for, say, a series of 10 m × 10 m sample plots) are compiled in a table in such a way that species occurring together are grouped, revealing the associations. The hierarchy is then produced by further grouping into higher units (also defined by character species) respectively termed alliance, order, class and division. It becomes possible to use the various hierarchical levels as mapping units which are surrogate measures of environmental characteristics such as soil moisture or fertility, and in the process aid land use planning. It is worth bearing in

Diagrammatic representation of the distribution of six hypothetical plant species along a moisture gradient and the manner in which such distributions form associations. For further explanation, see text.
*Source*: From R.H. Whittaker 1975: *Communities and ecosystems*. 2nd edn and originating in Westhoff and van der Maarel 1973: *Handbook of vegetation science* 5.

mind, however, that such classifications remain fundamentally human constructions, that is simplifications, of reality and that associations are really just convenient arbitrary units that only approximate the individuality of species and community composition. The observation that a particular association appears to parallel an environmental factor does not prove that the factor is functionally important. As pointed out by the American ecologist Robert Whittaker, classification is justified by usefulness, not necessarily by theory. Studies of the Quaternary palaeoecology of plant communities have also revealed that associations of particular species may be temporary and transient over time, dependent on the complex interplay between type and rate of climate change, availability of suitable habitat and dispersal capabilities of the species involved (see Bennett 1998).          MEM

**Reading and Reference**
Begon, M., Harper, J.L. and Townsend, C.R. 1996: *Ecology: individuals, populations, communities.* 3rd edn. Oxford: Blackwell Science. Pp. 679–92. · Bennett, K. 1998: *Evolution and ecology: the pace of life.* Cambridge: Cambridge University Press. · Whittaker, R.H. 1975: *Communities and ecosystems.* 2nd edn. New York: Macmillan.

**asthenosphere**   A zone within the earth's upper MANTLE, extending from 50 to 300 km from the surface to a depth of around 700 km, characterized by a lower mechanical strength and lower resistance to deformation than the regions above and below it. It is approximately, though not exactly, equivalent to the zone in the mantle which transmits seismic waves at a low velocity, due to its partially melted state. In the PLATE TECTONICS model the asthenosphere is regarded as the deformable zone over which the relatively rigid LITHOSPHERE moves.          MAS

**Reading**
Davies, P.A. and Runcorn, S.K. eds 1980: *Mechanisms of continental drift and plate tectonics.* London: Academic Press. · Mörner, N.-A. ed. 1980: *Earth rheology, isostasy and eustasy.* Chichester: Wiley. · Wyllie, P.J. 1976: The earth's mantle. In J.T. Wilson ed., *Continents adrift and continents aground.* San Francisco: W.H. Freeman. Pp. 46–57.

**astrobleme**   A term put forward by Dietz (1961) meaning 'star wound' and referring to the erosional remnant or scar of a structure of extra-terrestrial origin produced before the Pliocene by the impact of a meteorite on the earth's surface.          ASG

**Reading**
Dietz, R.S. 1961: Astroblemes. *Scientific American* 205, pp. 51–8.

**asymmetric valley**   A river valley or glacial valley of which one side is inclined in a different angle to the other. Such valleys are a feature of periglacial areas where differences in aspect cause considerable differences in the strength of frost weathering and solifluction, but they can also be caused by structural circumstances (see UNICLINAL SHIFTING).          ASG

**Reading**
Churchill, R.R. 1982: Aspect-induced differences in hillslope processes. *Earth surface processes and landforms* 7, pp. 171–82.

**asymmetrical fold**   A fold in geological strata which has one side dipping more steeply than the other.

**Atlantic coastlines**   Coastlines where the trend of the mountain ranges is at right angles or oblique to the coastline (e.g. south-west Ireland), whereas the Pacific (or concordant) type of coastline is parallel to the general trend-lines of relief (e.g. the Adriatic coast of former Yugoslavia).          ASG

**atmometer**   An instrument for measuring the rate of evaporation.

**atmosphere**   The gaseous envelope of air surrounding the earth and bound to it by gravitational attraction. Up to a height of about 80 km the relative proportions of the major constituent gases (apart from water vapour) are more or less constant, as given in the table. The percentage volume of water vapour varies between less than 1 per cent and more than 3 per cent. Only carbon dioxide, ozone and water vapour vary locally in concentration. The atmosphere has been divided into vertical regions based on

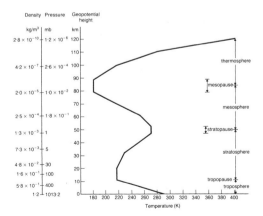

Average temperature structure of the atmosphere from 0 to 120 km.
*Source: Meteorological glossary.* 1972.

temperature as shown in the diagram. The atmosphere also contains natural pollutants in the form of AEROSOLS, dust and smoke from volcanoes, forest fires, soil erosion etc., and man-made pollutants such as sulphur dioxide, smoke, nitrogen dioxide, nitric oxide, ozone, lead and carbon monoxide.                    JET

**atmospheric composition** A unit mass of dry air is made up of 75.5 per cent nitrogen ($N_2$), 23.2 per cent oxygen ($O_2$), 1.3 per cent argon (A), 0.01 per cent carbon dioxide ($CO_2$) and smaller proportions of gases such as neon and helium. At a height of about 100 km molecular diffusion becomes comparable with mixing due to air motion and the light gases tend to float upwards. Also at these heights diatomic molecules (specially $O_2$) become split into their atomic components. Atmospheric air contains water vapour from about 4 per cent by mass at temperatures of 30 °C such as found in the tropics to 1 per cent at temperatures of 0 °C such as found in subpolar latitudes and in the middle TROPOSPHERE. Many molecules are spectacularly peculiar. Water has a large latent heat and changes state readily, consequently it plays a major role as a store of latent energy. Its high solvency encourages chemical reactions in the sea and in living matter. Carbon dioxide dissolves readily in sea water where it can be utilized by living matter ultimately to form carboniferous rocks. Otherwise it would clog up wavelengths through which terrestrial RADIATION escapes to space and lead to a hot 'run away GREENHOUSE' atmosphere like that of Venus. Ozone (at 0.0001 per cent by mass) removes virtually all solar radiation with wavelength less than 0.3 $\mu$m which would otherwise destroy living matter.                    JSAG

Composition of dry air

| | Molecular weight (12C = 12.000) | Volume (%) | Weight (%) |
|---|---|---|---|
| Dry air | 28.966 | 100.0 | 100.0 |
| Nitrogen | 28.013 | 78.09 | 75.54 |
| Oxygen | 31.999 | 20.95 | 23.14 |
| Argon | 39.948 | 0.93 | 1.27 |
| Carbon dioxide | 44.010 | 0.03 | 0.05 |
| Neon | 20.183 | 0.0018 | 0.0012 |
| Helium | 4.003 | $5.2 \times 10^{-4}$ | $7.2 \times 10^{-5}$ |
| Krypton | 83.800 | $1.0 \times 10^{-4}$ | $3.0 \times 10^{-4}$ |
| Hydrogen | 2.016 | $5.0 \times 10^{-5}$ | $4.0 \times 10^{-6}$ |
| Xenon | 131.300 | $8.0 \times 10^{-6}$ | $3.6 \times 10^{-5}$ |
| Ozone | 47.998 | $1.0 \times 10^{-6}$ | $1.7 \times 10^{-6}$ |

*Source: Meteorological Glossary 1972. London: HMSO.*

**Reading**
Goody, R.M. and Walker, J.C.G. 1972: *Atmospheres*. Englewood Cliffs, NJ: Prentice-Hall.

**atmospheric energetics** Concerns the energy content of the atmosphere and how it is changed from one form to another. Almost all energy is heat, latent heat and POTENTIAL ENERGY. THERMODYNAMIC DIAGRAMS relate these. A small fraction of the total of sensible, latent and potential energy (known as available potential energy) can be converted into the KINETIC ENERGY of the WIND.                    JSAG

**Reading**
Lorenz, E.N. 1967: *The nature and theory of the general circulation of the atmosphere*. Geneva: World Meteorological Organization.

**atmospheric instability** In general a system is unstable if an introduced disturbance increases in magnitude through time. Conversely the system is stable if the introduced disturbance is damped out through time. In the atmosphere this idea is applied at two main scales: cyclone-scale and cloud-scale.

Cyclone-scale instability is exemplified by the growth of cyclones within the extratropical westerlies. They are a manifestation of baroclinic instability. Cloud-scale instability results in vertical displacement of air parcels in an atmosphere initially in hydrostatic equilibrium. Cumulus clouds frequently result from this type of instability. (See also VERTICAL STABILITY/INSTABILITY.)                    BWA

**Reading**
Atkinson, B.W. 1972: The atmosphere. In D.Q. Bowen ed., *A concise physical geography*. Amersham: Hulton Educational. Esp. pp. 45–9. · —1981: Atmospheric waves. In B.W. Atkinson ed., *Dynamical meteorology: an introductory selection*. London and New York: Methuen. · Barry, R.G. and Chorley, R.J. 1992: *Atmosphere, weather and climate*. 6th edn. London and New York: Routledge.

**atmospheric layers** These are principally the TROPOSPHERE (or overturning layer) 0–10 km above the earth's surface, the STRATOSPHERE (or layer of constant temperature) 10–25 km, the ozonosphere (warmed through photochemistry involving oxygen) 25–60 km, the MESOSPHERE (some similarity with troposphere) 60–100 km and between 100 and 500 km the thermosphere (molecular conductivity balancing energy input), the IONOSPHERE (electrical charge on particles significant) and the exosphere (molecules liable to escape into orbit). Similar layers occur at similar values of the pressure (which is related to height) in atmospheres of other planets. Near the ground, there is a hierarchy of boundary layers: convective (containing the active regions of cumulus-scale motion) about 1 km deep, mechanical or Ekman (mixing due to mechanical stirring) about 300 m deep, logarithmic or constant flux about 10 m deep, and finally an unnamed layer penetrated by material objects such as trees, grass and waves, that interfere with the flow of air and transfer momentum, heat, moisture, salt, pollen, etc. into the atmosphere.                                    JSAG

**Reading**
Goody, R.M. and Walker, J.C.G. 1972: *Atmospheres*. Englewood Cliffs, NJ: Prentice-Hall. · McIntosh, D.H. and Thom, A.S. 1969: *Essentials of meteorology*. London: Wykeham Publications.

## atmospheric waves
An abstraction convenient for describing some phenomena closely related to the physical processes responsible for propagating characteristics through the atmosphere.

### Elastic waves
Propagate at the speed of sound (330 ms$^{-1}$) and transmit pressure pulses. They are usually of tiny amplitude but represent the practical limiting signal velocity (playing a role like that of the speed of light in classical physics).

### Gravity (-inertial) waves
Represent the action of restoring forces (gravity and Coriolis) acting because the atmosphere is stably stratified. They are analogous to the waves on the interface between two immiscible liquids when the lower is slightly denser than the upper. Short waves (gravity waves) have a characteristic period of about 600 s which is evident in many natural phenomena such as LEE WAVES and overshoot of cumulus tops. Very long waves (gravity-inertia waves) are dominated by Coriolis accelerations (GEOSTROPHIC WIND) with an upper bound to the duration of the period of $2\pi$/Coriolis parameter ($\simeq$12h); they are exemplified by cloud bands caused by mountain chains.

### Cyclone waves
Eddies in the western circumpolar flow around a hemisphere with horizontal dimensions of at most a few thousand kilometres. Although frequently quasi-circular in plan view, vertical cross-sections reveal wave forms in the temperature and pressure distribution. These waves, known also as baroclinic waves, lie within and are inextricably linked to ROSSBY WAVES.

### Rossby waves
These are very large (both wavelength and amplitude of several thousand kilometres) perturbations in the extra-tropical high atmosphere. They are usually two to six in number, encircle the globe from west to east in each hemisphere, contain the jet streams and are vital to the formation of extratropical cyclones and anti-cyclones and hence extra-tropical climate.                                    JSAG

**Reading**
Atkinson, B.W. 1981: Atmospheric waves. In B.W. Atkinson ed., *Dynamical meteorology: an introductory selection*. London and New York: Methuen. Pp. 110–15. · Eliassen, A. and Kleinschmidt, E. 1957: Dynamic meteorology. In S. Flügge ed., *Encyclopedia of physics*. Berlin: Springer-Verlag. Vol. 48, pp. 1–154.

## atoll
An annular form of CORAL ALGAL REEF consisting of an irregular elliptical reef, often breached by channels, around a central lagoon. There are over 400 atolls recorded in the world, most of which are found in tropical waters of the Indo-Pacific. Atolls vary greatly in size and shape, as well as in the depth of the central lagoon. The largest atoll is Kwajalein in the Marshall Islands ($120 \times 32$ km). As with all coral algal reefs, atolls are sensitive to fluctuations in relative sea level. Some atolls, for example Aldabra Atoll in the Seychelles archipelago, are now elevated by a few metres above present sea level; others have become drowned as their growth has failed to keep pace with changing sea level (e.g. Saya de Malha, Indian Ocean). Low, sandy islands (called cays) may form on the reef rim of atolls. Micro-atolls are rounded forms found often on reef flats, usually single colonies of massive corals less than 6 m in diameter with a flat or concave upper surface devoid of living coral. They grow preferentially where water is ponded at low tide.                           HAV

**Reading**
Guilcher, A. 1988: *Coral reef geomorphology*. Chichester: Wiley. · Woodroffe, C.D. and McLean, R. 1990: Micro-atolls and recent sea level change on coral reefs. *Nature* 244, pp. 531–4.

## atterberg limits
The results of tests (Index Tests) which, arbitrarily defined, show the properties of soils which have COHESION in that they represent changes, in state or water content, from solid to plastic to liquid materials. The Plastic Limit (PL) is the minimum moisture content at which the soil can be rolled into a thread 3 mm in diameter without breaking. The Liquid Limit (LL) is the minimum moisture content at which the soil can flow under its own weight. These are the ones most commonly used, but the Shrinkage Limit (SL) is the moisture content at which further loss of moisture does not further decrease the volume of the sample. The PL and LL are often combined to give the Plasticity Index (PI) from PI = LL–PL and the Liquidity Index (LI) from LI = (100$m$–PL)/(LL–PL),

where *m* is the natural moisture content of the soil. A chart of PI as ordinate, plotted against LL, is often used for comparing different types of soil and for classifying them. WBW

**Reading**
Mitchell, J.K. 1976: *Fundamentals of soil behaviour.* New York: Wiley. · Whalley, W.B. 1976: *Properties of materials and geomorphological explanation.* Oxford: Oxford University Press.

**aufeis** See ICING.

**auge** A hot, dry wind which blows from the south of France to the Bay of Biscay.

**aulacogens** can be thought of as a continental rifting system in which seafloor spreading proceeded for a while and then ceased. Such 'aborted pull-aparts' most commonly occur at places in continental lithosphere where three directions of sea-floor spreading are tending to occur at the same time. The common occurrence is for two of these directions to become predominant and for the third axis to become an aulacogen. ASG

**Reading**
Dewey, J.F. and Burke, K. 1974: Hot spots and continental break-up: implications for collisional orogeny. *Geology* 2, pp. 57–60.

**aureole, metamorphic** The zone of metamorphosed rock adjacent to an intrusion of igneous rock.

**aurora borealis** The 'Northern Lights'. Flashing white and coloured luminescence in the ionized layers of the earth's atmosphere about 400 km above the poles. The result of solar particles being trapped in the earth's magnetic field. The term 'aurora australis' has been applied to the phenomenon in the southern hemisphere.

**autecology** The ECOLOGY of individual organisms and of particular species. Originally used for the study of relationships between a single organism and its environment, the term is now equally widely used for the study of relationships between plants or animals of the same species, particularly species populations, and their environments (population ecology). Autecology provides the fundamental basis for understanding the distribution of organisms and their behaviour in communities, and involves the ecology of organisms at different stages of their life histories together with the environmental controls on germination, establishment, growth, reproduction, dispersal and survival. Modern autecology is an experimental science with important branches developed to physiological processes (physiological ecology) and genetics (evolutionary ecology). (See also SYNECOLOGY.) JAM

**Reading**
Bannister, P. 1976: *Introduction to physiological plant ecology.* Oxford and New York: Blackwell Scientific and Halsted-Wiley. · Daubenmire, R.F. 1974: *Plants and environment: a textbook of plant autecology.* New York and London: Wiley. · Macfadyen, A. 1963: *Animal ecology: aims and methods.* London and New York: Pitman. · Pianka, E.R. 1974: *Evolutionary ecology.* New York and London: Harper & Row. · Vernberg, F.J. and Vernberg, W. 1970: *The animal and the environment.* New York and London: Holt Rinehart & Winston.

**autochthonous** Matter which is formed or accumulates within a defined space (such as a lake or catchment) and which has not been subject to transport. The term is most frequently applied to sediments and, for example in the case of EVAPORITES, refers to the fact they accumulate *in situ* from substances available in the immediate locality. In aquatic communities, the autochthonous input of organic matter is that which is derived from photosynthesis of the locally occurring plants, either the benthic AQUATIC MACROPHYTES and ALGAE associated with the shallower water nearer the shoreline or the planktonic, mainly algal, flora of the open water. A substantial proportion of the organic matter accumulation in such situations is derived, however, from dead material formed outside the lake and transported into the system either by runoff or wind. This is known as allochthonous organic matter or sediment and its relative proportion depends on the dimensions of the water body and the kinds of terrestrial community associated with its catchment area. Some communities consist almost entirely of autochthonous material, for example the nutrient-deficient and mainly organic sediments of raised bogs (OMBROTROPHIC mires) are derived almost entirely from local sources because the level of the surface is such that inputs via runoff are not possible. Generally, the relative proportions of autochthonous to allochthonous material in a catchment increases downstream. The open ocean, for example, derives the majority of its organic and inorganic sediments from autochthonous sources. In tectonics, the term 'autochthonous' is used to describe rock formations in Alpine structures which have not been displaced by major thrusting, although they have been folded and faulted. MEM

**autocorrelation** The property of persistence in sequences of values measured over time or space. It is usually measured by comparing each value in the sequence either with its immediately previous (lag = 1) value or with the value at a fixed previous time or distance (lag > 1). For a sequence $x_1 \ldots x_n \ldots x_N$ the comparisons for lag $r (<= N)$ are between the $(N - r)$ pairs $x_n$ and $x_{(n-r)}$ as $n$ ranges from $r + 1$ to $N$. The pairs of values are then used to calculate a correlation coefficient which measures the degree of dependence for the given lag. The correlation coefficient is calculated as for a normal least-squares correlation, although in this context it is called the coefficient of autocorrelation. A correlogram may then be constructed in which the successive values of the coefficient for different lags is plotted against the lag. Comparison between correlograms for different types of sequence gives some idea of the type of persistence present, if any (for lag zero the coefficient is necessarily 1.0). A sequence of independent values clearly shows zero coefficients – that is, no autocorrelation. A simple Markov chain of values shows an exponential decline in the coefficients. For a sequence with a linear trend, the coefficient varies about a constant value; while for a regular wave form the coefficient also varies regularly over a range of positive and negative values. Linear or harmonic trends should be removed from a sequence to see whether other sources of persistence are present. Many if not most time and space sequences show some degree of autocorrelation, so care must be exercised in obtaining valid independent samples or applying parametric statistical tests.          MJK

**autogenic stream**  A stream whose flow is sustained by a continuously positive water balance along its course. Most perennial streams of humid areas are of this kind. The distinction is made between these streams and ALLOGENIC STREAMS which, for part or perhaps most of their course, traverse areas of negative water balance, and which are instead sustained by a water source in wetter headwaters. Derivation: terms from the Greek *autos* (self) and *allos* (other). Examples of the latter include the Darling River system in Australia and the lower Nile in Egypt.          DLD

**autogenic succession**  The process of community change (SUCCESSION) caused by the reaction of organisms, particularly plants, on their own environment. By the reaction mechanisms, organisms may so change their environment that other species are given a competitive advantage. The original organisms are eventually replaced, having brought about their own destruction. A.G. Tansley first used the term in the context of vegetation to distinguish this classic mechanism of succession from ALLOGENIC SUCCESSION,

which results from the action of external environmental factors independent of the organisms themselves. Consider, for example, a vegetation change in response to a change in soil pH. If this was caused by the *in situ* accumulation of acidic plant litter the succession would be autogenic, but if the cause was prolonged leaching due to heavy rainfall the succession would be allogenic. In reality the two concepts may be difficult, if not impossible, to separate. (See also COMPETITION; ECOSYSTEM.)          JAM

**Reading and References**
Botkin, D.B. 1981: Causality and succession. In D.C. West, H.H. Shugart and D.B. Botkin eds, *Forest succession: concepts and applications*. New York and Heidelberg: Springer. Pp. 36–55. · Connell, J.H. and Slatyer, R.O. 1977: Mechanisms of succession in natural communities and their role in community stability and organisation. *American naturalist* 111, pp. 1119–44. · Tansley, A.G. 1935: The use and abuse of vegetational concepts and terms. *Ecology* 16, pp. 284–307.

**autotrophic**  A descriptive term for those organisms which synthesize (usually, but not exclusively, through PHOTOSYNTHESIS) organic substances from simple inorganic source materials. Autotrophs are exemplified by the primary producer green plants in the assimilation of inorganic carbon dioxide and water into organic carbohydrates using solar radiation as the energy source to drive the process. By comparison HETEROTROPHS, consumers, must rely on organic molecules already synthesized. Autotrophs can be divided into two groups. Photoautotrophs are photosynthesizing and contain chlorophyll, the catalyst that facilitates the entrapment of energy from sunlight. Chemoautotrophs, mostly bacteria, are organisms in which energy is obtained from the oxidation of inorganic compounds *without* the use of light. Among the more remarkable of these organisms are the bacteria found in great concentrations in the ocean depths feeding, literally, on the hydrogen sulphide thrown out by hot volcanic vents. It has now been realized that there is a very substantial bacterial BIOMASS within the earth's crust as a whole, somewhat humorously accorded the acronym SLIME (subsurface lithoautotrophic microbial ecosystems) by Stevens and McKinley (1995). An autotrophic bacterial flora may well dominate the earth's biomass, and yet its existence was unheard of until the 1960s.          MEM

**Reference**
Stevens, T.O. and McKinley, J.P. 1995: Lithoautotrophic microbial ecosystems in deep basalt aquifers. *Science* 270, pp. 450–4.

**autovariation**  This term applied to the earth's environmental system has been invoked in some theories attempting to explain the causes of global CLIMATE CHANGE. As Goudie (1992, p. 272) notes 'there has been a variety of hypotheses in which it is envisaged that the atmosphere possesses a degree of internal instability which might furnish a built in mechanism of change'. Theories that suggest that positive feedback (see SYSTEMS) exist to influence global climate include the Ewing–Donn hypothesis relating to sea ice growth and albedo hypotheses whereby snowfalls persisting through the summer months increase albedo and enhance global cooling.                    DSGT

### Reference
Goudie, A.S. 1992: *Environmental change*. 3rd edn. Oxford: Clarendon Press.

**avalanche**   The sudden and rapid movement of ice, snow, earth or rock down a slope. Avalanches are an obvious and important mechanism of mass wasting in mountainous parts of the earth; they are also highly significant on subaqueous continental margins and deltas as well as in extra-terrestrial environments, for example on Mars. Avalanches occur when the shear stresses on a potential surface of sliding exceed the shear strength on the same plane. Failure is sometimes associated with increased shear stress in response to slope steepening or loading (for example, slope undercutting or snow or deltaic sediment accumulation), to reduced shear strength within the material (for example, increased PORE WATER PRESSURE or the growth of weak snow crystals), and sometimes to a combination of the two, especially when associated with an external trigger such as an earthquake.

Avalanches are commonly subdivided according to the material involved. *Snow avalanches* occur in predictable locations in snowy mountains and create distinctive ground features as they plunge down the mountain side (Rapp 1960). *Debris avalanches* involve the rapid downslope movement of sediment. On land they are commonly associated with saturated ground conditions. In subaqueous environments they reflect sediment overloading. One large example off the Spanish Sahara involved 18,000 km$^2$ of disturbance (Embley and Jacobi 1977). *Rock avalanches* are very rapid downslope movements of bedrock which become shattered during movement. These avalanches sometimes achieve velocities as high as 400 km per hour owing to the presence of trapped interstitial air; they can travel tens of kilometres from their source, sometimes with devastating effects on human life.                    DES

### Reading and References
Clapperton, C.M. and Hamilton, P. 1971: Peru beneath its eternal threat: analysis of a major catastrophe. *Geographical magazine* 43.9, pp. 632–9. (Huascaran rock avalanche.) · Embley, R.W. and Jacobi, R.D. 1977: Distribution and morphology of large submarine sediment slides and slumps on Atlantic continental margins. *Marine geotechnology* 2, pp. 205–28. · Nicoletti, P.G. and Sorriso-Valvo, M. 1991: Geomorphic controls of the shape and mobility of rock avalanches. *Bulletin of the Geological Society of America* 103, pp. 1365–73. · Perla, R.I. and Martinelli, M. 1976: *Avalanche handbook*. Agriculture handbook 489, (*Snow*). US Department of Agriculture, Forest Service. · Rapp, A. 1960: Recent development of mountain slopes in Karkevägge and surroundings, northern Scandinavia. *Geografiska annaler* 42A, pp. 71–200. · Williams, G.P. and Guy, H.P. 1971: Debris avalanches – a geomorphic hazard. In D.R. Coates ed., *Environmental geomorphology*. Binghamton: State University of New York.

**avalanche tarns**   Small water-filled depressions produced by repeated avalanche impact.

### Reading
Fitzharris, B.B. and Owens, I.F. 1984: Avalanche tarns. *Journal of glaciology* 30, pp. 308–12.

**aven**   A vertical passage or shaft which connects a cave with the surface or overlying chambers and passages.

**avulsion**   The diversion of a river channel to a new course at a lower elevation on its floodplain as a result of floodplain aggradation. It causes established meander belts to become abandoned and new ones to form.

### Reading
Smith, N.D., Cross, T.A., Dufficy, J.P. and Clough, S.R. 1989: Anatomy of an avulsion. *Sedimentology* 36, pp. 1–23.

**azimuth**   The arc of the sky extending from the zenith to the point of the horizon where it intersects at 90°.

**azoic**   Without life. Pertaining to the period of earth history before organic life evolved or to portions of the seas and oceans where organisms cannot exist.

**azotobacter**   The principal nitrogen-fixing bacteria. An aerobic bacteria which obtains energy from carbohydrates in the soil zone k.

# B

**backing wind**   See WIND.

**backshore**   The backshore of the coastal zone lies between the highest point reached by marine action and the normal high-tide level. On a low coast, the backshore zone is often in the form of a berm, above the normal reach of the tide. The berm often slopes gently landwards. On shingle BEACHES, the berm crest can attain 13 m above normal high-tide level, as on Chesil Beach, Dorset. On sandy coasts, foredunes may form in the backshore zone and washover fans are associated with barrier island backshore zones. On a steep coast, the backshore is that part of the platform and cliff foot affected by waves under storm conditions. A wave-cut notch is a common feature.

CAMK

**Reading**
Davies, J.L. 1980: *Geographical variation in coastal development*. 2nd edn. London: Longman.

**backswamp**   Low-lying marshy or swampy area on a FLOODPLAIN where overbank flood or tributary drainage water may become ponded between river levées and valley sides or other relatively elevated alluvial sediments. Such areas can be extensive where levées are well-developed and where near-channel deposits are aggrading. In settled areas such environments may be artificially drained leaving fine-grained soils which can be rich in organic materials and which may have characteristics developed under water-logged conditions.

JL

**backwall**   The arcuate cliffed head of a cirque basin or a landslide.

**backwash**   The return flow of water down a BEACH after a wave has broken. It is the return to the sea of the swash or uprush of the wave. It plays an important part in determining the gradient of the swash slope in association with the size of the beach material. On a coarse pebble beach the backwash is reduced in volume through percolation, so that a steeper slope is necessary to maintain equilibrium between swash and backwash. On a fine sand or wet beach the backwash is a large proportion of the swash, so a flat beach can remain in equilibrium. Rhomboid ripple marks are sometimes formed by backwash. Long waves and steep waves enhance the backwash and are associated with flatter swash slope gradients.

CAMK

**Reading**
Demarest, D.E. 1947: Rhomboid ripples marks and their relationship to beach slope. *Journal of sedimentary petrology* 17, pp. 18–22.

**backwearing**   The parallel retreat of a slope without a change in overall form or inclination. The term may be applied to escarpments and to side slopes: it is commonly used to contrast with down-wearing of a slope in which material is lost from the upper segments of the slope with consequent decreases in inclination. Parallel retreat implies that the resistance of rock and soil of the slope is constant into the hill which is being eroded, or that resistance does not control the slope form.

MJS

**badlands**   are generally regarded as the archetypal example of the effects of vigorous water EROSION, badlands can resemble miniature desert landscapes with weirdly shaped HOODOOS, barren steep and rounded slopes scarred by RILLS and GULLIES, and a maze of winding channels. The name likely derives from the early French explorers' expression *mauvaises terres à traverser* (meaning bad lands to cross) for such terrain in North America's western plains. While usually associated with dryland areas, badlands can form wherever weak, unconsolidated materials are exposed to periodic high intensity rainfall and rapid RUN-OFF. Badlands have formed on marine silts in Canada's high arctic and on deeply WEATHERED granites and basaltic LAVA flows in the humid tropics: they occur either naturally on weak mudstones and poorly cemented sandstones, or on industrial spoilheaps and unwisely used agricultural land. The rapidity of erosion, which can remove several millimetres of surface material in a single storm, coupled with the often infertile character of the material, inhibits plant growth, leaving the surface unprotected against heavy rains and thereby encouraging faster rates of water erosion. Badlands materials which contain swelling CLAY minerals develop DESICCATION cracks allowing percolation of water which generates subsurface flow to create an often complex network of PIPES, tunnels, and sinkholes formed by collapse. Some of these features resemble those found in limestone (KARST) areas but whereas limestone caves etc. form by chemical SOLUTION processes, in badlands they are the result of water erosion and are called PSEUDOKARST features to distinguish them from true karst landforms.                                                    IC

**Reading**

Bryan, R.B. and Yair, A., eds 1982: *Badland geomorphology and piping*. Norwich: Geo Books. · Campbell, I.A., 1997: Badlands and badland gullies. In D.S.G. Thomas ed., *Arid zone geomorphology*. Chichester: Wiley. Pp. 261–91. · Howard, A.D. 1994: Badlands. In A.D. Abrahams and A.J. Parsons eds, *Geomorphology of arid environments*. London: Chapman and Hall. Pp. 213–42.

**bajada**   Confluent alluvial and pediment fans at the foot of mountain and hill slopes encircling desert basins.

**ball lightning**   See LIGHTNING.

**bank erosion**   Removal of material from the side of a river channel. This may be accomplished by several processes: particle by particle removal following surface wash, frost heave, groundwater sapping and the dislodgement and fall of material and the subsequent entrainment of particles by flowing water; abrasion by transported ice; and mass failure of the bank. Bank erosion is usually associated with high flows, but may be at a maximum as water levels fall after a high-flow period and the lateral support provided by the deep water and the apron of previously slumped material are removed. Rates of erosion depend on prior conditions affecting bank strength, and in particular water content, which may vary seasonally. Bankside vegetation and root systems are also an important control on bank stability.

In detail bank erosion forms also reflect bank composition. In coarser materials erosion may be by particle entrainment with steep banks that have talus slopes at the base as long as such material is not picked up by the flowing water. In cohesive sediment bank failure may be in the form of shallow slips. In composite banks, with an upper cohesive fine layer above coarser materials, the process may be by removal of the underlying material accompanied by periodic collapse of overhanging 'cantilevers' of fine bank-top material. Until they are broken up and disaggregated, collapsed blocks may temporarily protect the bank base especially where given additional strength by vegetation and a root mat.

Bank erosion rates are highest where flow in river channels is asymmetric, as on the outer bank of meander bends. The exact location of such points may be a determinant in evolving channel patterns. For example, if located downstream of the bend apex, a meander loop may tend to translate downvalley rather than expand laterally. High rates of bank erosion may also relate to flow diverted by BARS in the channel, and the fact that non-cohesive banks may erode relatively rapidly may, among other factors, promote the formation of braided channel patterns (see BRAIDED RIVER).                            JL

**Reading**

Richards, K. 1982: *Rivers*. London and New York: Methuen. · Schumm, S.A. 1977: *The fluvial system*. New York: Wiley.

**bank storage**   Water retained in permeable channel-side deposits as soil and ground water. Drainage of such water by effluent seepage may contribute to the maintenance of river flows at low water, while under dry conditions there may be a downstream decrease in river discharge because of percolation into bank storage. This is particularly important in semi-arid environments, in allogenic streams, and on regulated rivers where low flow discharges are augmented by reservoir releases. A loss in the volume of water transmitted downstream has to be allowed for because of the bank storage factor.        JL

**bankfull discharge**   The river discharge which exactly fills the river channel to the bankfull level without spilling on to the floodplain, and which depends upon the definition of CHANNEL CAPACITY. Bankfull discharge has often been assumed to be a significant or critical channel-forming flow which is important in determining the size and shape of the river channel and it has therefore sometimes been regarded as equivalent to DOMINANT DISCHARGE. The frequency of occurrence of bankfull discharge varies considerably and has been quoted as ranging from 4 months to 5 years in Scotland, 1.3 to 14 years in lowland England and 1 to 30 years in twenty-eight basins in the western USA but care should be taken in quoting such recurrence intervals because at least sixteen different methods are available (Williams 1978) to calculate bankfull discharge.                                             KJG

**Reading and References**

Nixon, M. 1959: A study of bankfull discharges of the rivers of England and Wales. *Proceedings of the Institute of Civil Engineers* 12, pp. 157–74. · Williams, G.P. 1978: Bankfull discharge of rivers. *Water resources research* 14, pp. 1141–54.

**banner cloud**   Often seen to extend downward from isolated peaks rather like a flag. The classic example is that on the Matterhorn. The mechanism of the cloud is not yet understood.        BWA

**bar, nearshore**   A longitudinal sediment ridge developed in or adjacent to the surf zone as a result of WAVE action. Nearshore bars may occur individually, or in sets of up to thirty. Troughs separate them from the foreshore and from each other. Where multiple bars are present, their height and spacing usually increases with distance offshore. The presence of a nearshore bar is often associated with the development of a STORM BEACH, and the form may be ephemeral. On other coasts, however, bars may be in long term morphodynamic EQUILIBRIUM (Greenwood and Davidson-Arnott 1979), changing position and form only moderately through time.

There are three commonly recognized bar forms: parallel (to the beach) bars that are linear to sinuous; crescentic bars that are convex shoreward; and transverse bars that are attached to the foreshore. Inter-tidal ridge and runnel systems are also frequently described as a fourth type of nearshore bar. Longshore currents are common in troughs, and RIP CURRENTS are associated with transverse bar systems.

The location and geometry of subaqueous bars is dependent upon complex feedback between nearshore hydrodynamics, bar morphology, and sediment characteristics. Several models have been developed for the prediction of bar behaviour, but none have been found to be successful for all nearshore environments (Holman and Sallenger 1993). DJS

**References**
Greenwood, B., and Davidson-Arnott, R.G.D. 1979: Sedimentation and equilibrium in wave-formed bars. *Canadian journal of earth science* 16, pp. 312–32. · Holman, R.A., and A.H. Sallenger, Jr 1993: Sand bar generation: a discussion of the Duck Experiment series. *Journal of coastal research* SI 15, pp. 76–92.

**barchan** A crescent-shaped sand dune, in which the horns point down wind in the direction of dune migration. Barchans are a type of TRANSVERSE DUNE that tends to form where either sand transport rates are high and/or sand supply is limited, thus they frequently occur on the margins of sand seas and migrate over non-sandy surfaces. Barchan dunes are not necessarily a very common type of dune, but the frequent simplicity of form has made them the source of investigations into aeolian sand transport on dune forms. (See DUNE.) DSGT

**Reading**
Hastenrath, S. 1987. The barchan sanddunes of south Peru revisited. *Zeitschrift fur Geomorphologie* NF 31, pp. 167–78. · Howard, A.D., Morton, T.B., Gad-El-Hak, M. and Pierce, D.B. 1978. Sand transport model of barchan dune equilibrium. *Sedimentology* 25, pp. 307–38.

**barchanoid ridge** A type of TRANSVERSE DUNE that has the appearance of a series of linked BARCHAN dunes, perpendicular to the resultant sand transport direction. DSGT

**baroclinicity** A measure of the fractional rate of change of density in the horizontal. It is also frequently defined as a state where surfaces of equal pressure and surfaces of equal density intersect. In zones of large baroclinicity spontaneous generation of new motion systems is likely: primary depressions (see CYCLONE) in the baroclinic zone of middle ($30$–$60°$) latitudes, secondary depressions in the frontal zones of the primary depressions, and thunderstorms in the fronts of the secondary depressions. Large baroclinicity also defines the regions of transition between air masses, a concept which has lost favour in recent years with the recognition that these regions are often the result, rather than a cause, of the air motion in the air masses. JSAG

**barometer** An instrument for measuring atmospheric pressure. There are two types: mercury and aneroid. A mercury barometer works on the principle that atmospheric pressure is sufficient to support a column of mercury in a glass tube. As the pressure varies, so does the height of the column. An aneroid barometer comprises a series of vacuum chambers with springs inside to prevent total collapse. Atmospheric pressure changes are recorded by the compression or expansion of the chambers. This mechanism is frequently used in the barograph, which is a barometer that gives a graph of pressure changes through time. BWA

**Reading**
Meteorological Office 1956: *Handbook of meteorological instruments. Part I Instruments for surface observations*. London: HMSO.

**barranca** A steep-sided gully or ravine.

**barrier island** Barrier islands are elongated, mainly sandy features parallel to the coast and separated by a lagoon. They are not attached at either end, and are often separated by tidal inlets. Barrier islands make up 10–15 per cent of all the world coastlines. They normally consist of a berm in front of dune ridges, and are backed by washover fans and tidal marshes in the lagoon. They occur worldwide, but are particularly common in low to middle latitudes with low to moderate tidal range and swell wave regimes. They tend to migrate landwards on coasts where sea level is rising slowly, by washover under storm conditions. They are common on gently sloping coasts with wide shelves and coastal plains. The east coast of the USA provides good examples. CAMK

**Reading**
Hoyt, J.H. 1967: Barrier island formation. *Geological Society of America bulletin* 78, pp. 1125–36. · Oertel, G.F. and Leatherman, S.P. eds 1985: Barrier islands.

*Marine geology* 63, pp. 1–396. · Schwartz, M.J. ed. 1973: *Barrier islands*. Stroudsburg, Penn.: Dowden, Hutchinson and Ross. · Swift, D.J.P. 1975: Barrier island genesis: evidence from the central Atlantic shelf, eastern U.S.A. *Sedimentary geology* 14, pp. 1–13.

**barrier reef**   One of the whole suite of coral reef forms. It is characterized by the presence of a lagoon, or body of water, in between the reef and its associated coastline. This feature differentiates a barrier from a fringing reef. Charles Darwin identified barrier reefs as occupying the mid position of a developmental sequence of reefs caused by the submergence of the central land mass, or island. Fringing reefs form the start of this sequence and atolls the end. Like other coral reefs, barrier reefs are generally limited in occurrence to the tropical latitudes and are found on sediment-free coasts. Good examples of barrier reefs are found off the coast of Belize, and around the island of Mayotte, Comores Islands. Most barrier reefs are up to a kilometre in width and their lagoons are often up to a few kilometres wide. The Great Barrier Reef forms the largest stretch of reefs in the world, being almost 2000 km long, and is actually a compound form of many different reefs. It is backed by a lagoonal sea tens of kilometres wide.                                           HAV

**barrier spit**   The part of a barrier island system that is attached to an eroding headland. It is generally parallel to the coast and separated from it by a lagoon. An inlet, which is often tidal, separates a barrier split from a detached barrier island. One theory, first put forward by G.K. Gilbert, suggests that barrier spits are the first stage of barrier island formation. They are fed by longshore drift of material derived from the eroding headland, and are generally formed mainly of sand. Some barrier spits may be attached to older beach ridges, but one end of the elongated feature is always attached to the mainland, while the other is free but associated with a barrier island.                          CAMK

**Reading**
Carter, R.W.G. 1988: *Coastal environments*. London: Academic Press. · Gilbert, G.K. 1880: *Lake Bonneville*. US Geological Survey monograph 1, pp. 23–65. · Schwartz, M.L. 1971: The multiple causality of barrier islands. *Journal of geology* 79, pp. 91–4.

**barysphere**   The part of the earth's interior (i.e. core, mantle, and asthenosphere) which lies beneath the lithosphere.

**basal complex**   Describes the rocks which lie beneath the Pre-Cambrian shields that make up large portions of the continents.

**basal ice**   A relatively thin layer (with a vertical extent of up to tens of metres) of debris-rich ice present at the base of glaciers, which is produced at and interacts with the glacier bed. As a result of its different mode and environment of formation it may differ from 'normal' glacier ice in terms of its overall extent and properties (including its rheology), debris, solutes and gases.                                           ASG

**Reading**
Hubbard, B. and Sharp, M. 1989: Basal ice formation and deformation: a review. *Progress in physical geography* 13, pp. 529–58.

**basal sapping**   The process whereby a slope is undercut at the base. It can describe the retreat of a hillslope or escarpment by erosion along a spring line (mid-latitudes) by salt weathering (arid environments) or by glacial erosion (CIRQUE).

**basal sliding**   The process by which a glacier slides over bedrock. It is distinguished from processes associated with the internal deformation of ice within the body of a glacier. Two important processes of sliding are *enhanced basal creep*, where pressures caused by irregularities on the bed increase stresses within the ice and allow it to deform round the obstacles and *regelation*, where the pressures induced by the bed obstacles cause ice to melt on the upstream sides, flow round the obstacles as water, and refreeze on to the glacier on the downstream sides. Basal sliding is closely related to the presence and nature of water at the ice–rock interface. If no water is present basal sliding is restricted to enhanced basal creep and the ice–rock interface remains immobile. Water allows movement at the ice–rock interface and the higher the water pressure the quicker the rate of sliding. Glacier sliding takes place in a stick-slip fashion with periodic movements of a few centimetres, and probably relates to the impeding effect and sudden release of small frozen patches at the ice–rock interface. A recently recognized additional form of basal sliding incorporates the deformation of weak sediments beneath the weight of a glacier.                        DES

**Reading**
Paterson, W.S.B. 1981: *The physics of glaciers*. Oxford: Pergamon. Chapter 7.

**basalt**   A fine-grained and dark-coloured igneous rock. The lava extruded from volcanic and fissure eruptions. Rocks composed primarily of calcic plagioclase and pyroxene with or without olivine.

**base exchange**  (or CATION EXCHANGE) A vital soil reaction whereby bases or cations such as calcium, magnesium, sodium and potassium are made available as plant nutrients. These cations are usually loosely bonded on the surface of clay and organic colloidal particles in the soil complex, and cation exchange takes place when hydrogen cations, derived from organic decomposition and atmospheric sources, replace the metal cations, releasing the latter into the soil water around the root hairs, where the plant may absorb them.                                    KEB

**Reading**
Pears, N. 1985: *Basic biogeography*, 2nd edn. London and New York: Longman. · Trudgill, S.T. 1988: *Soil and vegetation systems*, 2nd edn. Oxford: Oxford University Press.

**base flow**  A term often used to describe the reliable background river flow component from a drainage basin. Base flow, or delayed flow, was originally considered to be the result of effluent seepage from groundwater storage but in some drainage basins this sustained component of flow is also produced by interflow from zones above the water table and particularly by throughflow or lateral drainage through the soil. The DEPLETION CURVE or base flow recession curve describes the way in which base flow discharge recedes at a particular site during a period without rainfall.                        AMG

**base level**  The lower limit to the operation of subaerial erosion processes, usually defined with reference to the role of running water. The level of the sea surface at any moment acts as a general base level for the continents, although there can be a wide range of local base levels, some above and some below sea level. The term was first used by J. W. Powell (1875) who defined it in a very broad fashion to include: first, the concept of the sea as 'a grand base-level, below which the dry lands cannot be eroded'; secondly, the existence of local or temporary base levels 'which are the levels of the beds of the principal streams which carry away the products of erosion'; and, finally, as 'an imaginary surface, inclining slightly in all its parts towards the lower end of the principal stream draining the area'. W.M. Davis (1902) considered that the breadth of Powell's definition had led to variety of practice and confusion. He therefore proposed that base level be restricted to 'simply...the level base with respect to which normal sub-aerial erosion proceeds'. This suggestion has largely been adopted.

Base levels can be identified on a number of different scales. A stream channel acts as the base level for processes on the adjacent slope. A major stream is the base level for the courses of its tributaries. A lake or a reservoir or a waterfall is the base level for the entire basin upstream. Clearly, these base levels can and will alter over time, with implications for the operation of subaerial erosional processes. By and large a fall in base level creates an increase in potential energy, by increasing the total relief, and may result in an acceleration of rates of erosion and down cutting. A rise in base level, on the other hand, reduces relief and potential energy and is frequently marked by aggradation.

One major feature of the Pleistocene has been the frequency with which the 'grand base-level' of the sea has fluctuated under the combined influence of EUSTASY and ISOSTASY. These fluctuations have had very complex repercussions in many drainage basins, even those where the effects of direct glaciation or tectonism are absent. An experimental study of the COMPLEX RESPONSE to a single change in base level is provided by Schumm and Parker (1973), who demonstrate that *two* sets of paired terraces may be produced in the lower reaches of a drainage basin. Such experiments must cast real doubt on many geomorphological studies which have considered that a separate base level change is required to account for each set of paired terraces observed.                        BAK

**Reading and References**
Davis, W.M. 1902: Base-level, grade and peneplain. *Journal of geology* 1, pp. 77–111. · Powell, J.W. 1875: *Exploration of the Colorado River of the West and its tributaries*. Washington: Government Printing Office. · Schumm, S.A. and Parker, R.S. 1973: Implications of complex response of drainage systems for Quaternary alluvial stratigraphy. *Nature (physical science)* 243, pp. 99–100.

**base saturation**  The condition arising when the cation exchange capacity of a soil is saturated with exchangeable bases – calcium, magnesium, sodium and potassium – when expressed as a percentage of the total cation exchange capacity.

**basement complex**  A broad term for the older rocks, usually Archean igneous and metamorphic rocks, which underlie more recent sedimentary rocks in any region. Such rocks are a feature of the ancient shield areas of the earth's surface.

**basic rocks**  Igneous rocks containing less than about 55 per cent silica, basalt a typical example (see ACID ROCKS).

**basin-and-range**  A type of terrain, as found in Utah and Nevada, where there are fault block mountains interpersed with basins. The basins often contain lakes, some of which may only fill up in pluvials.

**batholith** A large mass of intrusive igneous rock which extends to great depth and can cover or underlie very large areas.

**bathymetry** The measurement of water depth.

**bauxite** The main ore of aluminium. An impure aluminium hydroxide associated with the clay deposits of weathering zones, especially in tropical regions.

**beach** A coastal accumulation of varied types of sediment, usually of sand size or above. The sediment is derived from rivers and other sources and is moved by tides and waves to form a beach. Beaches have characteristic profile forms, which are determined by the steepness of the waves and the size of the sediments. These profiles may be highly changeable and many beaches display either swell or storm profiles at different times of year. Within the beach profile several different zones exist. These zones are called the breaker, surf and swash zones. These zones are related to the changes occurring to waves as they hit the shoreline. Within each zone are found different features, such as BERMS, BEACH RIDGES, and BARS (see Komar 1976 for a detailed description of different coastal profiles). Beaches also possess long profile shapes which are influenced by the nature of the coastline. As sea level changes, beaches may become elevated above sea level, producing RAISED BEACHES. HV

**Reading and Reference**
Carter, R.W.G. 1988: *Coastal environments*. London: Academic Press. · Hardisty, J. 1990: *Beaches form and process*. London: Unwin Hyman. · Komar, P.D. 1976: *Beach processes and sedimentation*. Englewood Cliffs, NJ: Prentice Hall.

**beach ridge** Accumulations of sediment forming a prominent feature on many beaches. Beach ridges can take a wide variety of forms.

Ridges may be formed near the top of shingle or sandy beaches (in sand these are called BERMS). Ridge features may also be formed in association with runnels near low-tide level on shallow gradient beaches. Multiple ridge forms often occur when sediment has been left by successive storm events. HV

**Reading**
Komar, P.D. 1976: *Beach processes and sedimentation*. Englewood Cliffs, NJ: Prentice Hall.

**beach rock** Beach rock is a sedimentary rock, or consolidated chemical sediment, which forms in the intertidal zone on beaches, most notably in the tropics. Beach rock may also develop along extra-tropical coastlines, such as around the Mediterranean and the Red Sea. It forms where a layer of beach sand and other material becomes consolidated by the secondary deposition of calcium carbonate at about the level of the water table. The cementing material may be aragonite or calcite and may come from groundwater or sea water (Stoddart and Cann 1965). Biochemical precipitation of calcium carbonate by micro-organisms may also be involved. Beach rock often forms relatively quickly, and is forming today in some areas as evidenced by the inclusion of recent human artefacts in the consolidated layer. Once formed, beach rock is relatively persistent, but often contains a suite of erosional features zoned according to their position relative to sea level. HAV

**Reference**
Stoddart, D.R. and Cann, J.R. 1965: Nature and origin of beachrock, *Journal of sedimentary petrology* 35, pp. 243–7.

**beaded drainage** A series of small pools connected by streams. The pools result from the thawing of ground ice and may be 1–3 m

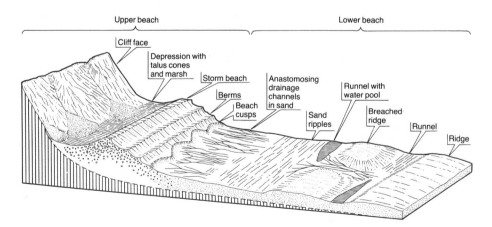

The idealized features of a sand and shingle beach.
*Source*: A.S. Goudie 1984: *The nature of the environment*. Oxford and New York: Basil Blackwell. Figure 8.4.

deep and up to 30 m in diameter (see THERMO-KARST).

**Reading**
Hopkins, D.M., Karlstrom, T.N.V., Black, R.F., Williams, J.R., Péwé, T.F., Fernald, A.T. and Muller, E.H. 1955: Permafrost and ground water in Alaska. *US Geological Survey professional paper* 264–F, pp. 113–47.

**bearing capacity** of a soil is the value of the average contact pressure between a foundation and the soil below it which will produce a (shear) failure in the soil. Strictly, this is the 'ultimate' bearing capacity; in soil mechanics a 'maximum safe' value is used, which is the ultimate value divided by a factor of safety. Various methods are available to enable calculation of appropriate values, according to the soil properties and the likely mode of failure.          WBW

**Reading**
Smith, G.N. 1974: *Elements of soil mechanics for civil mining engineers*. London: Crosby Lockwood Staples.

**Beaufort scale** Admiral Sir Francis Beaufort formalized a scale of WIND based on its effect on a man-of-war. The scale was later adapted for use on land. It was still internationally useful in 1946, when it was argued that the anemometers in common use were unable to register the shorter-term gust capable of destroying a ship.

JSAG

**Reading**
List, R.J. ed. 1951: *Smithsonian meteorological tables*. 6th revised edn. Washington: Smithsonian Institution. P. 119. (Contains full listing of the scale.)

**bed roughness** The surface relief at the base of a fluid flow, as on the bed of a river channel. This may consist of several elements: particle roughness, commonly defined with reference to size of larger particles in relation to the depth of fluid flow, form roughness produced by bedforms and the distorting effects of channel bends and 'spill' resistance, as where rapid

Reasonable current specifications for the Beaufort wind force scale

| Beaufort no. | Descriptive term | Windspeed (knots) | | Effect on sea surface |
| | | Mean | Limits | |
|---|---|---|---|---|
| 0 | Calm | 0 | <1 | Sea like a mirror. |
| 1 | Light air | 3 | 1–4 | Ripples with the appearance of scales, no foam crests. |
| 2 | Light breeze | 7 | 5–8 | Small wavelets, crests have glassy appearance and do not break. |
| 3 | Gentle breeze | 11 | 9–12 | Large wavelets, crests begin to break, perhaps scattered white horses. |
| 4 | Moderate breeze | 15 | 13–16 | Small waves becoming longer, fairly frequent white horses. |
| 5 | Fresh breeze | 19 | 17–21 | Moderate waves, many white horses, chance of some spray. |
| 6 | Strong breeze | 24 | 22–26 | Large waves form, white foam crests extensive, probably some spray. |
| 7 | Near gale | 29 | 27–31 | Sea heaps up and white foam from breaking waves blown in streaks. |
| 8 | Gale | 34 | 32–36 | Moderately high waves of great length, edges of crests begin to break into the spin-drift, foam blown in streaks. |
| 9 | Strong gale | 39 | 37–42 | High waves, dense streaks of foam, crests of waves begin to topple, tumble and roll over. |
| 10 | Storm | 45 | 43–48 | Very high waves with long overhanging crests, sea surface takes white appearance, visibility affected by spray. |
| 11 | Violent storm | 52 | 49–55 | Exceptionally high waves, sea completely covered by long white patches of foam, everywhere edges of wave crests blown into froth. |
| 12 | Hurricane | | >55 | Air filled with foam and spray, sea completely white with driving spray, visibility seriously affected. |

*Source*: Mather, J.R. 1987: Beaufort wind scale. In J.E. Oliver and R.W. Fairbridge eds, *The encyclopedia of climatology*. New York: Van Nostrand Reinhold. Table 2, p. 162.

changes occur where flow spills around protruding boulders. Velocity formulae (see FLOW EQUATIONS), relating flow velocity to the hydraulic radius and slope of river channels for UNIFORM STEADY FLOWS, incorporate a roughness coefficient (Manning's *n*), or related Darcy-Wiesbach friction coefficient or CHÉZY EQUATION. Sometimes roughness coefficients are estimated simply from the grain size on the bed, but it has to be remembered that this size may be variably effective at different depths and discharges, as may other forms of roughness or resistance.                                    JL

**Reading**
Graf, W.H. 1971: *Hydraulics of sediment transport.* New York: McGraw-Hill. · Richards, K. 1982: *Rivers.* London and New York: Methuen.

**bedding plane** The interface between two strata of sedimentary rocks, often a plane of weakness between two such strata.

**bedforms** Features developed by fluid flow over a deformable bed, as developed by wind on a bed of sand or streamflow over alluvial sediments. These forms may vary in size from small-scale ripples to larger BARS or DUNES. A hierarchy of forms may be present at any one place superimposed on one another and possibly related to different formative flows. The dimensions of bedforms may relate to flow magnitudes, as is particularly evident when comparing the giant 'ripples' tens of metres in length in gravel-sized material produced by the catastrophic draining of the Pleistocene ice-dammed Lake Missoula in North America with the ripples of dimensions in centimetres developed in sand in many shallow flows.

In flume experiments with sands, sequences of bedforms have been shown to develop with increasing stream power, bed shear stress and sediment transport rate. An initial PLANE BED with little sediment movement develops into one with ripples and then dunes of increasing dimension; there then follows re-establishment of a transitional plane bed phase under conditions of high sediment transport, finally with standing waves and antidunes developing into a chute and pool morphology. In detail the changes observed may be complex. Overlap of bedform types under given conditions of flow and sediment transport occurs. In part this may depend on the existing degree of development of bedforms of particular types. Under field conditions, while similar features and developments may be observed, things become even more varied, because of discharge variability and lag effects and variable flow conditions across channels, among other factors.

The existence of bedforms may be important in generating flow resistance and variability of channel flow conditions. This can lead, for example, to flow separation and the creation of localized areas of slackwater or upstream flow. These may be particularly important for flora and fauna, so that it may be desirable in channelized rivers to allow for or mimic such environments where bedforms are absent.                                    JL

**Reading**
Allen, J.R.L. 1970: *Physical processes of sedimentation.* London: Allen & Unwin. · — 1984: *Sedimentary structures.* Amsterdam: Elsevier. · Collinson, J.D. and Lewin, J. eds 1983: *Modern and ancient fluvial systems.* Special publication no. 6, International Association of Sedimentologists. Oxford: Blackwell Scientific.

**bedform reconstitution** The time that it takes for a bedform to move one wavelength in the direction of net transport (Allen 1974).
                                    DSGT

**Reference**
Allen, J.R.L. 1974: Reaction, relaxation and lag in natural sediment systems: general principles, examples and lessons. *Earth science reviews* 10, pp. 263–342.

**bedload, bedload equation** Fluid-transported sediment that moves along or in close proximity to the bed of the flow. This movement of the heavier and larger particles may be by rolling, sliding or saltation (the last being movement in a series of hops resulting from grains impacting on one another). For a given transporting flow the particles that move in this way are ones that are too heavy to be kept suspended in the flow itself since their fall velocity is greater than the upward velocity component of the turbulent fluid.

For rivers, bedload usually amounts to less than 10 per cent of total sediment transport, though higher proportions have been reported for mountain streams. Measurements available are, however, relatively few and possibly unreliable. The problem is that any device inserted in flowing water to trap just that sediment moving at or near the bed itself interferes with the pattern of fluid flow and sediment movement. Techniques include permanent traps or slots in the bed, both ones which are periodically emptied and therefore assess just the total amount of sediment transport in the sampling period, and ones which include a conveyor system or weighing device which samples sediment continuously (e.g. Leopold and Emmett 1976). Other portable basket-, bag- or pan-type samplers may be lowered onto the bed and then retrieved. These measurements are liable to varying and often unknown error (especially through faulty positioning on the bed) and trapping efficiency, but some long-term series of observations on European and North American rivers are now available. These show that even for a given discharge, bedload transport rates are unsteady and uneven across streams; it is also well known that bed material (which may include an additional component of suspension load) may accrete and move discontinuously in the form of migratory BARS. When bedload movement is in this form, it has been suggested that transport rates may be approximated by the volumetric transfer rate of such BEDFORMS over time.

Bedload movement has also been studied using tracers and the acoustic monitoring of inter-particle collisions; a particular aim may be to ascertain the hydraulic conditions at which bed material starts to move. Difficulties in measuring bedload transport, both practical and conceptual, have led to the development of a series of empirically calibrated bedload equations. Given certain information (for example, stream velocity, discharge or stream power and measures of bed material size and sorting), bedload transport rates may be calculated rather than directly measured. Some equations involve prediction of transport rates in terms of 'excess' shear or power above the threshold value at which transport starts. Some involve computation of total sediment transport rates rather than the bedload alone. From a practical point of view the problem is often that such equations may have to be used in conditions beyond those for which they have been designed and calibrated and, in practice, equations may give very different estimates from one another. None the less, properly used, such estimates have proved very useful as an aid to the design of reservoirs and other engineering works on rivers where bedload movement may be a practical problem.                                       JL

### Reading and Reference
Graf, W.H. 1971: *Hydraulics of sediment transport.* New York: McGraw-Hill. · Leopold, L.B. and Emmett, W.W. 1976: Bed load measurements, East Fork River, Wyoming. *Proceedings of the National Academy of Sciences* 73, pp. 1000–4. · Richards, K. 1982: *Rivers.* London and New York: Methuen.

**bedload pit trap** An excavation made in the bed of a stream in order to intercept part or the whole of the bedload delivered to the site, in order to be able to measure the size characteristics of the particles being carried and also the rate of sediment transport. The pit needs to be sufficiently long that grains cannot jump across it. Bedload traps may range from passive collecting pits lined with a durable material, and excavated during times of low flow to retrieve and

weigh the trapped material, to recording traps which may contain at their base a bladder which develops increasing internal pressure as bedload falls in upon it (Reid *et al.* 1980). This pressure can then be continuously recorded by running a small pressure tube to a transducer housed on the bank. Generally, pit traps only sample a width of channel of about 1 m, and the data derived from this are scaled-up to estimate the sediment load being carried within the whole bank-to-bank width of the stream.           DLD

### Reference
Reid, I., Layman, J.T. and Frostick, L.E. 1980: The continuous measurement of bedload discharge. *Journal of hydraulic research* 28, pp. 243–9.

**bedload tracers** are markings or other devices applied to bedload grains collected from a study stream, commonly pebbles and larger particles, which are then returned to the channel at known positions. Following a sediment transport event, the bed is searched for these marked grains, whose distance of travel can then provide useful information on stream competence and sediment transport mechanics (Laronne and Carson 1976). Tracers have included durable coloured paints, fluorescent paints searched for under ultraviolet light, and ferrous metal or ceramic magnets attached to the stone, or embedded within it, that are relocated using a metal detector. Some studies have employed many thousands of tracer particles that have been tracked for periods of years.                             DLD

### Reference
Laronne, J.B. and Carson, M.A. 1976: Interrelationships between bed material morphology and bed-material transport for a small, gravel-bed channel. *Sedimentology* 23, pp. 67–85.

**bedrock** The consolidated, unweathered rock exposed at the landsurface or underlying the soil zone and unconsolidated surficial deposits.

**Beer's law** establishes the linear relationship between the absorbency of light energy by a gas and the concentration of the gas, for a specific wavelength of energy. The equation is: $A(l) = a(l)lc$; where $A(l)$ is absorbency at a specific wavelength, $a(l)$ is the absorptivity coefficient for the gas at the specified wavelength, $l$ is the path length through the gas, and $c$ is the concentration of the gas. Beer's law is used by climatologists to estimate the transmittance of solar radiation through gases in the atmosphere for use in energy balance studies and climate modelling.                                   NJS

**benchmark** A surveyor's mark, usually cut on rock, a kerbstone or other relatively permanent structure, indicating a point of reference for levelling or surveying. Two types of benchmark can be defined; temporary benchmarks are often established when surveying a small area, but the height of these temporary benchmarks relative to the national height datum can only be accurately determined with reference to a national benchmark.                                        DJN

**benioff zone** The zone or plane of earthquake foci beneath some continental margins. The inclined plane dips deeper on the island of the continental margin.

**benthic** Pertaining to plants, animals and other organisms that inhabit the floors of lakes, seas and oceans.

**berg wind** A type of FÖHN wind blowing, mainly in winter, off the interior plateau of South Africa, roughly at right angles to the coast.

**Bergeron-Findeisen mechanism** See CLOUD MICROPHYSICS.

**berghlaup** An Icelandic term (literally, rock-leaping) which is sometimes used to refer to a large fallen earth or rock mass. No genetic interpretation is usually intended and it may be considered as an equivalent of bergsturz.
                                        WBW

**Bergmann's rule** states that, all other things being equal, an individual member of a species of warm-blooded creatures will have a greater body size in colder climates. Although not all species 'obey' this rule, it does appear to apply fairly widely amongst carnivores that have a wide geographical distribution (Klein 1984). Where this has been demonstrated through investigation of modern living animals, it has proved a useful palaeoenvironmental tool for studying temperature changes in the late Quaternary, where the remains of a particular species occur through sediments of different ages. In southern Africa, the principle of climatic controls on animal size variations has been extended for different species to identify links with not only temperature but precipitation changes (e.g. Avery 1983).                      DSGT

References
Avery, D.M. 1983. Palaeoenvironmental implications of the small Quaternary mammals of the fynbos region. In H.J. Deacon, Q.B. Hendry and J.J.N. Lambrechts eds, *Fynbos Palaeoecology: a preliminary synthesis*. SA National Scientific Progress Report 75, pp. 139–55. · Klein, R.G. 1984. The large mammals of southern Africa. In R.G. Klein ed., *Southern African prehistory and palaeoenvironments*. Rotterdam: Balkema. Pp. 107–46.

**bergschrund** The crevasse occurring at the head of a cirque or valley glacier because of the movement of the glacier ice away from the rock wall.

**berm** A ridge of sand parallel to the coast-line, commonly found on the landward side of steeply sloping beaches. It is a nearly horizontal feature formed by deposition at the upper limit of the swash zone. When steep beaches are transformed into more shallow gradient beaches, due to a change in wave regime, the berm is removed and a long-shore bar deposited just below low-tide level.                      HV

**Bernoulli's theorem and effect** describe the relationship between the pressure in a fluid flow in the direction of flow (dynamic pressure) and at right angles to it (static pressure).
   The theorem states that the sum of these two pressures is equal to the local HYDROSTATIC PRESSURE. The dynamic pressure exceeds the hydrostatic pressure by an amount $\rho u^2/2$ where $u$ is the local flow velocity and $\rho$ is the density of the fluid. The static pressure is thus less than the hydrostatic pressure by the same amount. The Bernoulli effect is produced where flow velocities differ laterally. The difference in static pressures tends to push objects towards the streams of more rapid flow, e.g. blowing between two oranges which are hanging so that they almost touch, causes them first to swing in to hit each other.                      MJK

**beta diversity** A measure of the degree of change in species composition of biological communities in relation to variations in local environments within a landscape. Beta or between-habitat diversity can be distinguished from alpha or within-habitat diversity, which measures the number of species within a single community. Landscapes in which the species composition changes rapidly in relation to variations in factors such as elevation, soil moisture, and degree of disturbance will have higher beta diversities than landscapes which have an absence of species changes along such habitat gradients.                      ARH

Reading
Whittaker, R.H. 1975: *Communities and ecosystems*. 2nd edn. London: Collier Macmillan. Chapter 4, pp. 111–91.

**Bhalme and Mooley drought index** One of many available numerical indexes designed for the assessment of DROUGHT duration and drought severity. The Bhalme and Mooley index was developed in the context of monsoonal conditions in India, but has been applied quite

| Bhalme and Mooley index value | Descriptive term |
|---|---|
| $\geq 4$ | Extremely wet |
| 3 to 3.99 | Very wet |
| 2 to 2.99 | Moderately wet |
| 1 to 1.99 | Slightly wet |
| 0.99 to −0.99 | Near normal |
| −1 to −1.99 | Mild drought |
| −2 to −2.99 | Moderate drought |
| −3 to −3.99 | Severe drought |
| $\leq -4$ | Extreme drought |

Source: Bhalme and Mooley 1980.

widely, and yields a score within the range +4 to −4, with negative values indicating drought and positive values indicating wetter than normal conditions (and hence, flood risk). This resembles the range of values generated by the Palmer index, another commonly used drought index. The values and associated descriptive terms are shown in the table.

The expression used to derive the index is

$$I_k = 0.50 I_{k-1} + (M_k/48.55)$$

where $I_k$ is the drought intensity of the $k$th month in a period of record, and $M_k$ is an index of moisture deficit (rainfall deficiency) calculated in relation to the long-term average rainfall and the COEFFICIENT OF VARIATION for the month $k$. For details of the calculation procedure, see Bhalme and Mooley (1980).     DLD

**Reference**

Bhalme H.N. and Mooley D.A. 1980: Large-scale droughts/floods and monsoon circulation. *Monthly weather review* 108, pp. 1197–211.

**biennial oscillation** See MACROMETEOROLOGY.

**bifurcation ratio** The ratio $(B_b = \sum n)/\sum(n+1)$ of number of streams of a particular order $(\sum n)$ to number of streams of the next highest order $\sum(n+1)$. It is therefore dependent upon techniques of stream ordering and gives an expression of the rate at which a stream network bifurcates. It has been correlated with hydrograph parameters and sometimes with sediment delivery factors. In a fifth Strahler order drainage basin, four values of $R_b$ could be calculated and so a weighted mean bifurcation ratio ($WR_b$) was recommended by Schumm (1956) as:

$$WR_b = \frac{\sum[R_{b_{n:n+1}} \times (N_n + N_{n+1})]}{N}$$

(See also ORDER, STREAM.)     KJG

**Reading and Reference**

Gregory, K.J. and Walling, D.E. 1976: *Drainage basin form and process*. London: Edward Arnold. · Schumm,

S.A. 1956: The evolution of drainage systems and slopes in badlands at Perth Amboy, New Jersey. *Bulletin of the Geological Society of America* 67, pp. 597–646.

**bioaccumulation** The build-up in concentration of a toxic substance in the body of an individual organism over time. Bioaccumulation occurs when an organism is unable to excrete the substance or break it down within the body, so that the substance remains in the body and its concentration increases each time an additional amount of the substance is taken in. Repeated exposure to toxic, conservative pollutants such as heavy metals and pesticide residues can result in numerous ill effects, including impaired reproductive capacity and birth defects, and sometimes ultimately in the organism's death. In many cases the effects may depend on the health of the animal. When chlorinated hydrocarbons such as pesticides and PCBs are bioaccumulated they tend to be concentrated in fatty tissues and released to circulate in the body when fat reserves are used during times of poor feeding.

The diet of animals that feed on bioaccumulators is enriched in these conservative materials and if, as is often the case, they too are unable to excrete them, they in turn acquire an even greater body concentration of the substance, a process known as biomagnification. Confusingly, some authors also use the term bioaccumulation for this process of biomagnification up a food chain.     NJM

**Reading**

McLachlan, M.S. 1996: Bioaccumulation of hydrophobic chemicals in agricultural food chains. *Environmental science & technology* 30, pp. 252–9. · Newman, M.C. and Jagoe, R.H. 1996: Bioaccumulation models with time lags: dynamics and stability criteria. *Ecological modelling* 84, pp. 281–6. · Woodwell, G.M., Wurster, C.F. and Isaacson, P.A. 1967: DDT residues in an east coast estuary: a case of biological concentration of a persistent insecticide. *Science* 156, pp. 821–4.

**biochemical oxygen demand (BOD)** The amount of oxygen used for the biochemical degradation of organic compounds by a unit volume of water at a given temperature and for a given time. The measure is commonly used as an index of organic pollution because the amount of dissolved oxygen in water has a profound effect on the plants and animals living in it. BOD is measured by storing a test sample in darkness at 20 °C, for three days (to give BOD3) or five days (to give BOD5), and the amount of oxygen taken up by the micro-organisms present is measured in milligrams per litre.

Water samples may also contain substances such as sulphides, which are oxidized by a purely chemical process, and in these cases the oxygen absorbed may form part of the BOD result. For this reason, the BOD test alone is no longer considered an adequate criterion for judging the presence or absence of organic pollutants, although it is still widely used particularly to assess the polluting capacity of sewage. Other aquatic pollutants, such as pesticides, dissolved salts and heavy metals, cannot be assessed using the BOD test. NJM

**Reading**
Clark, R.B. 1997: *Marine pollution*. 4th edn. Oxford: Clarendon Press. · Hoch, B., Berger, B., Kavka, G. and Herndl, G.J. 1995: Remineralization of organic matter and degradation of the organic fraction of suspended solids in the River Danube. *Aquatic microbial ecology* 9, pp. 279–88.

**biocides** See PESTICIDES.

**bioclastic** Pertaining to rock composed of fragmented organic remains.

**biocoenosis** (biocenose) The mixed community of plant and animals (a biotic community) in particular a HABITAT. It may be artificially partitioned into three components: a plant community (phytocoenosis), an animal community (zoocoenosis) and a community of microorganisms (micro-biocoenosis). First introduced by K. Möbius in 1877, the word is normally used in the sense of an actual community in the landscape rather than in the sense of an abstract concept or community type. The physical environment of a biocoenosis is its ECOTOPE; the biocoenosis with its ecotope make up a BIOGEOCOENOSIS. JAM

**Reference**
Möbius, K. 1877: *Die Auster und die Austernwirtschaft*. Berlin: Wiegundt, Hempel & Parey.

**biodegradation** The decomposition of organic substances by micro-organisms. The metabolism of aerobic (oxygen-utilizing) bacteria is primarily responsible for this breakdown.

The death and decay of organic matter is essential for the replenishment of raw materials in the biosphere. Reactions involved in the degradation of organic compounds are frequently those of their synthesis in reverse (Horne 1978). Catalysts for the degradation reactions are produced by micro-organisms, and are often identical or similar enzymes to those employed in synthesis (Bailey *et al.* 1978). Natural toxins and some complex structures are resistant to breakdown. For example, parts of the intricate chlorophyll molecule can be found fossilized in ancient sedimentary rocks. Also, organic compounds synthesized by indus-

trial chemistry (plastics, and chlorinated hydrocarbons, for example) are not readily biodegradable, and may persist in the environment for a considerable time (see BIOLOGICAL MAGNIFICATION). The presence of biodegradable material in bodies of water can lead to a significant reduction in the dissolved oxygen content of the water, as oxygen is needed for micro-organic activity to take place (BIOCHEMICAL OXYGEN DEMAND). RLJ

**References**
Bailey, R.A., Clark, H.M., Ferris, J.P., Krause, S. and Strong, R.L. 1978: *Chemistry of the environment*. New York and London: Academic Press. · Horne, R.A. 1978: *The chemistry of our environment*. New York and Chichester: Wiley.

**biodiversity** Biological diversity – 'an enormous cornucopia of wild and cultivated species, diverse in form and function, with beauty and usefulness beyond the wildest imagination' (Iltis 1988). This has recently become a major environmental issue because environments are being degraded at an accelerating rate, much diversity is being irreversibly lost through the destruction of natural habitats, and science is discovering new uses for biological diversity (Wilson 1988). The number of species of organism on earth is still imperfectly understood as is the rate at which they are being lost, but particular fears are expressed about the loss of species caused by rainforest exploitation.

Biodiversity has five main aspects:

1 The distribution of different kinds of ecosystems, which comprise communities of plant and animal species and the surrounding environment and which are valuable not only for the species they contain, but also in their own right.
2 The total number of species in a region or area.
3 The number of endemic species in an area.
4 The genetic diversity of an individual species.
5 The subpopulations of an individual species which embrace the genetic diversity. ASG

**Reading and References**
Iltis, H.H. 1988: Serendipity in the exploration of biodiversity – what good are weedy tomatoes? In E.O. Wilson ed., *Biodiversity*. Washington, DC: National Academy Press, pp. 99–105. · Wilson, E.O. ed. 1988: *Biodiversity*. Washington, DC: National Academy Press.

**biogeochemical cycles** The cycling, at various scales, of minerals and compounds through the ecosystem. The cycles (see CARBON CYCLE and NITROGEN CYCLE) involve phases of weathering of inorganic material, uptake and storage by organisms, and return to the pool of the soil, the atmosphere or ocean sediments. An increasing amount of research has been focused on the working out of the details of such cycles during the past decade as a result of concern over global environmental change and nutrient budgets. Classic studies, such as that of the Hubbard Brook Experimental Forest by Likens and associates, are often cited (Begon *et al.* 1990, pp. 688–91) as examples of well-documented small-scale studies. Diagrammatic representations of major element cycles are common in textbooks but there are fewer *quantified* studies on a global scale (Mannion 1991). The biogeochemistry of carbon has attracted particular attention because of the concern over GLOBAL WARMING and the GREENHOUSE EFFECT.     KEB

**Reading and References**
Begon, M., Harper, J.L. and Townsend, C.R. 1990: *Ecology*, 2nd edn. Oxford: Blackwell Scientific Publications. · Degens, E.T., Kempe, S. and Richey, J.E. eds 1991: *Biogeochemistry of major world rivers*. Chichester: Wiley. · Mannion, A.M. 1991: *Global environmental change*. Harlow, Essex: Longman Scientific and Technical.

**biogeocoenosis** (biogeocenose) A combination on a specific area of the earth's surface of a particular BIOCOENOSIS (biotic community) and its ECOTOPE (physical environment), e.g. a forest, a peat bog or an oyster bank. Introduced by V.N. Sukachev in 1944, the term was widely used in the former USSR as an equivalent to the western term ECOSYSTEM; a biogeocoenosis type being equivalent to an abstract ecosystem type. A biogeocoenosis is generally considered to possess a degree of homogeneity in its structure and a certain coherence in its functioning.     JAM

**Reading and Reference**
Sukachev, V.N. and Dylis, N. 1964: *Fundamentals of forest biogeocoenology*. Edinburgh and London: Oliver & Boyd. · Troll, C. 1971: Landscape ecology (geoecology) and biogeocenology – a terminological study. *Geoforum* 8, pp. 43–6.

**biogeography** The science of biological distribution patterns, and a discipline examining the characteristic spatial and temporal occurrence of the earth's organisms. As the name suggests, there are two components to this form of natural science. The biological component entails that its objects of study are biological entities, from species up through higher orders of taxonomic classification. The geographical component embodies the identification of distributional patterns and the search for an explanation as to the factors that underlie them. Depending on the level of taxonomy in question and the spatial scale of interest, there may be a strong degree of overlap between biogeography and ECOLOGY, since ecologists are also interested in organism distributions. The interdisciplinary science of biogeography has several elements which tend to distinguish it from ecology however, especially its focus on generally larger-scale spatial distributions (commonly regional, continental or even global) and also its frequent concern with higher taxonomic orders such as families. A further distinguishing factor is the biogeographical interest in evolutionary development over time and the dynamics of distribution patterns. Biogeographers also focus on the nature of the relationship between humans and organisms, although such an approach is not their sole preserve.

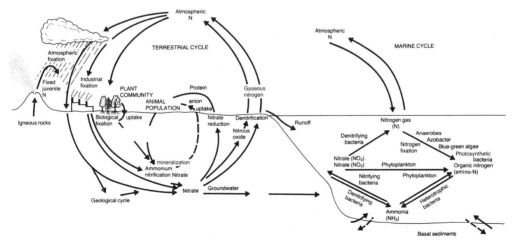

**Biogeochemical cycles**. Generalized diagrams of the terrestrial and marine nitrogen cycles.
*Source*: P. Furley and W. Newey 1983: *Geography of the biosphere*. London: Butterworths. Figure 3.7.

Biogeography has a distinguished history, although its identity as a named science is more recent so that some of its earlier and more famous proponents would not have described themselves as biogeographers *per se*. Perhaps the first biogeographical studies are the inventories of plant species generated for particular geographical areas (one of the oldest emanates from fourth-century China). The German polymath, Alexander von Humboldt, is arguably the first biogeographer as, in his seventeenth-century explorations of South America, he collated enormous numbers of plant specimens together with observations on their distribution patterns. The realization soon followed that species maintained characteristic distributions constrained by their adaptations to the physical environmental characteristics (especially climate) in their respective ranges. Both Charles Darwin and Alfred Russell Wallace developed a keen interest in the distribution patterns of organisms and their adaptational relationships to the environment. Such information was the key to the establishment of the notion of natural selection, although it was the German A.W.F. Schimper who formalized ideas on the nature of vegetation–climate interactions in his 1903 *Plant geography on a physiological basis*. The twentieth-century history of biogeography reflects several diverse and distinctive developments in the discipline and influenced by advances made in associated natural scientists such as botany, ecology and even geology.

Several distinctive forms of biogeography may be recognized in the contemporary literature. Prominent among them is a focus on distribution patterns, often of plant taxa (phytogeography), and their evolution over geological time. Few would argue that, ecologically, the rainforests of Africa and South America are functionally similar, but the taxonomic identities of the two regions are markedly distinct so that taxa in similar habitats are different because they have different histories. The sub-discipline of vicariance biogeography arose out of the recognition that history, by way of continental drift and other large-scale geological and climatic changes, has played the dominant role in determining taxonomic distribution patterns at the global scale. CLADISTIC biogeography goes one stage further in recognizing that the precise evolutionary relationships between taxa are a direct function of the particular sequence of events in their respective geographical areas of distribution, i.e. the relationships between geographical areas can be reconstructed from knowledge of the taxa occurring there.

Ecological ideas, in particular the development of the ECOSYSTEM concept, have proved central, especially in the contemporary geographers' (as opposed to biologists') approach to biogeography. This spawned the consideration of major vegetation formations, or BIOMES, which could be seen as an expression of the environmental relationships at the global scale. Another expression of ecological ideas, in this case the branch of population ecology, arose out of MacArthur and Wilson's (1967) ideas on ISLAND BIOGEOGRAPHY, which viewed species complements on islands, microcosms of larger scale ecosystems, as the product of an equilibrium between colonizing species and those becoming extinct. The potential for this idea to help resolve the growing CONSERVATION crisis was soon realized and, although direct application of the equilibrium approach has proved fruitless, it encouraged critical thinking as to the mechanisms underlying accelerated human-induced extinction. Indeed, biogeography exhibits a strong degree of involvement in that area of science dealing more broadly with human–environment relationships. Biogeographers, particularly those with training as geographers, have also made significant contributions in the field of Quaternary palaeoecology, where a dynamic approach to the rather more static large-scale biome studies has been fostered.

MEM

**Reading and Reference**
Brown, J.H. and Lomolino, M.V. 1998: *Biogeography*. 2nd edn. New York: Sinauer. · Cox, C.B. and Moore, P.D. 1993: *Biogeography: an ecological and evolutionary approach*. Oxford: Blackwell Scientific Publications. · MacArthur, R.H. and Wilson, E.O. 1967: *The theory of island biogeography*. Princeton, New Jersey: Princeton University Press. · Myers, A. and Giller, P.S. 1988: *Analytical biogeography: an integrated approach to the study of animal and plant distributions*. London: Chapman and Hall. · Tivy, J. 1993: *Biogeography: a study of plants in the ecosphere*. 3rd edn. Harlow: Longman Scientific and Technical.

**biogeomorphology** encapsulates concisely a developing and previously much neglected approach to geomorphology which explicitly considers the role of organisms. Two main foci are (*a*) the influence of landforms/geomorphology on the distribution and development of plants, animals and micro-organisms; and (*b*) the influence of plants, animals and micro-organisms on earth surface processes and the development of landforms. ASG

**Reading**
Viles, H.A. ed. 1988: *Biogeomorphology*. Oxford: Basil Blackwell.

**bioherm** *a*. An ancient coral reef.
*b*. An organism which plays a role in reef formation.

**biokarst** A KARST landform, usually small in scale, produced mainly by organic action. Strictly speaking, the term PHYTOKARST should be restricted to phenomena produced by plants alone. Biokarst features can either be erosional (as where organisms bore into or abrade carbonate rock surfaces) or constructional (as in the case of certain tufas and reef forms). ASG

**Reading**
Viles, H.A. 1984: Biokarst: review and prospect. *Progress in physical geography* 8, pp. 523–43.

**bioleaching** The process of using microorganisms to liberate metals from mineral ores. Direct leaching is when microbial metabolism changes the redox state of the metal being harvested, rendering it more soluble. Indirect leaching includes redox chemistry of other metal cations that are then coupled in chemical oxidation or reduction of the harvested metal ion and microbial attack upon and solubilization of the metal matrix in which the metal is physically embedded. Microbial mining of copper sulphide ores has been practised on an industrial scale since the late 1950s and since then bioleaching has also become common at uranium and gold mines. Bioleaching technologies have been developed by the mining industry in response to the lower grades of accessible ores and lower operating costs. Microbial mining saves energy, creates minimal pollution and recovers resources from tailings that would otherwise be wasted. An allied development in mining biotechnology is the use of bacterial cells to detoxify waste cyanide solution from gold mining operations (Agate 1996). NJM

**Reference**
Agate, A.D. 1996: Recent advances in microbial mining. *World journal of microbiology and biotechnology* 12, pp. 487–95.

**biological control** (biocontrol) The control of pestilential organisms such as insects and fungi through biological means rather than the application of man-made chemicals. This can include breeding resistant crop strains, inducing fertility in the pest species, disruption of breeding patterns through the release of sterilized animals or spraying juvenile hormones to interrupt life cycles, breeding viruses that attack the pests, or the introduction or encouragement of natural or exotic predators to control pest outbreaks. ASG

**biological magnification** The increased concentration of toxic material at consecutive TROPHIC LEVELS in an ecosystem. Toxins, such

as persistent pesticides and heavy metals, are incorporated into living tissue from the physical environment. Physical, chemical and biological processes operate to amplify the harmful substances in food chains by concentrating the quantities in individual organisms. The effect of these substances on organisms varies. In general, reproductive capacity is impaired at low concentrations and death occurs at high ones, sometimes via disease. Approximately 10 per cent of food at one trophic level is transferred to the next, the remainder being removed by respiratory and executive activity, but as toxic materials are not so readily broken down as other components of organic tissue (BIODEGRADATION), their transfer efficiency is higher. Thus a build-up of them occurs at successive trophic levels (Woodwell 1967).

DDT (PESTICIDES) residues are an example of accumulation in the physical environment. They have a low biodegradability, and are excreted slowly from organisms because they become dissolved in fatty tissues. Woodwell reports (in a study undertaken with Wurster and Isaacson) DDT concentrations of up to $36 \, kg \, ha^{-1}$ after two decades of application of the pesticide to a New York marsh ecosystem. Marsh plankton contained 0.04 ppm, minnows 1 ppm and a carnivorous gull 75 ppm of DDT in their tissues. RLJ

**Reading and Reference**
Jorgensen, S.E. and Johnsen, I. 1981: *Principles of environmental science and technology*. Amsterdam, Oxford and New York: Elsevier. · Odum, E.P. 1975: *Ecology: the link between the natural and social sciences*. 2nd edn. London and New York: Holt, Reinhart & Winston. · Woodwell, G.M. 1967: Toxic substances and ecological cycles. *Scientific American* 216, pp. 24–31.

**biological productivity** The rate at which organic matter accumulates over time within any given area. The term is usually applied at the ECOSYSTEM level, but may also be used with reference to an individual organism or population to account for growth, or the increase in BIOMASS over time. Biological productivity is a composite term and is used to refer generally to productivity that may be considered separately as NET PRIMARY PRODUCTIVITY, gross primary productivity and secondary productivity (of consumers), all of which involve the transfer of energy within an ecosystem. Biological productivity is measured in units of weight accumulating per unit area over time, for example as $g \, m^{-2} day^{-1}$, or $t \, ha^{-1} yr^{-1}$. There are defined relationships between gross primary productivity (GPP) and net primary productivity (NPP) in an ecosystem. GPP represents the total rate at which green plants convert solar radiation into carbohydrates via the process of PHOTOSYN-THESIS. NPP is the net rate of organic matter accumulating after allowance is made for the fact that the green plants themselves need to utilize some of the assimilated energy in order to exist and that some energy is lost to the system by the death or herbivory of the photo-synthesizing plants. Human populations are dependent for their existence on biological pro-ductivity, albeit often in an artificial and manipulated form as agricultural production. Humans, as consumers, rely on both net primary productivity of agricultural crops and the secondary productivity of herbivores. MEM

**bioluminescence** The light produced by some living organisms or the process of produc-ing that light, characteristic of glow-worms and some marine fish.

**biomantle** A layer or zone in the upper part of a soil that has been produced primarily by BIOTURBATION. Where present, biomantles gen-erally extend from the soil surface to the depth reached by burrowing invertebrates such as worms, ants and spiders, or small vertebrates (moles, gophers, etc). Over extended periods of time, the excavation of fresh burrows by organisms can progressively affect all of the upper part of the soil column. This may do one of two things: either, it will overturn and homogenize the layer that is being worked, and so produce a distinctive soil fabric, or, if there are stones in the soil too large to be moved, these will gradually accumulate as a layer at the base of the worked layer. The texture of the biomantle layer can be affected, as well as its structure. If the burrowing organisms build mounds of excavated spoils at the surface, then once again through extended periods of time all transportable materials in the worked layer will be brought to the surface before they are again mixed downward. While at the surface, the spoils are exposed to erosional processes (both wind and water), which may progressively winnow out finer, more readily moved grains. Thus, continued bioturbation can lead to a tex-tural coarsening of the biomantle layer of the soil. Such processes can be triggered in other ways, as when tree-throw brings up subsoil attached to the roots and exposes this to erosional sorting. Progressively, this too could result in the development of a distinctive bio-mantle whose depth would correspond to the rooting depth of the trees. DLD

**Reading**
Johnson, D.L. 1990: Biomantle evolution and the redis-tribution of earth materials and artifacts. *Soil science* 149,

pp. 84–102. · Nooren, C.A.M., van Breemen, N., Stoor-vogel, J.J. and Jongmans, A.G. 1995: The role of earth-worms in the formation of sandy surface soils in a tropical forest in Ivory Coast. *Geoderma* 65, pp. 135–48.

**biomass** The mass of biological material pre-sent per plant or animal, per community, or per unit area. On a world scale, and largely because of the efficiency with which vegetation colonizes land, most of the biomass is found in terrestrial rather than oceanic environments, in a ratio of several hundred to one. This is particularly the case for phytomass (the mass of growing and dead plant material). Whittaker and Likens (1975) placed the total continental phytomass at $1837 \times 10^9$ tonnes dry weight and that in oceans and estuaries at only $3.9 \times 10^9$ tonnes dry weight. On the other hand, the balance between terrestrial and oceanic animal biomass, which is at much lower levels all round, is much more even: similar estimates (Whittaker and Likens 1973) have set the former at $1005 \times 10^6$ tonnes, and the latter at $997 \times 10^6$ tonnes, both dry weights.

On land, about 90 per cent of phytomass is located in the world's forests. Tropical rain for-ests produce a mean phytomass of $45 \, \text{kg m}^{-2}$, other tropical and temperate forests one of $30$–$35 \, \text{kg m}^{-2}$, and boreal forests one of $20 \, \text{kg m}^{-2}$. Natural grasslands and tundra have phytomasses of $0.5$ to $5 \, \text{kg m}^{-2}$; and desert phytomasses range from $0.7 \, \text{kg m}^{-2}$ to much less. Within all vegetation communities a good deal of phytomass is accounted for by root growth: thus, roots give rise to over 50 per cent of the phytomass in many prairie grasslands of North America, and to an even higher per-centage in most deserts. Animal biomasses are greatest per unit area on land in tropical forests, and secondarily in the savannah grass-lands of Africa; and in oceanic environments in tropical reef zones, estuaries and on continen-tal shelves. DW

**References**
Whittaker, R.H. and Likens, G.E. 1973: The primary production of the biosphere. *Human ecology* 1, pp. 299–369. · —1975: Net primary production and plant bio-mass for the earth. In H. Reith and R.H. Whittaker eds, *The primary production of the biosphere*. New York: Springer-Verlag. Pp. 305–28.

**biome** A mixed community of plants and ani-mals (a biotic community) occupying a major geographical area on a continental scale. Usually applied to terrestrial environments, each biome is characterized by similarity of vegetation struc-ture or physiognomy rather than by similarity of species composition, and is usually related to climate. Within a particular biome the plants and animals are regarded as being well adapted

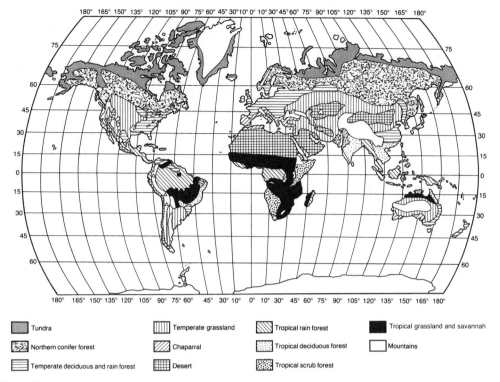

| | | | |
|---|---|---|---|
| Tundra | Temperate grassland | Tropical rain forest | Tropical grassland and savannah |
| Northern conifer forest | Chaparral | Tropical deciduous forest | Mountains |
| Temperate deciduous and rain forest | Desert | Tropical scrub forest | |

**Major biomes** of the world as if unaffected by human activity.
*Source*: I. Simmons 1979: *Biogeography, natural and cultural*. London: Edward Arnold. Figure 3.3.

to each other and to broadly similar environmental conditions, especially climate. Both CLIMAX VEGETATION and SERAL COMMUNITIES are represented. A group of biomes in which the plant and animal communities exhibit similar adaptations form a biome type. Thus the tropical rain forests of the Congo Basin and Amazonia are two biomes within the tropical rain forest biome type. Other biome types respectively include tundra, taiga, savannah and hot desert biomes (see LIFE FORM). Needless to say, large areas of many biomes have undergone transformation by human societies and small-scale maps that appear in texts and atlases are misleading (Holzner *et al.* 1983).                JAM

**Reading and Reference**
Brewer, R. 1979: *Principles of ecology*. Philadelphia and London: W.B. Saunders. · Carpenter, J.R. 1939: The biome. *American midland naturalist* 21, pp. 75–91. · Furley, P.A. and Newey, W.W. 1983: *Geography of the biosphere*. London: Butterworths. · Holzner, W., Ikusima, I. and Werger, M.J.A. eds 1983: *Man's impact on vegetation*. The Hague, Boston and London: W. Junk. · Kendeigh, S.C. 1974: *Ecology with specific reference to animals and man*. Englewood Cliffs, NJ: Prentice-Hall. · Shelford, V.E. 1963: *The ecology of North America*. Urbana: University of Illinois Press. · Walter, H. 1979: *Vegetation of the earth*. 2nd edn. Berlin: Springer-Verlag.

**biometeorology** The study of the effects of weather and climate on plants, animals and man. The International Society of Biometeorology, founded in 1956, has classified the subject into six main groups: phytological, zoological, human, cosmic, space and palaeo. Human biometeorology includes the study of the influence of weather and climate on healthy man and on his diseases and the effect of microclimates in houses and cities on health. Although Hippocrates discussed some of these topics over 2000 years ago it is only in the second half of the twentieth century that the main developments in this science have taken place. In some countries the old name of bioclimatology is still used.                DGT

**Reading**
Tromp, S.W. 1980: *Biometeorology*. London: Heyden.

**biosphere** The zone, incorporating elements of the hydrosphere, lithosphere and atmosphere, in which life occurs on earth. The term is occasionally used to refer only to the living component alone, although it is more commonly conceived as a zone of interaction between the other 'spheres'. This is appropriate, because life is dependent upon energy, processes and materials which are located in all three of the earth's other conceptual spheres, to the extent that the scheme is often represented as a series of overlapping circles with the biosphere in the nodal position. Used in this way, biosphere is synonymous with ECO-SPHERE. Implicit in the concept is the fact that life within the biosphere is organized hierarchically. The hierarchy has a fundamentally cellular basis and is structured through progressively larger and more complex agglomerations of organs, organisms and populations. By incorporating elements of the physical environment also, the higher levels of organization in the biosphere are represented by its ECOSYSTEMS and BIOMES.

In proportion to the dimensions of the earth and its atmosphere as a whole, the biosphere occupies a remarkably thin and, arguably, fragile layer. It extends a maximum of a few hundred metres above the land surface and penetrates relatively little into the crust, although recent research suggests that the microbial biomass may be considerably larger (see ANAEROBIC bacteria). Certainly the most conspicuous organisms are found within a few metres above and below the earth's surface. Approximately two-thirds of the biosphere is characterized by water, as oceans.

Crucial in the evolution of the biosphere has been the emergence of photosynthesizing green plants and the consequent modification of the atmosphere that appears to have facilitated the familiar ecosystem processes of the contemporary earth. Indeed the prominence of atmospheric oxygen in its role in ecological processes is a direct result of the evolutionary development of photosynthesizing AUTOTROPHS. In essence, the earth's contemporary atmosphere is a product of the biosphere.

Two fundamental kinds of processes characterize the biosphere, viz.: ENERGY FLOW and nutrient recycling, both of which involve functional linkages and exchanges between the various spheres in the earth system. Most of the energy which activates the biosphere is derived from solar radiation and becomes ecologically important via its assimilation through photosynthesis. Subsequently, this energy is utilized by organisms in successive TROPHIC LEVELS as part of the FOOD CHAIN. Progressively smaller absolute quantities of energy are available along the length of the food chain so that consumer biomass, for example herbivores and primary and secondary carnivores, is less than for the autotrophs. The incoming energy driving the entire system is eventually dissipated as respirational or heat loss but is, of course, kept topped up by the constancy of solar inputs. Nutrients within the biosphere, however, are not renewed in the same way and, instead, are recycled through the active components of the earth system over time. Hydrogen, carbon, oxygen and nitrogen, together with some forty or so other essential mineral elements recycle within the biosphere and maintain its integrity. Carbon, for instance, recycles through the atmosphere, biosphere, lithosphere and hydrosphere and, in an ideal, efficient BIOGEOCHEMICAL CYCLE, is consistently replenished so that circulating quantities remain stable over time. Due to excessive human exploitation of resources in the biosphere, however, imbalances in some of the nutrient cycles are becoming increasingly apparent and express themselves, for example, as environmental problems such as EUTROPHICA-TION and GLOBAL WARMING.

Arising out of the biosphere concept, in which the earth may be envisaged as a single, self-contained ecosystem, is the idea that actions on any scale anywhere on earth may have a cumulative, by implication negative, effect on the whole system. James Lovelock in 1974 took this idea to what he considered to be its logical conclusion, i.e. that the earth is in effect a superorganism and that it strives to maintain a regulated atmospheric composition and temperature condition. The so-called GAIA hypothesis is controversial but has served to focus attention on the nature of the biosphere and the human interaction with it.     MEM

**Reading**
Bradbury, I.K. 1998: *The biosphere*. 2nd edn. Chichester: Wiley.

**biostasy**   A term that was applied by Erhart (1956) to periods of soil formation, with rhexistasy referring to phases of denudation. In periods of biostasy there is normal vegetation, while in phases of rhexistasy there is dying out or lack of vegetation as a result of soil erosion resulting from climatic changes, tectonic displacement, etc. The period of rhexistasy is characterized by mechanical reworking, whereas biostasy is characterized by chemical decomposition.     ASG

**Reference**
Erhart, H. 1956: *La genèse des sols en tant que phénomène géologique*. Paris: Masson.

**biota**   The entire complement of species of organisms, plants and animals, found within a given region.

**biotechnology**   The use of microbial, animal or plant cells or enzymes to synthesize, break down or transform materials (Smith 1988). However, as Bull *et al.* (1982) and Bu'Lock (1987) note, alternative definitions of the term have arisen as a result of the philosophies and interests of the practitioners of a rapidly developing and sometimes controversial subject. It involves the interaction of biology and engineering to their mutual benefit, and is an interdisciplinary science concerned with a collection of technologies relevant to a number of parts of industry: the production of food and medicine, and the treatment of organic waste, for example. Biotechnological principles (for instance, in the manufacture of alcohol from grain, and cheese from milk) have been applied for centuries, but major advances in bioengineering have been made in recent decades. RLJ

**Reading and References**
Bull, A.T., Holt, G. and Lilly, M.D. 1982: *Biotechnology: international trends and perspectives*. Paris: OECD. · Bu'Lock, J.D. 1987: Introduction to basic biotechnology. In J. Bu'Lock and B. Kristiansen eds, *Basic biotechnology*. London and New York: Academic Press. Pp. 3–10. · Ginzburg, L.R. 1991: *Assessing ecological risks of biotechnology*. Boston and London: Butterworth-Heinemann. · Smith, J.E. 1988: *Biotechnology*. 2nd edn. London: Edward Arnold.

**biotic isolation**  The isolation of organisms by hereditary mechanisms which cause the restriction or elimination of interbreeding through processes wholly internal to the organism. The isolation results not from geographical or ecological isolation but from the incompatibility of reproductive structures and genetic isolation. Thus, although populations of the song sparrow (*Melospiza melodia*) and Lincoln's sparrow (*M. lincoleni*) live side by side over considerable areas of the USA, they do not integrate and remain distinct from each other. Most species are isolated by several kinds of mechanisms which are in turn controlled by the action of many different genes. (See also SYMPATRY.) PAS

**Reading**
Jones, S.B. and Luchsinger, A.E. 1979: *Plant systematics*. New York: McGraw-Hill. · Stebbins, G.L. 1977: *Processes of organic evolution*. 3rd edn. Englewood Cliffs, NJ: Prentice-Hall.

**biotic potential**  The total reproductive potential of an individual organism or a population; an important concept in the study of plant and animal population dynamics. Essentially it is a measure of the reproductive rate of a given species taking into account the inherited sex ratio, the number of young per female, and the number of generations per unit of time. The biotic potential of many species (e.g. small mammals or bacteria) is enormous and such species would soon swamp the earth if there were no environmental checks on reproduction and on the survival of offspring. PAS

**Reading**
Silvertown, J. 1982: *Introduction to plant population ecology*. London: Longman. · Solomon, M.E. 1976: *Population dynamics*. 2nd edn. London: Edward Arnold.

**biotope**  The HABITAT of a BIOCOENOSIS, or a micro-habitat within a biocoenosis. In the first sense the word is synonymous with ECOTOPE, the effective physical environment of a biocoenosis or biotic community. In the second sense it refers to a small, relatively uniform habitat within the more complex community, e.g. although a forest community occupies its own habitat, each layer or stratum within the forest may be regarded as a separate biotope. (See also NICHE.) JAM

**Reading**
Allee, W.C., Emerson, A.E., Park, O., Park, T. and Schmidt, K.P. 1951: *Principles of animal ecology*. Philadelphia and London: W.B. Saunders. · Daubenmire, R. 1968: *Plant communities: a textbook of plant synecology*. New York and London: Harper & Row.

**bioturbation**  The mixing of soil or surficial layers of terrestrial, marine or lacustrine sediments by the physical action of organisms. A familiar example would be the mulching of topsoil by the action of earthworms, termites or moles, which activity results in the organic and clastic soil particles being selectively reworked and sorted. Ploughing and mulching are forms of intentional human-induced bioturbation in soils aimed at mixing the topsoil and optimizing drainage and nutrient conditions. In the case of marine or lacustrine sediments, BENTHIC micro- and macrofauna, through their activities at the sediment surface and near-surface, may overturn the accumulating deposits such that apparent inconsistencies in chronological order of sedimentation may result. This is potentially problematic if the sediments are the object of palaeolimnological or palaeoceanographic study, for the microfossils and other sediment characteristics of interest may become displaced. Such a situation is most serious under circumstances of very slow rates of sediment accumulation, where physical disturbance over a few centimetres can re-sort materials deposited over several hundreds or thousands of years, hence impacting on the level of temporal resolution possible in the palaeoenvironmental reconstruction. It is for this reason that deep ocean sediments are unable to resolve environmental events of less than one thousand years duration. MEM

**bise, bize**  A cold, dry northerly to north-north-easterly wind occurring in the mountains of Central Europe during the winter months.

**black box** The term given to any model or treatment of a SYSTEM where the internal components of the system are ignored or excluded from direct treatment, by only considering the relationship between inputs to the system and outputs from the system. A good example might be a statistical treatment of the relationship between rainfall and runoff, which does not consider the operations of processes within the system (e.g. soil infiltration, initial soil moisture conditions, interception of rainfall by vegetation and subsequent evaporation). In practice, most models have an element of black box treatment, as there are normally always some processes operating at scales smaller than the model. These may be dealt with using statistical treatments which may have only a poor resemblance to the actual way in which they work.    SNL

**blanket bog** A type of bog, often composed of peat, which drapes upland terrain and infills hollows, in areas of high precipitation and low evapotranspiration.

**blind valley** A steep-sided, river-cut valley which terminates in a precipitous cliff. Although blind valleys can be found in any terrain near the source of a river, they are usually produced when a river flows onto limestone bedrock and sinks into subterranean passages, enabling downcutting to occur in the active river valley. This has the effect of depressing the river valley relative to the upstanding limestone rock, thus producing a steep cliff. Blind valleys are favoured sites for cave exploration: caves such as the famous Swildon's Hole, Mendip, south-west England are entered from a blind valley.    PAB

**blizzard** A snow storm, either of falling snow or deflated snow, usually accompanied by low temperature and high winds.

**block faulting** The process whereby large regions are tectonically disrupted to form complex systems of troughs and ridges or basins and block mountains. The result of tectonic uplift and subsidence of adjacent blocks of the earth's crust following faulting and fracturing on a grid pattern.

**block fields, block streams** Spreads or lines of boulders, generally angular, formed by *in situ* shattering by frost of a bedrock surface. They may surround features such as tors and nunataks and other landforms subjected to severe periglacial processes.

**blocking** An extreme state in which the tropospheric circulation takes the form of large-amplitude stationary waves. Such flow is frequently manifest in stationary anomalies in the weather which may have significant economic repercussions, such as the 1976 drought in the UK. Prediction of the initiation and persistence of blocks is difficult. Hydrology, transfer of energy by both solar and terrestrial radiation, interaction of synoptic-scale weather systems and the larger scale mean flow, and resonance of stationary waves are all possible relevant factors.    JSAG

**blow-hole** Vertical shaft leading from a sea cave to the surface. Air and water may be forced through the hole with explosive force as a rising tide or large waves cause large changes in pressure in the underlying cave.

**blowouts** Erosional hollows, depressions, troughs or swales within a dune complex (Carter *et al.* 1990). They form readily in vegetated dunes for a variety of reasons: shoreline erosion and/or washover, vegetation die-back and soil nutrient deficiency, destruction of vegetation by animals and fire, and human recreational activities. However, blowout topography need not necessarily arise from erosional processes; it may develop as areas of non-deposition between mobile dune ridges or as gaps in incipient foredunes. Two basic blowout morphologies have been identified: saucer blowouts (shallow, ovoid and dish-shaped with a steep marginal rim) and trough blowouts (deep, narrow, steep-sided with more marked downwind depositional lobes, and marked deflation basins).    ASG

**Reference**
Carter, R.W.G., Hesp, P.A. and Nordstrom, K.F. 1990: Erosional landforms in coastal dunes. In K.F. Nordstrom, N.P. Psuty and R.W.G. Carter eds, *Coastal dunes: form and process*. Chichester: Wiley. Pp. 217–50.

**bluehole** A circular, steep-sided hole which occurs in coral reefs. The classic examples come from the Bahamas (Dill 1977), but other examples are known from Belize and the Great Barrier Reef of Australia (Backshall *et al.* 1979). Although volcanicity and meteorite impact have both been proposed as mechanisms of formation, the most favoured view is that they are the product of karstic processes (e.g. collapse dolines) which acted at times of low sea level when the reefs were exposed to subaerial processes.    ASG

**References**
Backshall, D.G., Barnett, J. and Davies, P.J. 1979: Drowned dolines – the blue holes of the Pompey Reefs, Great Barrier Reef. *BMR journal of Australian geology and geophysics* 4, pp. 99–109. · Dill, R.F. 1977: The blue holes

– geologically significant sink holes and caves off British Honduras and Andros, Bahama Islands. *Proceedings of the 3rd International Coral Reef Symposium, Miami* 2, pp. 238–42.

**Blytt–Sernander model**  Provides the classic terminology of the HOLOCENE. It was established by two Scandinavians, A.G. Blytt and R. Sernander, who, in the late nineteenth and early twentieth centuries, undertook various palaeobotanical investigations that revealed vegetational changes from which climatic changes were inferred. They introduced the terms Boreal, Atlantic, Sub-Boreal, and Sub-Atlantic for the various environmental fluctuations that took place. Modern workers recognize that because factors other than climate affect vegetation change (e.g. man's intervention, soil deterioration through time etc.) the model may be simplistic.                               ASG

The classic European Holocene sequence

| Period | Zone number | Blytt–Sernander zone name | Radiocarbon years BP |
|--------|-------------|---------------------------|----------------------|
| Post Glacial | IX | Sub-Atlantic | post 2450 |
| | VIII | Sub-Boreal | 2450–4450 |
| | VII | Atlantic | 4450–7450 |
| | VI | Late Boreal | 7450–8450 |
| | V | Early Boreal | 8450–9450 |
| | IV | Pre-Boreal | 9450–12,250 |
| Late Glacial | III | Younger Dryas | 10,250–11,350 |
| | II | Allerød | 11,350–12,150 |
| | Ic | Older Dryas | 12,150–12,350 |
| | Ib | Bølling | 12,350–12,750 |
| | Ia | Oldest Dryas | |

**bodden**  A type of irregularly shaped coastal inlet brought about by the transgression of the sea in an area of undulating terrain, as along the Baltic coastline of eastern Germany.         ASG

**bog**  In the widest sense, the term bog may be taken to include any nutrient-poor, acidic peatland community with a distinctive plant community of sphagnum mosses, ericaceous sedges and coniferous trees. This broader use of the term includes poor fen. Bog is used in the restrictive sense to refer to those peatlands where the bog surface is irrigated more or less exclusively by precipitation inputs and is unaffected by minerotrophic groundwater. Another term for bog is ombrotrophic MIRE. The nutrient supply to the bog is low although nitrogen may be supplemented by atmospheric enrichment derived from industrial, urban and agricultural

sources. Bog biodiversity is low. Typically the pH is < 4.5 compared to fens where the pH range is 4.5–7.5.                               ALH

**Reading**
Lewis, W. M. ed. 1995: *Wetlands: characteristics and boundaries*. California: National Research Council.

**bog bursts**  The sudden disruption of a bog so that there is a release of water and peat which may then flow over a considerable distance.

**bogaz**  Narrow, deep ravines and chasms in karst areas, the products of limestone solution along bedding planes, joints and fissures.

**Bølling**  A short-lived LATE GLACIAL INTERSTADIAL dated at about 12,350–12,750 BP.

**bolson**  A low-lying trough or basin surrounded by high ground and having a playa at its lowest point to which all drainage trends.

**Bond cycles**  Groups of very rapid global temperature changes during the QUATERNARY, individually called DANSGAARD–OESCHGER (D–O) EVENTS, they are known as Bond cycles after Bond *et al.* (1992 and 1993). Bond and co-workers demonstrated that the D–O events, identified in ICE CORE records, were also present in the deep sea core record of ocean sediments. This suggests a very strong likelihood of these events having a climatic causation, and also indicate complex ice–ocean–atmosphere linkages.                               DSGT

**Reading and References**
Bond, G. and 13 others, 1992: Evidence for massive discharges of icebergs into the North Atlantic Ocean during the last glacial period. *Nature* 360, pp. 143–7. · Bond, G.C., Broecker, W.S., Johnsen, S., McManus, J., Labeyrie, L., Jouzel, J. and Bonani, G. 1993: Correlation between climate records from North Atlantic sediments and Greenland ice. *Nature* 365, pp. 143–7. · Williams M.A.J., Dunkerley, D., de Deckker, P., Kershaw, P., Chappell, J. 1998: *Quaternary environments*. 2nd edn. London: Arnold.

**bora**  From Latin *boreas*, the north wind. A dry, cold, gusty, north-east wind which affects the northern part of the Adriatic Sea and the Dalmatian coast (see Jurcec 1981). Peak frequency occurs during the months from October to February and a maximum gust speed of 47.5 m s$^{-1}$ has been recorded at Trieste. Two types of bora are recognized: anticyclonic and cyclonic, according to the atmospheric pressure field. The term bora is now applied to similar winds in other parts of the world, e.g. Oroshi in central Japan (Yoshino 1976).         AHP

**Reading and References**

Atkinson, B.W. 1981: *Mesoscale atmospheric circulations.* London and New York: Academic Press. · Jurcec, V. 1981: On mesoscale characteristics of bora in Yugoslavia. In G.H. Liljequist ed., *Weather and weather maps.* Stuttgart: Birkauser Verlag. · Yoshino, M.M. 1975: *Climate in a small area.* Tokyo: University of Tokyo Press. · —1976: *Local wind bora.* Tokyo: University of Tokyo Press.

**bore**  A tidal wave which propagates as a solitary wave with a steep leading edge up certain rivers. Formation is favoured in wedge-shaped shoaling estuaries at times of spring tides. Other local names include *egre* (England, River Trent), *pororoca* (Brazil), *mascaret* (France).          DTP

**boreal climate**  In Köppen's classification scheme a climate which is characterized by a snowy winter and warm summer, with a large annual range of temperature, such as obtains between 60° and 40° north.

**boreal forest**  The northern coniferous zone of the Holarctic. The most northerly section, transitional to the TUNDRA, is synonymous with TAIGA. The southerly sections are frequently made up of dense forest with closed canopies permitting little light to reach the floor, which possesses a variable cover of lichens, mosses and herbaceous plants. Typical hardy and undemanding trees include spruce, fir, and hemlock, with pine providing more open formations and having a denser ground cover. The forest is bounded approximately to the north by the 10 °C July average isotherm and to the south by areas with more than four months above 10 °C.          PAF

**Reading**

Larsen, J.A. 1980: *The boreal ecosystem.* New York: Academic Press.

**boreal (period)**  Period of time which, according to the BLYTT–SERNANDER division of European peat stratigraphy, occurred during the earlier part of the Holocene (approximately 8500 BP to 7500 BP) and was associated with somewhat warmer and drier conditions than occurred either immediately before or afterwards. It is inferred that the relatively high level of humification of peats associated with this period is indicative of both higher temperatures and lower precipitation levels. The interpretation of regionally contemporaneous climatic phases in the European Holocene is based on characteristic raised bog stratigraphy and associated pollen sequences, although the rigid zonation implied by the scheme has now been largely abandoned due to the recognition that relationships between peat stratigraphy, pollen zonation and climate change are more complex.          MEM

**bornhardt**  Distinctive hill with bare rock surfaces, dome-like summit and steep sides, described from east Africa by German geologist Bornhardt (1900) as INSELBERG now often take his name. Usually in granite or gneiss, and found on CRATONS and in deeply eroded OROGENS. Notable groups include the 'sugar loaves' of Rio de Janeiro, Idanre Hills of Nigeria; similar to exfoliation domes of Yosemite. Low porosity, from metasomatic infusion of granite or gneiss, may cause resistance to DENUDATION. Often surrounded by deeply weathered or less resistant rocks, they emerge as ground surface is lowered (Falconer 1912; Willis 1936; Ollier 1960; Thomas 1978). Range from low shield inselberg to high mountains, which could be formed by repeated phases of weathering and excavation, though King (1975) emphasized the role of rock mechanics and the retreat of marginal slopes by spalling of rock sheets. High bornhardts are probably of great geological age. Bornhardt rock surfaces are diversified by etch pits and TAFONI.          MFT

**References**

Bornhardt, W. 1900: *Zur Oberflächengestaltung und Geologie Deutsch-Afrikas,* Berlin. · Falconer, J.D. 1912: The origin of kopjes and inselbergs. *Transactions Section C, British Association for the Advancement of Science* 476. · King, L.C. 1975: Bornhardt landforms and what they teach. *Zeitschrift für Geomorphologie, N.F.* 19, pp. 299–318. · Ollier, C.D. 1960: The inselbergs of Uganda. *Zeitschrift für Geomorphologie N.F.* 4, pp. 43–52.· Thomas, M.F. 1978: The study of inselbergs. *Zeitschrift für Geomorphologie, Supplementband* 31, pp. 1–41. · Willis, B. 1936: East African plateaus and rift valleys: studies in comparative seismology. *Carnegie Institute, Washington, publication* 470, p. 358.

**boss**  A small batholith or any dome-shaped intrusion of igneous rock, especially one exposed at the surface through erosion of the less resistant host rocks.

**botryoidal**  Having a form resembling a bunch of grapes, often applied to aggregate minerals.

**bottom-sets**  Beds of stratified sediment that are deposited on the bottom of the lake or sea in advance of a delta.

**Bouguer anomaly**  A measure of the gravitational pull over an area of the earth after the Bouguer correction to a level datum, usually sea level, has been applied.

**boulder clay**  See TILL.

**boulder train** A stream of boulders derived by glacial transport from a specific and identifiable bedrock source, and carried laterally in a more or less straight line, thereby permitting former directions of ice movement to be inferred.

**boulder-controlled slopes** First described by Bryan (1922, p. 43) from the Arizona deserts. Bryan recognized slopes formed on rock with a veneer of boulders and assumed that the angle of the bedrock surface had become adjusted to the angle of repose of the 'average-sized joint fragment'. Measurements have subsequently shown that boulder-covered slopes exist over the range of angles up to about 37° and that the existence of a boulder cover may be due to a number of causes. Melton (1965) suggested that steep angles of the debris (34°–37°) may occur at the angle of static friction of that debris, and that angles of around 26° may be related to its angle of sliding friction. Where the boulders are core stones left by removal of fine-grained saprolite of a former regolith the boulders may lie at any angle up to about 37° and they do not exert any control on the bedrock slope which is essentially a relict weathering front from the base of the regolith (Oberlander 1972). Boulders may also lie upon bedrock which has resulted from the development of a slope angle in equilibrium with the mass strength of the rock (Selby 1982). The idea that the boulders control the angle of the bedrock slope on which they lie is, therefore, open to question in many cases.                MJS

**Reading and References**
Bryan, K. 1922: Erosion and sedimentation in the Papago country, Arizona, with a sketch of the geology. *US Geological Survey bulletin* 730-B, pp. 19–90. · Melton, M.A. 1965: Debris-covered hillslopes of the southern Arizona Desert – consideration of their stability and sedimentation contribution. *Journal of geology* 73, pp. 715–29. · Oberlander, T.M. 1972: Morphogenesis of granitic boulder slopes in the Mojave desert, California. *Journal of geology* 80, pp. 1–20. · Selby, M.J. 1982: Rock mass strength and the form of some inselbergs in the central Namib Desert. *Earth surface processes and landforms* 7, pp. 489–97.

**boundary conditions** Many physical phenomena can be modelled by mathematical deduction leading to generalized equations; in order to obtain simplified specific solutions to these equations their applicability is deliberately constrained by the definition of particular circumstances known as 'boundary' or 'initial' conditions (Wilson and Kirkby 1975, pp. 206–7 and 222–5). In particular the solution of differential equations requires definition of boundary conditions so that expressions can be found for the arbitrary constants resulting from integration of the equations.

An example is the theoretical derivation of the logarithmic velocity profile in turbulent flow. The rate of change of velocity ($v$) with height above the bed ($y$) – the 'velocity shear' – itself decreases with height, according to the differential equation:

$$dv/dy \cdot K/y$$

where $K$ is a constant incorporating the bed shear stress. The variation of velocity with height is obtained by integrating this equation, which gives:

$$v = K \ln y + C.$$

Here, $C$ is a constant of integration. It can be evaluated by specifying the boundary condition that $v = 0$ when $y = y_0$, so that:

$$0 = K \ln y_0 + C$$

and therefore:

$$C = -K \ln y_0.$$

This is actually an initial condition, since the height is defined at which the velocity is zero, and negative velocities are assumed not to occur. The velocity profile equation can now be simplified by inserting the expression for $C$:

$$v = K \ln y - K \ln y_0$$
$$= K \ln (y/y_0).$$

This describes a curve plotting as a straight line on a graph with a logarithmic height axis and an arithmetic velocity axis; the intercept on the height axis where $v = 0$ is $y_0$, and the gradient of the line is $K$ (Richards 1982, pp. 69–70).

Theoretical models may require multiple boundary conditions, some of which are necessarily dynamic in order to provide realistic solutions. The slope profile shape characteristic of different types of slope process is modelled by a partial differential equation in which the rate of change of local slope sediment transport with distance from the divide ($x$) equals the rate of change of local slope surface elevation ($y$) with time ($t$). To solve this equation and derive the characteristic form of profile as a graph of $y$ against $x$, the initial conditions are: (a) the divide is fixed at $x = 0$, and sediment transport is zero at the divide; and (b) an initial slope shape ($y = f(x)$) is defined from which the characteristic form evolves.

A boundary condition then defines the slope foot; this base level is fixed at $x_1$, and although a fixed base level elevation ($y = 0$ at $x = x_1$) may be set, a dynamic boundary condition may be established in which $y(x_1) = f(t)$. Quite different characteristic profiles will emerge from the solution of the initial general equation according

59

to the nature of this boundary condition, which models the behaviour of basal erosion at the slope foot.                                KSR

**Reading and References**
Richards, K.S. 1982: *Rivers: form and process in alluvial channels*. London: Methuen. · Sumner, G.N. 1978: *Mathematics for physical geographers*. London: Edward Arnold. · Wilson, A.G. and Kirkby, M.J. 1975: *Mathematics for geographers and planners*. Oxford: Oxford University Press.

**boundary layer**   When a fluid and a solid are in relative motion the boundary layer is the zone in the fluid closest to the solid surface within which a velocity gradient develops because of the retarding frictional effect of contact with the solid. The fluid is at rest relative to the solid immediately adjacent to the surface, but with distance from the surface the frictional effect diminishes and velocity increases, at a rate dependent on the local flow characteristics. The velocity gradient of the boundary layer occurs in overland flow on hillslopes, river flow in channels, the swash and backwash of beaches, airflow over a desert dune, but also immediately adjacent to a sand grain falling through the water or air, where it is the *relative* motion producing the boundary layer, rather than the fluid motion over a static solid surface. The diagram illustrates the development of a boundary layer over a surface parallel to the direction of motion within deep water flow. At the entry point where flow begins over the surface, a laminar boundary layer forms, but at some distance downstream, this is replaced by a turbulent boundary layer if the flow conditions are appropriate; that is, if the REYNOLDS NUMBER exceeds about 2000. The boundary layer is 'fully developed' if the velocity profile extends to the surface of the flow, which is normal in rivers. In a deep fluid layer, however, the motion at some distance from the surface is unaffected by the boundary influence and the velocity is that of a free, or external, fluid stream.

Within a laminar boundary layer viscous forces within the flow are pronounced, and adjacent fluid layers are affected by the molecular interference of the fluid viscosity. The velocity increases with distance from the solid in an approximately parabolic curve (Allen 1970, pp. 36–9). In a turbulent boundary layer, the pattern of velocity increase with distance from the bed is very complex. Close to the bed the fluid is sufficiently retarded for viscous effects to be pronounced and laminar flow occurs; this is the very thin 'laminar sublayer'. If grains on a sedimentary surface are smaller in diameter than the thickness of the laminar sublayer the flow is 'hydrodynamically smooth', and the grains are

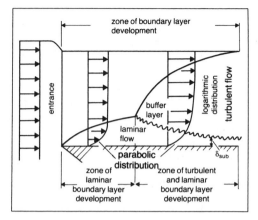

protected against entrainment. In the HJUL-STRÖM CURVE, threshold velocities are seen to be higher for silt and clay sizes than for sand sizes. Above the laminar sublayer is a buffer zone before the true turbulent velocity profile is reached. In the turbulent boundary layer interference between fluid elements occurs at a scale controlled by the depth of eddy penetration, and measured velocity profiles indicate that velocity increases with the logarithm of distance from the surface (Richards 1982, pp. 68–72). Under the BOUNDARY CONDITIONS defined above it is shown that this relationship takes the form:

$$v = K \ln \left( y/y_0 \right)$$

where $v$ is velocity, $y$ is distance from the surface, $y_0$ is the height where the velocity is zero, and $K$ is a constant which is equal to $v_*/K$. Here, $K$ is the von Karman constant (0.4) and $v_*$ is the 'shear velocity', defined as $\sqrt{\tau_0/\rho_w}$. This is a measure of the steepness of the velocity profile which is dependent on bed shear stress and water density. In hydrodynamically rough conditions where grains are large relative to the thickness of the laminar sublayer, $y_0$ equals one-thirtieth of the $D_{65}$ grain diameter. If the above equation is converted to common logarithms and the expressions for $K$ and $y_0$ are inserted, the equation for the logarithmic velocity profile in a hydrodynamically rough, turbulent boundary layer, becomes:

$$v/v_* = 5.75 \, log \left( y/D_{65} + 8.5 \right)$$

This equation may be used to fit to velocity data from the lower 10–15 per cent of the flow in order to project the curve to the bed. The local bed shear stress can then be estimated, as well as the velocity close to the bed sediment at heights where measurement is impractical, especially in the field. Note that both the turbulent fluctuations and the rapid increase of velocity above the bed material, causing strong lift forces, which occur under hydrodynamically rough bed conditions in turbulent flow, are important factors in the entrainment of sediment by the flow.　　　　KSR

### Reading and References

Allen, J.R.L. 1970: *Physical process of sedimentation*. London: Allen & Unwin. · Leeder, M.R. 1982: *Sedimentology: process and product*. London: Allen & Unwin. Pp. 47–66. · Open University 1978: Oceanography Unit 11 *Introduction to sediments*. Milton Keynes: Open University. Pp. 10–16. · Richards, K.S. 1982: *Rivers: form and process in alluvial channels*. London and New York: Methuen.

**bounding surface** is used to describe a break between different primary sedimentary structures. Bounding surfaces may occur in a hierarchy with major surfaces representing hiatuses in deposition due to factors that include climate change, and others representing hiatuses in episodic deposition and/or the erosion of one unit prior to deposition of the next. The term is particularly used in describing aeolian deposits.　　　　DSGT

### Reading

Brookfield, M.M 1977: The origin of bounding surfaces in ancient aeolian sandstones. *Sedimentology* 24, pp. 303–32. · Kocurek, G. 1988: First order and super bounding surfaces in eolian sequences: bounding surfaces revisited. *Sedimentary geology* 56, pp. 193–206.

**bourne** A stream or stream channel on chalk terrain that flows after heavy rain.

**Bowen ratio** is named after American astrophysicist Ira S. Bowen (1898–1978). The ratio of heat energy used for sensible heating (conduction, convection) to the heat energy used for latent heating (evaporation of water, sublimation of snow). Applicable to any moist surface, the Bowen ratio ranges from near zero for ocean surfaces to greater than two for desert surfaces, with negative values possible. The Bowen ratio is often employed in the surface energy balance equation in order to estimate fluxes of latent heat in the 'Bowen ratio–energy balance' method.　　　　AWE

**Bowen's reaction series** A series of minerals which crystallize from molten rock of a specific chemical composition, wherein any mineral formed early in the chain will later react with the melt, forming a new mineral further down the series; the minerals formed under decreasing temperatures of crystallization are more stable in the weathering environment.　　　　ASG

**brackish** Pertaining to water which contains salt in solution, usually sodium chloride, but which is less saline than seawater.

**braided river** A river whose flow passes through a number of interlaced branches that divide and rejoin. The term has been applied both to short reaches where a river splits around an island and to very extensive river networks on valley bottoms or alluvial plains, the whole of which may be criss-crossed by rapidly shifting channels with freshly deposited sediment between them. Braiding may be more apparent at some low levels than at others. For example, single channels of low sinuosity at high flows may assume a braided pattern as channels thread their way between sets of emergent BEDFORMS at low flow. By contrast, single channels may become multiple ones at high flows as inactive channels are reoccupied and developed.

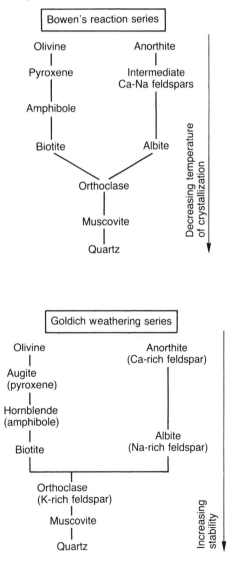

The term 'braided' is applied in a general sense to a whole family of multiple channel river patterns some of which have recently been given separate names. The term applies particularly to 'anastomosing' and 'wandering' rivers. The former, as identified by D.G. Smith in Canada, is a type of stable multi-channel system developed under aggrading conditions with levées and backswamps. They closely resemble deltaic distributary channel patterns, though they are found in some inland valleys. It is confusing that 'anastomosing' was earlier also used as an alternative for braiding in the general sense. 'Wandering' rivers, first identified in this sense by C.R. Neill, also in Canada, may consist of alternate stable single channel reaches and unstable multi-channel 'sedimentation zones'. The term has also been used as an alternative for patterns that are transitional between meandering and braided. Examples of these several variants or relations to braided channel patterns are discussed in chapters by D.G. Smith, M. Church, R.I. Ferguson and A. Werritty in Collinson and Lewin (1983).

Braided river patterns – in the general sense of multi-channel systems – appear to be created in various ways. Mid-channel bar development may lead to division of the channel and enlargement of the bar by accretion, possibly with the development of a vegetated island. Alternatively, migratory bars, exposed only at low flows, may simply be exposed bedforms continually shifting at high flows by erosion and accretion. Scour at channel junctions may be important for the local entrainment of sediment which is then redeposited as flows diverge again down-channel. Overbank flood flows also scour out new chute channels.

Particularly in view of the various channel-dividing or multiplying processes involved, not to mention the different kinds of pattern and pattern change, it is not surprising that various conditions and environments have been identified as conductive to the development of braided rivers. These include high energy environments (with steep-gradient channels and high or variable discharges) and high rates of sediment transport. It has been suggested that the braided pattern is one form of channel adjustment to prevailing hydraulic conditions, though it is not yet possible to predict exactly when and why braiding will occur. In general braiding may be found in contemporary ice-marginal gravel rivers (where fluctuating high discharges, high sediment supply rates and steep gradients may be combined) and in some semi-arid sandy rivers (where at least the first two may be common). Localized braiding, in the form of semi-permanent islands, may be widely found. Many of the world's largest rivers have braided lower courses. These include the Amazon, the Brahmaputra and the Hwang He. In North America, braided rivers are found in the arid south-west, on the Great Plains, and in glacier marginal environments in the Rockies and Coast Ranges of Western Canada. In Britain, braided rivers are rare, with a few good examples in Scotland but almost none elsewhere (see CHANNEL CLASSIFICATION).    JL

**A braided river** channel in the Rakaia valley, New Zealand. Such multi-thread channels tend to develop in areas with coarse debris, relatively steep slopes and variable discharge.

**Reading and References**
Collinson, J.D. and Lewin, J. eds. 1983: *Modern and ancient fluvial systems*. Special publication no. 6, International Association of Sedimentologists. Oxford: Blackwell Scientific. · Richards, K. 1982: *Rivers*. London and New York: Methuen.

**brash**   A mass of fractured rock that has been weathered *in situ*, also applied to a mixture of shattered rock or ice.

**Braun-Blanquet scale**   Standard means of measuring plant cover in a sample quadrat. The phytosociological scheme was developed by Josias Braun-Blanquet, a Swiss ecologist (1884–1980), who devised this widely used method for the quantitative description of vegetation COMMUNITIES and published it in his 1928 text book *Pflanzensoziologie*. The method is based on a five-point scale accounting for plant species cover abundance as follows:

| | |
|---|---|
| 5 | Cover over 75% |
| 4 | Cover 50–75% |
| 3 | Cover 25–50% |
| 2 | Cover 5–25% |
| 1 | Cover 1–5% |
| + | Species present but with negligible cover |

When cover values are computed for all the species in a quadrat, and for a number of quadrats in a community, it is possible to arrange the data in such a way as to identify typically recurring groups of species, or plant ASSOCIATIONS. Although not widely utilized by English-speaking ecologists, the Braun-Blanquet scheme was adopted in Europe and was influential in the development of the theoretical principles around the identification of plant associations and communities. Braun-Blanquet developed further five-point scales to account for, for example, the constancy (presence) of species within communities. Constancy is a measure of the evenness or otherwise of species distribution and may be expressed as the frequency of occurrence of particular species either in sample quadrats within a community or in stands between different communities. The scale is as follows:

| | |
|---|---|
| 5 | Frequency 81–100% |
| 4 | Frequency 61–80% |
| 3 | Frequency 41–60% |
| 2 | Frequency 21–40% |
| 1 | Frequency 1–20% |

Species classified as '1' under this scheme are regarded as 'rare', whereas those with a score of '5' are 'constant'. (See COVER, PLANT.)   MEM

**breccia**   A rock that has been greatly fractured into angular fragments, generally less than 2 mm in diameter, by tectonic activity, volcanism or transport over short distances.

**brodel**   A highly contorted and irregular structure in soils which have been subjected to churning by frost processes.

**brousse tigrée**   Vegetation banding, which may include grassland patterns but which generally consists of bands of more closely spaced trees alternating with bands of sparser vegetation. Its nature and origin have been well described thus by Mabbutt and Fanning (1987, p. 41): 'All are developed in arid or semi-arid areas, in open low woodlands or tall shrublands, with average annual rainfalls of between 100 and 450 mm; they occur on slopes of the order of 0.25%, too gentle for the development of drainage channels, but steep enough to maintain organized patterns of sheetflow; these slopes are mantled with alluvium or colluvium and the patterns are independent of bedrock. The associated soils are earths, and sandier crests or clay flats in the same areas do not have tree bands. The bands of denser vegetation, termed "vegetation arcs" run close enough to the contour to serve as form lines; hence they tend to be convex downslope on interfluves and convex upslope in shallow drainage ways. In drier areas the banding may be restricted to the better-watered depressions, but it is commonly best-developed on low interfluves, with the intervening depressions marked by uniformly dense tree cover. Such tracts of more concentrated sheetflow have been named "water lanes".

The bands commonly occur in fairly regular sequences or ladder-like "tiers" downslope, the tiers being bounded by water lanes. Tree bands may extend up to a kilometre, or more along the contour, but in detail they are commonly slightly irregular, "burgeoning here and becoming attenuated there; dying out and succeeding one another *en echelon*".

The downslope distance between bands ranges from 70–500 m, although it is mainly between 100 and 250 m and the interband intervals are commonly between two and four times as wide as the bands.'   ASG

**Reference**
Mabbutt, J.A. and Fanning, P.C. 1987: Vegetation banding in western Australia. *Journal of arid environments* 12, pp. 41–59.

**Brückner cycle**   A series of cold, wet seasons followed by a series of hot, dry ones which recur regularly over a period of about 35 years.

**Brune curve** The empirically based formula published by Brune (1953) linking the size of a reservoir to its efficiency in trapping sediment carried into it by streamflow. Brune used data from 44 reservoirs of varying size, all in the USA. He established a curve linking sediment trap efficiency to the ratio of dam storage volume C to the annual inflow volume I. This showed that reservoirs having I/C ratios of around 0.1 have trap efficiencies approaching 90 per cent for particles of silt size and larger. The ratio used by Brune is proportional to the average retention time of the impoundment, and thus reflects opportunities for sediment particles to settle. DLD

**Reference**
Brune G.M. 1953: Trap efficiency of reservoirs. *Transactions of the American Geophysical Union*, 34, pp. 407–18.

**Brunhes–Matuyama** The magnetic polarity epoch boundary (see PALAEOMAGNETISM) marking the major change or reversal in the Earth's magnetic field which occurred at around 730,000 years (determined by *K–Ar* dating) or 780,000 years before present (based upon dating using the oxygen-isotope record preserved within sea-floor sediments). The Brunhes–Matuyama boundary marks the change between the Matuyama epoch when the magnetic field was reversed and aligned south to north and the present day Brunhes polarity epoch where the magnetic field is aligned 'normally' (i.e. north to south). DJN

**brunizem** A prairie soil developed under grassland in temperate latitudes. Characteristically a brown surface zone overlies a leached horizon which grades into a brown subsoil on non-calcareous bedrock.

**Bruun rule** An empirical equation designed to predict absolute horizontal shoreline recession (*r*) arising from absolute sea level rise (*s*).

$$r = ls/h$$

where *l* and *h* are the length and height of the equilibrium cross-shore profile from beach crest to offshore. Shoreline recession is caused by the upward and landward movement of the cross-shore profile as it moves to readjust to the disequilibrium caused by a rise in sea level. Measurements of *l* and *h* are difficult as the seaward edge of the equilibrium profile has to be established. This position (closure depth) should relate to the start of onshore sediment transport by waves, but is usually regarded as variable depending on arbitrary definitions of the maximum wave causing such transport. The closure depth has in recent years come to

be defined by some multiple of significant storm-wave height associated with a return period of *n*-years (Hands 1983). The original study (Bruun 1962) related to measured profile recession of Florida barrier-island shorelines over twenty years. Schwartz (1967) thought that the equation had universality sufficient to indicate the status of a rule, but subsequent work indicates that this is an overstatement. Widespread use of the rule to establish building set-back lines on eroding coasts by coastal managers has been controversial. Doubts have been expressed about the universal validity of such cross-shore profile analysis when beach changes are dominated by long-shore sediment supply. The rule has been championed by workers with experience mainly of the open barrier beaches (i.e. USA). Recession of gravel-dominated beaches on closed or crenellate coasts does not conform to this absolute rule, though *r* and *s* are positively correlated when cited as rates of change (Orford *et al.* 1991). It is important to realize that sea level *per se* does not cause recession; it is merely the datum upon which waves and tides, which do the work of profile alteration, operate. JO

**References**
Bruun, P. 1962: Sea level rise as a cause of shore erosion. *Journal of the Waterways and Harbour Division of the American Society of Civil Engineers* 88, pp. 117–30. · Hands, B. 1983: The Great Lakes as a test model for profile responses to sea level changes. In P.D. Komar ed., *Handbook of coastal processes and erosion*. Boca Rotan, Florida: CRC Press, pp. 167–89. · Orford, J., Carter, R.W.G. and Forbes, D.L. 1991: Gravel barrier migration and sea level rise: some observations from Story Head, Nova Scotia, Canada. *Journal of coastal research* 7, pp. 477–88. · Schwartz, M.L. 1967: The Bruun theory of sea level rise as a cause of shoreline erosion. *Journal of geology* 73, pp. 528–34.

**Bubnoff units** provide a means for quantifying rates of slope retreat or ground loss (perpendicular to the ground surface). A unit equals 1 mm per 1000 years, equivalent to $1\,m^3\,km^{-2}$ (Fischer 1969). ASG

**Reference**
Fischer, A.G. 1969: Geological time-distance rates: the Bubnoff unit. *Bulletin of the Geological Society of America* 80, pp. 594–652.

**buffer** A solution to which large amounts of acid or alkaline solutions may be added without markedly altering the original hydrogen ion concentration (pH).

**buffer strip** A belt of vegetated land, generally running continuously along the banks of a stream, and maintained with the intention of protecting the stream habitat from disturbance related to land use, perhaps agriculture or forest logging, occurring beyond the buffer strip. Buffer strips are commonly required by the legislation that governs timber harvesting in forests that are also important for water supply. The intention is that runoff water carrying eroded soil will be slowed, and some of the eroded particles trapped in the buffer strip, before reaching and polluting the stream. Nutrients in agricultural runoff may also be held in the soils of the buffer strip. Strips of preserved forest are commonly 20–100 m in width, and may be required along all perennial water courses in the area being logged. Buffer strips are also used to trap soil eroded from tilled fields, and may take the form of grassed or wooded zones at the hillslope foot (Daniels and Gilliam 1996). Buffer strips may also serve as corridors of preserved habitat for wildlife, and by shading of the water course may limit any rise in water temperature that would occur following exposure to solar heating.

Much attention has been paid to the width required for efficient protection of the stream habitat. Experimental work generally suggests that a buffer of 30–50 m offers useful protection.                                                         DLD

**Reference**
Daniels R.B. and Gilliam J.W. 1996: Sediment and chemical load reduction by grass and riparian filters. *Soil Science Society of America journal* 60, pp. 246–51.

## buffering capacity
A buffer is a chemical compound that has the capacity to absorb or exchange hydrogen or hydroxide ions, and that allows the system to assimilate a limited amount of these ions without changing appreciably in pH. Buffering capacity is the quantitative ability of a solution to absorb hydrogen or hydroxide ions without undergoing a pronounced change in pH.

Soils are more strongly buffered in comparison with precipitation or freshwater. Different buffering systems come into play at particular ranges of soil pH. Carbonate minerals buffer soil pH within the range > 8 to 6.2; silicates from 6.2 to 5.0; cation exchange capacity from pH 5.0 to 4.2; aluminium from pH 5.0 to 3.5; iron from pH 3.8 to 2.4; and humic acids from pH 6 to 3.     ALH

## bulk density
The relationship of the mass of a soil or sediment to its volume, typically expressed in g cm$^{-3}$, using either a naturally damp or oven-dried sample (from which the wet and dry bulk density, respectively, can be calculated). Bulk density is measured in the laboratory from a sample which has been extracted using an open-ended metal cylinder of known volume. The cylinder is driven into

the ground, carefully removed, has caps placed on either end, and is then stored in a polythene bag. If the wet bulk density is required, the mass of the sample is determined by subtracting the weight of the cylinder plus caps from the total weight of the sample, cylinder and caps. For dry bulk density, the sample is carefully removed from the cylinder, dried at 105 °C, and weighed. In both cases, the volume of the sample can be determined by measuring the radius and length of the sampling cylinder. The bulk density can then be determined by dividing the appropriate weight by the original volume of soil. Dry bulk density is a parameter used, along with the density of particles within the sample, in determining the porosity of a soil or sediment, with wet bulk density used when estimating soil moisture content.                                                    DJN

**Reading**
Rowell, D.L. 1994: *Soil science: methods and application.* Harlow: Longman. Pp. 67–9.

## bush encroachment
is the process whereby shrubs come to dominate areas that previously were largely open grassland or mixed grass and woodland. It has been described from southern, eastern and Sahelian Africa, Australia and South America. The term is largely applied to SAVANNA environments and has widely been attributed to disturbances caused by the impact of marked grazing pressure by domestic livestock (e.g. Walker *et al.* 1981). Bush encroachment is seen as a major threat to sustainable productivity in dryland pastoral systems, but its occurrence and persistence can prove difficult to assess given the inherent temporal and spatial variability of disequilibrium savanna ecosystems (Behnke *et al.* 1993).

Several models have been expounded to explain the process of bush encroachment. Soil-based models (e.g. Walker and Noy-Meir 1982) assume that grasses have a competitive advantage over shrubs and trees in tapping rainfall that infiltrates into the ground, the so-called 'two-layer' model). Changes within soil resources, in terms of the vertical distribution of moisture and nutrients, may alter the balance in favour of shrub species. Ecological models (e.g. Westoby *et al.* 1989) place an emphasis on changes in grazing strategies and in the occurrence of fire, the latter seen as a natural mechanism that suppresses the potential for shrubs to become dominant. Clearly, soil and ecological elements are interlinked and are both affected by grazing patterns and natural events such as droughts. Recent empirical research is suggesting that soil water and nutrient distribution changes are not evident in areas where bush encroachment has occurred in central southern Africa (Dougill *et al.* 1999), and that interactions between fire regimes, grazing levels and rainfall distributions are the key to understanding the occurrence and potential persistence of bush encroachment.                                    DSGT

### References

Behnke, R.H., Scoones, I. and Kerven, C. 1993: *Range ecology at disequilibrium: new models on natural variability and pastoral adaptation in African savannas*. London: ODI. · Dougill, A.J., Thomas, D.S.G. and Heathwaite, A.L. 1999: Environmental change in the Kalahari: integrated land degradation studies for nonequilibrium dryland environments. *Annals, Association of American Geographers* 89, pp. 420–42. · Walker, B.H. and Noy-Meir, I. 1982: Aspects of the stability and resilience of savanna ecosystems. In Huntley, B.J. and Walker, B.H. eds, *Ecology of tropical savannas*. Berlin: Springer-Verlag. Pp. 556–90. · Walker, B.H., Ludwig, D., Holling, C.S. and Peterman, R.S. 1981: Stability of semiarid savanna grazing systems. *Journal of ecology* 69, pp. 473–98. · Westoby, M., Walker, B.H. and Noy-Meir, I. 1989: Opportunistic management for rangelands not at equilibrium. *Journal of range management* 42, pp. 266–74.

**bushveld**   The savanna lands of sub-Saharan Africa, ranging from open grassland, through parkland with scattered trees to dense woodland.

**butte**   A small, flat-topped and often steep-sided hill standing isolated on a flat plain. Often attributed to erosion of an older landsurface, the butte representing a remnant or outlier.

**Buys Ballot's law**   An observer in the northern hemisphere, standing with his back to the wind, will have low pressure to his left and high pressure to his right; the converse is true in the southern hemisphere. This law was formulated in 1857 by the Dutch meteorologist Buys Ballot. (See also CORIOLIS FORCE; GEOSTROPHIC WIND.)
                                    BWA

**bypass flow**   The movement of water through the soil along a pathway other than that provided by the microscopic pore spaces within the soil matrix. Among the alternate pathways that may be available are shrinkage cracks, faunal burrows, and voids left following the decay of plant roots. These may be classified as 'macropores', if > 1 mm in diameter. In many cases, macropores occupy a few per cent of the total soil volume, and a somewhat larger fraction of the total void space present in the soil. The significance of bypass flow is that the larger conduits that carry the flow allow much faster flow than is possible in the laminar conditions arising in the ordinary network of soil micropores that are only $\mu$m in diameter. Consequently, water may travel downslope towards streams much more rapidly when the soil matrix is bypassed. This allows rain water to provide a greater contribution to channel flow than would be possible if all flow were through the soil matrix. Furthermore, there is less opportunity for rapidly delivered water to have its properties, such as acidity, moderated within the soil, since the area of contact with the soil is small, and the period of residence short. In areas affected by acid rain, a major contribution to streams via bypass flow may result in a rapid drop in pH that is taxing for aquatic biota. Vertical bypass flow may carry water downward through the soil column rapidly, and result in very rapid water table fluctuations. If there is a significant volume of macropores that are not interconnected, the filling of these may delay the onset of runoff.

An important feature of bypass flow is that it can transmit water rapidly through soils whose matrix is not saturated. Normally, the hydraulic conductivity of soils declines markedly as they fall below saturation, since the largest pores drain first, leaving only small pores to transmit flow. The rapid transmission of water under unsaturated conditions can also be achieved in materials lacking macropores, through what has been termed *preferential flow* (e.g. Stagnitti *et al.* 1995). This may involve concentrations of flow along restricted paths through a medium that is relatively homogeneous, which arise from instabilities in the wetting front that may preferentially wet-up 'fingers' of soil extending to considerable depth.                                    DLD

### Reading and Reference

Beven K. and Germann P. 1982: Macropores and water flow in soils. *Water resources research* 18, pp. 1311–25. · Chen C. and Wagenet R.J. 1992: Simulation of water and chemicals in macropore soils. Part 1. Representation of the equivalent macropore influence and its effect on soilwater flow. *Journal of hydrology* 130, pp. 105–26. · Dingman S.L. 1994: *Physical hydrology*. New Jersey: Prentice-Hall. · Stagnitti F., Parlange, J.-Y., Steenhuis T.S., Boll J., Pivetz B. and Barry, D.A. 1995: Transport of moisture and solutes in the unsaturated zone by preferential flow. In V.P. Singh ed., *Environmental hydrology*. Dordrecht: Kluwer. Pp. 193–224.

**bysmalith**   A plutonic plug or mass of igneous rock which has been forced up into the overlying rocks causing them to dome up and fracture.

# C

caatinga   A form of thorny woodland found in areas such as north-east Brazil, and characterized by many xerophytic species.

caballing   The mixing of two water masses of identical *in situ* densities but different *in situ* temperatures and salinities, such that the resulting mixture is denser than its components.   ASG

caesium-137 analysis   The use of spatial patterns of accumulation or depletion of the anthropogenic isotope caesium-137 to determine rates of sedimentation and erosion. Caesium-137 ($^{137}$Cs), a fallout product from thermonuclear explosions, does not occur naturally. It decays with a half-life of 30 years. $^{137}$Cs analysis provides a means of rapidly assessing sediment loss or accumulation rates through the decades since 1954, when the isotope was first released from weapons testing.

$^{137}$Cs is delivered to the surface via precipitation and gravitational settling, where it is adsorbed strongly on clay and silt sized particles. Thereafter, the isotope is redistributed as soil particles are transported by erosion (Martz and DeJong 1987). Assuming that the initial distribution of $^{137}$Cs across the landscape is relatively uniform, areas with higher $^{137}$Cs contents indicate depositional sites, while those with low caesium-137 content are sites where erosion processes are active. Stable sites should have $^{137}$Cs contents that reflect the initial input of the radio-isotope, less the loss owing to radioactive decay, and are used as reference locations. Consequently, samples are collected for $^{137}$Cs analysis, and their caesium-137 contents related to that of the reference site in order to examine whether erosion or deposition has taken place.

The unit of radioactivity is the becquerel (Bq), which is defined as one nuclear transition per second, and $^{137}$Cs concentrations are commonly expressed as milli becquerels per gram of sample (mBq g$^{-1}$) or converted to the area function of milli becquerels per square centimetre of sample (mBq cm$^{-2}$).   KB

**Reading and Reference**
Campbell, B.L., Loughran, R.J. and Elliot, G.L. 1988: A method for determining sediment budgets using caesium-137. In *Sediment budgets*. Proceedings of the Porto Alegre Symposium, December. IAHS Publication no. 174, pp. 171–9. · Martz, L.W. and DeJong, E. 1987: Using caesium-137 to assess the variability of net soil erosion and its

association with topography in a Canadian prairie landscape. *Catena* 14, pp. 439–51.

Cainozoic   (Cenozoic) A geological era spanning the Palaeocene, the Eocene, the Oligocene, the Miocene, the Pliocene and the Pleistocene. It was a time of climatic decline, possibly associated with the breaking of the super-continent of Pangaea into the individual continents we know today, which moved into high latitudes so that ice caps could develop.   ASG

calcicole   A plant which flourishes with a large amount of exchangeable calcium in the soil. Examples include wood sanicle (*Sanicula europaea*) and traveller's joy (*Clematis vitalba*). Plants which clearly cannot tolerate such conditions are calcifuge; examples include common heather (*Calluna vulgaris*) and most other ericaceous plants. The effect of pH on mineral nutrition appears to be the operative factor.   KEB

**Reading**
Crawley, M.J. ed. 1986: *Plant ecology*. Oxford: Blackwell Scientific.

calcifuge   Any plant which grows best on acidic soils, e.g. bracken.

calcite compensation depth   is the critical depth in the oceans below which the rate of solution of the calcite crystalline form of calcium carbonate exceeds the rate of deposition. Calcite is chemically more stable than aragonite so that the latter is markedly more soluble at given temperatures and partial pressures of carbon dioxide. The two crystal forms, therefore,

Subdivisions of the Cainozoic era

|  | Date of beginning in millions of years |
| --- | --- |
| Pleistocene | 1.8 |
| Pliocene | 5.5 |
| Miocene | 22.5 |
| Oligocene | 36.0 |
| Eocene | 53.5 |
| Palaeocene | 65.0 |

*Source*: Berggren, W.A. 1969: Cainozoic stratigraphy, planktonic foraminiferal zonation and the radiometric time-scale. *Nature* 224, pp. 1072–5.

have differential solubilities in ocean water. Accordingly, the depth below which calcite dissolves, its compensation depth, is somewhat deeper than that of aragonite. (See CARBONATE COMPENSATION DEPTH.)                          MEM

**calcite saturation index** Measure of the degree to which water contains dissolved calcium carbonate in the calcite crystal form in relation to the amount it contains at saturation level. Saturation is, in turn, defined as the calcite concentration above which point it comes out of solution and is deposited. The index is calculated as the absolute concentration of dissolved calcite in a sample of water, divided by the saturation concentration for a given condition of water temperature, pressure, pH and partial pressure of carbon dioxide.                          MEM

**calcrete** A type of near-surface DURICRUST, predominantly composed of calcium carbonate, which occurs in a range of forms from powdery to nodular to highly indurated. It results from the cementation and displacive introduction of calcium carbonate into soil profiles, sediments and bedrock, in areas where vadose and shallow phreatic groundwater becomes saturated with respect to calcium carbonate. The term is synonymous with caliche and kunkur but distinct from other calcium carbonate cemented materials including cave deposits (such as SPELEOTHEMS), lacustrine STROMATOLITES, spring deposits (such as TUFA or travertine), marine deposits (such as BEACH ROCK) or cemented dune sand (AEOLIANITE).

Calcretes are generally white, cream or grey in colour, though mottling and banding is common. They are widespread in semi-arid areas such as the High Plains of the USA, the Kalahari Desert, southern Africa, in north Africa, Rajasthan, India and in western Australia, where evapotranspiration is in excess of precipitation. They can form by either pedogenic (soil-forming) processes or by direct precipitation from groundwater. Pedogenic calcretes develop through a variety of stages from powdery varieties to fully indurated hardpans, and typically form at the land surface. In contrast, groundwater calcretes may form at depths of several metres, with sites of carbonate precipitation linked to the water table position and to the presence of landscape depressions such as ephemeral lakes or palaeochannels. Carbonate source materials can be distant or local, can be moved either laterally or vertically to site of precipitation, and include weathered bedrock, surface runoff, dust, ground- and surface-water. Important mechanisms leading to the precipita-

tion of $CaCO_3$ include evapotranspiration, water uptake by soil organisms and shifts in environmental pH to above pH 9.0. The mean global chemical composition is 78% $CaCO_3$; 12% silica, 3% $MgO$; 2% $Fe_2O_3$; and 2% $Al_2O_3$, although variations in chemistry may occur dependent upon host sediment characteristics, cement composition, the presence of authigenic silica and silicates, the mode of origin and stage of development.                          DJN

**Reading**
Watson, A. and Nash, D.J. 1997: Desert crusts and varnishes. In D.S.G. Thomas ed., *Arid zone geomorphology: process, form and change in drylands*. Chichester: John Wiley. Pp. 69–107. · Wright, V.P. and Tucker, M.E. 1991: Calcretes: an introduction. In V.P. Wright and M.E. Tucker eds, *Calcretes*. Oxford: International Association of Sedimentologists/Blackwell Scientific. Pp. 1–21.

**caldera** A large, roughly circular, volcanic depression. Calderas usually have a number of smaller vents and can also contain a large crater lake. The distinction between volcanic craters and calderas is essentially one of size, one to two kilometres being the lower limit for the diameter of a caldera. Maximum diameters are in excess of 40 km. Calderas probably form in a variety of ways, but most proposed mechanisms attribute a primary role to collapse or subsidence, which may be related to explosive eruptions.                          MAS

**Reading**
Francis, P. 1983: Giant volcanic calderas. *Scientific american* 248, pp. 60–70. · Williams, H. and McBirney, A.R. 1979: *Volcanology*. San Francisco: Freeman, Cooper.

**calms** Winds with a velocity of less than one knot and which are represented by a force of zero on the BEAUFORT SCALE.

**calving** The breaking away of a mass of ice from a floating glacier or ice shelf to form an iceberg or brash ice (small fragments). Large tabular icebergs calve from ice shelves while smaller icebergs and brash ice are commonly produced by valley or outlet glaciers. Most calving is induced by stresses set up within the floating ice mass by ocean swell.                          DES

**cambering** The result of warping and sagging of rock strata which overlie beds of clay. The plastic nature of the clays causes the overlying rocks to flow towards adjacent valleys, producing a convex outline to the hill tops. Classic examples of cambering occurred in the Pleistocene when, under periglacial conditions, great rafts of limestone or sandstone subsided over lias and other clays along the escarpments of southern England. Cambering is often associated with the development of VALLEY BULGES. ASG

**canopy** Usually taken to be the uppermost stratum of woodland vegetation, the tree-top layer, though the term may also be used for any extensive above-ground leaf-bearing parts of plants. Despite the obvious importance of this zone in the interception of light and precipitation, and in the production of flowers, relatively little work has been reported, presumably due to practical difficulties. The role of the canopy in the woodland light climate, and therefore in tree regeneration and ground flora, is a vital one: some 80 per cent of incoming radiation may be intercepted in this zone and 10 per cent reflected from the upper leaves and twigs.

KEB

**Reading**
Crawley, M.J. ed. 1986: *Plant ecology.* Oxford: Blackwell Scientific. · Packham, J.R. and Harding, D.J.L. 1982: *Ecology of woodland processes.* London: Edward Arnold.

**capacity / non-capacity load** A classification of the load of sediment being carried by a stream, according to whether the stream has excess (unsatisfied) capacity, in which case it carries a *non-capacity load,* or is moving as much material as available stream energy permits (a *capacity* load). The figure shows that in general sediment transport capacity declines for increasingly large (and therefore heavy) particles. This is because a stone resting on the bed experiences an entraining drag force exerted by the water that is proportional to the area of the stone facing into the current. But the same stone experiences a retaining force (its weight) that is proportional to its volume. Given that for increasingly large stones, area increases as $(\text{diameter})^2$ while weight increases as $(\text{diameter})^3$, retaining forces outstrip drag. The figure also shows a notional curve of sediment supply, that suggests a similar decline in the supply of materials of increasingly large size. The bulk of the particles fed to most streams come from soil erosion in the catchment and along the channel banks, and are mostly fine. Fine particles are light, and can easily be carried in turbulent streamflow. The enormous capacity for the transport of fine sediment is exemplified by rivers carrying HYPERCONCENTRATED FLOWS. Thus, most streams carry a non-capacity load of fine sediments; this is clearly because, despite their available capacity, they are only fed limited quantities of fine materials by runoff from the

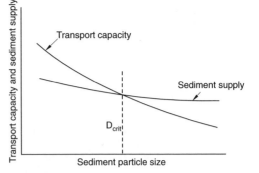

catchment. Thus, the supply curve crosses the capacity curve at some diameter $D_{crit}$. For particles larger than this, which are harder to transport, capacity generally lies below supply. The size corresponding to this diameter is approximately the silt/sand boundary (0.063 mm).

Many streams flow on beds of pebbles, cobbles, and other coarse particles. The way in which these materials move generates a number of sedimentological features that further contribute to their resistance to motion, and to their involvement in non-capacity transportation. These include ARMOURING, and cluster bedforms, in which smaller pebbles come to rest in the quieter water just downstream of larger, stationary, ones. These lee-side clusters of particles may be matched by upstream or stoss-side accumulations that come to rest as the flow passes up and over the obstacle clast that has triggered the accumulations. Both lee-side and stoss-side accumulations are protected from the force of the water and thus are not entrained. Imbrication, the geometric packing of platy particles so that they rest one upon the other, like a series of books that has been pushed over, also contributes stability to the particles concerned. DLD

**Reading**
Shen, H.W. 1971: Wash load and bed-load. In H.W. Shen ed., *River mechanics.* 2 vols. Fort Collins: H.W. Shen. Chapter 11, pp. 11–1 to 11–30.

**capillary forces** Essentially SURFACE TENSION and adsorptive forces. Water will rise up a narrow (capillary) tube as a result of adsorptive forces between the water and the tube surface and tension forces at the water surface. These forces bind soil moisture to the soil particles so that it is held in an unsaturated soil at less than atmospheric pressure. This is often called a SUCTION or tension and its strength may be determined using a TENSIOMETER.

AMG

**Reading**
Baver, L.D., Gardner, W.H. and Gardner, W.R. 1991: *Soil physics* 5th edn. New York: Wiley. · Smedema, L.K. and Rycroft, D.W. 1983: *Land drainage.* London: Batsford.

**cap-rock** A stratum of hard, resistant rock which overlies less competent strata and protects them from erosion.

**capture** (or river capture) The capture of part of one drainage system by another system during the course of drainage pattern evolution. Interpretation of drainage networks in terms of river capture was an integral feature of the Davisian CYCLE OF EROSION and the distance to base level or sea level, the exposure of easily eroded rocks, or the effects of discharge increase following climatic change could all be reasons why one river was able to erode more rapidly and so capture the headwaters of another. The beheaded stream becomes a misfit stream as it is now too small for the valley. River capture has certainly featured prominently in the evolution of world river systems and, for example, the easternmost tributary of the Indus was captured by the Ganges in geologically recent times, transferring drainage from a large area of the Himalayas from Pakistan to India. Knowledge of the sequence of river capture is sometimes necessary in the location of placer deposits which are alluvial deposits containing valuable minerals. Placer deposits from ore deposits may occur in an area no longer directly connected to the drainage system with ores outcropping in the headwaters (Schumm 1977). KJG

**Reference**
Schumm, S.A. 1977: *The fluvial system.* Chichester: Wiley.

**carapace** *a.* The upper normal limb of a recumbent fold.
*b.* A soil crust which is exposed at the surface, especially a surficial calcrete.

**carbon cycle** The 'life' cycle, carbon being one of the three basic elements (with hydrogen and oxygen) making up most living matter. Over 99 per cent of the earth's carbon is locked up in calcium carbonate rocks and organic deposits such as coal and oil, both being the result of millions of years of carbon fixation by living organisms on land and in the oceans. The biotic cycle is similarly split into terrestrial and oceanic subsystems. Photosynthesis by pigmented plants fixes the carbon dioxide from air and water; almost half is returned by plant respiration, the rest builds up as plant materials. The carbon is then returned to the atmosphere via animal respiration or plant decomposition. Fossil fuel consumption has increased atmospheric $CO_2$ fairly dramatically in the past few decades and is the basis of current concern over a GREENHOUSE EFFECT. KEB

**Reading**
Bach, W., Crane, A.J., Berger, A.L. and Longhetto, A. eds 1983: *Carbon dioxide.* Dordrecht, Holland: D. Reidel. · Bradbury, I.K. 1991: *The biosphere.* London: Belhaven Press. · Goudie, A.S. 1993: *The human impact.* 4rd edn. Oxford: Basil Blackwell.

**carbon dating** is the most widely used technique for dating carbon-bearing materials in the age range 0–40,000 years, and is therefore used in studies of late QUATERNARY palaeoenvironmental changes. It is based on measuring the relative abundance of radioactive carbon ISOTOPE ($^{14}C$) in comparison to a stable carbon ISOTOPE ($^{12}C$). $^{14}C$ is continuously formed at low approximately equilibrium levels in the atmosphere from the interaction of cosmic-ray neutrons with a stable isotope of nitrogen ($^{14}N$), where it is oxidized into carbon dioxide. Variations in the atmospheric concentration of $^{14}C$ are related to solar and other modulations of the cosmic ray flux. Interactions in the atmosphere-earth systems results in the fixation of carbon into a variety of biogenic and inorganic forms from where the radioactive decay of $^{14}C$ takes place. The radioactive decay of $^{14}C$ occurs, via emission of a beta (b-particle), as an exponentially declining trend with a HALF-LIFE of 5730 years.

The actual measurement procedure comprises either the counting of radioactive (b-particle) decay events (commonly referred to as conventional radiocarbon dating), or direct counting of abundances of stable ($^{12}C$) and unstable carbon atoms by accelerator mass spectrometric (AMS) methods. A major disadvantage of conventional radiocarbon dating is that only about 1 per cent of the $^{14}C$ atoms in a sample will emit a b-particle in about 80 years – only a very small portion of the total sample carbon content is therefore measured. As all the carbon atoms present in a sample are counted in the AMS method, the required sample size is less by several orders of magnitude. Not only does the AMS method's small sample size requirements (minimum size *c.*100 micrograms) provide access to a wider range of sample types than conventional methods (e.g., individual seed grains, included organic debris from rock varnish layers), but it also allows determinations to be made on separate chemical components from within samples – some of which may be more reliable for dating than others.

Ultimately both methods are limited by the half-life of $^{14}$C.

A variety of materials are datable by the radiocarbon methods, many of which are in fact relatively low in carbon content. While typically scarce, charcoal, wood and macrofossil fragments (including insects, chronomids and other aquatic organisms, bones, molluscs and snails) constitute the sample of preference for dating. Additional datable animal remains include faecal pellets, ivory, horn, hair and egg shells.

Application of radiocarbon methods to soil and sediment components has been an area or considerable investigation and debate. This relates to the open system nature of soils and the corresponding high likelihood of introduction of old or young carbon-bearing compounds and solutions via percolating groundwaters or modern root penetration. Application of radiocarbon methods to soil carbonates, and discrete CALCRETE and TUFA deposits have generally resulted in erroneous radiocarbon ages. This has been attributed to the incorporation of old carbonate in percolating groundwaters.

While the basic assumptions of the radiocarbon method are sound, a number of complicating factors must be considered in order to guarantee the generation of accurate absolute dates. Isotopic fractionation resulting from metabolic processes causes different kinds of samples to exhibit slightly different initial $^{14}$C activities to that of the atmospheric carbon dioxide reservoir from where it was derived. As the fractionation of the stable carbon isotopes ($^{13}$C/$^{12}$C) is proportional to that of the ratio $^{14}$C/$^{12}$C, it is possible to use the ratio of the more abundant and more easily measured stable isotopes to estimate, and correct for, the degree of fractionation within any given dating.

A further complication to radiocarbon dating lies in the need for calibration of the dates produced to determine absolute ages. This is necessary because the assumed constant $^{14}$C production rate (and therefore also the atmospheric ratio of $^{14}$C/$^{12}$C) is only approximately true due to subtle changes in the cosmic ray flux. Calibration is achieved by comparison to absolute chronologies generated from tree-rings, VARVES and from URANIUM SERIES dates on CORAL ALGAL REEFS.

In addition to correction for fractionation and calibration it is necessary to consider non-systematic factors which may influence radiocarbon evaluations for a given suite of samples. These include contamination from intrusion of younger material (e.g. roots, humic acids, bacteria), or the introduction of older materials; the 'hard-water effect' whereby ecological and hydrological systems fed by groundwater containing dissolved fossil carbonates produce organic materials (e.g. sapropels, shells) which are effectively depleted in $^{14}$C; and, the young age range of the method burning of $^{14}$C-free fossil fuels and the generation of $^{14}$C by atmospheric testing of nuclear devices has resulted in competing effects which limit the generation of radiocarbon dates over the last few centuries.
ss

**Reading**
Gehy, M. and Scheicher, H. 1990: *Absolute age determination: physical and chemical dating methods and their application.* New York: Springer-Verlag.

**carbon dioxide problem** Although the GREENHOUSE EFFECT and its enhancement by human actions is widely viewed as a major environmental issue, the so-called carbon dioxide problem which it is commonly associated with may also be viewed as a construct of computer-driven GENERAL CIRCULATION MODELLING. Some of these predict that a doubling of the atmospheric $CO_2$ concentration will enhance the planet's natural greenhouse effect to such an extent that it will lead to catastrophic global warming, which could melt the polar ice caps, flood coastal lowlands, produce simultaneous floods and droughts, cause havoc with agriculture, and lead to all manner of economic, social and political instability. Because of these dire potential consequences, $CO_2$ is often portrayed as a pollutant in the popular media, as on television, for example, where global warming is frequently discussed against a backdrop of industrial smokestacks spewing out ominous columns of smoke and ash. But, is there any justification for this verbal and visual abuse of $CO_2$, and is there any alternative perspective?

As defined in *Webster's New Universal Unabridged Dictionary*, a pollutant is something that makes things foul or unclean; i.e., pollutants taint, contaminate and defile things. In the case of carbon dioxide, however, we have a colourless odourless gas that cannot be seen nor smelled. It currently comprises only 0.036% of the air we breathe (Houghton *et al.* 1996); and although there is very little of it in the atmosphere, it is vital for nearly all forms of life. Without it, in fact, we would not be here, as carbon dioxide is the principal 'food' that plants use to construct their tissues (Wittwer 1995), which we either consume directly or indirectly when we eat animals that have fed upon them. Consequently, it is clear that carbon dioxide does not contaminate or defile things; it actually enhances and makes possible the very existence of life on earth. By all counts, then, $CO_2$ would seem to be just the opposite of a pollutant. Indeed, it is a vital atmospheric ingredient that makes our planet liveable and a veritable biological jewel in an immense sea of abiotic interstellar space (Idso 1995).

In view of these facts, it should not be surprising that not all scientists are overly concerned about the '$CO_2$ problem', for there are no hard data to confirm the ominous predictions of the climate models. Furthermore, the air's $CO_2$ content is known to have been several times higher than it is today in times past, without causing deleterious climatic effects. In fact, past climatic warm regimes have typically been referred to as 'optimum' periods of earth's climate history. In contrast to this muddled state of affairs on the climate side of the issue, there is no shortage of empirical data to demonstrate the direct biological benefits of increasing atmospheric $CO_2$ on vegetation. Kimball (1983a, 1983b), for example, produced two large reviews of this phenomenon which revealed that, on average, most agricultural and horticultural plants exhibit a 33% increase in productivity for an atmospheric $CO_2$ enrichment of 300 ppm. And in reviewing the plant science literature of the subsequent decade, Idso (1992) analysed the results of 342 peer-reviewed scientific journal articles that contained 1087 individual observations of plant responses to $CO_2$ enrichment, finding that a 300 ppm increase in atmospheric $CO_2$ enhanced plant productivity by an average of 52%.

In conclusion, although some scary theoretical scenarios involving $CO_2$-induced global warming have been produced by computer-driven climate models, the bulk of the hard scientific evidence would suggest that the increasing level of atmospheric $CO_2$ is not a problem, as $CO_2$ certainly is not a pollutant, but the primary – indeed, essential – raw material that is responsible for supporting nearly all life on earth. Perhaps the real carbon dioxide problem is that there still is not enough of this biological elixir in the air to bring the productivity of earth's vegetation back to its historic optimum of bygone years.                                    KEI

**References**

Houghton, J.T., Meira Filho, L.G., Callander, B.A., Harris, N., Kattenberg, A. and Maskell, K. 1996. *Climate change 1995: the science of climate change.* Cambridge: Cambridge University Press. · Idso, K.E. 1992. Plant responses to rising levels of atmospheric carbon dioxide: a compilation and analysis of the results of a decade of international research into the direct biological effects of atmospheric $CO_2$ enrichment. *Climatol. Pub. Sci. Pap.* 23.

Off. of Climatol.: Ariz. State Univ., Tempe. · Idso, S.B. 1995: *$CO_2$ and the biosphere: the incredible legacy of the industrial revolution.* University of Minnesota, St Paul, Minn.: Dept of Soil, Water & Climate. · Kimball, B.A. 1983a: Carbon dioxide and agricultural yield: an assemblage and analysis of 330 prior observations. *Agron. J.* 75, pp. 779–88. · — 1983b: *Carbon dioxide and agricultural yield: an assemblage of 770 prior observations.* Phoenix, Ariz.: US Water Conservation Lab. · Wittwer, S.H. 1995. *Food, climate, and carbon dioxide: the global environment and world food production.* Boca Raton, Fla.: CRC Press.

**carbonate compensation depth**   The critical depth in the oceans below which the rate of calcium carbonate solution exceeds that of deposition. It has long been observed that calcium carbonate remains of planktonic organisms are restricted to ocean floors above a certain depth. Below around 4000 to 5000 m in the Pacific, and somewhat deeper in the Atlantic and Indian Oceans, the calcium carbonate tests of plankton are dissolved. Controlling factors include pressure, partial pressure of dissolved carbon dioxide, temperature and pH. The solubility of calcium carbonate increases as the water temperature drops and, since ocean bottom water is relatively cold (2 to 3 degrees Celsius) and has higher pressures and carbon dioxide concentrations and correspondingly lower pH, this means that, in general, calcium carbonate goes into solution at such depths. (See also CALCITE COMPENSATION DEPTH.)                                    MEM

**carbonation**   The reaction of minerals with dissolved carbon dioxide in water. The process is dominant in the weathering of limestone, since rainwater contains a small proportion of carbon dioxide (0.03 per cent by weight) and thus acts as weak acid dissolving limestone rock. The conventional chemical reaction is shown in the following formula:

$$CO_2 + H_2O + CaCO_3 \rightleftharpoons Ca(HCO_3)_2$$

The $Ca(HCO_3)_2$ molecules have never been detected in solution and, while the product of carbonation is well known, the chemical process is not fully explained by this conventional equation (Picknett *et al.* 1976).                                    PAB

**Reference**

Picknett, R.G., Bray, L.G. and Stenner, R.D. 1976: The chemistry of cave water. In T.D. Ford and C.H.D. Cullingford eds, *The science of speleology.* London: Academic Press. Pp. 213–66.

**carnivore**   An animal-eating mammal of the order Carnivora, which depends solely on other carnivores or HERBIVORES for its food, and which is located in the higher TROPHIC LEVELS of ecological systems. Carnivores may be predators (e.g. the lion or wolf among the large land animals, many species of beetles, molluscs, centipedes and mites among the smaller); scavengers, such as jackals and seagulls; or animal parasites, including a wide range of bacteria, protozoa, nematodes and winged insects. Excepting the parasites, most carnivores are not restricted to a single species for their food supply; their ranges accordingly tend to be larger than those of the animals on which they depend.

DW

**carrying capacity** Represents the population size which the resources of an environment can just maintain without a tendency to decrease or increase. Begon *et al.* (1986, p. 209) explain it thus: 'As population density increases, the per capita birth rate eventually falls and the per capita death rate eventually rises. There must, therefore, be a density at which these curves cross. At densities below this point, the birth rate exceeds the death rate and the population increases in size. At densities above the cross-over point, the death rate exceeds the birth rate and the population declines. At the cross-over density itself, the two rates are equal and there is no net change in population size. This density therefore represents a *stable equilibrium*, in that all other densities will tend to approach it.

In other words, intraspecific competition, by acting on birth rates and death rates, can *regulate* populations at a stable density at which the birth rate equals the death rate. This density is known as the *carrying capacity* of the population and is usually denoted by *K*.'

ASG

**Reference**
Begon, M., Harper, J.L. and Townsend, C.R. 1986: *Ecology: individuals, population and communities.* Oxford: Blackwell Science.

*carse* A flat area of alluvium adjacent to an estuary.

**cascading systems** See SYSTEMS.

**case hardening** The feature or process of formation of a hard, resilient crust on the surfaces of boulders and outcrops of soft, porous rock through the filling of voids with natural cement. The cement may consist of a range of different materials, including iron and manganese oxides, silica and calcium carbonate. Beneath the hard surface the rock may be weakened, so that if the crust is breached, cavernous weathering may occur.

ASG

**cataclasis** The process of rock deformation accomplished by the fracture and rotation of mineral grains, as in the production of a crush breccia.

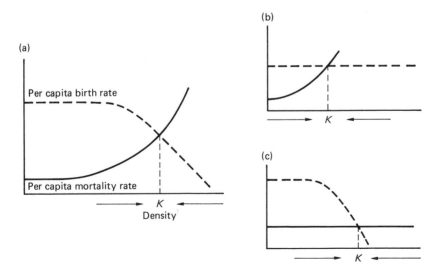

**Carrying capacity**. Density-dependent birth and mortality rates lead to the regulation of population size. When both are density-dependent (a), or when either of them is (b and c), their two curves cross. The density at which they do so is called the carrying capacity (K). Below this the population increases, above it the population decreases: K is a stable equilibrium. But this figure is a mere caricature of real populations.
*Source*: Begon *et al.* 1986, p. 210. Figure 6.5.

**cataclinal** Pertaining to a stream or river which trends in the same direction as the dip of the rocks over which it flows.

**catastrophe/catastrophism** In general use, the word catastrophe can be applied to any major, normally short-lived and sudden, misfortune leading to widespread change. Catastrophes can be found in both the physical and human environments, physical examples being storm surges, floods and hurricanes. In terms of catastrophe theory, however, catastrophes are more precisely defined events which affect systems and cause their organizations (Thom 1975). This varied use of the word catastrophe springs in part from the changing development of ideas, especially of those relating to earth history.

Catastrophism is a mode of thought that ascribes important change in the physical environment to the action of catastrophic events. It is often placed in direct opposition to the doctrine of UNIFORMITARIANISM which basically ascribes change in the physical environment to small-scale, commonly acting processes (see Gould 1984 for a more comprehensive description of the varied meanings of uniformitarianism).

The origins of catastrophism have often been traced to Baron Georges Cuvier (1769–1832). Cuvier was primarily a palaeontologist and he brought catastrophic ideas to the attention of his fellow geologists. The use of catastrophic episodes to explain earth history was necessitated by current religious and scientific views which held that the history of both rocks and the living world should be subsumed within biblical history. Two important data levels were revealed from the Bible, i.e. the Deluge and the Creation. Cuvier came to the conclusion that in order to incorporate the events revealed in the stratigraphic record, given a relatively short ordained time span of 75,000 years for the total history of the earth as accepted by most scientists at that time, sudden catastrophic changes needed to be invoked. Extinctions and structural discordances could only be explained in these terms. Cuvier's ideas were therefore at variance with those of James Hutton (1726–1797) who is regarded as one of the fathers of uniformitarian views.

Charles Lyell (1797–1875) somewhat unjustly saw catastrophism as being unscientific in its approach (Benson 1984). He was the first main exponent and propagator of uniformitarian views in the geological world and as a result of his persuasive arguments and those of others, uniformitarian views have dominated the study of geology and other earth sciences for over 100 years. Darwin's acceptance of uniformitarian views had great effect on his ideas on the progress of evolution.

Catastrophism and uniformitarianism can therefore be seen in their extreme forms to be two ends of a spectrum of approaches to explaining the physical environment. The development of ideas on the environment is, in part, affected by human history and cultural events which affect the way in which people think and view the world. When war and revolution are common, for example, catastrophic explanations of a variety of phenomena are likely to come into vogue.

In more recent years catastrophism has once again become accepted as a valuable explanatory tool. In palaeontology and biology, for example, the idea of 'punctuated equilibria' (i.e. discrete, sudden changes in species as opposed to gradual evolution) has gained support (Gould and Eldridge 1977) and there have been many explanations of past mass extinctions invoking catastrophic events.

Many geomorphological features can be most satisfactorily explained by recourse to catastrophic ideas (Dury 1980). A classic example of this is provided by the work of Bretz on the channelled scablands of eastern Washington. Bretz suggested in 1923 that these scablands could best be explained by the action of a single gigantic flood over a period of only a few days. Bretz's views did not achieve much recognition when they were published but they have since been shown to be broadly correct. Analogous features have recently been discovered on the surface of Mars and similar catastrophic explanations have been put forward to account for these.

Neocatastrophic views in the earth sciences have been strengthened by the development of catastrophe theory (Thom 1975). This complex, mathematical theory accounts for sudden changes in systems and may be used for modelling. Its potential has been recognized by several geomorphologists (e.g. Graf 1988), but it has received much criticism from mathematicians and its complexity has baffled many earth scientists.

Huggett (1990) provides a useful review of the history of catastrophism and its importance to biology, geology and geomorphology. It is clear that most environmental systems are affected by both catastrophic and gradual changes and the ideas of NON-LINEAR SYSTEMS recently introduced to the earth sciences attempt to include all such changes. Indeed, Huggett goes as far as to suggest that 'over the next few years catastrophic and gradual change will be unified by the theory of non-linear dynamics' (Huggett 1990, p. 200).    HAV

**Reading and References**
Benson, R.H. 1984: Perfection, continuity and common sense in historical geology. In W.A. Berggren and J.A. Van Couvering eds, *Catastrophes and earth history*. Princeton, NJ: Princeton University Press. Pp. 35–76. · Berggren, W.A. and Van Couvering, J.A. eds 1984: *Catastrophes and earth history*. Princeton, NJ: Princeton University Press. · Bretz, J.H. 1923: The channeled scablands of the Columbia plateau. *Journal of geology* 31, pp. 617–49. · Dury, G.H. 1980: Neocatastrophism? A further look. *Progress in physical geography* 4, pp. 391–413. · Gould, S.J. 1984: Toward the vindication of punctuational change. In W.A. Berggren and J.A. Van Couvering eds, *Catastrophes and earth history*. Princeton, NJ: Princeton University Press. Pp. 9–34. · —and Eldridge, N. 1977: Punctuated equilibria: the tempo and mode of evolution reconsidered. *Paleobiology* 3, pp. 115–51. · Graf, W.L. 1988: Applications of catastrophe theory in fluvial geomorphology. In Anderson, M.G. ed., *Modelling geomorphological systems*. Chichester: Wiley. · Huggett, R. 1990: *Catastrophism. Systems of earth history*. London: Edward Arnold. · Thom, R. 1975: *Structural stability and morphogenesis: an outline of a general theory of models*. Reading, Mass.: Benjamin.

**catchment control** The adjustment and arrangement of land use in a catchment so that as far as possible an appropriate quality and quantity of water suitable for distribution throughout the year can be ensured at minimum cost to the community.                    ASG

**Reading**
Newson, M. 1991: Catchment control and planning: emerging patterns of definition, policy and legislation in UK water management. *Land use policy* 8, pp. 9–15.

**catena** A sequence of contrasting soils formed along a topographic slope, which have acquired their different characteristics from differences in soil drainage, leaching, MICROCLIMATE erosion and depositional processes, and other factors which vary with slope position. The term was introduced by Milne (1936). In a catena, soils on steep upper sites may be well-drained and better aerated, and may lose easily erodible particles by water transport to flatter sites downslope where they come to rest. Footslope conditions may be generally wetter, less well drained and aerated, and the soils perhaps deeper owing to deposition of eroded materials, though very fine particles may leave the slope altogether. Soil differences of this kind can be further accentuated by associated differences in plant cover, soil biota, biological mixing, and in the supply of organic materials to the soil. The differentiated soils that make up a catenary sequence are often derived from the same parent materials and developed under the same climatic conditions. There may be differences in features like stone content, and in the sizes of rock particles found within the soils, if these materials are sorted during downslope migration from an outcrop near the crest. Along the catena, the upper layers of soils are most mixed by biota, and most susceptible to slope wash. Thus, along lower parts of the catena, the soils may be composed of relatively immobile lower layers derived primarily from the underlying bedrock together with mobile upper layers in which the material is actually undergoing progressive downslope movement.

DLD

**Reading and Reference**
Gerrard, J. 1992: *Soil geomorphology*. London: Chapman & Hall. Chapter 3, pp. 29–50. · Milne, G. 1936: Normal erosion as a factor in soil profile development. *Nature* 138, pp. 548–9.

**cation exchange** The interchange of cations between ADSORPTION sites on CLAY particles or organic matter, and the soil water. Clay particles generally carry a negative surface charge, which holds cations adjacent to the surface; many ions are also held in the interlayer spaces between the sheets which make up clay minerals. There are similar sites on organic macromolecules where cations can be held. Since most soils contain both clays and organic materials, cation exchange is often thought of as arising on the 'clay–humus' complex. The enormous surface area of clays, reaching hundreds of square metres per gram, is the other reason that clays are so important in retaining cations in the soil. The most important ions held are $Na^+$, $K^+$, $Ca^{2+}$ and $Mg^{2+}$, these being important plant nutrients, and the amounts held are expressed by the CATION EXCHANGE CAPACITY (CEC). In the soil, roots, root hairs, and MYCORRHIZAL FUNGI extract ions from the soil water. This increases the concentration gradient near the soil exchange complexes and additional ions are exchanged into solution, often with H+ taking their place. In soils affected by acidic precipitation, cation exchange is involved in the neutralization process, but this depletes the store of macronutrient cations available in the exchange complex.                    DLD

**Reading**
Cresser, M., Killham, K. and Edwards, T. 1993: *Soil chemistry and its applications*. Cambridge: Cambridge University Press.

**cation exchange capacity** A measure of the ability of the soil exchange complex, which consists of clays and organic matter (see CATION EXCHANGE), to supply the ions $K^+$, $Na^+$, $Ca^{2+}$ and $Mg^{2+}$ to the soil water. Cation exchange capacity (CEC) is determined in the laboratory by chemically extracting all of the available cations. It normally increases with the clay content of the soil, and is also influenced by the soil pH, not least because in acid soils, $H^+$ occupies some exchange sites. Commonly, amounts of the main cations held in exchange sites amount to 10–100 me (milli-equivalents) per 100 g of dry soil. Frequently, somewhat larger amounts are additionally held on the soil organic matter. Soils where the CEC is dominated by sodium are termed sodic soils and can be dispersible and unstable when wet. High CEC is associated with elevated soil mechanical strength.     DLD

**Reading**
Brady, N.C. and Weil, R.R. 1996: *The nature and properties of soils.* 11th edn. New Jersey: Prentice-Hall.

**cation-ratio dating** Biogenic ROCK VARNISH coatings or patinas formed on surface boulders, rock engravings and surface artefacts provide an opportunity to obtain chronological control on the timing of exposure of underlying surfaces. Such surface varnishes are common in DRYLANDS and may exhibit a detailed microstratigraphy when examined in thin section. As the ages thus established relate to the timing of COLONIZATION of exposed substrates they represent minimum ages of surface formation. The method is based on the observation that the ratio of certain cations ([potassium and calcium]/titanium) in varnish decreases with age. It is generally accepted that this is the result of preferential leaching of the more mobile potassium and calcium. The minimum age range of the method relates to the time required for initiation of a visible rock varnish. Visible varnishes form over periods of the order of a few thousand years while incipient varnish development may be observed microscopically over as little as a hundred years.

While the cation-ratio method requires calibration to other techniques to derive absolute age estimates, the relatively low costs (compared to AMS RADIOCARBON DATING or COSMOGENIC methods), its high speed of analysis, and ability to generate relative or absolute chronologies in varnishes which do not contain sub-varnish organic matter mean that the method has considerable utility despite its numerous limitations.     SS

**Reading**
Dorn, R.I. 1994: Surface exposure dating with rock varnish. In Beck, C. ed., *Dating in exposed and surface contexts.* Albuquerque: University of New Mexico Press. Pp. 77–114.

**causality** The relationship between events in which a second event or configuration (*B*) can be seen as the product of a prior event (*A*): in other words *A* is cause and *B* is effect. A simple causal relationship is one where *B* is only and always the result of *A*: an obvious example is the reaction of litmus paper to the application of an acid solution.

In physical geography – and in historical science in general – causality can rarely be established in a simple experimental fashion, but has to be inferred by repeated observations of *A* and *B*. Several problems arise. First, the joint occurrences of *A* and *B* may be fortuitous and there may be no physical connection between them. Secondly, both *A* and *B* may be responses to some other, truly causal event or variable, *C*, and the apparently direct causal link between them misleading. Thirdly, *A* may be a necessary but not a sufficient cause of *B*, i.e. some further agency or group of agencies is involved.

It is particularly difficult to infer causality with certainty when observations are spatially contiguous or coincident, although similar problems arise with temporal sequences.

**causse** A term synonymous with karst, derived from the name of the limestone landscape of the Central Massif of France.

**cave** A natural hole or fissure in a rock, large enough for a man to enter. Although caves can be found in any type of rock, they are most common in limestone regions and are formed by solutional processes of joint enlargement. Caves can be either horizontal or vertical in general form; the latter are usually termed potholes. Those produced by solutional processes are normally initiated (i.e. by joint enlargement) in the saturated or phreatic zone. Lowering of the water table allows normal stream or vadose conditions to cut canyons in the more circular phreatic cave tubes. Thus, compound cave cross-sections can result: in this specific case a keyhole-shaped passage is produced. (Indeed, the 20 km cave Agen Allwedd in South Wales is named from the Welsh: Keyhole Cave.) Solutional processes alone do not account for all limestone cave systems; often, when the water table lowers, the overburden of rock, now no longer supported by a water-filled cavity, collapses, producing extensive boulder falls in cave passages.

The general pattern of a cave system depends not only on the processes which have led to its formation but also on the regional jointing, folding and faulting. Caves develop along lines of weakness and the structural geology of the area will dictate the plan and depth of a cave almost as much as fluctuations in the water table.

Solutional caves can also form in rock salt, although such cavities usually form as isolated chambers rather than integrated cave passage networks. Ice too can provide solutional cave systems; some systems can be very long lasting (Bull 1983).

Although caves can also be produced by tectonic activity (which is regularly referred to in textbooks as a viable mechanism of cave formation), in practice they are few and far between. They form as cavities on the limbs and crests of tightly folded rocks but normally only very small recesses are formed, never long cave systems.

The largest of the cave systems formed in non-karstic rocks (PSEUDOKARST) are found in lava. Well-documented, long cave systems exist in Hawaiian lava flows (Wood 1976), sometimes exceeding 10 km in passage length. They are not of course the results of solutional processes but rather products of heat loss at the edges of lava flows, with corresponding continual flowing of molten lava in the core of the flow. Repeated eruptions utilize the same passages to transport their lava along these gently dipping tubes, perpetuating the lava cave system.          PAB

### References
Bull, P.A. 1983: Chemical sedimentation in caves. In A.S. Goudie and K. Pye eds, *Chemical sediments in geomorphology*. London: Academic Press. Pp. 301–19. · Wood, C. 1976: Caves in rocks of volcanic origin. In T.D. Ford and C.H.D. Cullingford eds, *The science of speleology*. London: Academic Press. Pp. 127–50.

**cavern**   See CAVE.

**cavitation**   Occurs in high velocity water (above $8$–$16\,\mathrm{m\,s^{-1}}$) in irregular channels, when local acceleration causes pressure to decrease to the vapour pressure of water and airless bubbles form. Subsequent local deceleration and increased pressure result in bubble collapse. This process is a manifestation of the conservation of energy; increased kinetic energy during flow acceleration is balanced by decreased pressure energy (see GRADUALLY VARIED FLOW). The bubble implosion generates shock waves that erode adjacent solid surfaces like hammer blows. Cavitation erosion occurs in waterfalls, rapids, and especially in subglacial channels where velocities of $50\,\mathrm{m\,s^{-1}}$ have been observed (Barnes 1956). Typical erosional products are pot-holes and crescent-shaped depressions called *sichelwannen*.          KSR

### Reference
Barnes, H.L. 1956: Cavitation as a geological agent. *American journal of science* 254, pp. 493–505.

**cellular automata**   An approach to distibuted numerical modelling in physical geography which explicitly recognizes that in many SYSTEMS, feedbacks between components of a system are spatially distributed. This follows from basic conservation laws, and notably the CONTINUITY EQUATION. For example, if one area of a bed of a stream erodes then the eroded material may well be available for deposition further downstream. Cellular automata have a number of key aspects: (1) process rules that apply to each cell; (2) strong dependence of these cell properties; (3) feedback between cell properties and process operations at the cell level; and (4) process rules that connect cells to one another. Despite widespread application in the physical and biological sciences in general, applications in physical geography are more unusual. A good example is provided by Murray and Paola (1994) who developed a cellular automata model for a BRAIDED RIVER. Their model divides the river into cells in the cross-stream and downstream directions and has: (1) a discharge-based BEDLOAD EQUATION; driven by (2) the DISCHARGE passing through each cell; which causes (3) the bed elevation of the cell to increase (where more sediment is supplied to the cell than can be transported) or decrease (where less sediment is supplied to the cell than can be transported); and where the discharge and sediment leaving any one cell is routed downstream, whilst being distributed laterally according to prevailing bed slope. The model produced sensible behaviour when tested using a range of specially developed quantitative methods. Cellular automata reflect one of the basic principles of any numerical model: they seem to have just enough process representation to capture the fundamentals of system behaviour. However, some view them as just another form of distributed numerical modelling, whilst others are concerned about the simplicity of process representation, and recognize the possibility that these models produce reasonable representations of real systems, but not necessarily for the right reasons. Nevertheless, these models may have much potential in any dynamic environmental system with strong spatial connectivity between system components.
          SNL

### Reading and Reference
Murray, A.B. and Paola, C., 1994: A cellular model of braided rivers. *Nature* 371, pp. 54–7. · —— 1998: Properties of a braided stream model. *Earth surface processes and landforms* 22, pp. 1001–25.

**Cenozoic**   See CAINOZOIC.

**cerrado**   A form of savanna vegetation, comprising grasses, small trees and tangled undergrowth, found in Brazil.

**channel capacity**   The size of the river channel cross-section to bankfull level, usually expressed as the cross-sectional area in square metres. Various possible definitions of the bankfull capacity have been used (Williams 1978), referring to: the heights of the valley flat, the active floodplain, the benches within the channel, the highest channel bars, the lower limit of perennial vegetation, the upper limit of sand-sized particles in the boundary sediment or the elevation at which the WIDTH–DEPTH RATIO becomes a minimum, where there is a first maximum of the bench index (Riley 1972), or the relation of cross-sectional area to top width changes. Williams concludes that the bankfull level to the active floodplain level is the most useful to the fluvial geomorphologist, whereas the banks of the valley flat are the most important to engineers. Many river channels have a compound cross-section in which it is difficult to determine exactly which of the several levels signifies the capacity level which is equivalent to bankfull stage defined elsewhere, and the sharp limits of lichen growth have been shown to allow the consistent identification of channel capacities from one area to another (Gregory 1976). Because some channels are incised into their floodplains and some are compound in cross-section it has been suggested that an active channel can be recognized within the bankfull capacity channel (Osterkamp and Hedman 1981). The upper limit defining the active channel is a break in the relatively steep bank to the more gently sloping surface beyond the channel edge and this break in

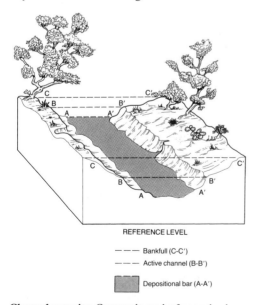

REFERENCE LEVEL

– – – Bankfull (C-C')

—— Active channel (B-B')

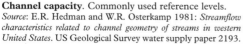  Depositional bar (A-A')

**Channel capacity**. Commonly used reference levels.
*Source*: E.R. Hedman and W.R. Osterkamp 1981: *Streamflow characteristics related to channel geometry of streams in western United States*. US Geological Survey water supply paper 2193.

slope normally coincides with the lower limit of permanent vegetation. The section of channel within the active channel is actively, if not totally, sculptured by the normal process of water and sediment discharge.

A definition of channel capacity is necessary before measurements of cross-sectional area can be related to values of drainage area or discharge. The size of channel at a particular location is related to a range of channel-forming discharges which have often been approximated by a single BANKFULL DISCHARGE or DOMINANT DISCHARGE value. Most recent research has shown how a range of flows acting upon the locally available bed and bank sediment will determine the size of channel capacity. In New South Wales it has been shown that a range of flows (recurrence interval 1.01 to 1.4 years on the annual series) affects the bedforms in the channel and a less frequently occurring range of flows (recurrence interval 1.6 to 4 years) is responsible for the channel capacity (Pickup and Warner 1976). Andrews (1980) has defined effective discharge as the increment of discharge that transports the largest fraction of the annual sediment load over a period of years, and from fifteen gauging stations showed such effective discharges to be equalled or exceeded between 1.5 and 11 days per year and to have recurrence intervals on the ANNUAL SERIES ranging from 1.18 to 3.26 years.

**Reading and References**

Andrews, E.D. 1980: Effective and bankfull discharges of streams in the Yampa river basin, Colorado and Wyoming. *Journal of hydrology* 46, pp. 311–30. · Gregory, K.J. 1976: Lichen and the determination of river channel capacity. *Earth surface processes* 1, pp. 273–85. · —ed. 1977: *River channel changes*. Chichester: Wiley. · Osterkamp, W.R. and Hedman, E.R. 1981: Perennial-streamflow characteristics related to channel geometry and sediment in Missouri River Basin. *US Geological Survey professional paper* 1242. · Pickup, G. and Warner, R.F. 1976: Effects of hydrologic regime on magnitude and frequency of dominant discharge. *Journal of hydrology* 29, pp. 51–75. · Richards, K.S. 1982: *Rivers: form and process in alluvial channels*. London and New York: Methuen. · Riley, S. 1972: A comparison of morphometric measures of bankfull. *Journal of hydrology* 17, pp. 23–31. · Williams, G.P. 1978: Bankfull discharge of rivers. *Water resources research* 14, pp. 1141–54.

**channel classification**  An approach to the description of river channels based upon classification of characteristic morphological (e.g. SINUOSITY, degree of braiding or ANABRANCHING, relative stability) or sedimentological (e.g. type of sediment LOAD) parameters. The most basic form of channel classification is based upon planform morphology and typically involves three types of channel: straight; MEANDERING and BRAIDED. However, actual observation of the range of fluvial forms seen in a variety of different environments has resulted in a number of additional classificatory variables. For instance, Schumm (1977) suggests the introduction of a sediment load parameter to recognize that sediment load controls the relative stability of a channel, and that this stability is important in terms of the types of channel pattern that one observes: SUSPENDED LOAD channels are relatively stable; mixed load channels are intermediate; and BEDLOAD channels are relatively unstable. Even then, additional classificatory variables, and even approaches, may be required (e.g. Rust 1978). For instance, ANASTOMOSING channels look like braiding channels, but are actually quite stable. The difficulty of fitting them into some channel classifications resulted in a different two-parameter classification (Rust 1978), based upon: (1) degree of sinuosity; and (2) degree of division.

Whilst these classifications are largely qualitative and descriptive, attempts have been made to discriminate channels quantitatively, both in terms of morphological properties and possible explanatory variables. In terms of quantification of channel patterns, this may involve indices of sinuosity (e.g. Hong and Davies 1979; Richards 1982) and braiding (e.g. Rust 1978; Friend and Sinha 1993). All of these seek to express the morphological property (e.g. total active channel length) as a function of the property that would be expected in a straight channel (e.g. total valley length). Relatively more attention has been given to identification of possible variables that might explain why different channels have different planforms. The earliest attempts to do this were by Leopold and Wolman (1957) and Lane (1957). Leopold and Wolman, whilst recognizing a continuum of river channel patterns from braiding through meandering to straight, and the role of other environmental controls, made a graphical distinction (figure) between the braiding and meandering state on the basis of just 2 variables, bankfull DISCHARGE ($Qb$) and channel slope ($s$):

$$S = 0.013Q_b^{-0.44} \qquad (1)$$

In a study of 36 field cases and 9 laboratory models, all with sandy beds, Lane (1957) identified two thresholds (2) and (3) separating three classes: meandering; transitional meandering-braiding; and braiding.

### Meandering-transitional threshold

$$S_b = 0.007Q_m^{-0.25} \qquad (2)$$

### Transitional-braiding threshold

$$S_b = 0.0041Q_m^{-0.25} \qquad (3)$$

where $Q_m$ is the mean annual discharge. Ferguson (1987) shows how it is possible to obtain theoretical support for the model developed by Leopold and Wolman (1957), and that the threshold identified in (1) is actually a threshold of specific STREAM POWER. These simple thresholds have had to be modified to recognize that discharge and channel slope are rarely the only controls upon river channel morphology. For instance, Carson (1984) argues that sandbed rivers braid at lower slopes than gravel-bed rivers for a similar discharge regime, and this will follow from the differences in the dominant modes of transport and also differences in bank stability (e.g. Kellerhals, 1982). Sand-bedded channels have more erodible banks, and hence braid more easily. This ties in with the observation of Murray and Paola (1994) that braiding is the fundamental instability of laterally unconstrained free-surface flows over COHESIONless beds and that meandering results from the partial suppression of braiding due to lateral channel constraints or transport of sediment in suspension. In essence, all straight channels that become unstable will braid, unless general environmental characteristics result in them meandering. Similarly, upstream sediment supply may be important. Carson (1984) argues that if the combined load from upstream and from bank erosion is greater than the transporting capacity, then shoaling and braiding will result due to local mid-channel incompetence leading to the progressive growth of mid-channel bars.

A number of important issues arise from attempts to classify river channels (e.g. Ferguson 1987). First, the determination of river channel pattern is not necessarily straightforward. For instance, some channel planforms are stage dependent and a braided channel exposed at low flows may appear straight at high flows. This was illustrated by Leopold and Wolman (1957) for Horse Creek, Wyoming, where upon closer examination the channel did not appear to be really braiding, but was a meandering river with too slow a sediment removal in dead zone areas and too rapid a rate of migration that made it appear braided. Similarly, when can a sinuous channel be labelled a meandering channel given that we may choose to associate particular processes (e.g. the RIFFLE-POOL sequence) with our definition of meandering (Richards 1982)?

Second, the need to include additional classificatory variables is a recognition that different planform morphologies are in fact part of a continuum of channel patterns, and this is explicit in the ideas of both Schumm and Rust. Even within a particular type of river channel (e.g. one that is braiding) it may be possible to identify a range of braiding characteristics (e.g. braiding intensity, relative bar stability) that relate to the particular characteristics of the environment through which the river is flowing. This is not just a problem of a continuum of pattern, but also of process (e.g. Richards, 1982). There is not much evidence that processes in single and multithread streams are particularly different, as was implicit in the reference to Murray and Paola (above). For instance, channel pattern may be viewed as an expression of excess energy. More complicated channel patterns result from the expenditure of excess energy, which serves to create erosion and hence increase total boundary resistance such that there is greater frictional dissipation of energy. If there is a continuum, it is important to ask if there is a fundamental building block to this continuum. One suggestion has been the riffle-pool or bar-pool system (e.g. Ferguson 1987), with the arrangement, density and behaviour of these units determining the nature of river channel pattern. Thus the critical question becomes one of when such units take on particular arrangements and densities, and hence produce particular channel patterns.

Third, the discriminatory functions that are used to separate channel pattern may be problematic if they vary over short time-scales. Discharge is important in this respect as development of a discriminatory function based upon discharge requires some assumption as to which discharge provides the best discriminator (bankfull discharge, mean annual discharge etc.). This is not simply a question of practicality, but basic philosophy, in that acceptance of any discriminator requires an assumption that channel pattern is adjusted to that discriminator to produce a particular EQUILIBRIUM state. Thus, channel classification could be criticized in the same way that REGIME THEORY is criticized: it fails to consider the way in which river channel morphology conditions the exact nature of river channel change. This means that it is important to distinguish between two questions: (a) when is a river channel likely to have a particular channel planform; and (b) under what conditions might a river channel change its characteristic planform. This latter question is about the dynamic behaviour of river channels. Channels can undergo changes in planform over a wide variety of temporal and spatial scales. These may be long-term, in response to secular change in external factors (e.g. tectonic-related uplift, climate). They may also be short-term. Harvey (1987), for instance, describes how a hillslope failure event in the Howgill Fells, northern England,

caused a change in river channel pattern from meandering to braided. However, simple divisions between time-scales in this manner ignores the role of a particular river channel planform in determining the way in which it responds to secular changes in external factors. For instance, it has been argued that the braiding process may help to maintain a channel in its braided state because of the effects of braiding upon the redistribution of sediment, whereby local deposition within the channel creates the topographic complexity that drives the braiding process (e.g. Lane and Richards 1997). If a channel is dependent upon braiding to maintain it in its braided state, then it may be more sensitive to changes in external factors that reduce or eliminate the braiding process. Thus, an important distinction for any channel type may be between braiding and braided and meandering and meandered, to recognize that an actively changing or dynamic channel may be more likely to undergo some sort of major change in channel pattern. Nevertheless, the basic discriminatory functions (e.g. (1)) may still provide a useful indicator of the probability of change between system states: the closer to the THRESHOLD or discriminating condition, the more likely that a river will change its pattern (Ferguson 1987) as is implicit in the transitional classification identified by Lane (1957).                    SNL

### Reading and References
Bridge, J.S., 1993: The interaction between channel geometry, water flow, sediment transport and deposition in braided rivers. In J.L. Best and C.S. Bristow, eds, *Braided rivers*. Geological Society Special Publication 75, pp. 13–71. · Carson, M.A., 1984: The meandering-braided river threshold: a reappraisal. *Journal of hydrology* 73, pp. 315–34. · Ferguson, R.I., 1987. Hydraulic and sedimentary controls of channel pattern. In K.S. Richards ed., *River channels: environment and process*. Oxford: Blackwell. Ch 6, pp. 129–58. · Friend, P.F. and Sinha, R., 1993: Braiding and meandering parameters. In J.L. Best and C.S. Bristow eds, *Braided rivers*. Geological Society Special Publication 75, pp. 105–11. · Harvey, A., 1987: Sediment supply to upland streams: influence on channel adjustment. In C.R. Thorne, J.C. Bathurst, and R.D. Hey eds, *Sediment transport in gravel-bed rivers*. Chichester: Wiley. Pp. 121–50. · Hong, L.B. and Davies, T.R.H., 1979: A study of stream braiding. *Geological Society of America Bulletin* 79, pp. 391–4. · Kellerhals, R., 1982: Effect of river regulation on channel stability. In R.D. Hey, J.C. Bathurst and C.R. Thorne eds, *Gravel-bed rivers*. Chichester: Wiley. Pp. 685–705. · Lane, E.W., 1957: A study of the shape of channels formed by natural streams flowing oevr erodible material. *US Army Corps of Engineers, Missouri Rivers Division*. Sediment Series 9. · Lane, S.N. and Richards, K.S., 1997: Linking river channel form and process: time, space and causality revisited. *Earth surface processes and landforms* 22, pp. 249–60. · Leopold, L.B. and Wolman, M.G., 1957: River channel patterns: braided, meandering and straight. *Professional Paper of*

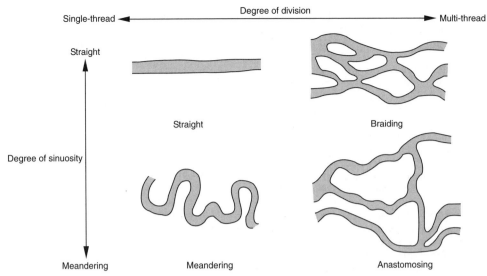

**Channel classification**. A classification of channels based on the parameters of sinuosity and division. *Source*: Adapted from Rust 1978.

*the U.S. Geological Survey* 282B. · Murray, A.B. and Paola, C., 1994. A cellular model of braided rivers. *Nature* 371, pp. 54–7. · Richards, K.S., 1982. *Rivers: form and process in alluvial channels*. London: Methuen. · Rust, B.R., 1978. A classification of alluvial channel systems. In A.D. Miall ed., *Fluvial sedimentology*. Canadian Society of Petrology and Geology Memoir 5, pp. 187–98. · Schumm, S.A., 1977. *The fluvial system*. New York: Wiley. · Thorne, C.R., 1997. Channel types and morphological classification. In C.R. Thorne, R.D. Hey and M.D. Newson, eds, *Applied fluvial geomorphology for river engineering and management*. Chichester: Wiley. Chapter 7, pp. 175–222.

**channel resistance**   Water flowing in a river channel encounters various sources of resistance which oppose downstream motion and result in energy loss. The potential energy of the water is converted to kinetic energy, and thence to work in overcoming frictional resistance and generating heat, as well as in transporting sediment. Channel resistance, or ROUGHNESS, includes grain resistance controlled by bed material size, internal distortion resistance which encompasses the form resistance of bedforms and flow separation at bends, and spill resistance caused by local acceleration at obstacles. Irregularity of channel form and bank vegetation add to flow resistance. The combined effect of these resistances is summarized by the composite roughness coefficients in the MANNING and CHÉZY EQUATIONS.                    KSR

**channel storage**   The volume of water that can be stored along a river channel because of the variations in channel morphology. As a flood HYDROGRAPH travels along a river channel the shape of the hydrograph will change as a result of the storage of water in the channel. Prediction of the character of the hydrograph along the channel is called FLOOD ROUTING.                    KJG

**channelization**   (or river channelization) The modification of river channels for the purposes of flood control, land drainage, navigation, and the reduction or prevention of erosion. River channels may be modified by engineering works including realignment or by maintenance measures by clearing the channel. Channelization can influence the downstream morphological and ecological characteristics of river channels through channel erosion giving larger channels, deposition of the sediment released and change in the river ecology. Because of these consequences downstream from channelization schemes and because of the effects that channelization measures can have on the landscape by aesthetic degradation, alternative methods of stream restoration or stream renovation have been suggested.                    KJG

**Reading**
Brookes, A. 1988: *Channelized rivers – perspectives for environmental management*. Chichester: Wiley. · —Gregory, K.J. and Dawson, F.H. 1983: An assessment of river channelization in England and Wales. *Science of the total environment* 27, pp. 97–111. · Keller, E.A. 1976: Channelization: environmental, geomorphic and engineering aspects. In D.R. Coates ed., *Geomorphology and engineering*. Binghampton NY: State University of New York Press. Pp. 115–40.

**chapada** A wooded ridge or elevated plateau in the savanna areas of South America, especially Brazil.

**chaparral** A vegetation type encountered in areas experiencing Mediterranean climates, characterized by evergreen shrubs with small leathery leaves. (See also MATTORAL.)

**chattermarks** Crescent-shaped gouges found on the surfaces of rocks and rock particles (even sand grains) either as individual features or as trails. They can be produced either by the grinding of rock-armoured basal ice riding over a rock outcrop to produce crescentic gouge trails (Chamberlain 1888 on rock; Gravenor 1979 on sand grains) or by impaction of subrounded grains on other grains in wind or water environments. These latter forms are termed Hertzian cracks in engineering science. Chattermarks on sand grains may also be produced by chemical etching, particularly in a beach environment (Bull *et al.* 1980). PAB

**References**
Bull, P.A., Culver, S.J. and Gardner, R. 1980: Chattermark trails as palaeoenvironmental indicators. *Geology* 8, pp. 318–22. · Chamberlain, T.C. 1888: The rock scorings of the great ice invasions. *US Geological Survey seventh annual report*, pp. 147–248. · Gravenor, C.P. 1979: The nature of the Late Paleozoic glaciation in Gondwana. *Canadian journal of earth sciences* 16, pp. 1137–53.

**cheiorographic coast** The characteristic coastline of areas which have experienced complex, tectonic uplift and subsidence, being made up of alternating deep bays and promontories.

**chelation** The chemical removal of metallic ions in a rock or mineral by biological weathering. The term derives from the Greek *chela* meaning claw and reflects the process by which the metallic ion is sequestered, held between a pincher-like arrangement of two atoms (a ligand). These ligands most frequently attach themselves to the metal ion through nitrogen, sulphur or oxygen atoms. Ligands are produced by organic molecules of plant, animal and microbial origin and are important, and much neglected, processes of rock disintegration. PAB

**Reading**
Ehrlich, H.L. 1981: *Geomicrobiology*. New York: Marcel Dekker.

**cheluviation** Results when water containing organic extracts combines with soil cations to form a chelate. This solution then moves downwards in the soil profile by a process of eluviation, transferring aluminium and iron sesquioxides into lower horizons. ASG

**chemosphere** A term sometimes applied to the region of the atmosphere, mainly between 40 and 80 km in altitude, in which photochemical processes are important.

**chenier ridge** A beach ridge which is surrounded by low-lying swamp deposits, and which tends to be made of sand or shell debris. Classic examples occur downdrift from the Mississippi delta, where individual cheniers are up to 3 m high, 1000 m wide, and 50 km long. They are generally slightly curved, with smooth seaward margins but ragged landward margins due to washovers. Conditions conducive to their formation include low wave energy, low tidal range, effective longshore currents and a variable supply of predominantly fine-grained sediment. ASG

**Reading**
Augustinus, P.G.E.F. ed. 1989: Cheniers and chenier plains. *Marine geology* 90, pp. 219–351. · Hoyt, J.H. 1969: Chenier versus barrier; genetic and stratigraphic distinction. *Bulletin of the American Association of Petroleum Geologists* 53, pp. 299–306.

**chernozem** A black soil rich in humus and containing abundant calcium carbonate in its lower horizons. A soil type characteristic of temperate grasslands, notably the Russian steppes.

**chert** A cryptocrystalline variety of silica, e.g. flint, or more specifically a limestone rock in which the calcium carbonate has been replaced by silica.

**Chézy equation** A FLOW EQUATION developed by Antoine Chézy in 1769 and experimentally tested using data from the River Seine. Its derivation assumes UNIFORM STEADY FLOW in which no acceleration or deceleration occurs along a reach, and the resistance to flow must therefore balance the component of the gravity force acting in the direction of the flow (Sellin 1969). The equation is:

$$\nu = C\sqrt{Rs}$$

where $\nu$ is mean velocity, $R$ is the HYDRAULIC RADIUS (often taken to be the mean depth in wide, shallow channels) and $s$ is energy or bed slope. $C$, the Chézy coefficient, is essentially a measure of 'smoothness' or the inverse of channel resistance, and is therefore inversely related to the coefficient in the similar MANNING EQUATION. KSR

**Reference**
Sellin, R.H.J. 1969: *Flow in channels*. London: Macmillan.

**Chicxulub impact** The event commonly attributed to be the trigger for mass extinctions about 65 million years BP, at the Cretaceous–Tertiary (K–T) boundary. The Chicxulub Crater is located on Mexico's Yucatan Peninsula, buried beneath approximately 1 km of shallow water carbonates, and it may be as much as 300 km in diameter. The impact was by a meteorite or asteroid at least 10 kilometres in diameter, travelling at a speed of about 30 km per second. The meteorite and 200,000 $km^3$ of crust were vaporized and ejected into the atmosphere. Widespread iridium and tektite deposits formed from the ejecta, and the global dust cloud caused the environmental changes that are linked to the extinctions.                    DJS

**Reading**
Sharpton, V.L., Dalyrmple, G.B., Marin, L.E., Ryder, G., Schuraytz, B.C., and Urrutia-Fucugauchi, J. 1992: New links between the Chicxulub impact structure and the Cretaceous/Tertiary boundary. *Nature* 359, pp. 819–21.

**chine** A small ravine or canyon which reaches down to the coast, especially in southern England. Chines are well developed in sandstones near Bournemouth.

**chinook** A warm, dry wind which blows down the eastern slopes of the Rocky Mountains of North America. It is warmed adiabatically during its descent and produces marked increases in temperatures, especially in the spring months. It has some similarity to the föhn winds of Europe.

**chlorofluorocarbons** (CFCs) Organic compounds of human origin derived from hydrocarbons. Large-scale production of CFCs did not occur until the 1950s, though they were invented in the 1920s. CFCs contain chlorine, fluorine and carbon atoms arranged in a chemically stable (inert) structure. This compound is nonflammable, nontoxic, noncorrosive and unreactive with most other substances. This allowed for the widespread use of CFCs as coolants, insulators, foaming agents, solvents and aerosols. CFCs are extremely stable in the lower atmosphere. The molecules do not dissolve in water or break down in biological processes, so that CFC molecules eventually drift up into the stratosphere, where the sun's electromagnetic radiation in the upper atmosphere breaks apart the CFC molecules, freeing chlorine (Cl) atoms. The excess chlorine atoms in the stratosphere would react with ozone ($O_3$) molecules producing chlorine monoxide (ClO) and oxygen ($O_2$): in essence destroying the ozone molecule. The chlorine monoxide molecule can combine with an oxygen (O) atom producing an oxygen molecule and freeing the chlorine atom to begin the process all over again. One chlorine atom could destroy up to 100,000 ozone molecules. Another by-product of chlorine in the upper atmosphere is the long residence time of the chlorine atoms (40–100 years). Because of the relationship between ozone destruction and chlorine the Montreal Protocol was passed in 1987 (revisions in 1989, 1990, 1992 and 1995) which diminished and eventually eliminated the production of CFCs. However, even with the Montreal Protocol Amendments and Adjustments, the levels of CFCs in the stratosphere will not return to the pre-1980 levels until the year 2050.         JAS

**chott** A seasonal lake, often very saline, flooded only during the winter months. Applied especially to the tectonically formed lake basins of North Africa.

**chronosequence** A sequence of soils, each having undergone weathering and soil development (pedogenesis) for a different period of time. If they are located on similar parent materials, and have experienced the same climatic and other influences, the soils of a chronosequence can be used to study the rates and mechanisms of soil formation. Various sites may host chronosequences, including flights of river terraces of varying age, coastal terraces produced by tectonic processes, multiple glacial till sheets, and successive lava flows, as well as sites disturbed by human activity such as rehabilitated mine sites of varying age. Chronosequences have been used to document rates of horizon development, accumulation of organic carbon, soil carbonates, clay enrichment, development of hardpans, and many other soil features.                    DLD

**Reading**
Eash, N.S. and Sandor, J.A. 1995: Soil chronosequence and geomorphology in a semi-arid valley in the Andes of southern Peru. *Geoderma* 65, pp. 59–79.

**chute** A narrow channel with a swift current, applied both to rivers and to the straits between the mainland and islands.

**circadian rhythm** The approximately 24-hour rhythm of activity exhibited by most living organisms: humans, higher animals, insects and plants. The cycle is to some degree independent of day and night cycles and seems to be an important organizing principle in animal and plant physiology. Organisms isolated from external stimuli will continue to display circadian rhythms of temperature, respiration, hormone levels etc. for some time, but may get 'out of phase' and need a diurnal cycle to reset their 'internal clocks'. KEB

**Reading**
Guthrie, D.M. 1980: *Neuroethology: an introduction.* Oxford: Blackwell Scientific. · Luce, G.G. 1972: *Body time*. London: Temple Smith.

**circulation index** A numerical measure of properties or processes of the large-scale atmospheric circulation. Indices have been devised to measure the strength of the east–west component of the circulation in middle latitudes – the zonal index, and the north–south component – the meridional index. The indices are usually in terms of differences in the mean pressures of two specified latitudes. The mean pressures are calculated along each of the latitudes. Lamb (1966) suggests that indices can usefully express circulation vigour if measured at points where the main air streams are most regularly developed. Indices were first used in statistical investigation connected with long range forecasting, e.g. Walker's North-Atlantic Oscillation (see Forsdyke 1951) but are now used more widely in studies of climatic change and the general circulation. AHP

**References**
Forsdyke, A.G. 1951: Zonal and other indices. *Meteorological magazine* 80, pp. 156–60. · Lamb, H.H. 1966: *The changing climate*. London and New York: Methuen.

**cirque** (also corrie, cwm) A hollow, open downstream but bounded upstream by an arcuate, cliffed headwall, with a gently sloping floor or rock basin. The cirque floor is eroded by glacier sliding while the backwall is attacked by BASAL SAPPING and subaerial rock weathering. Cirques are common in formerly glaciated uplands and have long caught the imagination of physical geographers. They were originally thought to have been formed during the waxing and waning of ice sheet glaciations, but few were occupied by active glaciers during ice sheet withdrawal. Instead it seems likely that they represent many stages during the past few million years when marginal glaciation affected mid-latitude uplands. Most mid-latitude cirques show a preferred orientation towards the north-east in the northern hemisphere and towards the south-east in the southern hemisphere, reflecting mainly the effect of shade in protecting the glacier from the sun but also the effect of wind-drifted snow accumulated by predominantly westerly winds. Preferred orientation is less important in polar and tropical mountains. Cirque altitude is an indication of former snow lines and it is common for basin altitudes to increase with distance from a coast. Cirques have

attracted much morphometric analysis and, although the main controls on their morphology remain unclear, it seems that they tend to become more enclosed and deeper with time. DES

**Reading**
Derbyshire, E. and Evans, I.S. 1976: The climatic factor in cirque variation. In E. Derbyshire ed., *Geomorphology and climate*. New York: Wiley. · Gordon, J.E. 1977: Morphometry of cirques in the Kintail-Affric-Cannich area of north-west Scotland. *Geografiska annaler* 59A. 3–4, pp. 177–94.

**cirrocumulus** See CLOUDS.

**cirrostratus** See CLOUDS.

**cirrus** See CLOUDS.

**cladistics** (or phylogenetic systematics) The elucidation of the evolutionary history of groups of organisms. Hennig (1966) was responsible for its initial development. Subsequently, it has been refined into a method with more general properties than those stated by Hennig and having a much wider application than he intended (Humphries and Parenti 1986).

Biological systematics seeks to describe and to classify the variation between organisms. Such classifications are often hierarchical and reveal patterns of association (TAXONOMY). Resemblances among the intrinsic properties of organisms (encompassing everything from their chemistry to conduct) can be classified to reveal phylogenetic patterns. As Eldredge and Cracraft

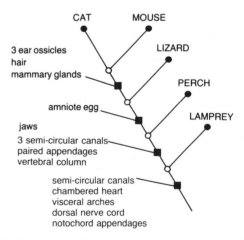

**A cladogram** for five kinds of vertebrates. Each level of the hierarchy (denoted by branch points) is defined by one or more similarities interpreted as evolutionary novelties. *Source*: N. Eldredge and J. Cracraft 1980: *Phylogenetic patterns and the evolutionary process*. New York: Columbia University Press. Figure 2.1.

(1980) point out, the pattern of similarity of features in organisms has resulted from either evolution by descent or special creation (CREATIONISM). If evolution by ancestry and descent is accepted, together with the fact that novel characteristics appear in organisms at different times during its course, a hierarchical phylogenetic pattern of clustered intrinsic resemblances (akin to similarities in taxa) can be formulated.

According to Hennig (1966), it is possible to distinguish two principal categories of resemblance in a monophyletic taxon (one having two or more species and including an ancestor and descendants). First, true evolutionary similarities. These may be of two kinds, either those acquired from a distant common ancestor and retained by one or more of its descendants, or those possessed solely by a particular group of organisms which acquired them from a recent common ancestor. The second category is that of false or misleading resemblance and derives from adaptations made during parallel and convergent EVOLUTION. Cladists contend that the intermittent retention of resemblances acquired from an ancient source renders such traits only useful as clustering agents where they can claim novel status. Their methodology also indicates that because ancestors do not appear to have special assemblages of novelties, their definition is difficult.

Accordingly, phylogenetic systematists concentrate upon recently acquired evolutionary resemblances in an attempt to ascertain patterns of novelties and establish proximate relationships between taxa. The intrinsic characteristics of organisms are examined in detail. An attempt is made to establish whether these characteristics are primitive or derived. Widespread traits within groups are thought to have originated there, while characters occurring in only a few instances are considered as derived. Significant patterns of phylogenetic resemblance have also emerged after the analysis of homologous characteristics at various stages in the life cycle of organisms. The taxon being investigated is also compared in detail with its nearest relative (sister), as part of the search for character distributions. The aim is to find a universal set (one or more) of similarities in order that a group in which they are all present may be defined. Resemblances which are common to only some of the group can also be used to categorize subsets. Inferred inherited similarities that define subsets at some level within a universal set are known as homologies. Thus homology is synonymous with synapomorphy and refers to novel, shared resemblance endowed by a recent common ancestor (Eldredge and Cracraft 1980).

Data from such analyses are presented as branching diagrams or cladograms, upon which are illustrated clusters of common resemblances regarded as evolutionary innovations. It is often possible to map more than one cladogram from one data set. When compared, these exhibit different patterns that require further investigation. An example provided by Eldredge and Cracraft (1980) will illustrate the principles and problems of cladistic analysis (see p. 91). The cat, mouse, lizard, perch and lamprey are vertebrates with a universal set of resemblances which include semicircular canals in their heads and a chambered heart. Subgroups can be made of the perch and lamprey as they lack an amniote egg, and the lizard, perch and lamprey which do not possess hair or mammary glands. The apparent conflict in the relationships between these organisms can be clarified with reference to the stages in their life cycles. Each has a substantial amount of cartilage in at least one stage of their ontogeny. Cartilage is superseded by bone in the advanced stages of development of the cat, mouse, lizard and perch. Thus bone is diagnostic of a subgroup (each member of which is also part of a bigger group including the lamprey), whose members possess a vertebral column, paired appendages, jaws and semicircular canals. A further subgrouping emerges with reference to an amniote egg, which is possessed by the cat, mouse and lizard. Consideration of earlier developmental stages of the organisms indicates that all five then had a general vertebrate egg, an amniotic membrane subsequently emerging in the cat, mouse and lizard. The cat and mouse can be combined in isolation from the lizard as they are endowed with hair, mammary glands and three ear ossicles, these again being adaptations from more widespread vertebrate traits. Such relationships can be expressed in the cladogram shown here.

The next stage in the cladistic analysis of these data is to look outside the group of five organisms for others with equivalent shared resemblances. In this context, two other chordates, the lancelet amphioxus and the tunicates are relevant. They, however, do not have the characteristics of the cat, mouse, lizard and perch grouping, nor of the large group which includes the lamprey. Thus the amphioxus, tunicates and lamprey comprise outgroups and the cat, mouse, lizard and perch a subgroup. The latter can be further subgrouped in that cats, mice and lizards possess amniote eggs but perch do not, their eggs being analogous to those of the outgroups. Finally, cats and mice are the exclusive possessors of hair, mammary glands and three ear ossicles, qualifying as a subgroup by virtue of these evolutionary novelties.

A progression in this type of analysis, whereby more complex and accurate hypotheses concerning ancestor–descendant relationships can be formulated, comes via additional information and takes the form of a phylogenetic tree, which can also be represented diagrammatically. At the end of this sequence is the phylogenetic scenario which consists of a tree with an overlay of adaptational narrative (Eldredge 1979).

Cladistics has caused considerable controversy and acrimony among systematists. Some cladistic viewpoints conflict, often sharply, with certain of those of the other major biological taxonomists – the evolutionary systematists and the numerical taxonomists. Eldredge and Cracraft (1980) state that cladistic strategy involves investigation of the structure of nature, and the formulation of hypotheses concerning this which contain the basis for their own evaluation. This approach differs from that of the Darwinian school of evolutionary taxonomy which, they contend, invokes acknowledged processes in the explanation of form, such that its conclusions are not amenable to disproval. The cladists do not claim to prove anything, but to carry out pattern analysis without fixed ideas concerning causal processes, and within a framework that accepts that evolutionary patterns and processes are related. They also accept that morphological and taxonomic diversity are related but not synonymous, stating that evolutionary systematists believe that they are equivalent. The principles of cladistics also have a bearing on speciation. Cladists recognize species as distinctive entities to the extent that the emergence of a new one interferes with the pattern of ancestry and descent. Darwinists explain long-term intrinsic modifications in species populations by means of natural selection and adaptation. Cladism leads to the notion that such gradual speciation is unlikely. Cladists believe that species remain unmodified for a period, and are quite suddenly replaced by others. Speciation occurs in areas of various dimensions. When these areas are defined, they can be related to the phylogenetic histories of the relevant species as depicted by cladograms (VICARIANCE BIOGEOGRAPHY).

Cladists disagree with numerical taxonomists who, they contend (mainly on the basis of computerized cluster analysis), feel that evolutionary history defies reason. They are at variance with both evolutionary systematists and numerical taxonomists over a fundamental tenet of their method. This relates to cladistic assemblages which comprise novelties rather than retentions, the latter being the usual characteristics employed by alternative approaches.

In consequence, the cladists' interpretation of the evolutionary status of certain taxa differs from that of their fellow systematists.     RLJ

**Reading and References**
Eldredge, N. 1979: Cladism and common sense. In J. Cracraft and N. Eldredge eds, *Phylogenetic analysis and paleontology*. New York and Guildford: Columbia University Press. Pp. 165–97. · —and Cracraft, J. 1980: *Phylogenetic patterns and the evolutionary process*. New York and Guildford: Columbia University Press. · Hennig, W. 1966: *Phylogenetic systematics*. Urbana: University of Illinois Press. · Humphries, C.J. and Parenti, L.R. 1986: *Cladistic biogeography*. Oxford: Clarendon Press.

**clast**  A coarse sediment particle, usually larger than 4 mm in diameter; clast sizes include pebbles, cobbles and boulders.

**clastic**  Composed of or containing fragments of rock or other debris that have originated elsewhere.

**clay**  Term applied in the textural classification of soils and sediments to particles having diameters of less than 2 $\mu$m, as well as in soil chemistry to a wide range of crystalline hydrous layer silicate minerals (the 'clay minerals'). The clay minerals, of which there are many varieties, are derived from the weathering of primary silicate minerals, and are made up of stacked layers of tetrahedrally and octahedrally structured sheets of oxygen atoms bound to cations, notably $Si^{4+}$, $Al^{3+}$, $Fe^{3+}$, $Fe^{2+}$ and $Mg^{2+}$. These materials ordinarily occur as small crystalline fragments less than a few $\mu$m in size. Whilst many clay particles, *texturally* defined, do consist of such clay mineral fragments, clay-sized particles of other materials such as quartz also occur. Furthermore, clay mineral particles, which carry a negative surface charge because of ionic substitution within their crystal lattice (say, $Si^{4+}$ replaced by $Al^{3+}$, leaving an unbalanced negative charge), are often held together in clumps or flocs, and when held in this way the composite particle composed of many clay particles can itself be of much larger size. (See also PARTICLE SIZE.)     DLD

**Reading**
Brady, N.C. and Weil, R.R. 1996: *The nature and properties of soils*. 11th edn. New Jersey: Prentice-Hall.

**clay dune**  A dune (often LUNETTE) where a significant proportion of the constituent sediments (up to 30 per cent) comprise clay minerals. These are likely to have been transported to the dune, via saltation and creep, as CLAY PELLETS, which may have become broken down, or further aggregated, after deposition. Clay dunes usually border lacustrine basins and result from the deflation of basin floor sediments during PLAYA or PAN stages. Clay dunes occur in parts of the Kalahari, Australia, North Africa and Texas, USA.                                    DSGT

**Reading**
Bowler, J.M. 1973. Clay dunes: their occurrence, formulation and environmental significance. *Earth science reviews* 9, pp. 315–38.

**clay–humus complex**  Consists of a mixture of clay particles and decaying organic material that attracts and holds the cations of soluble salts within the soil profile.

**clay pellet**  A particle of sediment that, although SAND or SILT sized, is in fact composed of aggregated CLAY size particles. Pellet formation may occur through the process of FLOCCU-LATION or through the breakdown of clay layers through salt efflorescence. Sand-size clay pellets will be transported through the processes of SALTATION, CREEP and REPTATION. On deposition, pellets may become deflocuated. If a particle size analysis of such a sediment is conducted, the resulting distribution may over-represent clay in respect of the transport processes and mechanisms through which the deposit was achieved. One environment where clay pellet formation may be important is on the floor of PANS and PLAYAS, with pellets transported by the wind to an adjacent LUNETTE DUNE.                                    DSGT

**claypan**  A stratum of compact but not cemented clayey material found within the soil zone.

**clay-with-flints**  An admixed deposit of clay and gravel, predominantly flint, occurring locally in depressions in the chalk uplands of southern England. Probably the insoluble components of the chalk and/or reworked Tertiary deposits.

**clear water erosion**  The erosion caused by rivers whose sediment load has been removed by the construction of a dam and reservoir. With a reduced sediment load, incision occurs rather than aggradation. This can create serious problems for bridges and other man-made structures downstream.

**cleavage**  Of minerals, the plane along which a crystal can be split owing to its internal molecular arrangement. Also those planes of weakness in fissile rocks which are not related to jointing or bedding.

**CLIMAP**  In 1971 a consortium of scientists from many institutions was formed to study the history of global climate over the past million years, particularly the elements of that history recorded in deep-sea sediments. This study is known as the CLIMAP (Climate, Long-range Investigation, Mapping and Prediction) project. One of CLIMAP's goals is to reconstruct the earth's surface at particular times in the past, and a good example is contained in CLIMAP Project Members (1976).                       JGL

**References**
CLIMAP Project Members 1976: The surface of the ice-age earth. *Science* 191, pp. 1131–7. · Gates, W.C. 1976: Modelling the ice-age climate. *Science* 191, pp. 1138–44.

**climate**  The long-term atmospheric characteristics of a specified area. Contrasts with weather. These characteristics are usually represented by numerical data on meteorological elements, such as temperature, pressure, wind, rainfall and humidity. These data are frequently used to calculate daily, monthly, seasonal and annual averages, together with measures of dispersion and frequency.

Climatic statistics have been published for a huge number of stations, and there are some useful publications which have summarized tables of major climatic phenomena. Among the most useful are Landsberg (1969), Wernstedt (1972), Müller (1982) and Pearce and Smith (1984).                                    ASG

**References**
Landsberg, H.E. ed. 1969 onwards: *World survey of climatology*. 15 vols. Amsterdam: Elsevier. · Müller, M.J. 1982: *Selected climatic data for a global set of standard stations for vegetation science*. The Hague: Junk. · Pearce, E.A. and Smith, C.G. 1984: *The world weather guide*. London: Hutchinson. · Wernstedt, F.L. 1972: *World climatic data*. Lemont, Penn.: Climatic Data Press.

**climate change**  The long-term variability associated with the earth–ocean–climate system. 'Long-term' in this instance refers to time periods greater than the day-to-day variability of weather. Consequently, climate change can refer to changes in average weather conditions from one year to the next or even to the changes associated with the transitions from the climate of the last GLACIAL to the present climate. The study of climate change has been classified into a number of smaller areas of investigation, including climate reconstruction, climate change theory and climate modelling.

Climate reconstruction or PALAEOCLIMATO-LOGY relates to the study of the earth's past climates and determination of those past climates. Records of actual measures of climate, such as temperature and precipitation, are often only available for limited places and for limited times of the past. Modern weather service observations, for example, have only been available for the last century and half and then primarily only in western Europe and North America. Consequently, other measures of climate variability, termed climate proxies, are needed to extend our understanding of past climates both geographically and temporally.

For the time associated with human civilizations, many historical records, diaries, planting records and even clothing and building materials can supply useful information on climate conditions for a given civilization. More detailed and geographically diverse information, however, can be obtained from various geological and biological indicators, termed physical climate proxies. Analyses of such biologic indicators as tree rings, lake varves, vegetation growth boundaries, and pollen distributions have provided detailed climate information such as the precipitation and temperature regimes of relatively remote regions of the earth several thousand years into the past.

Other techniques can be used to evaluate the climates of far earlier times in the earth's past. Isotope analysis can be applied to glacial ice cores or to deep-sea marine sediment cores to gain data on past temperature and global ice volumes (see OXYGEN ISOTOPE). The biogenic material found in a marine core may also be analysed for proxy climate data, since certain species display identifiable temperature-dependent characteristics. For example, the coiling direction of some kinds of foraminfera is dependent on water temperature, which in turn is influenced by atmospheric temperatures.

Abiotic proxy evidence can be garnered from such sources as the analysis of isotopic concentrations in calcium carbonate, such as the well-dated core taken from Devil's Hole, Nevada; the analysis of the extent of ancient desert sediments; and the analysis of fossil pollen accumulated within sediments. In all cases however, consideration has to be given to the relative influence of climatic and other environmental factors that have affected the magnitude and distribution of living species and the operation of ecological, sedimentary and geomorphological processes. Bradley (1985) gives an excellent summary of various means of climate reconstruction.

A second major component in the field of climate change is the formulation and evaluation of climate change theories. Such theories are constructed to explain either the various shifts in climate revealed through climate reconstruction and study of secular climate records or to account for anticipated changes in global or regional climate. Although hundreds of climate change theories have been hypothesized in the last century or so, two have reached great prominence and acceptance among climatologists.

The astronomical theory of climate change, often referenced by the name of one of its leading early researchers, Multin Milankovitch, states that periodic changes in the earth's climate are linked to cyclic orbital variations in the earth. In particular, the climatic changes of the last two million years have been statistically linked to variations in three of the earth's orbital mechanisms. Obliquity, changes in the axial tilt of the earth, operates with a periodicity of 41,000 years. Precession of the longitude of the equinoxes, the wobble in the earth's orbit which changes the timing in which perihelion (the time at which the earth is closest to the sun) occurs in relation to the equinoxes, occurs with variability of 21,000 years. The third mechanism, eccentricity, the variability in the shape of the earth's orbit, has a periodicity of 100,000 years. The cycles associated with these mechanisms have been identified in the analyses of deep-sea marine sediment cores across the globe.

A second climate change theory, the GREENHOUSE EFFECT (see also CARBON DIOXIDE PROBLEM) has received significant scientific and public attention in recent decades. In brief, carbon dioxide and other gases, including water vapour, act to absorb terrestrial radiation in the lower atmosphere. Such absorption acts as an extra source of heat to the surface. For example, a given locale is warmer under a cloudy sky than under a clear sky because of the greenhouse characters of the water vapour associated with clouds. Consequently, it is hypothesized, when all other factors are held constant, that increasing amounts of various greenhouse gases will lead to general global warming. This theory has been used to explain the abnormal global warmth revealed in reconstruction evidence from the age of the dinosaurs (the Cretaceous). Greenhouse theory has also been used to produce future scenarios of the earth's climate given the increasing amount of human-produced greenhouse gases emitted into the atmosphere. Two current concerns associated with this theory are the identification of potential feedback mechanisms that may serve to mitigate the global warming, and the identification of associated climate impacts beyond global warming, such as changes in regional precipitation.

A third subdiscipline in climate change studies is the creation, testing and application of numerical climate models. Climate models have become a major tool of investigation for climate change theories, as well as a means for evaluating regional differences in climate derived from reconstruction proxies and secular climate records. A hierarchy of climate models exists based on the complexity of their physical relationships and the number of their spatial dimensions.

Radiative-convective models (RCMs) are generally one-dimensional models which focus on mathematical expression of the radiative processes in the atmosphere, particularly as influenced by atmospheric water vapour and aerosols. Energy balance models (EBMs) have been formulated using a single point determination (zero-dimensional), with latitudinal dependence (one-dimensional), with latitude and vertical components (two-dimensional) and with all three spatial dimensions (three-dimensional). EBMs are constructed around a primary equation that addresses the various energy transfers (such as by advection of sensible heat, latent heat or by ocean currents) that occur in the earth's climate system.

The most complex of numerical climate models are those known as GCMs (see GENERAL CIRCULATION MODELLING) and incorporate many of the features of both RCMs and EBMs. GCMs are mathematical computer models describing the primary controls (such as radiative fluxes), energy transfers, circulations and feedbacks existing in the earth–ocean–climate system. GCMs have their roots in the early numerical weather models developed by J.G. Charney and the so-called 'father' of modern computing, J. von Neumann, in the late 1940s. Most GCMs are based on seven fundamental (or 'primitive') equations of the atmosphere.

Climate models have been used to study the interrelationships between climate processes (such as albedo and temperature), evaluate and explain reconstruction proxy data (such as the GCM simulations of the Cretaceous and Pleistocene climates), and to address possible effects of increased atmospheric carbon dioxide and other greenhouse gases. Trenberth (1992) gives an excellent review of both fundamentals of computer modelling and the applications to which they have been applied.

Climate change study has become one of the most important sciences in recent years. Appreciating the variability inherent in our climate system is becoming increasingly critical to both understanding the growth and decline of past human civilizations and, indeed, other species on this planet, as well as the formulation of future public policy regarding all elements of human life. Climate change study can best be characterized as a young, constantly changing discipline whose members are at the vanguard of the frontiers of science. RSC

**Reading and References**
Berger, A., Imbrie, J., Hays, J., Kukla, G. and Saltzman, B. 1984: *Milankovitch and climate*. 2 vols. Dordrecht: Reidel Publishing. · Bradley, R.S. 1985: *Quaternary paleoclimatology: methods of paleoclimate reconstruction*. London: Unwin Hyman Ltd. · Houghton, J.T., Meira Pilho, L.G., Callander, B.A., Haris, N., Kattenburg, A. and Maskell, K. 1996: *Climate change 1995*. Cambridge: Cambridge University Press. · Trenberth, K.E. ed. 1992: *Climate system modelling*. Cambridge: Cambridge University Press.

**climate modelling**   See GENERAL CIRCULATION MODELLING.

**climate modification**   See WEATHER MODIFICATION.

**climatic classification**   A systematic partitioning of the earth's climates based on the averages of weather elements, the variations of weather elements, and the effects of weather elements on the land surface environment. Due to some degree of regularity in the general circulation of the atmosphere, there exist broad patterns of climates across the earth which are modified by such additional controls as topography and proximity to large water bodies. An ideal classification of the climates should reveal all of the climates that occur on the earth, should show the relationships among climates, should allow for subdivisions down to local climates, and should demonstrate the controls on each climate. However, the ideal climate classification will never exist due to the facts that the overall climate system of the earth is too complex and the individual climates themselves are not spatially finite. Therefore, climate classification schemes are generally developed and utilized as dictated by their particular use.

The first attempt at climate classification was made by early Greek civilizations who used only latitude to subdivide the earth's global climate into three distinct zones: torrid, temperate, and frigid. Further attempts at classifying the earth's climates were not made until the end of the nineteenth and the beginning of the twentieth centuries. In the late 1800s, German plant physiologist Wladimir Köppen (1846–1940) began to recognize that a plant or plant communities at a particular place reflected a synthesis of the weather elements of that location. Comparing the global distribution of vegetation to his own data on world distributions of temperature and precipitation, Köppen identified a correlation between the atmosphere and the biosphere, and he subsequently began to distinguish climates based on those interrelations. The first version of the Köppen Climate Classification System, based solely on temperature, was published in 1884. After modifying the system to include climatic factors as a whole, Köppen published his second version in 1900 (Köppen 1900). Five significant modifications by Köppen himself occurred between 1918 and 1936, while several scientists have made further modifications since then.

The Köppen climate classification system itself uses letter symbols to distinguish different characteristics of each climate. Six major climate groups are subdivided once, and in some cases twice, based on specific temperature and moisture characteristics. Relative magnitudes of precipitation and its predominant timing within the year, as well as data on the warmest and coldest months of the year provide a finer definition of the climate groups. Largely based on the distribution patterns of the earth's vegetation, many of the climate groups derived from the Köppen system bear names associated with the type of vegetation that they support, such as the tropical rain forest, desert, and tundra climates.

Several scientists of the early to mid-twentieth century opposed Köppen's system, the most notable of whom was C. Warren Thornthwaite (1899–1963). Thornthwaite argued that the classification of the earth's climates required a greater understanding of spatial variations in the climatic water budget, not just simply variations in temperature and moisture (Thornthwaite 1948). Accordingly, the 1948 climate classification system published by Thornthwaite is based on the concept of potential evapotranspiration, or the amount of evapotranspiration that would occur from a well-watered land area. This maximum amount of water loss through evapotranspiration is a function of air temperature and vegetation cover. Thornthwaite's method differs from that of Köppen in that it expresses the effectiveness of precipitation and the efficiency of air temperature in the process of evapotranspiration, both of which indicate the degree to which vegetation can be supported. Similar to the method of Thornthwaite, Mikhail Budyko later developed an approach in which, instead of temperature, the net radiation available for evaporation from a moist surface is related to the amount necessary to evaporate the mean annual precipitation (Budyko 1974). As a reflection of the general climate, Budyko used that ratio to classify vegetation zones.

Due to its simplicity and usefulness, the Köppen system has become the most widely used climate classification system, and it continues to be of pedagogical value in modern geography,

particularly on scales larger than those of regional study. However, with many potential uses to the modern geographer, many methods of climate classification have been devised. Despite the attraction of using qualitative and subjective input to the process of climate classification, contemporary geographers realize the demand for a quantitative method. As such, the process of climate classification today begins much the same way as it did for Wladimir Köppen over a century ago with average values of weather elements such as temperature and precipitation.                                    AWE

**References**
Budyko, M.I. 1974: *Climate and life*. New York: Academic Press. · Koppen, W. 1900: Versuch einer Klassifikation der Klimatre, Vorzugweise nach ihren Beziehungen zur Pflanzenwelt. *Geographische Zeitschrift* 6, pp. 593–611. · Thornthwaite, C.W. 1948: An approach towards a rational classification of climate. *Geographical review* 38, pp. 55–94.

**climatic geomorphology** This subject developed during the period of European colonial expansion and exploration at the end of the nineteenth century, when unusual and spectacular landforms were encountered in newly discovered environments like deserts and the humid tropics. In the USA, W.M. Davis recognized 'accidents', whereby non-temperate climatic regions were seen as deviants from his normal cycle and introduced, for example, his arid cycle (Davis 1905). Some regard Davis as one of the founders of climatic geomorphology (see Derbyshire 1973), though the French (e.g. Tricart and Cailleux 1972) have criticized him for his neglect of the climatic factor in landform development. Much important work on dividing the world map into climatic zones (morphoclimatic regions) with distinctive landform assemblages was attempted both in France (e.g. Tricart and Cailleux 1972) and in Germany (e.g. Büdel 1982). This geomorphology described itself as geographical (Holzner and Weaver 1965).

In recent years certain limitations have become apparent:

1 Much climatic geomorphology has been based on inadequate knowledge of rates of processes and on inadequate measurement of process and form.
2 Many of the climatic parameters used for morphoclimatic regionalization have been seen to be meaningless or crude.
3 The impact of climatic changes in Quaternary times has disguised the climate–landform relationship.
4 Climate is one step removed from process.

5 Many supposedly diagnostic landforms are either relict features or have a form which gives an ambiguous guide to origin.

6 Climate is but one factor in many that affect landform development.

7 Climatic geomorphology tends to concentrate on the macroscale rather than investigating the details of process.

8 Macroscale regionalization has little inherent merit.                                         ASG

**Reading and References**

Birot, P. 1968: *The cycle of erosion in different climates*. London: Batsford. · Büdel, J. 1982: *Climatic geomorphology*. Princeton, NJ: Princeton University Press. · Davis, W.M. 1905: The geographical cycle in an arid climate. *Journal of geology* 13, pp. 381–407. · Derbyshire, E. ed. 1973: *Climatic geomorphology*. London: Macmillan. · — 1976: *Climate and geomorphology*. London: Wiley. · Holzner, L. and Weaver, G.D. 1965: Geographical evaluation of climatic and climato-genetic geomorphology. *Annals of the Association of American Geographers* 55, pp. 592–602. · Stoddart, D.R. 1969: Climatic geomorphology. In R.J. Chorley ed., *Water, earth and man*. London: Methuen. Pp. 473–85. · Tricart, J. and Cailleux, A. 1972: *Introduction to climatic geomorphology*. London: Longman.

**climatic hinge** An imaginary line in the tropical zone which separates areas receiving the bulk of precipitation in the northern summer from those affected by summer rainfall in the southern hemisphere. Although this currently equates to the position and movement of the Inter Tropical Convergence Zone, the concept has been used in palaeoclimatic studies, particularly in Africa, to identify the precipitation equator once it became apparent that long-term changes in the two hemispheres were neither identical nor synchronous. For example, during the late Glacial in East Africa the climatic hinge is thought to have been in the vicinity of southern Lake Malawi, some $13°$ south of the equator, as evidenced by low lake levels in the East Africa lakes, but high water levels in the Okavango Delta and associated lakes in the Kalahari Desert.                                         PSh

**climatic optimum** See ALTITHERMAL.

**climatic sensitivity parameter** A numerical index of the amount by which the mean surface temperature of the earth would change in response to a unit change in radiative forcing. The radiative forcing is primarily the solar radiation received by the earth, but also includes energy trapped within the earth–atmosphere system, and resulting fluxes of long-wave radiation. For example, greenhouse gases in the atmosphere trap heat energy, and the resulting radiant heat flux is estimated to increase radiative

forcing by about $4\,W/m^2$ for a doubling of $CO_2$. The climate sensitivity parameter, $l$, links this change in radiative forcing, $DQ$, to the rise in mean surface temperature, $D, T_s$, as follows

$$D(T_s = l = \times D \times Q)$$

The value of $lh$ is $< 1K\,m^2\,W^{-1}$ but its value is difficult to determine exactly. This is because many feedback processes, such as changes in snow, ice, and cloud cover, and the water vapour content of the atmosphere, are triggered by a change in $T_s$, and so affect the value of $l$.                                         DLD

**Reading**

Houghton, J.T., Jenkins, G.J. and Ephraums, J.J. eds 1990: *Climate change: the IPCC scientific assessment*. Cambridge: Cambridge University Press.

**climato-genetic geomorphology** An attempt to explain landforms in terms of fossil as well as contemporary climatic influences. Büdel (1982) recognized that landscapes were composed of various relief generations and saw the task of climato-genetic geomorphology as being to recognize, order and distinguish these relief generations, so as to analyse today's highly complex relief.

**Reference**

Budel, J. 1982: *Climatic geomorphology*. Princeton, NJ: Princeton University Press.

**climatology** Traditionally concerned with the collection and study of data that express the prevailing state of the atmosphere. The word study here implies more than simple averaging: various methods are used to represent climate, e.g. both average and extreme values, frequencies of values within stated ranges, frequencies of weather types with associated values of elements. Modern climatology is concerned with explaining, often in terms of mathematical physics, the causes of both present and past climates.                                         JGL

**Reading**

Barry, R.G. and Chorley, R.J. 1992: *Atmosphere, weather and climate*. 6th edn. London: Routledge. · Lockwood, J.G. 1979: *Causes of climate*. London: Edward Arnold.

**climax vegetation** Vegetation COMMUNITY regarded as existing in a state of equilibrium with given climatic and edaphic conditions and capable of self-perpetuation. The idea was initially developed by the American biologist, F.E. Clements in 1916, who hypothesized that, following colonization of a new surface (e.g. a lava flow or bare sand dune), ecological SUCCESSION would proceed until a final stage was reached in which the community was in harmony with prevailing environmental conditions. Climate was considered to be the dominant ecological determinant and the end product of succession was therefore referred to as the climatic climax community, for example, deciduous woodland in the cool temperate climates of northwest Europe and eastern North America. The climax community is perceived as having greater species diversity than earlier successional stages and is a correspondingly more stable community under conditions of higher levels of complexity and interdependence. Intuitively attractive though the notion is, it is now realized that the dynamics of ecosystems are such that equilibrium, implying stasis, is unlikely to be achieved due to spatial and temporal variability in environmental determinants, imposing ecological stresses of varying degrees of magnitude upon communities. Change, as opposed to stability, in community composition is probably the norm and the term 'mature community' is a more appropriate one to describe ecosystems eventually developing when the degree of stress is low.                                                    MEM

**Reading**
Burrows, C.J. 1990: *Processes of vegetation change.* London: Unwin Hyman.

**climbing dune**   A TOPOGRAPHIC DUNE that has developed on the windward side of a topographic obstacle, as airflow and sand transport is disrupted by the barrier. Where the hill slope is less than 30° sand transport is not disrupted, but at 30–50° a dune develops, up to twice as long as the height of the hill (Cooke *et al.* 1993). If the barrier is a discrete hill rather than an escarpment, the dune may eventually wrap around the obstacle. See also ECHO DUNE.
                                                    DSGT

**Reference**
Cooke, U., Warren, A. and Goudie, A.S.G. 1993. *Desert geomorphology.* London: UCL Press. Pp. 353–4.

**climogram**   Usually refers to two types of climatic diagram, a climograph and a hythergraph, which were introduced by T. Griffith Taylor in 1915. The diagrams depict graphically the annual range of climatic elements at a particular location, emphasis being placed on the way in which these elements affect the comfort of man. In a climograph mean monthly WET-BULB TEMPERATURE is plotted as the ordinate and mean monthly relative humidity as the abscissa. The points are then joined to form a 12-sided polygon. In a hythergraph mean monthly temperature is plotted against mean monthly precipitation. Data for several stations, representing different climatic zones, are usually plotted on a single climogram, for comparative purposes. Many other types of climogram have been devised (Monkhouse and Wilkinson 1971).                                          DGT

**Reference**
Monkhouse, F.J. and Wilkinson, H.R. 1971: *Maps and diagrams.* London and New York: Methuen. Pp. 246–50.

**cline**   A gradation in the range of physiological differences within a species which result from the adaptions to different environmental conditions.

**clinometer**   An instrument for measuring angles in the vertical plane, particularly those of dipping rocks and hillslopes.

**clinosequence**   A group of related soils that differ in character as a result of the effect of slope angle and position.

**clint**   The ridge or block of limestone between the runnels (grikes) on a rock outcrop. Clints form on remnant features caused by the solution of the limestone by water under soil or drift cover, etching weaknesses or joint patterns in the rock. Clint is an English term (helk is also used) but the features are called KARREN in German. The most famous clints and grikes can be found at Malham Tarn, Malham, Yorkshire.
                                                    PAB

**clisere**   The series of climax plant communities which replace one another in a particular area when CLIMAX VEGETATION is subjected to a major change in climate.

**clitter**   Massive granite boulders, especially those found on Dartmoor. Clitter probably represents a type of blockstream or blockfield which may have resulted from the fashioning of tors.

**cloud dynamics**   The role played by air motions in the development and form of clouds. Extensive sheets or layers of stratiform cloud are an expression of the gentle, uniform and widespread ascent of deep layers of moist air. In contrast, localized convective or cumuliform cloud bears witness to the presence of more vigorous but much smaller-scale ascent of moist bubbles of air.

Widespread ascent is associated with the extensive low-level convergence and frontal upgliding that characterizes middle latitude depressions, and involves vertical velocities of a few $cm\,s^{-1}$ which can reach $10\,cm\,s^{-1}$ in the vicinity of the fronts. Cirrostratus, altostratus and stratus are products of this kind of ascent, although they can be modified if they occur in thin layers and are not shielded by any higher level cloud. A stratus layer cools at its top by radiation to space and is warmed at its base by absorbing infrared radiation emitted from below. This process destabilizes the layer, leading to overturning and a dappled appearance as the stratus is gradually transformed into stratocumulus. Cirrocumulus and altocumulus can also form in this way.

Localized ascent is associated generally with strong surface heating in an unstable atmosphere such as occurs occasionally over summertime continents or in polar air outbreaks across oceans. The rising convective bubbles transport heat upwards in a central core of air which becomes visible as a cumulus cloud once condensation occurs. When these clouds occur on settled days in the form of fair weather cumulus they tend to be only 1–2 km deep and display updraughts of up to $5\,\mathrm{m\,s^{-1}}$. In unsettled weather convection can extend to the tropopause and these cumulonimbus clouds exhibit updraughts often stronger than $5\,\mathrm{m\,s^{-1}}$ and exceptionally up to $30\,\mathrm{m\,s^{-1}}$. These convective clouds are surrounded by clear sky in which the air is sinking gently.

Cumulonimbus clouds go through a unique dynamical life cycle which is most developed at maturity with an organized up- and downdraught, the latter of which flows out at the surface as a gust front. The dissipating stage is dominated by downdraughts associated with extensive rainshafts and evaporative cooling of the raindrops.                                        RR

## cloud forest

**cloud forest** Occupies those zones in mountainous terrain where clouds occur sufficiently regularly to provide moisture to support the growth of forest, which is often of broad-leaved evergreen type.

## cloud microphysics

**cloud microphysics** Deals with the physical processes associated with the formation of cloud droplets and their growth into precipitation particles. Liquid water droplets and ice crystals in clouds all possess a nucleus; the former have hygroscopic nuclei (e.g. dust, smoke, sulphur dioxide and sodium chloride) and the latter freezing nuclei which form most commonly between $-15\,°C$ and $-25\,°C$ (for example very fine soil particles). The hygroscopic nuclei vary in diameter between $4 \times 10^{-7}\,\mathrm{m}$ and $2 \times 10^{-5}\,\mathrm{m}$ and exhibit concentrations of about $10^{8}\,\mathrm{m^{-3}}$ in oceanic air and $10^{9}\,\mathrm{m^{-3}}$ in continental air. Microphysics is concerned with explaining how cloud droplets which have diameters from less than $2 \times 10^{-6}\,\mathrm{m}$ up to $10^{-4}\,\mathrm{m}$ grow to precipitation particle diameters of up to $5 \times 10^{-4}\,\mathrm{m}$ (drizzle) and above (rain).

It is observed that at a given temperature below $0\,°C$, the relative humidity above an ice surface is greater than over liquid water and that saturation vapour pressure over water is greater than over ice, notably between $-5\,°C$ and $-25\,°C$. The result is that in clouds where ice crystals and supercooled water droplets coexist, the crystals grow preferentially by SUBLIMATION at the expense of the droplets. This Bergeron–Findeisen mechanism is important in precipitation production especially in the extra-tropics where a good deal of clouds have substantial layers colder than $0\,°C$. This means that most rain and drizzle in these areas starts off as ice crystals or snowflakes high in the parent cloud.

In natural clouds the droplet size varies and those that have grown on giant nuclei are large and exhibit the greatest fall speeds. These droplets grow by colliding with slower moving smaller droplets and this process of coalescence is efficient enough to produce drizzle size or sometimes raindrop size particles. It is most favoured in deep clouds with prolonged updraughts and is fairly common in clouds of tropical maritime origin. Cloud electrification increases the efficiency of coalescence.                          RR

**Reading**
Mason, B.J. 1962: *Clouds, rain and rainmaking.* Cambridge: Cambridge University Press.

## cloud streets

**cloud streets** Rows of cumulus or stratocumulus clouds lying approximately along the direction of the mean wind in the layer they occupy. A single cloud street may form downwind of a persistent source of thermals, but extensive areas of parallel streets are common, being associated with a curvature of the vertical profile of the horizontal wind. The axes of adjacent rows are separated by distances which are approximately three times the depth of the layer of convection. They are especially frequent over the oceans during outbreaks of cold arctic or polar air. The air motion within the cloud streets is one of longitudinal roll vortices, with air undergoing spiralling motions.            KJW

**Reading**
Agee, E.M. and Asai, T. eds 1982: *Cloud dynamics.* (Advances in earth and planetary science.) Tokyo: Terra Scientific; Dordrecht: D. Reidel. · Atkinson, B.W. 1981: *Mesoscale atmospheric circulations.* London and New York: Academic Press. · Stull, R.B. 1988: *An introduction to boundary layer meteorology.* Amsterdam: Kluwer.

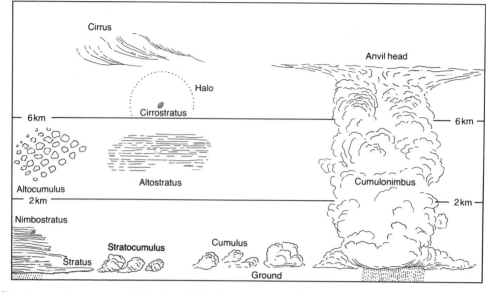

**Cloud genera** showing typical heights in middle latitudes.
*Source*: J.G. Harvey 1976: *Atmosphere and ocean*. London: Artemis. Figure 3.5.

**clouds** Both the most distinctive feature of the earth viewed from space (see SATELLITE METEOROLOGY) and the most transient element of the climate system. At any time about half of the globe is covered by clouds composed of water droplets or ice crystals and occurring at altitudes throughout the TROPOSPHERE. The dynamic nature of cloud processes (see CLOUD DYNAMICS) is most readily viewed on a bright summer day when vertical development is vigorous. Clouds are formed when air is cooled below its saturation point (see CLOUD MICROPHYSICS). This cooling is usually the result of ascent accompanied by ADIABATIC expansion. Different forms of VERTICAL MOTION give rise to different cloud types: vigorous local CONVECTION causes convective clouds; forced ascent of stable air (usually over an adjacent AIR MASS) produces layer clouds and forced lifting over a topographic feature gives rise to orographic clouds, which may be convective or stable in character. Clouds can also form as a result of other processes, such as the cooling of the lowest layers of the atmosphere in contact with a colder surface in, for instance, the formation of radiative FOG as a result of radiative cooling of the surface at night and advective fog resulting from movement of warm, moist air across a cold surface. The International Classification scheme (WMO 1956) groups clouds into four basic categories and uses the latin names given to them by an English chemist Luke Howard in 1803; *cumulus* – a heap or pile; *stratus* – a layer; *nimbus* – rain; and *cirrus* – a filament of hair. Additionally the

height of the cloud can be indirectly identified by the use of the terms: *strato* – low level; *alto* – middle level; and *cirro* – high. Thus, middle level, cumuliform cloud, often known as a 'mackerel sky', is called altocumulus. Clearly many different combinations of character and height descriptors are possible. The passage of a warm sector depression system often includes rain from nimbostratus clouds while conditions in summer frequently lead to cumuliform development, perhaps culminating in a fully fledged cumulonimbus cloud and a THUNDERSTORM.

AH-S

**Reference**
WMO 1956: *International cloud atlas*. Geneva: World Meteorological Organization.

**cluse** Originally a French term, now in general use to describe specifically a steep-sided valley which cuts through a mountain ridge in the Jura mountains.

**cluster bedform** A particular type of BEDFORM associated with gravel-bed rivers, and comprising an aggregation of individual grains, often of varying sizes. Cluster bedforms can be classified (e.g. Richards and Clifford 1991) by their orientation with respect to the flow: either transverse or parallel. Transverse bedforms can be classified into three scales of feature: (1) TRANSVERSE RIBS; (2) transverse CLAST dams; and (3) STEP-POOL SYSTEMS. Transverse ribs are the smallest features, comprising regularly spaced pebble, cobble or boulder ridges formed when rolling or sliding grains stall against already stationary grains. Transverse clast dams are larger in scale, often associated with a wider particle size distribution, with larger clasts forming the building blocks of the dams. It is not uncommon to observe infilling of fines between these dams. Step-pool systems are associated with particularly steep, coarse-bedded channels, with a low depth to grain size ratio. They are stable at most discharges, and the time-scales at which these three types of bedform adjust to flow and sediment transport is one of their major characteristic features. Ribs and dams are more dynamic than step-pool systems because they generally involve smaller clasts and are therefore moved at a much wider range of discharges. The main type of bedform that is parallel to the flow is the pebble cluster. This comprises a large core clast, with larger particles that have stalled on its STOSS side, and smaller particles that accumulate in the flow separation zone in its lee. These are exceptionally common in all gravel-bed channels and are also very mobile. As a group, cluster bedforms are now recognized to play a critical part of the dynamics of gravel-bed rivers. The can significantly enhance ROUGHNESS and create important flow structures, both of which have implications for the BEDLOAD transport process. In general, they increase roughness, reduce sediment transport rates, and may be a critical part of the sediment sorting process.                    SNL

**Reading**
Bluck, B.J. 1987. Bedforms and clast size changes in gravel-bed rivers. In K.S. Richards ed., *River channels: environment and process*. Oxford: Blackwell. · Richards, K.S. and Clifford, N.J., 1991. Fluvial geomorphology: structured bedas in gravelly rivers. *Progress in physical geography* 15, pp. 407–22.

**coastal dunes**   Deposits of AEOLIAN sand adjacent to large bodies of water. Coastal dunes may form in marine, LACUSTRINE or fluvial environments, with most forms similar to those found in arid environments. Foredunes are the characteristic coastal dune, as their form usually displays some influence of the adjacent water. Large dune complexes are most common along low relief coasts, adjacent to rivers, on BARRIER ISLANDS and spits, or along arid coastlines. Coastal dunes are also distinguished from dunes in other environments by the morphological signatures of hydrodynamic erosion, especially dune scarping, and by the importance of vegetation in controlling dune development. The sediment source for most coastal dunes is an adjacent beach, where aeolian erosion and transport provides the linkage of the systems. Along many coasts, sand stored in the dunes represents a significant fraction of the material in the sediment budget.

Coastal dune systems often require careful management. In many environments, these dunes provide critical habitat for plants and animals, and they may serve as important, local sources of fresh water. The dunes also offer important protection against coastal erosion and flooding. They act as a sediment reservoir to feed the nearshore system during erosive events, and when water levels are elevated, linear foredune systems act as barriers to flooding. For these reasons, coastal dunes are often socially valuable landforms.          DJS

**Reading**
Nordstrom, K.F., Psuty, N.P., and Carter, R.W.G eds 1990: *Coastal dunes: form and process*. Chichester: Wiley. · Sherman, D.J., and Bauer, B.O. 1993: Dynamics of beach-dune systems. *Progress in physical geography* 17, pp. 413–47.

**coccoliths**   The mineralized (calcite) components of single-celled green ALGAE called coccolithophores. They are generally spherical or oval in shape and are usually smaller than 20 $\mu$ms and are often referred to as nanofossils. Although a few species are adapted to either fresh or brackish water, coccolithophores are super-abundant in the world's oceans, and are one of the principal components of modern deep-sea sediments. Studies have shown that many species are associated with distinct thermal characteristics and they provide an important means of determining ocean palaeo-temperatures and hence past climates.     SLO

**cockpit karst**   The scenery produced by the solution of limestone resulting in a hummocky terrain of conical residual hills surrounded by DOLINES or SINKHOLES. They are very characteristic of Jamaican limestone scenery and are often associated with humid tropical karst landscapes. The hills are also called KEGELKARST or MOGOTES.                    PAB

**Reading**
Sweeting, M.M. 1972: *Karst landforms*. London: Macmillan.

**coefficient of permeability**   See PERMEABILITY.

**coefficient of variation (CV)**   is a relative measure of the variability within a group of observations or other numerical data. It is defined as the standard deviation divided by the mean, and expressed as a percentage. That is,

$$CV = s/x \,.\, 100$$

For example, the numbers 0, 1, ... 10 have a mean of 5 and a standard deviation of 3.3, and hence a CV of 66.3%. A useful property of the CV is that it is not affected by the units of measurement. Thus, using the CV, the variability in yearly rainfalls over a catchment, scaled in mm, can be compared directly to the variability in resulting river flows measured in ML, since the variability is expressed as a dimensionless percentage of the mean. For the numbers 0, 1, ... 100, both the mean and the standard deviation are 10 times larger than for the numbers 0, 1, ... 10 (50 and 33.2 respectively) but the CV is the same (66.3%).                    DLD

**Reading**
Clark, W.A.V. and Hosking, P.L. 1986: *Statistical methods for geographers*. New York: Wiley.

**coevolution** The evolutionary interaction between two species so that over a long period of time they become co-adapted to each other through a series of selective influences. The predator–prey relationship in animals provides some of the clearest examples. A predator such as a cheetah has evolved to outrun prey such as gazelles, whose defence is to run faster themselves, and to evolve more sensitive detection mechanisms, forcing the cheetah to stalk more quietly and unobtrusively. Coevolution between insects and plants has been the subject of much recent work. Besides the numerical relationship between the ecological 'conspicuousness' of plants such as the oak tree in England, with about 300 insect species associated with it, contrasting with the ash, with less than fifty associated insects, there are some very precise and beautiful examples which are less obvious. Chemical defences may be used by plants against particular insects – e.g. oak produces leaves with more tannin as a defence against defoliating caterpillars – or else mimicry may be employed. An example of this is the extraordinary way in which the leaves of *Passiflora* species in central America mimic the leaves of other rain-forest species in an attempt to deceive *Heliconiid* butterflies who otherwise may lay their eggs on the leaves. Some *Passiflora* species even produce small swellings on the leaves to mimic butterfly eggs and hence escape predation! Coevolution must also have occurred between larger herbivores and grazed vegetation: the production of spines, of tough unpalatable leaves and of basal shoots as in grasses, are all responses to grazing pressure.

Coevolution can also lead to SYMBIOSIS or mutualism, as in lichens, and other relationships which may be looked upon as controlled or 'beneficial' parasitism such as that between mycorrhizal fungi and their host plants. The whole field of pollination ecology also throws up many examples of coevolution between insects and plants.                    KEB

**Reading**
Howe, H.F. and Westley, L.C. 1986: Ecology of pollination and seed dispersal. In Crawley, M.J. ed, *Plant ecology*. Oxford: Blackwell Scientific. · Krebs, C.J. 1985: *Ecology: the experimental analysis of distribution and abundance*. 3rd edn. New York: Harper & Row. · Putman, R.J. and Wratten, S.D. 1984: *Principles of ecology*. Beckenham, Kent: Croom Helm.

**cohesion** The attraction of particles to each other (usually clay minerals in soils) which is not governed directly by a FRICTION law (i.e. it is independent of STRESS) but does provide a measure of the strength of a material. Thus sands do not exhibit cohesion but what is termed (intergranular) friction, while clays, or soils which contain clays, show cohesion. The strength is supplied by the structure of the clay minerals and the way in which chemical bonding is produced in these structures. It can be measured, as in soil mechanics, by the MOHR-COULOMB EQUATION.                    WBW

**col** A pass or saddle between two mountain peaks or a narrow belt of low pressure separating two areas of high pressure.

**cold front** A frontal zone in the atmosphere where, from its direction of movement, rising warm air is being replaced by cold air. It has an average slope of about 1 in 60 over the lowest several hundred metres rather like the forward bulge of a density bore. Above this level both warm and cold fronts have similar slopes. A cold front is often found to the rear of an EXTRA-TROPICAL CYCLONE marking a sudden change from warm, humid and cloudy weather to clear, brighter weather. Satellite images usually show cold fronts clearly, as a thick band of cloud spiralling away from the depression centre. Where uplift within the warm air is rapid, rainfall may be heavy along the cold frontal zone. As it passes, temperatures fall suddenly together with a rapid veering of the wind.                    PS

**cold pole** Can be defined as the location of lowest mean monthly temperature, or of lowest mean annual temperature, or of the coldest air in the TROPOSPHERE. The last case is usually indicated by the area of lowest THICKNESS on the chart of 1000–500 mb thickness, and in January is found in the northern hemisphere over north-east Siberia, and in July over the north pole. In the southern hemisphere the lowest 1000–500 mb thickness is found over the Antarctic continent. The lowest surface air temperatures in the northern hemisphere are found in north-east Siberia at Verhojansk and Oymjakon, where they have reached −68 °C. The lowest surface air temperatures in the southern hemisphere are recorded in Antarctica, where they have reached −88 °C.          JGL

**coleoptera**    Beetles, one of several orders of insects. In Quaternary studies the remains of coleoptera have proved especially valuable, for the following reasons: (1) unlike other orders, their hard chitinous exoskeletons, especially wing cases, withstand burial and preservation in sediments; (2) the order is one of the largest in the insect kingdom; (3) remains are well documented, facilitating relatively easy identification to the species level; (4) many species are stenotypic – i.e. have distinct environmental controls on their distribution; and (5) they respond relatively rapidly in their distribution in response to environmental changes. The use of coleoptera remains in Quaternary studies has especially been developed by G.R. Coope, who has noted that, due to the above reasons 'Coleoptera [are] one of the most climatically significant components of the whole terrestrial biota' (Coope 1977).          MEM

**Reference**
Coope, G.R. 1977. Quaternary Coleoptera as aids in the interpretation of environmental history. In F.W. Shotton ed., *British Quaternary studies: recent advances*. Oxford: Oxford University Press.

**colk**    A pot-hole in the bed of a river.

**colloid**    Any non-crystalline, partially solid substance. Often applied to individual, lens-shaped particles of clay.

**colluvium**    Material that is transported across and deposited on slopes as a result of wash and mass movement processes. It is frequently derived from the erosion of weathered bedrock (ELUVIUM) and its deposition on low-angle slopes, and can be differentiated from ALLUVIUM, which is deposited primarily by fluvial agency. Colluvium may be many metres thick, often contains fossil soil layers which represent halts in deposition, shows some crude bedding downslope, and is generally made of a large range of grain sizes. Cut and fill structures may represent phases when stream incision has been more important than colluvial deposition, and in southern Africa (Price Williams *et al.* 1982) many colluvial spreads have suffered from intense DONGA (gully) formation.          ASG

**Reference**
Price Williams, D., Watson, A. and Goudie, A.S. 1982: Quaternary colluvial stratigraphy, archaeological sequences and palaeoenvironment in Swaziland, Southern Africa. *Geographical journal* 148, pp. 50–67.

**colonization**    The occupation by an organism of new areas or habitats thereby extending either its geographical or its ecological range. For colonization to succeed three main phases are usually necessary; effective dispersal or migration, germination or breeding and establishment, and, finally, survival in the new site through time. Most colonizing species are subject to *R*-SELECTION. Species which are able to increase their numbers exponentially on arrival in a new area have strong advantage (e.g. annual and biennial plants), while animals exhibiting behavioural adaptability and a marked ability to learn new behaviour tend to be natural colonizing species. Man has proved a potent aid to many colonizers, spreading them to new areas where they have subsequently flourished (e.g. *Opuntia stricta* in Australia) or opening up new habitats ripe for colonization. (See also ALIENS; DISPERSAL; *R*- AND *K*-SELECTION.)          PW

**Reading**
Elton, C.S. 1958: *The ecology of invasions by animals and plants*. London: Methuen. · Krebs, C.J. 1978: *Ecology: the experimental analysis of distribution and abundance*. New York: Harper & Row. · MacArthur, R.H. 1972: *Geographical ecology*. New York: Harper & Row.

**combe, coombe**    A small, often narrow valley. Frequently applied to the dry or seasonal stream valleys of chalk country.

**comfort zone**    The range of meteorological conditions within which the majority of the population, when not engaged in strenuous activity, will feel comfortable. It is often expressed in terms of effective temperature (ET), which combines dry-bulb and WET-BULB TEMPERATURES and air movement into a single biometeorological index. The comfort zone varies with climatic zone and season of the year. Personal factors such as age, clothing, occupation and degree of acclimatization are also important. In the tropics, the comfort zone is often taken to range from 19 to 24.5 °C ET. In the UK, comparable values are 15.5–19 °C ET in summer and 14–17 °C ET in winter (Air Ministry 1959). The effects of direct exposure to solar radiation are not normally included in a consideration of the comfort zone.          DGT

**Reading and Reference**
Air Ministry 1959: *Handbook of preventive medicine*. London: HMSO. P. 164. · Terjung, W.H. 1966: Physiologic

climates of the conterminous United States: a bioclimatic classification based on man. *Annals of the Association of American Geographers*, 56, pp. 141–79.

**commensalism** The weakest type of association of species living together in SYMBIOSIS. One or more 'guest' species benefits from living in association with a 'host' species, but the latter neither benefits nor is harmed by the presence of the guests. Mites which live in human hair represent a case in point, as also does the pitcher plant (*Nepenthes*), which can host up to nineteen species. On a world scale commensalism is most common in oceans, especially in the Indian and West Pacific oceans, where at the generic level it attains 59 per cent among caridean shrimp populations (Vermeij 1978). DW

**Reference**
Vermeij, G.T. 1978: *Biogeography and adaption: patterns of marine life*. Cambridge, Mass. and London: Harvard University Press.

**comminution** The reduction of a rock or other substance to a fine powder, often as a result of abrasion.

**community** Any assemblage of populations of living organisms in a prescribed area or habitat. The term 'community' is an ecological unit used in a more general, broad, collective sense to include groups of various sizes and degrees of integration. The community comprises a typical species composition that has resulted from the interaction of populations over time.

Botanists and zoologists have defined the term community in widely differing ways. Three main ideas are involved in community definitions, which claim to have one or more of the following attributes:

co-occurrence of species,
recurrence of groups of the same species,
homeostasis or self-regulation.

Two opposing schools have developed in ecology over the question of the nature of the community. The *Organismic school* suggests that communities are integrated units with discrete boundaries. The *Individualistic school* suggests that communities are not integrated units but collections of populations that require the same environmental conditions. AP

**Reading**
Krebs, C.J. 1985: *Ecology: the experimental analysis of distribution and abundance*. New York: Harper and Row.

**compaction** In engineering terms, the expulsion of air from the voids of a soil. This is usually achieved by artificial means, e.g. with various types of roller in road and dam construction. It differs from CONSOLIDATION which is usually a natural process, although the latter may involve some air expulsion. The aim of compaction is to achieve a high bulk density or lower VOID RATIO, so improving the overall strength of the soil. WBW

**compensation flows** Designated river discharges that must continue to flow in a river below a direct supply reservoir or abstraction point to allow for other riparian activities and interests downstream. Such flows are legally established in different ways in different countries and at different times, and may be necessary to maintain water quality, to satisfy the needs of wildlife or recreation, or to provide water for other abstractors or users. JL

**competence** The competence of flowing water is the maximum particle size transportable by the flow. A related concept is the threshold flow, which is the minimum flow intensity capable of initiating the movement of a given grain size. When the largest grains on a stream bed are just mobile the flow is the threshold (critical) flow for those grains, and their size represents the competence of the flow. It is possible to predict theoretically the threshold flow conditions for non-cohesive grains coarser than 0.5–0.7 mm, by balancing moments caused by fluid drag and particle immersed weight (Carson 1971, p. 26; Richards 1982, pp. 79–84). The threshold shear stress increases linearly with the grain diameter, while threshold velocity increases with the square root of grain diameter or the one-sixth power of its weight (the so-called 'sixth-power law'). These relationships can be used to define the competence of a given shear stress or velocity (see HJULSTRÖM CURVE). However, the competence to *maintain* particle motion is usually greater than that to *initiate* motion, so large grains may be carried *through* a reach in which they cannot be entrained.

In flowing water grain motion is normally caused by fluid stresses. In air, however, an impact threshold also occurs. The kinetic energy of a sand grain travelling by SALTATION in air allows it to move grains up to six times its size. The concept of fluid competence is therefore less applicable to aeolian transport. KSR

**References**
Carson, M.A. 1971: *The mechanics of erosion*. London: Pion. · Richards, K.S. 1982: *Rivers: form and process in alluvial channels*. London and New York: Methuen.

**competition** The inevitable interactions between members of the same and different species attempting to secure limited resources from the environment, and leading to increases and decreases in the populations of more and less successful organisms. Competition takes many forms: between plants for light, water, space and nutrients; between animals for mates, food, shelter and for social position within a hierarchy. While many of the principles are the same for both plants and animals, there are several obvious practical differences, such as the ability of plants to spread and reproduce asexually. Competition is also a very complex phenomenon, with many of the precise mechanisms by which it occurs being poorly understood. The literature tends, therefore, to be over-reliant upon a relatively small number of classic case studies. More recently, there has been increasing recognition of the role of other factors in the distribution of species and the structure of plant and animal communities, particularly herbivory, predation and habitat disturbance.

The classic experiments concluded that species grown together in the same environment did not do as well as when grown separately, for one species tended to triumph and the other became extinct. This led to the 'competitive exclusion principle' and quantitative expressions such as the Lotka–Volterra equations (Begon *et al.* 1990, pp. 247–51), based on the logistic curve of POPULATION DYNAMICS. All discussion of competition also involves consideration of the concept of the NICHE, the position of an organism within the community and its interrelationships with the organisms around it, as extensively reviewed by Putman and Wratten (1984).

Reviews of the influence of competition on community structure tend to stress the importance of other factors, while not doubting that competition has a major role (e.g. Begon *et al.* 1990), though some see the predictions of a simple theory of competition for essential resources as the main explanation of the observed patterns (Tilman 1986), at least in plant communities.                    KEB

**Reading and References**
Begon, M., Harper, J.L. and Townsend, C.R. 1990: *Ecology*. 2nd edn. Oxford: Blackwell Scientific. · Putman, R.J. and Wratten, S.D. 1984: *Principles of ecology*. Beckenham, Kent: Croom Helm. · Tilman, D. 1986: Resources, competition and the dynamics of plant communities. In M.J. Crawley ed., *Plant ecology*. Oxford: Blackwell Scientific.

**complex response**  The term introduced by Schumm (1973) to describe the variety of linked changes which may occur within a drainage basin in response to the single passage of a geomorphical THRESHOLD. These changes may well

be both spatially and temporally separated and it may, in consequence, be difficult *ex post facto* to establish that all are ultimately due to a single event.                    BAK

**Reference**
Schumm, S.A. 1973: Geomorphic thresholds and the complex response of drainage systems. In Marie Morisawa ed., *Fluvial geomorphology*. Publications in geomorphology. Binghampton: State University of New York. Pp. 299–310.

**compressing flow**  Compressing and extending flows describe longitudinal variations in velocity in a moving medium. The concept is widely used in glacial geomorphology to describe longitudinal variations in flow along the length of a glacier (Nye 1952). Compressing flow refers to a reduction in the length of a unit of glacier ice in a downstream direction, and extending flow refers to an increase in length in the same direction. Such changes must be accompanied by a corresponding variation in glacier width or height. As valley walls often prevent lateral expansion the variations in a glacier are commonly accompanied by a change in depth, with compressing flow associated with thickening and extending flow with thinning. Any such thickening must be associated with an upward component of ice flow while thinning is associated with a downward component of ice flow. When such vertical flow is superimposed on normal down-glacier flow it can have important geomorphological effects by redistributing rock debris. Compressing flow can bring basal debris to the surface and its later deposition by the melting of underlying ice can lead to complex hummocky moraine landforms. Extending flow can carry surface and englacial material to the glacier bottom, thus replenishing the supply of erosive tools. Such vertical movements are usually taken up by ice deformation or CREEP, but where stresses are sufficiently high thrusting and faulting may occur. The association of compressing and extending flows with changing ice thickness leads to predictable spatial associations. Extending flow is characteristic of the accumulation zone where the glacier mass increases downstream, while compressing flow is common in the ABLATION zone where the glacier thins downstream, and especially near the snout where thinning is most marked. Extending flow is also associated with a bed convexity which steepens downstream, as in the case of an ICE FALL, while compressing flow occurs as the glacier crosses a bed concavity. Variations in the rate of BASAL SLIDING along the length of a glacier can also induce compressing and extending flows. For example, a change from warm- to cold-based thermal regime or a thinning of the basal water layer in a downstream direction favours compressing flow, while the opposite conditions favour extending flow.                DES

**Reference**
Nye, J.F. 1952: The mechanics of glacier flow. *Journal of glaciology* 2.12, pp. 82–93.

**compressional** Pertaining to descending atmospheric air masses, which as a result of pressure increases are warm and dry. Also pertaining to geological faults and fractures which are the product of lateral increases of pressure and to earthquake waves, specifically P-waves.

**conchoidal fracture** A fracture in a rock or mineral which is shaped like a shell, i.e. concave down. The characteristic fracture of siliceous rocks and minerals.

**concordant** Within the same plane. Applied particularly to the mountain summits of a region, the concordance of summits attesting to the existence of a plateau prior to incision and dissection of the landsurface.

**concordant coast** A coastline where the general orientation is parallel with, and controlled by geological structures. The western coastlines of the Americas are usually concordant, exhibiting control by adjacent faults. The DALMATIAN COAST of Yugoslavia is concordant. The east coast of the United States is discordant, with Cape Cod and Long Island representing orientations perpendicular to the general coastal trend.                DJS

**concretion** A solid lump or mass of a substance, or aggregates of these, incorporated within a less competent host material.

**condensation** The process of the formation of liquid water droplets from atmospheric water vapour mainly onto 'large' hygroscopic nuclei with diameters between $2 \times 10^{-7}$ m and $10^{-6}$ m. The cooling necessary to produce condensation in the atmosphere is effected either by ADIABATIC expansion during ascent, by contact with a colder surface or by mixing with colder air. The LATENT HEAT released during this process can be a significant source of heat in weather systems.                RR

**conductance, specific** A measure of the ability of an aqueous solution to conduct an electrical current. It is reported in units of microsiemens per centimetre at $t\,°C$ ($\mu S\,cm^{-1}t\,°C$). Pure water has a very low specific conductance, but the conductance will increase with an increasing concentration of charged ions in solution. Individual ions exhibit different conductance values and the level of conductance recorded for a given total ion concentration will therefore vary according to its ionic composition. For most dilute natural waters an increase in temperature of $1\,°C$ will increase the conductance by approximately 2 per cent and values of specific conductance are normally corrected to a reference temperature of 20 or $25\,°C$.

Because specific conductance (SC) measurements are simple and rapid to make, and reflect the total ion concentration, they have been widely used as a means of estimating the total dissolved solids (TDS) concentration of water samples. The relationship between SC and TDS for waters of different chemical composition is generally determined empirically and commonly takes the form:

$$TDS = KSC$$

where $K$ varies between 0.55 and 0.75. This relationship may be complicated by changing ionic composition and by the presence of non-ionized material in solution (e.g. $SiO_2$ and dissolved organic matter) which will contribute to TDS but not SC.                DEW

**Reading**
Foster, I.D.L., Grieve, I.C. and Christmas, A.D. 1982: The use of specific conductance in studies of natural waters and solutions. *Hydrological sciences bulletin* 26, pp. 257–69.

**conductivity, hydraulic** The term given to the parameter $K$ in the equation defining DARCY'S LAW. It is concerned with the physical properties of both the fluid and the material through which it flows, reflecting the ease with which a liquid flows and the ease with which a porous medium permits it to pass through it. Hydraulic conductivity has the dimensions of a velocity and is usually expressed in m s$^{-1}$. Since hydraulic conductivity may vary according to direction, $K_x$, $K_y$ and $K_z$ can be used to represent the hydraulic conductivity values in the $x$, $y$ and $z$ directions. Hydraulic conductivity should be distinguished from the term PERMEABILITY (or INTRINSIC PERMEABILITY) $k$ which refers only to the characteristics of the porous medium and not to the fluid which passes through it.

The saturated hydraulic conductivity of soils and other sediments may be measured in the laboratory with a PERMEAMETER. If samples are essentially undisturbed, the results are point values representative of the field conditions. However, the hydraulic conductivity of an aquifer (see GROUNDWATER) is best determined by field methods. Piezometer tests can be used to determine *in situ* $K$ values in a porous material around a piezometer tip. Pumping tests at a well provide measurements representative of a much larger aquifer volume (Freeze and Cherry 1979). PWW

**Reference**
Freeze, R.A. and Cherry, J.A. 1979: *Groundwater*. Englewood Cliffs, NJ: Prentice-Hall.

**cone of depression** The shape of the depression in the WATER TABLE surface around a well that is being actively pumped for GROUNDWATER. Pumping results in a lowering (or DRAW DOWN) of the water table that is greatest in the well itself, but reduces radially with distance from the well. In a rock with horizontal PERMEABILITY that is uniform in every direction, the cone of depression will be symmetrical about the well. But where groundwater flow occurs more readily in one direction than in another (perhaps because of the influence of dominant joints), the cone of depression will be asymmetrical and elongated in the direction of greatest permeability. Cones of depression can develop in wells pumping both unconfined and confined aquifers. PWW

**Reading**
Freeze, R.A. and Cherry, J.A. 1979: *Groundwater*. Englewood Cliffs. NJ: Prentice-Hall. · Linsley, R.K., Kohler, M.A. and Paulhus, J.L.H. 1988: *Hydrology for engineers* 3rd edn. New York: McGraw-Hill.

**confined groundwater** See GROUNDWATER.

**congelifluction** A general term applied to the movement of rock and earth, usually down hillslopes, as a result of freezing and thawing of ice, especially in permafrost regions.

**congelifraction** The weathering of rock by freeze–thaw process.

**congeliturbation** A general term for the heaving of the ground surface as a result of the freezing and thawing of ice.

**conglomerate** A rock which is composed of or contains rounded or water-worn pebbles and cobbles more than about 2 mm in diameter.

**connate water** Water trapped in the interstices of sedimentary rocks at the time of their deposition. It is usually highly mineralized and is not involved in active GROUNDWATER circulation, although connate waters may be expelled from their original location by compaction

pressure and migrate, accumulating in more permeable formations. (See also JUVENILE WATER; METEORIC WATER.) PWW

**conscious weather modification** See WEATHER MODIFICATION.

**consequent stream** A stream which flows in the direction of the original slope of the land-surface.

**conservation** Used in its more general sense to mean environmental conservation, the term relates to the sustainable utilization, through intervention and management, of the entire range of environmental resources and processes. Conservation is distinguished from preservation insofar as it incorporates the element of sustainable utilization, whereas preservation implies protection from any form of use. Thus, conservation refers to the active maintenance of elements of the environment, for example, soil, water, biological diversity and other natural resources in association with the recognition that exploitation of such resources is essential to human existence. The broader use of the term is, however, relatively recent; historically conservation has been synonymous with the establishment of set-aside areas, so-called nature reserves, for the preservation of, in the main, wild animals. More recently, the need for conservation of the spectrum of environmental processes, both within and beyond reserve boundaries, has been recognized. Many politicians have jumped on the 'conservation bandwagon' and the increase in environmental awareness that has marked the 1990s, following the Rio Earth Summit has led to the wider application of conservation, i.e. sustainable utilization, principles. From being a rather negative concept dealing with the preservation of existing environmental processes, conservation now invokes active and positive intervention as an integral component of economic development. The traditional 'conservation versus development' debates are placed in perspective by the acceptance of an expanded view of conservation.

The most common context for conservation, nevertheless, remains the goal of maintaining the earth's biological diversity, a branch of science which is referred to as conservation biology. Different levels of biological conservation are apparent, ranging from entire ecosystem conservation to the conservation of communities, habitats, species, populations or even gene pools. Biological conservation is based on sound scientific principles emanating from the fields of, in particular, genetics and population ecology. With respect to conservation of endangered plant and animal species within reserve areas, significant advances were made in the 1960s following the development of ISLAND BIOGEOGRAPHY and nowadays theory and practice around the management of so-called minimum viable populations represents a key element of conservation biology. Conservation using these principles is both *in situ* (i.e. in the natural environment) and *ex situ* (for example in botanical and zoological gardens).

Much of the debate around conservation remains focused on the protected area network. The World Conservation Strategy, established in 1980, formally introduced international policy aimed at conserving the environment through the maintenance of essential ecological processes and the preservation of genetic and species diversity. The call for each country to conserve a minimum of 10 per cent of its geographical area was made, although there is no scientific basis for such an arbitrary value and it by no means guarantees the meaningful protection of most of the plant and animal species within any country. In any case, very few nations would be able to boast a conservation estate of 10 per cent of their areas. In practice, politics and economics as opposed to science, have tended to dictate which areas are managed for conservation purposes. Numerous criteria can be put forward to evaluate an area, such as total species diversity, the presence of rare species and the *naturalness* or degree to which a site has been modified by human activities. By employing these and other conservation criteria, it is hoped that land-use planning decisions which minimize the environmental impact of developments will emerge.

To most, including politicians, biological conservation has an undeniable value, although it has proved remarkably difficult to convert such value into the kind of economic entity that influences hard decisions about where to spend taxpayers' money. Of late, attempts to answer the question 'why conserve nature?' have taken on a less abstract form with the development of ecological economics (Costanza 1991). Although it is a difficult process, it is possible to sum the various ethical, aesthetic and utility values of conserving, for example, biological diversity to construct a more persuasive, economically based argument in favour of conservation.

Increasing concern with the degree of SOIL EROSION, especially in seasonal rainfall climates, has resulted in better soil conservation practices with the intention of maintaining soil structure and fertility. Soil management actually has a long history in the early agricultural areas around the Mediterranean, the Middle and Far East, as reflected in the adoption of, for example, terraces and mulching techniques. Since 1934 in the United States, the Soil Conservation Service has offered advice to farmers about methods of soil tillage and management aimed at minimizing soil loss by wind and water, a movement which has since been emulated in many other countries.                    MEM

**Reading and Reference**
Costanza, R. ed. 1991: *Ecological economics: the science and management of sustainability*. New York: Columbia University Press. · Morgan, R.P.C. 1995: *Soil erosion and conservation*. 2nd edn. London: Longman. · Shafer, C.L. 1990: *Nature reserves: island theory and conservation practice*. Washington: Smithsonian Institute Press. · Spellerberg, I.F. and Hardes, S. 1992: *Biological conservation*. Cambridge: Cambridge University Press.

**consociation**  A natural vegetation community which is dominated by one species.

**consolidation**  In engineering terms, the expulsion of water from the void spaces (see VOID RATIO) of a soil. This is achieved by natural burial processes as sediment accumulates or when a structure is erected on a soil. A normally consolidated soil (usually clay) is one which has never been subjected to an overburden pressure (i.e. load on top) greater than that which it currently has. An over-consolidated soil is one which has had a greater overburden pressure. This last effect is usually achieved by denudation of the overlying material: lodgement TILL is over-consolidated because of the pressures previously applied by the glacier above. Over-consolidated clays have higher strengths than an otherwise equivalent normally consolidated clay, measured by an over-consolidation ratio.

                                                            WBW

**constant slope**  A term used by Wood (1942) to define the straight part of a hillside surface, lying below the free face, and having an inclination determined by the angle of repose of the TALUS material forming it. The constant slope extends upwards until it buries, or replaces the free face and so eliminates the supply of fresh talus debris. The WANING SLOPE of lower inclination is formed of weathered talus and eventually extends upwards and replaces the constant slope. The idea of a 'constant' slope is a theoretical construct with constancy of slope length and slope angle being of limited duration in a geological time span.                    MJS

**Reading and Reference**
Wood, A. 1942: The development of hillside slopes. *Proceedings of the Geologists' Association* 53, pp. 128–40. · Young, A. 1972: *Slopes*. Edinburgh: Oliver & Boyd.

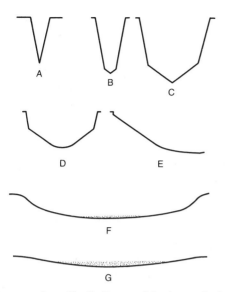

**Constant slope**. Wood's diagram of the slope cycle. A = free face only; B and C = constant slope forms; D and E = waning slope develops; F = waxing slope forms, waning slope rises up side of constant slope, alluvial filling represented by dots; G = constant slope has been consumed, alluvial fill deepens, slopes gradually flatten and approach a peneplain.
*Source*: Wood 1942.

**continental climate** A type of climate characteristic of the interior of large land masses of middle latitudes. The major effects of continentality are to produce extreme seasonal variations of temperature and to depress the mean annual temperature below the latitudinal average. Since the oceans are the main sources of atmospheric moisture the continental interiors also tend to be dry, allowing the subtropical deserts to extend into temperate latitudes. The most extreme form of continental climate is found in eastern Asia, where there is an annual range of temperature of about 60 °C.                    JGL

**continental drift** The movement of continents relative to each other across the earth's surface. Although it was a subject of speculation by numerous early workers, the first comprehensive case for continental drift was presented by Alfred Wegener in 1912. Wegener cited various lines of evidence in support of his notion of a super-continent (Pangaea) which gradually separated into a northern (Laurasia) and a southern (Gondwanaland) landmass before finally splitting into the continents of the present day. The evidence included the matching configuration of opposing continental coastlines, the similarity of geological structures on separate continental masses, the anomalous location of ancient deposits, indicating specific climatic conditions, and the distribution of fossil species through time. The theory was rejected at first by most geologists and geophysicists, largely because of a lack of a viable mechanism, but further support for continental drift was provided during the 1950s and 1960s through evidence from PALAEOMAGNETISM. In the late 1960s the idea was incorporated into the PLATE TECTONICS model.                    MAS

**Reading**
Du Toit, A.L. 1937: *Our wandering continents*. Edinburgh: Oliver & Boyd; New York: Hafner. · Hallam, A. 1973: *A revolution in the earth sciences: from continental drift to plate tectonics*. Oxford: Clarendon Press. · McElhinny, M.W. ed. 1977: *Past distribution of the continents*. Amsterdam: Elsevier. · Wegener, A. 1966: *The origin of continents and oceans*. Translated by John Biram from the fourth revised German edition. New York: Dover. · Wilson, J.T. ed. 1976: *Continents adrift and continents aground*. San Francisco: W.H. Freeman.

**continental freeboard** The average elevation of the continents relative to MEAN SEA LEVEL. Continental freeboard changes as a result of expansion and contraction of the volumes of the ocean basins relative to changes in continental volumes. It is an indicator of the ratio of the volume of ocean basins (or the volume of water in the basins) to the volume of the continents. If the ratio increases, there is EUSTATIC fall of sea level, and freeboard rises. Conversely, if the ratio decreases, sea level rises and freeboard falls. The present freeboard is approximately 750 m. Cogley (1984), including large, submerged regions of continental crust, estimated freeboard to be 126 m.

According to Wise (1974), the average freeboard has been relatively stable over the last 2.5 billion years, at a level about 20 m higher than present. Since the Precambrian, variation in freeboard has been within a range of $+/- 60$ m of the mean for more than 80 per cent of the time. He attributed this to ISOSTATIC equilibria in the LITHOSPHERE, and described the feedback mechanisms responsible. Galer and Mezger (1998) have formalized this concept as:

$$\Delta f = \left(\frac{\rho_o}{\rho_m} - 1\right)\Delta d_r + \left(\frac{\rho_b}{\rho_m} - 1\right)\Delta d_b - \left(\frac{\rho_c}{\rho_m} - 1\right)\Delta d_c$$

where $f$ is freeboard, $\rho$ is density and the subscripts $o$, $m$, $b$, and $c$ refer to ocean, mantle, oceanic crust, and continental crust, $d_r$ is reference water depth, and $d_b$ and $d_c$ are the thicknesses of oceanic and continental crusts.

Changes in freeboard are estimated using palaeogeographic maps of the area of continents, where inundation is interpreted from marine deposits in the geological record. The areal estimates are combined with HYPSOMETRIC

CURVES to reconstruct the continental elevation distributions. DJS

### References
Cogley, J.G. 1984: Continental margins and the extent and number of the continents. *Reviews of geophysics and space physics* 22, pp. 101–22. · Galer, S.J.G. and Mezger, K. 1998: Metamorphism, denudation and sea level in the Archean and cooling of the earth. *Precambrian research* 92, pp. 389–412. · Wise, D.U. 1974: Continental margins, freeboard and the volumes of continents and oceans through time. In C.A. Burk and C.L. Drake eds, *The geology of continental margins*. New York: Springer-Verlag. Pp. 45–58.

**continental islands** occur in close proximity to a continent, to which they are also geologically related, and which are detached from the mainland by a relatively narrow, shallow expanse of sea. They were formerly united to the mainland. Oceanic ISLANDS, on the other hand, are and have been, geographically isolated and rise from the floors of the deep ocean basins.

**continental shelf** A portion of the continental crust below sea level, consisting of a very gently sloping, rather featureless, surface forming an extension of the adjacent coastal plain and separated from the deep ocean by a much more steeply inclined continental slope. The gradient of most continental shelves is between 1 and 3 m km$^{-1}$. Some shelves reach a depth of over 500 m but 200 m is often conveniently used as a depth limit. Their mean width is around 70 km but there is a marked variation between different coasts. It is estimated that continental shelves cover approximately 5 per cent of the earth's surface. MAS

### Reading
Kennett, J.P. 1982: *Marine geology*. Englewood Cliffs, NJ and London: Prentice-Hall. · Shephard, F.P. 1973: *Submarine geology*. 3rd edn. New York: Harper & Row.

**continental slope** Lies to the seaward of the CONTINENTAL SHELF and slopes down to the deep sea floor of the abyssal zone.

**continuity equation** A statement that certain quantities such as mass, energy or momentum are conserved in a system, so that for any part of the system the net increase in storage is equal to the excess of inflow over outflow of the quantity conserved.

The most familiar example of a continuity equation is perhaps the STORAGE equation in hydrology which describes the conservation of mass for water. Continuity equations are equally applicable to the conservation of total mass of earth materials, or of mass for particular chemical elements or ions. In these contexts the continuity equation is one fundamental basis for models of hillslope or soil evolution (e.g. Kirkby 1971). The equation is usually used in these models as a partial differential equation, which formally connects rates of change at a site over time to rates of change over space at a given time. One simple form of the continuity equation is:

$$\partial s / dt + \partial Q / dx = a$$

where $s$ is the amount stored (e.g. as elevation of rock and soil materials), $Q$ is the rate of flow (e.g. as total transport of earth materials downslope), $a$ is the net rate of accumulation (e.g. as wind-deposited dust), and $x$ and $t$ are respectively distance downslope and time elapsed.

An equation of this type cannot normally be solved without specifying the relevant processes which control the rates of flow, accumulation, etc. Continuity equations are equally relevant to the conservation of energy in micro-meteorological studies and in many other physical and chemical systems. The concept of continuity dates back to Leonardo da Vinci and remains a fundamental principle underlying our understanding of the physical world. MJK

### Reading and Reference
Davidson, D.A. 1978: *Science for physical geographers*. London: Edward Arnold. · Kirkby, M.J. 1971: Hillslope process-response models based on the continuity equation. *Transactions of the Institute of British Geographers, special publication* 3, pp. 15–30.

**contour** A line on a map which joins areas of equal height or equal depth.

**contrail** An abbreviation for condensation trail which is a thin line of water droplets or ice crystals that condense after emission in the exhaust gases of aircraft engines. At a given height each engine type has a critical air temperature above which a contrail will not form; in the case of the British Isles they are seldom observed below about 6000 m in winter and 8500 m in summer. Once formed they generally spread in width and if persistent indicate slow EVAPORATION in a relatively humid upper TROPOSPHERE. RR

**contributing area** The area of a catchment which is, or appears to be, providing water for storm run-off.

The term was first used by Betson (1964) and may be calculated as the stream discharge divided by the rainfall (or net rainfall) intensity, usually expressed as a proportion or percentage of total catchment area. This ratio is meaningless when it is not raining, and is usually calculated over a total storm period, in the form total storm run-off (normally calculated as QUICK-FLOW) divided by total storm rainfall. For many well-vegetated small catchments, the contributing area is less than 5 per cent of the catchment, but is commonly many times greater where vegetation is sparse (within the area of storm rainfall). MJK

**Reference**

Betson, R.P. 1964: What is watershed runoff? *Journal of geophysical research* 69, pp. 1541–52.

**convection** In general, mass movement within a fluid resulting in transport and mixing of properties of that fluid. In the atmosphere a class of fluid motion in which warmer air goes up while colder air goes down. Unfortunately fluid dynamicists sometimes use the word in place of ADVECTION. In the case of fair-weather convection, air at low levels is slowly warmed by sunshine absorbed at the ground until it can lift away from the surface to be replaced suddenly by comparatively cool air from above. If the air is sufficiently moist cumulus cloud may form above the updraught as an indicator of the process. *Cumulonimbus* convection is more complex (see CLOUDS). Air rising from the surface is usually slightly cooler than its immediate environs but it is very moist, so that when it has risen above cloud base and made use of the latent energy of the water vapour by condensing to cloud droplets it becomes very buoyant. This air then accelerates vigorously, dragging the reluctant air near the surface after it. Slantwise convection occurs when the warm and cold air are initially side by side. This mechanism accounts for sea breezes, valley winds and weather systems (see MESOMETEOROLOGY). JSAG

**Reading**

Ludlam, F.H. 1980: *Clouds and storms*. University Park, Pa.: Pennsylvania State University Press.

**convergence** See DIVERGENCE.

**coordination number** Refers to the packing of objects around a given body. The packing of sand particles can be represented by a coordination number which refers to the number of contacts a given particle makes with other particles. The higher the coordination number the greater the interparticle friction and interlocking and the higher the SHEAR STRENGTH of the material. The value also affects (but is not a substitute for) POROSITY and PERMEABILITY of a material.

In chemical structures the coordination is also important in determining the properties of materials; this is especially true with clay miner-als. There are two basic building blocks in clays: first a silicon atom with four (larger) oxygen atoms around it in the form of a tetrahedron to give four-fold coordination; secondly, an atom of magnesium or aluminium surrounded by six hydroxyl radicals in an octahedron to give six-fold coordination. WBW

**Reading**

Whalley, W.B. 1976: *Properties of materials and geomorphological explanation*. Oxford: Oxford University Press.

**coprolite** Animal dung or excrement preserved or fossilized over time. Such deposits, particularly if preserved in sediments suitable for absolute or relative dating, provide valuable palaeoenvironmental evidence. The chemical conditions within the coprolite may facilitate the preservation of other fossils, for example, plant remains, including pollen grains, fish scales or parts of other devoured animals. Subsequent analysis may then allow for the reconstruction, either of the diet of the animal that produced the excrement, or of the surrounding vegetation types at the time of its production. Several types of coprolite have proved useful in this way, for example, hyena dung deposited in caves may preserve pollen blown into the cave. Excreta of the packrat (in North America) or hyrax (in Africa), both of which develop dung middens over extended periods of time, have been utilized for POLLEN ANALYSIS, since the sticky surface of fresh deposits acts as an excellent pollen trap. MEM

**coquina** A carbonate rock which consists largely or wholly of mechanically sorted, weakly to moderately cemented fossil debris (especially shell debris) in which the interstices are not necessarily filled with a matrix of other material. ASG

**coral algal reef** A marine structure, containing colonies of scleractinian corals. Normally composed mainly of calcium carbonate, REEFS are also complex and productive ecosystems. Skeletal carbonates provide much of the reef framework and organisms also assist in reef erosion and sediment production. The distribution of coral algal reefs is controlled by environmental factors, notably water temperature, clarity and salinity. Most reef-forming corals prefer sea temperatures between 17 and 33 °C, salinities of between 30 and 38 parts per thousand, and clear water. Light is also important, and coral growth is usually restricted to the upper 25 or 30 metres. Because of these factors, coral algal reefs are found mainly between latitudes

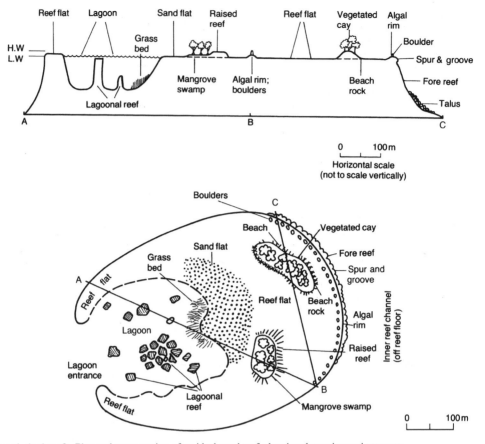

**Coral algal reefs**. Plan and cross-section of an ideal coral reef, showing the major environments.

30 °N and S on mud-free coastlines, particularly in western parts of the Pacific, Indian and Atlantic Oceans.

There are three main types of coral algal reef, i.e. fringing reefs, BARRIER REEFS and ATOLLS. Bank reefs and ridge reefs (found in the Red Sea) have also been described. Reefs formed on continental coastlines are often complex, multiple features which are difficult to categorize. Several theories have been proposed to explain the genesis of barrier reefs and atolls, invoking subsidence, sea-level change and sea-floor spreading.

The coral algal reef ecosystem has very different environmental conditions from the surrounding sea, encouraging a flourishing of life. Zooxanthellate scleractinian corals form much of the primary reef framework. These corals live in symbiotic association with unicellular algae, the *Zooxanthellae*, and are colonial organisms with the ability to reproduce sexually or asexually. Calcareous, encrusting organisms (such as coralline algae, corals, bryozoans, gastropods and serpulid worms) attach themselves to cav-

ities within the primary reef framework, forming a secondary structure.

Reef ecosystems are vulnerable to catastrophic events, such as hurricanes and bleaching episodes, which may cause mass mortality of corals. Human stresses, such as pollution and increase of sediment load, have damaged reefs in many areas. Recently, there has been much speculation over the future impacts of climatic warming on coral reefs (see Stoddart 1990). The geomorphology of coral algal reefs is controlled by the interplay of growth and erosion, producing a reef front (often with spur and groove topography) on the ocean side, a reef flat, and a back reef zone.                                    HAV

**Reading and Reference**
Guilcher, A. 1988: *Coral reef geomorphology*. Chichester: Wiley. · Hopley, D. 1982: *The geomorphology of the Great Barrier Reef*. New York: Wiley Interscience. · Stoddart, D.R. 1990: Coral reefs and islands and predicted sea-level rise. *Progress in physical geography* 14, pp. 521–36.

**coral bleaching** An unexplained phenomenon which threatens to destroy coral reefs.

Corals are bleached when the colourful symbiotic algae they house are lost. The algae can re-enter their hosts if conditions are favourable, and bleached reefs have recovered. When the algae are absent for any length of time, the coral dies. The extent of bleaching varies with depth: the shallower the water, the worse the bleaching. Scientists have hypothesized that the cause of the bleaching is stress, brought on by unusually warm water, changes in salinity, excessive exposure to ultraviolet radiation or extreme climatic changes. Most corals thrive when the water is between 25 and 29 °C, and it is believed that algae die when water temperatures exceed the upper limit.                     ASG

**Reading**
Brown, B.E. ed. 1990: Coral bleaching. *Coral reefs* 8, pp. 153–232. · —and Ogden, J.C. 1993: Coral bleaching. *Scientific American*, 268, pp. 44–50.

**core**   The intensely hot (2700 °K) inner part of the earth. It begins at around 2900 km from the surface at the Gutenberg Discontinuity. The outer portions of the core may be liquid and the inner solid.

**corestone**   A cobble or boulder of relatively unweathered rock which is or has been incorporated within the weathered rock which surrounds it.

**Coriolis force**   Also known as the geostrophic force. An apparent force on moving particles in a frame of reference which itself is moving, usually rotating. Such a force is required if Newton's laws of motion (see EQUATIONS OF MOTION) are to be applied in the rotating framework. The Coriolis force is of major importance to the movement of both oceanic waters and air. In meteorology the Coriolis force per unit mass of air arises from the earth's rotation and is equal to $-2\Omega \times V$, where $\Omega$ and $V$ are vectors representing respectively the angular velocity of the earth and the velocity of the air relative to the earth. In practical terms the force 'deflects' air particles to the right in the northern hemisphere and to the left in the southern hemisphere. It affects only the direction, not the speed of the wind. (See also GEOSTROPHIC WIND and WIND.)
                                                    BWA

**Coriolis parameter**   Twice the component of the earth's angular velocity ($\Omega$) about the local vertical, $2\Omega \sin \phi$, where $\phi$ is latitude.

**cornice**   An overhanging accumulation of wind-blown snow and ice found on a ridge or a clifftop, usually on the lee side.

**corniche**   An organic protrusion growing out from steep rock surfaces at about sea level, and providing a narrow pavement or sidewalk-like path at the foot of sea cliffs. Comparable rock ledges caused by erosional processes and coated with organic material are termed trottoirs. Corniches are often formed of calcareous algae. They are largely intertidal, being best developed in the inlets of exposed coasts and generally protrude about 0.2–2.0 m. Vermetids and serpulids may contribute to their development.
                                                    ASG

**cornucopian**   The NATURAL RESOURCE perspective that scarcity will not occur, either because of natural abundance, or because of technological and economic growth. It is an optimistic world view based upon faith in human ingenuity. The cornucopian argument opposes the neo-Malthusian view of resource limited growth. J.L. Simon (1996) was a notable cornucopian proponent. He attracted international attention by winning a famous bet with population biologist Paul Erlich concerning the price of metals, taken as an indicator of increasing or decreasing supply. The outcome of the wager was meant to vindicate the cornucopian argument.                                        DJS

**Reference**
Simon, J.L. 1996: *The ultimate resource*. Vol. 2. Princeton, NJ: Princeton University Press.

**corrasion**   Mechanical erosion of rocks by material being transported by water, wind and ice over and around them.

**corrie**   See CIRQUE.

**corrosion**   The process of solution by chemical agencies of a rock in water as distinct from the mechanical wearing away of rock by water or its bedload (CORRASION).

**cosmogenic isotope**   Radiogenic and stable isotopes may be created by the interaction of extraterrestrial cosmic rays and terrestrial atoms and are cosmogenic isotopes. The best known of such isotopes is radiocarbon ($^{14}C$). Their formation may be atmospheric (of meteoric origin) as is the case for $^{14}C$, or result from interactions with terrestrial rocks in the upper few metres of earth regolith (of hypogene origin). The advent of accelerator mass spectroscopy (AMS) (see also CARBON DATING) has allowed accurate analysis of cosmogenic isotope concentrations, which are typically low (at best a few atoms.$g^{-1}$.$year^{-1}$) in whole rock samples and individual minerals and precipitates. As the isotopes exhibit a wide range of HALF-LIVES, they provide a potential for establishing ages over a considerable range of time: from a few thousand to in excess of 5 million years.

Central to dating via cosmogenic isotopes is the assumption of a constant flux of cosmic rays and their secondary products for any given area, and therefore a constant (and known) rate of cosmogenic isotope production at source through time. *In situ* (hypogene) accumulation of cosmogenic isotopes has been described for exposed country rock, alluvial fan surfaces, flood and shoreline deposits and lava flows. Much of the atmospheric cosmogenic $^{36}Cl$ production is transported by the hydrological system to the oceans, while some remains trapped in evaporitic deposits associated with closed basins. ss

**Reading**
Stokes, S. 1997: Dating of desert sequences. In D.S.G. Thomas ed., *Arid zone geomorphology*. 2nd edn. Chichester: Wiley. Pp. 607–37.

**cost–benefit ratio** The ratio of the expected costs of an anticipated project to the expected benefits expressed in monetary terms. Projects are usually considered warranted if the cost-benefit ratio exceeds unity. All costs, including values assigned to qualitative concepts, that can be reasonably anticipated are compared to all reasonably expected benefits calculated in a similar manner (Sewell 1975, p. 10).

Costs in a cost–benefit analysis usually include material, labour, and associated direct costs of construction. For many long-term capital projects a major consideration is the amount of interest cost incurred on construction funds. Land acquisition costs also tend to be an important component, especially in urban areas or for water projects requiring extensive reservoirs. The costs associated with environmental degradation are usually included, but they are frequently difficult to quantify and are controversial items. Social costs including dislocation of homes and businesses and lost opportunities are also difficult to assess but none the less important to the decision-maker.

Benefits to be compared to costs usually account for increased productivity and increased land values near to the project. More difficult to define but of equal importance are enhanced environments that are more useful for human purposes than the original natural environment and increased opportunities in a social sense.

Although the mathematics of cost-benefit analysis are clearly defined (Thuesen *et al.* 1971), the approach suffers from two major weaknesses. First, the process gives little attention to those aspects which are difficult to convert to monetary values, yet to the decision-maker these may be the most important considerations. Secondly, the process tends to emphasize the short-term considerations because they are the most easily assessed, while the long-term considerations may be the most important.

WLG

**Reading and References**
Sewell, G.H. 1975: *Environmental quality management*. Englewood Cliffs, NJ: Prentice-Hall. · Thuesen, H.G., Fabrycky, W.J. and Thuesen, G.J. 1971: *Engineering economy*. 4th edn. Englewood Cliffs, NJ: Prentice-Hall.

**coulée** A flow of volcanic lava which has cooled and solidified.

**couloir** A deep gorge or ravine on the side of a mountain, especially in the Alps.

**Coulomb equation** See MOHR–COULOMB EQUATION.

**cover, plant** The proportion, or percentage, of ground occupied by the aerial parts of a plant species or group of species. With the overlapping of species in most plant communities, the combined percentage will nearly always exceed 100, except in very open vegetation. Special scales have been devised for estimating both the degree and character of plant cover in quadrats of a chosen size, such as the DOMIN SCALE and the BRAUN-BLANQUET SCALE. Some methods involve sampling, like the number of points touching a given species, whereas other systems are simply estimates by eye. PAS

**Reading**
Kershaw, K.A. 1973: *Quantitative and dynamic plant ecology*. 2nd edn. London: Edward Arnold. · Willis, A.J. 1973: *Introduction to plant ecology*. London: Allen & Unwin. Especially ch. 11.

**coversand** A term originally applied to any aeolian sand which masked older sediments but this generic definition has led to its indiscriminate use. Coversands are typically thin sandy deposits of cold-climate aeolian origin formed into a flat spatially continuous relief of uniform thickness. Only in valleys and against topographic barriers do they increase in thickness. DUNES directly associated with coversands are relatively rare and often indistinct. Whilst coversands are predominantly aeolian they do not preclude incorporation of sand or subsequent reworking by other processes. Use of the orientation coversand of dune morphology, bedding inclination and unit thickness has enabled the reconstruction of regional palaeowind directions (Bateman 1998).

Coversands are widely found on the lowlands of Europe with occurrences in Britain, The Netherlands, Germany, Denmark, and Poland. The northerly coversand limits coincide with the maximal position of the late PLEISTOCENE ice-sheet. Coversands occupy similar ice maximal positions in North America.

European coversands are all relict features with two main phases of coversand deposition reported; one around 18,000–15,000 years ago (Older Coversand) and another between 13,000–11,000 years ago (Younger Coversand) (Koster 1988; Bateman 1998). On the basis of their stratification they have been classified into two main FACIES. One, associated with the Younger Coversand phase, is typically unimodal, well sorted with parallel laminations and a large sand component derived from local sources. This type has little evidence of PERMA-FROST and indicates increasingly dry conditions. The other facies, associated with the Older Coversand phase, is characterized by an alternation of well sorted parallel-laminated beds of loam and sand with evidence of deposition under wet/moist conditions. Cryogenic deformation and FROST WEDGE casts are found only in the lower part of the facies (Older Coversand I).      MDB

**Reading and References**
Bateman, M.D. 1998: The origin and age of coversand in N. Lincolnshire, UK. *Permafrost and Periglacial Processes* 9, pp. 313–25. · Koster, E.A. 1988: Ancient and modern cold-climate aeolian sand deposition: a review. *Journal of*

*Quaternary Science* 3, pp. 69–83. · Kasse, C. 1997: Cold-climate Aeolian sand-sheet formation in north-western Europe (*c.*14–12.4 ka): a response to permafrost degradation and increased aridity. *Permafrost and Periglacial Processes* 8, pp. 295–311.

**crab-holes**   Small abrupt depressions in the ground surface which vary in diameter from a few centimetres to more than a metre, and in depth from *c.*5 to 60 cm. They occur in sediments which are prone to vertical cracking and horizontal piping.      ASG

**Reading**
Upton, G. 1983: Genesis of crabhole microrelief at Fowlers Gap, western New South Wales. *Catena* 10, pp. 383–92.

**crag and tail**   A landform consisting of a small rock hill and tapering ridge which is produced by selective erosion and deposition beneath an ICE SHEET. They range in scale from tens of metres to kilometres in length, with the tail pointing in the down-ice direction. The hill, or crag, is usually of strong rock that has resisted glacial erosion and forms an obstruction to the ice producing a 'pressure shadow' in its lee. This extends in a down-ice direction in proportion to the ice velocity and thickness and creates a gradually tapering zone of minimal erosion or even a cavity. Although often similar in appearance there are two types of crag-and-tail dependent upon the composition of the tail and processes

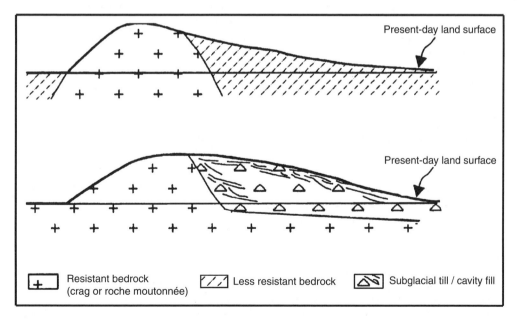

**Crag and tail forms**. Top (erosional), bottom (depositional) ice flow from left to right.
*Source*: Benn and Evans 1998.

that led to its formation. Erosional crag-and-tails consist of a highly resistant rock crag that protected less resistant bedrock in its lee from the full force of glacial erosion. The tail in this type consists of bedrock. Depositional crag-and-tails were formed by the inflow of glacial sediments into a cavity produced in the lee of the rock obstruction, and hence have tails composed of unconsolidated sediments. These tend to be smaller in scale. In practice it is hard to differentiate between these two types as they may both have glacial sediments at the surface of the tail. The significance of these landforms is that in common with SUBGLACIAL BEDFORMS they record former directions of ice flow and indicate that the ice was at pressure melting point, which permitted sliding at the bed.                                              CDC

**Reading**
Benn, D.I., and Evans, D.J.A. 1998: *Glaciers and glaciation*. London: Arnold.

**crater**  A depression at the crest or on the flanks of a volcanic cone where a pipe or vent carrying gases and lava reaches the surface. Also the impact scar left by a meteorite when it impacts on the surface of a planet or moon.

**craton**  A continental area that has experienced little internal deformation since the Precambrian (about 570 million years ago). These areas can be divided into a very stable core, known as a SHIELD, and a marginal platform zone where gently tilted or flat sedimentary rocks bury the Precambrian basement. Crustal movement in cratons is largely vertical and results in the formation of broad domes and basins. The term craton is also used in an alternative sense to refer to the very ancient core of shield.                                              MAS

**Reading**
Spencer, E.W. 1977: *Introduction to the structure of the earth*. 2nd edn. New York: McGraw-Hill. · Windley, B.F. 1984: *The evolving continents*. 2nd edn. London and New York: Wiley.

**creationism**  The theory that attributes the origin of all species of organisms (and indeed all matter) to special creation as opposed to evolution. Creationists maintain that plants and animals were brought into existence, in their present form, by the direct intervention of divine power. Some authorities, such as John Ray (1627–1705) and William Paley (1743–1805), emphasized the adaptation of organisms to their environment as evidence of the 'wisdom of God'. Creationists usually suggest that all organisms were created at the same time, or

over a very short period (as in the Genesis accounts), although Louis Agussiz (1807–1873) entertained the possibility of 'multiple creations'.

'Creation science' or 'scientific creationism' attempts to demonstrate the literal truth of accounts of the creation, such as that in the book of Genesis, using the techniques of modern science. Bone beds, such as that of the Rhaetic in Britain, are attributed to a catastrophic destruction of organisms caused by the deluge. Creation scientists have drawn attention to imperfections in the fossil record, and the rarity within it of links between the major groups of organisms, arguing that these invalidate the theory of evolution. They have stated that there are problems in explaining the evolution of complex structures, such as the vertebrate eye, by natural selection, for, creation scientists argue, an 'incompletely evolved' eye would be of no value to the organism. They have also taken certain recent scientific doubts about natural selection as a *principal* mechanism of evolution as calling into question the whole of evolutionary doctrine.

In certain jurisdictions, creation scientists have sought (and on occasions won) the right to place their ideas alongside those of evolutionary theory before school students. Unquestionably, creationists have caused textbook writers to be much more careful in their wording, but most evolutionists would argue that many of their ideas are based upon the misunderstanding, or in some cases, the deliberate distortion, of scientific evidence.                                              PHA

**Reading**
Baker, S. 1976: *Is evolution true?* Welwyn, Herts. and Grand Rapids, Mich.: Evangelical Press. · Ruse, M. 1982: Creation science: the ultimate fraud. In J. Cherfas ed., *Darwin up to date*. London: New Science Publications. Ch. 2, pp. 7–11.

**creep**  The gradual downslope movement of soil and rock debris on hillsides or glacier ice under the influence of gravity.

**crevasse**  A chasm or deep fissure in the surface of an ice sheet or glacier. Also a breach in a levée along the bank of a river.

**critical load**  A concept in pollution studies which involves the idea that there is a certain pollution load level above which harmful effects on biological systems, such as decline and disappearance of fish populations, will occur.  ASG

**Reading**
Brodin, Y-W. ed. 1992: The critical-load concept: an instrument to combat acidification and nutrient enrichment. *Ambio* 21, pp. 332–87.

**critical velocity (or critical erosion velocity)** is the flow velocity required to initiate the ENTRAINMENT of sediment particles by wind or water, e.g. in the study of fluvial sediment transport, the critical velocity of streamflow is that velocity which, if increased slightly, would trigger the beginning of motion in the bed materials. Thus, it indicates the threshold velocity below which the bed is stable, and above which particles on the bed would begin to move along in the flow. The critical erosion velocity is primarily a function of the mean particle size of the sediment in question. Generally, larger particles have a higher entrainment threshold. Particles < 0.06 mm (silt/clay), however, may also require faster fluid flows for entrainment because they tend to have additional molecular and electrostatic forces of cohesion, often become protected from erosion by larger particles, or lie below the ROUGHNESS HEIGHT in the zero-velocity layer. In aeolian systems, small particles such as these also have an affinity for the retention of moisture, thus increasing the forces resistant to entrainment.

There are two critical velocity thresholds: fluid (or static) and dynamic. The fluid threshold refers to the critical velocity where sediment is entrained only by the drag and lift forces of the moving fluid. However, one of the most significant influences on the entrainment of sediment is the existence of grains already in motion. In particular, the impact of grains in SALTATION on a surface often leads to the splashing of other grains into the fluid flow which may, in turn, also be entrained by the flow and undergo saltation. Such a saltating system carries inertia and often continues even when the flow velocity is reduced to below that of the fluid critical erosion velocity. Great efforts have been made to develop both empirical and theoretical formulae to enable the critical velocity to be calculated for various kinds of stream bed sediment. In general, these have greatest success in sands, and when the material concerned is well sorted. The values of critical erosion velocity for grains of known size for both wind and water are commonly determined by wind tunnel or flume experiments. Thresholds of motion on natural sediment beds are also influenced by factors such as sediment size mixtures, surface crusting, surface slope, moisture and vegetation.

A diversity of factors influences the value of critical velocity, including flow properties (depth, water temperature) and particle properties (density, size, shape, inter-particle cohesion). In addition, aspects of the sedimentology of the stream bed are very important, especially in gravel streams and where the bed material involves poorly sorted sediments (see CAPACITY/ NON-CAPACITY LOAD). Under these conditions, particle exposure and friction between particles lodged tightly together can greatly affect the critical velocity. Critical velocity tends to be lowest for fine to medium sands, with particle diameters of about 0.1–0.5 mm. Particles of this size can be entrained by flows whose mean forward velocity is about 0.5 metres per second. It increases for larger particles because of their greater weight, but also for finer particles (silts and clays) because attractive forces cause these particles to cohere. For example, cobbles are generally only entrained when the mean flow velocity exceeds several metres per second.

Various forms of critical velocity relation have been developed. One difficulty that arises in the use of velocity is that this parameter varies continuously from the bed to the water surface. Some critical velocity equations employ the *bottom velocity*, while others are based on the *mean velocity* of the flow. An empirical curve developed by Hjulström relates particle diameter to critical velocity expressed as the mean. However, the critical velocity is known to vary with water depth and with the slope of the channel, so that empirical relations provide only a general guide to entrainment conditions.

Given that streamflow is turbulent, it can be imagined that near the critical velocity, some of the smaller particles that are swept by powerful eddies would begin to move slightly, perhaps at locations scattered over the surface of the bed. As flow velocity increased, this motion would become both more vigorous, and more widespread across the bed. Thus, the notion becomes slightly blurred and involves a degree of uncertainty. As stream velocity increases past the critical velocity for even the largest particles in the bed, all grains would be involved in active transport, a condition known as the active bed state. Even under these conditions, particle collisions, and temporary burial and entrapment within the upper, mobile part of the bed ensure that particle motion is often discontinuous. Therefore, it must also be recalled that the critical velocity does not separate the resting condition from continual movement. Criteria other than velocity have been investigated for the establishment of stability/motion thresholds. These include the *critical discharge* and the *critical bed shear stress*, both defined analogously to the critical velocity. A widely used form of critical shear stress relation was developed by Shields (see SHIELDS PARAMETER). The use of shear stress to predict the critical conditions for sediment motion has the advantage that this parameter is directly related to the forces exerted on the bed. DLD/GFSW

**Reading**
Graf, W.H. 1971: *Hydraulics of sediment transport*. New York: McGraw-Hill. · Knighton, D. 1998: *Fluvial forms and processes: a new perspective*. London: Arnold. · Wiggs, G.F.S. 1997: Sediment mobilisation by the wind. In D.S.G. Thomas, ed., *Arid Zone Geomorphology*, 2nd edn. London: Wiley. Pp. 351–72. · Yalin, M.S. 1972: *Mechanics of sediment transport*. Oxford: Pergamon.

**cross-bedding** The arrangement of laminae and beds in sedimentary strata at different angles from the principal planes of stratification. The pattern of cross-bedding provides evidence of the environment and modes of deposition.

**cross-lamination** Thin layers of sediment, often only a few millimetres thick, that dip obliquely to the main bedding plane. The cross-laminae represent individual sedimentation units resulting from small-scale fluctuations in velocity and in rates of supply and deposition of silts and sands forming the lee, stoss and trough laminae of migrating ripples.                    JM

**cross-profile, valley/river channel** A profile may be surveyed at right-angles to the river flow direction across a river channel or across the valley in which the river channel occurs. Information from contours or topographic maps is often sufficiently detailed to draw valley cross-profiles. (See also CHANNEL CAPACITY.)                    KJG

**crumb structure** Pertains to those soils which have their fine particles accumulated in the form of aggregates or crumbs, so that they have a more open, coarser and workable texture.

**cryergic** Periglacial in its broadest sense. Pertaining to the periglacial features and processes which occur in those areas not immediately adjacent to glaciated areas.

**cryoplanation** The flattening and lowering of a landscape by processes related to the action of frost. (See also ALTIPLANATION.)                    ASG

**Reading**
Priesnitz, K. 1988: Cryoplanation. In M.J. Clarke ed., *Advances in periglacial geomorphology*. Chichester: Wiley. Pp. 49–67.

**cryostatic pressures** Freezing-induced pressures thought to develop in the ACTIVE LAYER in pockets of unfrozen material which are trapped between the downward migrating freezing plane and the perennially frozen ground beneath. Although recorded in experimental studies, the existence of substantial cryostatic pressures in the field has yet to be convincingly demonstrated. Generally, the presence of voids in the soil, the occurrence of frost cracks in winter, and the weakness of the confining soil layers lying above, prevent pressures of any magnitude from forming. Nevertheless, cryostatic pressures are often invoked to explain various forms of patterned ground and mass displacements (cryoturbations) in the active layer.                    HMF

**Reading**
French, H.M. 1976: *The periglacial environment*. London and New York: Longman. Esp. pp. 40–4. · Mackay, J.R. and Mackay, D.K. 1976: Cryostatic pressures in nonsorted circles (mud hummocks), Inuvik. *Canadian journal of earth sciences* 13, pp. 889–97. · Washburn, A.L. 1979: *Geocryology: a survey of periglacial processes and environments*. New York: Wiley. Esp. p. 167.

**cryoturbation** The process whereby soils, rock and sediments are churned up by frost processes to produce convolutions or involutions. The process is especially active in the zone above permafrost which is subject to seasonal freezing and thawing – the active layer. (See also FROST HEAVE.)                    ASG

**Reading**
Vandenberghe, J. 1988: Cryoturbations. In M.J. Clarke ed., *Advances in periglacial geomorphology*. Chichester: Wiley. Pp. 179–98.

**cryovegetation** Consists of plant communities comprising such types as algae, lichens and mosses, which have adapted to life in environments where there is permanent snow and ice.

**cryptovolcano** A small, roughly circular area of greatly disturbed strata and sediments which though suggestive of volcanism does not contain any true volcanic materials.

**cuesta** A ridge which possesses both scarp and dip slopes.

**cuirass** An indurated soil crust which mantles the landsurface protecting the underlying unconsolidated sediments from erosion.

**cumec** A measure of discharge, being an abbreviation for cubic metre per second.

**cumulative soil profiles** receive influxes of parent material while soil formation is still going on; i.e. soil formation and deposition are concomitant at the same site. Their features are thus partly sedimentological and partly pedogenic. Among topographic sites favourably situated for their formation are areas receiving increments of loess, river floodplains, and colluvial and fan deposits at the base of hillslopes.

                    ASG

**cumulonimbus** See CLOUDS.

**cumulus** See CLOUDS.

**cupola** A dome-shaped mass of igneous rock which projects from the surface of a batholith.

**current bedding** Layering produced in accumulating sediment by fluid flow which is oblique to the general stratification. Laminae (less than 1 cm) or thicker strata in tabular or wedge-shaped units may comprise near-parallel sets bounded by plane, curved or irregular surfaces, with boundaries representing accretionary limits or erosional truncations. Such features are produced through the development of fluid BED-FORMS varying in size from CURRENT RIPPLES to BARS, and their form may give indications of the direction of current flow, flow regime and sediment supply. Also called cross-bedding (a term now preferred by most sedimentologists and subdivided, into cross-lamination and cross-stratification) or false bedding, this phenomenon may help in the identification of sedimentary environments. For example, large-scale current bedding may relate to delta growth, the progradation of aeolian dune slip faces or to river channel point bar sedimentation. Smaller-scale features derive from ripple development, and different kinds of ripples (e.g. straight-crested or linguoid) may produce contrasted current bedding patterns.          JL

**Reading**
Allen, J.R.L. 1970: *Physical processes of sedimentation*. London: Allen & Unwin. · —1982: *Sedimentary structures*. Vol. 1. Amsterdam: Elsevier. · Collinson, J.D. and Thompson, D.B. 1982: *Sedimentary structures*. London: Allen & Unwin.

**current meter** An instrument for measuring the velocity of flowing water in freshwater and marine environments. Many principles have been employed and available types include rotating, electromagnetic, optical and pendulum current meters. The rotating current meter is the most widely used for river measurements and consists of a propeller (horizontal axis type) or a rotor formed by a series of cups (vertical axis type) which rotates at a speed proportional to the flow velocity. The revolutions are counted over a fixed period of time, and velocity is computed from calibration data. The meter body may be mounted on a wading rod or suspended on a cable.          DEW

**Reading**
Buchanan, T.J. and Somers, W.P. 1969: Discharge measurements at gaging stations. *US Geological Survey techniques of water resources investigations*. Book 3, ch. A8.

**current ripples** Small-scale wave-like undulations developed by fluid flow over a sandy or coarse silty bed. Their spacing or wavelength is usually less than 50 cm and the height difference between trough and crest is seldom more than 3 cm. Larger sand features may be called dunes or sandwaves and larger dynamically related features in coarser sediment may be termed BARS.

Current ripples may be described in terms of their plan and profile characteristics. They may be straight, sinuous or indented (linguoid, cuspate, lunate) in plan and may have peaked, rounded and asymmetrical crests. Ripple development occurs through flow separation from the bed; dimensions may be related to applied fluid shear stress.          JL

**Reading**
Allen, J.R.L. 1968: *Current ripples*. Amsterdam: North-Holland.

**currents, nearshore** In the nearshore environment water particles beneath wind-driven waves develop elliptical orbits when the water depth is less than half the wave length, and an oscillatory bed current develops normal to the wave front. This increases in velocity as water depth decreases, and gains in geomorphological effectiveness when it is competent to move sediment. The wave approach is often oblique to the shore at angles of 10–20°, so littoral drift of sediment occurs in the swash and backwash when the wave breaks. The wave-normal current may be resolved into shore-normal and longshore component current (Pethick 1984, pp. 34 ff). The longshore current is accentuated by the need to satisfy continuity, since an outward flow is required to balance the shoreward flow of water. This takes the form of a rip current in a clearly defined channel running parallel to the beach, then forming a localized stream passing through the breaker zone at speeds of up to 1 m s$^{-1}$ to disperse eventually about 300 m offshore. Rip currents form part of a series of cell-like circulation systems related to EDGE WAVES.          KSR

**Reference**
Pethick, J.S. 1984: *An introduction to coastal geomorphology*. London: Edward Arnold. Pp. 34–5.

**currents, ocean** Drift currents in the oceans are driven by the major global wind systems, and therefore contribute to the net poleward energy transfer necessitated by latitudinal imbalance in solar radiation receipt. Near the equator dominantly east-west currents are driven by the trade winds, and an easterly directed

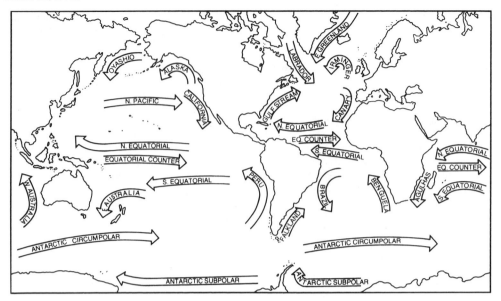

**Surface currents** of the world during the northern hemisphere winter.

countercurrent at the equator completes a circulation known as 'gyre'. Within the ocean basins currents follow the continental margins. In the North Atlantic a clockwise circulation includes the eastern and western 'boundary currents', the Canaries Current and the Gulf Stream. Drift currents may have clearly defined temperature and density boundaries with the surrounding water, and migrating meanders which intensify then break away to form cells or rings. With increasing depth beneath the surface successive deflection of the current occurs relative to that above, to the right in the northern hemisphere and to the left in the southern, because of the Coriolis force which results from the earth's rotation. This increasing deflection is the EKMAN SPIRAL.                                    KSR

**currents, river**  River currents vary both across the section within which the flow is confined, and with depth because of the bed friction that controls the vertical velocity profile. In addition, SECONDARY FLOWS occur with components directed across the channel. The main longitudinal current is generally fastest in the deepest part of the cross-section (the thalweg), but shifts position with changes in river discharge. In a meander bend at low discharge a high sinuosity current (the 'pool current') follows the thalweg closely, but at high discharge a 'bar-head current' of lower sinuosity and steeper slope develops across the point bar. Sediments deposited by river currents have grain sizes which reflect current strength, and structures

which record the current direction. Coarse sediments and their structures tend to display more consistent directional properties, whereas finer sediments are deposited by currents of more varied direction which are forced to conform to the bed topography created by the higher flows.                                    KSR

**currents, tidal**  Tidal currents are horizontal water movements asociated with the tidal rise and fall of the sea surface. In the open ocean standing wave tides are associated with continuous currents of about 0.3–0.5 m s$^{-1}$ which change direction through 360° in one tidal cycle (clockwise in the northern hemisphere). At the coast, and especially in estuaries, the tide is more like a progressive wave, and is associated with stronger, reversing, currents which flood as the tidal wave crest approaches the land, and ebb as the tidal wave trough approaches. Slackwater periods occur at high and low tides, and maximum tidal current velocities occur at mid-tide. In an estuary, as the tide wave progresses inland its amplitude is decreased by frictional effects and it becomes asymmetric as the crest of the tidal wave entering the deeper water at the mouth and travelling faster catches the slower moving trough ahead of it. Ebb and flood current velocities are influenced by the tidal range and asymmetry, and by the estuary morphology and freshwater discharge (Pethick 1984, pp. 59 ff). The flood is often shorter than the ebb, so higher velocities are needed to discharge the same volume of water through the same cross-section. More sediment may thus be moved landward than seaward, and the head of an estuary becomes a sediment sink. KSR

**Reference**

Pethick, J.S. 1984: *An introduction to coastal geomorphology.* London: Edward Arnold. Pp. 58–63.

**cusp, beach** A three-dimensional, scallop-shaped BEACH form occurring in rhythmic sets along the shore. Beach cusps can form through erosion, accretion and are most common on REFLECTIVE BEACHES. The regular spacing of beach cusps has made them a focus of considerable research attention. The EDGE WAVE hypothesis and the system self-organization hypothesis are two competing models. The former requires that beach cusp spacing be tied to edge wavelengths. The latter requires that spacing be a function of run-up height. Attempts to discern the appropriate model remain unsuccessful because of a lack of appropriate field data. DJS

**Reading**

Coco, G., Huntley, D.A., and O'Hare, T.J. 1999: Beach cusp formation: analysis of a self-organization model. In N.C. Kraus and W.G. McDougal eds, *Coastal sediments*

'99. Reston, Va.: American Society of Civil Engineers. Pp. 2190–205.

**cutan** A thin coating of clay on soil particles or lining the walls of a void in the soil zone.

**cut-off** An abandoned reach of river channel, produced particularly where a meander loop has become detached from the active river channel because the neck of the loop has been breached. The abandoned reach may be occupied by an OXBOW lake which gradually fills with sediment. Different cut-off processes are possible: with highly sinuous channels adjacent reaches may impinge on each other directly (neck cut-offs), while in other cases the scouring out of longer short-circuiting channels across the inside of meander bends during floods may produce cut-offs at much lower sinuosities (chute cut-offs). The accumulation of sediment in BARS may also lead to the detachment of channel reaches. The term is sometimes applied to the new channel itself as well as to the abandoned one or to the process in general. JL

**cutter** Linear slots cut in bedrock by solution along a guiding structural element, they are

**Cut-off** on the Darling River, New South Wales, Australia.

115

equivalent to the British term, grike (White 1988).                                                    ASG

**Reference**
White, W.B. 1988: *Geomophology and hydrology of karst terrain*. New York: Oxford University Press.

**cyanobacteria** Photosynthetic, single-celled organisms lacking an enclosed nucleus or other specialized cell structures. Sometimes referred to blue-green algae, they are prokaryotes and more commonly classified in the present day as bacteria. Cyanobacteria contain chlorophyll, although it is distributed throughout the cell rather than confined in cell organelles as in the green plants. Other pigments in many species impart a bluish or reddish tinge. Most species are colonial or form filaments. Cyanobacteria are found abundantly in a wide range of habitats, for example, on tree bark, rocks and in moist soil, where they are responsible for nitrogen fixation; still others symbiotically coexist with fungi to form lichen. In fresh waters subject to EUTROPHICATION, cyanobacteria may form toxic blooms. In the tropics, mats of cyanobacteria filaments grow into humps called stromatolites.                                                     MEM

**cycle, biochemical** See BIOGEOCHEMICAL CYCLES.

**cycle, climatic** A recurrent climatic phenomenon. The term is best reserved for changes of strictly periodic origin such as the annual temperature cycle, but it is often used loosely to describe many changes of climate which occur at approximately fixed intervals.

The cycle of major glaciations with a peak about every $10^5$ years is a good example. Various short period cycles have been described by numerous authors, but they are often the subject of some controversy. Examples are the supposed climatic cycles related to solar activity in general and SUNSPOTS in particular. (See MILANKOVITCH HYPOTHESIS.)                                                JGL

**cycle of erosion** The sequence of denudational processes and forms which, in theory, exist between the initial uplift of a block of land and its reduction to a gently undulating surface or peneplain close to BASE LEVEL. The concept was codified and popularized by the American geographer, William Morris Davis (1850–1934), who termed the sequence the geographical cycle (1899).

In reality, the process described is not a cycle at all, but a one-way movement of mass from higher to lower elevation. The true cycle involved is that first described by James Hutton (1788) as the sequence of uplift, denudation, sedimentation and lithification, followed by renewed uplift. This *geological* cycle is the most basic of the numerous biogeochemical cycles in which the earth's atoms are repeatedly recombined into new chemical, geological and biological compounds. In all these cycles, there are many different pathways which matter may follow over time periods of enormously different lengths.

The Davisian concept of the cycle of erosion is thus mis-named. It is also over-simple, since it emphasizes a generally relentless progression of material from highlands to the oceans or enclosed basins, and the creation of a characteristic sequence of landforms during the process. The early, intermediate and late stages of the cycle are termed youth, maturity and old age. While it was recognized that changes in base level might interrupt the progression, leading to rejuvenation, the general tenor of Davis's ideas led to the belief that such interruptions were unusual.

Davis worked originally upon the sequence of forms produced by fluvial action in areas of humid climate: this was termed the 'normal' cycle. Later, supposedly distinctive, sequences were added by Davis and others to represent the development of arid, glacial, karst, coastal and periglacial regions, together with a recasting of Darwin's theory of coral reef development (1842) in a more explicitly cyclic form.

The reality and the theoretical utility of the cycle of erosion as a concept have been increasingly challenged. In particular, the frequent and intense changes in global climate which appear to have occurred during the Pleistocene – to say nothing of associated fluctuations in base level – render it improbable that any area will contain the simple sequence of forms and, even if these occur, it is difficult to see how they can be the products of any simple, unidirectional set of processes. Therefore, while it is true that 'what goes up must come down', it does not seem useful to ascribe the descent to the operation of Davisian cycles of erosion.                      BAK

**Reading and References**
Chorley, R.J. 1965: A re-evaluation of the geomorphic system of W.M. Davis. In R.J. Chorley and P. Haggett eds, *Frontiers in geographical teaching*. London: Methuen. · Darwin, C.R. 1842: *The structure and distribution of coral reefs*. London: Smith, Elder. · Davis, W.M. 1899: The geographical cycle. *Geographical journal* 14, pp. 481–504. · Hutton, J. 1788: Theory of the earth: or an investigation of the laws observable in the composition, dissolution and restoration of land upon the globe. *Transactions of the Royal Society of Edinburgh* 1, pp. 209–304.

**cyclone** (or depression) A region of relatively low atmospheric pressure, typically one to two thousand kilometres across, in which the low level winds spiral counter-clockwise in the northern hemisphere and clockwise in the southern. Cyclones are common features of surface weather maps and are frequently associated with windy, cloudy and wet weather.

Cyclogenesis (or the formation of cyclones) occurs in preferred areas and is usually most vigorous in wintertime; such areas in the northern hemisphere are the western North Atlantic, western North Pacific and Mediterranean Sea. The birth and subsequent movement of cyclones is closely linked to the presence of the planetary scale ROSSBY WAVES in the atmosphere. They form frequently in the downstream or eastern limb of these large-scale troughs and move as features embedded in the deep generally poleward flow.

The inward spiralling air ascends within the system and flows out in the upper troposphere. In any atmospheric column, if more mass is exported aloft than is imported at low levels, the surface pressure will fall, while surface pressure will rise and the low will fill if there is a net gain of mass in the column.

Cyclones are highly transient features which are associated with disturbed weather and, if frontal, with strong horizontal gradients of TEMPERATURE and HUMIDITY, and sharp changes in cloud cover and type. Across the extra-tropical ocean basins they carry out very important heat transport in a meridional direction which offsets to some extent the persistent equator to pole imbalance in the RADIATION budget. In frontal cyclones warm air is transported poleward and cooled while cold air moves towards the equator and is warmed. RR

**Reading**
Palmén, E. and Newton, C.W. 1969: *Atmospheric circulation systems*. New York and London: Academic Press.

**cyclostrophic** A term which relates to the balance of forces in atmospheric systems in which the flow is tightly curved, e.g. near the centre of a hurricane. In this case the centrifugal force is substantially larger than the CORIOLIS FORCE and the cyclostropic wind ($V$) is:

$$V = \frac{P_n^{1/2}}{R_T}$$

where $P_n$ is the horizontal pressure gradient force and $R_T$ the local radius of curvature of the isobars. RR

**cyclothem** A uniform sequence of sedimentary strata repeated several times through a stratigraphic succession and indicative of repetitive cycles of sedimentation under similar environmental conditions.

**cymatogeny** The warping of the earth's crust over horizontal distances of tens to hundreds of kilometres with minimal rock deformation, producing vertical movements of up to thousands of metres. The term, introduced by L.C. King in 1959, describes crustal movements intermediate between EPEIROGENY and OROGENY and applies not only to the formation of broad domal uplifts but also to the linear vertical movements represented by mountain ranges such as the Andes. Uplift is assumed to be induced by vertical movements associated with processes active within the earth's MANTLE and not to arise from the large-scale horizontal movements proposed in the PLATE TECTONICS model. MAS

**Reading and Reference**
King, L.C. 1959: Denudational and tectonic relief in southeastern Australia. *Transactions of the Geological Society of South Africa* 62, pp. 113–38. · —1967: *The morphology of the earth*. 2nd edn. Edinburgh: Oliver & Boyd; New York: Hafner.

# D

**dalmatian coast** A coastline characterized by chains of islands running parallel to the mainland, deep bays and steep shorelines, being the product of subsidence of an area of land with mountain ridges running parallel to the coast.

**Daly level** Named after R.A. Daly, a student of coral reef development who first put forward the theory that coral reefs were affected by cold periods during the Pleistocene when colder temperatures prevailed and there were lower sea levels. Daly suggested that during these phases the coral would be destroyed and wave-cut platforms, formed by marine planation, would develop on the dead coral. These platforms are, according to Daly, evidenced by present-day lagoon floors. Subsequent sea-level rise would encourage regrowth of corals at the edges of these platforms. If Daly's theory is correct lagoon floors should be flat and occur at similar levels throughout the world.          HAV

**Reading**
Stoddart, D.R. 1973: Coral reefs: The last two million years. *Geography* 58, pp. 313–23.

**dambo** A Bantu word describing a shallow valley or depression which, though seasonally waterlogged, frequently contains no stream channel. The flanking slopes are gentle and they are associated with land surfaces of very low relief. Dambos were first described from present-day Zambia by Ackermann in 1936, and the term now embraces similar landforms elsewhere (e.g. *vlei* in S. Africa, *fadamas* in N. Nigeria). Dambos are typical valley forms of the SAVANNA. In Zambia, interfluve areas are covered by miombo woodlands containing *Brachystegia*, *Julbernadia* and *Isoberlinia*, which give way abruptly on lower slopes to *Ludetia simplex* grasslands and then to *Andropogon* and *Hyparrhenia* grasses bordering the inner valley floor which contains plants such as *Hyparrhenia bracteata* (Mäckel 1974; Thomas and Goudie 1985). While upper slopes may be underlain by *ferralsols* (*ultisols*) the valley floors are associated with *vertisols*.

Many dambos broaden out upstream, becoming bottle or hammer shaped. This suggests that they may drain from former enclosed hollows or *pans*, and/or that they become widened by chemical sapping of surrounding slopes. The importance of chemical DENUDATION processes (*etchplanation*), is indicated by the occurrence of core boulders, and the formation of 2:1 smectite clays in the central seepage zone where divalent ions $(Ca^{++}, Mg^{++}, Fe^{++})$ liberated by rock weathering become fixed. However, most dambos are also erosional forms and show evidence of buried stream channels, infilled with sands and clays. Late Quaternary environmental change was probably responsible for many of these features.

Dambo margins are prized by farmers for their moist, productive soils, and dambo floors are valued for dry season grazing. These wetlands are delicate ecosystems, deriving their distinctive features from a combination of geomorphic setting, environmental history, and a seasonal climate with an intense dry season.
                                                                          MFT

**Reading and References**
Ackermann, E. 1936: Dambos in Nordrhodesien. *Wissenschaft Veröff. Dt. Mus. Länderkde*, Leipzig, NF 4, pp. 147–57. · Boast, R. 1990: Dambos: a review. *Progress in physical geography* 14, pp. 153–77. · Mäckel, R. 1974: Dambos: a study in morphodynamic activity on the plateau regions of Zambia. *Catena* 1, pp. 327–65. · Thomas, M.F. and Goudie, A.S. eds 1985: Dambos: small channelless valleys in the tropics. *Zeitschrift für Geomorphologie*, *Supplementband* 52, p. 222.

**dams** The damming of rivers by artificial structures to create reservoirs has been one of the most dramatic and widespread deliberate impacts that humans have had on the natural environment. Such structures change river hydrology, sediment loads, riparian vegetation, patterns of aggradation and erosion, the migration of organisms, seismic activity, etc.          ASG

**Reading**
Petts, G.E. 1984: *Impounded rivers – perspectives for ecological management*. Chichester: Wiley.

**Dansgaard–Oeschger (D–O) events** The high resolution of climatic change proxies preserved in ICE CORE evidence, particularly in the form of OXYGEN ISOTOPE records, has allowed identification of rapid CLIMATE CHANGE during the QUATERNARY period, particularly the last GLACIAL cycle. For example, Dansgaard *et al.* (1989) identified evidence of very rapid warming, of up to 7 °C in *c*.50 years at the end of the YOUNGER DRYAS. Other evidence has shown that INTERSTADIALS during the last Glacial began equally rapidly. Such rapid temperature

118

changes are known as Dansgaard–Oeschger (D–O) events. Their cause is not fully understood.

DSGT

**Reference**

Dansgaard, W., White, J.W.C. and Johnsen, S.J. 1989: The adrupt termination of the Younger Dryas event. *Nature* 339, pp. 532–4.

**Darcy's law**  Defines the relationship between the discharge of a fluid through a saturated porous medium and the gradient of HYDRAULIC HEAD (figure 1). It is the most important law of GROUNDWATER hydrology.

Henri Darcy was a French engineer. In 1856 he published the results of an experiment undertaken to determine the nature of water flow through sand (see figure 2). He found the outflow discharge from the sand to be directly proportional to the loss in hydraulic head (*h*) after a given length (*l*) of flow through the sand. This relationship may be written as:

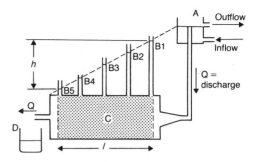

**Darcy's law: 1.** Apparatus showing the relationships expressed in Darcy's law. A, constant head device; B, manometers; C, porous medium being measured; D, outflow with discharge Q. Note the change in head h with distance l.

*Source*: D.I. Smith, T.C. Atkinson and D.P. Drew 1976. In T.D. Ford and G.H.D Cullingford eds, *The science of speleology*. London and New York: Academic Press. Figure 6.2.

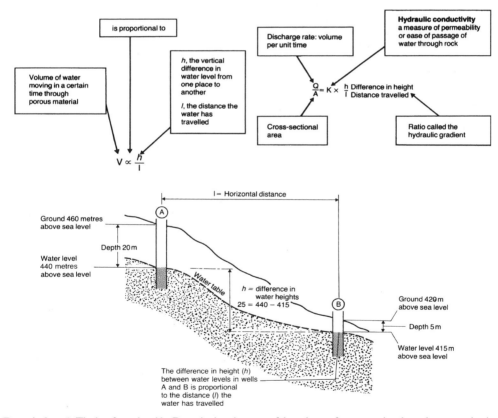

**Darcy's law: 2.** The law formulated by Darcy is given in terms of the volume of water moving through any opening in a given amount of time, essentially a velocity term, and the geometry of the general flow, or the ratio of the vertical to the horizontal distance. Darcy reasoned that the permeability of a rock is what more or less slowed down the flow for a given drop of height (h) in a certain distance (l) and so made this into an equation by multiplying the right-hand side by a proportionality factor, called K. Darcy identified K as a measure of the permeability of the rock, or in other words, how easily it transmits water. From this equation, we can either determine the velocity of flow, or knowing the velocity, the hydraulic conductivity.

*Source*: F. Press and R. Siever 1978: *Earth*. San Francisco: Freeman. Box 6–1.

$$\nu = -K\frac{dh}{dl}$$

where $\nu$ is the specific discharge, $dh/dl$ is the hydraulic gradient, and $K$ is a constant of proportionality known as the hydraulic CONDUCTIVITY. Thus if $dh/dl$ is held constant, $\nu \propto K$.

The specific discharge is sometimes termed the macroscopic velocity, the Darcy velocity or Darcy flux. It is the volume rate of flow through any cross-sectional area perpendicular to the flow direction (Freeze and Cherry 1979). It has the dimension of a velocity, but it should be clearly distinguished from the microscopic velocities of water passing through individual inter-granular spaces during flow through the porous medium.

Darcy's law is valid for groundwater flow in any direction, including when it is being forced upwards against gravity in circuitous groundwater flow paths. It may also be used to describe the flow of moisture in soil. However, there are limits to its validity. Freeze and Cherry (1979) point out that if it were universally valid, a plot of specific discharge $\nu$ against hydraulic gradient $dh/dl$ would reveal a straight line relationship for all gradients between zero and infinity. This is not the case. There appear to be both upper and lower limits to its validity.

Darcy's law is only applicable under conditions of laminar flow. But at the upper limits to laminar flow where the REYNOLDS NUMBER is in the range of 1 to 10, the law breaks down. Hence in the non-linear laminar flow regime, Darcy's law is not valid. Evidence is less conclusive at the lower limit, but some work suggests that there may be a threshold hydraulic gradient below which flow does not occur.

The above equation describing Darcy's law is one-dimensional in form, but it may also be developed to describe three-dimensional flow (Freeze and Cherry 1979). In three-dimensions, specific discharge $\nu$ is a vector with components $\nu_x, \nu_y$ and $\nu_z$. Hydraulic conductivity $K$ may not be the same in each direction. Thus a three-dimensional generalization of Darcy's law may be written as:

$$\nu_x = -K_x\frac{dh}{dx}$$

$$\nu_y = -K_y\frac{dh}{dy}$$

$$\nu_z = -K_z\frac{dh}{dz}$$

PWW

**Reading and Reference**

Castany, G. 1982: *Principes et méthodes de l'hydrogéologie.* Paris: Dunod. · Freeze, R.A. and Cherry, J.A. 1979: *Groundwater.* Englewood Cliffs, NJ: Prentice-Hall. ·

Todd, D.K. 1980: *Groundwater hydrology.* 2nd edn. New York and Chichester: Wiley.

**Darwinism** The biological theory of evolution by natural selection as propounded by Charles Darwin (1809–1882) and set out in *Origin of Species* (1859) and *The Descent of Man* (1871). In demonstrating the mutability of species Darwin succeeded in setting the whole idea of scientific enquiry free from theological constraints. Thus the biblical notion of successive deluges modifying the landscape (CREATIONISM) was superseded by a theory of 'evolution' whereby random variations in fauna and flora would be selectively preserved and inherited by subsequent generations. Darwinism is sometimes taken to refer to any kind of evolution (the term did not appear in *Origin of Species* until the fifth edition), indeed to any kind of evolutionary theory which relies on the natural selection principles but rejects the doctrine of inheritance of acquired characteristics as suggested by Lamarck. Specifically, however, the idea of adaptation to the environment and selective change at the species level revolutionized the scientific community of the nineteenth century so much that the development of geography as a science became possible. Darwin's theory allowed the disciplines geomorphology, pedology, ecology and natural history more generally to calculate more *time* in which the sequential progression of development occurred; dispelled were the time-limited ideas of Bishop Ussher and the Anglican Church that the world was a divine creation formed 4004 BC. The pre-Darwinian ideas of landform studies in the Linnaeus taxonomic form produced only confusion in the early nineteenth-century scientific community. This was due in the main to the failure of taxonomic organization to provide any unifying principle which would allow scientists to order the myriad of landform types which had been recognized.

W.M. Davis, utilizing the Darwinian principle of *evolution* through time, provided a cycle of landscape evolution from youth to maturity and finally to old age. This was a direct analogue of the Darwinian idea of a plant or animal undergoing sequential change through time. Such change embodies a second important quality of Darwinism, the idea of organization in change. The importance of this in nineteenth-century scientific thinking was the rejection of the preconception of Platonic thinking of immutability of form. Dominant in the rationale of geologists at this time was the idea of changeless ideas or forms. Indeed it was generally held that change and variation were no more than illusions and that genuine reality was of fixed types permanently distinguished from one another. Such ideas delayed the recognition of evolution in nature and specifically prevented the science of ecology from developing. Indeed the idea of the inter-relationship between fauna and flora and their environment is a basic tenet of ecological understanding. Darwinism became the underlying principle of the subject.

Darwinism also provides geography with the idea of struggle and selection. Cause–effect relationships preoccupied pre-Darwinian thought at the expense of *process*. Subsequent Darwinian disciples have stressed the importance of the environmental influence. In a broad sense these 'Darwinian impacts' which include Social Darwinism, Darwinism and its influence upon politics, theology, philosophy, psychology, anthropology, literature and even music all affect our general viewpoint within the specific field of physical geography. All these facets of the 'Darwinian revolution', as it is now called, relied initially (as did geology and geomorphology) on the idea of evolution. Darwin's important message, which lay relatively unheeded until the 1930s, was the idea of randomness or chance variation in nature. Indeed natural selection of species was effected (according to Darwin) by this mechanism of chance.

Many objections to Darwinism have been raised; the most serious concerned the method of inheritance. In 1867 Jenkin pointed out that favourable variations in species would soon disperse when interbred with the 'normal' non-variant types within that species. This damaging comment could, however, have been answered with reference to the work of Mendel who recognized the structure of inheritance, but Darwin however, reverted to a Larmarckian stance, a theory which he termed Pangenesis.

Darwinian ideas of evolution were in general widely accepted as a doctrine, but the natural selection idea was rejected. By the end of the nineteenth century only two important scientists, August Weismann and Alfred Russel Wallace, believed in the random or chance variation idea. Weismann argued that inheritance of acquired characteristics was wrong, indeed impossible and Wallace, the often overlooked co-originator of Dawinism, maintained an unswerving loyalty to the doctrine. With Mendel's work and subsequent development of genetics, Darwin's theory remains the only plausible explanation of life. In the geographical sciences, however unintended, Darwinism provides evolution through time, organization and relationships between plants and animals and the environment, struggle and selection and, to a limited extent, chance and randomness. Generally, Darwinism provides the science of geography with a framework free of the static view of nature and landforms, free of the Linnaean taxonomic viewpoint and free of the Greek view of changeless ideas and forms. Alfred Russel Wallace wrote of Darwin in *Natural selection and tropical nature* (1895):

Nature and Nature's laws lay hid in night. God said, 'Let Darwin be' and all was light.     PAB

### References

Darwin, C.R. 1859: *On the origin of species by means of natural selection; or, the preservation of favoured races in the struggle for life.* London: John Murray. · —1871: *The descent of man and selection in relation to sex.* London: John Murray. · Green, J.C. 1980: The Kuhnian paradigm and the Darwinian revolution in natural history. In G. Gutting, ed., *Paradigms and revolutions.* Notre Dame, Indiana: University of Notre Dame Press. · Oldroyd, D.R. 1983: *Darwinian impacts.* Milton Keynes: Open University. · Stoddart, D.R. 1966: Darwin's impact on geography. *Annals of the Association of American Geographers* 56, pp. 683–98. · Wallace, A.R. 1895: *Natural selection and tropical nature.* London: Macmillan.

**data logger** An instrument which records environmental data (e.g. temperature, windspeed, river stage) from external monitoring equipment so allowing an observer to be absent from the apparatus during data collection.

Data loggers are particularly useful where data has to be collected over long time periods in harsh terrain (e.g. deserts or arctic environments) or where data are required at different places at the same time over spatially extensive areas. A good example of the historic use of data loggers is in the recording of air PRESSURE on BAROMETERS using an ink line drawn on a chart attached to a continuously rotating and clockwork drum.

Today, digital data loggers are driven by electrical power and can be programmed by computer to accept data from a wide range of measuring probes recording environmental variables from wind speed and humidity to river velocity and bank erosion. With the introduction of solar power and satellite technology it is now possible for data loggers set-up in remote locations to record and then transfer data to base stations perhaps many thousands of kilometres away. Such an approach means that the data logger and measuring equipment may only have to be visited on an annual basis for routine maintenance.

Data loggers are also now able to accept very high frequency measurements of environmental data (> 10 Hz). Whilst this can prove very useful in certain experimental situations it can also be problematic in terms of the amount of data that can be collected in a short space of time which then needs to be stored and analysed.

GFSW

**Reading**
Whalley, W.B. 1990: Measuring and recording devices. In A. Goudie ed., *Geomorphological techniques*. 2nd edn. London: Unwin Hyman. Pp. 186–91.

**daya** A small, silt-filled solutional depression found on limestone surfaces in some arid areas of the Middle East and North Africa.

**Reading**
Mitchell, C.W. and Willimott, S.G. 1974: Dayas of the Moroccan Sahara and other arid regions. *Geographical journal* 140, pp. 441–53.

**dead ice topography** See STAGNANT ICE TOPOGRAPHY.

**débâcle** The breaking up of ice in rivers in spring.

**deciduous forest** An area in which the dominant life form is trees, whose leaves are shed at a particular time, season or growth stage. Deciduousness is a protective mechanism against excessive transpiration and represents an alternative strategy to being evergreen. Leaf shedding occurs most commonly in either cold or dry conditions, when the availability of soil water to roots is reduced. If transpiration were to continue unchecked during these periods, deciduous trees with a full leaf canopy would suffer serious water deficiencies. During the resting season buds are enclosed in tough, protective scales, but the timing of bud formation varies for each different tree species.     PAF

**Reading**
Cousens, J. 1974: *An introduction to woodland ecology.* Edinburgh: Oliver & Boyd. · Reichle, D.E. ed. 1973: *Analysis of temperate forest ecosystems.* New York: Springer-Verlag.

**décollement** A feature resulting from the detachment of strata from underlying beds during folding, with the result that the upper strata slip forward.

**decommissioning** The process of dismantling and disposing of old nuclear reactors and other integral components of nuclear plant following final reactor shutdown. As the first generation of nuclear reactors (e.g. from nuclear submarines and Britain's Magnox power stations) come to the end of their operational lives, the costs and problems of decommissioning take on increasing significance. The actual process of dismantling reactors will be delayed by the presence of high levels of radioactivity, so that one solution is to entomb reactors under a mound of sediment until radioactivity levels have decayed to a point where dismantling can more safely take place. Large amounts of nuclear waste will need to be disposed of. The events involved in decommissioning are as follows (from Mounfield 1991, p. 368):

(*a*) defuelling and shipment off the site of all spent fuels, a major task that could take several years for a gas-cooled reactor;
(*b*) decontamination (chemical cleaning) of accessible pipework and, if necessary, of parts of the site;
(*c*) packaging and removal of operational solid radwastes;
(*d*) dismantling and removal of all buildings and structures outside the biological shield, including contaminated boilers, circulatory systems and fuel ponds;
(*e*) 'mothballing' or 'entombment' of the reactor for a period of time (determined partly by the decay rate of cobalt-60);
(*f*) dismantling and removal of the reactor itself and the biological shield;
(*g*) site decontamination to enable free entry and reuse.     ASG

**Reference**
Mounfield, P. 1991: *World nuclear power.* London: Routledge.

**decomposer** An organism which helps to break down dead or decaying organic material, and so aid the recycling of essential nutrients to plant producers. Bacteria and small fungi are the major decomposer groups. Digestive enzymes are released from the bacterial cells or fungal filaments, which turn the organic material into a soluble form capable of being ingested, and at the same time $CO_2$ and $H_2O$ are released back into the environment. In natural ecological systems, many thousands of decomposer organisms ensure the efficient operation of the detritus FOOD CHAIN.     DW

**deductive** (science) Also known as critical rationalism, this employs a research method based on logical structures and systematic explanations. Scientific explanations are based on judgements derived from the analysis and assessment of observations and measurements (Popper 1972). In a formal sense, a HYPOTHESIS is established and tested, though MULTIPLE WORKING HYPOTHESES may be established and tested against each other. The hypothesis is accepted or refuted on the basis of the balance of the evidence obtained by data analysis. A key component of deductive science is that nothing is ever proved, since further data collection and analysis could always provide evidence that contradicts the accepted hypothesis; thus acceptance and rejection are asymmetrical alternatives. The deductive approach is widely used in physical geography, even if not in a rigid formal sense. Its use and basis is analysed in a geographical context by Haines-Young and Petch (1986). (See also INDUCTIVE.)          DSGT

**References**
Haines-Young, R.H. and Petch, J.R. 1986: *Physical geography: its nature and methods.* London: Paul Chapman. · Popper, K.R. 1972: *The logic of scientific discovery.* London: Hutchinson.

**deep weathering**  A term widely used to denote the existence of a chemically weathered layer or mantle (see also WEATHERING PROFILE) exceeding the depth of the soil profile. It describes the deep penetration of chemical decay into susceptible rocks, often producing a *saprolite* with significant clay content. In extra-glacial areas deep weathering is widespread in feldspathic rocks (esp. granite, but also gneiss and many volcanic rocks), in the humid climates of temperate and tropical regions. Often seen as typical of tropical cratons (shields) such as West and central Africa, western Australia and South America, where profiles can exceed 100 m, but thick saprolites are found in Palaeozoic mountains of temperate areas such as Appalachia, western and central Europe and eastern Australia, and in Mesozoic rocks within the humid tropics.

Deep weathering is associated with the concentration of oxides and saprolites are mined for economic minerals (e.g. gold, aluminium, copper, nickel). In geomorphology the stripping of weathering covers by erosion is thought to reveal bedrock forms (see BORNHARDT; ETCHPLAIN; TOR), and some glaciated landscapes may be little more than exposed portions of this bedrock relief. Spatial patterns of deep weathering indicate thickening beneath summits in humid areas; deep troughs following shatter belts and fault zones, and a strong influence from fracture (joint) patterns in the bedrock (Thomas 1966). Described by Gerrard (1988), Ollier (1984) and Thomas (1994).          MFT

**References**
Ollier, C.D. 1984: *Weathering.* 2nd edn., London: Longman. · Gerrard, A.J. 1988: *Rocks and landforms.* London: Unwin Hyman. · Thomas, M.F. 1966: Some geomorphological implications of deep weathering patterns in crystalline rocks in Nigeria. *Transactions of the Institute of British Geographers* 40, pp. 173–93. · — 1994: *Geomorphology in the tropics.* Chichester: Wiley.

**deflation**  The process whereby the wind removes fine material from the surface of a beach or a desert. It is a process which contributes to the development of LOESS, DUST storms and some STONE PAVEMENTS.          ASG

**deforestation**  The removal of trees from a locality. This removal may be either temporary or permanent, leading to partial or complete eradication of the tree cover. It can be a gradual or rapid process, and may occur by means of natural or human agencies, or a combination of both.

Spurr and Barnes (1980) enumerate the major causes of deforestation. Natural tree removal is of relatively little significance on a global scale. Its mechanisms often lead to partial and temporary clearance that is followed by secondary succession, as a result of which forest develops again. The major natural cause of tree removal is fire resulting from lightning strike. Such burns are an essential part of certain forest ecosystems (e.g. in some types of pine forest). Gales may cause trees to be broken or uprooted in what is termed windthrow. Disease can also lead to the elimination of forest trees. Native animals (e.g. elephants in savanna woodland) can also damage trees by removing foliage and bark, and by trampling and uprooting them. Temporary severe weather (such as cold or drought) can lead to tree death (DIE-BACK), while secular modifications of climate may contribute to deforestation. In the latter instance, several millennia of reduced temperatures and increased precipitation may, for example, accelerate the accumulation of soil organic matter and inhibit forest growth by preventing tree regeneration.

The principal cause of deforestation is human activity. Forests are often permanently cleared on a large scale for a variety of agricultural and urban-industrial purposes using cutting and burning techniques. Some 33 per cent of the biosphere (*c*.$4 \times 10^9$ ha) is at present forested, compared with approximately 42 per cent about a century ago. Although this represents a pronounced recent decline, it is merely part of a process which has been operational in certain parts of the world for millennia.

As Lamb (1979) notes, while some trees continue to be removed, major deforestation is no longer a feature of most developed countries. There are some exceptions (e.g. parts of the boreal coniferous forests), but, in general, these areas have already been effectively deforested, and their emphasis is now upon woodland preservation. In north-west Europe, for instance, postglacial pollen records indicate the presence of an extensive mixed deciduous forest from *c*.7–$5 \times 10^3$ years BP, after which time it was progressively cleared by human activities. The forest soils, also found fossil (PALAEOSOLS), were fertile, and hence enticing to prehistoric and later agriculturalists.

Some developed countries have a considerably shorter history of major deforestation.

These (such as the USA, Australia and New Zealand) were settled by Europeans, who, together with their introduced animals, brought about substantial tree losses in hundreds rather than thousands of years. This is illustrated by New Zealand which was over two-thirds forested at the time of European colonization, and now has less than a 20 per cent tree cover.

The most significant deforestation at present is in the less-developed countries. Tropical tree cover has been removed for millennia by native inhabitants as an aid to hunting, for fuel, itinerant agriculture and settlement. As population increases, the cleared area is getting larger, and secondary succession to forest rarer, with lasting changes being brought about in the vegetation. Tropical rain forest is also being subjected to the constant, organized, commercial removal of its hardwood timber for export to developed countries. If, as is increasingly the case, such felling is entire (clear), regeneration of similar forest vegetation is unlikely.

Forests are complex ecological structures. They represent the optimum sites for photosynthesis within the biosphere, and thus contain a substantial proportion of the earth's biomass. Moreover, they possess considerable biotic diversity, which makes them important gene pools (see GENECOLOGY). Their photosynthetic activity means that they assimilate a considerable amount of atmospheric $CO_2$, the concentration of which is increasing as a result of fossil fuel burning. Increased $CO_2$ concentration may be a contributor to a global warming of climate (GREENHOUSE EFFECT). Forest removal could therefore amplify this trend, because less $CO_2$ will be taken up by the trees. Additionally, the burning of wood releases $CO_2$ to the atmosphere. Tree burning also depletes atmospheric oxygen, and destroys an important source of oxygen (see PHOTOSYNTHESIS).

As well as destroying trees, deforestation eliminates dependent animal habitats in the forest ecosystem. Deforestation involving fire may also kill animal as well as plant life. As Tivy and O'Hare (1981) observe, tree removal causes changes in the light, temperature, wind and moisture regimes of an area. A greater quantity of light is able to reach the ground, where temperatures will also be increased. A consequence of this is an accelerated rate of organic matter decomposition. Both temperature ranges and windspeeds are greater after forest clearance. There is also a higher intensity of rainfall, and a lower evapotranspiration rate. Hence, more run-off occurs, as does enhanced leaching and soil erosion. Studies at the Hubbard Brook Experimental Catchment in the USA (Likens

*et al.* 1977) have revealed details of the biogeochemistry (BIOGEOCHEMICAL CYCLES) resulting from deforestation. For example, compared with a forested area, run-off from one that was deforested increased, and had a higher concentration of dissolved matter. The increased run-off was principally the result of the diminution in evapotranspiration, while the greater quantity of dissolved matter was derived mainly from that not taken up by plants as nutrients, and from substances released by the higher rate of organic matter breakdown on the former forest floor.                                                                    RLJ

### Reading and References
Lamb, R. 1979: *World without trees.* London: Wildwood House. · Likens, G.E., Bormann, F.H., Pierce, R.S., Eaton, J.S. and Johnson, N.M. 1977: *Biogeochemistry of a forested ecosystem.* New York, Berlin and Heidelberg: Springer-Verlag. · Richards, J.F. and Tucker, R.P. eds 1988: *World deforestation in the twentieth century.* Durham, North Carolina: Duke University Press. · Spurr, S.H. and Barnes, B.V. 1980: *Forest ecology.* 3rd edn. New York and Chichester: Wiley. · Tivy, J. and O'Hare, G. 1981: *Human impact on the ecosystem.* Edinburgh and New York: Oliver & Boyd.

**deformation**   A general geological term for the disruption of rock strata, folding, faulting and other tectonic processes.

**De Geer moraines**   Landforms of glacial deposition consisting of 'swarms of small ridges orientated perpendicular to the ice flow direction' (Larsen *et al.* 1991, p. 263). They are thought to form either marginally to an ice body by glacier pushing at the grounding line or subglacially by material being squeezed up into basal crevasses.                             ASG

### Reference
Larsen, E., Longva, O. and Follestad, B.A. 1991: Formation of De Geer moraines and implications for deglaciation dynamics. *Journal of Quaternary science* 6, pp. 263–77.

**deglaciation**   The process by which glaciers thin and withdraw from an area. The usual cause is climatic amelioration which reduces snow accumulation or increases ABLATION, but sea-level rise relative to the land can also increase calving and thus the rate of ablation. The literature is full of arguments about whether the dominant process of deglaciation is thinning (downwasting) or snout retreat (back-wasting). Since both must occur together it is probably helpful to regard the relative dominance of one or other as varying according to glacier type, location and the nature of the climatic or sea-level change. (See also STAGNANT ICE TOPOGRAPHY.)                             DES

**degradation** The lowering, and often flattening, of a landsurface by erosion.

**degree day** For a given day the difference between the mean temperature and a given threshold, normally 15.5 °C for a heating degree day:

$$\text{degree days} = 15.5 - \frac{\left(\begin{array}{c}\text{maximum} \\ \text{temperature}+ \\ \text{minimum} \\ \text{temperature}\end{array}\right)}{2}$$

If the mean temperature is greater than 15.5 °C, negative degree days can be used to estimate air conditioning needs. For space heating, two other formulae have been devised for the days when the maximum temperature is above 15.5 °C.

If the daily maximum temperature is above 15.5 °C by a lesser amount than the daily minimum temperature is below 15.5 °C, then:

degree days $= \frac{1}{2}$ (15.5 − minimum temperature) $-\frac{1}{4}$ (maximum temperature − 15.5).

If the daily maximum temperature is above 15.5 °C by a greater amount than the daily minimum temperature is below 15.5 °C, then:

degree days $= \frac{1}{4}$(15.5 − minimum temperature).

Generally speaking, degree days are used for heating purposes and ACCUMULATED TEMPERATURE is used for agricultural purposes.　　JET

**delayed flow** The part of the streamflow which lies below an arbitrary cut-off line drawn on the hydrograph, representing the more slowly responding parts of the catchment.

The division between QUICKFLOW and delayed flow is usually made by a line which rises from the start of the hydrograph rise at a gradient of $0.55 \ 1 \ s^{-1}/km^2$. h until the line meets the falling limb of the hydrograph. This procedure was suggested by Hewlett (1961) as an arbitrary but objective basis of separating the hydrograph peaks associated with each storm. It was earlier proposed as a replacement for older methods of hydrograph separation (e.g. by Linsley *et al.* 1949) which were considered to have only a spurious physical basis (see BASE FLOW).　　MJK

**References**

Hewlett, J.D. 1961: Soil moisture as a source of base flow from steep mountain watersheds. *US Department of Agriculture: Southeastern Forest Experimental Station paper* 132. · Linsley, R.K., Kohler, M.A. and Paulhus, J.L.H. 1949: *Applied hydrology*. New York: McGraw-Hill.

**dell** A small, well-wooded stream or river valley.

**deltas** Accumulations of river-derived sediment deposited at the coast when a stream enters a receiving body of water, which may be an ocean, gulf, lagoon, estuary or lake. Deltas result from the interaction of fluvial and marine (or lacustrine) forces. Sediment accumulations at the mouths of rivers are subject to reworking by waves and tidal currents. Development of deltas involves the progradation of river mouths and delta shorelines producing a subaerial deltaic plain surmounting delta-front deposits which have accumulated to seaward.

Deltas exhibit appreciable variability of morphology and depositional patterns, reflecting global, regional and local variations in such controlling factors as river discharge, marine energy, tidal regime and geological structure and dynamics (Coleman and Wright 1975; Wright 1978, 1982). Although processes such as ocean currents may be important in a particular case, it is generally agreed that the interaction of river, wave and tide regimes is the major factor influencing delta morphology and sediment types (Galloway 1975). The diagram shows a tripartite classification using selected major deltas to illustrate variations in deltaic characteristics associated with the relative contribution of each process.

Deltas consist of subaqueous and sub-aerial regions. The former constitutes the foundation of a delta and can be subdivided into a prodelta sedimentary unit and a delta-front sedimentary unit. The prodelta is typically fine grained whereas the delta front often possesses a bar morphology consisting of coarser-grained material (Wright 1982). These bars may result from a number of processes operating at or adjacent to river mouths (Wright 1978). The subaerial portion can be divided into two subregions: a lower delta plain which extends to the inland limit of tidal influence, and an upper delta plain dominated by fluvial processes. Distributary channels, natural levees, overbank crevasse splays, tidal channels and flats, interdistributary depressions (lakes, swamps, marshes, lagoons and bays), beaches, cheniers and dunes are landforms which may be found on many deltas of the world. Over time the active portion of a delta dominated by riverine processes may be abandoned because of the development of a new distributary system. In the abandoned sector marine, estuarine and paludal processes will modify the primary delta landscape.　　BGT

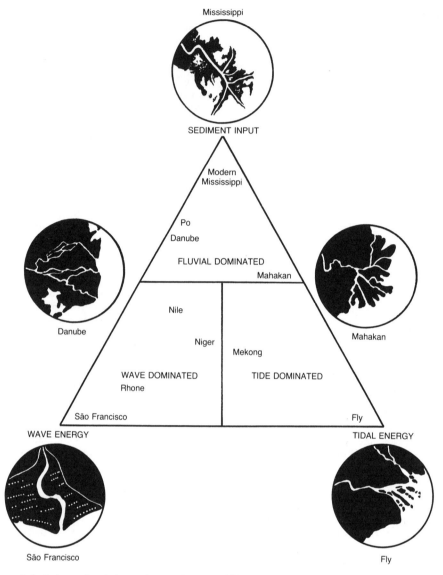

Mississippi

SEDIMENT INPUT

Modern
Mississippi

Po
Danube

FLUVIAL DOMINATED

Mahakan

Nile

Niger

Mekong

Danube

WAVE DOMINATED
Rhone

TIDE DOMINATED

Mahakan

São Francisco

Fly

WAVE ENERGY

TIDAL ENERGY

São Francisco

Fly

**Delta** morphological types in relation to three environmental factors.
*Source*: Galloway 1975.

**References**
Coleman, J.M. and Wright, L.D. 1975: Modern river deltas: variability of processes and sand bodies. In M.J. Broussard ed., *Deltas: models for exploration*, Houston, Texas: Houston Geological Society. · Galloway, W.E. 1975: Process framework for describing the morphologic and stratigraphic evolution of deltaic depositional systems. In M.L. Broussard ed., *Deltas: models for exploration*. Houston, Texas: Houston Geological Society. · Wright, L.D. 1978: River deltas. In R.A. Davis Jr ed., *Coastal sedimentary environments*. New York: Springer-Verlag. · — 1982: Deltas. In M.L. Schwartz ed., *The encyclopedia of beaches and coastal environments*. Stroudsburg: Hutchison and Ross.

**demoiselle** A pillar of earth or other unconsolidated material that is protected from erosion by a capping boulder.

**dendroecology** The study of the width of the annual growth rings of trees in order to interpret specific ecological events that resulted in changes in a tree's ability to photosynthesize and fix carbon.

Today studies involve a variety of coniferous and deciduous tree species, along with some shrubs, in a variety of areas, particularly in arctic and alpine areas where temperature limits tree growth, and in specific localities where events such as floods or glacial movements also restrain growth. The isotopic composition of the wood, rather than the width of the ring, has recently yielded information about past temperatures. Data provided by dendroecology are used by all sciences involved in palaeoecological reconstruction, e.g. geography, geology, archaeology, hydrology, forestry, biology, limnology, and ecology.                                        RHS

**Reading**
Douglas, P.E. 1919: Climatic cycles and tree growth. Vol. 1. *Carnegie Institute of Washington Publication* 289. · Fritts, H.C. 1976: *Tree rings and climate*. New York: Academic Press.

**denitrification**   The removal of nitrogen from a reservoir, commonly groundwater enriched in N from agricultural landuse, or industrial flue gases containing $NO_x$. Nitrogen is one of the components of organic matter (e.g. in amino acids and proteins), and occurs naturally in the environment as well as at elevated levels in contaminated soils and soil water. Within the soil, especially in waterlogged reducing environments, bacteria are able to convert nitrates ($NO_3^-$) into nitrogen gas, which then escapes from the system. Rates of gas release of $> 50$ mg $N_2^{-1}m^2$ day$^{-1}$ have been observed from riparian soils carrying forest cover, and the ability of the soil zone to retain and process nitrogen compounds before they reach the stream environment is an important goal in the preservation of streamside BUFFER STRIPS. Soil denitrification is exploited in the treatment of wastewater (e.g. from tertiary sewage treatment plants) that is used in irrigation. Industrial denitrification is the removal of $NO_x$ compounds that are involved in the acid rain phenomenon from the flue gases released by industries using fossil fuels. Denitrification of this kind employs a catalytic reduction process that yields nitrogen gas.                                        DLD

**Reading**
Pinay, G. and Decamps, H. 1988: The role of riparian woods in regulating nitrogen fluxes between the alluvial aquifer and surface water: a conceptual model. *Regulated rivers: research and management* 2, pp. 507–16.

**density**   The frequency of a phenomenon or mass of a substance per unit or volume. Stream density, for example, refers to the number of stream channels per unit area of a landscape.

**density current**   A descending body of air or water with a high suspended sediment load. Turbidity currents on lake-bed and sea-floor slopes, clouds of falling volcanic ash and dust storms are all examples of density currents.

**density dependence**   The action of environmental factors controlling the growth of populations of organisms, which vary with the density of the population. In contrast, factors that affect population size independently of the number of individuals present are described as density independent. Climatic and physical factors, such as floods or earthquakes, are usually classed as density independent factors. Resources (food, shelter) are often considered to be density dependent, as are biotic factors such as competition and parasites.

The relative importance of density dependent and independent factors in regulating population numbers has been a subject of considerable debate. Andrewartha and Birch (1954) argued that natural populations are controlled primarily by density independent factors and emphasized the ultimate role of severe climatic conditions. An alternative view stressed the importance of density dependent factors involving competition (Lack 1954). These differences in viewpoint may also be influenced by contrasts in the ecosystems analysed, which range from small organisms, such as insects, in arid areas in the case of Andrewartha and Birch, to larger organisms, such as birds, in temperate landscapes in the case of Lack.

The regulation of populations may often involve the interaction of density independent and dependent factors. Moreover, density dependent factors may sometimes vary in their effectiveness from one population density to another. In some cases predation may control population size at low prey densities but become ineffective at high densities.                                        ARH

**Reading and References**
Andrewartha, H.G. and Birch, L.C. 1954: *The distribution and abundance of animals*. Chicago: University of Chicago Press. · Lack, D. 1954: *The natural regulation of animal numbers*. Oxford: Oxford University Press. · Odum, E.P. 1971: *Fundamentals of ecology*. 3rd edn. Philadelphia and London: W.B. Saunders. Ch. 7, pp. 162–233.

**denudation**   Literally, the laying bare of underlying rocks or strata by the removal of overlying material. It is usually defined as a broader term than 'erosion', to include weathering and all processes which can wear down the surface of the earth. While some denudational processes may continue to operate below sea level, most discussions are concerned with subaerial denudation. Studies of other planets – notably Mars – make it clear that denudation is also a factor to be considered in the explanation of non-terrestrial topography (Francis 1981).

The principal agent of subaerial denudation on the earth is water, as Playfair clearly recognized as long ago as 1802. Other important agents are related to stresses generated by pressure (atmospheric, gravitational and crustal); and the actions of organisms, including man. The impact of bodies such as meteorites, although of limited importance on the modern earth, assumes great significance on Mercury and the moon.

The rate at which denudation proceeds is fundamentally dependent upon the intensity with which the different agents operate, singly or collectively; and the ability of the ground surface and the underlying materials to withstand the stresses generated. It is this intimate connection between the nature of the applied force and the actual resistance of earth materials which makes it difficult to produce realistic generalizations about denudation rates, either over space or through time. Surface geometry, the chemical constituents of rocks, tectonic setting and climate are key factors in determining the denudation environment.                        BAK

**Reading and References**
Fournier, F. 1960: *Climat et érosion: la relation entre l'érosion du sol par l'eau et les précipitations atmosphériques.* Paris: Presses Universitaires de France. · Francis, P. 1981: *The planets.* London: Penguin Books. · Langbein, W.B. and Schumm, S.A. 1958: Yield of sediment in relation to mean annual precipitation. *Transactions of the American Geophysical Union* 39, pp. 1076–84. · Ollier, C.D. 1981: *Tectonics and landforms.* London: Longman. · Playfair, J. 1802: *Illustrations of the Huttonian theory of the earth.* London: Cadell & Davies.

**denudation chronology**   An attempt by geomorphologists to reconstruct the erosional history of the earth's surface. The original definition of geomorphology that emerged in the US Geological Survey in the 1880s saw the new science of geomorphology as being in all essentials equivalent to what is now termed denudation chronology.

In the mid-nineteenth century planed-off surfaces had been identified by British geomorphologists in areas of complex structure and lithology, such as mid-Wales, while, with the exploration of the walls of the Colorado Canyon by Major Powell and his colleagues, great unconformities were recognized, leading to the concepts of BASE LEVEL and peneplain. In Europe, Suess postulated that planation surfaces might be susceptible to correlation on a worldwide basis as a result of worldwide (eustatic) changes of sea level in the geological past. Denudation chronology therefore arose as a prime focus of geomorphology, the aim of which was to use the study of erosional remnants to reconstruct the history of the earth where the stratigraphic record was interrupted or unclear. Techniques were developed to help in the identification of

such erosional remnants, including superimposed contours, altimetric frequency curves, etc. (Richards 1981) and particular energy was expended on trying to fathom out whether surfaces were the product of marine or subaerial denudation. Crucial in such an analysis was the degree of adjustment of streams to structure; streams on subaerial peneplains were thought to be better adjusted than those developed on marine planation surfaces. Much of the evidence for denudation chronology was morphological, with all the implications that this has for its reliability (Rich 1938). The tectonic warping of small remnants rendered height correlation difficult. Supposedly accordant summits might have been greatly lowered by erosion, areas of flat ground might be susceptible to a whole range of different interpretations of their origin, and adjustments of streams to structure might be affected by a variety of tectonic factors, including antecedence (Jones 1980). The more successful attempts at denudation chronology were able to supplement the morphological evidence with information gained from deposits resting on the planation surfaces. In southern England, for example, Wooldridge and Linton (1955) were able to use the Lenham Beds to establish the presence of the supposed marine Calabrian Transgression of Plio-Pleistocene times. Other notable studies include those of Baulig (1935) in France, D.W. Johnson (1931) in the USA, and E.H. Brown (1960) in Wales. In the 1960s, as geomorphology became less concerned with evolution and more concerned with process studies, morphometry, and systems, considerable dissatisfaction was expressed about denudation chronology as a basis for the discipline (Chorley 1965) but with developments in plate tectonics, in our knowledge of the importance of Pleistocene events, in the amount of information that can be gained from a study of submarine deposits in basins like the North Sea, and with improvements in dating techniques, it remains a viable branch of study.                        ASG

**References**
Baulig, H. 1935: *The changing sea level.* London: Philip. · Brown, E.H. 1960: *The relief and drainage of Wales: a study in geomorphological development.* Cardiff: University of Wales Press. · Chorley, R.J. 1965: The application of quantitative methods to geomorphology. In R.J. Chorley and P. Haggett eds, *Frontiers in geographical teaching.* London: Methuen. Pp. 148–63. · Johnson, D.W. 1931: *Stream sculpture on the Atlantic Slope: a study in the evolution of Appalachian rivers.* New York: Columbia University Press. · Jones, D.K.C. ed. 1980: *The shaping of southern England.* London: Academic Press. · Rich, J.L. 1938: Recognition and significance of multiple erosion surfaces. *Bulletin of the Geological Society of America* 49, pp. 1695–722. · Richards, K.S. 1981: Geomorphometry and geochronology. In A.S. Goudie ed., *Geomorphological tech-*

*niques*. London: Allen & Unwin. Pp. 38–41. · Wooldridge, S.W. and Linton, D. 1955: *Structure, surface and drainage in south-east England*. 2nd edn. London: Philip.

**denudation rates**  Provide a measure of the rate of lowering of the landsurface by erosion processes per unit time and are expressed in millimetres per 1000 years (mm 1000 year$^{-1}$) or in the direct equivalent of cubic metres per square kilometre per year (m$^3$ km$^{-2}$ year$^{-1}$). These rates are commonly calculated using information on sediment (physical denudation) and solute yields (chemical denudation) from drainage basins, coupled with an estimate of soil or rock density. As such they are an index of the rate of denudation in the upstream catchment area. Maximum reported denudation rates are probably those for the island of Taiwan which exceed 10,000 m$^3$ km$^{-2}$ year$^{-1}$ (Li 1976).

There are, however, several important problems in the derivation and interpretation of denudation rates based on measurements of river loads. For example, in the case of suspended sediment yields, it must be recognized that only a proportion of the eroded sediment will be transported to the basin outlet and that the associated denudation rate may be an underestimate. With the dissolved load, however, a large proportion may reflect non-denudational sources and should not be included in the calculation (Janda 1971). Furthermore, it may be unrealistic to convert values of river load to a uniform rate of surface lowering over the entire catchment and to assume that current river loads are representative of past conditions (Meade 1969).                                     DEW

**References**

Janda, R.J. 1971: An evaluation of procedures used in computing chemical denudation rates. *Bulletin of the Geological Society of America* 82, pp. 67–80. · Li, Y.H. 1976: Denudation of Taiwan Island since the Pliocene epoch. *Geology* 4, pp. 105–7. · Meade, R.H. 1969: Errors in using modern stream-load data to estimate natural rates of denudation. *Bulletin of the Geological Society of America* 80, pp. 1265–74.

**deoxygenation**  The depletion of the amount of oxygen dissolved in a reservoir such as a surface water body (stream, lake or artificial impoundment). Deoxygenation can result from the oxidation of organic material, as may happen in the lower water column of a large reservoir where oxidation of the remains of the former land vegetation can progressively consume available oxygen until reducing conditions set in. Deoxygenation in surface water can also result from EUTROPHICATION or excessive enrichment of the available nutrient supply, which may result in seasonal blooms of algae. Oxida-

tion of the remains of these microorganisms may again lead to progressive deoxygenation of the waterbody.                                     DLD

**depletion curve**  (or recession curve or base-flow recession curve) Represents gradual drainage of water from storage in a drainage basin and it is often possible to construct a master depletion curve for a particular site on a river system. This depletion curve will usually provide a very reliable means of predicting the decline of base-flow discharge at a site during dry conditions, although some drainage basins may exhibit seasonal variations in the form of the curve as a result of differences in loss of stored water through evapotranspiration (Federer 1973).

The depletion curve may be derived by producing a composite curve from the recession limbs of storm HYDROGRAPHS at a gauging site (see diagram) and one of a number of functions may be fitted to the depletion curve to describe it and to allow quantitative comparison of curves from different sites within a drainage basin, different drainage basins or to check the consistency of curves produced from different hydrograph recession limbs at the same site. One widely applied depletion function is:

$$Q_t = Q_o e^{-\alpha t}$$

where $Q_o$ is flow at any time in the period of depletion or base flow, $Q_t$ is flow after time $t$ from flow $Q_o$, $\alpha$ is the recession coefficient, and $e$ is the base of natural logarithms.

This function will plot as a straight line on semi-logarithmic graph paper and a modification of the method of depletion curve construction shown in the diagram is to plot the recession limbs on semi-logarithmic graph paper so that they define a straight line. However, in practice a perfect straight line is rarely found and this is probably partly a result of the fact that a number of stores are contributing to the depletion curve, all of which produce a recession flow at different rates.

The sources of discharge contributing to the depletion curve are likely to be very different in different drainage basins. Traditionally, it has been assumed that depletion flow is derived from effluent seepage from an aquifer and so the recession coefficient ($\alpha$) has often been called the aquifer coefficient, implying that it is a simple parameter of the drainage characteristics of a single aquifer. However, in practice such a simple situation is highly unlikely and the number of stores contributing to base flow will vary with the structure and size of the catchment. Many drainage basins are not underlain by efficient aquifers and yet they exhibit depletion flow which is largely generated from soil

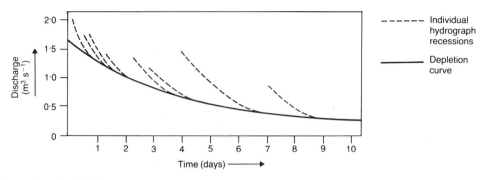

Construction of a depletion curve.

moisture storage. Higher discharges on a depletion curve are almost certainly produced by drainage from soil moisture and from the unsaturated zone of aquifers as well as from the saturated zone. Drainage basins underlain by more than one aquifer may experience effluent seepage from different aquifers at different points on the drainage network and may even lose flow by influent seepage at some locations. As a result, the depletion curve should not be expected to have a simple form because every store contributing to flow should have its own depletion curve which should produce a complex composite curve (see RECESSION LIMB OF HYDROGRAPH). AMG

**Reading and References**
Bako, M.D. and Owoade, A. 1988: Field application of a numerical method for the derivation of baseflow recession constant. *Hydrological processes* 2, pp. 331–6. · Federer, C.A. 1973: Forest transpiration greatly speeds streamflow recession. *Water resources research* 9, pp. 1599–605. · Hall, F.R. 1968: Base-flow recessions – a review. *Water resources research* 4, pp. 973–83. · Linsley, R.K., Kohler, M.A. and Paulhus, J.L.H. 1988: *Hydrology for engineers*. 3rd edn. New York: McGraw-Hill.

**depression** See CYCLONE.

**depression storage** Consists of water trapped in small surface depressions or hollows during a rainfall event. Depression storage must be filled before overland flow can occur and this component of the total volume of rainfall will eventually be evaporated or will infiltrate the soil. (See also SURFACE DETENTION; SURFACE STORAGE.) AMG

**Reading**
Sneddon, J. and Chapman, T.G. 1989: Measurement and analysis of depression storage on a hillslope. *Hydrological processes* 3, pp. 1–13.

**depth–duration curve** Relates the magnitude of rainfall to its duration. This type of curve is usually constructed to relate the magnitude of extreme rainfall events to their duration at a single site or over an area. AMG

**Reading**
Niemczynowicz, J. 1982: Areal intensity–duration–frequency curves for short term rainfall events in Lund. *Nordic hydrology* 13, pp. 193–204. · Shaw, E.M. 1988: *Hydrology in practice*. 2nd edn. New York: Van Nostrand Reinhold.

**desalinization** (or desalination) The production of freshwater from saline brines, especially seawater, by distillation or any other process.

**desert** Scientific definitions of deserts have been based on a range of criteria including the nature and development of drainage systems, the types of rock weathering processes that operate, ecological communities, and the potential for crop growth. Most definitions of deserts are however related to moisture deficiency, and the term is generally synonymous with DRYLANDS. True desert conditions are most clearly represented by the HYPER-ARID and arid components of drylands. Deserts do experience significant rainfall events, but these are either unreliable, irregular, or confined to only part of the year. Some deserts also have perennial rivers, with sources in wetter regions, flowing through them, for example the Nile which waters extremely low rainfall areas of Egypt and the Sudan. Deserts are rarely totally devoid of plant and animal life. Desert species have adaptive strategies that permit the accumulation and retention of available moisture, and/or physiologies that allow biological functions to be slowed or shut down at times of acute moisture stress.

Deserts embrace a wide range of landscape systems (see table) that reflects their tectonic and continental settings and the interplay between different geomorphological processes. SAND SEAS are not dominant at the global scale

Arid zone landscapes in different regions (expressed as a percentage of area[a])

|  | SW USA | Sahara | Libya | Arabia | Australia[b] |
|---|---|---|---|---|---|
| Mountains | 38.1 | 43 | 39 | 47 | 16 |
| Low angle bedrock surfaces | 0.7 | 10 | 6 | 1 | 14 |
| Alluvial fans | 31.4 | 1 | 1 | 4 | |
| River plains | 1.2 | 1 | 3 | 1 | 13 |
| Dry watercourses | 3.6 | 1 | 1 | 1 | |
| Badlands | 2.6 | 2 | 8 | 1 | — |
| Playas | 1.1 | 1 | 1 | 1 | 1 |
| Sand seas | 0.6 | 28 | 22 | 26 | 38 |
| Desert flats[c] | 20.5 | 10 | 18 | 16 | 18 |
| Recent volcanic deposits | 0.2 | 3 | 1 | 2 | — |

Notes
[a] Percentages given are only approximate, with the degree of accuracy differing between areas.
[b] From Mabbutt (1977). Categories used by Mabbutt do not necessarily coincide with those used in other areas: included for comparison only. The remaining data are from a study by Clements et al. (1957) for the US Army.
[c] Undifferentiated: includes areas bordering playas.

despite popular images of deserts and early views that suggested aeolian processes were the dominant influence on desert landscapes (see AEOLATION). The major climatic changes that have affected the earth in the Quaternary period are known on theoretical and empirical grounds to have impacted on the distribution and extent of desert conditions. The rock record has also been used to identify the sporadic existence of desert sand seas and dune deposits as far back as the Proterozoic (Glennie 1987). The distribution and occurrence of such deposits reflects not simply climatic changes that may have affected the earth in earlier geological times but the changing location of land masses relative to aridity-inducing climate systems, as a function of plate tectonic movements.

At the scale of Quaternary glacial–interglacial cycles, changes in the extent of deserts and drylands can be expected to have occurred due to changes within the partitioning of moisture in the global hydrological cycle, and alterations to the positioning of major climatic systems. Evidence for dryland changes comes from deep-sea core sediments, sections of which may contain dust and aeolian sands derived from the increased operation of aeolian processes under conditions of expanded aridity, or changes in fluvially derived sediments indicative of changed catchment weathering environments, and from terrestrial sediments and landforms. Terrestrial evidence not only points towards larger deserts at times within the Quaternary (e.g. Stokes et al. 1997) but, through the existence of palaeolake shorelines, significant fluvial deposits and cave speleothems, times of reduced desert extent as

well. Human actions may also influence the extent of desert-like conditions through the impact of DESERTIFICATION. DSGT

### Reading and References
Blume, H.-P. and Berkowicz, S.M. 1995: Arid ecosystems. Advances in Geoecology 28, Catena Verlag, Cremingen. · Clements, T., Merriam, R.H., Stone, R.O., Mann, J.F. Jr and Eymann, J.L. 1957. A study of desert surface conditions. Headquarters Quartermaster Research and Development Command, Environmental Protection Research Division Technical report EP53. · Cooke, R.U., Warren, A., and Goudie, A.S. 1993: Desert geomorphology. London: UCL Press. · Glennie, K.W. 1987: Desert sedimentary environments, present and past: a summary. Sedimentary geology 50, pp. 135–66. · Mabbutt, J.A. 1977: Desert landforms. Canberra: ANU Press. · Stokes, S., Thomas, D.S.G. and Washington, R., 1997: Multiple episodes of aridity in southern Africa since the last interglacial period. Nature 388, pp. 154–8.

**desert pavement** See STONE PAVEMENT.

**desert varnish** A thin patina (just a few microns thick) of shiny dark stain that develops on rocks in arid areas and some other environments. It is composed primarily of iron and manganese oxides which may either be derived from in situ weathering of the underlying rock or a result of inputs (probably aeolian) of materials from outside the rock. Although geomorphologically insignificant, it may have some potential for dating purposes (See ROCK VARNISH.) ASG

### Reading
Whalley, W.B. 1983: Desert varnish. In A.S. Goudie and K. Pye eds, Chemical sediments and geomorphology. London: Academic Press.

**Desert varnish** developed on a rock slope in the dry Hunza valley in the Karakorams of Pakistan. Note the contrast in tone between the varnish and the rock underlying it as revealed by the petroglyphs.

**desertification** (also desertization) A term coined by the French forester Aubreville in 1949 to describe land degradation. Since then there have been over 100 published definitions of this controversial term which gained prominence after the 1977 UN Conference on Desertification, itself prompted by social and environmental concerns in the sub-Saharan Sahel region of Africa. The term has been confused with DROUGHT in some circles (though there are clear links with drought) and its social impacts have sometimes been associated with famine that may in fact have non-environmental causes (Ollson 1993). Desertification has consequently been heavily critiqued in recent years in both the environmental and social sciences (see e.g. Thomas and Middleton 1994 and Stiles 1995). In 1995 The UN Convention to Combat Desertification (CCD) was signed; it has since been ratified by the governments of over 150 countries. Coincident with the CCD have been attempts to clarify the scientific and social dimensions of this environmental issue, and to establish an agreed definition.

In the CCD, desertification is defined as *land degradation in arid, semi-arid and dry-subhumid areas resulting from various factors including climatic variations and human activities*. The problem is therefore confined to the susceptible DRYLANDS, with land degradation regarded as soil erosion, internal soil changes, depletion of groundwater reserves and irreversible changes to vegetation communities (see also SOIL ERO-

SION, SALINIZATION, BUSH ENCROACHMENT). In many respects, desertification is no different than land degradation occurring world-wide, except that it specifically refers to its occurrence within dryland areas. Whether there is anything special about desertification, in environmental process terms, is a moot point which has been considered by writers such as Mainguet (1991). It is perhaps political and social dimensions of the problem (including its apparent severe occurrence in some of the world's poorest nations, in Africa and Asia) that have given credence to desertification as an environmental issue worthy of special and urgent consideration (see Stiles 1995; Thomas 1997).

In environmental terms, the soil degradation component of desertification comprises water and wind erosion, and physical and chemical changes within the soil. These factors reduce the potential productivity of the land. A systematic global survey of soil degradation, commissioned by the United Nations Environment Programme (UNEP) and conducted in the early 1990s, estimated that soil degradation and erosion affects up to 1035 million ha of drylands world-wide, though gained reliable and agreed assessments of the full extent of desertification have been difficult, and sometimes controversial (Thomas and Middleton 1994). Overall, the survey suggested that water erosion was the dominant form of soil desertification in 48% of affected areas, wind erosion in 39%, chemical changes (that include nutrient depletion and salinization) in 10% and physical changes (including soil crusting and compaction) in 4%. The full data set is explored in Middleton and Thomas (1997); it can however be additionally noted that the level of desertification was only extreme or severe in 4% of all affected lands. Even given the precision limitations of the global survey, it serves to identify the diversity and insidiousness of desertification, and adds further weight to dispelling images of sand dunes advancing over productive land, a common media misrepresentation, as the common face of the problem. As Toulmin (1995, p. 5) has concisely noted, 'dryland degradation does not involve moving sand dunes. Rather, it concerns the gradual impoverishment of agricultural and pastoral systems, which makes them less productive and more vulnerable to drought'. The status of vegetation degradation as a form of desertification is even more difficult to assess given the natural dynamics of dryland ecosystems and debates over how permanent vegetation changes really are (see e.g. BUSH ENCROACHMENT).

Though the CCD definition includes possible multiple causes for desertification, it is undoubtedly the case that the principal agent of degradation is human actions. Though desertification has often been viewed as a particular problem of the twentieth century, the ability of humans to detrimentally alter drylands is not new. Salinization in Mesopotamia around 2500 BC has been attributed to agriculture, while in central Mexico, prehispanic societies caused severe soil erosion (O'Hara et al. 1993). In the twentieth century it is the growth and changing distributions of human populations in drylands, and the spread of technological advances, that have increased the propensity for dryland degradation. Several issues make drylands especially susceptible to degradation by humans. These do not necessarily relate to the sometimes supposed fragility of their ecosystems – which as Behnke et al. (1993) show, is disputable for many dryland areas – but to the nature of environmental systems and human activities.

1 The inherent natural variability of dryland climates and ecosystems tend to be overridden today when humans use technologies and methods imported from more consistent temperate environments.
2 Geomorphic processes in drylands tend to be characterized by significant periods of quiescence, punctuated by abrupt episodes of activity. Thus areas cleared of natural vegetation by grazing pressures, or bare in the period immediately following harvesting, are particularly susceptible to rapid erosion from high intensity rain storms or windy conditions.
3 The rapid growth of urban areas in deserts and drylands since the Second World War has placed significant pressures on limited water resources and options for waste disposal.
4 In dryland areas of the developing world, declining rural populations, in response to migration to urban centres, can lead to the failure of traditional soil and vegetation conservation techniques. Tiffen et al. (1994) have shown how more people on the land can in certain circumstances halt and reverse environmental degradation.

DSGT

**References**
Behnke, R.H., Scoones, I. and Kerven, C. 1993: *Range ecology at disequilibrium: new models on natural variability and pastoral adaptation in African savannas.* London: ODI. · Mainguet, M., 1991: *Desertification.* Berlin: Springer Verlag. · Middleton, N.J. and Thomas, D.S.G. 1997: *World atlas of desertification.* 2nd edn. London: UNEP/ Edward Arnold. · O'Hara, S.L., Street-Perrot, F.A. and Burt, T.P. 1993: Accelerated soil erosion around a Mexico highland lake caused by prehispanic agriculture. *Nature* 362, pp. 48–51. · Ollson, L. 1993: On the causes of famine: drought, desertification and market failure in the Sudan. *Ambio* 22, pp. 395–403. · Stiles, D. ed. 1995: *Social aspects of sustainable dryland management.* Chichester: John Wiley. · Thomas, D.S.G. 1997: Science and the desertification debate. *Journal of arid environments* 37, pp. 599–608. · —and Middleton, N.J. 1994: *Desertification: exploding the myth.* Chichester: John Wiley. · Tiffen, M., Mortimore, M. and Gichuki, F. 1994: *More people, less erosion: environmental recovery in Kenya.* Chichester: John Wiley. · Toulmin, C. 1995: The convention to combat desertification: guidelines for NGO activity. *IIED dryland paper* 56. London: IIED.

**desiccation** This is both a concept and a process. The concept of progressive desiccation has existed since at least the mid-nineteenth century, and is the belief that parts of the world are getting drier, through reduced rainfall levels and depleted groundwater reserves, giving rise to the spread of deserts. It has been attributed to both climatic change and human mismanagement (Goudie 1972). Warren and Khogali (1992) have used the term to describe the reduction of moisture in drylands that results from a dry event at the scale of decades – i.e. longer than a DROUGHT. Desiccation is also used to describe the process of the drying out of an individual water body, for example a lake during a drought period or at the end of the rainy season. (See also DESERTIFICATION; DRYLANDS.)

DSGT

**References**
Goudie, A.S. 1972: The concept of post-glacial desiccation. *Oxford school of geography research paper series* 4. ·

Warren, A. and Khogali, M. 1992: *Assessment of desertification and drought in the Sudano-Sahelian region*. New York: UNSO/UNDP.

**design discharge** The discharge which a structure or development is designed to resist or to cope with. A dam across a river must be designed to retain a flood of a particular size and a flood prevention scheme will also be designed to convey a flood of a particular magnitude. Although very large recent events may provide the experience against which schemes may be designed, a design discharge is usually selected by reference to a specific RECURRENCE INTERVAL and this has to be such that it will provide a reasonable expectation of protection and yet not be too costly. Therefore, for land drainage works or for flood prevention schemes, estimated recurrence intervals in the range 50–75 years, or occasionally 100 years, may be used.          KJG

**desquamation** Onion-weathering. The disintegration of rocks, especially those in desert areas, by peeling of the surface layers.

**desulphurasation** The removal of sulphur compounds, primarily sulphur dioxide ($SO_2$) from the flue gases of coal-fired power stations and other industrial sources in order to prevent their escape into the atmosphere, where they may contribute to the acid rain phenomenon. Gas desulphurasation processes include water scrubbing, and absorption using lime or active carbon. Commonly used processes achieve reductions in sulphur emissions of 50–99 per cent. Desulphurasation is also applied to heavy fuel oils in order to reduce sulphur emissions associated with their use.          DLD

**Reading**
Department of the Environment 1991: *Manual of acidic emission abatement technologies. Volume 1: coal-fired systems*. London: HMSO.

**devensian** See late GLACIAL.

**dew, dewpoint** Dew is formed as a condensate from moist air which is cooled by contact with a surface which loses heat by RADIATION. It occurs once the chilling has lowered the air's TEMPERATURE to its dewpoint which is the temperature at which an air sample becomes saturated by cooling at constant PRESSURE and absolute HUMIDITY. The numerical value of dewpoint is obtained by using humidity tables in conjunction with measurements of dry-bulb and wet-bulb temperature.          RR

**Reading**
Monteith, J.L. 1957. Dew. *Quarterly Journal of the Royal Meteorological Society* 83, pp. 322–41.

**diabatic** See ADIABATIC.

**diachronous** Pertaining to a sedimentary unit of a single facies which belongs to two or more units of geological time.

**diaclinal** Describes those rivers whose courses cross the strike of geological structure at right angles.

**diagenesis** Post-depositional changes which have altered a sediment, particularly cementation and compaction.

**diamictite, diamicton** Terms proposed by the American geologist R.F. Flint for non-sorted terrigenous deposits and rocks containing a wide range of particle sizes, regardless of genesis. Examples of diamictites include till and mudflow deposits.

**diapir** An anticlinal fold which has resulted from the upward movement of mobile rocks, such as halite, lying beneath more competent strata. Sometimes a surface dome produced by such movements.

**diastrophism** Tectonic processes which produce dramatic changes in the shape of the earth's surface, such as orogenies, faulting and folding.

**diatoms** These are microscopic, unicellular ALGAE (*Bacillariophyceae*) with a shell, called a frustule, which is made of silica. The frustule, which is etched by rows of tiny holes, is composed of two valves that are held together by siliceous belts called girdle bands. Round *et al.* (1990) recognized three principal types of diatom pattern: (1) centric, (2) simple pennates and (3) raphid pennates. Centric diatoms are radially symmetrical and are generally planktonic (floating) while the pennates are bilaterally symmetrical. The shape, size and style of ornamentation on the shell vary from one species to another and are used to help identify the range of diatoms present in a sample. The frustule of all except the most fragile of diatoms is generally well preserved in sediments and allows the identifications of diatoms down to species and in exceptional cases sub-species level. Diatoms are very small, ranging in size from about 5 $\mu$m to 2 mm although most species fall into the size range 20–200 $\mu$m. A very high powered light microscope is required to identify and count the different species with the diatom magnified between 200 and 1000 times.

Diatoms can be found in almost every aquatic environment where there is sufficient light for photosynthesis and distinct diatom communities exist within a wide range of microhabitats. Diatoms are commonly preserved in LAKE sediments and useful palaeoenvironmental indicators for several reasons. If well preserved and abundant the frustules may be readily identified and counted. Furthermore, many species live in very defined ecological conditions, thus providing important information on pH, salinity and mineral conditions. Unlike POLLEN which gives a regional picture of environmental change, the results from diatom analysis generally relate to the lake being studied, providing a more detailed view of change on a local scale.    SLO

### Reference

Round, F.E., Crawford, R.M. and Main, D.G. (1990). *The diatoms*. Cambridge: Cambridge University Press.

**diatreme** The general term for vents and pipes which have been forced through sedimentary strata by the forces of underlying volcanism. Kimberlite pipes are examples of diatremes.

**die-back** Mortality beginning at the extremities of plants. One or more phenomena lead to stress which manifests itself in the decline and death of leaves, shoots and roots. Plants susceptible to die-back are often woody and possess restricted environmental tolerances. In deciduous trees, a characteristic first sign of it is the premature discoloration and fall of leaves on outermost branches, with no subsequent leaf regrowth in that area (Norton 1985).

There appears to be a number of possible causes of die-back. Extreme and lasting drought may initiate it, as seems to have occurred in the shallow-rooting sugar maple (*Acer saccharum*) in the eastern United States during pronounced dry spells of the first, third and sixth decades of this century (Westing 1966). Recent die-back of sugar maple groves in Ontario, Quebec and Vermont could be linked to ACID PRECIPITATION (Norton 1985). Drought may also heighten the risk of disease which leads to die-back. For instance, fungi which do not harm white ash (*Fraxinus americana*) growing under normal conditions have probably been responsible for canker formation during dry spells in New York State over the past half-century (Silverborg and Ross 1968). Water abundance may be related to die-back. For example, salt-marsh substrate in southern England possessed reducing conditions that caused the build-up of sulphide, the toxic effect of which may have led to die-back in cord-grass (*Spartina townsendii*) (Goodman and Williams 1961).

Progressive die-back lowers the resistance of plants to other environmental factors (insect attack, for example) unfavourable to their survival, and thus can combine with such factors to bring about death (Norton 1985).    RLJ

### Reading and References

Goodman, P.J. and Williams, W.J. 1961: Investigations into 'die-back' in *Spartina townsendii* agg. III. Physiological correlates of 'die-back'. *Journal of ecology* 49, pp. 391–8. · Norton, P. 1985: Decline and fall. *Harrowsmith* 9, pp. 24–43. · Silverborg, S.B. and Ross, E.W. 1968: Ash dieback disease development in New York State. *Plant disease reporter* 52, pp. 105–7. · Westing, A.H. 1966: Sugar maple decline: an evaluation. *Economic botany* 20, pp. 196–212.

**diffluence** The process of a glacier overflowing its valley into an adjacent one. Also the separation in airflow which occurs when an airstream decelerates.

**diffusion equation** A flow equation for *transient* flow through a saturated homogeneous, porous rock in which hydraulic CONDUCTIVITY is the same in each direction. *Steady* flow through a porous medium requires the rate of fluid flow into a given volume of the rock to be equal to the rate of flow out of it. This is expressed mathematically by the equation of continuity:

$$-\frac{dv_x}{dx} - \frac{dv_y}{dy} - \frac{dv_z}{dz} = 0$$

where $v_x, v_y, v_z$ are specific discharges in directions $x$, $y$, and $z$. If the porous rock is homogeneous and hydraulic conductivity is the same in each direction, then another equation incorporating DARCY'S LAW can be written to express the steady state saturated flow through it:

$$\frac{d^2h}{dx^2} + \frac{d^2h}{dy^2} + \frac{d^2h}{dz^2} = 0$$

where $h$ is the HYDRAULIC HEAD and $x$, $y$, and $z$ are a coordinate system defining position. This partial differential equation is known as *Laplace's equation*. It is incorporated into another equation termed the *diffusion equation*, in order to describe *transient* saturated flow through a porous medium with similar homogeneous properties, as follows:

$$\frac{d^2h}{dx^2} + \frac{d^2h}{dy^2} + \frac{d^2h}{dz^2} = \frac{pg(\alpha + n\beta)}{K}\frac{dh}{dt}$$

where $p$ is density, $g$ is gravitational acceleration, $\alpha$ is the vertical compressibility of the aquifer, $\beta$ is the compressibility of water, $n$ is porosity, and $K$ is hydraulic conductivity.

Since $p\,g\,(\alpha + n\beta) = $ specific storage $S_s$, the right-hand side of the equation may be simplified to $S_s/K\,dh/dt$.

The specific storage is the volume of water that a unit volume of aquifer releases from storage per unit decline in $h$. The solution $h$ $(x, y, z, t)$ describes the value of the hydraulic head at any point in a flow field at any time (Freeze and Cherry 1979). PWW

**Reference**
Freeze, R.A. and Cherry, J.A. 1979: *Groundwater.* Englewood Cliffs, NJ: Prentice-Hall.

**digital elevation model** See DIGITAL TERRAIN MODEL.

**digital image processing** A REMOTE SENSING technique involving the handling and modification of images that are held as discrete units, e.g. the sampling, correction and enhancement of a Landsat/MSS image can be achieved by this method.

A discrete image comprises a number of individual picture elements known as pixels, each one of which has an intensity value and an address in two-dimensional image space. The intensity value of a pixel, which is recorded by a digital number (DN), is dependent upon the level of ELECTROMAGNETIC RADIATION received by the sensor from the earth's surface and the number of intensity levels that have been used to describe the intensity range of the image. There are three stages in the processing of a discrete image. First, the images which are stored on computer-compatible tapes are read into a computer; next, the computer manipulates these data; then results of these manipulations are displayed. These three stages can be performed using a range of computers; workstations and the larger personal computers are currently the most popular among physical geographers.

There are many techniques for the processing of digital images, and physical geographers tend to concentrate on six of them: image restoration and correction, image enhancement, data compression, colour display, image classification and the development of geographic information systems.

1 Image restoration and correction form the first stage in any image-processing sequence and include, first, the restoration of the image by the removal of effects whose magnitudes are known, like the non-linear response of a detector or the curvature of the earth and, secondly, the correction of the image by the suppression of effects whose magnitudes can only be estimated, such as atmospheric scatter or sensor wobble.

2 Image enhancement involves the 'improvement' of an image in the context of a particular application. The most popular image enhancements are the selective increase in image contrast (stretching), ratioing wavebands against each other to display differences between wavebands, and digital filtering to smooth or sharpen edges within an image.

3 Data compression involves the reduction of many images into one image for ease of interpretation.

4 Colour display involves the combination of images with colour, again for ease of interpretation.

5 Image classification can be achieved by several techniques, notably the density slicing of one image or the supervised classification of several images. Density slicing involves the grouping of image regions with similar DN, either automatically or interactively. Supervised classification involves the careful choice of wavebands, the location of small but representative training areas, the determination of the relationship between object type and DN in the chosen wavebands, the extrapolation of these relationships to the whole image data set and the display and accuracy assessment of the resultant images.

6 Geographic information systems involve the combination and use of any spatial data that can be referenced by geographic coordinates. The three processing steps for this operation are, first, data encoding where spatial data are broken into polygons or grids; secondly, data management, where these data are spatially filed and, thirdly, data manipulation where these data are retrieved, transformed, analysed, measured, composited or modelled.

PJC

**Reading**
Curran, P.J. 1985: *Principles of remote sensing.* Harlow: Longman Scientific and Technical. · Mather, P.M. 1987: *Computer processing of remotely-sensed images: an introduction.* Chichester: Wiley. · Moik, J.G. 1980: *Digital processing of remotely-sensed images.* Washington, DC: National Aeronautics and Space Administration. · Richards, J. 1986: *Remote sensing digital image analysis: an introduction.* Berlin: Springer-Verlag.

**digital terrain model (DTM)** A model of the land surface used in a GEOGRAPHIC INFORMATION SYSTEM. Another land surface model, the digital elevation model or DEM, simply models height, whereas a DTM also models terrain shape. In practice, the terms are often used interchangeably. The commonest form consists of point estimates of elevation located on a regular grid. This can be used in standard RASTER GIS and DTMs in this form are available for many countries and areas. The other common format is the

Triangulated Irregular Network (TIN) where points of known height are connected into a series of triangles covering the land surface.

Applications include: (1) calculation of basic terrain parameters, such as slope steepness; (2) hydrological analyses, including estimating the CATCHMENT of any point on a stream; (3) determining the visibility between points (viewshed analysis) used in military planning and visual impact assessment.          SMW

**Reading**
Weibel R. and Heller M. 1991: Digital terrain modelling. In D.J. Maguire, M.F. Goodchild and D.W. Rhind eds, *Geographical information systems*, Vol. 1. Harlow: Longman. Pp. 269–97.

**dikaka**  Accumulation of dune and sand covered by scrub or grass vegetation, extended to include plant-root cavities in dune sediments (calcified root tubules).

**dilation**  (or dilatation) Describes the action of PRESSURE RELEASE in a rock mass by the removal of overlying material by erosional processes. Severe glaciation may cause dilation joints to open in glaciated terrain, while in granite areas the opened joints on many INSELBERGS may be a result of pressure release following the removal of the overlying sedimentary or metamorphosed rocks. The joints often approximately parallel the ground surface configuration.          ASG

**dilution effect**  A term used to describe the behaviour of those solute concentrations in a stream which decrease during a storm run-off event. This decrease is ascribed to the dilution of solute-rich base flow by additional inputs of storm run-off which, in view of its shorter residence time within the drainage basin, possesses lower solute concentrations. Some solute concentrations in a stream may, however, increase during periods of storm run-off.          DEW

**dilution gauging**  A method of measuring river discharge by introducing a tracer into the river channel and timing the passage of the tracer over a known length of channel. (See also DISCHARGE.)

**Reading**
Water Research Association 1970: *River flow measurement by dilution gauging*. Water Research Association technical paper TP74. Medmenham: Water Research Association. · White, K.E. 1978: Dilution methods. In R.W. Herschy ed., *Hydrometry: principles and practices*. Chichester: Wiley.

**diluvialism**  The belief in the role of Noah's flood, as reported in the book of Genesis, in shaping the landscape. Before the true origin of glacial drift was recognized such materials were ascribed to a great deluge, when 'waves of translation' covered the face of the earth. The heterogeneous and unsorted character of the drift seemed ample proof that it had been laid down in the turbulent waters of a universal flood, and Dean Buckland of Oxford termed such material 'diluvium' to distinguish it from the 'alluvium' formed by rivers. By the 1830s recognition of the often complex stratigraphy of the drift, and the Ice Age, greatly weakened the diluvial viewpoint. As the catastrophic interpretation of landscape and geological history gave way to UNIFORMITARIANISM, diluvialism became obsolete.          ASG

**Reading**
Davies, G.L. 1969: *The earth in decay*. London: Macdonald.

**dimensionless number**  A dimensionless number is one which is scaled by a parameter that: (1) has identical dimensions to it; and (2) has basic theoretical or empirical bases for being used in such a way. The need for dimensionless numbers arises from the fact that the magnitude of a variable will tend to change as the scale of the investigation changes. Thus, by scaling the variable's magnitude, with respect to the experiment, it becomes possible to compare information acquired from a range of different experiments, even where the experiments are conducted at different scales. For instance, in BOUNDARY LAYER flows, we tend to express the height above the boundary (units of length) as a proportion of the total boundary depth (also units of length). Use of depth means that identical dimensions are being used. The theoretical basis of this scaling is also strong, as we know that boundary layer characteristics do change as a function of depth, so by scaling using depth, we can compare situations with different depths. It follows that dimensionless numbers should be central to the representation of most empirical evidence in PHYSICAL GEOGRAPHY.          SNL

**dimethylsulphide (DMS)**  The most abundant volatile sulphur compound in seawater is produced by planktonic algae and bacterial decay. It oxidizes in the atmosphere to form a sulphate aerosol that is a major source of cloud-condensation nuclei. Because of this, DMS may have an important climatic impact through its impact on cloud albedo and the earth's radiation budget (Charlson *et al.* 1987). Increasing cloud production and albedo caused by increased planktonic productivity resulting from global warming could act as a negative feedback in the climate system. This is a hypothesis that merits careful consideration. ASG

**Reference**

Charlson, R.J., Lovelock, J.E., Andreae, M.O. and Warren, S.G. 1987: Oceanic phytoplankton, atmospheric sulphur, cloud albedo and climate. *Nature* 326, pp. 655–61.

**dip**   The angle between the inclination of sedimentary strata and horizontal.

**dipslope**   The more gentle slope of a cuesta; the slope of the landsurface that approximates the dip of the underlying sedimentary rocks.

**dipwell**   A device designed to measure the position of the water table below the ground surface. The water table position is measured at atmospheric pressure. Thus a dipwell differs from a PIEZOMETER which measures pressure head in the saturated zone and a TENSIOMETER which measures matric potential in the unsaturated zone. A dipwell may be constructed and installed very simply using plastic drain pipe *c*.40–60 mm internal diameter drilled (2–3 mm holes) along the length of the pipe at regular intervals and sealed at the base with a rubber bung. A loose cap should be fitted to the top of the dipwell above the ground surface to stop direct precipitation input. The length of the dipwell depends on the estimated water table depth over the hydrological year. Continuous water table measurements can be obtained by installing a calibrated pressure transducer inside the dipwell and linking this to a data logger.

ALH

**dirt cone**   A conical hill or dome of ice that is completely mantled by till or rock fragments. It owes its existence and form to the mantle of debris which retards the rate of surface lowering resulting from the melting of the ice.

**discharge**   The volume of flow of water or fluid per unit time. It is usually expressed in cumecs which are cubic metres per second ($m^3 s^{-1}$) but for small discharges it may be more conveniently expressed as litres per second ($1 s^{-1}$). In Imperial Units the cusec (cubic feet per second) was originally used and is still employed in the USA. It is necessary to measure discharge and to obtain continuous records of discharge variation for the investigation of the HYDROLOGICAL CYCLE.

Discharge may be measured in a number of different ways and the method adopted at a particular gauging station will depend upon the size of the river, the stability of the channel, the variability of the flow and of the sediment trans-ported, and the length and accuracy of the record required. The major methods of discharge measurement are:

1   *Volumetric gauging* involves collection of the total volume of flow over a specified period of time. It is the most accurate method but can only be used where it is easy to collect the discharge in a large container and to time the increases in water level. Can therefore be used to measure flow from small plots or experimental areas.

2   *Control structures* are structures installed in the cross-section of the stream channel which includes weirs and flumes. Both types of control structure have a formula which relates depth of water to discharge. A weir is a structure placed across the channel and may be sharp crested, in which case the plate inserted in the channel cross-section has a sharp edge on the V notch and the angle at the centre of the V may be 90° or 120° or other angles. Alternatively, the weir may be broad crested and this is preferred for large basins and has various forms which include flat-V and Crump types and often need a rating curve established to relate the depth of water above the weir and the velocity of water. Where the gauging station needs to measure a range of flows it may be necessary to construct a compound weir where a V notch may occur in the centre of a rectangular cross-section, for example. The flume is an artificial channel constructed by raising the channel bed into a hump or by contracting the sides of the channel, or by combining both. The cross-section of the flume is adapted to suit the range and magnitude of discharge and it may be rectangular, triangular or trapezoidal. Flumes have the advantage that silt and debris are easily carried through whereas it could collect upstream of a weir. Several types of flume exist and in a standing wave flume or critical depth flume there is a direct relationship between depth of water upstream of the throat of the flume and discharge. In other cases (Parshall and Venturi flumes) head is measured within and upstream of the throat and the difference between the two values is directly related to discharge.

3   *Velocity-area technique* is the most frequently used method for discharge measurement and depends upon the fact that discharge is:

$$Q = Va$$

where $V$ is velocity and $a$ is water cross-sectional area. The velocity is usually measured by current meter and this is used in each of several verticals across the channel and the spacing between the verticals should not exceed 5 per cent of the channel width. At many gauging stations the velocity and hence discharge is measured at a range of flows, and a rating curve is constructed relating discharge to depth of water or stage. This rating curve can then be the basis for converting continuous records of river stage into discharge values.

4 *Dilution gauging* is a method of discharge measurement depending upon calculation of the degree of dilution by the flowing water of an added tracer solution which may be sodium chloride ($NaCl$) or sodium dichromate ($Na\text{-}Cr_2O_3$). The tracer may be injected either at a constant rate or by gulp injection. In the latter case an amount of the tracer solution is introduced instantaneously into the stream and the passage of the 'Ionic Wave' or slug past a downstream site at a known distance downstream is measured usually using a conductivity meter.

5 *The slope-area method* of estimating discharge is effected by using a flow equation whereby velocity ($V$) can be estimated by surveying water surface slope ($S$), hydraulic radius ($R$) and estimating roughness ($n$) using an equation such as the Manning equation in which:

$$V = \frac{R^{\frac{2}{3}}S^{\frac{1}{2}}}{n}$$

6 *Electromagnetic gauging* is particularly useful where there is no stable-discharge relationship or where weed growth impedes flow. An electric current passed through a large coil buried beneath the river bed induces an electromotive force in the water and the force recorded by probes at each side of the river is directly proportional to the average velocity through the cross-section.

7 *Ultrasonic gauging* can be used where water flow is not hampered by vegetation or sediment. The time taken for acoustic pulses beamed from transmitters on one side of the river to travel to sensors on the other side is recorded and gives mean velocity at a specified depth.

Most of the above methods of discharge measurement will provide a value for a single moment. To obtain continuous records of discharge it is usual to employ a stage recorder which will give continuous records of water depth which can subsequently be converted to discharge values.

The discharge record obtained from a gauging station has to be expressed in a form which can be analysed in relation to controlling parameters. This can be done by establishing the general character of the discharge record and by calculating daily, monthly or annual flows for a specific period, usually a year. A further way is to calculate the total run-off volume for a specified period and this is usually expressed as depth of run-off from the entire catchment area and calculated by dividing the total volume of water which passes the gauging station by the surface area of the drainage basin. The run-off ($R$) can then be compared directly with precipitation ($P$) over the basin during the same time period and the ratio gives the run-off percentage ($R/P \times 100\%$). If the average flow is plotted for the year a diagram can be drawn to show the river regime. Such regime diagrams reflect the broad influence of climate and some river regimes will show a major concentration during a short period, for example in an area affected by snowmelt, whereas other areas will have a fairly uniform distribution of discharge throughout the year.

A discharge record may also be analysed for a short period and individual HYDROGRAPH events may be identified either by extracting specific parameters from a single hydrograph or by generalizing the hydrographs as UNIT HYDROGRAPHS.

Variation in river discharge reflects the pattern of climate over the basin, particularly the incidence and intensity of precipitation, the drainage basin characteristics and also the effects of change in the drainage basin including LAND USE changes. KJG

**Reading**
Gregory, K.J. and Walling, D.E. 1976: *Drainage basin form and process*. London: Edward Arnold. · Herschy, R.W. ed. 1978: *Hydrometry, principles and practices*. Chichester: Wiley.

**disclimax** A stable plant community resulting from the disturbance of CLIMAX VEGETATION. The word is normally used for communities disturbed to such an extent by the activities of humans or domesticated animals that the former climax has been largely replaced by new species. The species may even be introduced, as in the case of the prickly pear cactus, which has formed a disclimax over wide areas in Australia. It is similar to the term plagioclimax, which has been used for many plant communities of the English landscape produced by grazing, cutting or burning over many centuries. (See also SERE; SUBCLIMAX; SUCCESSION.)  JAM

**Reading**
Oosting, H.J. 1956: *The study of plant communities.* San Francisco: W.H. Freeman. · Tansley, A.G. 1949: *The British Islands and their vegetation.* Vol. I. Cambridge: Cambridge University Press. · Vogl, R.J. 1980: The ecological factors that produce perturbation-dependent ecosystems. In J. Cairns Jr ed., *The recovery process in damaged ecosystems.* Ann Arbor: Ann Arbor Science.

**disconformity**   An unconformity in a geological sequence which is not represented by a difference in the inclination of the strata above and below.

**discordance**   An unconformity. A difference in the inclination of two strata which are contiguous.

**disjunct distribution**   A geographical distribution pattern in which two or more populations of an organism are exceptionally widely separated for the organism concerned, thus creating a major discontinuity, and may involve now isolated relict populations, exceptional long-range migration or the separation of populations through continental movement. The southern hemisphere beeches (*Nothofagus*), for example, are thought to have been formerly linked on one continental landmass, Gondwanaland, but today they occur widely disjunct in South America, New Zealand, Australia and New Guinea.  PAS

**Reading**
Stott, P.A. 1981: *Historical plant geography: an introduction.* London: Allen & Unwin.

**dispersal**   The mechanism of migration, by which plants and animals are disseminated over the surface of the earth. In plants, dispersal involves the transport of any spore, seed, fruit or vegetative portion which is capable of producing a new plant in a new locality. These propagules or diaspores may be carried by air or water, on or inside animals, by the movement of soil or rock, or they may be exploded from the parent plant. In animals, dispersal depends on the mechanisms for movement, which may range from facilities for flying to swimming and running. For both plants and animals, man is a potent agent of dispersal. (See also ALIENS; MIGRATION.)  PAS

**Reading**
Seddon, B. 1971: *Introduction to biogeography.* London: Duckworth. Especially ch. 8.

**dissection**   The destruction of a relatively flat landscape through incision and erosion by streams.

**dissipative beach**   Beaches can be classified into two basic types: dissipative and reflective. A third type, termed intermediate, represents those beach states which contain elements of the two basic types (Wright *et al.* 1979, 1982).

Under dissipative conditions, incident waves break and lose much of their energy before reaching the beach face. Broken waves or dissipative bores form the resulting surfzone with bore height decreasing in amplitude towards the shore. Depending on incident wave height, the surfzone may be as wide as 500 m under fully dissipative conditions.

Dissipative beaches are characterized by a wide low gradient beach face extending from the foot of the foredune into the surfzone. Multiple bars or breaker zones may be present across the surfzone. Water circulation in the surfzone is dominated by strong onshore flow in the upper water column (alongshore if incident wave approach is oblique to the shoreline), and offshore towards the bed (Wright *et al.* 1982).  BGT

**References**
Wright, L.D., Chappell, J., Thom, B.G., Bradshaw, M.P. and Cowell, P. 1979: Morphodynamics of reflective and dissipative beach and inshore systems: Southeastern Australia. *Marine geology* 32, pp. 105–40. · Wright, L.D., Guza, R.T. and Short, A.D. 1982: Dynamics of a higher energy dissipative surfzone. *Marine geology* 45, pp. 41–62.

**dissolved load**   Material in solution transported by a river and including both inorganic and organic substances. There is no clear boundary between a true solution and the presence of material as fine colloidal particles and all material contained in a water sample which has been passed through a 0.45 $\mu$m filter is conventionally regarded as dissolved. Sources of the dissolved load include rock weathering, atmospheric fallout of aerosols of both oceanic and terrestrial origin, atmospheric gases, and decomposition and mineralization of organic material. In general the total concentration (mg l$^{-1}$) of material in solution in a river will decline as discharge increases due to a DILUTION EFFECT, but the load transported (kg s$^{-1}$) will increase.

On a global basis, the dissolved loads of perennial rivers and streams range from < 1.0 t km$^{-2}$ year$^{-1}$ to a maximum of about 500 t km$^{-2}$ year$^{-1}$, although even higher levels may exist in streams draining saline deposits. Meybeck (1979) has estimated the mean dissolved load transport by rivers from the landsurface of the globe to the oceans at 37.2 t km$^{-2}$ year$^{-1}$, and Walling and Webb (1983) have pointed out the importance of mean annual run-off and lithology in controlling the global pattern of dissolved load transport. DEW

**References**

Meybeck, M. 1979: Concentration des eaux fluviales en éléments majeurs et apports en solution aux océans. *Revue de géologie dynamique et de géographie physique* 21, pp. 215–46. · Walling, D.E. and Webb, B.W. 1983: The dissolved loads of rivers: a global overview. In *Dissolved loads of rivers and surface water quantity/quality relationships.* IAHS publication no. 141. Pp. 3–20.

**dissolved oxygen** Will be present in most natural waters, but interest in this water quality parameter has focused largely on rivers, because of its importance for fish and other aquatic life and the potential for significant spatial and temporal variation. The dissolved oxygen content of streamflow primarily reflects interaction with the overlying air, since oxygen from the atmosphere is dissolved in the water. Assuming equilibrium conditions, the dissolved oxygen concentration is essentially a function of the water temperature, which influences its solubility, and of the atmospheric pressure, which reflects the partial pressure of the gas. The solubility of oxygen at 0 °C and 760 mm atmospheric pressure is 14.6 mg $l^{-1}$ and this will decrease with increasing temperature to 7.6 mg $l^{-1}$ at 30 °C. Deviations from the equilibrium may occur in both a positive and a negative direction. Supersaturation can result from the production of oxygen within the water body through photosynthesis by macrophytes and algae during daylight hours. Oxygen will be consumed by the respiration of aquatic organisms and by the biochemical oxidation of organic material and pollutants, and reduction in dissolved oxygen concentration will occur if this consumption exceeds the rate of atmospheric reaeration. The potential oxygen consumption associated with an organic pollutant is expressed by its biochemical oxygen demand or BOD value.
DEW

**dissolved solids** The total concentration of dissolved material in water is frequently expressed as a total dissolved solids (TDS) concentration. This parameter can be determined by evaporating to dryness a known volume of water. The evaporation is normally carried out at a temperature of 103–105 °C, but some standard procedures specify a temperature of 180 °C. The value obtained should be viewed as only approximate, as there is a possibility of loss of dissolved material by volatilization or of incomplete dehydration of the residue. The residue will contain both inorganic and organic material. An alternative procedure involves summation of the results obtained from analysis of individual constituents, although these need to be expressed in terms of an anhydrous residue (e.g. bicarbonate will exist as carbonate) in order to ensure comparability. There may be poor correspondence between results from the two methods, through the problems outlined above and the general lack of data for dissolved organic material.

Because of their generalized nature, measurements of TDS concentration are of limited value in water quality assessment and pollution studies but they are frequently employed in investigations of DISSOLVED LOAD transport by rivers and in associated assessments of chemical DENUDATION RATES.

On a global scale, discharge-weighted TDS concentrations encountered in rivers range between minima of approximately 5–8 mg $l^{-1}$ recorded in several tributaries of the Amazon, and maxima of 5000 mg $l^{-1}$ or more associated with streams draining areas of saline deposits. Values are, however, typically in the range 30–300 mg $l^{-1}$ and Meybeck (1979) cites a mean inorganic TDS concentration of 99.6 mg $l^{-1}$ for world river water. Dissolved organic matter generally constitutes only a small proportion of the total dissolved solids, and concentrations commonly fall in the range 2–40 mg $l^{-1}$, with a mean of approximately 10 mg $l^{-1}$. The inorganic component is dominated by a limited number of major elements and $Ca^{2+}$, $Mg^{2+}$, $Na^+$, $K^+$, $Cl^-$, $HCO_3^-$, $SO_4^{2-}$ and $SiO_2$ generally account for 99 per cent of the material, with a number of lesser constituents comprising the remainder. Several workers have attempted to define a world average river composition in terms of major inorganic constituents and those provided by Livingstone (1963) and Meybeck (1979) are listed in the table. In the latter case, an attempt has been made to deduct anthropogenic contributions.
DEW

Average composition of world river water

| | Concentration (mg $l^{-1}$) | | | | | | | | |
|---|---|---|---|---|---|---|---|---|---|
| | $Ca^{2+}$ | $Mg^{2+}$ | $Na^+$ | $K^+$ | $Cl^-$ | $SO_4^{2-}$ | $HCO_3^-$ | $SiO_2$ | Total |
| Livingstone 1963 | 15.0 | 4.1 | 6.3 | 2.3 | 7.8 | 11.2 | 58.4 | 13.1 | 118.2 |
| Meybeck 1979 | 13.4 | 3.35 | 5.15 | 1.3 | 5.75 | 8.25 | 52.0 | 10.4 | 99.6 |

**References**

Livingstone, D.A. 1963: Chemical composition of rivers and lakes: data of geochemistry, Chapter G. *US Geological Survey professional paper* 440G. · Meybeck, M. 1979: Concentrations des eaux fluviales en éléments majeurs et apports en solution aux océans. *Revue de géologie dynamique et de géographic physique* 21, pp. 215–46.

**distributary**   A stream channel which divides from the main channel of a river. One of the channels a river subdivides into when it becomes braided or when it reaches its delta.

**distribution graph**   A graph first employed by Bernard in 1935 to represent the unit hydrograph (see UNIT RESPONSE GRAPH) in percentage form. In a distribution graph ordinates are expressed as the percentage of total run-off. According to unit hydrograph theory the storm discharge response is directly proportional to the effective rainfall, and therefore the percentage of run-off in each time period will remain the same regardless of effective rainfall amount if the rainfall duration is constant.   AMG

**Reading and Reference**

Bernard, M. 1935: An approach to determine stream flow, *Transactions of the American Society of Civil Engineers* 100, pp. 347–95. · Wilson, E.M. 1969: *Engineering hydrology*. London: Macmillan.

**diurnal tides**   Occur in a limited number of parts of the world and consist of only one high and one low tide in each 24 hour period. They only occur where the coastline has the correct configuration.

**divergence**   The phenomenon of air flowing outwards from an air mass being replaced by air descending from above, also the splitting of oceanic currents, often as a result of offshore winds, allowing upwelling of cold water from the depths.

Written as div $\vee$ it is given by:

$$\text{div } \vee = \frac{\partial u}{\partial x} + \frac{\partial v}{\partial y} + \frac{\partial w}{\partial z}$$

where $\vee$ is the three-dimensional velocity vector, having $x$, $y$ and $z$ components $u$, $v$, and $w$; units are $(\text{time})^{-1}$. Often in meteorology the term is used to mean horizontal divergence, being the instantaneous fractional rate of change of an infinitesimal horizontal area within the fluid, given by:

$$\text{div } \vee_H = \frac{\partial u}{\partial x} + \frac{\partial v}{\partial y}.$$

Typical values of large-scale horizontal divergence in the atmosphere are about $10^{-5}\text{s}^{-1}$. Negative divergence is called convergence.

Local variations of density in the atmosphere are relatively small, so that horizontal divergence and convergence usually result in VERTICAL MOTION.   KJW

**Reading**

Atkinson, B.W. ed., 1981 *Dynamical meteorology: an introductory selection*. London and New York: Methuen.

**diversity**   Usually the number and relative abundance of species in a biological community (see ECOSYSTEM). However, the resource and habitat diversity of ecological systems can also be measured (Magurran 1988). A small proportion of the species in a community usually accounts for a considerable number of its members (see DOMINANT ORGANISM), while a large proportion will be represented by relatively few individuals. Such a situation leads to the presence of a common and rare species component in a community, with the latter category mainly responsible for its diversity (Odum 1971). There are two types of species diversity. Local (alpha) diversity relates to species variety within a particular community, while regional diversity (see BETA DIVERSITY) is that which exists between communities in juxtaposition (Whittaker 1975; Ricklefs 1980).

Species diversity may be either totally enumerated or, more commonly, sampled. Its most basic measure is given by the ratio between the number of species present ($S$) and the number of individuals ($N$). The number of species per unit area can also be ascertained. Estimation of diversity from the number of individuals provides no insight into the commonness or rarity of species. Communities with the same $S/N$ ratio may possess major differences in species relative abundance. Among measures employed to accommodate both number and relative abundance, an index of diversity measured by the Shannon–Weiner function is favoured. This takes the form:

$$H = -\sum_{i=1}^{S} (p_i)(\log_2 p_i)$$

where $H$ = index of species diversity; $S$ = number of species; $p_i$ = proportion of the total sample belonging to the $i$th species (Krebs 1985).

While such measures provide indices of diversity, the calculated values do not relate to the frequency distribution of the sample. As Hutchinson (1978) states, four main distributions have been recognized in relation to numbers of species and individuals. The most commonly encountered seems to be log-normal, although declining geometrical, log-series and random distributions have been observed.

Important species diversity variations include those on the equatorial-polar latitudinal gradient (high to low respectively) and on islands of increasing size (low to high respectively). Possible causes of diversity differences are numerous, complex, interactive and imperfectly understood. They have been discussed by Pielou (1975) and Krebs (1985).

Greater speciation, hence a more diverse biota, seems to be aided by the combination of physical environmental equability and stability, together with habitat variety, which is characteristic of low rather than high latitudes. Competition for resources also appears to be greater in tropical than other ecosystems, so that there tends to be more niche differentiation in low latitudes. The latter, in turn, helps promote species diversity. On islands, species abundance appears to approximately double when insular area is ten times as great. Here, species diversity is also a function of parameters such as habitat variety and competition. The number of species present is determined by an equilibrium between immigration to, and extinction within, the area (see ISLAND BIOGEOGRAPHY).

Until recently, it was widely assumed that species diversity fostered stability in ecosystems. However, mathematical modelling of ecosystem components has prompted May (1981) to postulate the converse. He argues that high diversity levels lead to increased susceptibility to disruptive effects, so that a complex ecosystem will become unstable if its balance is disturbed.

Over the short term, communities appear to diversify with time. At present, it is usual for mature communities to contain more species than youthful ones. Longer-term patterns of diversity, deduced from fossil taxa, are more debatable. The uncertainty is largely due to the uneven distribution of geological evidence in both space and time. Gould (1981) discusses the two major hypotheses relating to long-term diversity trends. The first states that diversity has increased throughout geological time. The second contends that the available ecological niches were filled earlier in biospheric history. Since that time, there has been a steady-state in respect of diversity, with variations in the equilibrium number of species occurring in response to plate-tectonically controlled environmental fluctuations.

Species diversity measures can provide a pointer to the welfare of ecosystems, and are of potential use in conservation strategy and environmental monitoring (Magurran 1988). Species diversity is often regarded as synonymous with biodiversity. However, the latter also includes the assessment of genetic diversity, which fosters evolution, and of the diversity of ecosystems, within which there are links between biotic and abiotic components (World Conservation Monitoring Centre 1992).                                    RLJ

**Reading and References**
Gould, S.J. 1981: Palaeontology plus ecology as palaeobiology. In R.M. May ed., *Theoretical ecology: principles and applications*. 2nd edn. Oxford and Boston: Blackwell Scientific. Pp. 295–317. · Hutchinson, G.E. 1978: *An introduction to population ecology*. New Haven and London: Yale University Press. · Krebs, C.J. 1985: *Ecology: the experimental analysis of distribution and abundance*. 3rd edn. New York and London: Harper & Row. · Magurran, A.E. 1988: *Ecological diversity and its measurement*. London and Sydney: Croom Helm. · May, R.M. 1981: Patterns in multi-species communities. In R.M. May ed., *Theoretical ecology: principles and applications*. 2nd edn. Oxford and Boston: Blackwell Scientific. Pp. 197–227. · Odum, E.P. 1971: *Fundamentals of ecology*. 3rd edn. Philadelphia: Saunders. · Pielou, E.C. 1975: *Ecological diversity*. New York and London: Wiley. · Ricklefs, R.E. 1980: *Ecology*. 2nd edn. London: Nelson. · Whittaker, R.H. 1975: *Communities and ecosystems*. 2nd edn. New York and London: Macmillan. · World Conservation Monitoring Centre 1992: *Global biodiversity: status of the Earth's living resources*. London and New York: Chapman and Hall.

**diversivore** An animal which may vary its food intake to include both a plant-eating and an animal-eating pattern. The best examples are human beings for, although on a world scale humans are probably more herbivore than anything else, consuming large amounts of cereals in particular, they also ingest meats and fish (notably for protein), and choose to eat some decomposers too (e.g. fungi). There are several other well-known examples: in Welsh rivers the net-spinning caddis fly (*Hydropsyche*) feeds on detritus, green algae, mayflies and midges, and the stonefly (*Perla*) lives off a similar diet (Jones 1949). Such organisms do not fit easily into the 'normal' structure of FOOD CHAINS and TROPHIC LEVELS. In the real world, food chains with a high degree of diversivory are rare, except for insect–host parasite systems (Yodzis 1981).    DW

**References**
Jones, J.R.E. 1949: A further ecological study of a calcareous stream in the Black Mountain district of South Wales. *Journal of animal ecology* 18, pp. 142–59. · Yodzis, P. 1981: The stability of real ecosystems. *Nature* 289, pp. 674–6.

**doab** A term used in the Indian subcontinent to describe the low alluvial plain between two converging rivers.

**doldrums** A zone of light, variable winds, low atmospheric pressure, high humidity and temperature, and frequent cloudy and unsettled weather located near or slightly north of the equator. The doldrums are bounded to the north by the north-east trade winds of the northern hemisphere and to the south by the south-east trade winds of the southern hemisphere (see INTERTROPICAL CONVERGENCE ZONE). They shift north and south with the seasons, being farthest north in June to October. In a broader sense, the term doldrums refers to the lethargic, monotonous, warm, humid weather of summer and to the listless, often despondent, human response to it.                                                    WDS

**doline**  A roughly conical depression formed essentially by the solution and/or collapse of underlying limestone strata. Dolines are often, but not necessarily, sites of sinking water (hence the name SINKHOLE or swallow hole) and can be mantled by subsequent glacial drift deposits. When limestone dissolves and collapses beneath gritstone (as on the north crop of the South Wales limestone) dolines can form on rock other than limestone – they are then called interstratal karst dolines (Thomas 1974). They can be surface expressions of collapsed cave passages and are common sites for cave digging activities. The age of formation of many of the British dolines has been questioned; they may be considerably older than was once thought (Bull 1980).                                          PAB

**References**
Bull, P.A. 1980: The antiquity of caves and dolines in the British Isles. *Zeitschrift für Geomorphologie Supplementband* 36, pp. 217–32. · Thomas, T.M. 1974: South Wales interstratal karst. *British Cave Research Association* 3, pp. 131–52.

**dolocrete**  A form of calcrete in which magnesium carbonate is a major component, and which probably forms as a groundwater precipitate near the water table of a brackish water body.                                                    ASG

**Reading**
El-Sayed, M.I., Fairchild, I.J. and Spiro, B. 1991: Kuwaiti dolocrete: petrology, geochemistry and groundwater origin. *Sedimentary geology* 73, pp. 59–75.

**dome dune**  A low, circular or oval mound formed where dune height is inhibited by unobstructed strong winds. It generally lacks a slip face. (See ZIBAR.)

**domestication**  The manipulation of the life and reproduction of plants and animals by humans to produce genetic changes that result in new races. The domestication of biota represents a change in human economies from food collecting (e.g. hunting, fishing, gathering) to food production, sometimes called the Neolithic

revolution. The transition from hunting and gathering to food production must always have been a gradual process, although once inaugurated it may often have spread quickly. The recognition of the process of domestication in archaeological contexts is often difficult and may be apparent only when it is complete, so that the beginnings of such processes are often lost beyond recovery. The reasons why domestication started in such early foci as south-west and south-east Asia, and Meso-America are still under discussion: the innovation and stimulus derived from a city, climatic change and 'oasis' formation, ecosystem models, environmental potential models, and demographic pressure have all been invoked. Once established, food-producing societies raise the carrying capacity of the land (the CARRYING CAPACITY for late Pleistocene hunter-gatherers was 0.1 person $km \gg fb^2$, for early dry farms 1–2 persons $km \gg fb^2$, early irrigated lands 6–12 persons $km \gg fb^2$), encourage the development of sedentary societies, cause changes in the structure of society, allow craft specialization, the development of surpluses and very probably the growth of civilization as we usually define it. Whether food production allows rapid population growth or is called forth by it, there is clearly a close link between the two.       IGS

**Reading**
Bender, B. 1975: *Farming in prehistory: from hunter-gatherer to food-producer*. London: Arthur Barker. · Harris, D.R. Alternative pathways towards agriculture. In C.A. Reed ed., *Origins of agriculture*. The Hague: Mouton. Pp. 179–243.

**domin scale**  Plant abundance scale which is used to rate the relative vigour and cover of species within a sample quadrat. As with the BRAUN-BLANQUET scale, the scheme was developed to facilitate the quantitative analysis of co-occurrences of plant species and is a phytosociological tool used to identify plant ASSOCIATIONS. This is a 10-point scale, it is non-linear and, although in principle it is the same as the Braun-Blanquet method, it allows for rather more divisions:

10 species cover almost complete, 90 to 100% of sample area;

9 species covers approximately 75 to 90% of sample area;

8 species covers approximately 50 to 75% of sample area;

7 species covers approximately 33 to 50% of sample area;

6 species covers approximately 25 to 33% of sample area;

5 species covers approximately 20% of sample area;

4 species covers up to 10% of sample area;

3 species occurs as several individuals which cover less than 4% of sample area;

2 species occurs as several individuals, no measurable cover;

1 species occurs as one or two individuals with normal vigour, no measurable cover;

+ species occurs as a single individual with reduced vigour, no measurable cover.

(See COVER, PLANT.)                                    MEM

**dominant discharge**   The discharge to which the average form of river channels is related. The dominant discharge that determines the size of a river channel cross-section or the size of river channel pattern at a particular location will also depend on the character and quantity of sediment transported and also on the composition of the bed and bank materials. The dominant discharge will not be a single value but a range of flows. (See also CHANNEL CAPACITY.)
                                                       KJG

**dominant organism**   An organism of principal importance in either the whole or part of a community. Dominance may be physiognomic (of a particular form, such as a tree – see LIFE FORM), taxonomic (of an evolutionary category, such as a species), or ecological (of quantity and function, such as the amount of standing crop (see BIOMASS), the competitive ability of a producer or consumer). An organism can exhibit more than one type of dominance. This is especially so in the case of physiognomy and taxonomy (Shimwell 1971).

Both plants and animals may dominate. However, plants comprise most of the earth's biomass and are essential for the survival of animals (see ECOSYSTEM). Thus, strictly speaking, plants always dominate communities (Odum 1975). Nevertheless, as Daubenmire (1968) points out, there are instances (in agricultural ecosystems, for example) in which animals assume an important role. Similar regard may be given to the current ascendancy of man in the majority of world ecosystems.      RLJ

**Reading and References**
Daubenmire, R.F. 1968: *Plant communities: a textbook of plant synecology*. New York and London: Harper & Row. · Odum, E.P. 1975: *Ecology: the link between the natural and social sciences*. 2nd edn. London and New York: Holt Reinhart & Winston. · Shimwell, D.W. 1971: *Description and classification of vegetation*. London: Sidgwick & Jackson. · Willis, A.J. 1973: *Introduction to plant ecology*. London: Allen & Unwin.

**dominant wind**   The wind which plays the most significant role in a particular local situation, in contrast to the PREVAILING WIND.

**donga**   Derived from the Nguni word *Udonga*, meaning a wall, the term used in southern Africa to describe a gully or badland area caused by severe erosion. Dongas are especially prevalent in colluvium and in weathered bedrock in areas where the mean annual rainfall lies between 600 and 800 mm. Where the materials in which they are developed have high exchangeable sodium contents, they may have highly fluted 'organ pipe' sides (Stocking 1978). At Rorke's Drift they provided a snare for unwary troops in the battle between the British and Zulus.      ASG

**Reference**
Stocking, M.A. 1978: Interpretation of stone lines. *Southern African geographical journal* 60, pp. 121–34.

**dormant volcano**   A volcano which, though not currently or perhaps even recently active, is not extinct since it is likely to erupt in the future.

**double mass analysis**   A plot of cumulative values of one variable, or values from one site against cumulative values of another variable or from another site. It has often been used to plot the values of precipitation recorded at one station against the records of another station or against the average of several other stations. Unless the double mass curve which is plotted has a slope of 1.0 there is a difference between the sites and this may be interpreted in terms of instrument siting or of temporal variations which affect one variable but not the other.
                                                       KJG

**Reading**
Dunne, T. and Leopold, L.B. 1978: *Water in environmental planning*. San Francisco: W.H. Freeman.

**downwelling**   The process of accumulation and sinking of warm surface waters along a coastline. A change of airflow of the atmosphere can result in the sinking or downwelling of warm surface water. The resulting reduced nutrient supply near the surface affects the ocean productivity and meteorological conditions of the coastal regions in the downwelling area.      ASG

**draa**   A large-scale accumulation of aeolian sand dunes geographically distinct from other sand fields but smaller than a sand sea.

**drainage**   May refer either to the natural drainage of the landsurface or to the system of land drainage introduced by human activity. Natural drainage of the landsurface is organized in drainage basins which are those areas in which water is concentrated and flows into the DRAINAGE NETWORK. The drainage basin is usually defined by reference to information on surface elevation,

145

for example, from contours on topographic maps although the position of the WATERSHED on the ground surface, which is the line separating flow to one basin from that to the next, may not correspond to the PHREATIC DIVIDE beneath the surface. The pattern of natural drainage has been studied in relation to the DRAINAGE DENSITY and drainage basin characteristics which can be quantified and used in rainfall–run-off modelling and in the interpretation of river discharge; to the nature of the drainage network including the pattern of the drainage and also the stream ORDER; and to the evolution of the drainage pattern. In the course of drainage evolution the details of the several patterns such as trellis or rectangular (see diagram for DRAINAGE NETWORK) may be related to geological structure such as the alternation of hard and soft rocks or to the presence of joints or faults. Where drainage patterns are discordant with the structure and cross folds or faults, for example, it has been suggested that either the drainage has been superimposed from a cover rock that originally occurred above the rocks at present exposed in the landscape, or the drainage was antecedent and the drainage pattern was maintained as the structures were developed by endogenetic uplift giving the folded and/or faulted structures.                    KJG

**drainage density**  Is calculated by dividing the total length of stream channels in a basin ($\Sigma L$) by the drainage basin area ($A_d$) as:

$$D_d = \frac{\Sigma L}{A_d}$$

Drainage density is a very significant measure of drainage basin character because the values of $D_d$ reflect the climate over the basin and the influence of other drainage basin characteristics including rock type, soil, vegetation and land use, and topographic characteristics. Drainage density also has an important influence upon streamflow because water flow in channels is faster than water flow over or through slopes. The higher the drainage density the faster the hydrograph rise and the greater the peak discharge. Although drainage density is a very significant parameter it must be used with careful attention given to the extent of the drainage network which can be determined from topographic maps of different scales, from remote sensing sources, or from field survey; and also to the composition of the DRAINAGE NETWORK because networks are composed of channels which have flows for different periods of the year.                    KJG

**Reading**
Gregory, K.J. 1967: Drainage networks and climate. In E. Derbyshire ed., *Geomorphology and climate*. Chichester: Wiley. Pp. 289–318.

**drainage network**  The system of river and stream channels in a specific basin or area (see diagram). Whether a particular headwater stream is included in the drainage network depends upon channel type, and networks have been analysed qualitatively and quantitatively. Qualitative classifications of drainage networks depend upon the way in which drainage patterns reflect either underlying geological structure (rectangular, parallel, dendritic, trellised), prevailing regional slope (radial, centripetal), geomorphological history (deranged), or some combination of these (e.g. annular). Quantitative analysis has focused upon drainage NETWORK structure founded upon stream ORDER and upon DRAINAGE DENSITY. When viewed according to channel process the drainage network can be regarded as composed of PERENNIAL, INTERMITTENT and EPHEMERAL streams.                    KJG

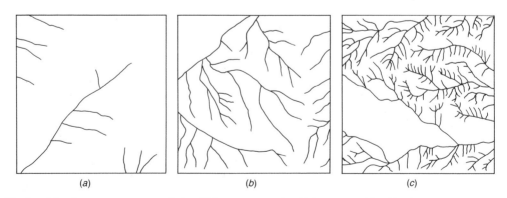

|  (a)  |  (b)  |  (c)  |

**Drainage density** measurements can distinguish between networks which are coarse (*a*), with a low to medium density (*b*) and to fine (*c*) which have a high drainage density value.

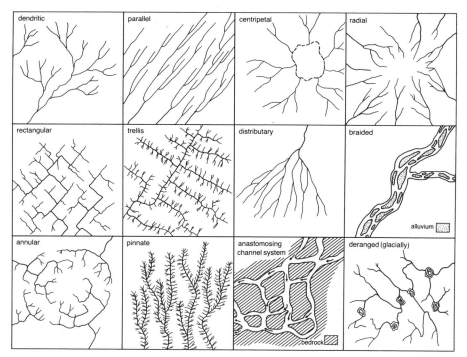

**Drainage network**. Types of drainage pattern.
*Source*: H.F. Garner 1974: *The origin of landscapes*. New York: Oxford University Press. Pp. 60–1.

**draw down** The extent to which the WATER TABLE is reduced in elevation as a result of pumping water from a well. The amount of draw down diminishes logarithmically with distance from the site of pumping, and this determines the shape of the CONE OF DEPRESSION in the water table. Under ARTESIAN conditions, draw down may also occur in the potentiometric surface should heavy pumping occur at a bored well. In a coastal AQUIFER, the reduction in HYDRAULIC HEAD caused by draw down will encourage salt water intrusion beneath the well (see GHYBEN–HERZBERG PRINCIPLE).        PWW

**Reading**
Freeze, R.A. and Cherry, J.A. 1979: *Groundwater*. Englewood Cliffs, NJ: Prentice-Hall. · Linsley, R.K., Kohler, M.A. and Paulhus, J.L.H. 1988: *Hydrology for engineers*, 3rd edn. New York: McGraw-Hill.

**dreikanter** A pebble found on the surface in desert regions; it has three distinct facets on its upper surface as a result of wind abrasion.

**drift potential** A term used to describe the *potential sand transport* due to aeolian processes in a particular wind environment. Three different types of drift can be calculated when a SAND ROSE is produced: the potential drift from each compass direction for which wind data are available, the total drift potential (DP), which is the sum of all individual directional potentials, and the resultant or net drift potential (RDP) which is the resolved value in the resultant drift direction (RDD).        DSGT

**dripstone** Any accumulation of water-soluble salts on the roofs, walls and floors of caves. A general term for stalactites and stalagmites.

**drop size** (and the distribution of drop sizes) This forms one of the major determinants of the effects of rain on exposed soils. Similarly, beneath plant canopies, the sizes of water drops that fall from the leaves and branches (termed *gravity drops*) is equally important to soil splash. In general, larger drops have higher TERMINAL VELOCITIES and deliver more energy to the soil surface, where it may be expended in breaking down soil aggregates or splashing grains. In rain, drop sizes are no smaller than about 0.1 mm diameter (smaller drops are kept aloft by atmospheric turbulence) and no larger than about 6 mm (larger drops are unstable and break up during their fall). The modal diameter is often about 2–2.5 mm. It tends to be larger in intense thunderstorms, when only larger and heavier drops are able to fall through the strong updrafts feeding moisture into the convective clouds, and smaller in low-intensity rain.        DLD

147

**Reading**
Bubenzer, G.D. and Jones, B.A. 1971: Drop size and impact velocity effects on the detachment of soils under simulated rain. *Transactions of the American Society of Agricultural Engineers* 14, pp. 625–8.

**drought** This is 'a rather imprecise term with both popular and technical usage' (Kemp 1994, p. 41). Droughts can occur in any environment, and the perception and therefore definition of a drought varies with the climate conditions that generally pertain in the environment concerned. While a drought linked to a rainfall deficit is strictly a meteorological phenomenon, drought may also be assessed relative to economic and social needs. To this end, an assessment solely linked to rainfall deficits might more appropriately be called a meteorological drought, to distinguish it from considerations related to other factors. In particular, an assessment of an even broader term might be: 'There are a plethora of purpose-specific subtypes of drought that have been proposed [see Agnew and Anderson 1992 for a review] which include water-supply drought, climatic drought, hydrological drought, socio-economic drought and economic drought, which means a water shortage that negatively affects the established economy of the country or region affected' (Sandford 1987). Some of these subtypes are not especially helpful in the analysis of water deficits, but the term AGRICULTURAL DROUGHT has relatively wide usage and may be helpful in the assessment of impacts upon crop production. In order to take non-crop producing areas into account, Rasmussen (1987) provides a useful wider socio-economic definition that refers to drought as an extended and significant negative departure of rainfall from the regime that society has adjusted to. The remainder of this entry focuses upon the meteorological focus of drought.

Meteorological drought is the condition of dryness resulting from a lack of PRECIPITATION. The deficit is commonly sufficiently persistent for EVAPORATION to lead to a substantial decrease in the moisture content of soils and in other hydrological parameters such as groundwater flow and streamflow. Official definitions exist in many countries. In the former USSR a drought was defined as a period of 10 days with total rainfall not exceeding 5 mm; in India it is when rainfall is less than 80 per cent of normal levels (both from Agnew and Anderson 1992). A number of indices exist for the assessment of drought from meteorological data, of which the PALMER DROUGHT SEVERITY INDEX is one widely used example.

In the United Kingdom, an *absolute drought* is a period of at least 15 consecutive days during which no day reports more than 0.2 mm of rain. A *partial drought* is a spell of at least 29 days during which there may be some days that experience slight rain but for which the mean daily rainfall does not exceed 0.2 mm. Absolute droughts occur about once a year on average in the lowlands of south-east England. In contrast, more prolonged drought is in a sense a regular annual feature of some tropical climates where one rainy season is separated from the next by a long dry season, found characteristically across areas at the limit of the poleward excursion of the INTERTROPICAL CONVERGENCE ZONE.

Droughts in the mid-latitudes are associated with the unusual persistence of anticyclonic conditions and especially with the presence of a BLOCKING anticyclone. Under such a regime the rain-bearing frontal systems are steered around the flanks of the stationary high pressure area so that anomalous dryness in one place is linked to anomalous wetness in others. For example, May 1975–August 1976 was a period of extreme drought stretching from Scandinavia to western France, with southern England recording only 50 per cent of the long-term mean precipitation for a 16-month period. In August 1976 the same area recorded less than 50 per cent of the mean precipitation, while at the same time Iceland to the north-west and the northern Mediterranean, to the south of the blocking high, reported over 150 per cent of the respective means. The Sahelian drought of subtropical Africa that commenced in 1969 and which still persisted through the 1970s is believed to be the result of various factors including the anomalous southward expansion of the Azores anticyclone and, through the operation of TELECONNECTIONS, the effects of EL NIÑO. This long period of rainfall depression may also be called a DESICCATION event. A detailed analysis of meteorological data by Agnew (1990) has shown that whilst the effects of the Sahel drought have undoubtedly been severe, its spatial distribution was markedly uneven from year to year reflecting the spottiness of rainfall in DRYLANDS. Consequently, some of the environmental and social consequences attributed to drought impacts may be due to other factors, including human actions, that form part of DESERTIFICATION.     DSGT/RR

**References**
Agnew, C, 1990: Spatial aspects of drought in the Sahel. *Journal of arid environments* 18, pp. 279–93. · —and Anderson, E. 1992: *Water resources in the arid realm.* London: Routledge. · Kemp, D.D. 1994: *Global environmental issues: a climatological approach.* London: Routledge. · Rasmussen, E.M. 1987: Global climate change and variability: effects on drought and desertification in Africa. In M. Glantz, ed., *Drought and hunger in Africa.*

Cambridge: Cambridge University Press. · Sandford, S. 1987: Towards a definition of drought. *Symposium on drought in Botswana, Gaborone.* Hanover NH: Clark University Press. Pp. 33–40.

**drumlin** An oval-shaped hill, largely composed of glacial drift, formed beneath a GLACIER or ICE SHEET and aligned in the direction of ice flow. There are no strict definitions relating to their size but they tend to be up to a kilometre long and 50 m in relief. They are widespread in formerly glaciated areas and are especially numerous in Canada, Ireland, Sweden and Finland. Drumlins are considered to be part of a family of related landforms including FLUTES, MEGA-SCALE GLACIAL LINEATIONS, and ROGEN MORAINE which are collectively referred to as SUBGLACIAL BEDFORMS. Their formation remains controversial (see below) but in spite of this they are extremely useful for reconstructing former ice sheets. The word drumlin is a derivation of a Gaelic word for a rounded hill.

Whilst there are many variations in shape, the 'classic' drumlin is a smooth, streamlined hill that resembles an egg half buried along its long-axis. They tend to exist as fields or swarms of landforms rather than as isolated individuals, with a typical swarm comprising tens to thousands of drumlins. Viewed en masse, drumlins within a swarm display a similar long-axis orientation and morphology to their neighbours, and are closely packed, usually within two to three times the dimensions of their drumlin length. The majority of drumlins in a swarm have their highest elevation and blunter end pointing in an upstream direction, with the more gently sloping and pointed end, or tail, facing down-ice. The upstream blunt end is called the stoss end and the downstream end called the lee. A common measure of their shape is the elongation ratio, which is the maximum drumlin length divided by maximum width. Typical elongation ratios are 2:1 to 7:1. Variations in drumlin shape include spindle-like forms, two-tailed forms that resemble a BARCHAN dune in plan view, and they also exist as perfect circular hills with an elongation ratio of 1:1. There is a whole branch of investigation, called drumlin morphometry, which uses measures of shape, size and spacing to try and develop or test theories for their formation.

The timing or synchroneity of drumlin formation within a field remains unknown. Some researchers deem drumlin formation as occurring close to ice margins and believe fields are built up incrementally as the margin retreats, whereas others believe that extensive patterns of drumlins may have formed approximately synchronously under wide swathes of an ice sheet. Drumlin patterns have been found to lie cross-cutting each other with some superimposed upon others. This demonstrates that older landforms can be preserved beneath ice flow and that more than one flow direction is recorded.

The internal composition of drumlins reveals a perplexing array of different sediment types and structures. Some have rock cores surrounded by a concentric sheath of TILL, but they are mostly filled with unconsolidated sediments that are poorly sorted, and may contain silts, sands, gravel and boulders. They may however also be found with fluvially sorted sediments at their core or in a lee-side position. Some interpret this as demonstrating that fluvial deposition was part of the drumlin formation process but others favour a two-stage process of fluvially deposited material which was later shaped into a drumlin. Tectonic structures such as thrusts and folds have also been found and have been taken to imply the sediments have been deformed during the drumlinization process.

Due to the inaccessibility of glacier beds, active drumlin formation has not been observed first-hand, so it is perhaps no surprise that their formation remains something of a mystery. Part of the difficulty is that a good theory must be capable of explaining the full range of observed drumlins and other subglacial bedforms and their wide variation in shapes, scales and internal composition. There have been many hypotheses and theories that attempt to explain their formation. Menzies (1979) and Patterson and Hooke (1995) provide good overviews of the 'drumlin problem'. Put simply, drumlins may have formed by a successive build of sediment to create the hill (i.e. deposition or accretion) or pre-existing sediments may have been depleted in places leaving residual hills (i.e. erosion), or possibly a process that blurs these distinctions. Hypotheses have been proposed for all these cases but most common have been those involving some form of sediment accretion. These, however, have difficulty explaining drumlins with cores of pre-existing fluvially sorted sediments.

Observations of the nature of the bed of contemporary ice sheets have revealed that the forward motion of ice can, in part, be accomplished by deformation of the soft sedimentary bed. This has led to the deforming bed model of glacier flow, which has become the most widely accepted, but still unproven, mechanism for drumlin formation. If the sediments of the bed are weak they may deform as a result of the shear stress imparted by the overlying ice. If parts of this deforming till layer vary

149

in relative strength, then the stronger, stiffer portions will deform less and remain static, whilst the intervening weaker portions will deform more readily and become mobile. The relative strength is thought to be controlled by grain size, with coarse grained sediments (e.g. gravels) remaining strong as they do not allow a build up of pore water pressures, and fine grained sediments as easily deformable. So a till layer with spatially variable strength will have static or slow moving strong patches, around which the weaker more deformable till will flow. This can explain the cores of drumlins (strong patches; rock-cored, coarse-grained or with preserved fluvially sorted sediments) surrounded by more easily deformed till which is responsible for the streamlining. It also explains the occurrence of folds and thrusts commonly observed in drumlins. In this deforming bed model of drumlin formation (Boulton 1987) the position of each drumlin is controlled by sediment inhomogeneities and the streamlined shape by deformation. As deformation continues, drumlins may be uprooted and become mobile.

An alternative to the above, is a model developed by Shaw *et al.* (1989) that views drumlins and other subglacial bedforms to be the result of meltwater erosion and deposition as a consequence of large floods beneath the ice. In this meltwater model, regional scale outburst floods from the central regions of the ice sheet produce sheet flows of water, tens to hundreds of kilometres wide and deep enough to separate the ice from its bed. Turbulent water during the floodstage erodes giant drumlin-shaped scours in the base of the ice, which are then infilled with sediment as the flood wanes and as the ice presses down onto its bed. This is the cavity-fill drumlin and explains how fluvially derived sediments may appear in drumlins. A related mechanism is also thought to operate whereby vortices in the flood water erode down into the till bed leaving intervening ridges of the original material, which are the second type of meltwater drumlins. These could therefore contain tills or almost any material, as the composition is unrelated to shaping event.

The difficulty in evaluating these theories arises from the fact that the deforming bed and meltwater models are each so comprehensive as to be able to predict the wide variety of observed drumlin characteristics. This makes it hard to use geomorphological observations to test between them. Also, both are still at the stage of qualitative theories rather than physically based models. Deforming beds have been observed to exist and so have subglacial floods, the question remains as to which are capable of producing drumlin forms and over the widespread patterns for which they are observed. The answers must surely lie in numerical modelling to examine plausible mechanisms tested against large-scale drumlin patterns.    CDC

**Reading and References**
Benn, D.I., and Evans, D.J.A., 1998: *Glaciers and glaciation.* London: Arnold. · Boulton, G.S. 1987. A theory of drumlin formation by subglacial sediment deformation. In J. Menzies and J. Rose, eds, *Drumlin symposium.* Balkema, Rotterdam, pp. 25–80. · Menzies, J. 1979: A review of the literature on the formation and location of drumlins. *Earth science reviews,* 14, pp. 315–59. · Patterson, C.J. and Hooke, R. LeB. 1995. Physical environment of drumlin formation. *Journal of glaciology* 41.137, pp. 30–8. · Shaw, J., Kvill, D. and Rains, B. 1989. Drumlins and catastrophic subglacial floods. *Sedimentary geology* 62, pp. 177–202.

**dry deposition**   The process by which pollutant gases or particles are transferred directly from the atmosphere on to liquid and solid surfaces. Particularly important for the deposition of $SO_2$, but deposition of $NO_2$ is slow and dry deposition is not a major removal process for atmospheric $NO_x$. Deposition is primarily through turbulent transfer and the velocity of deposition depends upon relative humidity, pH and, especially in urban areas, aerodynamic resistance, whereby surfaces exposed to the wind experience greatest deposition. Deposition is independent of rainfall, and maps of Britain produced by the Review Group on Acid Rain (1987) show dry deposition of sulphur to be more widespread and loadings to be considerably higher than those for wet-deposited non-marine sulphur. Particulate deposition tends to concentrate near to source areas and particulates deposited between rains can be concentrated in the early stages of surface run-off creating an acid surge. Dry deposition can be increased by land use changes, for example, afforestation, which increase aerodynamic resistance and/or surface area.    BJS

**Reference**
Review Group on Acid Rain 1987: *Acid deposition in the United Kingdom.* 2nd report. Warren Spring Laboratory. London: HMSO.

**dry valley**   A valley which is seldom, if ever at the present time, occupied by a stream channel. These valleys are widespread on a variety of rock types, including sandstones, chalk and limestone in southern England, but they are also known from many other parts of the world, including the coral reefs of Barbados. An enormous range of hypotheses has been put forward to explain why they are generally dry.

## Hypotheses of dry valley formation

### Uniformitarian

1 Superimposition from a cover of impermeable rocks or sediments
2 Joint enlargement by solution through time
3 Cutting down of major through-flowing streams
4 Reduction in catchment area and groundwater lowering through scarp retreat
5 Cavern collapse
6 River capture
7 Rare events of extreme magnitude

### Marine

1 Non-adjustment of streams to a falling Pleistocene sea level and associated fall of groundwater levels
2 Tidal scour in association with former estuarine conditions

### Palaeoclimatic

1 Overflow from proglacial lakes
2 Glacial scour
3 Erosion by glacial meltwater
4 Reduced evaporation caused by lower temperatures
5 Spring snowmelt under periglacial conditions
6 Run-off from impermeable permafrost.

The uniformitarian hypotheses require no major changes of climate or base level, merely the operation of normal processes through time; the marine hypotheses are related to base-level changes; and the palaeoclimatic hypotheses are associated primarily with the major climatic changes of the Pleistocene. Dry valleys show a considerable range of shapes and sizes, from mere indentations in escarpments, to great winding chasms like Cheddar Gorge in the Mendips.                                    ASG

### Reading

Goudie, A.S. 1993: *The nature of the environment*, 3rd edn. Sect. 4.15. Oxford and New York: Basil Blackwell.

## dry weather flow

A term used to describe low flow in a river as an alternative term to BASE FLOW. It is also used to refer to the total daily rate of flow of domestic and trade waste sewage in a sewer in dry weather. The daily total of sewage discharge is used to represent dry weather flow because both domestic and trade waste sewage vary greatly in quantity with 24 hour periods. The dry weather flow is the background flow in the sewer and may be 'measured after a period of seven consecutive days during which rainfall has not exceeded 0.25 mm' (Bartlett 1970).

### Reading and Reference

Bartlett, R.E. 1970: *Public health engineering-design in metric sewerage*. Oxford: Elsevier. · Linsley, R.K., Kohler, M.A. and Paulhus, J.H.L. 1982: *Hydrology for engineers*. 3rd edn. New York: McGraw-Hill.

## drylands

These cover *c*.47% of the earth's land surface and embrace HYPER-ARID, arid, SEMI-ARID and dry subhumid environments (Thomas 1997). Drylands are closely allied to the notion of *aridity*, which has four main causal factors: tropical and subtropical atmospheric stability; continentality; topographically induced rain shadows; and in some coastal situations such as the south-western coasts of South America and southern Africa, cold ocean currents that reduce evaporation from the sea surface.

Attempts to quantify aridity have focused on the balance between PRECIPITATION inputs, and moisture losses through EVAPOTRANSPIRATION. Recent attempts to establish the extent of drylands on the basis of moisture availability have used an aridity or moisture index in the form P/PET, where P = annual precipitation and PET = potential evapotranspiration. Meigs' (1953) moisture index used aggregated monthly to annual moisture surplus and deficit data instead of P, with PET calculated by the Thornthwaite method. Because of limitations in the availability of meteorological data from many desert and dryland areas, a recent widely available assessment (by Mike Hulme for the United Nations Environment Programme (UNEP), in Middleton and Thomas 1997) has used the simpler Penman method of calculating PET. To further rationalize values, Hulme also used meteorological data for a defined time period (1951–80) to calculate P/PET, rather than simply taking mean values from each station supplying data. This overcomes the problem that in some developing parts of the world mean values calculated from data runs of a few decades would be treated as equivalent to those produced from a century or more of data from parts of Europe. This approach also allows account to be given to a major climatological characteristic of drylands: high interannual and interdecadal climatic variability. This permits the construction of dryland climate surfaces for different decades, allowing spatial changes in their extent to be determined (Hulme 1996).

In terms of P/PET values, hyper-arid areas have values less than 0.05; arid = 0.05–< 0.20; semi-arid = 0.20–0.50 and dry-subhumid = 0.50–0.65 (Hulme 1996). The world's drylands support *c*.17% of the human population: in Africa, this rises to nearly 50% of that continent's population. The significance of this is that these people are both susceptible to the natural climatic variability inherent in drylands, and are potential agents of environmental change, including DESERTIFICATION. Arid, semi-arid and dry-subhumid areas have together been termed the susceptible drylands, because of their potential to be degraded by human actions. Indirect human impacts, through possible anthropogenic global warming, may also impact significantly on drylands. It has been estimated that drylands provide a net contribution to enhanced greenhouse gas levels of only 5-10% of the global total (Williams and Balling 1995), but many dryland areas are predicted to be amongst the biomes that respond most rapidly to global warming, through decreased soil moisture levels, a greater incidence of drought and higher mean temperatures. Significantly however, some view drylands as having the potential to play a critical role on the mitigation of global warming, through carbon sequestration. Opportunities exist for enhancing dryland soil and biomass carbon storage, not least because drylands are a major reserve (c.75%) of global soil carbonate carbon, which participates less in global carbon fluxes than organic carbon. If anti-desertification land restoration measures proposed by the UN were to be implemented, the sequestered carbon would be equivalent to 15% of current annual $CO_2$ emissions. This demonstrates clearly the global significance of dryland environments. DSGT

### Reading and References

Hulme, M. 1996: Recent changes in the world's drylands. *Geophysical research letters* 23, pp. 61–4. · Meigs, P. 1953: World distribution of arid and semi-arid homoclimates. In *Arid zone hydrology*. UNESCO Arid zone research series 1, pp. 203–9. · Middleton, N.J. and Thomas, D.S.G. 1997: *World atlas of desertification*. 2nd edn. London: UNEP/Edward Arnold. · Squires, E. Glenn 1997: Carbon sequestration in drylands. In N.J. Middleton and D.S.G. Thomas eds, *World atlas of desertification*. 2nd edn. London: UNEP/Edward Arnold. Pp 140–3. · Thomas, D.S.G. ed. 1997: *Arid zone geomorphology: process, form and change in drylands*. 2nd edn. Chichester: John Wiley. · Williams, M.A.J. and Balling, R.C. 1995: *Interactions of desertification and climate*. London: Edward Arnold.

**du Boys equation** In 1879 Paul du Boys developed one of the earliest bedload transport equations from consideration of the TRACTIVE FORCE of flowing water. The bedload transport weight per unit width of channel per unit time ($g_b$) is expressed as a function of the mean bed shear stress ($\tau_o$) in excess of the threshold shear stress ($\tau_{oc}$) required to initiate particle motion:

$$g_b = A(\tau_o - \tau_{oc})\tau_o$$

The constant $A$ and the threshold mean bed shear stress ($\tau_{oc}$) are both regarded as functions of sediment particle size alone. Derivation of the equation is simply based on the assumption that bed sediment moves in discrete layers whose velocity decreases linearly from the bed surface to zero at a depth dependent on the fluid shear stress at the bed (Embleton and Thornes 1979, pp. 238–9). KSR

### Reading and Reference

du Boys, P.F.D. 1879: Études du régime du Rhône et l'action exercée par les eaux sur un lit à fond de graviers indefiniment affouillable. *Annales des Ponts et Chaussées* Ser. 5.18, pp. 141–95. · Embleton, C. and Thornes, J.B. eds 1979: *Process in geomorphology*. London: Edward Arnold.

**dune** A dune is a subaerial accumulation of sediment, with particles usually sand-sized (2 mm or less in diameter). Dunes are formed by the wind, in aeolian environments, and in water, generated by TURBULENT FLOW STRUCTURES in fluvial environments. Active AEOLIAN dunes are aerodynamically shaped, with different forms being the result of the overall sand transporting wind environment, the supply of sediment and the characteristics of the terrain in which they occur. Dunes may vary in size from less than a metre to over 200 m high, they may be ephemeral forms or the result of sediment accumulation over many millennia. Dunes usually comprise quartz sand, but can also consist of carbonate sediments, gypsum, CLAY PELLETS (see also CLAY DUNE), and volcanic ash. The constituent particles of active dunes remain loose, but gypsum and carbonate sands may readily become cemented. Dunes most commonly occur in dryland and coastal environments, where suitable sediment is available for entrainment by the wind and where surface vegetation does not impede transport by SALTATION and CREEP. Approximately 20 per cent of the world's drylands are covered by aeolian deposits, including SAND SHEETS and dunes. Extensive areas of dunes and sand sheets are called SAND SEAS. Active dunes are defined as those that are presently undergoing aeolian processes, though rates of surface change may be slow and total dune mobility does not necessarily occur. Dunes that are stabilized, fixed and/or degraded, sometimes termed relict/relic dunes, occur when the total package of environmental factors favouring aeolian processes does not now exist. Relict dunes occur on the margins of many active sand seas and are testimony to expanded dryland conditions at various times during the Holocene and late Quaternary.

A TOPOGRAPHIC DUNE occurs where airflow and sediment transport is disrupted by hills, valleys or other structural features. Most dunes however are not influenced by topography, and occur in relatively low relief situations, where the initiation of dune forms is probably largely a function of changes in surface ROUGHNESS, which leads to sediment deposition and the formation of a sand patch. Following initiation, dune development results from the transport of sediment up the windward or *stoss* slope of the patch towards the *crest*. The growing dune increasingly modifies its own BOUNDARY LAYER, such that flow separation on the lee side allows a SLIP FACE to develop. Desert dunes commonly occur not as isolated forms but in dunefields where many individual dunes exist. Following McKee (1979), *simple dunes* occur where individual dunes are discrete forms, *compound dunes* occur when adjacent dunes of the same basic type coalesce or merge with each other, and *complex dunes* comprise superimposed forms of different dune types.

Different types of simple dune result primarily from variations in formative wind regimes, with marked morphological variability occurring within all dune types. TRANSVERSE, LINEAR and STAR dunes form respectively in unimodal, bimodal and multimodal wind regimes. A SAND ROSE may be constructed from wind data to assess the wind environments for different dune environments. The terminology used to describe dunes is wide ranging and reflects local usages; see, for example BARCHAN, LONGITUDINAL DUNE, SEIF, and RHOURD. Other types of dune result from the controlling influences of vegetation covers (see NEBKHA, PARABOLIC DUNE, BLOWOUTS), sediment type (see ZIBAR) and sediment supply (see AKLÉ, LUNETTE DUNE). Not all dune types are mobile and migrate, which is a function of the wind environments in which they occur (Thomas 1992). Transverse dunes are migratory forms, moving downwind in the overall direction of sand transport, linear dunes tend to extend downwind but may also experience slow lateral movement if one transport component of the bimodal regime is stronger than the other, and star dunes are accumulatory forms that may also experience a slow migration in one direction. With the exception of blowouts and parabolic forms, most coastal dunes – especially in temperate environments – are stationary, accumulating sand bodies, anchored by the vegetation that traps sand deflated from the beach zone. DSGT

**Reading and References**

Lancaster, N. 1995. *Geomorphology of desert dunes*. London: Routledge. · McKee, E.D. (ed.) 1979: *A study of global sand seas*. US Geological Survey Professional Paper 1052. · Thomas, D.S.G. 1992. Dune activity: concepts and significance. *Journal of arid environments* 22, pp. 31–8. · —1997: Sand seas and aeolian bedforms. In D.S.G. Thomas ed. *Arid zone geomorphology: process, form and change in drylands*. Chichester: Wiley. Pp. 373–412.

**dune memory** See MEMORY CAPACITY (DUNE).

**dune network** Sand dunes in coastal and continental settings do not always produce discrete forms, but may overlap or encroach upon each other to produce *compound* or *complex* forms. In extreme cases dune networks, sometimes called AKLÉ or reticulate dunes, may result, especially where the supply of sand of the thickness of the sand body is large. Dune networks may be distinguished from compound and complex dunes by the fact that the interdune areas are completely enclosed by dune bodies, creating interdune cells that are usually a few tens of metres across. Dune networks are usually formed by overlapping TRANSVERSE DUNE forms, in which elements with different orientations respond to sand transporting winds deriving from different directions. Dune networks may form relatively stable patterns, especially where the ridges are large and respond to different seasonal wind directions, or may comprise unstable, ephemeral patterns that respond to short-term wind directional changes. (See DUNE.) DSGT

**Reading**

Warren, A. and Kay, A.W. 1987. The dynamics of dune networks. In L.E. Frostick and I. Ried eds, *Desert sediments, ancient and modern*. Geological Society of London, special publication 35. Oxford: Blackwell. Pp. 205–12.

**durability** A term used with reference to building materials, which gives an indication of the service life of the material (e.g. a building stone). It is not an absolute quality nor can it be easily quantified, as it must be related to tests appropriate to the type of deterioration which the structure is likely to undergo (e.g. various weathering phenomena). Such tests can be standardized. WBW

**Reading**

Frohnsdorff, G. and Masters, L.W. 1980: The meaning of durability and durability prediction. *American Society for Testing and Materials special technical publication* 691, pp. 17–30.

**duricrust** A hard crust or nodular layer formed at or near the land surface by the processes of weathering. The crust may be composed dominantly of iron oxides (LATERITE, FERRICRETE), aluminum oxides (alcrete, bauxite), silica (SILCRETE) or calcium carbonate

(CALCRETE). Crusts may also form from more soluble minerals such as gypsum and halite, but are generally not classified as duricrusts as they do not persist in the landscape beyond rainfall events.

The formation of duricrusts requires stable, low gradient, landscapes and high temperatures which promote intense chemical activity. The crusts themselves form in a range of climatic regimes. Where rainfall is high and leaching intense residual accumulations of insoluble ions (Al-Fe-Ti) form the bulk of the material, whereas environments in which evaporation exceeds precipitation are dominated by absolute accumulation, as in the case of calcrete. Thus on a global scale alcretes and ferricretes are associated with humid tropical regions and calcretes with deserts. However local conditions frequently determine the duricrust type, particularly in semi-arid regions, and inter-grade crusts of two or more components are common. Silica mobility in particular is responsive to changes in pH and silcrete occupies a wide climatic range. The persistence of duricrusts in the landscape may also lead to survival of both climatic change and continental wandering. The bauxites of Le Baux (France) and the sarsen stones (silcretes) of southern England and the Paris Basin are relicts of warmer, wetter climates of the geological past.

Models of duricrust formation require movement of the ions through the near surface zone, either downwards from the atmosphere, upwards from weathering rock, or laterally through the regolith. Where translocation and precipitation occur in the soil profile the resulting duricrust can be described as pedogenic. Non-pedogenic crusts, sometimes termed groundwater duricrusts, have subsurface water as a transporting agent and frequently occur on the margins of valleys and ephemeral lakes. Precipitation may occur directly to form pisoliths and nodules, or may indurate pre-existing rock, sediment or soil, thereby preserving the host fabric.

Regional duricrust suites in ancient shield landscapes, such as India, Australia and southern Africa, have originated in the Tertiary era. As they are resistant to erosion they may form cap rocks on mesas and escarpments, suggesting relief inversion in the long term. Not all duricrusts, however, are of great antiquity: examples have been reported of ferricretes and calcretes containing twentieth-century artefacts, and of silcretes formed by bacteria in the space of a few years.                                        PSh

**Reading**
Ollier, C. and Pain, C. (1996) *Regolith, soils and landforms.* Chichester: Wiley.

**duripan**  A synonym for silcrete. An indulated and cemented silica-rich accumulation within the soil zone or zone of weathering.

**dust**  This consists of fine particles (commonly silica) of less than about 0.07 mm diameter (SILT size) which have undergone ENTRAINMENT by wind and are suspended in the atmosphere. Whilst dust sources may include fine-grained glacial deposits or even industrial or domestic AEROSOLS the greatest sources are the DESERT or semi-desert areas of the world where the lack of stable vegetated surfaces allows the EROSION of available fine sediment by strong winds. Where visibility is reduced to less than 1 km by suspended sediment a dust storm condition is said to exist, with visibility up to 2 km termed dust HAZE. Turbulent eddies within the atmosphere are capable of keeping fine sediment entrained in the atmosphere for many days and airborne dust can be transported thousands of kilometres from its source. Evidence for such long-range transport comes from reports of 'red rain' in the UK caused by Saharan dust and red dust settling on snow in New Zealand originating from the Channel Country of central Australia.

The dustiest places on earth tend to be in the SEMI-ARID regions with between 100–200 mm of annual precipitation, rather than the hyper-arid deserts. This is because the WEATHERING and FLUVIAL activity in these semi-arid areas helps to produce and concentrate fine sediment in dry lake beds or floodplains which may then be available for entrainment into the atmosphere. Pye (1987) also notes that human cultivation and activity in the semi-arid zone may disrupt stable sediment surfaces which may then be eroded. The Seistan Basin in Iran records 80 dust storms a year and is considered to be one of the dustiest regions on the globe (Middleton 1997).

The importance of dust has been generally underestimated both in its geomorphological significance and also its effects on human health and activities. More recently the significance of airborne dust has gained ground in both academic and applied research. It has been estimated that the Niger river, the most significant active river draining the Sahara, carries 15 million tonnes $yr^{-1}$ of sediment, whilst airborne dust accounts for between 60 and 200 million tonnes $yr^{-1}$ (Cooke *et al.* 1993). The vast LOESS deposits of the world are now also thought to be partly aeolian in origin with the deposition of airborne dust contributing significantly to both the 'glacial' and 'desert' loess theories.

Environmental hazards caused by dust are associated with the DEFLATION, transportation and deposition of fine sediment. In arid and semi-arid areas in particular the deflation of silty sediment from soils is very damaging as it is in the top few centimetres that most soil nutrients are accumulated. The removal of these soil nutrients by high winds can also cause severe problems downwind where dust deposition may bury plants or contaminate food and water supplies. It is, however, the transportation of dust in the atmosphere which has received the most attention in terms of effects on human health and activity. Dust entrained in a strong wind may be highly abrasive and major crop and livestock damage can occur. A lot of attention has also focused on fine sediment particles less than 10 microns in diameter (called PM10s). Such PM10s are easily inhaled into the deepest parts of the human lung and may contribute to lung damage and cancer. Problems including bronchitis, emphysema and conjunctivitis have also been linked to airborne dust. Visibility reduction on roads and runways during severe dust storms has also caused loss of life (Middleton 1997) whilst at a larger scale there is mounting concern as to the reflective properties of airborne dust reducing the input of solar radiation to the earth's surface and hence contributing to global cooling.

The Dust Bowl on the Great Plains of the USA in the mid-1930s is perhaps the best known example of how disruptive human activities and agricultural mismanagement may result in severe and life-threatening dust storms (Worster 1979). The environmental hazards associated with dust transport are beginning to be understood and a recognition of the consequences of poor management techniques on dust generation in hazardous areas is leading to improved management practices aimed at reducing the ERODIBILITY of agricultural surfaces and also reducing the EROSIVITY of the wind. There are still many areas, however, where airborne dust generation is a serious problem. One problem which is very difficult to control is the use of off-road vehicles (ORVs) which can lead to the loss of 8–80 kg/km of silt (Pye 1987). Urban activities around localized water supplies, building sites and military manoeuvres can all seriously disrupt naturally wind resistant surfaces and lead to an increase in dust storm potential (Pye 1987). Dust blown from the exposed former sea-bed of the Aral Sea (Gill 1996) which is rapidly desiccating due to water mismanagement in the surrounding basin, has been called one of the worst environmental disasters to date.

GFSW

**References**
Cooke, R.U., Warren, A. and Goudie, A. 1993: *Desert geomorphology*. London: UCL Press. · Gill, T. 1996: Eolian sediments generated by anthropogenic disturbance of playas: human impacts on the geomorphic system and geomorphic impacts on the human system. *Geomorphology* 17. 1–3, pp. 207–28. · Pye, K. 1987: *Aeolian dust and dust deposits*. London: Academic Press. · Middleton, N. 1997: Desert dust. In D.S.G. Thomas, ed., *Arid zone geomorphology*. 2nd edn. London: Wiley. Pp. 413–36. · Worster, D. 1979: *Dust bowl*. Oxford: Oxford University Press.

**dust storm** A large volume of dust-sized sediment blown into the atmosphere by a strong wind. The ENTRAINMENT of dust from the ground surface is controlled by the nature of the wind, the nature of the soil or sediment itself and the presence of any surface obstacles to wind flow. Dust storms can occur in any environment but are most common in drylands. Meteorologists define a dust storm as a dust-raising event that reduces horizontal visibility to 1000 m or less, as distinct from 'dust haze' which refers to dust that is suspended in the air but which is not actively being entrained within sight of the observer (it has been raised from the ground prior to the observation or at some distance away). Visibility in a dust haze may or may not be less than 1000 m. The other types of dust event are 'blowing dust', which is material being entrained within sight of the observer but not obscuring visibility to less than 1000 m, and the 'dust devil' which is a localized column of dust that neither travels far nor lasts long. McTainsh and Pitblado (1987) have further refined these definitions by including the relevant international meteorological observers' 'SYNOP' present weather (or 'ww') codes for each event type.

Different terrain types vary greatly in their susceptibility to dust storm occurrence. Although silts and clays occur widely in soils and sediments, a number of factors influence their susceptibility to the deflation of dust. Among these factors are the ratios of clay-, silt- and sand-sized particles, the moisture content of the soil, the presence of particle cements such as salts or organic breakdown products, the compaction of sediments and the presence of crusts or armoured surfaces. The most favourable dust-producing surfaces are areas of bare, loose and mobile sediments containing substantial amounts of sand and silt but little clay. Terrains that satisfy these conditions are most commonly found in geomorphologically active landscapes where tectonic movements, climatic changes and/or human disturbance are responsible for rapid exposure, incision and reworking of sediment formations containing dust (Pye 1987).

NJM

**Reading and References**
McTainsh, G.H. and Pitblado, J.R. 1987: Dust storms and related phenomena measured from meteorological

records in Australia. *Earth surface processes and landforms* 12, pp. 415–24. · Middleton, N.J. 1997: Desert dust. In D.S.G. Thomas ed., *Arid zone geomorphology: process, form and change in drylands*, 2nd edn. Chichester: John Wiley. Pp. 413–36. · Péwé, T.L. ed. 1981: Desert dust: origins, characteristics and effects on man. *Geological Society of America special paper* 186. · Pye, K. 1987: *Aeolian dust and dust deposits*. London: Academic Press. · Goudie, A.S. and Middleton, N.J. 1992: The changing frequency of dust storms through time. *Climatic Change* 20, pp. 197–225.

**dust veil index**   (DVI) A quantitative method developed by Lamb (1970) for comparing the magnitude of volcanic eruptions. The formulae use observations either of the depletion of the solar beam, temperature lowering in the middle latitudes, or the quantity of solid matter dispersed as dust. The reference dust veil index is 1000, assigned to the Krakatoa 1883 eruption, and the index is calculated using all three methods, where the information is available, for statistical comparison purposes.

ASG

**Reading**
Lamb, H.H. 1970: Volcanic dust in the atmosphere with a chronology and assessment of its meteorological significance. *Philosophical Transactions of the Royal Society of London, Series A* 266, pp. 425–533.

**dyke**   A sheet-like intrusion of igneous rock, usually orientated vertically, which cuts across the structural planes of the host rocks. The wall or trough formed by differential weathering of such an intrusion when exposed at the landsurface.

**dynamic equilibrium**   A term so widely used in physical geography that it has lost almost all strictness of definition and seems to mean a situation which is fluctuating about some apparent average state, where that average state itself is also changing through time. In Huggett's view (1980) the problems with the definition arise because an idea from thermodynamics (which allows for micro-scale chaotic movement of molecules while the whole system is in equilibrium) has been transposed to apply to a macro-scale. Following Huggett, one can agree that dynamic equilibrium in physical geography today 'is taken either as synonymous with "steady state" or with some false equilibrium in which the system appears to be in equilibrium (*qua* steady state) but in reality is changing very, very slowly with time'.

In this broad sense the term has been neatly substituted for a number of earlier, equally ill-defined concepts, most notably GRADE and climax. Unfortunately, such substitution cannot be said to have resulted in clarification. On the

macro-scale, the problem is a statistical one, which applies to any attempt to equate a belief that a long-term, directed change is occurring with observations which indicate departures from some theoretically persistent condition. In such a case, how can the fluctuations be separated from the underlying trend? Conversely, how should it be decided if some of the fluctuations are too great to be 'permissible' elements of a dynamic equilibrium, but instead indicate a departure from that equilibrium state and the onset of a relaxation period leading to some new equilibrium? The choice seems to be entirely a product of the spatial and temporal scale under discussion.

In practice it is virtually impossible to decide whether any part of the physical environment *is* in dynamic equilibrium, since it is equally difficult to decide that any part is *not* in that state. Unless the term is defined with clarity its use adds very little except confusion to any discussion of change over time.

BAK

**Reference**
Huggett, R. 1980: *Systems analysis in geography*. Oxford: Clarendon Press.

**dynamic source area**   The changing part of a catchment which is physically contributing overland flow at any time. The SOURCE AREA of a catchment is envisaged as changing in response to subsurface flow conditions before and during storms. This concept is important to an understanding of hillslope hydrological response.

MJK

**Reading**
Kirkby, M.J. ed. 1978: *Hillslope hydrology*. Chichester: Wiley.

**dynamical meteorology**   Study of the forces acting on, and the subsequent motion of air in weather systems; compare SYNOPTIC METEOROLOGY; SATELLITE METEOROLOGY. The control of most of the important physical processes in the atmosphere, such as condensation, depends critically on air motion (see WIND). Hence dynamical meteorology is seen by many to represent the core of meteorology.

JSAG

**Reading**
Aktinson, B.W. ed. 1981: *Dynamical meteorology: an introductory selection*. London and New York: Methuen.

**dynamics**   As distinct from statics, the study of the forces on, and accelerations of, moving bodies. The application of this study to the atmosphere gives rise to DYNAMICAL METEOROLOGY. The basic principles can, of course, be applied to all aspects of the natural environment which involve moving bodies, e.g. oceanography, hydrology, slope mechanics.          JSAG

**Reading**
Atkinson, B.W. ed. 1981: *Dynamical meteorology: an introductory selection*. London and New York: Methuen. Esp. pp. 1–20.

# E

earth hummocks  A form of patterned ground characterized by rounded hummocks which produce an irregular net pattern over the landsurface. (See MMA MOUNDS.)

earth pillar  A pinnacle of soil or other unconsolidated material that is protected from erosion by the presence of a stone or boulder at the top.

earthflow  See MASS MOVEMENT TYPES.

earthquake  A series of shocks and tremors resulting from the sudden release of pressure along active faults and in areas of volcanic activity. The shaking and trembling of the earth's surface associated with subterranean crustal movements. (See tables 1 and 2.)

Table 1  Modified Mercalli scale of earthquake intensity

| | |
|---|---|
| I | Not felt except by a very few under especially favourable circumstances. |
| II | Felt only by a few persons at rest, especially on upper floors of buildings. Delicately suspended objects may swing. |
| III | Felt quite noticeably indoors, especially on upper floors, but many people do not recognize it as an earthquake. Standing motor cars may rock slightly. Vibration like passing truck. |
| IV | During the day felt indoors by many, outdoors by few. At night some awakened. Dishes, windows, doors disturbed; walls make creaking sound. Sensation like heavy truck striking building. Standing motor cars rocked noticeably. |
| V | Felt by nearly everyone; many awakened. Some dishes, windows, etc. broken; a few instances of cracked plaster; unstable objects overturned. Disturbances of trees, poles, and other tall objects sometimes noticed. Pendulum clocks may stop. |
| VI | Felt by all, many frightened and run outdoors. Some heavy furniture moved; a few instances of fallen plaster or damaged chimneys. Damage slight. |
| VII | Everybody runs outdoors. Damage negligible in buildings of good design and construction; slight to moderate in well-built ordinary structures; considerable in poorly or badly designed structures; some chimneys broken. Noticed by persons driving motor cars. |
| VIII | Damage slight in specially designed structures; considerable in ordinary substantial buildings, with partial collapse; great in poorly built structures. Panel walls thrown out of frame structures. Fall of chimneys, factory stacks, columns, monuments, walls. Heavy furniture overturned. Sand and mud ejected in small amounts. Changes in well-water levels. Disturbs persons driving motor cars. |
| IX | Damage considerable in specially designed structures, well-designed frame structures thrown out of plumb; great in substantial buildings, with partial collapse. Buildings shifted off foundations. Ground cracked conspicuously. Underground pipes broken. |
| X | Some well-built wooden structures destroyed; most masonry and frame structures destroyed with foundations; ground badly cracked. Rails bent. Landslides considerable from river banks and steep slopes. Shifted sand and mud. Water splashed over banks. |
| XI | Few, if any, masonry structures remain standing. Bridges destroyed. Broad fissures in ground. Underground pipelines completely out of service. Earth slumps and land slips in soft ground. Rails bent greatly. |
| XII | Damage total. Waves seen on ground surfaces. Lines of sight and level distorted. Objects thrown upward into the air. |

*Source*: US Geological Survey.

Table 2  Energy equivalents of earthquakes compared with the Richter scale

| Earthquake magnitude | TNT equivalent |
|---|---|
| 1.0 | 0.17 kg |
| 1.5 | 0.9 kg |
| 2.0 | 5.9 kg |
| 2.5 | 28 kg |
| 3.0 | 179 kg |
| 3.5 | 450 kg |
| 4.0 | 5.5 t |
| 4.5 | 29 t |
| 5.0 | 181 t |
| 5.3 | 455 t |
| 5.5 | 910 t |
| 6.0 | $5.7 \times 10^3$ t |
| 6.3 | $14.4 \times 10^3$ t |
| 6.5 | $28.7 \times 10^3$ t |
| 7.0 | $181 \times 10^3$ t |
| 7.1 | $228 \times 10^3$ t |
| 7.5 | $910 \times 10^3$ t |
| 7.7 | $1811 \times 10^3$ t |
| 8.0 | $5706 \times 10^3$ t |
| 8.2 | $11,421 \times 10^3$ t |
| 8.5 | $28,711 \times 10^3$ t |
| 9.0 | $181,999 \times 10^3$ t |

**easterly wave** A shallow disturbance in the trade winds, which moves westwards causing convergence and associated storms.

**eccentricity** Literally non-concentric, it is widely used to describe an orbit that does not have its axis located centrally, i.e. the moving body does not always remain equidistant from the axial point. The earth has an eccentric orbit around the sun, and since the degree of eccentricity varies through time, this is one component of earth-orbital theories of the causes of CLIMATE CHANGE. DSGT

**echo dune** A TOPOGRAPHIC DUNE that is detached from the obstacle that induces its formation, due to the development of a reverse flow eddy that sweeps the area adjacent to the obstacle clear of sand. The reverse flow eddy develops when the upwind hill slope is steeper than 50°. An echo dune may form either upwind or downwind of the topographic barrier. DSGT

**ecliptic** The plane within which the earth orbits about the sun. The path the sun appears to take as it moves across the sky.

**ecological energetics** The study of energy fixation, transformation and movement within ecological systems. The three most important types of energy found are solar energy, chemically stored energy, and heat energy. Short-wave energy from the sun, in the visible light spectrum of from 0.36 to 0.76 $\mu$m, is fixed within the cells of green plants, blue-green and other algae, and phytoplankton, by transformation, through the process of PHOTOSYNTHESIS, into chemical energy, following which it is passed down either the grazing or the decomposer FOOD CHAIN. The long-wave heat energy associated with plant and animal RESPIRATION may leave ecological systems at any time. Heat is not directly available for reuse as an energy source by organisms. Both plant and animal communities at high altitudes (above 2000 m) may also receive small but significant amounts of ultraviolet radiation, with wavelengths of less than 0.36 $\mu$m, which is potentially damaging to cell tissue, and against which chemical and physical defences are required.

To avoid confusion, and since all forms of ecologically useful energy can be converted into heat, the calorie or kilocalorie is used as the standard unit of measurement in ecological energetics. One calorie is equivalent to the amount of heat required to raise the temperature of 1 g of water by 1 °C; and 1 kilocalorie (kcal) = $10^3$ calories. The joule (J), a measure of

mechanical or work energy, is now increasingly utilized: the amount of work needed to raise 1 g weight against gravity to a height of 1 cm = 981 erg; $10^7$ erg = 1 J; 4.2 J = 1 cal. The conversion factor between mechanical and heat energy was not discovered until the mid-nineteenth century.

Studies in ecological energetics are inevitably complex, not least because of the large numbers of different organisms found within most biological systems, the uncertainty as to the precise roles of some of these in respect of energy transfer, and the often complicated technology required to unravel the intricacies of energy flow, and their changing patterns in time and space. Even so, some understanding of the essential energy controls of individual land-based organisms had been gained from A.L. Lavoisier's studies in France on respiration, as early as 1777. And from the 1920s, investigations into the energy flow of biologically simple lake systems were under way, particularly in North America.

But the major theoretical step forward was provided by R.L. Lindemann in 1942. It was he who first formalized the concept of TROPHIC LEVELS, and the energy relationships between them. He suggested that the standing crop (BIOMASS) of each trophic level might be measured, and the result then translated into energy equivalents. He hypothesized that the mean energy flow between different levels could be represented by simple equations, which took into account the ultimate controls of the first and second laws of thermodynamics. If the energy content of any trophic level is given as $\Lambda$ along with a subscript to indicate the food-chain role of the organisms within it ($\Lambda_1$ = producers; $\Lambda_2$ = herbivores; $\Lambda_3$ = carnivores, etc.), and the passage of energy between any two levels is designated by $\lambda$, then that which is transferred from $\Lambda_n$ to $\Lambda_{n+1}$ will be $\lambda_{n+1}$. Further, the heat loss in respiration may be termed $R_1$, $R_2$, etc, according to the particular trophic level under review. The heat loss $R$, coupled with the quantity of energy in transfer between any two trophic levels may jointly be described as $\lambda_n$, and is the amount which, at the lower trophic level, does not go into biomass production. From this, the rate of energy exchange in any trophic level may be expressed as:

$$\frac{\Delta\Lambda_n}{\Delta_t} = \lambda_n + \lambda_n',$$

or in other words, the rate at which energy is taken up by that level ($\lambda_n$) to form new biomass, minus the rate at which energy is lost from it ($\lambda_n'$).

159

In many respects Lindemann's theory marked a turning point in the study of ecological energetics, for it provided a framework for all later research. It encouraged Slobodkin (1959) to undertake a series of laboratory experiments on *Daphnia* populations, from which he deduced that the mean transfer of energy from one trophic level to another would be *c*.10 per cent (it is now known that it can vary widely from this: see TROPHIC LEVELS). Others (e.g. Odum 1971) have accumulated a vast amount of data on the patterns of energy transfer in both natural and agricultural systems. Still others (e.g. Kozlovsky 1968) have suggested alternative methods of examining the efficiency of energy transfer between trophic levels, as for example that related to consumption efficiency:

Consumption efficiency =

$$\frac{\text{intake of food at trophic level } n}{\text{net productivity at trophic level } n-1} \times 100$$

This may increase slightly from the lowermost trophic levels, but seems in general to fall between 20 and 25 per cent; and the majority of the remainder goes into the decomposer food chain.

Overall, ecological energetics has brought home the importance of small and microscopic organisms in the patterns of energy transfer everywhere, and the need to maintain the presence of these in balanced systems. A good deal is now known about the fixation of solar energy into plant cells at the primary producer level, but much still remains to be learnt about energy transfer and fixation further down the food chain (see BIOLOGICAL PRODUCTIVITY; NET PRIMARY PRODUCTIVITY). DW

**Reading and References**
Kozlovsky, D.G. 1968: A critical evaluation of the trophic level concept. I. Ecological efficiencies. *Ecology* 49, pp. 48–60. · Lindemann, R.L. 1942: The trophic-dynamic aspect of ecology. *Ecology* 23, pp. 399–418. · Odum, E.P. 1971: *Fundamentals of ecology*. 3rd edn. Philadelphia: W.B. Saunders. · Phillipson, J. 1966: *Ecological energetics*. London: Edward Arnold. · Slobodkin, L.B. 1959: Energetics in *Daphnia pulex* populations. *Ecology* 40, pp. 232–43.

**ecological explosions** Ecological events marked by an enormous increase in the numbers of some kind or kinds of organism. The term was carefully defined by Elton (1958) and was employed to indicate the bursting out from control of populations that were previously held in restraint by other forces. Classic examples of such explosions are found in the epidemics of infectious viruses and bacteria, such as influenza and bubonic plague, and in the rapid spread of the North American grey squirrel (*Sciurus*

*carolinensis*) after its introduction to Britain in the nineteenth century.

Many organisms subject to such population outbursts are serious agricultural pests, such as the desert locust (*Schistocerca gregaria*). In most locust species it is possible to identify two types of distributional area, namely a highly localized *outbreak area*, where the species persists permanently and where swarms form, and an *invasion area*. The causes of the devastating plagues appear to be related to weather conditions, particularly to moisture, operating in and through the process of phase transformation, in which the locusts exhibit polymorphism, changing from solitary forms (*solitaria*) to gregarious or swarming forms (*gregaria*). The causes of many ecological explosions remain far from clear and it is interesting to observe that many species which are rare in their normal habitat experience such bursts of population when spread by man to new areas and environments. PAS

**Reading and Reference**
Drake, J.A., Mooney, H.A., di Castri, F., Groves, R.H., Kruger, F.J., Rejmánek, M. and Williamson, M. eds 1989: *Biological invasions: a global perspective*. Chichester: Wiley. · Elton, C.S. 1958: *The ecology of invasions by animals and plants*. London: Methuen. · Krebs, C.J. 1978: *Ecology: the experimental analysis of distribution and abundance*. 2nd edn. New York: Harper & Row. Esp. ch. 16.

**ecological transition** A concept (Bennett 1976) which concerns the reduction in the farmer's dependence on his land that often accompanies his incorporation into the cash economy. The economic opportunities of industry or urban life gradually provide viable social alternatives to rural life and individual farmers can afford to be less concerned about the possibility of long-term decline in the productivity of their land. Over-cultivation may result from this reduced ecological sensitivity in rural populations. ASG

**Reference**
Bennett, J.W. 1976: *The ecological transition: cultural anthropology and human adaptation*. Oxford: Pergamon.

**ecology** The scientific study of the interactions between organisms and their environment. Ecology examines the relationships between organisms belonging to both the same and different taxonomic groups, and between those organisms and their physical environmental circumstances. The word was first used by Ernst Haeckel in 1869, but many of its concepts are much more recent. The ECOSYSTEM concept is central to an understanding of the nature of ecological relationships and dates from the 1930s syntheses of Arthur Tansley. Still other associated concepts, such as those developed in relation to population ecology, came much later, so that ecology can be considered as a relatively new scientific discipline. Essentially, according to Begon *et al.* (1996), ecology operates at three levels: the organism, the population and the community. At the organism and population level (autecology), ecology is concerned with the way in which individuals are influenced by the physical environment and the presence or absence of particular species, while at the community level (synecology), ecology is concerned with the composition of assemblages of different species. A number of concepts integral to ecology have emerged due to the focus on the ecosystem as the fundamental unit of study, in particular the ideas of NICHE, BIOLOGICAL PRODUCTIVITY, SUCCESSION and CLIMAX VEGETATION. Ecology has had a profound influence on the understanding of relationships between humans and their environment. The ongoing search for a more complete understanding of the role of humans in the BIOSPHERE is really a function of intellectual developments within the science of ecology.

MEM

**Reading and Reference**
Begon, M., Harper, J.L. and Townsend, C.R. 1996: *Ecology: individuals, populations and communities*. 3rd edn. Oxford: Blackwell Science. · Bush, M.B. 1997: *Ecology of a changing planet*. New Jersey: Prentice-Hall.

**ecosphere** A synonym for biosphere, but also used as a definition of the zone between Venus and Mars where life as we know it may be possible.

**ecosystem** The North American ecologist E.P. Odum defines an ecosystem as 'any unit that includes all of the organisms in a given area interacting with the physical environment so that a flow of energy leads to . . . exchange of materials between living and non-living parts of the system'. This amplifies the earlier definition given by A.G. Tansley, who coined the term in 1935, and confirms the concept of the ecosystem in an aggregative hierarchy (see Tansley 1946). Individuals aggregate into populations, populations come together in communities, and a community plus its physical environment comprise an ecosystem. In many ways the concept is independent of scale, for the definition is valid for a drop of water with a few micro-organisms in it or for the whole of planet earth, but in usual practice the term is used for units below the scale of the major world BIOMES. Although an ecosystem may be characterized in synoptic terms, i.e. by an inventory of its components, both biotic and physical, the essential features of

the term are (*a*) that it implies a functional and dynamic relation between the components, going beyond a frozen mosaic of species distribution, and (*b*) that it is holistic, implying that the whole possesses emergent qualities which are not predictable from our knowledge of the constituent parts. The study of functional relationships in ecosystems has usually concentrated on phenomena which can be accurately measured and which are common to both biotic and abiotic parts of the system: energy, water, and mineral nutrients are frequent examples. Energy flow through the various TROPHIC LEVELS of a system and its dissipation into heat can be used to see how the system has in its evolution partitioned the energy, and how efficiently it is passed from level to level. Studies of nutrients have often revealed mechanisms for keeping them within the ecosystem: under natural conditions relatively little of the nutrient capital of the system is lost in run-off or animal migration. In arid areas, the use of water by the system may be similarly conserved by a variety of adjustments within the ecosystem, as well as the physiological responses of individual plants and animals. The temporal dimensions of the system are also amenable to study; for example, population numbers through time are often collected. For most species, each ecosystem has a CARRYING CAPACITY, an optimum level for a particular population, which may be a simple number or subject to fluctuations of various kinds. Again, the changes in species composition and physiognomy of an ecosystem through time may be studied, as in the SUCCESSION from bare ground left by a glacier through various types of vegetation to a stable, self-reproducing forest. When succession has apparently terminated at an ecosystem type which sustains itself and gives way, under natural conditions, to no other then this is said to be a mature or climax ecosystem. The applied side of the concept is evident in the idea of BIOLOGICAL PRODUCTIVITY, which is the rate of organic matter production per unit area per unit time, a rate which can be used to compare natural ecosystems with those affected by human activity or indeed totally human-made. The concept of STABILITY is important in the human–biophysical interface because it relates to the resilience of an ecosystem to human-induced perturbation. If we perform a particular act of environmental manipulation, will an ecosystem recover its former state (given the cessation of the impact) or will it perhaps break down irreversibly? The concept itself may also apply to ecosystems with a large human-directed component, such as agriculture (the term agro-ecosystem is sometimes used), pastoralism, fisheries, forestry and even cities themselves, as in the work on Hong Kong by K. Newcombe and colleagues (Boyden *et al.* 1981), in which the urban area is seen as a functional ecosystem with inputs and outputs of energy and matter.

IGS

**Reading and References**
Boyden, S., Miller, S., Newcombe, K. and O'Neill, B. 1981: *The ecology of a city and its people: the case of Hong Kong.* Canberra: ANU Press. · Odum, E.P. 1969: The strategy of ecosystem development. *Science* 164, pp. 262–70. · Putman, R.J. and Wratten, S.D. 1984: *Principles of ecology.* London: Croom Helm. · Tansley, A.G. 1946: *Introduction to plant ecology.* London: Allen & Unwin.

**ecotone** The transition on the ground between two plant communities. It may be a broad zone and reflect a gradual blending of two communities or it may be approximated by a sharp boundary line. It may coincide with changes in physical environmental conditions or be dependent on plant interactions, especially COMPETITION, which can produce sharp community boundaries even where environmental gradients are gentle. It is also used to denote a mosaic or interdigitating zone between two more homogeneous vegetation units. They have special significance for mobile animals through edge effects (such as the availability of more than one set of HABITATS within a short distance). JAM

**Reading**
Allee, W.C., Emerson, A.E., Park, O., Park, T. and Schmidt, K.P. 1951: *Principles of animal ecology.* Philadelphia and London: W.B. Saunders. · Daubenmire, R. 1968: *Plant communities: a textbook of plant synecology.* New York and London: Harper & Row.

**ecotope** The physical environment of a biotic community (BIOCOENOSIS). It includes those aspects of the physical environment that are influences on or are influenced by a biocoenosis. Together with its biocoenosis, the ecotope forms an integral part of a BIOGEOCOENOSIS. There are two major component parts of the ecotope: the effective atmospheric environment (climatope) and the soil (edaphotope). (See also BIOTOPE; HABITAT.) JAM

**ecotoxicology** 'The science which includes all studies carried out with the intention of providing information to further our understanding of the effects that chemicals and radiations (that become bio-available as a direct or indirect result of man's activities) exert on organisms in their natural habitats' (Depledge 1990, p. 251).

ASG

**Reading and Reference**
Depledge, M.H. 1990: New approaches in ecotoxicology: can individual physiological variability be used as a tool to investigate pollution effects? *Ambio* 19, pp. 251–2. ·

Levine, S.A., Harwell, M.A., Kelly, J.R. and Kimball, K.D. 1988: *Ecotoxicology: problems and approaches.* Berlin: Springer Verlag.

**ecotype** Coined by Turesson (1922) to describe populations of organisms within a single species that exhibit genetically produced differences in morphology or physiology which have adapted to a particular habitat, but can interbreed with other ecotypes (ecospecies) of the same species without loss of fertility. Well-known examples come from plants that, because of their low mobility, exhibit evolutionary isolation of subpopulations on small geographic scales. In many cases, these differences amongst habitats result from developmental responses, the phenotypes of populations are fixed but vary among individuals from place to place, which may limit the exchange of genes. Such region and habitat differences in adaptations broaden the ecological tolerance ranges of many species by dividing them into smaller sub-populations, each differently adapted to consistent local environmental conditions. AP

**Reading**
Krebs, C. 1985: *Ecology: the experimental analysis of distribution and abundance.* 3rd edn. New York: Harper and Row. · Ricklefs, R.E. 1990: *Ecology.* 3rd edn. San Francisco: W.H. Freeman. · Turesson, G. 1922: The genotypical response of the plant species to the habitat. *Hereditas* 6, pp. 147–236.

**ecozone** Large-scale ecological unit of classification with broadly consistent landscape and vegetation characteristics and where organisms and their physical environments endure as a system. The scale is such that an ecozone defines an area of at least $100,000 \, km^2$ and which is, therefore, mappable at a scale of 1:3,000,000. Diagnostic features include landforms, soils, water features, vegetation and climate; in essence an ecozone is a large-scale ECOSYSTEM. The term is somewhat loosely and variously applied as the approximate equivalent to BIOME, zonobiome or ecoregion. Schulze (1995) recognizes nine terrestrial ecozones on the global scale as follows: polar and sub-polar zone, boreal zone, humid mid-latitude zone, arid mid-latitude zone, tropical and sub-tropical arid zone, mediterranean zone, seasonal tropical zone, humid sub-tropical zone and humid tropical zone. The correspondence of this scheme with numerous biome-scale global ecological treatments is striking. Some countries, for example Canada and the United States, use the ecozone concept as a mapping unit in the development of large-scale land use and ecological planning frameworks. MEM

**Reading and Reference**
Bailey, R.G. 1996: *Ecosystem geography*. New York: Springer. · Schulze, J. 1995: *The ecozones of the world: the ecological divisions of the geosphere*. Hamburg: Springer.

**edaphic** A term referring to environmental conditions that are determined by soil characteristics. Edaphic factors include the physical, chemical and biological properties of soils such as pH, particle size distribution and organic matter content. Plants and animals may be influenced by soil characteristics, although in many instances these edaphic factors interact with other aspects of the HABITAT.

Areas underlain by serpentine rock provide a well-defined example of the influence of edaphic factors on plant distribution. Serpentine soils are low in major nutrients but contain very high levels of chromium, nickel and magnesium. These sites are occupied by distinctive plant communities adapted to these unusual soil conditions.                                             ARH

**Reading**
Watts, D. 1971: *Principles of biogeography*. New York and London: McGraw-Hill. Ch. 4, pp. 175–84.

**edaphology** The science that deals with the influence of soils on living things, particularly plants, including human use of land for plant growth.

**eddies** In any model or data set we tend to see a slowly varying signal (the 'drift' or the 'climate') plus resolved fluctuations (eddies) and unresolved fluctuations (TURBULENCE and perhaps noise). In a useful record or model these signals and fluctuations define distinct physical processes. When visualizing the nature of the eddy flow it is desirable to subtract the mean flow by plotting the flow relative to the eddy. (Imagine trying to describe the mechanism of a motor-car engine relative to coordinates fixed in a pedestrian.) When this is done there are usually some closed STREAMLINES which are taken by some authors to define an eddy. There is also a tendency to notice eddies only in flow plotted relative to the ground. This makes the definition more pictorial and less mechanistic. Many eddies are noticed because they are the finite-amplitude form of mathematically well-defined solutions to a linearized equation, e.g. Kelvin–Helmholtz. A popular class of eddies is found to the lee of, or propagating downwind of, bluff bodies like steep mountains or islands, such as Jan Mayen or Madeira, where accompanying cloud features can be seen on satellite pictures.       JSAG

**Reading**
Atkinson, B.W. 1981a: *Dynamical meteorology; an introductory selection*. London and New York: Methuen. · —1981b: *Meso-scale atmospheric circulations*. London and New York: Academic Press.

**eddy diffusivity** (eddy viscosity) A conceptual device designed to represent the mixing effect of eddies. By analogy with the kinetic theory of gases the flux of a quantity is supposed to be proportional to the mean gradient of the quantity, the distance over which the air moves and the speed with which it does so. Molecular mean free path of $10^{-7}$ m multiplied by molecular speed of $300\,\mathrm{m\,s^{-1}}$ gives the kinematic molecular diffusivity of $3 \times 10^{-5}\mathrm{m^2\ s^{-1}}$. Eddy radius of $10\,\mathrm{m}$ multiplied by a fluid velocity of $1\,\mathrm{m\,s^{-1}}$ gives kinematic eddy diffusivity of $10\ \mathrm{m^2\ s^{-1}}$.       JSAG

**Reading**
McIntosh, D.H. and Thom, A.S. 1969: *Essentials of meteorology*. London: Wykeham Publications.

**edge waves** Gravity or infragravity WAVES with crests oriented parallel to the shoreline, generated by reflection of incident waves and subsequent REFRACTION of the reflected wave. Their amplitude is a maximum at the water line. The wave length, $\lambda$, is controlled by its period, $T_e$, modal number, $n$, and beach slope, $\beta$:

$$\lambda = \frac{g}{2\pi} T_e^{2}(2n + 1)\tan\beta$$

where $g$ is gravitational acceleration. $T_e$ may be synchronous or subharmonic with incident wave period. The mode describes the offshore structure of the edge wave relative to the mean water level. Standing edge waves may be the mechanism for the formation of rhythmic BEACH topography including beach CUSPS, rip current channels, and nearshore BARS.       DJS

**Reading**
Holman, R. 1983: Edge waves and the configuration of the shoreline. In P.D. Komar ed., *Handbook of coastal processes and erosion*. Boca Raton, Fla.: CRC Press. Pp. 21–33.

**EDM** See ELECTROMAGNETIC DISTANCE MEASUREMENT.

**effective precipitation** Actual PRECIPITATION minus the amount lost back to the atmosphere by EVAPORATION or EVAPOTRANSPIRATION. It is therefore the amount that enters a soil, plus run-off.

**effective rainfall** (ER) The proportion of the total PRECIPITATION that is available for a specific purpose. This definition can be refined only when the scale and object of a particular study are known. An agroclimatologist (see AGROMETEOROLOGY), for instance, calculating the availability of water for plant growth, may simply assume that it equals the amount reaching the soil surface and that intercepted (see INTERCEPTION) by vegetation,

$$ER = S + ITC$$

or they may take into account RUN-OFF and drainage losses, i.e.

$$ER = S + ITC - DRA - RO.$$

A hydrologist examining discharge may consider it necessary to include run-off and GROUNDWATER contributions, i.e.:

$$ER = RO + GWflow$$

whereas a hydrogeologist may be interested only in the amounts of water remaining in aquifers, i.e.:

$$ER = GWstor \qquad \text{CTA}$$

**effective stress** A concept related to total stress in the MOHR–COULOMB EQUATION, which takes the effects of pore water pressure into account when the strength of a soil is calculated. The pressure exerted by water in soil pore spaces acts upon the grains, tending to force them apart when saturated (positive pore water pressure) or pull them together when the pore spaces have only a small amount of water in them (negative pore water pressure). The normal stress ($\sigma$) is modified by the subtraction of the pore water pressure ($u$) so that $\sigma' = (\sigma - u)$, where $\sigma'$ is the effective normal stress. The Mohr–Coulomb equation or failure criterion then becomes:

$$s = c' + \sigma'\tan\phi$$

where the primes denote that effective stresses have been considered. Positive values of $u$ thus tend to decrease the soil strength but negative values (suction) increase it. WBW

**Reading**
Mitchell, J.K. 1976: *Fundamentals of soil behaviour.* New York: Wiley. · Smith, G.N. 1974: *Elements of soil mechanics for civil and mining engineers.* London: Crosby Lockwood Staples. · Statham, I. 1977: *Earth surface sediment transport.* Oxford: Clarendon Press. · Whalley, W.B. 1976: *Properties of materials and geomorphological explanation.* Oxford: Oxford University Press.

**effluent** The polluted water or waste discharged from industrial plants.

**effluent stream** A stream which flows from a lake or the small distributary of a river.

**egre** A tidal bore.

**Eh** See REDOX POTENTIAL.

**Ekman layer** The transition layer in the atmosphere above the surface ($c.10$ m deep) layer where the friction of the earth's surface causes winds to blow across the isobars at an angle; and below the geostrophic wind level (about 1 km above the surface) where there is no friction and the pressure gradient force is balanced by the CORIOLIS FORCE, causing winds to blow parallel to the isobars. In the Ekman layer, eddy viscosity and density are assumed to be constant. This layer was originally theorized by V.W. Ekman in 1902 to describe wind-driven ocean currents. It is also known as the Ekman spiral due to the spiralling directional shift of the winds with altitude clockwise (counter-clockwise) in the northern (southern) hemisphere.

In the ocean, the Ekman spiral describes the currents generated by surface winds. At the surface the frictional effect is maximum and the currents move at about 45° from the wind's direction. As depth increases, current speed decreases and direction becomes increasingly opposed to the surface wind direction. The depth at which the current direction and surface wind direction are 180° apart is called the depth of frictional influence. NJS

**Ekman spiral** An idealized mathematical description of wind distribution in the atmospheric BOUNDARY LAYER. It is an equiangular spiral which forms the locus of the end points of the wind vectors as a function of height, all having a common origin. The GEOSTROPHIC WIND is its limit point. The assumed conditions under which this description is valid are that EDDY DIFFUSIVITY and density are constant within the layer, and the geostrophic wind is constant and unidirectional.

If the $x$-direction is taken along the geostrophic wind direction, the equations for the components of the wind vectors, $u$ and $v$, in the $x$- and $y$-directions, respectively, at any level $z$ are:

$$u = v_g(1 - e^{-z/\beta}\cos^{z/\beta}); v = v_g e^{-z/\beta}\sin^{z/\beta}$$

where $v_g$ is the geostrophic wind speed and $\beta = (2K_M/f)^{\frac{1}{2}}$ where $K_M$ is the eddy viscosity and $f$ is the CORIOLIS PARAMETER.

Through most of the layer the wind blows across the isobars towards lower pressure at an angle which is maximum at the surface at 45°.

The theory of the spiral was developed by Ekman in 1902 for the variation of wind-driven ocean current with depth below the surface. KJW

**Reading**
Holton, J.R. 1972: *An introduction to dynamic meteorology.*
New York and London: Academic Press.

## elastic rebound theory

The contention that the faulting of rocks results from the sudden release, through movement, of the elastic energy that has accumulated owing to pressure and tension in the earth's crust. Sudden movement dissipates the energy.

## electromagnetic distance measurement

A highly accurate method for determining the distance between two points based on measuring the transit time of electromagnetic waves between an emitting instrument (commonly mounted on a theodolite), a reflector and back again. Most often using infrared radiation over distances of the order of 1 km, instruments based on microwave emission (EDM) can operate over much longer distances.  GFSW

**Reading**
Wilson, J.P. 1983: *Land surveying.* Plymouth: MacDonald & Evans.

## electromagnetic radiation

Energy that is propagated through space or through a material in the form of an interaction between electric and magnetic wave fields. This is the link between the earth's surface and the majority of sensors used in REMOTE SENSING. The three measurements used to describe these waves are: wavelength ($\lambda$) in micrometres ($\mu$m), which is the distance between successive wave peaks; frequency ($\gamma$) in Hertz (Hz), which is the number of wave peaks passing a fixed point in space per unit time; and velocity ($c$) in m s$^{-1}$, which within a given medium is constant at the speed of light. As wavelength has a direct and inverse relationship to frequency, an electromagnetic wave can be characterized either by its wavelength or by its frequency.

Electromagnetic radiation occurs as a continuum of wavelengths and frequencies from short wavelength, high-frequency cosmic waves to long wavelength, low-frequency radio waves.

The wavelengths that are of greatest interest in remote sensing are visible and near infrared radiation in the waveband 0.4–1 $\mu$m, infrared radiation in the waveband 3–14 $\mu$m and microwave radiation in the waveband 5–500 mm.

PJC

**Reading**
Curran, P.J. 1985: *Principles of remote sensing.* Harlow: Longman Scientific and Technical. · Drury, S. 1990: *A guide to remote sensing: interpreting images of the Earth.* Oxford: Oxford University Press. · Monteith, J.C. and Unsworth, M.H. 1990: *Principles of environmental physics.* 2nd edn. London: Edward Arnold.

## electron spin resonance (ESR)

A phenomenon, relating to the presence of a paramagnetic effect in insulating minerals, that is exploited to determine ages for bones, teeth, precipitated carbonates, shells, corals and volcanic materials. The effect is measured by placing a sample within an external magnetic field and exposing it to microwaves. The age produced relates to the time of precipitation of the specific mineral used. Materials may be dated back to millions of years. The ESR signal measured relates to a population of trapped charges within the crystalline lattice of some minerals. Hydroxyapatite is the most common mineral exploited when measurements are made on bones and tooth enamel. The trapped charges come about from the interaction of electrons within the crystal and energy deposited following natural radioactive decay of uranium, thorium and potassium in the environment and a small contribution from cosmic rays. Additionally, uranium incorporated into enamel and bone may also make an internal dose contribution. Displaced electrons are transferred to the conduction band and after a short period of diffusion are lodged in traps between the conduction and valence bands. These traps come about due to ionic defects and impurities within the atomic structure. The trapped electron population grows larger at a rate which is directly related to the level of environmental radiation, and for paramagnetic centres. When a sample containing such populated paramagnetic centres is brought into an external magnetic field and exposed to microwaves the magnetic moment of the centre changes relative to the magnetic field. This occurs only for specific magnetic field strengths and microwave frequencies. At such times the centre is said to undergo resonance and its frequency is characteristic for a given trap type. By convention, ESR spectra are plotted as the first derivative of ESR intensity versus $g$-value (the ratio of the microwave frequency to the magnetic field strength). The suitability of a signal for dating relates to the following criteria:

1 There must have been a zeroing effect or process which deletes all previously accumulated populated traps
2 The ESR signal must grow in proportion with radiation dose
3 The signal must be stable over geological time-scales
4 The number of trapping sites is fixed or changes systematically
5 The ESR signal is not influenced by sample preparation methods, anomalous fading or secondary influences (such as superimposed signals from organic radicals).

As in LUMINESCENCE DATING the age of a sample is calculated by establishing the total amount of radiation damage, the sensitivity of the sample to radiation damage, and the rate of radiation dose. The age equation is

$$\text{Age}(ka) = \frac{\text{Equivalent dose}(Gy)}{\text{Dose rate}(Gy \cdot Ka^{-1})}$$

The palaeodose is frequently also termed the accumulated dose or equivalent dose. Additional radiation dose is applied to samples in the laboratory to ascertain the sample sensitivity to radiation and hence the palaeodose. Unlike luminescence-based techniques the trapped charge responsible for producing the ESR signal is not destroyed during the measurement process. As such, ESR measurements are repeatable and may be measured to a high degree of precision. To simplify the determination of the radiation dose level during antiquity (the dose rate), it is common to remove the outer 2 mm of material from a specimen, in doing so removing the effects of beta irradiation from the surrounding environment. A complication however is the need to model the timing of incorporation of any uranium which is present within the sample. For this reason sites which produce samples with low natural uranium levels are preferred.

ESR, in combination with THERMOLUMINES-CENCE, has played a key role in providing dates for archaeological sites in the Levant which are considered to be the earliest known localities of Homo Sapiens Sapiens. A recent application of the method relates to dating fault movements.

SS

**Reading**
Grun, R. and Stringer, C.B. 1991: Electron spin resonance dating and the evolution of modern humans. *Archaeometry* 31.2, pp. 153–99.

**ellipsoid** Closed surface, symmetrical about three mutually perpendicular axes, in which all plane sections are either ellipses or circles. The name is also given to the solid contained by such a surface. The ellipsoid is used to represent the earth's shape (see GEOID) in geodetic calculations. The ellipsoidal earth is slightly flatter at the poles, the difference between polar and equatorial radii being approximately 21 km. In a geological context, the triaxial ellipsoid may be used to represent various optical properties of crystals or to denote values of stress or strain within materials. MEM

**El Niño** The name translates from Spanish as 'the boy child'. Peruvian fishermen originally used the term – a reference to the Christ child – to describe the appearance, around Christmas, of a warm ocean current off the equatorial South American coast of Peru, Ecuador and Chile. This is normally a region of upwelling of cold, nutrient-rich water brought up from the lower depths of the ocean to replace surface water driven westward by the TRADE WINDS. During typical episodes, from late December to March, the upwelling usually weakens and is replaced to some extent by warm water moving in from the west and north.

El Niño events occur irregularly at intervals of 2–7 years, although the average has been, until recently, about once every 3–4 years and lasting 12–18 months. El Niño events often result in flooding in California and parts of the midwestern United States, while the southern half of the United States experiences cooler than normal winters. Winters are generally warmer than normal in the northern half of the United States. During El Niño years, there are fewer hurricanes in the Atlantic.

Changes to the atmosphere and ocean circulation during El Niño events include: (1) warmer than normal ocean temperatures across the central and eastern tropical Pacific Ocean; (2) increased convection or cloudiness in the central tropical Pacific Ocean – the focus of convection migrates from the Australian/Indonesian region eastward towards the central tropical Pacific Ocean; (3) weaker than normal easterly trade winds; and (4) low (negative) values of the Southern Oscillation Index (SOI).

Until the 1997–8 El Niño, the 1982–3 episode had been the strongest and most devastating of the twentieth century. During that period, the trade winds not only collapsed; they reversed and caused weather-related disasters on nearly every continent. Australia, Africa and Indonesia suffered droughts while Peru received the heaviest rainfall on record. The 1982–3 El Niño was blamed for thousands of deaths and billions of dollars in damage to property and livelihoods worldwide. The damage from the 1997–8 event may far exceed that amount.

MLH

**Reading**
Enfield, D.B. 1989: El Niño, past and present. *Review of geophysics* 27, pp. 159–87. · Pickard, G.L. 1979: *Descriptive physical oceanography*. New York: Pergamon.

**eluviation** The downward transportation by percolating water of certain soil materials from upper layers of the soil towards lower layers, where they may be set down. Both materials in solution and fine solid particles may undergo eluviation. (The corresponding term for the arrival of such materials and their accumulation in deeper layers is ILLUVIATION.) Also used to refer to the lateral transport of soil materials that takes place in seepage water moving down the topographic slope (*lateral eluviation*). Evidence includes loss of fine clays and soluble materials from the upper soil HORIZONS (though this can be explained in other ways) and in the presence of accumulations of these materials in the subsoil. Depositional clay skins (CUTANS) surrounding larger grains, or lining the walls of void spaces, and more abundant in the lower soil than in the upper parts, provide one line of evidence for the eluviation of materials.　　DLD

**Reading**
Brady, N.C. and Weil, R.R. 1996: *The nature and properties of soils*. 11th edn. New Jersey: Prentice-Hall.

**eluvium**　The material which is produced through the rotting and weathering of rock in one place. *In situ* weathered bedrock.

**Emerson test**　A test that determines the stability of soil aggregates when in contact with water. Small air-dry soil aggregates, 3–5 mm in diameter, are placed in beakers of distilled water, and their condition observed for up to 12 hours (Emerson 1967). The sample may slake (slowly or rapidly disintegrate into separate grains that lie on the bottom of the beaker, but leaving the water clear), partially disperse (indicated by a turbid cloud of fine particles surrounding the test aggregate) or remain stable. The rate and extent of these processes are scored, and aggregates that remain stable can be subjected to mechanical stirring to inspect for dispersion such as might result from raindrop impact or soil tillage.　　DLD

**Reference**
Emerson, W.W. 1967: A classification of soil aggregates based on their coherence in water. *Australian journal of soil research* 5, pp. 47–57.

**endangered species**　One of a group of terms used to describe the status of wildlife. The following definitions are those employed by the Species Survival Commission of the World Conservation Union (IUCN) and are accepted for use by international bodies such as the Convention on Trade in Endangered Species of Flora and Fauna.

*extinct*: species not definitely located in the wild during the past fifty years.
*endangered*: species in danger of extinction, whose survival is unlikely if the causal factors continue operating.
*vulnerable*: species believed likely to move into the 'endangered' category in the near future if the causal factors continue operating.
*rare*: species with small world populations that are not at present 'endangered' or 'vulnerable'.

*indeterminate*: species known to be 'endangered', 'vulnerable' or 'rare', but not enough information is available to determine which category is appropriate.
*insufficiency known*: species that are suspected of belonging to one of the above categories but are not definitely known to be, due to lack of information.
*threatened*: a general term used to denote species which are in any of the above categories.
　　ASG

**endemism**　The confinement of plant and animal distributions to one particular continent, country or natural region. Thus the white spruce (*Picea glauca*) is endemic to North America, the coast redwood (*Sequoia sempervirens*) to coastal California and Southern Oregon, and the genus *Dasynotus* to a few square kilometres of Idaho. Organisms confined to one small island, one mountain range or just a few restricted localities are termed 'local' or 'narrow' endemics, whereas those with more substantial distributions are called 'broad' endemics.

The number of endemics in the northern hemisphere is lower than that in the southern hemisphere, although their count is reduced in both hemispheres where the lands were occupied by the cap of the continental glacier during the Pleistocene. Endemism is most marked on islands, such as Darwin's famous Galápagos Islands, particularly in the warmer regions of the world. No less than 90 per cent of the native plants of the Hawaiian Islands are endemic to the island group. Endemism is also marked on isolated mountain tops, especially in the tropics. There is usually a close relationship between the number and type of endemics and the geological age of the habitats they occupy. Naturally, the study of endemism is one of the main ways of characterizing the different faunal and floristic regions of the world. (See also ISLAND BIOGEOGRAPHY; REFUGIA; VICARIANCE BIOGEOGRAPHY.)
　　PAS

**Reading**
Daubenmire, R.F. 1978: *Plant geography, with special reference to North America*. New York and London: Academic Press. · Richardson, I.B.K. 1978: Endemic taxa and the taxonomist. In H.E. Street ed., *Essays in plant taxonomy*. London and New York: Academic Press. Pp. 245–62. · Stott, P. 1981: *Historical plant geography*. London: Allen & Unwin. · Williamson, M. 1981: *Island populations*. Oxford: Oxford University Press.

**endogenetic**　Pertaining to the forces of tectonic uplift and disruption originating within the earth and to the landforms produced by such processes (see EXOGENETIC).

**endoreic** Drainage systems which do not reach the oceans, terminating in inland locations. They are therefore a feature of some DRYLANDS, where, due to insufficient flow from wetter (allogenic) headwater areas and the excess of EVAPOTRANSPIRATION over local desert (endogenic) precipitation, rivers may terminate before draining to the coast. De Martonne and Aufrere (1928) classified the world's drylands on the basis of endoreism and areicism (no drainage at all).

The Okavango in central southern Africa is an example of such a river, terminating in an extensive inland delta or fan. Tectonic and structural factors may also contribute to endoreism. The Okavango is the last remaining major river from a more extensive endoreic system that developed with the division of Gondwanaland and the creation of a tectonic rim (or hingeline) around southern Africa. The endoreic system has now largely been captured by more aggressive ALLOGENIC rivers that have cut back through the hingeline and captured endoreic channels (Thomas and Shaw 1988).                      DSGT

**References**
de Martonne, E. and Aufrere, L. 1928: Map of internal basin drainage. *Geographical review* 17, p. 414. · Thomas, D.S.G. and Shaw, P. 1988. Late Cainozoic drainage evolution in the Zambesi basin: geomorphological evidence from the Kalahari rim. *Journal of African earth sciences* 7, pp. 611–18.

**endrumpf** A peneplain. A landsurface that has been reduced to a flat plain or gently undulating landscape by erosive processes.

**energy flow** May be loosely defined as the energy transformations which occur within the planet earth system. Energy is the capacity to do work and, for present purposes, includes mechanical, chemical, radiant and heat energy. Mechanical energy may be further subdivided into kinetic and potential energy. Kinetic energy or free 'useful' energy is possessed by a body by virtue of motion and is measured by the amount of work required to bring the body to rest. Potential energy is stored and becomes useful only when converted into the free form and can do work. This includes movement, friction, and the expenditure of heat. On earth, energy sources are solar radiation, rotational energy of the solar system and radiogenic heat involving geothermal heat flow. It is no exaggeration to claim that physical geography is the science of energy flow from these sources through the atmospheric, oceanic, aeolian, fluvial (hydrological cycle), glacial, biological, human, tectonic and geothermal systems as described by the laws of thermodynamics and as defined by complex circulation patterns and thermal gradients. The main circulations are atmospheric circulation, heat and moisture balance, photosynthesis and ecological energetics, the hydrological cycle and tectonics. The important gradients are latitudinal, altitudinal, seasonal, daily, heat gradients and across system contrasts (e.g. land–sea).   DB

**Reading**
Bloom, A.L. 1969: *The surface of the earth*. Englewood Cliffs, NJ: Prentice-Hall. · Caine, N. 1976: A uniform measure of sub aerial erosion. *Bulletin of the Geological Society of America* 87, pp. 137–40. · Chapman, D.S. and Mach, A.N. 1975: Global heat flow: a new look. *Earth and planetary science letters* 28, pp. 23–32.

**energy grade line** Water flowing down a channel loses energy because of the work done in overcoming FRICTION, both internally and with the channel perimeter, and in transporting sediment. The rate of energy loss per unit length of channel is measured by the energy gradient, which is the slope of the energy grade line. This line may be plotted above the water surface at a distance equal to the velocity head ($v^2/2g$), and therefore measures the variation of the total energy (potential and kinetic) of the flow (see GRADUALLY VARIED FLOW). In UNIFORM STEADY FLOW the energy grade line, water surface and channel bed are parallel. In gradually varied flow, the three slopes differ but the energy grade line always slopes downwards in the direction of flow.                                    KSR

**englacial** Describes conditions within the body of a glacier and is therefore to be distinguished from the *subglacial* environment which is beneath a glacier, the *supraglacial* environment on the glacier surface and the *proglacial* environment in front of the glacier margin.

**ENSO** An abbreviation for the EL NIÑO-SOUTHERN OSCILLATION phenomenon. This phenomenon is a natural variation in the ocean-atmosphere system that occurs every 3 to 7 years. Though it takes place in the tropical south Pacific, ENSO affects the weather worldwide, and next to the seasons themselves it is the most important variation in the global climate system.

ENSO is a deviation from the normal ocean-atmosphere system in the tropical south Pacific. Under 'normal' conditions, atmospheric pressure is higher in the western Pacific than in the eastern Pacific, and as a result surface flow is from east to west. This easterly flow piles up warm surface water in the western Pacific, where sea surface heights are about a half metre higher than in the east. The easterly flow also causes an upwelling of cold water along the coast of South America, and thus sea surface temperatures there are about eight degrees Celsius cooler than in the west.

Dramatic changes occur during an ENSO event, which typically evolves over the course of an eighteen-month period. Atmospheric 'mass' seesaws from the eastern to the western Pacific, and the easterly trade winds weaken and recede eastward. This recession reduces the coastal upwelling and allows the warm waters amassed in the western Pacific to 'slosh' back toward the east. Precipitation follows the warm water eastward, and the typically arid regions of coastal South America receive heavy rainfall. Indonesia and Australia, in contrast, experience drought. The eastward displacement of the warmest water also has TELECONNECTIONS outside the tropical Pacific; drought in north-eastern Brazil and south-eastern Africa, for example, usually accompany ENSO, while the western US experiences wet winters.          RSW

**entrainment**  The process by which surface sediment is incorporated into a fluid flow (such as air or water) as part of the operation of ERO-SION. Sediment is entrained into a flow when forces acting to move a stationary particle overcome the forces resisting movement.

In water and air sediment is subjected to three major forces of erosion: lift, surface drag and form drag. Lift is a result of the fluid flowing directly over the particle forming a region of low pressure (the Bernoulli effect), in contrast to the relatively high pressure beneath the particle. The particle then gets 'sucked' into the flow. Surface drag is the SHEAR STRESS on the particle provided by the flow velocity and the form drag relates to the pressure differences upstream and downstream of the particle. When these movement forces overcome particle cohesion, packing and weight, the particle tends to shake in position and then lift off, spinning into the flow. The velocity required to initiate this entrainment is termed the CRITICAL EROSION VELOCITY and is primarily a function of mean particle size.

Once entrained into the fluid, sediment may move in a number of ways. Coarse particles tend to slide or roll along the bed whilst smaller particles may undergo SALTATION. Very fine particles (silt or clay) tend to be transported by SUSPENSION where the TURBULENT FLOW STRUC-TURES within the fluid carry the particles without impacting the surface.

In GLACIAL environments sediment may become entrained into moving ice by falling onto the ice surface from above, or by joint block removal which involves the 'plucking' of joint-separated rock by glacial ice.          GFSW

**Reading**
Thorne, C.R., Bathurst, J.C. and Hey, R.D. eds. 1987: *Sediment transport in gravel-bed rivers*. Chichester: Wiley. · Wiggs, G.F.S. 1997. Sediment mobilisation by the wind. In D.S.G. Thomas ed., *Arid zone geomorphology*, 2nd edn. Chichester: Wiley. pp. 351–72.

**entropy**  A concept in thermodynamics which describes the quantity of heat supplied at a given temperature. The entropy of a unit mass of substance remains constant in the process of adiabatic expansion, i.e. when no additional heat is supplied to the mass. The term has been transferred by analogy to other fields (notably information theory) and introduced into geomorphology by Chorley (1962) and Leopold and Langbein (1963) as a measure of energy unavailable to perform work (high entropy equals greater unavailability of energy). This analogue use has been widely criticized as inappropriate.          BAK

**References**
Chorley, R.J. 1962: Geomorphology and general systems theory. *United States Geological Survey professional paper* 500-B. · Leopold, L.B. and Langbein, W.B. 1963: The concept of entropy in landscape evolution. *United States Geological Survey professional paper* 500-A.

**envelope curve**  A curve drawn on an X–Y scatterplot that delimits the part of the graph within which points representing a particular phenomenon occur. When the relationship between the two variables involves a great deal of scatter, techniques like the fitting of a least-squares regression line become inappropriate. It is under these conditions that an envelope curve may prove to be a useful analytical tool. The figure (from Enzel *et al.* 1993) shows an example in which the sizes of floods are plotted in relation to the river catchment area. The envelope curve shown on p. 170 delimits the largest floods occurring for any basin area, and its form suggests that there is an approach toward an upper limit on flood sizes for even the largest basins.

          DLD

**Reference**
Enzel, Y., Ely, L.L., House, P.K., Baker, V.R. and Webb, R.H. 1993: Paleoflood evidence for a natural upper bound to flood magnitudes in the Colorado River basin. *Water resources research* 29, pp. 2287–97.

**environmental assessment**  An attempt objectively to evaluate the quality of the environment in terms of biophysical attributes and/or aesthetic value. It is used to give the natural environment credibility and comparability with socioeconomic data in planning decisions. The concept of environmental assessment has been important since the 1960s with the gradual

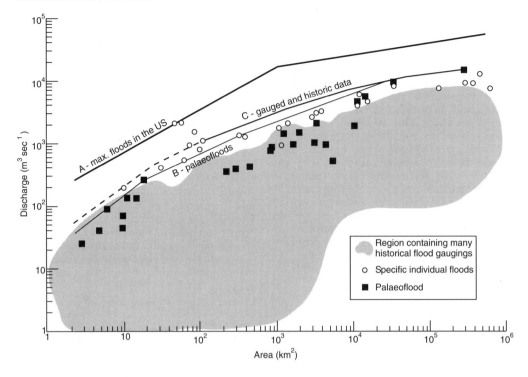

recognition that problems such as loss of habitat and genetic variety, pollution, population growth and an increasing reliance on non-renewable energy and mineral resources cannot be solved by economic growth or technology and that finite limits exist within the environment.                                    SMP

**environmental economics**   The aim to conserve, maintain, use and reuse natural resources so that the quality of life is retained without excessive waste. It recognizes that the major cause of environmental problems in market systems is failure of the incentives generated by these markets to lead towards efficient use of resources.

At the beginning of the twentieth century Alfred Marshall suggested that the basis of orthodox economics – the market – presented an oversimplified idea of reality and introduced the concept of external costs and benefits. These describe instances where one fiscally independent economic unit directly affects another, without intervention of the market. On the cost side the instance of a locomotive igniting an adjacent field is commonly quoted; a good example of a beneficial externality is that of bees pollinating an orchard owner's apples.

The 1960s and 1970s witnessed the rapid development of these ideas owing to an increas-

ing awareness of environmental problems, particularly pollution, and the notion that both free enterprise and Marxist economics encourage the squandering of natural resources. A profound asymmetry had developed in the effectiveness and efficiency of the market system  which works well in stimulating the exploitation, processing and distribution of basic resources, but almost completely fails in the efficient disposal of residuals to common property assets.

The natural environment has clearly been largely ignored in the conventional economic account and commonly the earth is regarded as a bottomless rubbish dump. Boulding (1966, 1970) describes this as the 'cowboy economy', in which success is measured in terms of the amount of material turned over by the factors of production. He compares this to the 'spaceship economy' in which maintenance of existing capital stocks within limits is the criterion for success.

The practice of economics is certainly still the single most important feature governing the relationship between man and the environment. However, environmental economics acknowledges its own limitations for regulating current human resource requirements. Price cannot always be equalled with value; for instance, unspoilt countryside may have intrinsic value but cannot have an accurate price fixed upon it. The interest of the individual in the market is often not the same as the general interest (Hardin 1968), indeed some resources (e.g. clean air) are not within the market so are not subject to the choices normally available. Furthermore, market economies are not well suited to respond to problems which suddenly force themselves upon resource managers or the public (e.g. disease causing a harvest to collapse) nor to cope with the long timelags within which complex technology needs to develop substitutes.

The ultimate goal of environmental economics is to reach the steady state with little or no economic growth in the industrialized nations (Boulding 1966). In so doing it attempts to investigate that which now governs price and supply of materials and, in particular, the role of energy, the economic relations of rich and poor countries and the construction of new measures of human welfare in terms of the whole resource process, rather than just a portion of it, which has for so long been the sole concern of economists.                                               SMP

**Reading and References**

Boulding, K.E. 1966: Ecology and economics. In F. Fraser Darling and J.P. Milton eds, *Future environments of North America*. New York: Natural History Press. · — 1970: The economics of the coming spaceship earth. In H. Jarrett ed., *Environmental quality in a growing economy*. Baltimore: Johns Hopkins University Press. · Cottrell, A. 1978: *Environmental economics*. London: Edward Arnold; New York: Halsted Press. · Hardin, G. 1968: The tragedy of the commons. *Science* 162, pp. 1243–8. · Knese, A.V. 1977: *Economics and the environment*. London and New York: Penguin. · Marshall, A. 1930: *Principles of economics*. 8th edn. London: Macmillan.

**environmental impact**  A net change, either positive or negative, in man's health and well-being and in the stability of the ecosystem on which man's survival depends. The change may result from an accidental or a planned action and can affect the change in balance either directly or indirectly.

Direct impacts are generally premeditated and planned and are commonly felt soon after environmental modification. The effects are often long term but normally reversible and include alterations such as land use changes, various constructional and excavational activities, the direct ecological impact of agricultural practices and the direct effects of weather modification programmes.

In contrast, indirect effects are normally unplanned and are often socially, if not economically, undesirable. Effects are often delayed until well after the original impact and depend upon the sensitivity of the system to change, the existence of threshold conditions, and interaction between different side-effects of the initial impact. Many impacts are long term, cumulative and irreversible, difficult to identify and almost impossible to predict, and include the introduction of DDT and other toxic elements into the environment and the subsequent accumulation of those into food chains over a wide area, triggering long-term and possibly long-range climatic modifications by particulate and gaseous pollution and indirect local climatic effects associated with changed land surface configuration or material composition.

Many impacts are caused by pressures related to the rapid increase in population growth, especially on physical resources – land, food, water, forests, metals – and on the biological environment, whose ability to remove and recycle human waste and provide an important set of functions such as pest control or fish production is being severely strained. In addition there are pressures on society's ability to dispense services – education, medicine and law administration – and in personal values such as privacy, freedom from restrictive regulations and the opportunity to chase a lifestyle.

It is not only population growth that causes these pressures. The consumption of materials and energy per person have grown simultaneously, linked with the type of technology that is being employed to facilitate consumption, and the economic, political and social forces influencing decision-making are contributory factors. The ability of both individuals and governments to react has not kept pace.

The environmental impact may be the final stage of a basic resource process, whereby humankind takes a resource from the environment, uses it in some fashion and then returns it to the environment. Human actions, whether they be legislative proposals, policies, programmes or operational procedures may well set in motion or accelerate environmental effects – which are, for example, dispersal of pollutants, soil erosion or displacement of persons – unless some form of preventative management is initiated. If management at this stage is ineffectual or avoided, however, an environmental impact will probably occur, and any subsequent management may be too late, too expensive or merely palliative. Unfortunately most knowledge is still concerned with what is put into the system rather than with how the environment responds to it, and it is this response that defines the nature and magnitude of the environmental impact.

If, at a later stage, humans complete their action the response may be a lessening of the environmental impact (figure 1) owing to natural

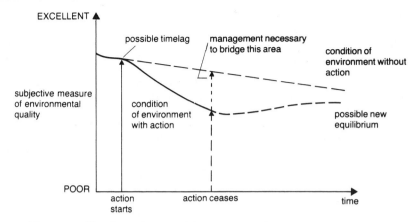

Environmental impact: 1. Changing environmental impact resulting from human activity.

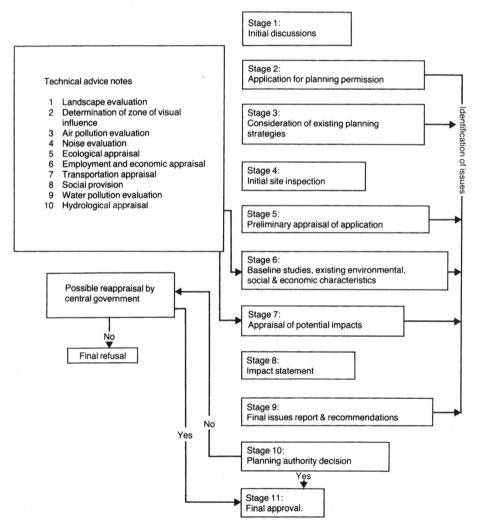

Environmental impact: 2. An approach to project appraisal under existing development control procedures.
*Source*: Clark 1976.

homeostasis in the system. A good example is a colliery spoil heap which gradually revegetates sometime after mining has ceased. It is rare, however, for the new equilibrium even to approach the original environmental quality. There are two main approaches to the assessment of environmental impacts. In the first the problem of resources receives particular attention and some of the more subjective ecological and aesthetic aspects of the environment, although recognized as important, are not included. One attempt (Ehrlich *et al.* 1977) tried to relate the problem in terms of population and consumption:

environmental impact = population × consumption of goods per person × environmental impact per quantity of goods consumed; or
environmental impact = population × affluence × technology.

However most forms of consumption give rise to many forms of environmental impact; changes in technology might reduce some impacts but increase others and different impacts associated with alternative technologies (e.g. oil spills and coal mining) are difficult to compare, with the result that the above measure is almost impossible to quantify.

The second approach considers the probable consequences of human intervention in the natural environment with the goal of minimizing environmental damage, while developing resources. Known as environmental impact assessment (EIA) it was developed by ecologists and economists in the USA and became a legal requirement under S102(2)(C) of the Natural Environmental Policy Act 1969 for all major federal actions affecting the human environment.

In Europe, EIA is used early in existing planning procedure and does not replace it. It identifies adverse effects, suggests alternatives to proposals and considers short-term local environmental use in relation to long-term productivity. It imposes limitations, therefore, on the magnitude and impact of resource processes.

The flowchart (figure 2) outlines a general impact evaluation methodology. The critical part is stage five – preliminary appraisal – as the value of any assessment of impact will depend heavily on the thoroughness with which the appraisal work is undertaken. Both the specific characteristics and the potential interactions between the existing local situation and that in the development proposal are evaluated and are normally identified using an impact matrix, which acts as a framework and checklist.

Most commonly used is the Leopold matrix which lists all possible environmental, social and economic parameters and attempts to evaluate the importance and magnitude of each impact identified. Others include the component interaction matrix, normally used for biological, physical and climatic dependencies, the disruption matrix, which measures environmental disturbance for each alternative proposal and the Sorenson matrix, a three-stage computer analysis which may also suggest corrective action or control mechanisms.

Baseline studies (stage six) are used to provide information for the matrix and both are considered to appraise the potential impacts, and produce an ENVIRONMENTAL IMPACT STATEMENT. SMP

### Reading and References

Clark, B. 1976: Evaluating environmental impacts. In T. O'Riordan and R. Hey eds, *Environmental impact assessment.* Farnborough: Saxon House. · Ehrlich, P.R., Ehrlich, A.H. and Holdren, J.P. 1977: *Ecoscience.* San Francisco: W.H. Freeman. · O'Riordan, T. 1981: *Environmentalism.* 2nd edn. London: Pion. · Park, C.C. 1980: *Ecology and environmental management.* Studies in physical geography 3. Folkestone, Kent and Dawson, Colorado: Westview Press. · Simmons, I.G. 1981: *The ecology of natural resources.* 2nd edn. London: Edward Arnold.

**environmental impact statement** (EIS) The summary of all the information gathered on each potential ENVIRONMENTAL IMPACT that might be realized for a given development proposal. It is an integral part of the procedure of environmental impact analysis.

The EIS discusses as succinctly as possible (1) a brief description of a proposed action; (2) the likely impact of the proposed action on the environment; (3) any predicted adverse or beneficial effects as a result of the proposal; (4) whether the impacts are likely to be long term or short term; (5) whether the impacts will be reversible or irreversible; (6) the range of direct and indirect impacts associated with the proposed action; and (7) whether the impacts are likely to be of local and/or national strategic significance. In addition, the prospects for the area under consideration if the development does not take place are outlined so that decision-makers can compare the potential effects of approving the application with the implications of the 'no change' alternative.

Most EISs do not need non-environmental questions (such as economic impact) to be included, although many agencies voluntarily include such information. The aim of the EIS is to identify and develop methods and procedures and use empirical information to ensure that unquantified environmental amenities and values are given appropriate consideration in decision-making along with economic and technical considerations. An EIS may, therefore, include rankings or hierarchies to differentiate the degree of importance or magnitude of impacts, although this procedure can lead to a false sense of objectivity. SMP

**Reading**
O'Riordan, T. and Hey, R. eds 1976: *Environmental impact assessment*. Farnborough: Saxon House. · Wathern, P. ed. 1990: *Environmental impact assessment: theory and practice*. London: Routledge.

**environmental issue** A concern that has arisen as a result of the human impact on the natural environment and/or the ways in which the natural environment affects human society. Such issues include DEFORESTATION, DESERTIFICATION, SOIL EROSION, POLLUTION, CLIMATE CHANGE, the OZONE HOLE and ACID RAIN. Research into, and teaching of, environmental issues is not the domain of the physical geographer alone since the word 'issue' inherently implies that human values play a key role in defining the subject matter (Middleton 1999), and consequently some issues, for example the CARBON DIOXIDE PROBLEM, are highly controversial. A physical geographer may approach these applied issues from a pure stance, bringing a knowledge of physical environmental processes and their rates of change, while a human geographer's approach may involve perceptions, cultural contexts and political viewpoints. Both approaches are important. A full understanding of desertification, for example, requires both an appreciation of the physical processes operating in the dryland environment and a grasp of the social, economic and political factors which affect the everyday lives of desert inhabitants (Thomas and Middleton 1994). In the study of soil erosion it has become clear that we are well aware of how soil erosion works as a set of processes, and physical scientists have developed many methods for combating the soil erosion problem. Nevertheless, soil erosion problems still plague many world regions, suggesting that the emphasis on the technical is not sufficient. The answers to the question 'why is soil erosion still a problem?' lie in an appreciation of the way local inhabitants view the problem (some may not view it as such at all) and the factors which force or encourage people to abuse their soil resource (Blaikie 1985). Hence, a full appreciation of any environmental issue must consider it as a set of physical processes viewed in a social context.

This understanding of what makes an issue an issue reflects the need for a holistic approach to the subject matter and environmental issues have been advocated as a rallying point for physical geography, to offset the subject's tendency towards fragmentation through the increasing development of specialisms (e.g. Stoddart 1987). According to this view, physical geography should focus on both the way the natural environment works and how this knowledge can be used to improve our understanding of human–environment relations, including the synergy that occurs between the two sets of forces, to concentrate on applied problems. Some see this role of applied environmental management as requiring a broader base for physical geography, involving further integration with the social science aspects of geography as a whole, although a division between such 'environmentalists' and the pure school of 'reductionist' physical geography continues to be apparent (Newson 1992). NJM

**References**
Blaikie, P. 1985: *The political economy of soil erosion*. London: Longman. · Middleton, N.J. 1999: *The global casino: an introduction to environmental issues*. 2nd edn. London: Arnold. · Newson, M. 1992: Twenty years of systematic physical geography: issues for a 'New Environmental Age'. *Progress in physical geography* 16, pp. 209–21. · Stoddart, D.R. 1987: To claim the high ground: geography for the end of the century. *Transactions of the Institute of British Geographers* NS 12, pp. 226–327. · Thomas, D.S.G. and Middleton, N.J. 1994: *Desertification: exploding the myth*. Chichester: Wiley.

**environmental lapse rate** See LAPSE RATE.

**environmental magnetism** Thompson and Oldfield (1986) who produced an accessible synthesis of magnetic theory, measuring techniques and the applications of magnetic measurements in a wide range of environmental contexts first coined the term environmental magnetism. Earlier studies of the natural magnetic properties of crustal rocks showed how certain rock minerals enabled the recording of the earth's palaeomagnetic field which eventually led to the confirmation of plate tectonics as the mechanism of continental drift (Vine and Mathews 1963). And the French geophysicist Le Borgne (1955) carried out simple magnetic studies of soil and deduced that strongly magnetic minerals formed as a result of pedogenic processes and fire. These studies subsequently triggered many applications of magnetic measurements in geomorphology, pedology, climate change and pollution studies. The dichotomy of palaeomagnetism and mineral magnetism, though far from being mutually exclusive, still represents the main magnetic framework today. Both divisions exploit the magnetic properties of Fe-bearing minerals, particularly those which exhibit ferrimagnetic (e.g. magnetite) and imperfect antiferromagnetic (e.g. haematite) behaviour. These minerals retain a measurable magnetization after a magnetic field has been removed, termed remanent magnetization.

Natural remanent magnetization (NRM) is induced by the earth's magnetic field as the minerals cool in igneous rocks from molten lava, precipitate from solution as structured crystals or physically align with the field in the surface of lake and marine sediments. The NRM of a sample can be analysed to reconstruct the original inclination (vertical) and declination (horizontal) angle components, and often the intensity of the magnetic field. When viewed over geological timescales, the data provide a magnetostratigraphic record showing periods when the polarity of the earth's magnetic field was either the same as today's (normal) or when it was reversed. Comparison of the palaeomagnetic pole positions between rocks of different ages allows reconstruction of polar wander paths and the direction of continental drift. Over the past few million years, igneous rock records are well dated using K–Ar ages and the record shows three major polarity zones in the last 3.4 million years: normal Brunhes (0–0.78 Ma), reversed Matuyama (0.78–2.60 Ma), and normal Gauss (2.60–3.40 Ma), with a larger number of shorter but nevertheless worldwide polarity intervals. The polarity record provides the means for dating samples, correlating different sediment and rock sequences and examining the mechanisms of the earth's interior. In sediments deposited over the past few thousand years, the polarity remains normal but there are significant fluctuations in the magnetic field caused by secular variation in the position of the geomagnetic pole. Records of secular variation in recent lake sediment records compare well with the laboratory records of declination and inclination kept since the sixteenth century in Europe and USA, and are used as a sediment dating technique. The NRM preserved in fired artefacts such as tiles and hearths is often used to help date archaeological contexts. Recently scientists have even succeeded in matching the NRM of red paint pigments from museum paintings to historical records of secular variation!

In mineral magnetic studies, measurements such as remanent magnetization induced in the laboratory and magnetic susceptibility are used to identify Fe-bearing minerals, the processes of their formation and their origins (see also MAGNETIC SUSCEPTIBILITY). Primary minerals, such as titanomagnetite, are found in igneous rocks and derived sediments, while secondary ferrimagnetic minerals (SFMs) are formed as a result of biogeochemical processes in soils and sediments, fire and combustion of fossil fuels. The formation of SFMs in soil often leads to magnetic enhancement of surface or sub-surface horizons, a phenomenon used in geomorphology to help trace the erosion and transport of surface soil from slopes to river. In some landscapes, the distribution of both primary and secondary minerals enables a variety of sources to be classified and their sediments to be traced through stream networks. Several studies have extended sediment-source tracing to lake and estuarine sediment records to enable reconstruction of the history of erosion and have shown clearly the long-term impact of human activities. In contrast, SFMs formed in thick loess deposits, such as those on the Chinese Loess Plateau, are sensitive to climatic conditions, particularly rainfall, and have been used to reconstruct glacial-interglacial sequences of climate for the region. These long terrestrial records parallel magnetic-based climate records from marine sediments (see Maher and Thompson 1999). In the North Atlantic, the presence of primary minerals in the sediments has been linked to the southward movement of icebergs and the release of minerogenic detritus following their melting. In warmer oceans, the magnetic records respond variously to the glacial-interglacial shifts in wind-blown dusts, biological productivity and the deposition of carbonate, and the presence of magnetite formed by bacteria. A major source of human magnetite is fossil fuel combustion, especially fly ash from coal. Magnetic measurements also help to discriminate between metallurgical dusts, rust particles and vehicle emissions. Measurements of recent lake sediments and peat profiles record the timing and trends of magnetite-pollution from many countries. Magnetic properties of atmospheric dusts, road dusts and river sediments in urban areas help to identify sources of pollution, and in some studies have provided an estimation of pollution by heavy metals and polycyclic aromatic hydrocarbons. During the 1990s the subject of environmental magnetism has expanded greatly. As a result there is now a better appreciation of mineral dissolution, iron sulphide formation, and magnetotactic bacteria, a stronger marriage between theory and measurement, and a more comprehensive use of both room temperature and thermal magnetization and remanent properties.　　　　　JAD

**Reading and References**

Dearing, J.A. 1999: *Environmental magnetic susceptibility*. Kenilworth: Chi Publishing. · Le Borgne, E. 1955: Susceptibilité magnétique anormale du sol superficiel. *Ann. geophys*, 11, pp. 399–419. · Maher, B.A. and Thompson, R. 1999: *Quaternary climates and magnetism* Cambridge: Cambridge University Press. · Thompson, R. and

Oldfield, F. 1986: *Environmental magnetism*. London: George Allen and Unwin. · Vine, F.J. and Mathews, D.H. 1963: Magnetic anomalies over ocean ridges. *Nature* 199, pp. 947–9.

**environmental management** Provides resources from the bioenvironmental systems of the planet but simultaneously tries to retain sanative, life-supporting ecosystems. It is therefore an attempt to harmonize and balance the various enterprises which humans have imposed on natural environments for their own benefit. To achieve this, long-term strategies are evolved, based on reducing stress on ecosystems from contamination or over-use. In addition, environmental management pursues short-term strategies that are sufficiently flexible to preserve the long-term options: in other words, no resource process that brings about irreversible environmental changes should be allowed to develop.

This temporal scale is of major importance in environmental management; for example, engineering solutions may be necessary in the short term to check localized coastal erosion, but in the long term conservation of the entire coast with acceptance of slow erosion, accretion or movement may be required. Extended time perspectives may also have to accommodate extreme events.

Different approaches are similarly evident with the spatial element. Using the coastal example again: management may consider a single beach profile or a complete sedimentary cell involving supply of material from a river mouth, movement, storage in beaches and abstraction to dunes or marine deeps. In a predominantly agricultural area the maintenance of a wild population of predatory birds for scientific interest or pest control is not possible with an isolated nature reserve, but requires a whole network of protected areas.

Environmental management is, therefore, dealing with the rationalization of the resource process – the flows of material (and energy) from natural states through a period of contact with man to their ultimate disposal. It has much in common with environmental planning and those two are sometimes used interchangeably. Strictly, planning approaches the natural resource problem with a cultural or demand bias, whereas in environmental management the emphasis is on the resources themselves. It is a dynamic discipline which does not urge the preservation of resources at all costs but attempts to identify or specify major groups of resources, consider the way each changes, evaluate and resolve conflicting demands upon them and finally conserve the resources. It

believes that the very process by which renewable resources are produced can be manipulated, but that the production of non-renewable resources is virtually unmanageable because of the time-scale under which they develop.

Traditionally, environmental management has considered that, generally speaking, environmental problems need more adjustment to socioeconomic systems and has been more concerned with the maintenance of the ecological and geomorphological balance – for instance the study of and control of movements of pollutants and pesticides in food chains, overtrampling of ecologically interesting swards – leading to the preparation of conservation strategies. However, it is becoming more socially aware as renewable resource systems become so thoroughly altered by man that they can seldom be left to produce or even function in a stable condition without interdisciplinary management involving biological and physical, economic and political, and scientific and aesthetic approaches.

In terms of values environmental management is rather ambivalent towards economic growth, recognizing that there is an absolute limit to materials and the surface area of the planet, but seeing little reason to prevent resource use unless ecological stability is threatened. In any natural ecosystem the overall limiting factor must be the amount of incident solar energy, but within this context other boundaries may operate. Humans may alleviate a critical unit, e.g. by using chemical fertilizers, or may introduce a new lower limit, e.g. untreated sewage in coastal waters may decrease the light reaching littoral and sublittoral vegetation, limit productivity and so reduce the recreational potential of the system.

Different societies have differing attitudes to environmental management determined by their own order of priorities. In the USA and Europe the dominant purpose of environmental management in the past has been to obtain useful materials, an emphasis which is decreasing in favour of more concern about the life-supporting role of the ecosystem and the aesthetic value of the environment given greater impacts upon it. In contrast, the struggle to obtain food has always dominated environment management in countries such as Egypt, where the gathering of useful materials has taken a secondary role and is largely regarded only as a basic development aimed at export markets, and the care of wildlife and aesthetic preservation of the environment is of peripheral interest with little value.

It is unrealistic to pretend that totally successful environmental management is currently much more than a concept, except perhaps in the relatively simple situation of Antarctica, where resource processes are readily identifiable and human intrusion is limited. Some process response reactions, e.g. the avoidance of flood hazard, might be regarded as environmental management at a local scale, but there are critical problems within the concept. One of the most important is the dualism between ecology and economics which have completely different resource and time perspectives. It is also quite difficult in many cases to distinguish between changes which have been brought about exclusively by man and those which are at least partly natural. A further difficulty is that problems of environmental management are not the same worldwide on account of differing attitudes of wealth and its distribution; sociopolitical systems, population growth rates, and the implementation of western 'developed' culture leading to rapid urbanization and industrialization, which takes no account of natural environmental processes. Indeed, in many instances environmental problems are caused by unbalanced or over-rapid development rather than a complete disregard for environmental management. SMP

**Reading**
Blacksell, M. and Gilg, A.W. 1981: *The countryside: planning and change.* Resource Management series 2. London and Boston: Allen & Unwin. · Douglas, I. 1983: *The urban environment.* London: Edward Arnold. · Edington, J.M. and Edington, M.A. 1977: *Ecology and environmental planning.* London: Chapman & Hall; New York: Wiley. · Goldsmith, F.B. and Warren, A., 1993 *Conservation in progress.* London and New York: Wiley · O'Riordan, T. 1981: *Environmentalism.* 2nd edn. London: Pion. · Park, C.C. 1980: *Ecology and environmental management.* Studies in physical geography 3. Folkestone, Kent and Dawson, Colorado: Westview Press. · Simmons, I.G. 1981: *The ecology of natural resources.* 2nd edn. London: Edward Arnold.

**epeiric sea** A shallow body of marine water on the continental shelf which is connected with an ocean.

**epeirogeny** The warping of large areas of the earth's crust without significant deformation. It can be contrasted with OROGENY, which is associated with linear zones of uplift. Epeirogenic uplift can affect regions thousands of kilometres across and is the major form of uplift in most CRATONS. The causes of the predominantly vertical movements associated with epeirogeny are uncertain but may be related to expansion resulting from localized heating within the crust or at the base of the LITHOSPHERE, possibly in conjunction with phase changes in the MANTLE. MAS

**Reading**
Crough, S.T. 1979: Hot spot epeirogeny. *Tectonophysics* 61, pp. 321–33. · Ollier, C.D. 1981: *Tectonics and landforms.* London: Longman.

**ephemeral plant** A plant, generally found in arid and semi-arid regions, in which the life cycle is completed within a very short period, perhaps only a few weeks. According to RAUNKIAER'S LIFE FORMS classification, these are therophytes. Seed germination in such species is triggered by a particular combination of environmental conditions, usually involving substantial rainfall and, as a consequence, although they are nominally 'annuals', they do not necessarily reappear annually. These species have a large reservoir of seeds capable of surviving many years in the dormant state before germination is triggered by moisture inputs. Seed germination and growth is then exceptionally rapid, followed by flowering and seed production. Because the life cycle is completed in such a brief period following rain, it is unusual for these species to have other adaptations to drought, although succulence is a feature of a significant proportion of ephemerals in the Namib Desert. The diversity of ephemeral species is especially great in the winter-rainfall desert regions of the world, such as in the western South American and south-western African coastal arid zones. In the semi-arid parts of the south-western Cape of South Africa, spectacular floral displays of ephemeral plants, normally following spring rains, regularly occur, in which there is mass flowering of plant species, many of which belong to the *Asteraceae* (daisy) family. Most of the plants are insect pollinated and the annual display of brightly coloured flowers represents an important component of the ecological dynamics of these regions in supporting an impressive diversity of wasps, beetles and other pollinating insects whose populations are correspondingly ephemeral. MEM

**Reading**
Inouye, R.S. 1991: Population biology of desert annual plants. In G.A. Polis, ed., *The ecology of desert organisms.* Tucson: University of Arizona Press. Pp. 25–54.

**ephemeral stream** A stream which is often one of the outer links of a DRAINAGE NETWORK and which contains flowing water only during and immediately after a rainstorm which may be fairly intense. As the water flows along the ephemeral stream channel it may infiltrate into the channel bed as a transmission loss by influent seepage and therefore the peak discharge may decrease downstream along the ephemeral channel by as much as 5 per cent per km of channel. In arid and semi-arid areas of the world ephemeral streams are very extensive and represent the major channel type. KJG

**Reading**
Renard, K.G. and Laursen, E.M. 1975: Dynamic behaviour model of ephemeral streams. *Journal of the Hydraulic Division of the American Society of Civil Engineers* 101, pp. 511–28. · Thornes, J.B. 1977: Channel changes in ephemeral streams: observations, problems and models. In K.J. Gregory ed., *River channel changes*. Chichester: Wiley. Pp. 317–35.

**epicentre** The point on the earth's surface which lies directly above the focus of an earthquake.

**epiclimate** (or nanoclimate) The climate on the surface of leaves, in the air cavities in litter, along the slopes of an ant hill, or in the fissures in rocks. Epiclimate extends vertically a few centimetres, or perhaps a decimetre and extends horizontally for centimetres. Studies of epiclimate are important for ecophysiology and population ecology of very small organisms. RCB

**epilimnion** The surface layer of water of a lake or sea. The water which lies between the surface and the thermocline.

**epipedon** A diagnostic surface horizon which includes the upper part of the soil that is darkened by organic matter, or the upper eluvial horizons, or both.

**epiphyte** A plant which grows upon the surface of another plant but does not obtain food from the host plant.

**epoch** A unit of geologic time equivalent to a series, a division of a period.

**equation of state** A relationship between properties of a material, and/or forces acting on it, in a state of equilibrium. The most familiar examples of equations of state in physical geography refer to the balance of forces acting on a body or within a material in equilibrium. Stability analysis for landslides is, for instance, based on such a balance. Equations of state may also refer to other properties, and the term has a special significance in thermodynamics as the unique relationship between temperature, pressure and volume for a body of fluid. MJK

**equations of motion** Expressions governing the motion of a body or a material under the action of a force or forces. The equations most commonly take the form:

$$\text{force} = \text{mass} \times \text{acceleration}$$

in either its linear form or as moments (torques) about a centre. Since force is a vector, the equation is a vector equation, and may be resolved to give up to three component equations in directions which are mutually at right angles. Equations of motion are one example of expressions which control the rate of a process, usually subject to the constraints of the CONTINUITY EQUATION. The term was originally used in the context of solid bodies, but may also be used to describe the motion of a fluid such as water or air, either travelling with a physical body of fluid or describing motion at a fixed point. MJK

**equatorial rain forest** A lowland evergreen TROPICAL FOREST lying approximately 5° north and south of the equator in near-continuous rainfall climates, over 2000 mm year$^{-1}$, and not limited by low temperatures. The forests are multilayered, over 30 m tall, shallow rooted and often buttressed, containing a profusion of climbing plants and epiphytes and the greatest diversity and abundance of plants and animals of any terrestrial biome. Their BIOLOGICAL PRODUCTIVITY also heads the league for terrestrial biomes. Although the main global formations are comparable in structure, life forms and animal adaptations, the biological evolution and species composition is profoundly different in each area. PAF

**Reading**
Golley, F.B., Lieth, H. and Werger, M.J.A. eds 1982: *Tropical rain forest ecosystems*. Amsterdam: Elsevier. · Longman, K.A. and Jenik, N. 1987: *Tropical forest and its environment*. 2nd edn. London: Longman.

**equatorial trough** A narrow, fluctuating belt of unsteady, light, variable winds, low atmospheric pressure, and frequent small-scale disturbances. It is located near the equator between the trade wind belts of the two hemispheres. However its position, breadth, and intensity are constantly changing. From time to time it disappears completely, especially over the continents. Along its meandering position occur most of the frequent, heavy showers for which the tropics are so well known. The equatorial trough includes the prevailing calms of the DOLDRUMS and is frequently referred to as the INTERTROPICAL CONVERGENCE ZONE (ITCZ). WDS

**Reading**
Rumney, G.R. 1968: *Climatology and the world's climates*. New York: Macmillan.

**equifinality** Arises when a particular morphology (e.g. a landform) can be generated by a number of alternative processes, process assemblages, or process histories. Under such circumstances the morphology alone cannot be used as a basis for reconstructing the process of origin of a feature. For example, a central assumption in CLIMATIC GEOMORPHOLOGY is that landforms differ significantly between climatic zones because of variation in the climatic factors that control weathering, run-off, erosion and deposition. However, specific landforms may originate in different ways, and are therefore not restricted to single climate zones. U-shaped valleys are characteristic of glaciated highlands, but also occur in high-relief sub-tropical areas where basal sapping maintains steep valley sides after intensive chemical weathering at the water table (Wentworth 1928). Tors are also features produced by quite distinct sets of processes in different areas; both deep chemical weathering with subsequent stripping of the weathered mantle and frost-shattering with mass movement generate similar morphological features. Thus the supposed characteristic forms of particular process assemblages and climatic regimes may in fact have diverse origins which display equifinality, and a simple correlation between form, process and climate cannot be demonstrated.                                    KSR

#### Reference
Wentworth, C.K. 1928: Principles of stream erosion in Hawaii. *Journal of geology* 36, pp. 385–410.

### equilibrium
A concept commonly applied to environmental open systems, that is, systems in which the quantities of stored energy or matter are adjusted so that input, throughput and output of energy or matter are balanced. For example, the earth receives short-wave solar radiation at the top of the atmosphere. Of the total receipt ($263 \, kcal \, cm^{-2} \, year^{-1}$), 31 per cent are reflected and 69 per cent ($181 \, kcal \, cm^{-2} \, year^{-1}$) are absorbed. Input and output are balanced by the earth maintaining an equilibrium mean temperature such that it emits $181 \, kcal \, cm^{-2} \, year^{-1}$ of long-wave radiation. In a river system equilibrium is often defined as a balance of erosion and deposition. This is achieved by morphological adjustments which maintain sediment transport continuity. If a short river reach experiences more bedload input from upstream than output downstream, the excess is deposited and the slope is steepened and the cross-section shallowed. The bedload input can then be transported through the reach more effectively, and the output is increased to balance the input. Any local particle detachment (bed erosion) is balanced by deposition. This equilibrium is maintained by negative feedback; the deposition of excess load changes reach morphology so that the transport capacity increases and further

deposition is prevented. The inputs to an open system vary through time, for example on a seasonal basis, but so long as average annual input is constant the system state is constant, and the equilibrium is one in which the relationship between form and process is stationary. This is a STEADY STATE (Chorley and Kennedy 1971, pp. 201–3). If the annual average input is changing through time sufficiently slowly for the system to adjust the condition is a DYNAMIC EQUILIBRIUM. Technically, however, there is always a lag between the change in the process input variable and the internal morphological adjustment of the system, so the term *quasi-equilibrium* is sometimes used in this case.   KSR

#### Reference
Chorley, R.J. and Kennedy, B.A. 1971: *Physical geography: a systems approach*. London: Prentice-Hall.

### equilibrium line
A notional line describing some sort of balance between process and form. The notion can be applied widely, e.g. to the profile of a slope or plan of a beach, and is closely bound up with concepts of EQUILIBRIUM in natural systems. (See DYNAMIC EQUILIBRIUM; EQUILIBRIUM SHORELINE.)

### equilibrium line of glaciers
A notional altitudinal line on a glacier where ABLATION balances accumulation. In most situations this means that the summer is just warm enough to melt the snow and ice that has accumulated during the previous winter. Above the equilibrium line on a glacier is the accumulation zone where accumulation exceeds ablation each year, while below the equilibrium line is the ablation zone where ablation exceeds accumulation each year. The amount of snow and ice melted at the equilibrium line each year is a measure of the activity of a glacier with high values implying high velocities (Andrews 1972). Glaciers are most active in mid-latitude, temperate areas and become less active towards continental interiors and the poles. The equilibrium line altitude varies in a similar way. Where there is high winter accumulation the summer temperature must be high in order to melt the snow and ice. Thus equilibrium line altitudes tend to be low near maritime coasts and to rise towards continental interiors in response to precipitation gradients. This pattern is brought out by the distribution of both present-day mountain glaciers and abandoned glacial CIRQUES.

In temperate environments where the glacier ice is at the pressure melting point the equilibrium line may coincide with the FIRN line which marks the line separating bare ice from

179

snow at the end of the ablation season. But on cold glaciers snowmelt may percolate down and freeze onto the glacier as SUPERIMPOSED ICE. Under these circumstances the positions of the firn line and equilibrium line on a glacier may differ. DES

**Reading and Reference**
Andrews, J.T. 1972: Glacier power, mass balances, velocities and erosion potential. *Zeitschrift für Geomorphologie* NF 13, pp. 1–17. · Paterson, W.S.B. 1981: *The physics of glaciers*: Oxford and New York: Pergamon.

**equilibrium shoreline** A hypothetical state that actual shorelines may or may not approximate. It is a dynamic state in which the geometry of the beach reflects a balance between materials, processes and energy levels (climate). The ideal EQUILIBRIUM beach has curvature and sand prism characteristics which are adjusted so closely that the energy available transports the detritus supplied, over a period to be measured in years rather than months, days or seconds. ASG

**Reading**
Tanner, W.F. 1958: The equilibrium beach. *Transactions of the American Geophysical Union* 39, pp. 889–91.

**equipotentials** Lines on a GROUNDWATER map joining points of equal fluid potential (or HYDRAULIC POTENTIAL). Fluid potential at any point is the product of HYDRAULIC HEAD and acceleration due to gravity. Consequently, since gravitational acceleration is practically constant, a WATER TABLE contour map with equal contour intervals is a potentiometric map in the horizontal plane. The distance between the contours (equipotentials) depicts the gradient of the potential. Hence hydraulic gradient varies inversely with contour spacing (Ward and Robinson 1990). In accordance with DARCY'S LAW, water always flows in a down gradient direction perpendicular to the equipotentials, the path followed by a particle of water being known as a streamline. A mesh formed by a series of equipotentials and corresponding streamlines is known as a flow net.

Equipotentials may also be constructed in the vertical plane. If fluid potential increases with depth, groundwater flow will be towards the surface, but if it decreases vertically flow will be downward (Hubbert 1940). PWW

**Reading and References**
Freeze, R.A. and Cherry, J.A. 1979: *Groundwater*. Englewood Cliffs, NJ: Prentice-Hall. · Hubbert, M.K. 1940: The theory of groundwater motion. *Journal of geology* 48, pp. 785–944. · Ward, R.C. and Robinson, M. 1990: *Principles of hydrology*. 3rd edn. Maidenhead, Berks: McGraw-Hill.

**era** The largest unit of geological time, being a span of one or more periods.

**erg** A sand desert. A desert area characterized by sand sheets and dunes. A sand sea.

**ergodic hypothesis** As used in geomorphology, suggests that under certain circumstances sampling in space can be equivalent to sampling through time. Geomorphologists have sometimes sought an understanding of landform evolution by placing such forms as regional valley-side slope profiles and drainage networks in assumed time sequences. The concept of the cycle of erosion was based to a large extent on ergodic assumptions, as was Darwin's model of coral reef development. Chorley *et al.* (1984, p. 33) point to certain dangers in ergodic reasoning: landforms may be assembled into assumed time sequences simply to fit preconceived theories of denudation; there is always a risk of circular argument; and form variations may result from factors other than their position in time. ASG

**Reference**
Chorley, R.J., Schumm, S.A. and Sugden, D.E. 1984: *Geomorphology*. London and New York: Methuen.

**erodibility** The susceptibility of a surface on sediment to EROSION. Erodibility is a function or factors such as rock hardness and strength, the particle size distribution of a consolidated or unconsolidated material, the degree and nature of organic content and plant cover, etc. See also EROSIVITY. DSGT

**erosion** The group of processes whereby debris or rock material is loosened or dissolved and removed from any part of the earth's surface. It includes weathering, solution, corrasion and transportation.

**erosion pin** Any rod, usually metal and ranging from a stake to a nail, that is fixed into the ground surface and used to monitor surface level changes (both losses and gains). Pins may initially be applied flush to the ground surface, to minimize intrusion into sediment movement processes, but as such are susceptible to burial. Normally they are left protruding and surface change measured against the initial known length of the protruding part. Pins should be as thin as possible, but need to be sufficiently robust, and sufficiently embedded into the ground surface, to resist disturbance. Pins have been used to monitor, for example, slope erosion and surface change on sand dunes (e.g. by Wiggs *et al.* 1995). DSGT

**Reference**
Wiggs, G.F.S., Thomas, D.S.G., Bullard, J.E. and Livingstone, I.P 1995: Dune mobility and vegetation cover in the southwest Kalahari Desert. *Earth surface processes and landforms* 20, pp. 515–29.

**erosion surface** A term commonly used in Britain to describe a flattish plain resulting from erosion. Since, strictly speaking, erosion surfaces may be far from flat, it is probably more helpful to use the term *planation* surface instead.

Planation surfaces assume a central role in a geomorphological approach concerned with the evolution of landscape since they are generally regarded as the end product of either a cycle of erosion – the peneplain in the Davisian sense (see CYCLE OF EROSION) or of a particular blend of surface processes, for example the pediplain in a semi-arid environment (see PEDIMENT), the etchplain in a humid tropical environment, or the wave-cut platform in the coastal environment.

In the first half of this century, when the study of landscape evolution was the prime goal of geomorphology, much attention was devoted to the identification of present and relict planation surfaces as the key to understanding DENUDATION CHRONOLOGY (King 1950; Wooldridge and Linton 1955). In many parts of Britain relict planation surfaces (summit planes) are common and comprise master features of the total landscape (see, e.g., Brown 1960; Sissons 1967). A major difficulty was the problem of dating, and the evolutionary approach in geomorphology became unfashionable in the 1960s and 1970s. Now, with new forms of radiometric dating and stratigraphic evidence available from offshore sediments, there are signs of a new lease of life for studies of landscape evolution and real prospects of understanding the significance of relict planation surfaces (e.g. Lidmar-Bergstrom 1982). (See also BASE LEVEL.)　　　　DES

**Reading and References**
Brown, E.H. 1960: *The relief and drainage of Wales.* Cardiff: University of Wales Press. · —and Clayton, K., eds. *The geomorphology of the British Isles.* (A series of regional volumes; varying dates.) London: Methuen. · King, L.C. 1950: The study of the world's plain lands. *Quarterly journal of the Geological Society* 106, pp. 101–31. · Lidmar-Bergstrom, K. 1982: Pre-Quaternary geomorphological evolution in southern Fennoscandia. *Sveriges Geol Unders. Series C*, 785. · Ollier, C.D. 1979: Evolutionary geomorphology of Australia and Papua-New Guinea. *Transactions of the Institute of British Geographers* ns 4.4, pp. 516–39. · Sissons, J.B. 1967: *Evolution of Scotland's scenery.* Edinburgh: Oliver & Boyd. · Wooldridge, S.W. and Linton, D.L. 1955: *Structure, surface and drainage in south-east England.* 2nd edn. London: George Philip.

**erosivity** A measure of the potential ability of a soil to be eroded by a given geomorphological agency. For given soil and vegetation conditions the effects of a storm, for example, can be compared with another storm quantitatively and a scale of erosivity values can be produced. The ERODIBILITY of a soil is the vulnerability of a soil to erosion; for given rainfalls one soil can be compared with another and a scale of values produced. Thus EROSION can be considered as a function of both erosivity and erodibility and they are related in the universal soil loss equation, details of which are given in Morgan (1986).　　　　WBW

**Reading**
Morgan, R.P.C. 1986: *Soil erosion and conservation.* London: Longman. · Selby, M.J. 1993: *Hillslope materials and processes.* 2nd edn. Oxford: Oxford University Press.

**erratic** (glacial) A rock or boulder that has been carried to its present location by the action of a glacier. In the English-speaking world in the first half of the nineteenth century erratics were commonly attributed to ice rafting in the universal flood, although in Switzerland, Germany and Norway their true origin had already been appreciated. The significance of erratics in demonstrating former widespread glaciers in Britain was shown by Agassiz (1840).

**Reference**
Agassiz, J.L.R. 1840: Glaciers, and the evidence of their having once existed in Scotland, Ireland and England. *Proceedings of the Geological Society* 3, pp. 321–2.

**eruption** A discharge or eruption of volcanic material, either gaseous, liquid or solid, at the earth's surface.

**escarpment** The steeper slope of a cuesta. Often used as a synonym for a cuesta.

**esker** A sinuous ridge of coarse gravel representing the deposits of a MELTWATER stream normally flowing subglacially. Eskers may be hundreds of kilometres in length and 100 m in height. In many situations they are beaded, which means that mounds occur along their length, particularly at points where they change direction. It is common to find that eskers form complex patterns of tributaries and distributaries and that sometimes ridges are discontinuous or linked by rock-cut meltwater channels. Most eskers are the channel deposits of subglacial meltwater rivers and their orientation is usually parallel to that of overall ice flow (Shreve 1972). The deposits are related to both closed channel flow and open channel flow within a conduit (Bannerjee and McDonald 1975; Saunderson 1977). DES

**References**
Bannerjee, I. and McDonald, B.C. 1975: Nature of esker sedimentation. In A.V. Jopling and B.C. McDonald eds, *Glaciofluvial and glaciolacustrine sedimentation.* Society of Economic Paleontologists and Mineralogists special publication 23. Tulsa, Oklahoma. Pp. 132–54. · Saunderson, H.C. 1977: The sliding bed facies in esker sands and gravels: a criterion for full-pipe (tunnel) flow? *Sedimentology* 24, pp. 623–38. · Shreve, R.L. 1972: Movement of water in glaciers. *Journal of glaciology* 11, 62. pp. 205–14.

**ESP**  See EXCHANGEABLE SODIUM PERCENTAGE.

**ESR**  See ELECTRON SPIN RESONANCE.

**estuary**  The section of a river which flows into the sea and is influenced by tidal currents. Estuaries form transition zones between freshwater rivers and salt-water oceans, with fluctuations in water level, salinity, temperature and velocity. They are constantly modified by erosion and deposition, resulting in tidal flats and salt marshes, deltas, spits and lagoons. A funnelling of tidal currents may produce powerful periodic waves or estuarine bores. Shallow sedimentary estuaries are rich in nutrients and very high in BIOLOGICAL PRODUCTIVITY, providing nurseries for fish and other animals. Deep estuaries such as fiords are colder, less productive and less biologically diverse.

Estuaries can be classified into a number of types on the basis of chemical characteristics (Dyer 1986): salt wedge estuaries in tideless seas, partially mixed estuaries where there are appreciable tidal movements, and well-mixed estuaries where the strength of the tidal currents is strong relative to river flow. They can also be classified on the basis of their tidal range. This determines the tidal current and residual current velocities and therefore the amount and source of sediments. *Microtidal estuaries* occur where tidal range is less than 2 m and so are dominated by freshwater inflow upstream of the mouth and by wind-driven waves seaward of the mouth. They often contain a fluvial delta and spits and bars at the seaward margin. In *mesotidal estuaries* (tidal range *c.*2–4 m) tidal currents are of greater importance but because

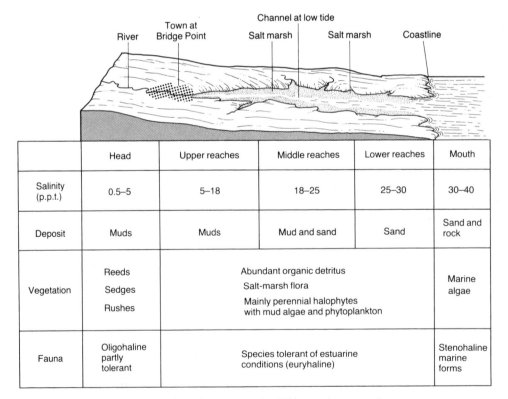

| | Head | Upper reaches | Middle reaches | Lower reaches | Mouth |
|---|---|---|---|---|---|
| Salinity (p.p.t.) | 0.5–5 | 5–18 | 18–25 | 25–30 | 30–40 |
| Deposit | Muds | Muds | Mud and sand | Sand | Sand and rock |
| Vegetation | Reeds Sedges Rushes | Abundant organic detritus Salt-marsh flora Mainly perennial halophytes with mud algae and phytoplankton | | | Marine algae |
| Fauna | Oligohaline partly tolerant | Species tolerant of estuarine conditions (euryhaline) | | | Stenohaline marine forms |

**The estuarine environment**. Zones of transition separate the different environments shown.
*Source*: P. Furley and W. Newey 1983: *Geography of the biosphere*. London: Butterworth. Figure 13.12.

of the still somewhat modest tidal range tidal flow does not extend very far upstream. Thus most mesotidal estuaries are relatively stubby. In the case of *macrotidal estuaries*, tidal ranges in excess of 4 m produce a situation where tidal influences extend far inland. Such estuaries have long, linear sand bars parallel to the tidal flow, but their most distinguishing characteristic is their trumpet-shaped flare. The Severn Estuary in Britain, the Delaware Estuary in the USA, and the Plate Estuary in Latin America are prime examples of this type.                   PAF/ASG

**Reading and Reference**
Dyer, K.R. 1986: *Coastal and estuarine sediment dynamics*. Chichester: Wiley. · Ketchum, B.H. ed. 1983: *Estuaries and enclosed seas*. Amsterdam: Elsevier. · McLusky, D.S. 1981: *The estuarine environment*. Glasgow: Blackie.

**etchplain** A term first used by Wayland (1933) to describe erosional plains in east Africa, where crystalline rocks are weathered to tens of metres depth. This SAPROLITE is removed by erosion during uplift and dissection, and renewed during periods of stability. Willis (1936) described Tanganyika Plateau as an 'etched peneplain' extensively mantled by regolith, and the rock floor as an 'etched surface'. Büdel (1957, 1982) developed similar concept of double PLANATION SURFACES (wash surface and basal weathering surface or WEATHERING FRONT), and explained inselberg-studded plains ('Rumpfflächen') by stripping of saprolite. Linton's (1955) theory of TOR formation is similar. Etchplains exhibit saprolite 3-30 m thick; commonly a sharp transition to fresh rock indicating intense weathering, but often a zone of less advanced decomposition or GRUS 100-150 m thick. Some regard true etchplains as only those etched rock surfaces which are widely exposed (Ollier 1984); others include stripped rocky landsurfaces and the weathered plains from which they are derived (Thomas 1994; Willis 1936). Associated with a 'cratonic regime' (Fairbridge and Finkl 1980), etchplains result from prolonged weathering during $10^7$-$10^8$ y, interrupted by shorter ($10^5$-$10^6$ y) periods of saprolite stripping and pediplanation. Described from east Africa and Tamil Nadu (southern India) in seasonal tropics, etchplains are also described from humid tropics of west Africa and Guyana, sub-tropics of south-western Australia, and as ancient palaeo-forms in Scandinavia, Britain and Palaeozoic massifs in Europe and North America.

Etchplains are often viewed as special forms of PENEPLAIN, a term often used generically for any surface of extensive planation, and they may contain *pediments* cut across the soft saprolite (see Adams 1975).                   MFT

**References**
Adams, G. ed. 1975: Planation surfaces. *Benchmark papers in geology*, 22. Penhsylvania: Dowden, Hutchinson and Ross. · Büdel, J. 1957: Die 'doppelten Einebnungsflächen' in den feuchten Tropen. *Zeitschrift für Geomorphologie*, NF 1, pp. 201-88. · —(1982): *Climatic geomorphology*. Trans. L. Fischer and D. Busche. Princeton: Princeton University Press. · Fairbridge, R.W. and Finkl, C.W., Jr 1980: Cratonic erosional unconformities and peneplains. *Journal of Geology* 88, pp. 69-86. · Linton, D.L. 1955: The problem of tors. *Geographical journal* 121, pp. 470-7. · Ollier, C.D. 1984: *Weathering*. 2nd edn. London: Longman. · Thomas, M.F. 1994: *Geomorphology in the tropics*. Chichester: John Wiley & Son. · Wayland, E.J. 1933: Peneplains and some other erosional platforms. *Annual report and bulletin, protectorate of Uganda geological survey*, Department of Mines. Note 1, pp. 77-9. · Willis, B. 1936: *East African plateaus and rift valleys: studies in comparative seismology*. Carnegie Institute, Washington, Publication, 470.

**etesian wind** The prevailing wind over the Aegean sea in summer (from the Greek *etesios*, annual). It blows steadily from the north with moderate force, bringing dry cold continental air and clear sky from mid-June to the beginning of October.                   JSAG

**Reading**
Meteorological Office. 1962: *Weather in the Mediterranean*. Vol. I. London: HMSO.

**eugeogenous rock** Rock which produces a large amount of debris when it is decomposed by weathering.

**eulittoral zone** The portion of the coastal zone which extends seawards from high-water mark down to the limit of attached plants (generally at a depth of 40–60 m).

**euphotic zone** The surface layer of a body of water in which photosynthesis can take place because of the availability of light.

**eustasy** A term which embraces sea-level changes of a worldwide nature. Local changes of sea level complicate the eustatic pattern and are caused by local factors such as ISOSTASY, OROGENY and EPEIROGENY. During the first decades of the twentieth century, following on in part from the work of Suess, a number of workers including De Lamothe, Deperet, Baulig, and Daly, proposed that most sea-level oscillations and strandlines of the Quaternary were glacio-eustatic (see Guilcher 1969). They believed, correctly, that sea level oscillated in response to the quantity of water stored in ice caps during glaciations and deglaciations.

They suggested that there was a suite of characteristic levels in Morocco and elsewhere

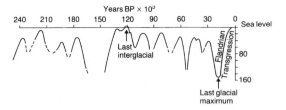

**Eustasy**. The nature of world sea-level change over the past quarter of a million years.
*Source*: A.S. Goudie 1984: *The nature of the environment*. Oxford and New York: Basil Blackwell.

around the Mediterranean which could be related to different glacial events:

1 Sicilian (80–100 m);
2 Milazzian (55–60 m) – between the Günz and the Mindel;
3 Tyrrhenian (30–35 m) – between the Mindel and the Riss;
4 Monastirian (15–20 and 0–7 m) – between the Riss and the Würm; and
5 Flandrian – the present post-Würm transgression.

The transgressions of the interglacial were succeeded by regressions during glacials and the height of the various stages declined during the course of the Pleistocene. Total melting of the two main current ice caps – Greenland and Antarctica – would raise sea level a further 66 m if they both melted. Deep-sea core evidence, however, does not suggest that in previous interglacials of the Pleistocene these two ice caps did disappear, and without a general melting of them, sea level would only have been a few metres higher than now in the interglacials. This fact does not tie in too happily with the simple glacio-eustatic theory of progressive sea-level decline during the Pleistocene. Some factors other than glacio-eustasy must be responsible for the proposed high sea levels of early Pleistocene times. Because other, sometimes local, factors have played a role, few people now seriously believe that through height alone one can correlate shorelines over wide areas on the basis of a common interglacial age.

Nevertheless, low Quaternary sea levels brought about by the ponding up of water in the ice caps were quantitatively extremely important. Donn *et al.* (1962) on the basis of theoretical considerations from known ice volumes, reckon that in the Riss, possibly the most extensive of the glaciations, sea levels might have been lowered by 137–159 m below current sea level. In the last glacial (Würm–Wisconsin–Weichsel) they give a figure for lowering of rather less: 105–123 m. On the basis of isotopic dates for coral and associated material in the Great Barrier Reef (Australia), California, and the south-east sea areas, Veeh and Veevers (1970) favour the conclusion that towards the end of the last glaciation, sea level dropped universally to at least −175 m, some 45 m deeper than hitherto suspected.

Although orogeny is normally regarded as being an essentially local factor of sea-level change, and eustasy as being worldwide, there is one class of process, *orogenic eustasy*, whereby a local change can have worldwide effects. It therefore acts as some sort of a link between these two main types of change, and is a worldwide change of sea level produced by changes in the volumes of the ocean basins resulting from orogeny (mountain building) (Grasty 1967).

In recent years the importance of a third type of eustatic change has been identified, notably by Mörner (1980). This is termed *geoidal eustasy*. The shape of the earth is not regular, and at present the GEOID (caused by the earth's irregular distribution of mass) has a difference between lows and highs of as much as 180 m. The ocean surface reflects this irregularity in the geoid surface, which varies according to forces of attraction (gravity) and rotation (centrifugal), and will respond by deformation to a change in these controlling forces. The possible nature of such changes is still imperfectly understood, but they include fundamental geophysical changes within the earth, changes in tilt in response to the asymmetry of the ice caps, changes in the rate of rotation of the earth, and redistribution of the earth's mass caused by ice-cap waxing and waning.

Although glacio-eustasy is the most important of the eustatic factors that have affected world sea levels during the course of the Quaternary, it is worth looking at some of the other minor eustatic factors which have played a role, especially over the long term. The infilling of the ocean basins by sediment, for example, would tend to lead to a sea-level rise. Higgins (1965) estimated that this could lead to a rise of 4 mm 100 years$^{-1}$. This is equivalent to a rise of 40 m in a million years. Two very minor factors are the addition of juvenile water from the earth's interior and the variation of water level according to temperature. The latter could raise the level of the sea by about 60 cm for each 1 °C rise in temperature of the seawater. The former could probably add about 1 m of water in a million years. The evaporation and desiccation of pluvial lakes, some of which had large dimensions, would be unimportant in affecting world sea levels, adding a maximum of about 10 cm to the level of the sea, if they were all to be evaporated to dryness at the same time (Bloom 1971).

Another cause of eustatic changes of sea level, especially in the Holocene, is the process called 'isostatic decantation'. Isostatic uplift in the neighbourhood of the Baltic basin and of Hudson Bay has led to a reduction of the volume of these seas; and the water from them has thus been decanted into the oceans to affect worldwide sea levels. A comparison of the area and volume of the late-glacial precursor of Hudson Bay with Hudson Bay itself suggests that the volume of water decanted could only be sufficient to cause a rise in world sea level of about 0.63 m. The contribution of the Baltic Sea would be even less. This factor can thus be largely ignored.                                    ASG

**Reading and References**
Bloom, A.I., 1971: Glacial eustatic and isostatic controls of sea level since the Last Glaciation. In K.K. Turekian, ed., *The Late Cenozoic glacial ages*. New Haven: Yale University Press. Pp. 355–79. · Donn, W.L., Farrand, W.R. and Ewing, M. 1962: Pleistocene ice volumes and sea level changes. *Journal of geology* 70, pp. 206–14. · Goudie, A.S. 1992: *Environmental change*. 3rd edn. Oxford: Clarendon Press. Ch. 6. · Grasty, R.L. 1967: Orogeny, a cause of world wide regression of the seas. *Nature* 216, p. 779. · Guilcher, A. 1969: Pleistocene and Holocene sea level changes. *Earth science reviews* 5, pp. 69–98. · Higgins, C.G. 1965: Causes of relative sea-level changes. *American scientist* 53, pp. 464–76. · Mörner, N.A. 1980: *Earth rheology, isostasy and eustasy*. New York: Wiley. · Veeh, H. and Veevers, J.J. 1970: Sea level at −175 m off the Great Barrier Reef, 13,600 to 17,000 years ago. *Nature* 226, pp. 526–7.

**eutrophic** Pertaining to lakes and other freshwater bodies which abound in plant nutrients and which are therefore highly productive. Lakes tend to become more eutrophic as they become older, and eutrophication can also result from the addition of nutrients as a result of pollution. This can cause phenomena such as algal blooms. (See also NUTRIENT STATUS.)

**eutrophication** The addition of mineral nutrients to an ECOSYSTEM, generally raising the NET PRIMARY PRODUCTIVITY. It is usually used of human-induced additions of elements such as nitrogen and phosphorus to salt and freshwater, which are naturally low in those elements, but it also occurs in terrestrial systems and may be a natural phenomenon. In current usage, it very often relates to the loads of $N_2$ and P in fresh and offshore waters heated by such effluents as sewage, fertilizer run-off and detergents. The effects are often algal blooms, deoxygenation of water through consequent bacterial activity and, in the sea, rapid growth of small organisms called dinoflagellates which are implicated in 'red tides'. Eutrophication is usually a local or regional problem at most,

though enclosed seas like the Mediterranean may be more vulnerable than open oceans. Its main drawback is the loss of expensively gained nutrients which then have to be replaced, since they cannot be economically retrieved from the water in which they have become diluted.    IGS

**Reading**
Whitton, B.A. ed. 1975: *River ecology*. Berkeley and Los Angeles: University of California Press.

**evaporation** The diffusion of water vapour into the atmosphere from freely exposed water surfaces. This includes water losses from lakes, rivers, even clouds and saturated soil and plant surfaces but it does not incorporate transpiration losses from plants. It is imperative therefore to distinguish between the process of evaporation which concerns only free-standing water bodies and that of EVAPOTRANSPIRATION. While the process of evaporation is generally understood, its accurate measurement has long proved difficult. Before considering the various instruments and formulae available for this task it is necessary to introduce the variables that govern rate of water loss.

The rate of evaporation is partly controlled by solar radiation, which supplies the energy required to transform liquid water into water vapour, i.e. the latent heat of vaporization or $2.44 \times 10^6 \text{J kg}^{-1}$ at a temperature of 25 °C. The proportion of net radiation received by the earth ($Q^*$) that is available for this process ($QLE$) depends not only upon the transmission, absorption and reflection of the earth's atmosphere and surface, but also the amounts employed for heating the atmosphere ($QH$), and for heating the ground ($QS$), i.e. (Oke, 1987):

$$Q^* = QH + QLE + QS$$

where $Q^*$ is the net radiation balance ($W \text{ m}^{-2}$), $QH$ is the sensible heat flux, $QLE$ is the latent heat flux, and $QS$ is the ground heat flux.

The humidity of the air above an evaporating surface will eventually increase until, when the air becomes saturated, evaporation will cease unless these layers are dispersed. The atmospheric 'demand' for moisture is therefore controlled not only by the radiation balance but also by humidity and windspeed. The rate of evaporation is influenced in addition by the characteristics of the water body itself, that is depth, extent and water quality (Ward and Robinson 1990). Understanding evaporation rates is further complicated by the need to distinguish between potential and actual evaporation.

Actual evaporation is the observed rate of water loss, whereas the potential rate is the evaporation that would occur from a free-standing freshwater surface due to atmospheric demand if there were no other limiting factors. These two rates could be equal where the water supply is plentiful, e.g. a large freshwater lake, but potential rates may not be reached when the evaporating surface does not have the same characteristics, e.g. over soils. In addition potential rates may be exceeded over small water bodies when relative humidities are low. Potential evaporation is therefore a theoretical concept referring only to the atmospheric demand for moisture over a large freshwater evaporating surface.

Several instruments have been designed to measure the volume of water loss directly from saturated surfaces and thus provide an estimate of potential rates. Atmometers are commonly employed because they are cheap and simple to use. The device consists of an inverted measuring cylinder with a porous plate at its base. As most of the energy used in evaporating water is derived from the surrounding air rather than direct solar radiation they are over-sensitive to humidity and windspeed (Jackson 1989). The instrument in most widespread use is the evaporating pan, in effect a small reservoir of water approximately 1.2 m across and 25 cm high (US Weather Bureau class 'A'). It is simple and effective but the loss of water is influenced by its exposure, design, material and colour. In addition, estimation of the depth of water is liable to significant observer error.

Potential evaporation rates can be obtained by surrounding these instruments by extensive saturated surfaces or by using empirical coefficients to convert observations. Doorenbos and Pruitt (1975) for example, provide coefficients for evaporating pans but information is also required on the windspeed, relative humidity and exposure.

The measurements of actual evaporation from soils can be estimated by lysimeters, consisting of isolated columns of soil placed in large containers thus enabling the observation of changes in soil moisture and drainage. Evaporation rates can then be deduced by a tabulation of the water balance:

$$E = P - D - RO \pm \Delta S$$

where $E$ is evaporation, $P$ is precipitation, $D$ is drainage, $RO$ is run-off, and $\Delta S$ are changes in soil moisture. Unfortunately these devices are difficult to construct and maintain, and the container walls can distort lateral movements of soil moisture. Despite their potential high level of accuracy, lysimeters therefore tend to be used only in research studies.

The water balance approach can be used for water bodies of any size but an alternative

method is to measure directly the diffusion of water vapour through the atmosphere, rather than estimating the various inputs and outputs of the hydrological cycle. Rosenberg (1974) lists two main approaches, the eddy correlation approach and profile techniques. Eddy correlation is based on the instantaneous measurement of humidity and vertical windspeed at a selected height when at time ($t$):

$$QLE(t) = \rho L q(t) W(t)$$

where $QLE(t)$ is the evaporation flux (strictly the vertical transport of latent heat), $q(t)$ is the water vapour content, $W(t)$ is the instantaneous windspeed, $L$ is the latent heat of vaporization, and $\rho$ is the density of dry air.

The usefulness of this equation depends upon the instantaneous measurement of humidity and heat fluxes (Moore et al. 1976) resulting in expensive and highly sophisticated instrumentation. A device, the HYDRA, has been developed by the Institute of Hydrology, Wallingford and used in a study of evaporation at Lake Toba, Indonesia (Sene et al. 1991).

The profile method relies on either the measurement of vertical humidity gradients (Sellers 1965) or simultaneous changes in temperature and humidity. This approach is based on the equation:

$$QLE = \rho L K_\omega \left( \frac{\Delta q}{\Delta z} \right)$$

where $K_\omega$ is the eddy diffusivity of water vapour in the air, $\Delta q$ is the atmospheric humidity difference between two heights $\Delta z$ apart, and $\Delta z$ is the height increment.

Unfortunately the diffusivity of water vapour is a variable, especially when the air is buoyant above the boundary layers, and the measurements of humidity need to be precise. An alternative approach called Bowen's ratio is therefore often used, based on the assumption that the thermal diffusivity and the diffusivity of water vapour are equal, then:

$$QH = \rho c_p K_h \frac{\Delta T}{\Delta z}$$

where $QH$ is the sensible heat flux, $c_p$ is the specific heat of air at constant pressure, $K_h$ is the thermal diffusivity, and $\Delta T / \Delta z$ is the temperature gradient in the vertical.

Then Bowen's ratio is:

$$B = \frac{QH}{QLE} = \frac{c_p \Delta T}{L \Delta q}$$

and if

$$Q^* = QLE + QH + QS,$$

and

$$QLE = (Q^* - QS)/(1 + B).$$

Unfortunately the profile approach, like the eddy correlation technique, requires exact and instantaneous measurements, in addition to steady state conditions. Hence it is mainly used for research applications, where accurate results over short time periods are required (e.g. Stannard and Rosenberry 1991). Routine measurements of evaporation tend to rely upon the water balance approach but this method is limited by data reliability. Atmometers and pans are widely used because of low cost and simplicity but they both suffer from errors.                    CTA

### References

Doorenbos, J. and Pruitt, K.C. 1975: *Crop water requirements*. FAO Irrigation and Drainage Paper 24, Rome. · Jackson, I. 1989: *Climate, water and agriculture in the Tropics*. 2nd edn. New York: Longman. · Oke, T. 1987: *Boundary layer climates*. 2nd edn. London: Routledge. · Moore, C.J., McNeil, D.D. and Shuttleworth, W.O. 1976: *A review of existing eddy correlation sensors*. Institute of Hydrology Report 32, Wallingford. · Rosenberg, N.J. 1974: *Microclimate: the biological environment*. New York: Wiley. · Sellers, W.D. 1965: *Physical climatology*. London: University of Chicago Press. · Sene, K.J., Gash, J.H.C. and McNeil, D.D. 1991: Evaporation from a tropical lake: comparison of theory with direct measurements. *Journal of hydrology* 127, pp. 193–217. · Stannard, D.I. and Rosenberry, D.O. 1991: A comparison of short term measurements of lake evaporation using eddy correlation and energy budget methods. *Journal of hydrology* 122, pp. 15–22. · Ward, R.C. and Robinson, M. 1990: *Principles of hydrology* 3rd edn. London: McGraw-Hill.

**evaporite**   A water-soluble mineral (or rock composed of such minerals) that has been deposited by precipitation from saline water as a result of evaporation, especially in coastal sabkhas or in salt lakes of desert areas. Among the most common minerals are sodium chloride (halite) and calcium sulphate (gypsum and anhydrite).                    ASG

### Reading

Warren, J.K. 1989: *Evaporite sedimentology*. Englewood Cliffs, NJ: Prentice-Hall.

**evapotranspiration**   The diffusion of water vapour into the atmosphere from a vegetated surface. Evapotranspiration is influenced by climate, soil hydrology and plant characteristics and this term has been criticized for combining these elements when they can respond in different ways to energy and moisture variations. Nevertheless, it is convenient to consider a homogeneous vegetated surface for many hydrological applications and this concept is widely used.

Plant stomata remain open for the processes of photosynthesis and respiration, which are essential for growth. This exposes the moist interiors of the stomata and encourages the diffusion of water vapour, that is, transpiration. In addition, water can be lost through small openings in the corky tissue covering stems and twigs. This water loss establishes a water deficit within the plant that may be transmitted to the roots. Whether or not this occurs depends upon a number of factors, e.g. plant physiology, moisture content, size of deficit, etc. (Begg and Turner 1976). Providing water is not limiting, transpiration rates will be largely controlled by atmospheric conditions and the characteristics of the vegetated surface, i.e. extent, structure and stomatal behaviour. Hence the concept of potential evapotranspiration relates to a specific plant cover, in addition to climate. Several definitions have been proposed and Ward and Robinson (1990) quote an approach by Penman that has gained widespread acceptance: 'Evaporation from an extended surface of a short green crop, actively growing, completely shading the ground, of uniform height and not short of water.'

Potential evapotranspiration ($PE_t$) is frequently calculated for a grass cover, but these values can be converted to any cropped surface by the crop coefficients provided by Doorenbos and Pruitt (1975) and Wright (1982) where:

$$\frac{PE(crop)}{PE_t} = \text{crop coefficient.}$$

The crop coefficient is assumed to be a constant, depending upon ground cover, i.e. state of development of the plant and prevailing meteorological conditions.

Actual evapotranspiration ($AE_t$) rates may exceed potential rates because of an increase in the extent of the evaporating surface, e.g. in forested areas. The reverse frequently occurs, due to water deficits. Immediately after rain has fallen the evaporation component of evapotranspiration will be high, but it will decrease markedly as soil and plant surfaces dry out. Thereafter, transpiration will be the main contributor relying upon soil moisture supplies, but water may only be absorbed by the roots when the soil has a higher moisture potential. Hence transpiration rates will be reduced as soil moisture contents decrease, but the exact relationship between evapotranspiration and soil moisture is complex. Hagan and Stewart (1972) present a comprehensive tabulation of the soil moisture contents at which crops start to undergo stresses that will ultimately reduce yields. While useful for irrigation scheduling, this study does not indicate the rates at which transpiration will decrease beyond these threshold values. Jackson (1989) shows that $AE_t$ responses to a limiting moisture supply are complex, being influenced by $PE_t$ rates and plant physiological adaptations including the rooting system and stomatal opening.

Actual evapotranspiration is then a combination of atmospheric conditions, soil moisture content and plant characteristics. Rates of actual water loss can be measured directly by monitoring plant water status, soil moisture changes or by using a lysimeter (see EVAPORATION). The practical difficulties of routinely monitoring such moisture changes over an extensive area have resulted in studies inferring $AE_t$ rates from estimates of $PE_t$.

Various empirical formulae have been devised to measure potential evapotranspiration, based on the assumption that it is controlled by the earth's energy balance which is reflected in mean monthly temperatures, e.g. Blaney-Criddle and Thornthwaite. These formulae rely upon empirical coefficients that limit their applicability to regional monthly water balance computations. Penman (1948) devised an alternative formula by applying understanding of the diffusion of water vapour from a moist surface to incorporate both energy ($H$) and aerodynamic ($EA$) functions in the following form:

$$PE_t = \left(\frac{\Delta}{\gamma} H + EA\right) / \left(\frac{\Delta}{\gamma} + 1\right)$$

where $PE_t$ is the potential evapotranspiration (mm of water $day^{-1}$), $\Delta$ is the slope of the saturation vapour pressure curve at mean air temperature ($mb\,°C^{-1}$), and $\gamma$ is the psychometric constant ($0.66\,mb\,°C^{-1}$).

Observations of temperature, windspeed, vapour pressure and either duration of daylight or the net radiation balance are required on a daily basis. The equation does contain a large number of constants but fortunately values have been calculated for most parts of the world and hence it has been widely employed in numerous studies. Unfortunately, two of the assumptions upon which the formula is based, that the moisture and heat fluxes from the ground are constant to a height of 1 m and that the heat stored in the ground is negligible, may become violated for time periods of less than one day. Shuttleworth (1988) suggests that such semi-empirical equations are only reliable over periods of ten days or more.

The Penman formula has become the standard approach for the measurement of $PE_t$ for over forty years because of its general reliability and modest data requirements. However, in 1990 (Monteith 1991) the FAO decided to phase out the use of $PE_t$ and crop coefficients to calculate crop water requirements through dissatisfaction with this empirical approach. In the future the 1965 Penman–Monteith revision (Monteith 1981) will be used. Monteith adopts a more rational process approach by incorporating the aerodynamic resistances created by the vegetated surface with the surface resistances to the diffusion of water vapour exerted by the transpiring plant. Thus $AE_t$ rates can be obtained with knowledge of these resistances and this approach has been used in the MORECS (Thompson *et al.* 1981) system of $AE_t$ calculation in the UK.

An alternative approach, the CRAE, summarized by Morton (1983) has emphasized the importance of areal estimates and rejects the sole use of point meteorological measurements because of the complementary relationship between $PE_t$ and $AE_t$. That is, when water is scarce, $AE_t$ will be low, relative humidities will be low, temperatures will be high and consequently $PE_t$ will be high. Calculations of areal $AE_t$ are made using knowledge of water availability and $PE_t$ rates (Kovacs 1987), but criticisms have been levelled concerning processes in the convective boundary layer of the atmosphere. Lemeur and Zhang (1990) have compared estimates of $AE_t$ by the CRAE and Penman–Monteith methods and concluded that the latter was the most appropriate but there remains much debate over the most suitable method for $AE_t$ determination.          CTA

**References**
Begg, J.E. and Turner, N.C. 1976: Crop water deficits. *Advances in agronomy* 28, pp. 161–217. · Doorenbos, J. and Pruitt, W.O. 1975: *Crop water requirements*. FAO Irrigation and Drainage Paper 24, Rome. · Hagan, R.M. and Stewart, J.I. 1972: Water deficits, irrigation design and programming. *Proceedings of the American Society of Civil Engineers Irrigation and Drainage Division* 98, pp. 215–35. · Jackson, I. 1989: *Climate, water and agriculture in the Tropics*. 2nd edn. New York: Longman. · Kovacs, G. 1987: Estimation of average areal evapotranspiration. *Journal of hydrology* 95, pp. 227–40. · Lemeur, R. and Zhang, L. 1990: Evaluation of three evapotranspiration models in terms of their applicability for an arid region. *Journal of hydrology* 114, pp. 395–411. · Monteith, J.L. 1981: Evaporation and surface temperature. *Quarterly Journal of the Royal Meteorological Society* 107, pp. 1–27. · — 1991: Weather and water in the Sudano-Sahelian zone. *Proceedings of Niamey Workshop on Soil Water Balance in the Sudano-Sahelian Zone*. IAHS Publ. 199, pp. 11–29. · Morton, F.J. 1983: Operational estimates of areal evapotranspiration and their significance to the science and practice of hydrology. *Journal of hydrology* 66, pp. 1–76. · Penman, H.L. 1948: Natural evaporation from open water, bare soil and grass. *Proceedings of the Royal Society, Series A* 193, pp. 120–45. · Shuttleworth, J.W. 1988: Macrohydrology; the new challenge for process hydrology. *Journal of hydrology* 100, pp. 31–56. · Thompson, N. Barrie, I.A. and Ayles, M. 1981: The Meteorological Office rainfall and evaporation calculation system: MORECS. *Hydrological Memoir 45*. Meteorologi-

cal Office, Bracknell. · Ward, R.C. and Robinson, M. 1990: *Principles of hydrology.* 3rd edn. London: McGraw-Hill. · Wright, J.L. 1982: New evaporation crop coefficients. *Proceedings of the American Society of Civil Engineers Irrigation and Drainage Division* 108, pp. 57–74.

**evergreen** Plants which have long-lasting leaves. If the length of the photosynthetic or growing season is short, or if nutrients are in very short supply, evergreen forms are favoured since they do not have to use resources in building up their photosynthetic capacity each year. The relative dominance of evergreen trees and shrubs in high latitude and high altitude areas, and of evergreen dwarf shrubs in peatlands, are examples of this response to climate and nutrient availability. **KEB**

**Reading**
Crawley, M.J. ed. 1986: *Plant ecology.* Oxford: Blackwell Scientific. · Larsen, J.A. 1982: *Ecology of the northern lowland bogs and conifer forests.* New York and London: Academic Press.

**evolution** Cumulative development in the characteristics of species over time. Classically, evolution is regarded as the progressive change in features of populations occurring through the course of sequential generations brought about by the process of natural selection. In fact, evolution must also explain the diversification of organisms of presumed common ancestry through geological time and must, therefore, account for speciation and extinction as well as progressive change as a result of natural selection. Convergent evolution is the process by which unrelated or distantly related groups evolve similar morphologies or adaptational traits, e.g. the reduced limbs of whales and penguins. COEVOLUTION is the contemporaneous evolution of two ecologically linked taxonomic groups such as flowering plants and their pollinators.

Charles Darwin is regarded as the father of evolutionary theory, although natural scientists had been grappling with the problem of explaining the diversity of life well before his publication, in 1859, of the first edition of *On the origin of species by natural selection.* In 1832, the British geologist Charles Lyell applied the term evolution to the notion of organic transmutation over time. Indeed, Charles Darwin himself did not use the word extensively at all, preferring instead the phrase 'descent with modification'. Darwin's famous voyage, as ship's naturalist, aboard the British brig, HMS *Beagle*, undertaken between 1831 and 1836, was to be instrumental in formulating his ideas about evolution. His short visit to the Galapagos Islands, 800 km off the west coast of South America, appears to

have been especially influential. It was here that Darwin observed and reported on morphological variations between individuals belonging to the same species and was also struck by the degree of adaptation to the environment displayed by several apparently closely related different species of 'finch', all confined to different habitats or to completely different islands. These ideas were to prove crucial in the development of his idea of modification between generations, and of common ancestry – in short, of evolution.

It transpires that Darwin and a contemporary British zoologist, Alfred Russel Wallace, developed the idea of evolution by natural selection independently. Wallace penned a letter to Darwin in 1858 in which he outlined his own hypotheses concerning the development of species through time and which appear to have persuaded Darwin that it was time to go public with the theory. On 1 July 1858 a joint paper was presented, *in absentia*, to the Linnean Society in Piccadilly and *Origin* soon followed. Furore greeted its publication fuelled by the logical conclusion of common ancestry that humans were, according to Darwin and Wallace's ideas, descended from the apes, an idea which clearly affronted the dominant conservative Christian philosophy of the Victorian age. But acceptance did eventually come and, indeed, the theory profoundly influenced the development of thinking on a number of other aspects of physical geography, such as the CYCLE OF EROSION and SUCCESSION. It appears to have frustrated Darwin that the crux, for him at least, of the idea of evolution, that of natural selection (or 'survival of the fittest' as it has been commonly perceived), was relatively ignored in the heat of the debate that centred on common ancestry.

The essential mechanism of natural selection may be summarized as follows:

1 In most species, far more young are produced than can possibly survive to reproductive age.

2 Organisms vary: individuals in a population differ from each other in numerous minor morphological and behavioural ways. Some of these variations are heritable and may be passed on to the succeeding generation.

3 There is a 'struggle' for existence, competition between organisms in which those better suited to the prevailing environment are more likely to survive to reproductive age.

4 Individuals with characteristics best suited to survival are more likely to reproduce

successfully and will, ultimately, predominate in a population. Adaptive change is therefore a result of natural selection operating over long periods based on the gradual accumulation of small, favourable variations present in populations.

Acceptance of these ideas was slow in coming, not least because the mid-nineteenth century scientific understanding of the processes of 'heritability' was incomplete. Gregor Mendel's theory of 'particulate inheritance' put forward in the 1860s certainly provided a mechanism, although it was only eighty years later that the 'modern synthesis' emerged (Huxley 1942). According to this integration, evolution was interpreted as a combination of speciation operating in concert with natural selection. Processes other than natural selection, for example GENETIC DRIFT, may also play a role.

More recently, Darwin and Wallace's ideas have come under scrutiny, especially in regard to the pace of evolutionary change. Their view of the process of natural selection was one of incremental, gradual accumulation of 'successful' adaptations within a species over time. The fossil record, on the other hand (despite its almost notorious incompleteness and arguable unreliability) reveals no such gradual changes in morphology of species over time, but rather a history of relative stability interrupted by relatively rapid changes. In 1977 two American biologists, Stephen Jay Gould and Niles Eldredge, published their explanation of this observation. They argued that evolution is not a gradual process, but is instead characterized by relatively long periods of negligible or limited change (stasis) punctuated by short, intense periods of rapid development based on speciation events. Gould (1996) has further argued that apparently progressive adaptation to environmental conditions over time, as implied by natural selection, is probably an artefact, rather than fundamental, of the way the process operates.

The most recent interpretations of evolution described as a 'post-modern' synthesis (Bennett 1997) assess the relative importance of natural selection and speciation events in the evolutionary process. Bennett argues, on the basis of evidence from the Quaternary period, that the incremental changes brought about by natural selection are relatively unimportant in macroevolution and that the processes of migration, extinction and speciation prompted by environmental changes on $10^3$ to $10^4$ year time-scales are the real pacemakers of evolution.     MEM

**Reading and References**
Bennett, K.D. 1997: *Evolution and ecology: the pace of life.* Cambridge: Cambridge University Press. · Darwin, C.

1859: *On the origin of species by natural selection, or the preservation of favoured races in the struggle for life.* London: John Murray. · Desmond, A. and Moore, J. 1991: *Darwin.* London: Michael Joseph. · Gould, S.J. 1996: *Life's grandeur: the spread of excellence from Plato to Darwin.* London: Jonathan Cape. · —and Eldredge, N. 1997: Punctuated equilibria: the tempo and mode of evolution reconsidered. *Paleobiology* 3, pp. 115–51. · Huxley, J. 1942: *Evolution: the modern synthesis.* London: Allen and Unwin.

**evorsion**   The erosion of rock or sediment in a river bed through the action of eddies and vortices.

**exaration**   The process by which glaciers pluck or quarry bedrock. Abrasion is not involved.

## Exchangeable Sodium Percentage (ESP)

This measures the proportion of sodium among the major cations present in the soil exchange complex (see CATION EXCHANGE). It is given by:

$$ESP = \{[Na^+]/[Na^+ + K^+ + Ca^{2+} + Mg^{2+}]\} \times 100$$

ESP is significant because the abundance of sodium in the soil exchange complex strongly influences the behaviour of the soil when in contact with water. The sodium ion is small and carries only a single electrical charge. It is therefore very mobile in the soil environment. When soil with a high ESP is wetted by dilute rainwater, sodium rapidly moves into solution in the immediate vicinity of the soil clays where it ordinarily resides. This sets up strong concentration gradients between the soil water and the rainwater, and osmosis drives water molecules from the rainwater into the more saline water among the soil clays. This flow of water requires that there be expansion of the soil matrix, and this may disrupt the soil aggregates, leading to dispersive behaviour and the release of small, readily erodible particles. This unstable behaviour that is triggered by large amounts of sodium can be reduced if the soil solution carries large concentrations of other salts, whose presence restricts the entry of additional sodium. Alternatively, if very soluble materials are applied to the soil surface, so that infiltrating rainwater rapidly collects a load of dissolved ions, the damaging effects of osmotic breakdown can be curtailed.     DLD

**Reading**
Brady, N.C. and Weil R.R. 1996: *The nature and properties of soils* 11th edn. New Jersey: Prentice-Hall.

**exfoliation**   Onion-weathering or desquamation. The weathering of a rock by peeling off of the surface layers.

**exhaustion effects** Encountered frequently in detailed studies of the variation of suspended sediment concentrations in rivers and streams during storm run-off events, and attributable to a progressive reduction or exhaustion in the availability of sediment for mobilization and transport. They may occur during an individual event, where they will be reflected in the occurrence of maximum suspended sediment concentrations and loads before the hydrograph peak, or during a closely spaced sequence of hydrographs, where they will give rise to a progressive decrease in the sediment concentrations associated with similar levels of water discharge.
DEW

**exhumation** The exposure of a subsurface feature through the removal by erosion of the overlying materials.

**exogenetic** Pertaining to processes occurring at or near the surface of the earth and to the landforms produced by such processes. By contrast ENDOGENETIC processes originate within the earth.

**exotic** A term normally used to describe a plant or animal which is kept, usually in a semi-natural or artificial manner, in a region outside its natural provenance. A wide range of garden plants and aviary birds are exotics, having been introduced by horticulturalists and bird fanciers from widely differing regions of the world. The term tends to be used in a rather restricted sense to describe especially spectacular species collected from markedly different climatic regimes. Exotic should be contrasted with the term ALIENS, which is the commoner designation for introductions which are more generally naturalized in their new locations.     PAS

**Reading**
Whittle, T. 1970: *The plant hunters*. London: Heinemann.

**expanding earth** The idea that the physical size of the planet may have increased through geological time has a long history, having been raised periodically during the eighteenth and nineteenth and twentieth centuries. Various lines of evidence have been put forward in support of the idea. Carey (1975, 1976) performed exacting map reconstructions of the supercontinent of Pangaea in the configuration of about 200 Ma BP but found that there were always gaps, which he termed *gaping gores*, that were difficult to account for. He found that these could all be closed at once if the continental reassembly were carried out on a globe of smaller diameter than the present earth (70–80 per cent of its present size). Others (e.g. Steiner 1977) tallied the areas of new crust created by sea-floor spreading and consumed along SUBDUCTION ZONES, and argued that more crust had been created than consumed, therefore requiring the surface area of the globe to increase. Other geometric arguments have been raised, including the difficulty of reconciling the simultaneous expansion, since the break up of Pangaea, of the Pacific and Atlantic Oceans, both of which have evolved since that time. The length of the day has increased through geological time, as revealed by the number of diurnal growth bands per annual cycle in fossil corals. Earth expansion has been raised as a mechanism to account for the slowing rotation of the planet that is the cause of this day length variation, but it rather appears to be related to the steady retreat of the moon.

Diverse mechanisms have been proffered to account for the expansion of the planet. Ongoing planetary cooling, recrystallization and change in rock fabrics and overall density provide one category. Another involves the very slow temporal change in what are thought of as 'constants', like the gravitational constant G. Decline in parameters like this, involved fundamentally in determining the forces that bind matter together, are envisaged in a number of cosmological hypotheses, and may play a part in the observed expansion of the universe.     DLD

**References**
Carey, S.W. 1975: The expanding earth: an essay review. *Earth science reviews* 11, pp. 105–43. · —1976: *The expanding earth*. Amsterdam: Elsevier. · Steiner J. 1977: An expanding earth on the basis of sea-floor spreading and subduction rates. *Geology* 5, pp. 313–18.

**expansive soils** Soils whose volume can be increased substantially as they become wet, and which shrink as they dry out, leading to a range of what are termed 'shrink–swell phenomena'. The wetting and drying cycles are mostly seasonal. Because of the requirement that there be a large surface area across which water can be exchanged, expansive soils commonly have at least 30 per cent clay, often fine clay, and the most important swelling minerals are 2:1 layered smectite clays such as montmorillonite. Water and exchangeable cations enter the spaces between structural sheets within these minerals, causing a volume increase. As determined using a ribbon of soil whose length is measured when dry and when wet, expansion may amount to 20 per cent in some cases. Volume increase accounts for a number of distinctive features in expansive soils, including marked local variation in the soil caused by subsoil material being forced towards the surface (associated shearing of the soil may produce SLICKENSIDES), and by various kinds of buckling of the soil surface (see GILGAI). Drying of these soils may be associated with the opening of deep cracks in the soil, which may later conduct water and facilitate re-wetting of the subsoil. Expansive soils form the Vertisol order in the US Soil Taxonomy. They are associated with many management problems, including damage to foundations and roads, cracking of pipelines, and leakage from impoundments via tension cracks.     DLD

**Reading**
Coulombe, C.E., Wilding, L.P. and Dixon, J.B. 1996: Overview of vertisols: characteristics and impacts on society. *Advances in agronomy* 57, pp. 289–375.

**experimental catchment**  A small drainage basin used for detailed investigations of hydrological and geomorphological processes. During the UNESCO International Hydrological Decade (1965–74) the term was frequently used in a more limited context to refer to small ($< 4\,\mathrm{km}^2$) catchment studies, where the vegetation cover or land use was deliberately modified in order to study the hydrological impact of such changes. These studies were seen as experiments. Various strategies involving single, paired and multiple watershed experiments have been employed in order to decipher the nature and magnitude of changes in catchment behaviour.     DEW

**Reading**
Rodda, J.C. 1976: Basin studies. In J.C. Rodda ed., *Facets of hydrology.* Chichester: Wiley. · Toebes, C. and Ouryvaev, V. eds. 1970: *Representative and experimental basins: an international guide for research and practice.* Paris: UNESCO.

**experimental design**  The prior planning, in terms of methods, sampling, statistical replication, and validation, of a programme of data collection, experimentation or testing. This forms a vital initial phase of HYPOTHESIS testing in geographical work as well as across the disciplines generally. It is necessary to consider, in designing an experiment, the variability (temporal and/or spatial) of many phenomena, because this will determine where and when sampling would be required in order to derive representative values. Some phenomena, like distributions of soil moisture, vary diurnally, or seasonally, or are dependent on antecedent weather conditions. Often, there will be marked spatial variation also. In sampling such phenomena, the researcher has to consider the number of samples that would be required in order to estimate the value of the variable with an acceptable level of error, and whether the time and budget available would permit this. In studying a process like soil erosion where there are many

influential variables, a decision has to be made about which will be measured. The ones selected need to be physically informative, perhaps statistically independent, and adequate for resolution of the research problem being studied. In addition, the particular means by which a variable will be measured requires consideration. For instance, in describing plant cover on an experimental plot, a worker might employ projected canopy cover, or biomass, or might decide to tally separately the leaf area of individual species growing at different heights above the soil surface. Each description has some value, but the worker must evaluate the various options in advance of fieldwork, and select the means of description (termed *parameterization*) that seems likely to yield the most relevant data.

Anderson and McLean (1974) identified 11 stages in the conduct of an experiment:

1  recognition that a problem exists;
2  formulation of the problem;
3  selecting factors and treatments to be used in the experiment;
4  specifying the variables to be measured;
5  definition of the 'inference space' for the problem: i.e. the range of conditions to which it is hoped that the results will apply;
6  random selection of subjects or sites;
7  assignment of treatments to the subjects or sites;
8  deciding on the kinds of statistical analysis that will be used with the data;
9  collection of the data;
10  analysis of the data;
11  drawing up of conclusions.

Not all of these will be required in all geographical studies, but many will. Jumping from step 1 or 2 to step 9 is all too common, but can only result in mistakes being made and inefficiency in the collection and later attempted analysis of the results. Often, it is desirable to run a small pilot investigation after step 2, in order to become more familiar with the issues and locations involved, and to test data gathering equipment or methods. Especially if working in a remote area, or one that is costly to visit, a 'shakedown' trial will often prove helpful.     DLD

**Reading and Reference**
Anderson, V.L. and McLean, R.A. 1974: *Design of experiments.* New York: Marcel Dekker. · Kempthorne, O. 1975: *The design and analysis of experiments.* New York: Krieger.

**extending flow**  See COMPRESSING FLOW.

**extinction**  The disappearance of an individual organism, a group of organisms or a local population from existence. Although many recorded extinctions are directly attributable to the actions of man, the fossil record clearly reveals that nearly all lineages have become extinct without leaving descendants and that evolutionary conservatism will frequently lead to extinction in the face of significant environmental changes, such as climatic change, the development of new competitors, predators or diseases, or the loss of a major food supply. Many classic mass extinctions, such as that of the dinosaurs, remain enigmas unexplained by modern science, although there is never a dearth of speculation, scientific or otherwise, on the causes of their demise. The process of extinction continues, however, and over the past 200 years or so it is estimated that some 53 birds and 77 mammals have become extinct.

One particularly relevant principle in the discussion of extinction is that of 'competitive exclusion' or the 'ecological replacement principle'. When one group disappears, another appears to take its place in a corresponding or identical habitat. This theory therefore asserts that when an ecological resource is used simultaneously by more than one kind of organism and when this resource is insufficient to furnish all their needs, then the better or best adapted organism will eventually eliminate the competitors. It has even been argued that new forms must always replace old ones and that for every species that evolves, another becomes extinct. It is, however, clear that competition is not always present in the process of extinction, which may simply reflect evolutionary conservatism in relation to the speed and direction of environmental change. Indeed, competition may resolve itself by producing diverging adaptations, leading to overall diversification and not to extinction. Other factors involved in extinctions include overspecialization, reduced mutation, loss of behavioural versatility and changes in community patterns.

Whatever the causes, the palaeontological record clearly indicates that extinction has paralleled evolution throughout the entire history of the plant and animal kingdoms, a fact which has yet to be faced by the modern CONSERVATION movement, which so often aims to maintain all forms, whatever the cost and against all odds, in the battle between birth and death. (See also ISLAND BIOGEOGRAPHY.)                    PAS

### Reading

Cox, C.B. and Moore, P.D. 1993: *Biogeography: an ecological and evolutionary approach*. 5th edn. Oxford: Blackwell Scientific. · Krebs, C.J. 1978: *Ecology: the experimental analysis of distribution and abundance*. New York: Harper & Row. · MacArthur, R.H. 1972: *Geographical ecology*. New York: Harper & Row. · Stanley, S.M. ed. 1987: *Extinction*. New York: Scientific American Library.

**extra-tropical cyclone**  An area of low pressure which develops in the westerly wind belts of middle latitudes and is associated with characteristic weather patterns. Also known as a depression or low. At the surface the extra-tropical cyclone consists of an area of low pressure surrounded by winds blowing in an anticlockwise direction in the northern hemisphere and in a clockwise direction in the southern hemisphere. Central pressure values vary greatly from about 930 hPa to 1015 hPa. The most intense cyclones usually occur in winter and,

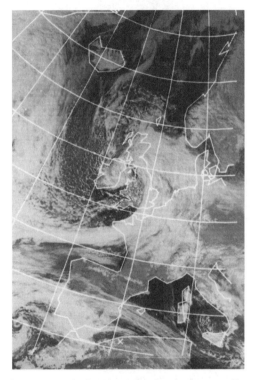

**An extra-tropical cyclone**  is centred over eastern Ireland with dense cloud forming the characteristic spiral away from the centre. Ahead of the system the cloud boundary is rather diffuse with evidence of some cirrus cloud over southern Norway. The sudden change in cloud type at the cold front is visible down the centre of England where convective shower clouds in the cold air behind the front contrast with the layer cloud ahead of the front. Over the Atlantic Ocean to the west of Ireland, patterns are clearly visible indicating variable convection. Another cyclone is seen to the west of Spain together with jet stream clouds over North Africa. The mountains of Norway, the Alps and Pyrenees stand out because of their snowcover. Photograph taken at 14.55 GMT on 17 March 1980.

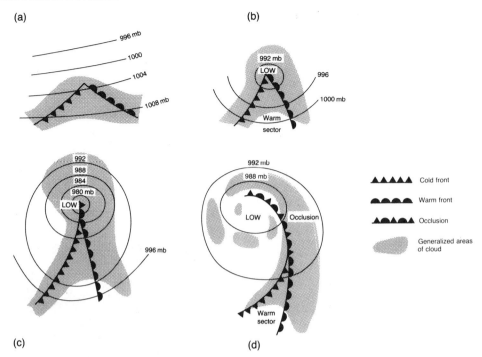

(a)

(b)

(c)

(d)

Cold front

Warm front

Occlusion

Generalized areas of cloud

on average, those of the southern hemisphere are deeper than those in the northern hemisphere. The diameter of the system may range from about 500 km to over 3000 km. The associated weather is unsettled with cloud, strong winds, periods of rain and sudden changes of temperature as the frontal zones pass. Satellite photographs show the beauty of the cloud patterns around the cyclones. They usually consist of a large 'comma-shaped' mass of cloud near the centre of the low with bands of frontal cloud near the warm and cold fronts streaming away towards the equator on a curved path (see illustration). Many differences of detail are found between cyclones but almost all follow this basic pattern. Over time the cyclones tend to move eastwards or north-eastwards, but some will remain almost stationary or follow unusual tracks depending upon air circulation in the middle atmosphere.

Cyclones are dynamic features of our atmosphere undergoing a sequence of changes from initiation to decay. The start of cyclone development involves a weak surface trough or front and a wave in the upper-level westerlies. Intensification of the surface low takes place as a result of divergence in the upper westerly wave which leads to the formation of a cyclonic circulation along the surface front. Establishment of the cyclonic circulation allows the flow of warm air to occur to the east of the cyclone centre and cold air transfer to the west of the centre. This temperature modification changes the distribution of uplift from one in which there is a general area of ascent and low-level convergence to one in which ascent takes place mainly east and poleward of the surface low and descent west and equator-ward of the low. The circulation associated with the cyclone causes the cold front to move eastward and equator-ward and the warm front to move eastward and poleward. This stage of development is shown in figure 1(a). Further advection of warm and cold air together with changes in the upper airflow produce a gradual (or sometimes rapid) change in the appearance of the surface low. By stage (d) the storm has reached its maximum intensity and started to occlude. Cold air is present at all levels over the cyclone centre. As the system decays, advection over the centre almost ceases. Surface friction and internal dissipation ensure that this eddy in the westerly circulation loses its identity.

In the northern hemisphere troughs in the westerlies have favoured positions of location because of the mountain ranges and the distribution of land and ocean. As a result cyclones have preferred areas of formation: off Newfoundland and north-east of Japan for example. In the southern hemisphere, the westerly flow is more symmetrical about the pole and longitudinal variations in cyclone formation are less pronounced. (See also CYCLONE.) PS

**Reading**
Carlson, T.N. 1988: *Mid-latitude weather systems*. London: Harper-Collins. · Palmén, E. and Newton, C.W. 1969: *Atmospheric circulation systems*. New York: Academic Press.

**extreme value theory**  A group of statistical methods that is widely applied in the study of natural phenomena such as floods. Extreme values (EVs) represent the outliers among populations, such as the shortest individuals, or the most intense storm events, or the largest flood experienced in each year of a period of record. Sets of such extreme values often follow a form of statistical distribution that is called an *extreme value distribution*. Many different mathematical functions have been employed to describe and analyse data for which some form of extreme value distribution is appropriate. These include the Gumbel, log Gumbel, Weibull, and others (Kinnison 1985). Distributions of extreme events can be investigated in two ways – in relation to the *numbers* of events of particular sizes, and the distribution of *sizes* of the events themselves. Different kinds of distributions apply to each form of investigation. If an appropriate EV function can be fitted to some observational flood size data collected over 30 years, for example, then it may be used to estimate the magnitude of the larger events that might have been encountered had a longer period of record, say 500 or 1000 years, been available.          DLD

**Reference**
Kinnison, B.R. 1985: *Applied extreme value statistics*. Columbus: Battelle Press.

**extrusion flow of glaciers**  A view that the lower layers of a glacier are squeezed out by the weight of the overlying ice and move faster than it. The concept has been abandoned as physically impossible.

**extrusion, volcanic**  A feature produced by rocks which have been deposited at the earth's surface after eruption, in a molten or solid state, from volcanic vents and fissures.

**exudation basin**  A depression occurring at the head of glaciers emanating from the Greenland ice cap.

**eye**  The centre of a cyclone where air descends from the upper atmosphere filling the low pressure zone. A calm area around which cyclonic winds blow.

**eyot**  An islet in a river or lake.

# F

**fabric** The three-dimensional arrangement of the particles of a sediment. It is a bulk property and can be specified at a variety of scales and in a number of ways. It is possible to refer to the 'clay fabric', the organization and arrangement of clays over a volume of a few millimetres. Most usually, clay fabrics are related to the engineering properties of the soil; especially to the response to STRAIN; this is similar to the analysis of strain and fabrics of metamorphic rocks. The clay fabric may have a rather different (microfabric) from the overall fabric, which takes much larger particles into account and covers a much greater volume. The volume taken is critical to measurement and interpretation of fabrics. The fabrics of stones (clasts) are often used as indicators of palaeocurrent direction, streams or ice sheet movement (TILL FABRIC ANALYSIS). They can also be used to help to identify a specific geomorphological mechanism, e.g. imbrication of pebbles in a stream, as well as to help explain the mass behaviour of a sediment, e.g. preferred permeability direction. Fabrics are sometimes determined in two dimensions but ideally three dimensions should be used. Where stones are employed, a preferred orientation is determined from analysis of a number (more than fifty) of individual measurements which are statistically analysed to give a measure of the fabric for that position. A clast long axis ($a$) needs to be determined and its orientation (or azimuth), with respect to, say, true north and its plunge (with respect to an angle below the horizontal) determined. These two values can be plotted on an equal area projection (or net) and contoured for all the clasts taken. Alternatively, the maximum projection ($a/b$) plane can be determined and the orthogonal to this (the $c$ axis) measured with respect to azimuth and plunge. This again can be plotted. Values of azimuth and plunge (of a line) or dip (of a plane) can be used to calculate mean orientations directly by various statistical measures and tested against several types of spherical distribution. WBW

**Reading**
Goudie, A.S. ed. 1990: *Geomorphological techniques*. 2nd edn. London: Allen & Unwin. · Whalley, W.B. 1976: *Properties of materials and geomorphological explanation*. Oxford: Oxford University Press.

**facet** A flat surface on a rock or pebble produced by abrasion.

**facies** The characteristics of a rock or sediment which are indicative of the environment under which it was deposited. A distinct stratigraphic body which can be distinguished from others on the basis of appearance and composition. Lateral variations in the nature of a stratigraphic unit.

**factor of safety** The ratio of RESISTANCE to FORCE: whenever this falls below 1.0, acceleration will take place. Most often this term is used in the context of slope stability; LANDSLIDE movement takes place when the sum of resisting forces falls below the sum of driving forces. Since the least resistant material is unlikely to have been sampled, and there is an error margin on the estimation of force, factors of safety below 1.3 are generally considered unstable; higher or lower threshold values may be appropriate. Since both resistance and force vary over time, the worst case expected should be used, e.g. zero cohesion or maximum pore-water pressure. Calculation of a factor below 1.0 means that some measurements or assumptions are inaccurate; otherwise movement would have taken place. IE

**falling dune** A TOPOGRAPHIC DUNE that has developed on the lee side of a topographic obstacle, as airflow and sand transport is disrupted by the barrier. A falling dune usually develops when a CLIMBING dune has reached the top of an obstacle, allowing sand to be transported over the barrier, permitting deposition on the lee side. (See also ECHO DUNE; LEE DUNE.) DSGT

**fallout** The descent of solid particles from the earth's atmosphere and/or the particles themselves. Finer, lighter particles tend to fallout further from source than coarser, heavier particles. The particles may be of natural origin, such as mineral dust from a dust storm, or from anthropogenic sources, such as sulphur particles emitted by an oil-fired power plant. Both can be construed as part of the flow of material through biogeochemical cycles. The measurement of caesium-137 deposited on soils after radioactive fallout from atmospheric weapons testing has been widely practised as a means to estimate erosion and deposition (e.g. Bajracharya *et al.* 1998). NJM

**Reference**
Bajracharya, R.M., Lal, R. and Kimble, J.M. 1998: Use of radioactive fallout cesium-137 to estimate soil erosion on three farms in west central Ohio. *Soil Science* 163.2, pp. 133–42.

**false-bedding**   The stratification of sediments in several units inclined to the general stratification. The product of fluvial, littoral and aeolian sedimentation.

**fan delta**   An alluvial fan extending into a body of water. The subaerial fan results from fluvial and slope processes, and shares many of the characteristics of alluvial FANS. The subaqueous fan is affected also by lake or marine processes, like a DELTA.                    IE

**Reading**
Schumm, S.A., Mosley, M.P. and Weaver, W.E. 1987: *Experimental fluvial geomorphology*. New York: Wiley. Pp. 351–65 and 372–4.

**fanglomerate**   Indurated alluvial fan gravel.

**fans**   Depositional landforms whose surface forms a segment of a cone that radiates downslope from the point where the stream leaves the source area. The coalescing of many fans forms a depositional piedmont that is commonly called a *bajada*. Each fan is derived from a source area with a drainage net that transports the erosional products of the source area to the fan apex in a single trunk stream. The plan view of the cone-shaped deposit is broadly fan-shaped with the contours bowing downslope. Overall radial profiles are concave and cross-fan profiles are convex. They vary greatly in size from less than 10 m in length to more than 20 km, and many large fans are thicker than 300 m. The debris that makes up fans decreases in size downfan, but is frequently coarse, and much of it has been transported by mudflow activity. Deposition is caused by decreases in depth and velocity where streamflow spreads out on a fan, and by infiltration of water into permeable superficial deposits.

Alluvial fans are widespread, especially in arid, mountainous and periglacial areas, but are especially notable in particular tectonic environments where there is a marked contrast between mountain front and depositional area. Uplift creates mountainous areas that provide debris and increased stream competence.                    ASG

**Reading**
Bull, W.B. 1977: The alluvial-fan environment. *Progress in physical geography* 1.2, pp. 222–70. · Rachocki, A. and Church, M.A. 1990: *Alluvial fans: a field approach*. Chichester: Wiley.

**fatigue failure**   Fracture as a result of repeatedly applied cyclic stresses, at levels far below the instantaneously determined strength of a material: it is widely recognized as a major contributor to failure in metals. In aircraft, regular vibrations have led to disastrous fracturing. Experimental studies of cyclic stresses in rocks have shown that fatigue fracturing occurs at 80–60 per cent of the ultimate strength and after a number of cycles which range from $10^3$ to $10^6$. The importance of fatigue failure in rocks exposed to repeated wetting and drying, thermal expansion and contraction, and salt crystallization and dissolution, is not known. It may, however, be a significant cause of several forms of rock fracturing which have not yet been explained satisfactorily.                    MJS

**Reading**
Selby, M.J. 1993: *Hillslope materials and processes*. 2nd edn. Oxford: Oxford University Press. Ch. 8.

**fault**   A crack or fissure in rock, the product of fracturing as a result of tectonic movement. The line along which displacement of formerly adjacent rocks has taken place as a result of earth movements. Normal faults develop under a pattern of stress which is predominantly tensional. A down-faulted block between a pair of more or less normal faults is known as a graben, and the up-faulted block is called a horst. By contrast, reverse faults are normally associated with zones of compression. Where the angle of dip is low the term thrust fault is used. When the mean compressive stress is vertical, strike-slip faults are formed (also called wrench faults and transcurrent faults). Where both horizontal and vertical movements are significant, the term oblique-slip fault is applied.                    ASG

**faunal realms**   The largest divisions into which the faunas of the world are customarily grouped; also called faunal kingdoms or empires. The earliest enduring classification of world faunas was drawn up by Sclater in 1858 and dealt with the distribution of birds, a group already well known by that date. This was then extended by Günther to include reptiles and finally consolidated into a widely accepted and applicable system by Alfred Russel Wallace, particularly in his early classic of zoogeography, *The geographical distribution of animals* (1876: see WALLACE'S REALMS). In this, he subdivided each of Sclater's original six regions into four subregions. The Sclater–Wallace division is still in use today, despite attracting continuous criticism, and it is constantly under revision. Smith (1983), for example, has attempted a logical and statistical derivation of world mammal faunal regions.

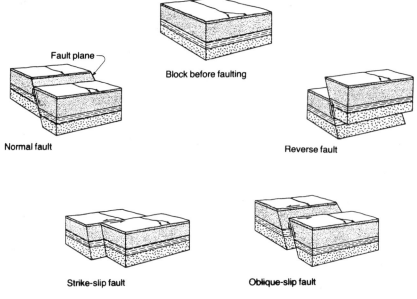

**Fault types**
*Source*: F. Press and R. Siever 1978: *Earth*. San Francisco: W.H. Freeman. Figure 20.33.

The main faunal realms normally recognized are:

1 The *Nearctic*, comprising virtually the whole of North America (e.g. the common raccoon, the American bison).
2 The *Neotropical*, including Central America and South America (e.g. sloths, opossums, vicuña, guanaco).
3 The *Palearctic*, embracing Europe and Asia north of the tropics (e.g. the European bison).
4 The *Ethiopian*, comprising Africa, mainly south of the Sahara (e.g. the aardvark, lemurs).
5 The *Oriental*, covering tropical Asia (e.g. the orang-utan, the tarsier, tree-shrews).
6 The *Australian*, a complex realm including parts of South-East Asia, Oceania, Australia and New Zealand (e.g. the Australian marsupials, such as the kangaroo, the Tasmanian wolf, etc.).

Because they have many closely related species or animals in common, the *Nearctic* and the *Palearctic* are usually linked together in one realm known as the *Holarctic* (e.g. lynx, wolf, fox, brown bear, weasel). This larger unit is, in turn, frequently joined with the *Ethiopian* and the *Oriental* realms to create one major single division known as *Arctogaea*, a concept first suggested by Huxley in his system. In Huxley's classification, the *Neotropical* realm was called *Neogaea* and the *Australian* realm *Notogaea*, terms still in use today.

The boundary zones between some of these realms are far from clear-cut and controversy has particularly surrounded that between the Australian and Oriental realms in the region known as Wallacea (see WALLACE'S LINE). Essentially, the realms reflect the present-day distribution of animals and are basically defined longitudinally by the great oceanic barriers and latitudinally by the subtropical warm temperate dry belt. It is also apparent that the different faunas are the product of continental drift, brought about through the processes of plate tectonics and sea-floor spreading, and that the distinctive marsupial fauna of the Australian realm, for example, survived because this land-mass was separated from the rest of the world before placental mammals were able to invade it and oust the more primitive forms. Left in isolation, the marsupials were able to radiate by ADAPTIVE RADIATION into a large number of different types. (See also FLORISTIC REALMS.)   PAS

**Reading and Reference**
Illies, J. 1974: *Introduction to zoogeography*. London: Macmillan. · Schmidt, K.P. 1954: Faunal realms, regions, and provinces. *Quarterly review of biology* 29, pp. 322–31. · Smith, C.H. 1983: A system of world mammal faunal regions. 1. Logical and statistical derivation of the regions. *Journal of biogeography* 10, pp. 455–66. · Udvardy, M.D.F. 1969: *Dynamic zoogeography*. New York: Van Nostrand Reinhold.

**feather edge**   The thin edge of a wedge-shaped sedimentary rock that tapers out and eventually disappears as it abuts an area where no deposition has taken place.

**feedback**  See SYSTEMS.

**feedback loops or homeostasis**  See SYSTEMS.

**feldspars**  Aluminosilicates of K, Na and Ca, and as such are important components of many rocks and rock weathering products. Feldspars have alkali and plagioclase types and in total frequency are the second most abundant mineral (after quartz) in SAND.

**felsenmeer**  See BLOCK FIELDS, BLOCK STREAMS.

**fen**  A low-lying area partially inundated by water which is characterized by accumulations of non-acid peat.

**feral relief**  Describes the landscape occurring in areas where the sides of the main valleys are dissected by insequent streams. It is a product of rapid run-off with intense dissection.

**ferrallitization**  A combination of intense weathering and efficient removal of the more soluble weathering products under warm, wet conditions. There are three basic aspects to the process: intensive and continuous weathering of parent material involving the release of iron and aluminium oxides and silica, as well as of bases; translation of soluble bases and silica; and formation of 1:1 kaolin-type clays.

**Ferrel cell**  A thermally indirect, weak, MERIDIONAL CIRCULATION in middle latitudes of the atmosphere, in which air rises in the colder regions around 60° latitude and descends in the warmer regions around 30° latitude (subtropic high pressure belts). This cell, in which the surface winds are predominantly westerly, forms part of a three-cell (in each hemisphere) mean meridional circulation pattern, the other two cells being the Hadley and polar cells.     KJW

**Reading**
Holton, J.R. 1992: *Introduction to dynamical meteorology*. New York and London: Academic Press.

**Ferrel's law**  States that all moving masses of air move to the right (as they proceed) in the northern hemisphere and to the left in the southern hemisphere. This is because air, in moving from higher to lower pressure, is subject to the effect of the earth's rotation (see CORIOLIS FORCE).     KJW

**Reading**
Palmén, E. and Newton, C.W. 1969: *Atmospheric circulation systems*. New York and London: Academic Press.

**ferricrete**  An iron-pan or a near-surface zone of iron oxide cementation. An accumulation of iron oxides and hydroxides within the soil zone as a result of weathering or soil-forming processes such as laterization. It is a type of duricrust and tends to be associated with deep weathering profiles.

**fetch**  The extent of the area of sea or ocean over which waves have developed through the effect of broadly unidirectional winds.

**fiard**  A river valley which has been drowned by the sea owing to a rise in sea level or subsidence of the land.

**field capacity**  The condition reached when a soil holds the maximum possible amount of water in its voids and pores after excess moisture has drained away. A measure of the volume of water a soil can hold under these conditions.

**field drainage**  The process of artificially accelerating the movement of water through soil. This is undertaken to lower levels of saturation, to reduce the extent of surface water ponding or run-off and to minimize the periods during which saturation or surface water exists on agricultural land. This process is alternatively called under-drainage, in contrast to the surface and arterial drainage which involve ditching and the modification of existing water courses to accelerate channelized run-off.

Field drainage may be implemented by excavating a network of trenches and inserting plastic piping or tubing, earthenware pipes or permeable rock fill, and then backfilling the trenches again. In heavy cohesive soils an extra network of mole drainage may be added. This is undertaken by drawing a former (the 'mole') through the soil at the required depth, the hole remaining open for a number of years afterwards. Deep ploughing ('subsoiling') may also accelerate field drainage.     JL

**Reading**
Green, F.H.W. 1978: Field drainage in Europe. *Geographical journal* 114, pp. 171–4. · Robinson, M. 1990: Impact of improved drainage on river flows. *Institute of Hydrology, Wallingford, Report*, 113.

**Finger Lakes**  A group of semi-parallel lakes in New York State. The name originated in the Indian legend of the Great Spirit placing his hand on the earth causing the finger-like indentations. An alternative explanation is that they were cut by glacial erosion.          DES

**Reading**
Clayton, K.M. 1965: Glacial erosion in the Finger Lakes region (New York State, USA). *Zeitschrift für Geomorphologie* 9, pp. 50–62. · Coates, D.R. 1966: Discussion of K.M. Clayton: Glacial erosion in the Finger Lakes region (New York State, USA). *Zeitschrift für Geomorphologie* 10, pp. 469–74. (Also in the same issue pp. 475–7, Clayton, K.M. Reply to Professor Coates.)

**fiord**  (fjord) A glacial trough whose floor is occupied by the sea. Fiords are common in uplifted mid-latitude coasts, for example in Norway, East Greenland, eastern and western Canada and Chile. Historically, the origin of fiords has generated a great deal of interest, for example an extreme view championed by Gregory (1913) that they are essentially tectonic forms. Today it seems easiest to think of them as essentially glacial with such features as steep sides, overdeepened rock basins and shallow thresholds at the coast as characteristic of erosion by ice streams or valley glaciers exploiting either pre-existing river valleys or underlying weaknesses in the bedrock (Holtedahl 1967).          DES

**Reading and References**
Gregory, J.W. 1913: *The nature and origin of fiords*. London: John Murray. · Holtedahl, H. 1967: Notes on the formation of fjords and fjord valleys. *Geografiska annaler* 49A, pp. 188–203. · Loken, O.H. and Hodgson, D.A. 1971: On the submarine geomorphology along the east coast of Baffin Island. *Canadian journal of earth sciences* 8.2, pp. 185–95. · Syvitski, J.P.M., Burrell, D.C. and Skei, J.M. 1987: *Fiords: processes and products*. New York: Springer-Verlag.

**fire**  This is a very important ecological factor, especially in regions with seasonally dry climates. Fire has played a significant role in the evolution of plant species in many different BIOMES, including boreal forests, Mediterranean-climate shrublands, SAVANNA and temperate grasslands and is a major determinant of species diversity. For plant species adapted to burning, so-called pyrophytes, fire is an integral and essential component of their life cycles, provided that the fire regime is appropriate. Fire-adapted species may survive fire either by resprouting, usually from underground parts, or by regeneration from soil or canopy-stored seed bank, in which case the adult plant is actually killed by the fire. Fire regime is accounted for by variations in the type, intensity, timing and frequency of fire, all of which are dependent on climate, availability of fuel and occurrence of ignition events. Fire has many positive implications for plant species with appropriate adaptations, since the combination of open space, increased nutrient availability and temporary reduction in numbers of potential seed predators is favourable for seedling establishment in the immediate post-fire environment. Fire may stimulate flowering, seed release or seed germination in such instances.

Humans, of course, have altered and manipulated the fire regime for a very long time, perhaps for more than one million years (Gowlett *et al.* 1981). Fire is utilized by people in a range of environments for a variety of reasons, for example, in clearing vegetation, attracting game, to deter pests and disease, to improve grazing for domestic stock and to make pottery etc. Given the virtual ubiquity of current human populations, it is difficult to know what would constitute a 'natural' fire regime. Most fire-prone vegetation types are now subject to anthropogenic burning regimes, either intentionally or otherwise and, indeed, it has been argued that some vegetation types, for example the savannas, may owe their distribution and characteristics to long-term human-induced fires.          MEM

**Reading and Reference**
Bond, W.J. and van Wilgen, B.W. 1996: *Fire and plants*. London: Chapman and Hall. · Gowlett, J.A.J., Hairns, J.W.K., Walton, D.A. and Wood, B.A. 1981: Early archaeological sites, hominid remains and traces of fire from Chesowanja, Kenya. *Nature* 294, pp. 125–9.

**firn**  The term generally applied to snow which has survived a summer melt season and which has not yet become glacier ice.

**firn line**  See EQUILIBRIUM LINE.

**fission track dating**  Based on the principle that traces of an isotope, $^{238}$U, occur in minerals and glasses of volcanic rocks, and that this isotope decays by spontaneous fission over time, causing intense trails of damage called tracks. These narrow tracks, between $5\mu$ and $20\mu$ in length, vary in their number according to the age of the sample. By measuring the numbers of tracks an estimate can be gained of the age of the volcanic minerals.          ASG

**Reading**
Miller, G.H. 1981: Miscellaneous dating methods. In A.S. Goudie ed., *Geomorphological techniques*. London: Allen & Unwin. Pp. 292–7.

**fissure eruption**  The eruption of volcanic gases, lavas and rocks through a large crack or chasm in the earth's surface rather than through a pipe or vent.

**Flandrian transgression**  Occurred as the ice sheets of the last glacial melted in the Holocene and Late Pleistocene. World sea levels rose,

**A fissure** developed in a lava platform in northern Iceland. Note the volcanic cone in the background. Many of the world's great spreads of lava were extruded from such fissures.

possibly by as much as 170 m. The continental shelves were flooded, and river and glacial valleys were transformed into rias and fiords respectively. This transgression was more or less complete by 6000 years ago, and may locally have reached a few metres above the present level. It caused the separation of Ireland from mainland Britain, and of Britain from the continent of Europe. In areas of very gentle slope, like the Arabian Gulf, the rapid lateral spread of the transgression (by as much as 100 m year$^{-1}$ at its peak) may have contributed to the biblical story of the flood.          ASG

**Reading**
Kidson, C. 1982: Sea-level changes in the Holocene. *Quaternary science reviews* 1, pp. 121–51.

**flash**  A local name from north-western England, especially Cheshire, for water-filled depressions produced by ground subsidence associated with salt dissolution. Many of them have resulted from salt mining and brine pumping.          ASG

**Reading**
Bell, F.G. 1975: Salt and subsidence in Cheshire, England. *Engineering geology* 9, pp. 237–47.

**flash flood**  Flash floods are local or rapid increases in DISCHARGE commonly, but not exclusively, in previously dry valleys. They are normally associated with rapid transport of large amounts of water and/or sediment. They may be either natural, such as due to a sudden rainstorm (e.g. Schick and Lekach 1987) or involve anthropogenically created hazards such as a dam break (e.g. Jarrett and Costa 1985).          SNL

**References**
Jarrett, R.D. and Costa, J.E., 1985: Hydrology, geomorphology and dam break modelling of the 15 July 1982, Law Lake and Cascade Lake Dam failures, Larimer County, Colorado. *US Geological Survey, open file report* 84–612. · Schick, A.P. and Lekach, J., 1987: A high magnitude flood in the Sinai desert. In L. Mayer and D. Nash eds, *Catastrophic flooding*: Massachusetts: Allen and Unwin. Chapter 18.

**flashiness**  The rapidity with which the stage discharge increases at a stream cross-section. The hydrograph (a graphical plot of water discharge versus time) of a flashy stream shows a rapid increase in discharge over a short period, with a quickly developed high peak in relationship to normal flow (Ward 1978, p. 10). On the other hand, a sluggish stream develops peaks more slowly, and the peaks are relatively lower than those developed on flashy streams.

The term flashy depends on personal judgement and is rarely quantified. It is frequently applied to streams in arid or semi-arid regions where the normal flow is zero. Flash floods cause waves or bores to descend the stream course, so that hydrographs show nearly instantaneous rises from zero to high values of discharge in extreme cases of flashiness (Schick 1970).

Flashiness results from the actions of several possible factors that intensify flood peaks, including accelerated run-off rates caused by land use changes, basin shapes that become progressively more narrow in the downstream direction, especially intense rainfall events, and rapid snowmelt.          WLG

**References**
Schick, A.P. 1970: Desert floods. In *Symposium on the results of research on representative and experimental basins*. International Association of Scientific Hydrology publication. No. 96. Pp. 479–95. · Ward, R. 1978: *Floods: a geographical perspective*. London: Macmillan; New York: Wiley.

**flatiron**  Steep triangular cliff facets resulting from the presence of a capping of rock resistant to erosion which protects the underlying more readily eroded rocks.          ASG

**Reading**
Schmidt, K-H. 1989: Talus and pediment flatirons – erosional and depositional features on dryland cuesta scarps. *Catena* Suppl. 14, pp. 107–18.

201

**float recorder**   A device for recording the water level in a river channel, lake, well or similar situation. A flexible cable or steel tape connecting a float and counterweight passes over a pulley which transmits vertical movements of the float to the recording mechanism. Several principles are employed. For example, in a vertical recorder the pulley moves the pen vertically across a moving chart, whereas with a horizontal recorder the pulley rotates the chart against the pen, which is driven across the chart by a clock. Digital recording devices also exist (see DIS-CHARGE).                                                    DEW

**floating bogs**   Refer to any lake-fill bog or kettle-water wetland with a quaking *Sphagnum* mat. Such bogs, which are common in parts of North America, are generally referred to as 'schwingmoor' or 'kesselmoor' in Europe.   ASG

**Reading**
Warner, B.G. 1993: Palaeoecology of floating bogs and landscape change in the Great Lakes drainage basins of North America. In F.M. Chambers ed., *Climate change and human impact on the landscape*. London: Chapman & Hall. Pp. 237–45.

**flocculation**   The process of aggregating into small lumps, applied especially to clays.

**flood**   A high water level along a river channel or on a coast that leads to inundation of land which is not normally submerged. River floods which involve inundation of the FLOODPLAIN may be caused by:

*Precipitation* when storm precipitation is very intense; when precipitation is very prolonged and follows a period of wetter than average conditions; when the snow-pack melts and snowmelt floods are an annual feature of many river regimes; when rain falls on snow and accelerates snowmelt; or when ice and snowmelt are combined and melting river ice produces breakup floods.

*Collapse of dams* which may be natural where a landslide temporarily blocks a river or a log jam dams the river channel; or which may occur when a man-made dam bursts and the impounded water flows down the river valley as a flood wave or when a landslide into a reservoir leads to a wave overtopping the dam wall.

*Drainage of ice-dammed lakes* which can lead to the release of great volumes of water (in Iceland JÖKULHLAUPS are flood waves which roar down-valley as a result of the failure of a lake dammed by, or contained within, a glacier).

In addition to being effected by the above causes coastal floods may also be produced by:

*High tides*: especially in combination with river floods.
*Storm surges*: abnormally high sea levels at about the time of spring tides.
*Tsunami*: large waves produced by submarine earthquakes, volcanic eruptions, landsliding or slumping.

The FLOOD FREQUENCY in a particular area may vary over time, especially as a result of human activity, and in some areas the damage produced by floods has increased despite expenditure on flood protection simply because there is a tendency to assume that flood protection measures have completely eliminated the flood hazard. Responses to flood hazard include:

*Flood protection* by structural measures along the river or coast to reduce the effects of flooding; these include walls and embankments, river diversion schemes, flood barriers.

*Flood reduction* by taking action in the drainage basin by afforestation, controlled vegetation or agricultural changes; or by the construction of small or large dams.

*Flood adjustment* by adjusting to the hazard by accepting it; by land use zoning; by taking emergency measures when floods occur; or by flood-proofing so that flooding will damage structures and buildings as little as possible.         KJG

**Reading**
Ward, R.C. 1978: *Floods: a geographical perspective*. London: Macmillan.

**flood frequency**   An analysis using data from the period of hydrologic records to establish a relationship between discharge and return period or probability of occurrence as a basis for estimating discharges of specific recurrence intervals. Several methods are available for calculating the recurrence intervals or return periods ($T$) but one most frequently employed ranks the $N$ years of record from the highest (rank $m = 1$) to the lowest (rank $m = N$) and uses:

$$T = \frac{N+1}{m}$$

To obtain the $N$ years of record use of the highest discharge from each year of record is the basis for the ANNUAL SERIES, whereas using the $N$ highest independent discharges provides the annual exceedance series. The recurrence intervals may be plotted against the discharges for a specific gauging station and the curve fitted by eye or by fitting an appropriate theoretical probability distribution. A range of distributions are available and the Gumbel extreme value type 1 based on GUMBEL EXTREME VALUE THEORY is one frequently used. The flood frequency plot of discharge against recurrence interval can be the basis for estimating the recurrence interval of discharge including the BANKFULL DISCHARGE and for giving the DESIGN DISCHARGE for a particular return period when a structure is being planned. When a regional flood frequency analysis is required using the data from a series of gauging stations within the region it is necessary to undertake a homogeneity test which identifies a homogeneous region according to the ten-year flood estimated from the probability curve for each gauging station.                                    KJG

**Reading**
NERC 1975: *Flood studies report*. 5 vols. London: Natural Environment Research Council. · Newson, M.D. 1975: *Flooding and flood hazard in the United Kingdom*. Oxford and New York: Oxford University Press. · Ward, R.C. 1978: *Floods: a geographical perspective*. London: Macmillan.

**floodout**   The part of an ephemeral desert stream system, located in the lowest reaches of the channel, where organized channel flow gives way to multiple distributary channels and ultimately to shallow non-channelized flow. Progressive transmission losses taking place along the channel from its headwater source areas progressively reduce the sediment-transporting capacity of the stream. Resulting deposition in the channel, together with a diminished ability to flush away wind-blown sands which may accumulate there, lead to falling channel capacity (Mabbutt 1977). Many floodouts are located where EPHEMERAL STREAMS first contact dunefields lying across their path. When large floods arrive at such a location, overbank flows spill from the channel and cut many small channels leading away from the trunk stream, perhaps draining along inter-dune corridors. Floodout locations appear to become semipermanent in this way, and frequently channels collapse at floodouts before reaching saline lake basins that would otherwise be their flow terminus.                                    DLD

**Reading and Reference**
Mabbutt, J.A. 1977: *Desert landforms*. Canberra: Australian National University Press. · Tooth, S. 1999: Floodouts in central Australia. In A.J. Miller and A. Gupta eds, *Varieties of fluvial form*. Chichester: Wiley. Pp. 219–47.

**floodplain**   The *hydraulic floodplain* is an engineering concept for any surface subject to river flooding within a given return period (say, every 100 years). Such surfaces may be formed of bedrock or Pleistocene sediment and only slightly modified by the contemporary river. The *genetic floodplain* is a geomorphological term for a relatively flat alluvial landform, constructed largely by the flow regime of the present river and subject to flooding. Commonly flood-

plains flank a clearly defined river channel, but some occur in valleys without channels, while others form downstream of channels (a *flood-out*). A genetic floodplain shows strong affinity to the river that has formed it, with sedimentology, stratigraphy and morphology reflecting flow and catchment conditions. With changes in flow regime, rivers may incise, leaving abandoned floodplains as RIVER TERRACES.

There are three main processes of floodplain formation: LATERAL ACCRETION, OVERBANK DEPOSITION and BRAIDED RIVER accretion. Three less common processes are oblique accretion (the onlapping of relatively steeply dipping convex-bank deposits, as the channel migrates), counterpoint accretion (sediment deposited against the concave bank of tightly curving bends) and abandoned channel (OXBOW) accretion. These processes are determined in individual reaches by variations in stream power, sediment character and local vegetation, making it possible to classify floodplains into a wide variety of types but essentially three broad classes: (1) high-energy, non-cohesive sediment; (2) medium-energy non-cohesive sediment; and (3) low-energy cohesive sediment floodplains.

In detail, floodplain surfaces consist of LEVÉES, scrolls (prior levees), abandoned channels and BACKSWAMPS that greatly influence the passage of overbank flow. Floodplains generally support fertile agricultural soils, but can be hazardous, rapidly changing geomorphological environments for human occupation.        GCN

**Reading**
Knighton, D. 1998: *Fluvial forms and processes*. London: Arnold. · Lewin, J. 1978: Floodplain geomorphology. *Progress in physical geography* 2, pp. 408–37. · Nanson, G.C. and Croke, J.C. 1992: A genetic classification of floodplains. *Geomorphology* 4, pp. 459–86.

**flood routing**   The technique of determining the timing and shape of the flood wave at successive points along a river channel. Techniques for flood peak estimation applicable to small areas are not appropriate for larger basins because as a flood wave is routed along the channel it will change in shape as a result of the storage offered by the channel and the characteristics of the channel morphology. Flood routing may be accomplished either by a theoretical method or by an empirical method of which the MUSKINGUM METHOD is well known.        KJG

**Reading**
Lawler, E.A. 1964: Flood routing. In V.T. Chow ed., *Handbook of applied hydrology*. New York: McGraw-Hill. Sect. 25–II.

**floristic realms**   The largest divisions into which the floras of the world are customarily

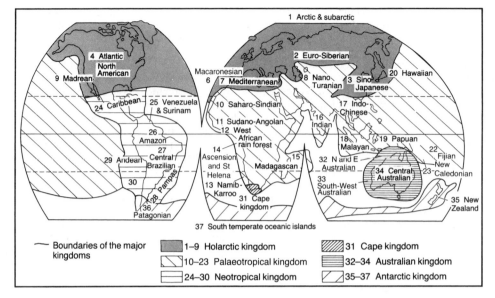

**Floristic realms**
*Source*: A.S. Goudie 1993: *The nature of the environment*. 3rd edn. Oxford and New York: Basil Blackwell.

grouped; also called floristic kingdoms or empires. In turn, these may be subdivided into floristic regions, provinces and districts. The earliest substantial system was that of Engler (1879–1882). Floristic realms are mainly characterized by the plant families or substantial sections of families which are endemic to that particular portion of the globe. The main realms normally recognized are: the *Holarctic* floral realm (e.g. most of the Pinaceae), which embraces North America, Greenland, Europe and the northern part of Asia; the *Palaeotropical* realm (e.g. the pitcher plant family, the Nepenthaceae); the *Neotropical* realm (e.g. the Bromeliaceae); and the *Austral* realm (e.g. the Proteaceae), which includes southern South America, southern Africa, Australia and New Zealand. The boundary between the holarctic and the tropical realm is, with a few exceptions, marked by the southward limit of the northern hemisphere coniferous family (Pinaceae) and the northward limit of the palms (Palmae or Arecaceae).

The evolution of these different realms reflects above all the geological history of the earth and the movement of continents brought about by the processes of plate tectonics and sea-floor spreading although, of course, the floristic realms are not confined to specific continents. (See also FAUNAL REALMS; WALLACE'S REALMS.) PAS

**Reading**
Good, R. 1974: *The geography of the flowering plants*. 4th edn. London: Longman. · Moore, D.M. ed. 1982: *Green planet: the story of plant life on earth*; Cambridge: Cambridge University Press. Esp. ch. 6.

**flow duration curve** Shows the percentage of time that given flows are equalled or exceeded at a particular site. They provide a useful summary of flow reliability at a site and, if the flow axis is standardized for variations in catchment area, the graph forms a very good basis for comparing the flow conditions in different drainage basins or at different sites within the same drainage basin. AMG

**Reading**
Dunne, T. and Leopold, L.B. 1978: *Water in environmental planning*. San Francisco: W.H. Freeman. · Gregory, K.J. and Walling, D.E. 1976: *Drainage basin form and process*. London: Edward Arnold.

**flow equations** Define the inter-relationships between velocity, depth, slope and boundary roughness or energy losses in open channel flow. A variety of flow conditions can be defined including UNIFORM STEADY FLOW and GRADUALLY VARIED FLOW. In the former the water surface is parallel to the bed, and width, depth and velocity are constant along a reach. Under such constrained circumstances the velocity formulae developed by Chézy and Manning (see CHÉZY and MANNING EQUATIONS) relate mean velocity to depth, slope and a roughness coefficient. These equations may be used to estimate flood velocities if flood depth and slope are measured and a suitable value of the roughness coefficient is identified. The equations which define point velocity variation with height above the channel bed – that is, the velocity profile shape within the BOUNDARY LAYER – also relate to uniform flow conditions. In gradually varied flow, which is more common in natural river channels, depth and velocity vary from section to section and bed and water surface slopes are not parallel. Under these conditions the flow properties must be modelled by an energy-balance equation based on the principle of conservation of energy between closely spaced sections. This is the BERNOULLI equation, which defines the total energy of the flow as the sum of potential energy, pressure energy, and kinetic energy.                KSR

**flow regimes**  Four flow regimes are defined, in terms of the depth and velocity of open channel flow, by the combination of criteria based on the Froude and Reynolds numbers. These describe the relationships between inertial, viscous and gravitational forces acting on the flow. The Froude number distinguishes subcritical ($F_r < 1$) from supercritical ($F_r > 1$) flow, and the Reynolds number identifies laminar ($R_e < 500$) and turbulent ($R_e > 2000$) flow with a transitional state between these limits. The flow regimes are laminar subcritical, laminar supercritical, turbulent subcritical, and turbulent supercritical. In river channels, turbulent subcritical conditions are normal, while overland flow on hillslopes is often laminar except when disturbed by rainsplash.

In a SANDBED CHANNEL, the progressive change of flow intensity during the passage of a flood along a reach is associated with a systematic sequence of changing bedforms. In the subcritical regime, as Froude and Reynolds numbers increase, small sand ripples up to 4 cm high and 60 cm long are replaced first by dunes with superimposed ripples and then by dunes up to 10 m high and 250 m long, depending on river size. At Froude numbers approaching unity as the flow passes into the upper regime supercritical state, dunes wash out to a plane bed with intense sediment transport, and then antidunes and standing waves form. Thus the flow regimes are closely related to bedform changes, which are in turn associated with changes of sediment transport rate and ROUGHNESS (Simons *et al.* 1961).                KSR

**Reading and Reference**
Richards, K.S. 1982: *Rivers: form and process in alluvial channels.* London and New York: Methuen. · Simons, D.B., Richardson, E.V. and Albertson, M.L. 1961: Flume studies using medium sand (0.45 mm). *United States Geological Survey water supply paper* 1498A.

**flow resources**  These may be defined as NATURAL RESOURCES that are 'naturally renewed within a sufficiently short timespan to be of relevance to human beings' (Rees 1990, p. 15).

They are also known as *renewable resources*. Two categories of flow resource can be discerned: those where flows are dependent on human activity, and whose future availability may be compromised by excessive use (see also SUSTAINED YIELD) and those where human usage has no impact on future availability. To this effect, Rees (1990) gave solar energy, air, water (at the global scale), wind and tidal energy as examples of the latter, termed *non-critical zone* flow resources, and fish, forests, soil and water in aquifers as examples of those in danger of losing their renewability through human actions. These can be termed *critical zone* flow resources, and in effect they can become STOCK RESOURCES if their ability to regenerate is compromised.                DSGT

**Reference**
Rees, J. 1990: *Natural resources: allocation, economics and policy.* 2nd edn. London: Routledge.

**flow till**  See TILL.

**fluid mechanics**  A branch of applied mathematics dealing with the motion of fluids and the conditions governing such motion. Mechanics generally includes kinematics and dynamics, which respectively cover the geometry of motion and the forces involved in motion. Thus the continuity equation which defines channel discharge as the product of cross-section area and velocity is a kinematic statement, while the theoretical velocity profiles of laminar and turbulent boundary layers are the product of dynamical analysis of forces acting upon and within moving fluids. HYDRAULICS deals specifically with the mechanics of *liquid flow*, particularly water.                KSR

**flume**  A structure built within a river channel in order to measure the flow. The structure, normally built of concrete, introduces a constriction or contraction into the channel. This produces critical flow conditions for which a unique and stable relationship will exist between the flow depth upstream of the constriction and discharge. The standing wave, critical depth or Parshall flume is the most common design, but there is a wide variety of specialized flume designs which have been produced for specific channel and flow conditions. A flume can provide very accurate measurements of discharge and may be preferred to a WEIR in many situations because sediment is able to pass through the structure. Calibration may involve formulae or field rating. (See also DISCHARGE.)                DEW

**Reading**
Ackers, P., White, W.R., Perkins, J.A. and Harrison, A.J.M. 1978: *Weirs and flumes for flow measurement.* Chichester: Wiley.

**flute** Also known as fluted moraine, these are elongated streamlined ridges of sediment produced beneath a GLACIER or ICE SHEET and which are aligned in the direction of ice flow. They are the smallest landform type in the family of SUBGLACIAL BEDFORMS, and typically occur as numerous parallel ridges, tens of centimetres to a few metres high and wide, and tens of metres in length. These rather delicate landforms are best observed close to recently retreated glacier margins, and due to their poor preservation potential are much less commonly found in relation to former ice sheets. CDC

**Reading**
Benn, D.I., and Evans, D.J.A. 1998: *Glaciers and glaciation*. London: Arnold.

**fluting** The production of flute marks on soft, cohesive sediment surfaces by flow-induced scour, often associated with TURBIDITY CURRENTS. Flute marks are usually streamlined, and characteristic shapes include bulbous, conical, and linguiform. Turbulent vortices cause the EROSION from either direct fluid stresses on the bed or by ABRASION from entrained particles. A bed defect acts as a locus for flow separation and vortex generation. The resulting erosion potential is greatest near the upstream end of the flute. Flute marks are often preserved as casts on the bottom of a subsequently deposited sand bed, where they are termed sole marks. DJS

**Reading**
Allen, J.R.L. 1982: *Sedimentary structures: their character and physical basis*, Vol. 2. Amsterdam: Elsevier Science.

**fluvial** This is the common usage of the term and refers to anything that is of or referring to a river, including things that grow or live in a river or that are produced by the action of a river. It is commonly combined as follows: (1) fluvial geomorphology, which refers to the study of the morphology of environments worked by rivers; (2) fluvial processes, which refer to flow processes and solute and sediment transport in rivers; (3) fluvial deposits or sediments, which refer to material worked or deposited in the landscape by rivers; and (4) fluvial erosion or denudation, which refers to erosion of a land surface by rivers. SNL

**fluvioglacial** See GLACIOFLUVIAL.

**fluviokarst** A type of landform developed in limestone areas by a combination of river action and of true KARST processes. Included in this category are gorges and DRY VALLEYS.

**flux** The rate of flow of some quantity. It is perhaps most easily perceived in the context of fluids (air, water) but it also applies to many other elements of the natural environment, such as mass and energy. Many apparently unchanging forms in the natural world ensue from different fluxes into and out of the system. As such, fluxes are a primary element in natural environmental processes. BWA

**flux divergence** The variation of fluxes through space. This relatively simple notion has profound implications in the natural environment. As W.M. Davis (1909) noted, landforms are a product of structure, process and stage. It takes but little thought to realize that probably everything in the universe is a function of structure, process and stage – hence the notion can be applied, at least, to all facets of physical geography. In this context we are most usually concerned with the structure and form of a constituent of the natural environment and, more particularly, the processes whereby the structure and form change through time.

In most natural systems fluxes are the essence of process: mass through geomorphological features; water through rivers, oceans, plants; chemicals through soils; air, energy and pollutants through the atmosphere. If the flux of a quantity through a system is constant everywhere, the size and form of the system is frequently unchanging. For example, if the flow of water into and out of a bath is equal, the depth of the water in the bath remains constant; the water itself, of course, is constantly changing. Casual inspection of the water body could easily give the impression that it is unchanging, simply because its depth is constant. In the same way, casual inspection of a scree could easily suggest that it is an unchanging feature whereas in reality significant mass fluxes occur. Clearly, the way to change the depth of the water in the bath is to ensure that input and output differ, that is, to ensure that there is flux divergence in the flow of water through the system. Hence we see that flux divergence is the way in which the state of a feature changes through time and is thus the essence of natural environmental processes.

For many years physical geographers monitored form, with little appreciation of process (and hence fluxes and flux divergences). In the past two decades more emphasis has been put on process and to some extent on fluxes (e.g. river discharge). But there remains much scope for better and more extensive measurements of fluxes and their divergences in the natural environment so as to increase our understanding of the mechanisms of environmental phenomena. BWA

**Reference**
Davis, W.M. 1909: *Geographical essays*. Boston: Ginn.

**flyggberg** An asymmetrical hill 1–3 km across and 100–300 m high shaped by the action of overrunning ice. Swedish derivation: *flygg*, a steep rock face; *berg*, mountain.

**flysch** Deposits of marine sandstones, shales, marls and clays produced during the initial uplift of the Alps by sedimentation and later deformation of materials eroded from the uplifted rocks. The word now refers to any thick succession of alternations of the rocks mentioned above, interpreted as having been deposited by turbidity currents or mass-flow in a deep water environment within a geosynclinal belt.

**fog** If the air near the ground is cooled enough and its moisture content is high CONDENSATION may occur and a fog will develop. By definition, a fog exists if the visibility in cloudy air is 1 km or less. There are several ways by which moist air may be cooled enough to form fog – loss of RADIATION at night (radiation fog), warm air passing over a cold surface (advection fog), and air flow up an incline (upslope fog). Fog may also form when warm rain falls through cold, but saturated, air (frontal fog) and when cold air passes over warm water (steam fog).          WDS

**Reading**
Neiburger, M., Edinger, J.G. and Bonner, W.D. 1982: *Understanding our atmospheric environment*. San Francisco: W.H. Freeman.

**föhn** (chinook in North America) The name given to a strong and gusty downslope (or 'fall') wind which occurs on the lee side of a mountain range when stable air is forced to flow over the mountains by the large-scale pressure gradient. The air at the foot of the mountains is characteristically dry and warm as a result of the ADIABATIC compression during its descent. In some cases the air may ascend over the range with cloud and precipitation occurring on the windward side, but in other cases a stable layer near the summit level blocks the cross-mountain flow and air descends from the summit level. Föhn onset can give rise to dramatic increases in temperature and such winds are important in accelerating snow-melt in spring along the northern flanks of the Alps and Caucasus and the eastern flanks of the Rocky Mountains.          RGB

**Reading**
Atkinson, B.W. 1981: *Meso-scale atmospheric circulations*. London and New York: Academic Press.

**fold** A bend in formerly planar strata of rock resulting from movement of the crustal rocks. If the strata are flexured in just one plane they are termed a monocline. The arching up of strata

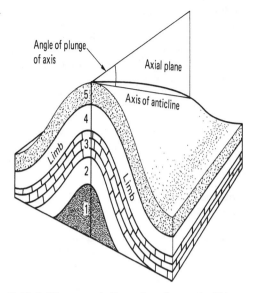

**Fold: 1**. Diagrammatic illustration of parts of a fold.

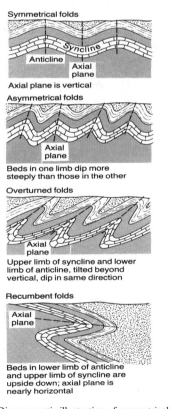

**Fold: 2**. Diagrammatic illustration of symmetrical, asymmetrical, overturned and recumbent folds.
*Source*: F. Press and R. Siever 1978: *Earth*. San Francisco: W.H. Freeman. Figure 20.23–4.

produces an anticline, while the depression of strata produces a syncline. In a recumbent fold the strata are overturned and both limbs of the folds are nearly horizontal. The horizontal compression that creates recumbent folds may lead to the shearing of the upper part of the fold along a thrust fault. The strata moved forward over a thrust fault form a nappe.          ASG

**foliation**  Fine layering of rocks, usually metamorphic and igneous rock, as a result of parallel orientation of minerals. Slates and schists are typical foliated rocks.

**food chain, food web**  Food chains represent the transfer of food (and, therefore, energy) from one type of organism to another, in sequence and in a linear relationship: for instance, in a marine environment: algae → small fish → squid → man. Normally, however, this would represent an extreme oversimplification of the feeding relationships present in any community, which are much better portrayed as a food web (see diagram).

There are two major types of food chain or food web, termed *grazing* and *detrital*. At the base of the grazing food chain or web are producer organisms which are autotrophs, i.e. they are able to fix incident light energy and so manufacture food from subsequent chemical reactions (see PHOTOSYNTHESIS). These may be green plants, blue-green or other algae, or phytoplankton. All other organisms within it are

dependent heterotrophs, i.e. they eat or rearrange existing organic matter. In the overall sequence of food consumption patterns, they may be classed as primary consumers (HERBIVORES), secondary consumers (CARNIVORES, which eat herbivores), tertiary consumers (top carnivores, which eat other carnivores), and so on.

Although essentially similar, detrital food chains begin at producer level with the breakdown of dead or decaying organic material by DECOMPOSER organisms. Consumers then take up the nutrients released by the decomposers, and some of the decomposers themselves; and they in turn are eaten by a range of carnivores. Relationships between the grazing and detrital food chains and webs are complex; in general one may say that much more energy passes through the latter than the former, often in a ratio of *c*.10 to 1 in organically rich communities.          DW

**Reading**
Pimm, S.L. 1982: *Food webs*. London: Chapman & Hall.

**foraminifera**  Often abbreviated to forams, these comprise a soft body enclosed within a shell that often resembles a gastropod or cephalopod in form. Depending on the species, the shell may be made of organic compounds, sand grains and other particles cemented together, or crystalline calcite. Fully-grown individuals range in size from about 0.4 mm to almost 20 centimetres in length. They are found in all

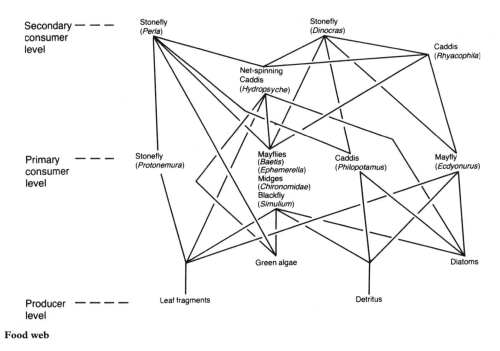

**Food web**

marine environments, from the intertidal zone to the deepest ocean trenches, and from the tropics to the poles, but species can be very particular about the environment where they live. Because different species are found in different environments their fossils can be used to determine past environments.                                SLO

**force** An agency changing the velocity of an object. Force is mass times acceleration, hence measured in dimensions of mass times length per the square of time: 1 Newton, N, is 1 kilogram-metre per second squared, $1\,kg\;m\;s^{-2}$. Like velocity and acceleration, force is a vector quantity: it has direction as well as magnitude. It is common to resolve force into components in 2 or 3 mutually perpendicular directions, e.g. normal to the surface and tangential to the surface: the latter is a shear force and the former contributes to RESISTANCE. Body forces (gravitational, magnetic) are produced without physical contact: surface forces result from physical contact with other bodies, leading to STRESS. All mechanical processes involve force, for which the energy is derived from a combination of gravity, climate and geothermal heat. For geomorphological work to be done, forces must overcome resistances.                                IE

**Reading**
Carson, M.A. and Kirkby, M.J. 1972: *Hillslope form and process*. Cambridge: Cambridge University Press. Ch. 3. · Chapman, R.E. 1995: *Physics for geologists*. London: UCL Press. Chs 1 and 2.

**forebulge** When land is isostatically depressed by the growth of an ice body (e.g. ICE SHEET), the deformation of the MANTLE includes a compensatory rise in surface level, the forebulge) at a distance from the location of depression.                                DSGT

**foredune** A distinctive COASTAL DUNE, with geometry controlled by wind, nearshore processes, sand and vegetation. Foredunes develop in locations at least partly protected from EROSION by nearshore processes. Vigorous plant colonization causes the dune to migrate toward the water, but hydrodynamic erosion moves the dune front landward. As a result, the location of the dune represents an EQUILIBRIUM between the two systems. The characteristic morphology of mature foredunes is a near-continuous ridge with an erosion scarp on the windward side and dune grass cover on the crest and lee slope. PARABOLIC DUNES may develop from BLOWOUTS through the foredune.                                DJS

**Reading**
Bauer, B.O. and Sherman, D.J. 1999: Coastal dune dynamics: problems and prospects. In A.S. Goudie, I. Livingstone, and S. Stokes eds, *Aeolian environments, sediments, and landforms*. Chichester: Wiley.

**foreset beds** Layers of sediment which have been laid down on the inclined surface of an advancing deltaic deposit or sand dune.

**forest** See BOREAL FOREST; DECIDUOUS FOREST; MONSOON FOREST; TROPICAL FOREST.

**forest hydrology** The branch of the science of physical geography (Linsley *et al.* 1988, p. 1) that is concerned with the 'study of how forests affect the hydrologic cycle, with particular reference to the regulation of streamflow, water supplies, and erosion control' (Storey *et al.* 1964). The relationships between water and forests have been the subject of scientific and political interest for several centuries. In 1215 King Louis VI of France set out a decree for the management of forests and related waters, and in Switzerland a law of 1342 protected forests from overcutting as an avalanche protection measure (Storey *et al.* 1964). The impacts of deforestation on accelerated soil erosion have been a subject of study and debate in the Mediterranean area for five hundred years. In the New World the US government issued its first report on the hydrological impacts of deforestation in 1849, and in 1891 began setting aside vast areas of forest for the purpose of watershed protection.

From the human perspective forests are considered to have beneficial effects on the hydrological cycle (Kittredge 1948). The forest foliage breaks the fall of raindrops, lessening the erosion caused by drop splash on the surface below. Large quantities of moisture are intercepted by the foliage, temporarily stored on leaves and branches, and then allowed slowly to drop to the surface. This process ensures that precipitation is more likely to be absorbed into the surface instead of being fed to the surface too rapidly for effective infiltration.

The overhead canopy of vegetation in the forest also reduces water loss from the surface by means of evaporation, although this saving is at some cost because the trees transpire large amounts of moisture through their leaves. The forest root system acts as an important soil binder and retards erosion. The litter on the forest floor, organic material that has fallen from the actively growing vegetation, absorbs large amounts of moisture that otherwise would quickly run off into channels. Finally, the forest protects snow cover with shade, preventing rapid melt and flash flooding in downstream areas. Melting takes place over a relatively longer period of time, releasing water later in

209

the melt season when it is more useful for irrigation.                                                    WLG

**References**
Kittredge, J. 1948: *Forest influences*. New York: McGraw-Hill. · Linsley, R.K., Kohler, M.A. and Paulhus, J.L. 1988: *Hydrology for engineers*. 3rd edn. New York: McGraw-Hill. · Storey, H.C., Hobba, R.L. and Rosa, J.M. 1964: Hydrology of forest lands and rangelands. In V.T. Chow ed., *Handbook of hydrology*. New York: McGraw-Hill.

**form line**  A contour line on a map, the precise position of which has not been accurately surveyed but interpolated.

**form ratio**  The ratio of the depth of a stream or river to its width.

**Fourier analysis**  (or harmonic analysis) A mathematical technique widely used to determine the periodic components of a wave form by a trigonometric series. It relies upon terms of a Fourier series that represents a repeating wave form as components of a fundamental sine wave and a series of added 'harmonics' of this frequency. This is the Fourier principle. A Fourier series is the sum of a number of sine and cosine terms of the series which describes a function $f(x)$ within the interval of width $2^1$. There are Fourier coefficients for the fundamental period and one each for each subsequent sine and cosine term (1 to $n$). In the series the importance of the component is measured by its amplitude. The general form of Fourier analysis is:

$$\hat{P} = \bar{P} + \sum_{r=1}^{N/2} A_r \cos(r\theta - \Phi_r)$$

where $\hat{P}$ is the estimated value for any of the $N$ observations, $\bar{P}$ is the mean of the observations, $A_r$ is the amplitude of a harmonic wave at frequency $r$, $\theta$ equals $2^1 x/p$ with $x$ representing the time and $p$ representing the fundamental period (e.g. 12 for monthly or 24 for hourly analyses), and $\Phi_r$ is the phase angle of the fitted wave. The first harmonic produces a fit with one maximum and one minimum occurring through the fundamental period and spaced exactly $p/2$ apart. The second harmonic has two evenly spaced maxima and two minima; $N/2$ harmonics are required to perfectly reproduce the original time series.

The technique can be used to detect important periodicities in a series (which may be a time series, showing a periodicity, e.g. rainfall maxima, sunspot cycles). However, the number of sampled points needs to be sufficient, at least twice the number of the highest value coefficient used, to avoid spurious results ('aliasing'). The technique has also been used to define a shape

and for this, the more points sampled on the line to be identified the more precise will be the definition. Particles, in two dimensions, can also be described in this way by 'unwrapping' the perimeter of the particle outline. In this case, polar coordinates are used to provide the sampled points. Values of the summed harmonics can be used to identify basic shapes as well as angularity and roughness components.                                    WBW/RCB

**fractal**  A geometric form which is infinitely repeated at all scales. This mathematical concept has been applied in geography (physical and human) by including repetition that is statistical rather than exact, by accepting that there are upper and lower limits to the scales involved, and sometimes by allowing the type of variability to change with scale. Thus geographers and earth scientists deal with *statistical pseudo-fractals* or *pre-fractals*, which are irregular over a broad range of scales. Interesting experiments on whether mapped surfaces are scale-free in this way, or scale-bound, can be performed if extraneous clues such as contour labels or anthropogenic features are excluded: can we estimate map scale and contour interval?

Fractal concepts have been applied with varying degrees of success to time series of processes, to turbulence-related features in the atmosphere and oceans, to fractured and other features in geology, to PARTICLE FORM, and to REMOTE SENSING images. One of the earliest contributions was the work of H.E. Hurst on the long-term record of DISCHARGE for the River Nile, which shows *persistence* of deviations above or below average. Here we concentrate on fractal aspects of three distinct features in geomorphology: the coastline, the land surface and the drainage network.

Indices of coastal irregularity have a long history, but the scaling properties of coastal length were noted by L.F. Richardson and developed by B. Mandelbrot. Many coastlines are *self-similar*: one part, magnified, statistically resembles the whole. Hence as measurements are made at higher resolution (e.g. shorter straight-line segments), the estimated length of a coast increases without limit. This increase may be linear on a log-log plot, for segments from around 100 m to 100 km. The rate of increase gives the fractal dimension, $D$, usually between 1.05 and 1.40. Thus the irregularity of coastlines gives them a fractal dimension greater than their conceptual Euclidean dimension (1.0), but less than that of Brownian motion (random increments: 1.5).

Similar relationships are found for topographic contours and for profiles, but the fractal dimensions estimated tend to differ. Fractal concepts can be applied to the land surface by scaling the vertical dimension by a different factor than the horizontal: this means it can be *self-affine* rather than *self-similar*. A smooth, differentiable surface has a dimension (Euclidean) of 2.0. The land surface tends to show fractal dimensions between 2.05 and 2.40, i.e. 1.0 more than coastlines. However, estimation is more difficult for surfaces and there is little agreement between the results of different techniques, among which box-counting and *variogram* analysis are to be preferred. In addition to sampling problems, these differences indicate that real land surfaces deviate from the assumptions (e.g. self-affinity, invertibility and anisotropy) underlying the techniques: even impossible values of $D$ are sometimes calculated. Surfaces simulated by approximations to *fractional Brownian motion* are useful as null hypotheses for comparison with real surfaces, or as initial rough surfaces on which geomorphologic processes can be simulated.

Most published results for land surfaces estimate fractal dimension over a limited scale range, between 10-fold and 100-fold (1 to 2 cycles of decimal logarithms). The concept of scale-independence is more valuable if it applies over at least 3 cycles of length, e.g. 100 m to 100 km, which is very demanding in terms of data. However, curvature or changes of slope are commonly observed leading to estimation of two or even three different dimensions, each over a limited scale range. More recent work has suggested that land surfaces exhibit *multifractality*, that is, continuous variation of fractal dimension with scale. Even so, there are many characteristics that are incompatible with either unifractality or multifractality.

The situation is different for DRAINAGE NETWORKS. Rain falls over the whole land surface and run-off produces a space-filling network of drainage. The fractal dimension of river networks in map view approaches the theoretical maximum value of 2.0. Models based on fractal-related processes, such as self-avoiding invasive percolation, can simulate realistic drainage basins. Realism is increased by use of multifractal models. Note that river channels have $D$ around 1.16, but river networks have $D$ around 1.9.                                    IE

**Reading**
Evans, I.S. and McClean, C.J. 1995: The land surface is not unifractal: variograms, cirque scale and allometry. *Zeitschrift für Geomorphologie* NF, *Supplement-Band* 101, pp. 127–47. · Gao, J. and Xia, Z. 1996: Fractals in physical geography. *Progress in physical geography* 20, pp. 178–91. · Lavallée, D., Lovejoy, S., Schertzer, D. and Ladoy, P. 1993: Nonlinear variability of landscape topography: multifractal analysis and simulation. In N.S.N. Lam and L. De Cola eds, *Fractals in geography*. Englewood Cliffs, NJ: Prentice-Hall. Pp. 158–92. · Mark, D.M. and Aronson, P.B. 1984: Scale-dependent fractal dimensions of topographic surfaces: an empirical investigation with applications in geomorphology and computer mapping. *Mathematical geology* 16, pp. 671–83. · Rodríguez-Iturbe, I. and Rinaldo, A. 1997: *Fractal river basins: chance and self-reorganization*. Cambridge: Cambridge University Press. · Xu, T.-B., I.D. Moore and J.C. Gallant 1993: Fractals, fractal dimensions and landscapes: a review. *Geomorphology* 8, pp. 245–62.

**fractal dimension**   (or fractal) A concept put forward by Mandelbrot (1982) which refers to the space filling of curves such that the dimension is between one and two; the trail of Brownian motion is a fractal. The length of a coastline can be measured in different ways which will give values greater than the linear distance between the end points. In extreme cases, the length is very large but it can be characterized by fractals. Complex lines can be analysed by using the ideas of fractal dimension. For a closed loop (such as the outline of a particle) with perimeter $P$, measured by stepping off a constant length $S$, a plot of log $P$ versus log $S$ tends to give a constant slope $b$. The fractal dimension ($D$) is equal to the slope term plus one ($D = b + 1$) and $\log P \propto (D - 1) \log S$.                                    WBW

**Reading and Reference**
Mandelbrot, B.B. 1982: *The fractal geometry of nature*. San Francisco: W.H. Freeman. · Orford, J.D. and Whalley, W.B. 1983: The use of the fractal dimension to quantify the morphology of irregular particles. *Sedimentology* 30, pp. 655–68.

**fracture**   The term given to the splitting of a material into two or more parts; the material is said to have failed. It usually refers to brittle fracture where the stressed body ruptures rapidly with the release of energy to form new surfaces by crack propagation after little or no plastic deformation. Brittle fracture may take place through crystals, along cleavage planes or between grains to give 'inter-granular' fracture. Brittle fracture occurs normal to the maximum applied tensile stress component. This type of failure is usual with hard rocks. Ductile failure is fracture which occurs after extensive plastic deformation and with slow crack formation; this is typical of clays.                    WBW

**Reading**
Whalley, W.B. 1976: *Properties of materials and geomorphological explanation*. Oxford: Oxford University Press. · —Douglas, G.R. and McGreevy, J.P. 1982: Crack propagation and associated weathering in igneous rocks. *Zeitschrift für Geomorphologie* 26, pp. 33–54.

**fractus**   Cloud having a broken or shattered appearance, perhaps a convective cloud growing in a place where the wind shear tears apart the incipient cloud, as in fractocumulus. It is a temporary phase of development in which buoyancy and shear are not yet reconciled.　　　JSAG

**Reading**
Ludlam, F.H. and Scorer, R.S. 1966: *Cloud study*. London: John Murray.

**fragipan** An acidic, cemented horizon between the base of the soil zone and the underlying bedrock or parent material. Fragipans are normally compact and brittle, and may often be bonded by clays. Other agents may also be involved in the bonding, including silica (Bridges and Bull 1983), iron, aluminium and organic matter. Many are the result of periglacial processes, and they are widespread in Europe and North America.　　　ASG

**Reading and Reference**
Bridges, E.M. and Bull, P.A. 1983: The role of silica in the formation of compact and indurated horizons in the soils of South Wales. *Proceedings of the International Symposium on Soil Micromorphology 1983*, pp. 605–13. · Grossmann, R.B. and Carlisle, F.J. 1969: Fragipan soils of the eastern United States. *Advances in agronomy* 21, pp. 237–79.

**frazil ice** Fine spicules of ice in suspension in water, commonly associated with the freezing of seawater.

**free face** The wall of a rock outcrop that is too steep for debris to rest upon it. It is the portion of a cliff that lies above a scree or talus, and from which, through rockfall and other processes, scree formation may take place.

**freeze–thaw cycle** A cycle in which temperature fluctuates both above and below 0 °C. The amplitude of the temperature change and the period of time over which the fluctuation occurs are important considerations since freezing does not occur instantaneously, nor does it always occur at 0 °C. A typical diurnal freeze–thaw cycle is one in which the temperature ranges from +0.5 °C to −0.5 °C within a 24 hour period.

Because of their supposed significance with respect to frost shattering of rock, the frequency and efficacy of freeze–thaw cycles have been the subject of both field and laboratory investigations. Field studies indicate that most freeze–thaw cycles per year ($\approx$ 40–60) occur in subarctic alpine regions which experience diurnal temperature rhythms. High latitudes experience few cycles on account of the seasonal temperature regimes. In all areas most cycles occur in the upper 0–5 cm of the ground

and only the annual cycle occurs at depths in excess of 20 cm. Laboratory studies, in which rock samples are subject to repeated freeze–thaw cycles of varying amplitude and intensity, suggest that the number of freeze–thaw cycles is more important than their intensity as regards rock shattering. If correct, the low frequency of freeze–thaw cycles recorded in present-day periglacial environments suggests that frost shattering may be overemphasized as a physical weathering process. Hydration shattering and cryogenic (i.e. frost) weathering in general may be equally if not more important.　　　HMF

**Reading**
Washburn, A.L. 1979: *Geocryology: a survey of periglacial processes and environments*. New York: Wiley.

**freezing front** The boundary between frozen or partially frozen ground and non-frozen ground. During freezing in permafrost regions freezing fronts move downwards from the ground surface and upwards from the permafrost table. In seasonally frozen ground only a downward moving freezing front exists. The freezing front is sometimes equated with the cryofront, the position of the 0 °C isotherm in the subsurface, forming the boundary between cryotic (i.e. temperature less than 0 °C) and non-cryotic (i.e. temperature more than 0 °C) ground. The permafrost base, the permafrost table, and the top and base of the cryotic portion of the active layer all constitute cryofronts, or freezing fronts. (See also PERMAFROST.)　　　HMF

**Reading**
van Everdingen, R.O. 1976: Geocryological terminology. *Canadian journal of earth sciences* 13, pp. 862–7.

**freezing index** A measure of the combined duration and magnitude of the below-freezing temperatures which occur in a freezing season. It is expressed in DEGREE DAYS.

**friction** A force resisting relative motion between two solids or between a solid and a fluid. It is a fundamental property in studies of sediment transfer or transport, since frictional resistance to motion must be overcome before masses or particles of sediment can be moved, and sedimentary landforms modified as a result.

The force required to move a solid block on a plane surface is its weight ($W$) multiplied by the static coefficient of friction ($\mu$) between the block and the underlying surface. If the surface is sloping at an angle $\beta°$, the normal force on the surface is $W \cos \beta$, and the force required to move the block is $\mu W \cos \beta$. The block will slide under its own weight when $\tan \beta = \mu$

(Statham 1977, pp. 12–14). The interlocking friction in a sediment between individual grains is measured by the angle of internal friction ($\phi$), which is, however, difficult to define and measure since it is dependent on density, water content, and test conditions (Statham 1977, pp. 41–9). Dry sediment on a slope will slide if the slope angle equals or exceeds the angle of internal friction, which is therefore a limiting or threshold slope angle in the landscape.

Friction also occurs between a solid bed surface and a fluid passing over it. This slows the immediately adjacent fluid to a standstill, but shear in the fluid above allows the development of a velocity gradient within the BOUNDARY LAYER, in which the frictional resistance is successively less effective with distance from the bed. In UNIFORM STEADY FLOW the friction exerted by the bed on the flow is equal and opposite to the drag of the fluid on the bed, and in loose, mobile sediment, this drag (the TRACTIVE FORCE) may overcome the resistance to motion of non-cohesive sediment grains, which is dependent on their weight and the frictional contact between them.     KSR

**Reference**

Statham, I. 1977: *Earth surface sediment transport*. Oxford: Clarendon Press.

**fringing reef**   See REEF.

**front**   A sharp transition zone separating air of different temperatures and origins. The term was introduced by the Bergen School of Meteorology in 1918 as part of their work on EXTRA-TROPICAL CYCLONE structure. The front has a three-dimensional form. It extends into the atmosphere as a gently sloping surface of about 1 in 100 so that the cold, denser air appears as a wedge shape beneath the warmer air. The front lies in a trough of lower pressure accompanied by changes in wind velocity, pressure and temperature as the front passes a site. The intensity of change varies greatly from one front to another.

Horizontal convergence and associated vertical motion are a necessary feature of a well-defined front. According to early ideas the rising of the warm air at a front took place along the frontal surface itself, but recent work, based on

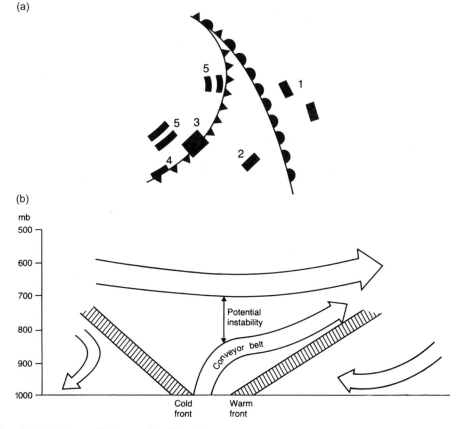

(a) Frontal rain band types. (b) Cross-section of frontal system.

Doppler radar, research aircraft, satellite images and rain-gauges, has shown a much more complex structure of air movements near fronts. Basic to the uplift of air at the warm front is a well-defined flow of low-level, moist, warm air within the warm sector which moves parallel to the cold front then ascends above the main warm frontal surface, eventually running parallel to the surface warm front but at a higher level (see figure (b)). Interaction between this flow and that at even higher levels (600 hPa) may trigger off potential instability which produces linear areas of heavier precipitation. Similar rain bands of subsynoptic scale, known as mesoscale precipitation areas (MPA), have been found associated with cold and occluded fronts. Five types of frontal rain bands have been identified: (1) warm frontal; (2) warm sector; (3) cold frontal – wide; (4) cold frontal – narrow; and (5) post frontal (see figure (a)). Aircraft observations have confirmed that clouds developing in frontal MPAs do have different liquid water contents and ice particle concentrations from other clouds associated with the fronts. Radar provides a continuous picture of the movement and patterns of the MPAs, but forecasting their initiation is difficult because of the lack of detailed information about atmospheric structure near fronts, and because the horizontal resolution of the present numerical weather prediction models is insufficiently fine to represent the features precisely.                    PS

**Reading**
Atkinson, B.W. 1981: *Meso-scale atmospheric circulations.* London: Academic Press. · Browning, K.A. 1982: *Nowcasting.* London: Academic Press. · Carlson, T.N. 1991: *Mid-latitude weather systems.* London: Harper-Collins.

**frontogenesis** The process of intensification of the thermal gradient at a frontal zone. It takes place mainly by horizontal confluence and convergence when the isotherms are suitably aligned. Since uplift must follow surface convergence, frontogenesis is helped if the upper atmospheric air movements favour the continuation of rising air. The reverse process is known as frontolysis. In this situation the thermal gradient becomes progressively less distinct and the cloud band eventually disperses.                    PAS

**frost action** The mechanical weathering process caused by alternative or repeated cycles of freezing and thawing of water in pores, cracks, and other openings, usually at the ground surface. The expansion of water upon freezing (approximately 9 per cent by volume) forces material, commonly rock, apart. Termed frost wedging, its efficacy largely depends upon the frequency of FREEZE–THAW CYCLES, the availability of moisture, and the lithological/strength characteristics of the material. Other terms for frost wedging include frost shattering, gelifraction and frost riving. The term 'frost action' is sometimes used to include a wider range of frost-related processes, such as frost heaving, frost creep, thermal contraction cracking and frost weathering. (See also FROST WEATHERING.)                    HMF

**frost creep** The downslope movement of the particles as a result of the frost heaving of the ground and subsequent settling upon thawing, the heaving being predominantly normal to the slope and the settling more nearly vertical. Although frost creep is commonly associated with GELIFLUCTION (and is usually included within it in rate measurements because of the difficulty of distinguishing between their contributions to total movement) it is a separate process. Movement associated with frost creep decreases from the surface downwards and depends upon frequency of FREEZE–THAW CYCLES, angle of slope, moisture available for heave, and frost susceptibility of soil. Studies in East Greenland indicate that frost creep exceeds gelifluction by not more, and probably less than 3:1 in most years, and frost creep usually resulted in 30–50 per cent of total annual movement on slopes. In the Colorado Rockies measurements indicate that solifluction is a more effective process than frost creep in the saturated axial areas of lobes, but less effective than frost creep at their edges.                    HMF

**Reading**
Benedict, J.B. 1970: Downslope soil movement in a Colorado alpine region: rates, processes and climatic significance. *Arctic and alpine research* 2, pp. 165–226. · Washburn, A.L. 1979: *Geocryology: a survey of periglacial processes and environments.* New York: Wiley.

**frost heave** The predominantly upward movement of mineral soil during freezing caused by the migration of water to the freezing plane and its subsequent expansion upon freezing. Frost heaving is usually associated with the active layer above permafrost or with seasonally frozen ground. As such, ICE SEGREGATION is an essential component of frost heave. Field studies indicate that heave occurs not only during the autumn freeze-back period but also during winter when ground temperatures are below 0 °C. Frost heaving processes include the upheaving of bedrock blocks, upfreezing of objects, tilting of stones, formation of NEEDLE ICE, and the sorting and migration of soil particles. Frost heaving presents important geotechnical problems in the construction of roads, buildings, pipelines, and airfields in cold environments.                    HMF

## Reading

Mackay, J.R. 1983: Downward water movement into frozen ground, western Arctic coast. *Canadian journal of earth sciences* 20, pp. 120–34. · Slusarchuk, W.A., Clark, J.I. and Nixon, J.F. 1978: Field tests of a chilled pipeline buried in unfrozen ground. *Proceedings of the Third International Conference on Permafrost*, 10–13 July 1978, Edmonton, Alberta. National Research Council of Canada, Ottawa, Vol. 1. · Washburn, A.L. 1979: *Geocryology: a survey of periglacial processes and environments*. New York: Wiley. Esp. pp. 79–96.

**frost smoke**  (also arctic smoke) A fog produced by the contact of cold air with relatively warm water. It is commonly associated with leads of open water which open up in a sea-ice cover but may also occur at the ice edge or over water which is beginning to freeze.

**frost weathering**  A general term used to describe the complex of weathering processes, both physical and chemical, which operate, either independently or in combination, in cold non-glacial environments. The most important physical weathering process is frost wedging which characteristically produces angular fragments of varying sizes. The predominant size to which rocks can be ultimately reduced by frost wedging is generally thought to be silt. Porous and well-bedded sedimentary rocks, such as shales, sandstones and limestones are especially susceptible to frost weathering. Features attributed to frost weathering include extensive areas of angular bedrock fragments (blockfields and blockslopes) and irregular bedrock outcrops termed tors. It has been estimated that frost weathering (i.e. rockfalls induced by frost wedging) over a 50 year period caused steep rock faces in Longyeardalen, Spitsbergen, to retreat at a rate of 0.3 mm year$^{-1}$.

Many aspects of cold climate weathering are not fully understood. For example, it has been suggested that hydration shattering may be responsible for the large blockfields and blockslopes characteristic of low temperature alpine and polar environments, but this has yet to be proved. Equally, experimental studies in the former USSR indicate that under cold conditions the ultimate size reduction of quartz (0.05–0.01 mm) is smaller than for feldspar (0.1–0.5 mm), a reversal of what is normally assumed for temperate environments. Finally, the current emphasis upon frost wedging in periglacial environments should not obscure the fact that chemical weathering can be significant. The dominance of physical weathering tends to mask chemical effects. In places, physical and chemical effects combine, as in salt wedging. Like frost wedging, this process breaks up rock into silt-size particles and is particularly effective in cold, arid regions, such as the ice-free areas of Antarctica. Solutional effects in limestone terrain may also be present and karst terrain exists in permafrost regions, further illustrating the inadequacy of a simplistic view of frost weathering.                                                      HMF

## Reading

van Everdingen, R.O. 1981: *Morphology, hydrology and hydrochemistry of karst in permafrost terrain near Great Bear Lake, NWT*. National Hydrology Research Institute, paper 11, Calgary, Alberta. · French, H.M. 1976: *The periglacial environment*. London and New York: Longman. · Jahn, A. 1976: Contemporaneous geomorphological processes in Longyeardalen, Vestspitsbergen (Svalbard). *Biuletyn peryglacjalny* 26, pp. 253–68. · Washburn, A.L. 1979: *Geocryology: a survey of periglacial processes and environments*. New York: Wiley. · White, S.E. 1976: Is frost action really only hydration shattering? A review. *Arctic and alpine research* 8, pp. 1–6.

**frost wedge**  See ICE WEDGE.

**Froude number**  ($F_r$) The dimensionless ratio of inertial to gravity forces in flowing water:

$$F_r = v/(\sqrt{gd}),$$

where $v$ is velocity, $g$ is gravitational acceleration and $d$ is depth. The term ($\sqrt{gd}$) is the velocity of a small gravity wave (a surface ripple) and, if $F_r < 1$, the flow is subcritical or tranquil and ripples formed by a pebble dropped into the water travel upstream because their velocity exceeds that of the stream. If $F_r > 1$ the flow is supercritical, and when $F_r = 1$ the flow is critical. Sudden spatial changes from supercritical to subcritical are HYDRAULIC JUMPS. In sandbed channels, temporal changes of FLOW REGIME during floods cause a consistent sequence of bedform changes, with sand dunes washing out to form antidunes at a local Froude number of approximately unity. However, *mean* Froude numbers in natural channels rarely exceed 0.4–0.5, because the associated rapid energy losses cause bank erosion, channel enlargement, and a reduction of flow velocity and Froude number – an example of negative feedback (see SYSTEMS).                     KSR

**fulgurite**  A tube in sand or rock produced by the fusing effects of a lightning strike (Withering 1790). Sand fulgurites are especially common in areas of dry, loose, quartz sand typical of deserts.                                                      ASG

## Reference

Withering, W. 1790: An account of some extraordinary effects of lightning. *Philosophical transactions of the Royal Society of London*, Series D. 80, pp. 293–5.

**fulje**  A depression between barchans or barchanoid sand ridges, especially where dunes are pressing closely on one another. In Australia the term may be used to describe a blowout or small parabolic dune.                                    ASG

**fumarole** A small, volcanic vent through which hot gases are emitted.

**fungi** A diverse grouping of organisms involved in nutrient cycling, lacking the ability to photosynthesize. An important group occurs as MYCORRHIZAE, a mutualistic (symbiotic) association with plant roots occurring in up to 80 per cent of terrestrial plant species (van der Heijden *et al*. 1998). Fungi are capable of breaking down a wide range of organic materials, including cellulose, both living and dead. They inhabit a broad range of environments, marine as well as terrestrial. Fungi display branching systems of filaments called *hyphae* which can extend in all directions from the location where a fungal spore germinates, and these can generate very large osmotic gradients that result in the transfer of nutrients from the medium in which the fungus grows. In association with an alga, fungi form the *lichens*, another very important group of non-vascular plants, especially in the soils of dry environments where they are able to survive severe moisture stress.                              DLD

**Reference**
Van der Heijden, M.G.A. Klironomos, J.N., Ursic, M., Moutoglis, P., Streitwolf-Engel, R., Boller, T., Wiemken, A. and Sanders, I.R. 1998: Mycorrhizal fungal diversity determines plant biodiversity, ecosystem variability and productivity. *Nature* 396, pp. 69–72.

**funnelling ratio** A means of describing the volume of STEM FLOW that is carried to the soil along the trunk of a tree or other large plant. It was defined by Herwitz (1982) as the ratio of the observed volume of stem flow to the volume that would be expected if the water was intercepted solely on parts of the plant directly above the tree base, and equal in area to the basal area of the trunk. A ratio $> 1$ indicates that water is being funnelled into the trunk from peripheral parts of the canopy. Using canopy dimensions of rain forest trees, and stem flow volumes measured using stem flow collars, Herwitz found funnelling ratios as high as 156. Funnelling of this intensity delivered sufficient stem flow volume to exceed the infiltration capacity of the soil around the tree, and to create local sources of Hortonian surface run-off.         DLD

**Reference**
Herwitz, S.R. 1982: The redistribution of rainfall by tropical rainforest canopy tree species. *Proceedings of the First National Conference on Forest Hydrology*. Melbourne: Institution of Engineers, Australia, National Conference Publication 82/6, pp. 26–9.

**fynbos** Heathlands found in the Mediterranean-climate region of south-western South Africa. Like the plant communities which characterize other parts of the world with similar fire-prone and temperate winter-rainfall climates (for example, macchia or maquis of the Mediterranean basin, and chaparral of southwestern USA), fynbos is dominated by sclerophyllous (i.e. evergreen, hard-leaved) shrubs. In the case of fynbos, plant diversity is exceptionally high and there may be as many as 120 different plant species within a 10 m × 10 m sample plot. Taxa typically include shrubs belonging to the families *Proteaceae* and *Ericaceae*, and a group of reed-like plants of the *Restionaceae* family.                              MEM

**Reading**
Cowling, R.M. ed. 1992: *The ecology of fynbos: nutrients, fire, diversity*. Cape Town: Oxford University Press.

# G

**gabbro** A basic igneous rock composed of calcic plagioclase and clinopyroxene with or without orthopyroxene and olivine. Usually coarse-grained and dark grey to black in colour.

**Gaia hypothesis** In 1979 James Lovelock published his so-called Gaia hypothesis, which proposed that the earth system was in some respects a self-regulating (via FEEDBACKS) single system (i.e. a 'living planet'). Gaia is named after the Greek goddess of the earth.

**Reference**
Lovelock, J.E. 1979: *Gaia: a new look at life on earth.* Oxford: Oxford University Press.

**gallery forest** Forest which lines the banks of a river in an area where away from the river's favourable hydrological circumstances such forest does not occur.

**GARP** (Global Atmospheric Research Programme) A project born out of the World Weather Watch in the late 1960s and early 1970s with the aim of studying the *global* atmospheric circulation by both observational and theoretical means. The observational programme has involved the use of satellites, balloons, radars and ocean-buoys as well as conventional surface and upper air observing systems. The First Garp Global Experiment (FGGE), conducted in the late 1970s was mankind's largest ever scientific experiment. These observational programmes have been complemented by programmes of numerical modelling of the GENERAL CIRCULATION of the atmosphere. BWA

**garrigue** Xerophytic, evergreen scrubland occurring on thin soils in areas with a dry 'Mediterranean type' of climate. Much of it may result from anthropogenic landscape degradation. It consists of low thorny shrubs and stunted evergreen oaks.

**gas laws** The thermodynamic laws applying to perfect gases. In particular they relate the pressure, density and temperature of gases in different ways.

The most important, dealing with 'perfect gases', are:

Boyle's law: 'The pressure of a given mass of gas, at constant temperature, is inversely proportional to its volume.'

This may be summarized as $pV =$ constant.

Charles's law: 'The volume of a given mass of gas, at constant pressure, increases by 1/273 of its values at 0 °C for every degree Celsius rise in temperature.'

These two laws can be related through the general equation of state for a perfect gas:

$$pV + RT$$

where $T$ is the absolute temperature, $p$ is the pressure and $V$ the volume, and $R$ is the gas constant.

Dalton's law of partial pressures: 'A mixture of gases has the same pressure as the sum of the partial pressures of its components.' WBW

**gauging stations** Sites at which river flow is determined. The gauging sites may only be equipped to provide point measurements in time or they may provide continuous measurements. The accuracy of the flow estimates will vary according to the gauging technique employed. (See also DISCHARGE; HYDROMETRY.) AMG

**GCM** See GENERAL CIRCULATION MODELLING.

**geest** Ancient alluvial sediments which still mantle the landsurfaces on which they were originally deposited.

**gelifluction** A type of solifluction occurring in periglacial environments underlain by permafrost. Suitable conditions for gelifluction occur in areas where downward percolation of water through the soil is limited by the permafrost table and where the melt of segregated ice lenses provides excess water which reduces internal friction and cohesion in the soil. Particularly favoured sites include areas beneath or below late-lying snowbanks. Rates of movement, which generally vary between 0.5 and 10.0 cm per year, usually decrease with depth. Frost creep is usually measured as a component of gelifluction. As with solifluction, features related to gelifluction include sheets, stripes and lobes. HMF

217

**Reading**
Washburn, A.L. 1979: *Geocryology: a survey of periglacial processes and environment.* New York: Wiley.

**gendarmes**  Pinnacles of rocks projecting vertically from a ridge.

**genecology**  The study of the genetics of populations in relation to habitat; the study of species (and other taxa) through a combination of the methods and concepts of both ecology and genetics. Some species display a range of ECOTYPES – genetic varieties existing in different environments; the investigation of these ecotypes constitutes a part of genecology.          PHA

**general circulation of the atmosphere**  Either the totality of atmospheric fluid motions or that part of the circulation associated with the synoptic – and planetary – scale horizontal wind field. The latter is driven either directly or indirectly by horizontal heating gradients in a stably stratified atmosphere and accounts for about 98 per cent of the atmospheric kinetic (wind) energy. The remaining kinetic energy is associated with small-scale motions driven by convective instability.

The energy associated with the planetary – and synoptic-scale – atmospheric disturbances is constantly being syphoned off by small-scale fluid motions which interact among themselves to transfer energy to smaller and smaller scales and ultimately down to random molecular motions. The large-scale circulation is maintained against this frictional dissipation by drawing on the reservoir of POTENTIAL ENERGY inherent in the spatial distribution of atmospheric mass resulting from the gradients of diabatic heating induced primarily by the sun. The conversion from potential to kinetic energy is achieved primarily through vertical motions in the atmosphere.

In the tropics most of the kinetic energy is associated with quasi-steady thermally driven circulations, which include the seasonally varying MONSOONS, the result of land–sea heating contrasts, and the large-scale HADLEY CELLS, especially prominent over the tropical Atlantic and Pacific Oceans. Within these cells the rising currents near the equator, forming the INTERTROPICAL CONVERGENCE ZONE, gain both potential energy and heat energy, the latter through condensation. This energy is carried poleward by the circulation in the upper TROPOSPHERE and lower STRATOSPHERE to latitudes of 30° or 40° where the now descending air is heated by ADIABATIC compression and the potential energy is converted into the kinetic energy of the low

latitude trade winds and, to some extent, the middle latitude westerlies.

In middle and high latitudes much of the kinetic energy is associated with moving disturbances called baroclinic waves, which develop within zones of strong horizontal temperature gradient along the POLAR FRONT. The wave energy is gained primarily through the vertical displacement of warm air masses by cold air masses, thus depleting the potential energy of the system. That the middle latitude westerlies can move from west to east faster than the rotating earth is due to the fact that these winds are maintained against frictional retardation primarily by the poleward gradient of temperature and potential energy.          WDS

**Reading**
Lorenz, E.N. 1967: *The nature and theory of the general circulation of the atmosphere.* Geneva: World Meteorological Organization. · Palmén, E. and Newton, C. W. 1969: *Atmospheric circulation systems.* New York: Academic Press. · Smagorinsky, J. 1972: The general circulation of the atmosphere. In D.P. McIntyre ed., *Meteorological challenges: a history.* Ottawa: Information Canada. Pp. 3–42. · Wallace, J.M. and Hobbs, P.V. 1977: *Atmospheric science.* New York: Academic Press.

**general circulation modelling**  One of the primary tools of investigation now used by climate researchers is the general circulation model (GCM). GCMs are mathematical computer models describing the primary controls (such as radiative fluxes), energy transfers, circulations and feedbacks existing in the earth-ocean-climate system. GCMs have their roots in the early numerical weather models developed by J.G. Charney and the so-called 'father' of modern computing, J. von Neumann, in the late 1940s. Most GCMs are based on seven fundamental (or 'primitive') equations. These equations are:

The equation of state (or ideal gas law). This basic physics principle mathematically describes the relationship between pressure, density and temperature.

The three equations of motion (or, when expressed in vector notation, often called 'the Navier–Stokes equation'). These equations describe the forces producing airflow in the three spatial dimensions. The forces mathematically included in those equations include gravity, the pressure gradient force, the Coriolis effect, and various acceleration terms.

The first law of thermodynamics. This principle links temperature changes to changes in energy inputs into the system and energy changes within the system.

An equation addressing conservation of mass. This mathematical relationship ensures that mass is neither created nor destroyed throughout the climate domain during the course of the simulation.

An equation addressing moisture distribution and phase through the model domain. Because water is of critical importance to the climate system, separate accounting of moisture (and its various phases of ice, snow, liquid water and water vapour) must be conducted within the GCM.

To ensure mathematical closure, these seven equations must be associated with seven field variables (variables expressed as functions of time and space). Consequently, the seven fundamental variables in a GCM are pressure ($p$), temperature ($T$), density (O), zonal ($E–W$) wind ($u$), meridional ($N–S$) wind ($v$), vertical wind ($w$), and specific humidity ($q$). In general, the mathematical expressions are solved such as the change in one of the variables over time is expressed as the result of the given variable's spatial variability. For example, all other factors being constant, a net convergence of moisture into a given area will lead to an increasing level of moisture at that location over time.

GCMs are constructed over spatial domains that can be expressed by cartesian or spatial coordinates (termed gridpoint models) or by Fourier transform functions (referred to as spectral models). The spatial resolution of these models continues to be refined. Early GCMs of the 1970s and 1980s commonly had spatial dimensions of 5 ¼ latitude × 5 ¼ longitude or even 10 ¼ × 10 ¼. Today, many GCMs operate with resolutions of 2.5 ¼ latitude by 2.5 ¼ longitude or have higher resolution submodels operating within larger spatial domains.

Temporally, the equations of a given GCM are solved over a fairly short time basis (on the order of minutes or hours) and the results are subsequently used as the new input to compute additional changes in the state variables over time. This means that an incredible number of calculations (over both space and time) must be made to produce a single GCM simulation. Normally, because of these heavy computational demands, GCM simulations are run only for a limited number of simulated years.

The first GCMs only addressed the fundamental principles of atmospheric circulation with very limited or extremely simplified inputs and outputs. Today GCMs are much more complex and include mathematical expressions of ocean circulation and interaction with the atmosphere, with land surface processes such as vegetation or cryosphere (snow and ice surfaces), and atmospheric chemistry (including radiative processes associated with ozone and carbon dioxide). Although inclusion of these processes and others has greatly improved GCMs, these models remain imperfect representations of the actual earth-ocean-climate system. Consequently, the results from climate simulations by GCMs should not be taken as true predictions of climate change but rather as either information used to improve our understanding of the climate system or as potential outlooks of future climate based only on our current understanding of climate.

Because GCMs have become such critical tools of climatic investigation, they have been used to study a number of important climate concerns. GCMs have been used as primary tools in the investigation of (1) the global climatic effects of increased atmospheric carbon dioxide; (2) the effects of deforestation on global climate; (3) the reconstruction of past climates (such as the Cretaceous time of the dinosaurs); and (4) studies of the global effects of the South Pacific phenomena known as El Niño–South Oscillation (ENSO).

Advances in numerical programming continue to focus on the inclusion of high-resolution spatial domains and of more complex processes into GCMs. However, because of the heavy numerical demands made by these inclusions, such advancement in GCMs is often limited by the size, speed and availability of modern computers.                                       RSC

**general system theory** Was largely developed by the biologist, Von Bertalanffy, whose basic statement appeared in 1950 and was substantially extended in 1962 (where the term becomes general systems theory). The fundamental proposition of the theory is that systems (defined as structured sets of objects and their attributes) may be identified in all studies of phenomena; and that more may be learnt by the comparison of the ways in which similar (isomorphic) systems function, than by the standard academic concentration on the distinctiveness of their component parts. Thus what is of interest in a comparison of – say – towns, drainage basins and bathtubs, is that they may be considered to share fundamental attributes which relate to their physical boundedness and to the transfers of energy and mass across those boundaries. These transfers serve to integrate the components of each system and this interrelatedness of parts is a diagnostic and important feature.

Since its introduction by Von Bertalanffy, general system theory (or GST) has become a field of study in its own right. The history of the ideas *per se* in physical geography is, however, less than clear and one is forced to admit that

GST (as opposed to general ideas of SYSTEMS) has been of little direct moment.

The first explicit use of Von Bertalanffy's ideas in geomorphology was made by Strahler (1950, p. 676), who stated 'A graded drainage system is perhaps best described as an open system in a steady state... which differs from a closed system in equilibrium in that the open system has import and export of components.' This and subsequent studies by Strahler and others were then taken up by Chorley (1960) in a major paper which apparently focused on the ideas of GST, rather than simply on concepts of physical systems. Chorley's discussion is perhaps remarkable in that it nowhere defines GST, although it introduces a series of the key concepts: notably, open and closed systems, entropy, steady state, self-regulation, equilibrium, hierarchial differentiation and organization. In later works (see Chorley and Kennedy 1971; Bennett and Chorley 1978), references to both GST and Von Bertalanffy are exiguous. Similarly Huggett (1980) makes no reference to either GST or its author. Only Chapman (1978, p. 404) is prepared to confess that he is making explicit use of GST.

While it may be argued that the *framework* provided by GST has been widely adopted in physical geography, it seems abundantly clear that the use made has been so derivative that the actual aims and aspirations of GST itself are largely unknown – and therefore irrelevant – to discussions of systems in physical geography.                                    BAK

**References**
Bennett, R.J. and Chorley, R.J. 1978: *Environmental systems: philosophy, analysis and control.* London: Methuen. · Chapman, G.P. 1978: *Human and environmental systems: a geographer's appraisal.* London: Academic Press. · Chorley, R.J. 1960: Geomorphology and general systems theory. *United States Geological Survey professional paper* 500–B. · —and Kennedy, B.A. 1971: *Physical geography: a systems approach.* London: Prentice-Hall International. · Huggett, R. 1980: *Systems analysis in geography.* Oxford: Clarendon Press. · Strahler, A.N. 1950: Equilibrium theory of erosional slopes approached by frequency distribution analysis. *American journal of science* 248, pp. 673–96; 800–14. · Von Bertalanffy, L. 1950: The theory of open systems in physics and biology. *Science* 111, pp. 23–9. · —1962: General systems theory, a critical review. *General systems* VII.

**genetic drift**  The effect of sampling error in causing random changes in the relative frequency of genes in a gene pool. Random change in gene frequency occurs in all populations and it is maintained that genetic drift provides a mechanism for evolution. The smaller the population size the greater the possibility of gene loss or fixation: a gene normally present in one in

10,000 individuals may not be present in a population of 100. Particularly important is the 'founder effect': when a small sample of an organism's population is isolated, for example on an island or mountain peak, it may have a different gene frequency from the parent group. This is one reason why organisms on remote islands are frequently ENDEMIC species or subspecies. An animal that has been reduced to a very small population by climatic catastrophe, disease or overhunting is likely to have an impoverished gene pool for some time after recovery.      PHA

**Reading**
Bonnel, M.L., and Selander, R.K. 1974: Elephant seals: genetic variation and near-extinction. *Science* 184, pp. 908–9.

**geo**  A deep, narrow cleft or ravine along a rocky sea coast which is flooded by the sea.

**geochronology**  A term used both to describe the chronologies of long-term environmental change, usually in the context of the QUATERNARY period, derived from the use of numerical age dating techniques (e.g. LUMINESCENCE DATING, CARBON DATING, URANIUM SERIES DATING), and to describe the techniques used to produce a chronology of change (i.e. a *geochronometric technique*).      DSGT

**geocryology**  The study of frozen, freezing and thawing terrain (but not glaciers) is known as permafrost science or more generally termed geocryology. This widespread term is usually associated with earth materials having a temperature below 0 °C. Geocryology seeks to promote an understanding of the dynamics of such environments, especially the study of the origins and history of permafrost.      AP

**Reading**
Washburn, A.L. 1979: *Geocryology: a survey of periglacial processes and environments.* London: Edward Arnold.

**geode**  A roughly spherical or globular inclusion within a mass of rock. Geodes are hollow and frequently exhibit mineral crystals growing into the central void.

**geodesy**  The determination of the size and shape of the earth by mathematical means and surveys.

**geo-ecology**  See LANDSCAPE ECOLOGY.

**geographic information system (GIS)**
Computer system for the storage, analysis and display of spatial data. Spatial data relates to features on the surface of the earth, at any

scale from the global to the local, and consists of two elements: location (where things are) and attributes (what things or places are like). There are two main ways of storing this information on the computer, giving two main types of GIS: (1) In VECTOR GIS location is stored by classifying objects as points (e.g. spot heights, buildings), lines (e.g. rivers, roads) and areas (e.g. woodlands, towns). Each object has attributes associated with it, normally stored in a database file. (2) In RASTER GIS location is stored by imposing a grid of square PIXELS over the area of interest – since the pixels are of fixed size, if the location of one point on the grid is known, the location of all other points can be calculated. Attributes are stored by recording a value in each pixel – these are usually presence/absence of an object, value in a classification (e.g. land use) or value of a geographical variable (e.g. elevation).

The input of data is an important and time consuming element of using GIS. The commonest sources of data are maps, remote sensing, surveys and address-based data. Maps are normally captured by digitizing. The paper map is attached to a digitizing table, which can record the location of points, lines and areas on the map under operator control. This produces vector data, but this can be converted to raster by software if needed. REMOTE SENSING data are already in a digital, raster format but raw satellite images must often be processed to generate useful information from the radiation values. Many large surveys, such as population censuses, provide data for small areas. If the boundaries of the areas are available in digital format, the survey information can be attached as attributes. Personal information often contains some form of postal address or postal code (e.g. postcode in the UK, zip code in the USA). This provides a point location which can be stored in a GIS, allowing the mapping of items such as crime incidence or cases of disease. Personal data can also be aggregated to provide a picture of the typical make-up of small areas. Such geodemographic information can then be used to target products and services more closely at potential customers.

Some GIS systems have developed from systems for computer cartography, and the production of paper maps from digital map data is still an important application of GIS. Advantages of digital methods include the ability to map only selected features, to experiment with the symbolism used and to produce maps unconstrained by traditional map sheet boundaries ('seamless mapping').

The major advantage of GIS systems lies in their ability to analyse data, which takes two main forms: (1) Analysis which uses simple queries of existing data. An important example is facilities management – the use of GIS for managing infrastructure such as cables and pipes. This is best suited to vector GIS, and is commonly used by utility companies. (2) Analysis which modifies or combines existing sets of data. This can be done using both vector and raster, although the types of data and analysis involved are often different.

Vector GIS is well suited to analysis involving discrete objects, such as the use of digitized road networks to estimate travel times between points. Raster GIS is well suited to the analysis of environmental data, where phenomena vary continuously across space. Such analysis commonly makes use of a DIGITAL TERRAIN MODEL (DTM) of the land surface. Both vector and raster can be used for cartographic modelling (Tomlin 1990) in which operations are used to modify or combine the original layers of data to answer queries. For instance to find potential sites for an agricultural crop, layers indicating where individual factors such as soil type, temperature, precipitation are suitable might be combined (using an overlay operation) to find those areas where all the factors are suitable.

In the past GIS software took the form of general purpose packages which required considerable training to use. Increasingly, some elements of GIS functionality are finding their way into more widely available software and onto the world wide web. For instance, route finding facilities are provided in simple PC packages and tourist information services may make use of a map interface, with a radial search to locate all attractions near to a given location.      SMW

**Reading**
Heywood I., Cornelius, S. and Carver, S, 1998: *An introduction to geographical information systems*. Harlow: Addison Wesley Longman. · DeMers, M.N. 1997: *Fundamentals of geographic information systems*. New York: Wiley. · Burrough, P.A. and McDonnell, R. 1998: *Principles of geographical information systems*. Oxford: Oxford University Press. · Longley, P., Maguire, D., Goodchild, M.F. and Rhind, D. 1999: *Geographical information systems: principles, techniques, management and applications*. 2 vols, 2nd edn. Cathedral City, Calif.: Adams Business Media. · Tomlin C.D. 1990: *Geographic information systems and cartographic modelling*. Englewood Cliffs, NJ: Prentice-Hall.

**geoid** The equipotential surface that would be assumed by the sea surface in the absence of tides, water-density variations, currents and atmospheric effects. It varies above and below the geometrical ellipsoid of revolution by as much as 100 m due to the uneven distribution of mass within the earth. The mean sea-level surface varies about the geoid by typically decimetres, but in some cases by more than a metre. DTP

**geological time-scale** The divisions of geological time as listed in the table.

Reading
Harland, W.B. *et al* 1990: *A geologic time scale.* Cambridge: Cambridge University Press.

**geomorphology** A term that arose in the Geological Survey in the USA in the 1880s, possibly coined by J.W. Powell and W.J. McGee. In 1891 McGee wrote: 'The phenomena of degradation form the subject of geomorphology, the novel branch of geology.' He plainly regarded geomorphology as being that part of geology which enabled the practitioner to reconstruct earth history by looking at the evidence for past erosion, writing: 'A new period in the development of geologic science has

dawned within a decade. In at least two American centres and one abroad it has come to be recognised that the later history of world growth may be read from the configuration of the hills as well as from the sediments and fossils of ancient oceans... The field of science is thereby broadened by the addition of a coordinate province – by the birth of a new geology which is destined to rank with the old. This is geomorphic geology, or geomorphology.'

Many scientists had studied the development of erosional landforms (see Chorley *et al.* 1964) before the term was thus defined and since that time its meaning has become broader. Many geomorphologists believe that the purpose of geomorphology goes beyond reconstructing earth history (see DENUDATION CHRONOLOGY), and that the core of the subject is the comprehension of the form of the ground surface and the processes which mould it. In recent years

Geological time-scale

| Era | Sub-era/period/ Subperiod/Epoch | | | Age (Ma BP) |
|---|---|---|---|---|
| CAINOZOIC | Quaternary | Holocene | | 0.01 |
| | | Pleistocene | | 1.64 |
| | Tertiary — Neogene | Pliocene | | 5.2 |
| | | Miocene | Late | 14.2 |
| | | | Early | 23.3 |
| | Tertiary — Palaeogene | Oligocene | | 35.4 |
| | | Eocene | | 56.5 |
| | | Palaeocene | | 65.0 |
| MESOZOIC | Cretaceous | Late | | 97.0 |
| | | Early | | 145.6 |
| | Jurassic | Late | | 157.1 |
| | | Middle | | 178.0 |
| | | Early | | 208.0 |
| | Triassic | | | 245.0 |
| PALAEOZOIC | Permian | | | 290.0 |
| | Carboniferous | | | 362.5 |
| | Devonian | | | 408.5 |
| | Silurian | | | 439.0 |
| | Ordovician | | | 510.0 |
| | Cambrian | | | 570.0 |
| PRECAMBRIAN | | | | |

*Source*: Harland *et al* 1990. (Note: the classification stands but the numbers have been revised from further determinations since 1989.)

there has been a tendency for geomorphologists to become more deeply involved with understanding the processes of erosion, weathering, transport and deposition, with measuring the rates at which such processes operate, and with quantitative analysis of the forms of the ground surface (morphometry) and of the materials of which they are composed. Geomorphology now has many component branches (e.g. ANTHROPOGEOMORPHOLOGY; APPLIED GEOMORPHOLOGY). ASG

**Reading and References**
Bloom, A.L. 1978: *Geomorphology: a systematic analysis of late Cenozoic landforms.* Englewood Cliffs, NJ: Prentice-Hall. · Butzer, K.W. 1976: *Geomorphology from the earth.* New York: Harper & Row. · Chorley, R.J., Dunn, A.J. and Beckinsale, R.P. 1964: *The history of the study of landforms.* Vol. 1. London: Methuen. · — Schumm, S.A. and Sugden, D.E. 1984: *Geomorphology.* London and New York: Methuen. · McGee, W.J. 1891: The Pleistocene history of northeastern Iowa. *Eleventh annual report of the US Geological Survey.* Pp. 189–577. · Rice, R.J. 1988: *Fundamentals of geomorphology.* 2nd edn. London: Longman. · Sparks, B.W. 1986: *Geomorphology.* 3rd edn. London: Longman. · Tinkler, K.J. 1985: *A short history of geomorphology.* London and Sydney: Croom Helm. · Twidale, C.R. 1976: *Analysis of landforms.* Sydney and New York: Wiley.

**geophyte** A herbaceous plant which has parts beneath the ground surface which survive when the parts above ground die back.

**geostatistics** This describes the special set of statistical methods which recognize that many phenomena of concern to geographers are spatially distributed, and that some classical statistics will result in information that represents this spatial distribution inadequately, or even incorrectly (e.g. see AUTOCORRELATION). Potentially, geostatistics is a very powerful tool for the physical geographer and allows: (1) description of the spatial features of a data set; and (2) estimation of key quantities, both for an area as a whole, or to provide spatially-variable local estimates. The latter is of particular importance as it provides a means of interpolating from a given set of data to areas where little or no information is available. A good example of a method that can be used to do this is *kriging* which, at its simplest, seeks to estimate local values of a variable using a model based on weighted linear combinations of available data. SNL

**Reading**
Isaaks, E.H. and Srivastava, R.M., 1989. *An introduction to applied geostatistics.* Oxford: Oxford University Press.

**geostrophic wind** The geostrophic wind is a wind with a velocity determined by an exact balance of the CORIOLIS FORCE and the horizontal pressure gradient force. This balance results in a configuration of velocity and pressure as described by BUYS BALLOT'S LAW, i.e. in the northern hemisphere, when one has one's back to the wind, low pressure lies on the left and high pressure on the right. The converse is true in the southern hemisphere. The geostrophic wind thus blows *along* the isobars and its magnitude is a direct function of the horizontal pressure gradient force, and an inverse function of height and latitude. It is not defined at the equator. At other latitudes the geostrophic wind is a reasonable approximation to the real wind. BWA

**Reading**
Atkinson, B.W. 1972: The atmosphere. In D.Q. Bowen, ed., *A concise physical geography.* Amersham: Hulton Educational. Pp. 1–76, esp. pp. 33–48. · Hess, S.L. 1959: *Introduction to theoretical meteorology.* New York: Henry Holt.

**geosyncline** A very large depression, perhaps several hundred kilometres across and up to 10 km deep, the terrestrial or marine floor of which is built up by sedimentation.

**geothermal energy** Gravitational collapse and radioactive breakdown generate heat in the earth's interior, which may be 'extracted' for human use by pumping water into the earth and extracting it as steam. DSGT

**Gerlach trough** A sediment trap designed to catch a sample of OVERLAND FLOW and the sediment it carries down a hillside. Troughs with a variety of shapes and sizes have been used to collect slope-wash sediment. They have an upslope lip which is either flush with the surface or inserted beneath the uppermost organic soil layers: disturbance from installing this lip is a major source of error. Sediment settles out and/or is filtered out of water which eventually overflows from the trough, sometimes through a total water or flow rate meter. The trough is protected by a lid. A second major source of error is the accurate delimitation of the area which yields sediment to the trough. This type of installation is an alternative to direct measurements of slope lowering by erosion pins, etc. MJK

**Reading**
Goudie, A. ed. 1990: *Geomorphological techniques.* 2nd edn. London: Unwin Hyman.

**geyser** A spring or fountain of geothermally heated water that erupts intermittently with explosive force as a result of increases in pressure beneath the surface.

**Ghyben–Herzberg principle**  This refers to the relationship between freshwater and saltwater in a coastal aquifer. Ghyben and Herzberg were two European scientists who independently investigated this relationship around the turn of the century. They found that since freshwater is less dense than seawater, it rises above underlying intruding saltwater. In unconfined aquifers beneath small islands, a lens of freshwater floats on seawater which surrounds and underlies it, whereas at the edges of larger landmasses there is a sloping interface with freshwater extending to the coast near the surface and seawater penetrating inland at depth (see diagram). The Ghyben–Herzberg principle can be expressed by the equation:

$$Z_s = \frac{\rho f}{\rho s - \rho f} Z_w$$

where $Z_s$ is the depth below sea level to the interface between fresh and saltwaters; $\rho f$ and $\rho s$ are the density of fresh and saltwaters respectively; and $Z_w$ is the elevation of the water table above sea level. Hence if the density of the freshwater is 1 and that of seawater is 1.025, then under hydrostatic equilibrium the depth $Z_s$ to the interface is 40 times the height of the water table above sea level. Consequently, if pumping from a well in a coastal aquifer results in a draw down of the water table by 1 m, then saltwater will intrude upwards beneath the well by a distance of 40 m (see diagram).

The Ghyben–Herzberg principle simplifies the relationship usually found in nature, because groundwater conditions are usually dynamic rather than static. As a result, the equation usually underestimates the depth to the interface with saltwater, which is commonly seaward of the calculated position.          PWW

**Reading**
Freeze, R.A. and Cherry, J.A. 1979: *Groundwater.* Englewood Cliffs, NJ: Prentice-Hall. · Hubbert, M.K. 1940: The theory of groundwater motion. *Journal of geology* 48, pp. 785–944. · Todd, D.K. 1980: *Groundwater hydrology.* 2nd edn. New York and Chichester: Wiley. · Ward, R.C. and Robinson, M. 1990: *Principles of hydrology.* 3rd edn. Maidenhead: McGraw-Hill.

**gibber**  A desert plain which is mantled with a layer of pebbles or boulders. It is a type of stone pavement.

**Gibbsian distribution**  A distribution used in statistical mechanics, that links, for a system capable of many different configurations, the probability of occurrence of any one state to the total energy associated with that state. This distribution has been employed in the study of the structure of stream networks, which topology shows can exist in principle in an extremely

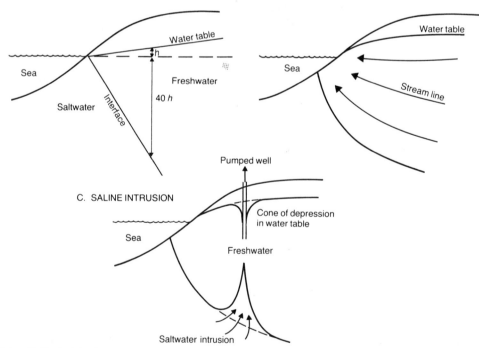

Stages of saltwater intrusion

large number of different forms, but which in nature display certain kinds of regularity (as exhibited in some of the empirically based 'laws' of drainage basin composition). The analysis of system energy using the Gibbsian distribution seeks to develop explanations for these observed regularities in the form of stream networks.                                    DLD

**Reading**
Troutman, B.M. and Karlinger, M.R. 1994: Inference for a generalized Gibbsian distribution on channel networks. *Water resources research* 30, pp. 2325–38.

**gilgai**   A class of soil surface forms, including various kinds of undulations and closed depressions. Derived from an Australian aboriginal word meaning 'waterhole'. Gilgai features include up to three components: the *depression* or low-lying part, the *mound* or elevated part, and sometimes a *shelf* or planar area lying at an elevation between that of depression and mound. The relative spatial extent of each component leads to four basic gilgai types: (1) mound and depression equally developed; (2) mound of greater extent than depression; (3) depression of greater extent than mound; (4) gilgai with mound, shelf, and depression (Paton 1974). The distance separating adjacent mounds or depressions is typically 1–20 m, and the amplitude from depression to mound 10–50 cm, but reaching 1–2 m in rare instances. The development of gilgai is ill understood but appears to be related to shrink-swell behaviour in the subsoil materials of EXPANSIVE SOILS. In some cases, the mound overlies a diapiric structure involving intense deformation, in which subsoil material approaches or reaches the soil surface.                                    DLD

**Reference**
Paton, T.R. 1974: Origin and terminology for gilgai in Australia. *Geoderma* 11, pp. 221–42.

**gipfelflur**   A plane within which uniform summit levels occur in a mountainous region, especially where the uniformity is neither structural nor the residual portion of a peneplain.

**glacial**   *a.* (adjective) Describes a landscape occupied by glaciers. In this usage the term is similar to *glacierized*, an alternative which has not found general favour. The term *glaciated* describes a landscape which has been covered by glaciers, but normally in the past.
   *b.* (noun) Those occasions during Ice Ages when ice sheets were expanded and average global climates were colder and drier than during the intervening INTERGLACIALS such as exists at present. During many of the seventeen or so

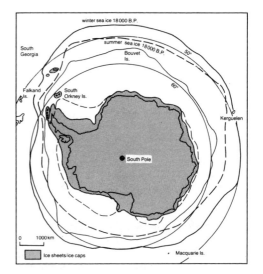

**Glacial: 1.** The Antarctic during glacial maxima. The thinner lines represent the modern equivalents. *Source*: Sugden 1982. Figure 7.2.

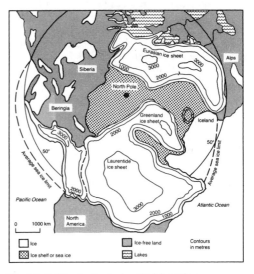

**Glacial: 2.** The Arctic during glacial maxima. *Source*: Sugden 1982. Figure 7.3.

PLEISTOCENE glacials ice sheets covered Canada and the northern USA, northern Europe, Britain north of the environs of London, and northwestern Siberia. In addition, the existing ice sheets of Greenland and Antarctica expanded offshore while mountain glaciers throughout the world extended into lower altitudes. Sea ice extended further towards the equator as global ocean temperatures fell (figures 1 and 2). Atmospheric and oceanic circulation was modified. It seems likely that the globe as a whole was drier with subtropical deserts extending

225

their equator margins and the equatorial rain forest being restricted to discrete islands by the spread of savannah conditions. Mid-latitude areas in the northern hemisphere saw increased wind action with extensive loess deposits in Europe, China and North America.          DES

## Reading

Bowen, D.Q. 1978: *Quaternary geology.* Oxford: Pergamon. · CLIMAP Project Members, 1976: The surface of the Ice-Age earth. *Science* 191, 4232, pp. 1131–7. · Goudie, A.S. 1983: The arid earth. In R. Gardner and H. Scoging eds, *Megageomorphology.* Oxford: Oxford University Press. Pp. 152–71. · Sugden, D.E. 1982: *Arctic and Antarctic.* Oxford: Basil Blackwell; Totowa, NJ: Barnes & Noble.

**glaciation level**   (also called glacial limit) The altitude above which mountain glaciers occur. Since glacier location is also influenced by topography, in particular by the need for sufficiently gentle slopes on which to form, the commonly used method of fixing the glaciation level is to take the altitude midway between the highest topographically suitable mountain without a glacier and the lowest topographically suitable mountain carrying a glacier (Østrem 1966). Delimited in this way, the glaciation level varies predictably over the globe. It rises from near sea level in high latitudes towards the equator in response to temperature changes, but superimposed on this trend are variations reflecting depression in humid areas, such as the mid-latitude and equatorial regions. The glaciation level also rises from maritime coastal locations towards continental interiors.          DES

## Reference

Østrem, G. 1966: The height of the glacial limit in southern British Columbia and Alberta. *Geografiska annaler* 48A.3, pp. 126–38.

**glacier**   A mass of snow and ice which, if it accumulates to sufficient thickness, deforms under its own weight and flows. If there is insufficient snow to maintain flow, as occurs in some dry polar areas, the glacier may be essentially stagnant and termed a *glacier réservoir* (Lliboutry 1965). If the snowfall is sufficient the snow is transformed to ice and flows from the accumulation zone to the ABLATION zone as a *glacier évacateur*.

There are three main types of glacier:

1 *Ice sheet or ice cap* where the ice builds up as a dome over the underlying topography. Such domes are often drained radially by outlet glaciers.
2 *Ice shelf* where the ice forms a floating sheet in a topographic embayment and flows towards the open sea.

3 *Mountain glaciers* which are constrained by the underlying topography of the mountains and form a wide variety of types, e.g. cirque, valley, piedmont glaciers.          DES

### Reading and Reference

Lliboutry, L. 1965: *Traité de glaciologie.* 2 vols. Paris: Masson. · Sugden, D.E. and John, B.S. 1975: *Glaciers and landscape.* London: Edward Arnold.

**glacier milk**   A popular name given to glacial meltwater with sufficient suspended sediment load to give it a milky-green colour.

**glacier table**   A stone resting on a pillar of ice which protrudes above a glacier surface. The ice has been protected from melting by the presence of the overlying stone.

**glacieret**   A small glacier, such as may develop from a SNOW PATCH.

**glacierization**   The process whereby a landscape is progressively covered by glacier ice.

**glacioeustasy**   See EUSTASY.

**glaciofluvial**   The activity of rivers which are fed by glacial MELTWATER. The main characteristics of such streams are the highly variable discharge and the high sediment loads. Discharge varies markedly on a wide variety of time-scales. Variations over a matter of seconds or minutes relate to the sudden release or closure of basal water pockets as a result of glacier sliding. Diurnal fluctuations respond to high rates of melting by day and produce high flows in the evenings. Fluctuations over a matter of days reflect prevailing weather patterns, whereas a strong seasonal summer flow reflects the effect of a glacier in storing winter precipitation only to release it in the ABLATION season. One particularly sudden seasonal peak in discharge may accompany the rapid emptying of a marginal or subglacial lake (see JÖKULHLAUP). The muddy colour of meltwater streams reflects their high suspended sediment loads and measurements as high as $3800 \, mg \, l^{-1}$ have been measured. In addition the bedload is high and may amount to 90 per cent of the suspended sediment load. Not surprisingly glaciofluvial landforms may reflect prodigious feats of erosion and sedimentation. Formerly glaciated areas, particularly in mid-latitudes, contain abundant erosional evidence in the form of deeply incised meltwater channels and giant pot-holes, subglacial channel courses such as ESKERS and KAMES, and extensive areas of proglacial OUTWASH and lake deposits (glaciolacustrine).          DES

A classification of glaciofluvial deposits

| Dominant sediment | Environment | General form | Relationship to ice | Genetic term |
|---|---|---|---|---|
| Ice-contact deposits Sand and gravel | Fluvial | Ridge | Marginal, Subglacial Englacial, Supraglacial | Esker |
| | | Mound | | Kame Kame complex |
| | | Spread with depressions | Marginal | Kettled sandur |
| Proglacial deposits sand and gravel | Fluvial | Spread | Proglacial | Sandur |
| Silt and clay | Lacustrine | | Proglacial/ marginal | Lake plain |
| Sand and gravel | | Terraces, ridges | | Beach |
| Clay, sand and gravel | | Terrace | | Kame delta |
| Silt and clay | Marine | Spread | | Raised mud flat |
| Sand and gravel | | Terraces, ridges | | Raised beach |
| Clay, sand and gravel | | Terrace | | Raised delta |

*Source*: Price, R.J. 1973: *Glacial and fluvioglacial landforms*. Edinburgh: Oliver & Boyd. Table 3, p. 138.

**Reading**
Elliston, G.R. 1973: Water movement through the Gornergletscher. *Symposium on the Hydrology of Glaciers, Cambridge 9–13 Sept. 1969*. International Association of Scientific Hydrology 95, pp. 79–84. · Østrem, G., Bridge, C.W. and Rannie, W.F. 1967: Glaciohydrology, discharge and sediment transport in the Decade Glacier area, Baffin Island, NWT. *Geografiska annaler* 49A, pp. 268–82.

**glacioisostasy** The state of balance that the earth's crust will attempt to achieve when loaded or unloaded by an ICE SHEET. ISOSTASY describes the equilibrium between a less dense and rigid lithosphere 'floating' on the dense and more mobile mantle. The addition of an ice sheet provides greater loading, which given enough time, will depress the lithosphere by an amount related to the thickness of ice and in proportion to the ratio of the densities of ice and the mantle.

Since the density of glacier ice is about one-third that of the crust / mantle, it follows that depression is about 0.3 times the ice thickness. This is a simplification because depression will be greatest under the central portion of the ice sheet, diminishing radially away from this. As the lithosphere possesses rigidity, some of the load is transmitted beyond its margins, which can lead to the depression extending as far away as 250 km.

The lithosphere can respond to a change in loading on a 1000 year time-scale but may take over 10,000 years to reach an equilibrium. As the ice sheets of the last Glacial disappeared between 10,000 and 7000 years ago the land surface in these areas is still adjusting to the unloading. This is often called glacioisostatic rebound and is as much as 1 cm per year for the central portion of the LAURENTIDE ICE SHEET with an estimated 150 m of uplift still to take place. In areas of close proximity to former ice sheets, RAISED BEACHES provide a partial record of the glacioisostatic rebound, although the effects of GLACIOEUSTASY and HYDROISOSTASY also need to be taken into account.                CDC

**Reading**
Benn, D.I., and Evans, D.J.A., 1998: *Glaciers and glaciation*. London: Arnold. · Sabadini, R., Lambeck, K. and Boschi, E. eds. 1991: *Glacial isostasy, sea level, and mantle rheology*. Dordrecht: Kluwer.

**glaciotectonism** Those structures and landforms (e.g. displaced megablocks) produced by deformation and dislocation of pre-existing soft bedrock and drift as a direct consequence of glacier ice movement.                ASG

**Reading**
Aber, J.S. 1985: The character of glaciotectonism. *Geologie en mijnbouw* 64, pp. 389–95.

**glacis** A gentle pediment slope, especially in arid and semi-arid regions.

**glei, gley** A clayey soil rich in organic material that usually develops in areas where the soil is waterlogged for long periods. Various component processes are involved: the reduction of ferric compounds, the translocation of iron as

227

ferrous compounds or complexes, and precipitation of iron as mottles and minor indurations.

**Glen's law** The relationship between the deformation of ice over time and SHEAR STRESS discovered by J.W. Glen (1955). It has the form:

$$\dot{\varepsilon} = A\tau^n$$

where $\dot{\varepsilon}$ is the strain rate (deformation rate), $\tau$ is the shear stress, A is a constant depending on ice temperature, crystal size and orientation, impurity content and perhaps other factors, and $n$ is a constant whose mean value is normally taken as equal to 3. The relationship models the secondary creep of ice which involves several separate processes, such as the movement of dislocations within crystals, crystal growth, the migration of crystal boundaries and recrystallization.

Glen's law is of fundamental importance in understanding glacier motion. It shows how sensitive glacier ice is to increasing shear stress and, for example, when the shear stress is doubled, the rate of deformation increases 8 times. This inherent sensitivity helps explain the characteristic shallow surface profile of glaciers. It also explains why most internal deformation occurs at the bottom of glaciers and it shows how glaciers move by internal deformation in the absence of BASAL SLIDING.                DES

**Reading and Reference**
Glen, J.W. 1955: The creep of polycrystalline ice. *Proceedings of the Royal Society of London*, Series A. 228, pp. 519–38. · Paterson, W.S.B. 1981: *The physics of glaciers*. Oxford and New York: Pergamon.

**glint line** The escarpment of Palaeozoic rocks which borders the Scandinavian and Laurentide shields and is associated with a line of lakes. Infrequently used today.

**global environmental change** There are two components to this (Turner *et al.* 1990): systemic global change and cumulative global change. In the systemic meaning, 'global' refers to the spatial scale of operation and comprises such issues as global changes in climate brought about by atmospheric pollution. In the cumulative meaning, 'global' refers to the areal or substantive accumulation of localized change, and a change is seen to be 'global' if it occurs on a worldwide scale, or represents a significant fraction of the total environmental phenomenon or global resource. Both types of change are closely intertwined. For example, the burning of vegetation can lead to systemic change through such mechanisms as carbon dioxide release and albedo change, and to cumulative change through its impacts on soil and biotic diversity.        ASG

**Reference**
Turner, B.L., Kasperson, R.E., Meyer, W.B., Dow, K.M., Golding, D., Kesperson, J.X., Mitchell, R.C. and Ratick, S.J. 1990: Two types of global environmental change. Definitional and spatial-scale issues in their human dimensions. *Global environmental change* 1, pp. 14–22.

**global ocean circulation** Surface currents of the oceans, driven by the prevailing winds, have their motion modified by the earth's rotation (see CURRENTS, OCEAN). In the Atlantic, Indian and Pacific Oceans surface circulation patterns are dominated by flow around the subtropical GYRES where the motion is in quasi-geostrophic balance between pressure gradient and Coriolis forces (see GEOSTROPHIC WIND). Here the velocities of western boundary currents, such as the Gulf Stream, can be as great as $3\,\mathrm{m\,s^{-1}}$. In equatorial regions a more complex pattern, which includes a powerful undercurrent flowing from west to east, results from the weakness of Coriolis effects together with the mean northward displacement of the INTERTROPICAL CONVERGENCE ZONE relative to the geographical equator. The directions of the currents in the northern Indian Ocean reverse seasonally under the influence of the monsoonal winds. At high latitudes in the Atlantic and Pacific Oceans cyclonic (counter-clockwise) subpolar gyres are present; more complex gyratory flows have been deduced for the Arctic. In contrast to the landlocked northern polar regions, a strong eastward-flowing current encircles Antarctica unimpeded by landmasses.

Whereas the surface circulation is relatively accessible to study, below the THERMOCLINE water moves some two or three orders of magnitude more slowly and on a much broader scale, hence less is known with any certainty. This deep or thermohaline circulation originates with the small differences in density between WATER MASSES from different source regions. At high latitudes, because of the low ambient temperatures and the addition of salts expelled in the formation of ice, the density of surface ocean water is increased to the point at which it naturally sinks to deeper levels. The principal source regions of deep ocean water are to be found in the Weddell and Ross Seas in the Antarctic and in the Norwegian Sea in the Arctic. Water from the former regions, known as Antarctic Bottom Water, is the densest of all water masses and flows northwards close to the bottom to cover much of the Indian Ocean and the Pacific and Atlantic Oceans to about 30°N. The presence of the Norwegian and Weddell Sea source regions at the extremities of the Atlantic Ocean, combined with its natural topography, gives rise to a

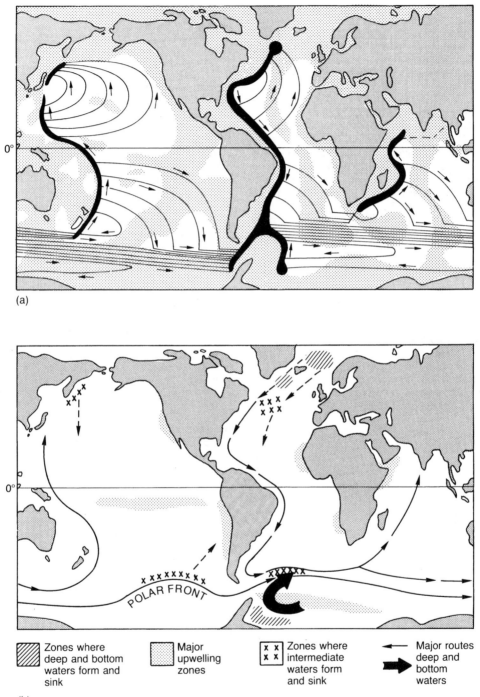

(a)

(b)

| | | | | | | |
|---|---|---|---|---|---|---|
| ▨ | Zones where deep and bottom waters form and sink | ▨ | Major upwelling zones | x x<br>x x | Zones where intermediate waters form and sink | ← ▶ Major routes deep and bottom waters |

**Global ocean circulation**: (a) Model of deep-water circulation of the world ocean with source regions in the north Atlantic and in the Weddell Sea. (b) Circulation within the deep ocean layers.
*Source*: Modified after H. Strommel in Tolmazin 1985, p. 143.

relatively strong meridional flow. The southward flow of Atlantic deep water from the Norwegian Sea is concentrated in a western boundary current moving at a few $cm\,s^{-1}$. Approaching the Southern Ocean it mixes with both Antarctic bottom water below and Antarctic intermediate water above before being carried eastwards in the strong circumpolar current. The resultant water mass flows northwards as a western boundary current into the Indian and Pacific Oceans, both of which lack significant source regions of deep water. Since there are no obvious zones of upwelling to match the large-scale downwelling of dense water, it is concluded that the return flow to the surface is principally by very slow upward diffusion through the thermocline. The circulation may not be completed until, after perhaps a thousand years, water is returned by wind-driven surface currents to the polar seas.     JEA

**Reading**
Pond, S. and Pickard, G.L. 1978: *Introductory dynamic oceanography*. Oxford: Pergamon. · Tolmazin, D. 1985: *Elements of dynamic oceanography*. Boston and London: Allen & Unwin.

## global positioning system (GPS)

A system for establishing accurate positions and heights anywhere on the earth using satellites. The best known system is run by the American military, and uses a series of 24 satellites. Their 12-hour orbits are designed so that at any point on the earth, between 5 and 8 satellites are always overhead. The satellites broadcast information on their current position and two types of time-related signal – the C/A and P codes which can be used by a receiver to estimate how far away each satellite is. Given information on the distance to at least 4 satellites whose position is known, the receiver can then estimate its own position by trigonometry.

The accuracy of this position depends on the accuracy of the signal from the satellites, and on the method of processing. The accuracy of the basic signal is deliberately degraded for non-military users, so that using a single GPS receiver, rapid readings will only give positions to a horizontal accuracy of about 100 m, sufficient for applications such as basic navigation.

Greater accuracy can be achieved in several ways: (1) Differential GPS. Two receivers are used, one at a fixed position of known location. This can transmit corrections to one or more other receivers, which can then record positions to 2–5 m accuracy. (2) Using the more accurate P code, taking uninterrupted readings over a longer period and post-processing the data. The maximum accuracy which can be achieved is 10–20 mm, suitable for high-precision topographic surveying.     SMW

**Reading**
Dodson, A.H. and Basker, G.A. 1992: GPS: GIS problems solved? *Proceedings of the AGI '92 conference*, 1.12.1–1.12.7, London: AGI.

## global warming

The possibility that the earth is warming or will warm up because of increasing concentrations of certain gases in the atmosphere (see GREENHOUSE EFFECT). Over the past century the earth's mean surface temperature has increased by perhaps 0.3–0.6 °C, and it is possible that it may increase by several more degrees during the next hundred years. The degree of warming may, however, be greatest in higher latitudes. The potential impacts of such warming are now the subject of considerable research and include cryospheric melting, sea-level rise, shifts in vegetation and precipitation belts, changes in hurricane characteristics, modifications to land use and human affairs, and changes in disease distribution.     ASG

**Reading**
Houghton, J.T., Callander, B.A. and Varney, S.K. 1992: *Climate change 1992: the supplementary report to the IPCC Scientific Assessment*. Cambridge: Cambridge University Press. · —, Jenkins, G.J. and Ephraums, J.J. 1990: *Climate change: the IPCC Scientific Assessment*. Cambridge: Cambridge University Press.

## Gloger's rule

The pigmentation of warm-blooded animal species tends to decrease away from the equator as mean annual temperatures decrease.

## GLOSS

(Global Sea-Level Observing System) A worldwide network of sea-level gauges, defined and developed under the auspices of the Intergovernmental Oceanographic Commission. Its purpose is to monitor long-term variations in the global level of the sea surface by reporting the observations to the Permanent Service for Mean Sea Level. The levels of the gauges are fixed by GPS: a satellite-based global positioning system, capable of accurately locating points in a three-dimensional geometric framework.     DTP

## gnammas

Holes produced in rock surfaces, especially igneous rocks and sandstones, by weathering processes. They are a type of rock basin.

## gneiss

A coarse-grained igneous rock, often a granite, that has been metamorphosed, producing a banded or foliated structure.

**goletz terrace** (also known as cryoplanation terrace or altiplanation terrace) A hillside or summit bench which is cut in bedrock and transects lithology and structure. It is confined to cold climates.

**Reading**
Washburn, A.L. 1979: *Geocryology*. London: Edward Arnold.

**Gondwanaland** See SUPERCONTINENT.

**gorge** A deep and narrow section of a river valley, usually with near vertical rock walls. More generally a narrow valley between hills or mountains.

**gouffres** Large pipes or vertical shafts that occur in limestone areas.

**graben** A valley or trough produced by faulting and subsidence or uplift of adjacent blocks (horsts).

**grade, concept of** One of the most confusing concepts in geomorphology, partly because of its inextricable relationship with gradient. Introduced by G.K. Gilbert in 1876, it relates the gradient of a channel to the balance between corrasion (erosion), resistance and transportation. The idea was adopted, adapted and debated by W.M. Davis, J.E. Kesseli and J.H. Mackin in particular and re-surfaces in the influential paper by S.A. Schumm and R.W. Lichty (1965) as the 'graded' time span. It is now generally assumed to be roughly equivalent to DYNAMIC EQUILIBRIUM and, in practical terms, should be viewed – by extension from work on regime theory in alluvial canals – as a state in which channel form is relatively constant despite variations in flow (usually 2–10 years). The application to hillslope form and materials is distinctly problematic.                          BAK

**References**
Gilbert, G.K. 1876: The Colorado Plateau province as a field for geological study. *American journal of science* 3rd series 12, pp. 85–103. · Schumm, S.A. and Lichty, R.W. 1965: Time, space and causality in geomorphology. *American journal of science* 263, pp. 110–19.

**graded bedding** Comprises sedimentary units that exhibit a vertical gradation in mean grain size. Normal grading is where a fining-upward sequence is present, and may result from deposition in a waning current, from a decrease in sediment supply, or from the progressive sorting or settling out of different size fractions. Inverse grading exhibits an upward-coarsening sequence, and may result from deposition in rising flow conditions or from an increase in sediment supply.                          JM

**graded slopes** Those possessing a continuous regolith cover without rock outcrops. The concept of grade was used by G.K. Gilbert (1876) to indicate a condition of balance between erosion and deposition, brought about by adjustments between the capacity of a stream to do work and the quantity of work that the stream has to do. This definition was formally introduced by W.M. Davis (1899, reprinted in 1954) and applied to hillslopes in the words: 'a graded waste sheet . . . is one in which the ability of the transporting forces to do work is equal to the work they have to do. This is the condition that obtains on those evenly slanting, waste-covered mountain sides which have been reduced to a slope that engineers call *the angle of repose*' (1954, p. 267). Rocky outcrops are not graded because waste can be removed from them faster than it is supplied by weathering. On slopes from which outcrops have been eliminated the 'agencies of removal are just able to cope with the waste that is there weathered plus that which comes from farther uphill' (1954, p. 268). Graded waste slopes decline in angle as the waste becomes finer in texture as a result of weathering 'so that some of its particles may be moved even on faint slopes' (1954, p. 269).

Because of the difficulty of determining the volumetric relationships between weathering and removal, and texture and removal processes, Young (1972, p. 100) has suggested that the term graded slope be used to indicate those lacking outcrops: this definition would make it equivalent to a soil or regolith-covered slope.                          MJS

**Reading and References**
Davis, W.M. 1899: The geographical cycle. *Geographical journal* 14, pp. 481–504. · —1954: The geographical cycle. In D.W. Johnson ed., *Geographical essays*. New York: Dover; London: Constable. · Gilbert, G.K. 1876: The Colorado Plateau province as a field for geological study. *American journal of science* 3rd series 12, pp. 16–24, 85–103. · Young, A. 1972: *Slopes*. Edinburgh: Oliver & Boyd.

**graded time** A time-span intermediate between the longer interval of 'cyclic time' and the shorter period of 'steady time' (Schumm and Lichty 1956). It is defined as 'a short span of cyclic time during which a graded condition or dynamic equilibrium exists' (1965, p. 114), with respect to the landforms and, by reference to Mackin's (1948) discussion of GRADE, it is implied that this 'short span' will be a 'period of years'. In Schumm and Lichty's view the chief practical considerations governing studies on

231

the graded time-scale are that time and initial relief become irrelevant to the enquiry, while the morphology of drainage networks and hillslopes and the hydrologic outputs of drainage basins are dependent variables, contingent upon the independent controls of geology, climate, vegetation, disposition of relief above base level, and the manner in which run-off and sediment are generated within the landscape.

The concept is discussed again by Schumm (1977, pp. 10–13) and, rather confusingly, it is there redefined (figures 1–5) as equivalent to steady-state equilibrium time, with a time span of 100–1000 years. It seems clear that Schumm intends the term to be used to imply an intermediate timescale in which the focus of investigations is on fluctuations in hydrologic outputs, channel morphology and hillslope form viewed as responses to spatial or temporal patterns of variation in the 'independent' variables listed in the 1965 paper.                    BAK

**Reading and References**
Mackin, J.H. 1948: Concept of the graded river. *Bulletin of the Geological Society of America* 59, pp. 463–512. · Schumm, S.A. 1977: *The fluvial system*. New York and London: Wiley. · — and Lichty, R.W. 1965: Time, space and causality in geomorphology. *American journal of science* 263, pp. 110–19.

**gradient wind**   Results from a balance of horizontal pressure gradient force, CORIOLIS FORCE and the centripetal acceleration (or centrifugal force) that exists when air moves in a curved path, such as occurs in a cyclone and an anticyclone. BUYS BALLOT'S LAW applies to this wind just as it does to the geostrophic wind, but because more forces are involved, the velocity of the gradient wind is different to that of the GEOSTROPHIC WIND. The one exception to this occurs when airflow is straight, giving a zero centripetal acceleration and hence geostrophic balance and wind.                    BWA

**Reading**
Atkinson, B.W. 1972: The atmosphere. In D.Q. Bowen ed., *A concise physical geography*. Amersham: Hulton Educational. Pp. 1–76, esp. pp. 33–48. · Hess, S.L. 1959: *Introduction to theoretical meteorology*. New York: Henry Holt.

**gradually varied flow**   In most natural river channels the flow is gradually varied because the cross-section and bed slope change downstream and the water surface is not parallel to the bed. Under these conditions the FLOW EQUATIONS for UNIFORM STEADY FLOW are not strictly applicable except locally, and a more detailed analysis of the flow energy is required. So long as the streamlines in a short reach are approximately parallel, and pressure within the flow is there-

fore HYDROSTATIC, the total energy of a unit mass of water at the bed is the sum of its potential energy, pressure energy, and kinetic energy:

$$E = \rho_\omega g z_1 + \rho_\omega g d_1 + \rho_\omega v_1^2 / 2$$

or

$$E = \rho_\omega g (z_1 + d_1 + v_1^2 / 2g)$$

where the term in brackets is the 'total head' $H_1$ and:

$$H_1 = z_1 + d_1 + v_1^2 / 2g$$

is the BERNOULLI equation. This can be used to define the conservation of energy between two adjacent sections. An energy balance equation between sections 1 and 2 of the diagram states that:

$$z_1 + d_1 + v_1^2 / 2g = z_2 + d_2 + v_1^2 / 2g + h_L$$

where $h_L$ is the head or energy loss between the sections.

It is clear that in gradually varied flow the water surface, bed slope and energy grade line are not parallel. The energy grade line always slopes downwards in the direction of flow and measures the rate of dissipation of energy (the energy loss) caused by flow resistance and sediment transport. If the water surface slope is gentler than the energy slope the kinetic energy term decreases downstream as the flow decelerates, whereas a steeper water surface slope would reflect accelerating flow and conversion of energy from potential to kinetic forms (Richards 1982, pp. 72–6). In a POOL AND RIFFLE stream the variations of velocity and depth from section to section reflect this continual conversion of energy from potential to kinetic forms, and vice versa, in response to the changing bed slope.                    KSR

**Reference**
Richards, K.S. 1982, *Rivers: form and process in alluvial channels*. London and New York: Methuen.

**granite**   A coarsely crystalline igneous rock composed predominantly of quartz and alkali feldspars. Additional constituents are commonly mica and hornblende.

**granulometry**   The process or method of PARTICLE SIZE determination of sediments, usually referring to those with a diameter of 2 mm or less. A number of granulometric methods exist including the use of settling tubes, by sieving or by laser defraction. Due to the different principles involved, data obtained by one method may not be directly comparable with those from another.                    DSGT

**grasslands** Regions in which the natural or the plagioclimax vegetation is dominated by grasses or grass-like plants and non-grass-like herbs. They include temperate grasslands of the steppes, prairies, pampas and veld, tropical grasslands or SAVANNAS, and smaller zones on mountains, in high latitudes and as patches within other plant formations resulting from fire, soil or drainage controls. Before man's modification of the natural plant and animal cover, grasslands probably occupied around 40–5 per cent of the land surface, a figure increased by the maintenance of grazing land and decreased by conversion to other forms of land use to around 25 per cent today. PAF

**Reading**
Coupland, R.T. ed. 1979: *Grassland ecosystems of the world*. Cambridge and New York: Cambridge University Press.

**gravel** Though frequently used to describe small, rounded, unconsolidated rock fragments in fluvial systems, gravel can also refer to any sediment comprising uncemented rock fragments coarser than SAND (2 mm diameter) and smaller than pebbles, with the upper size limit in the 10–60 mm range. Gravel particles may be angular or rounded, spheroid or rod-shaped, and may result from *in situ* weathering or rock breakdown during transport. DSGT

**gravimetric method** A means of soil moisture determination involving taking, weighing, oven drying and reweighing a soil sample and expressing the moisture content (or sample loss in weight) as a percentage of the oven dry weight of the sample. AMG

**Reading**
Reynolds, S.G. 1970: The gravimetric method of soil moisture determination. Parts I, II and III. *Journal of hydrology* 11, pp. 258–300.

**gravity** The force imparted by the earth to a mass which is at rest relative to the earth. All masses are attracted to each other according to Newton's law of universal gravitation, but the earth is also rotating and therefore a centrifugal force is also exerted on the mass in question. Hence the force observed, and commonly called gravity, is the combination of the true gravitational force and the centrifugal force. The standard acceleration of gravity at sea level at 45° latitude is $9.80665 \, \mathrm{m \, s^{-2}}$. BWA

**gravity faulting** An important process that operates in mountainous areas; high available relief enables major rock movements to occur under the influence of gravity, creating hilltop valleys and depressions, and sometimes double summits (*doppelgrate*). ASG

**Reading**
Beck, A.C. 1968: Gravity faulting as a mechanism of topographic adjustment. *New Zealand journal of geology and geophysics* 11, pp. 191–9. · Paschinger, V. 1928: Untersuchungen über Doppelgrate. *Zeitschrift für Geomorphologie* 3, pp. 204–36.

**gravity wave** A wave disturbance in which buoyancy acts as a restoring force on fluid parcels displaced from an equilibrium state. The restoring force acts only in the vertical, frequently producing simple harmonic motion around the equilibrium level. The wave ensues because this simple harmonic motion occurs in a horizontal flow. The resultant of the two components of velocity (vertical and horizontal) at any instant gives the velocity of the parcel of air under consideration. A sequence of such velocities throughout one cycle of simple harmonic motion describes the gravity wave form. Examples of gravity waves are lee waves in the atmosphere and water waves. Gravity waves are also known as buoyancy waves for obvious reasons. BWA

**great escarpment** An escarpment formed by the uplift associated with rifting that contributed to the breakup of Gondwanaland. Great escarpments are found, with various degrees of dissection and erosion, around the margins of southern hemisphere continents (Ollier 1985). The great escarpment of southern Africa extends more or less continuously, 150–200 km inland of the modern coastline, from Angola to southern Tanzania. DSGT

**Reference**
Ollier, C.D. 1985: Morphotectonics of continental margins with great escarpments. In M. Morisawa and J.T. Hack eds, *Tectonic geomorphology*. Binghampton Symposia in geomorphology, international series *Tectonic* 15, pp. 3–25. Boston: Allen and Unwin.

**great interglacial** A phase of Pleistocene history identified in the classic four-glacial model developed by A. Penck and E. Brückner (see PENCK AND BRÜCKNER MODEL) who believed that the interglacial between the Mindel and the Riss glacials lasted a longer time than any others. Studies of the Pleistocene record in ocean cores tend not to support this view. ASG

**greenhouse effect** The greenhouse effect is a natural phenomenon that occurs when short-wave solar radiation from the sun passes largely unattenuated through the earth's atmosphere, is absorbed at the planetary surface, is reradiated upward as longer-wavelength thermal radiation, and is absorbed by various atmospheric constituents and again reradiated. Since some of this latter radiation flux is directed downward, it results in a surface warming that would not occur in the absence of an atmosphere. And this extra warming is what is commonly called the greenhouse effect.

The most important 'greenhouse gases' are water vapour and carbon dioxide. Their presence in the atmosphere allows the earth to maintain an average temperature of approximately 15 °C. Without them, the surface temperature of the planet would be about −19 °C, and the earth could not support life. Consequently, it is clear that we owe our very existence to the greenhouse effect. With the onset of the Industrial Revolution, however, the $CO_2$ content of the air began to rise, and people began to worry that this phenomenon might have a 'dark side' we had not anticipated (see CARBON DIOXIDE PROBLEM).

Spurred on by this concern, scientists are attempting to determine whether increasing concentrations of atmospheric greenhouse gasses will have any effect on global temperature or climate patterns. Scientists who use computer models (see GENERAL CIRCULATION MODELLING) to study these questions commonly report that a temperature rise of 2 to 4 °C will result from a doubling of the atmosphere's $CO_2$ content, leading to catastrophic events such as the melting of earth's polar ice caps, which could produce widespread flooding and famine. However, scientists who feel that even our best computer models cannot yet accurately emulate the many interacting complexities of the natural world use empirical data to conclude that a global temperature increase up to 0.5 °C may be more likely, which is an order of magnitude lower than the warming predicted by current computer models. Consequently, the greenhouse effect has become a topic of both scientific and political debate; and it will probably remain so for years to come. KEI

**Reading**
Houghton, J.T., Meira Filho, L.G., Callander, B.A., Harris, N., Kattenberg, A. and Maskell, K. 1996. *Climate change 1995: the science of climate change*. Cambridge: Cambridge University Press. · Williams, M.A.J. and Balling, R.C. 1996: *Interactions of desertification and climate*. London: Edward Arnold. · Karl, T.R., Nicholls, N. and Gregory, J. 1997: The coming climate. *Scientific American* May, pp. 78–83. · Kemp, D.D. 1994: *Global environmental issues: a climatological approach*. London: Routledge.

**grey wether** See SARSEN.

**greywacke** A sediment composed of coarse fragments of quartz and feldspar, usually poorly sorted.

**grèzes litées** Bedded screes of angular rock fragments associated with cold climates and frost shattering. French examples have thicknesses up to 40 m, with layers dipping at as much as 40°, whereas Polish examples are generally only a few metres thick (Dylik 1960). The inclination of the layers parallels that of the slopes, and, in contrast to ordinary gravitational debris slides, the deposits show a striking predominance of fines in their distal parts. Snow patches may play a role in their formation (Guillen 1964) and downwash is an important process. The rhythmic nature of the sediments suggests that under cold conditions the following process occurs: first, freezing of rocks on a cliff face causes disintegration, and the coarse debris thus released slides downward over frozen subsoil; secondly, the following phase of thaw causes a mantle of half-fluid material rich in fines to spread over the stony layer. ASG

**References**
Dylik, J. 1960: Rhythmically stratified slope waste deposits. *Biulteyn Peryglacjalny* 8, pp. 31–41. · Guillen, Y. 1964: Les grèzes litées comme depôts cyclothémiques. *Zeitschrift für Geomorphologie Supplement-band* 5, pp. 53–8.

**grike** (gryke) The cleft or runnel in bare limestone pavements which separates the CLINTS. Grikes are formed when limestone is dissolved by water, probably under a soil cover (Trudgill 1972) normally along joint pattern weaknesses. Grikes are called *kluftkarren* in German. PAB

**Reference**
Trudgill, S.T. 1972: The influence of drifts and soils on limestone weathering in NW Clare. *Proceedings of the University of Bristol Speleological Society* 13, pp. 113–18.

**GRIP** Acronym for the Greenland ICE CORE Project.

**ground frost** Ground, but not necessarily air, below freezing. On cool dry nights there is not enough moisture in the air to stop terrestrial (heat) RADIATION escaping to space. When, as is usual, the earth is a good radiator but a poor conductor of heat, the ground may cool sufficiently to give CONDENSATION (dew). If the air is dry enough the moisture may sublimate into crystalline form to make frost. Such conditions are hazardous for plants whose cells are disrupted by ice crystals. Whether crystals form depends on the amount of 'antifreeze' (sugar and starch) in their tissues. JSAG

**Reading**
Monteith, J.L. 1973: *Principles of environmental physics*. London: Edward Arnold.

**ground ice** A body of more or less clear ice within frozen ground. It takes many forms, some of the more common being PORE ICE, SEGREGATED ICE, ice veins and ICE WEDGES, PINGO ice, and massive icy beds. Buried glacier ice, buried ICING ice, and buried snowbank ice are sometimes regarded as forms of ground ice even though they are of surface origin. In places, ground ice may constitute more than 50 per cent by volume of the upper 2–3 m of permafrost. Generally speaking, ground ice amounts decrease with increasing depth. Aggradational landforms associated with the formation of ground ice include open- and closed-system pingos, ice wedge polygons, palsas and peat plateaux, and seasonal frost mounds. The degradation of ice-rich permafrost causes THERMOKARST and results in thaw slumping, thaw depressions and lakes, and hummocky unstable topography. HMF

**Reading**
Mackay, J.R. 1972: The world of underground ice. *Annals of the Association of American Geographers* 62, pp. 1–22.

**ground moraine** See TILL.

**groundwater** Strictly defined, groundwater includes all sub-surface water in a solid, liquid or gaseous state, other than that which is chemically combined with any minerals present. In practice it is usual to include all sub-surface water that is part of the HYDROLOGICAL CYCLE, which encompasses all groundwater except CONNATE WATER which is prevented from active circulation and some frozen groundwaters which may have a very long turnover time. This definition is wide enough to encompass soil water components such as saturated THROUGHFLOW but it is more common to distinguish groundwater as a larger, deeper, more slowly moving body located in sediments or bedrock (AQUIFERS), as opposed to soil water.

Groundwater occurs within two vertically separated zones, a VADOSE (or aerated or unsaturated) zone beneath the soil which is situated above a PHREATIC (or saturated) zone. The boundary between the two is termed the WATER TABLE or PIEZOMETRIC SURFACE and marks the upper limit of saturation in an unconfined aquifer. The shape of the water table usually mimics that of the topography but in a much smoother, more subdued manner, with its depth below the surface varying spatially with climate and rock type and temporally due to seasonal variations in groundwater RECHARGE.

The upper vadose zone is often simply regarded as a zone in which water is in transit from the surface to the phreatic zone. However,

it holds around 1.5% of global non-frozen fresh water, and recent research in areas of non-porous fractured bedrock and karst terrain indicates that it can be an important store of water, sustaining base flow in subterranean streams above the water table. Groundwater in the phreatic zone accounts for around 95% of the global fresh, non-frozen water reserve. It can be considered as being either unconfined if it is stored near the surface within an aquifer which is situated above an impermeable *aquiclude* (a rock which holds and transmits little water) or semi-permeable *aquitard*, or confined if it is held within an aquifer between two aquicludes. Confined groundwater is often referred to as ARTESIAN, and may emerge at the surface at an *artesian spring* or well. An artesian spring is one in which pressure within the confined groundwater body forces the water upwards towards the *potentiometric surface* (i.e. the level which the phreatic surface would reach were the groundwater not confined, which would lie above the ground surface). PERCHED WATER TABLES are a variety of unconfined groundwater aquifer of relatively limited extent which stand above the main groundwater body and are formed either above an aquiclude of limited area or where percolation rates exceed the hydraulic conductivity of an aquitard. Groundwater has been encountered at depths of up to 4000 m during drilling and mining but, in general, it diminishes below a few hundred metres. The lower limit of groundwater penetration is determined by a variety of factors including the availability of voids in the rock and the hydrostatic pressure. At great depths there may be sufficient lithostatic pressure exerted by surrounding rocks that all voids are closed and the rock is effectively impervious.

Groundwater is an important contribution to base flow in some streams, especially those situated upon permeable aquifers, with water entering the channel by effluent seepage. This water may have taken a considerable time from falling as precipitation and contributing to recharge before exfiltrating, with phreatic groundwater flow-through times varying from 2 weeks to 10,000 years depending upon location. Flowthrough times are also significant in terms of human utilization of groundwater resources. Areas where groundwater is extracted at a higher rate than recharge is occurring are likely to experience falling water tables and, in exceptional cases, surface subsidence due to aquifer collapse. Excessive groundwater extraction in coastal zones may lead to the intrusion of saline water into the aquifer.

In addition to forming a major part of the hydrological system, groundwater can also act

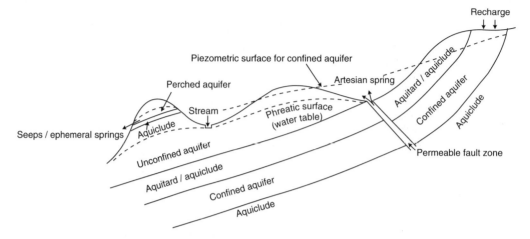

Forms of groundwater storage.
*Source*: Jones 1997.

as an important geomorphological agent. Sub-surface groundwater flow through fissures and joints in limestone bedrock can lead to the development of KARST landforms such as caves and subterranean channels by solutional processes. Emerging groundwater can contribute to the erosion of rock or sediment faces, and the development of scarps, DRY VALLEYS and some canyons, by spring SAPPING or seepage erosion. The height of the water can also act as an important control on the operation of specific processes, such as where the maximum extent of aeolian DEFLATION is limited by the depth to the regional water table and in PLAYA systems where fluctuations in water table height influence water chemistry and surface morphology. Groundwater can also act as an important medium for transporting minerals in solution which is important in the development of many near-surface duricrusts such as CALCRETES, SILCRETES and LATERITE. DJN

**Reading**
Brown, A.G. (ed) 1995: *Geomorphology and groundwater.* Chichester: John Wiley. · Fetter, C.W. 1994: *Applied hydrogeology.* 3rd edn. New York: Macmillan. · Jones, J.A.A. 1997: *Global hydrology: processes, resources and environmental management.* Harlow: Longman. · Price, M. 1996: *Introducing groundwater.* 2nd edn. London: Chapman & Hall.

**growan** Decomposed granite or related rock as found on Dartmoor, south-west England, and its environs. The growan may be produced by chemical weathering or metamorphic processes, and it has been suggested that the stripping of growan from corestones of sounder rock may play a role in the formation of tors.

**groyne** A man-made barrier running across a beach and into the sea which has been constructed to reduce the erosion of the beach by longshore currents. Although groynes may serve to reduce erosion at the place where they are constructed, by reducing the movement of material along the coastline by longshore drift they may cause beach starvation and erosion elsewhere.

**grumusol** Under the classification of the SEVENTH APPROXIMATION this may be regarded as a vertisol and is in effect a modern American term for a black cotton soil.

**grus** An accumulation of poorly sorted angular quartz grains and clayey material derived locally from weathered granite.

**guano** Thick accumulations of bird excrement, usually found on islands where the birds nest. The material is used as a fertilizer as it is rich in phosphates. In some caves (and belfries) there may be a substantial accumulation of bat guano.

**gull** A fissure or crack, sometimes sediment filled, which opens up on escarpments as a result of the tensions produced by CAMBERING.

**gully** A steep sided trench or channel, often deep (several metres), that is cut into poorly consolidated bedrock, weathered sediment or soil. The agent of gullying is ephemeral flow of running water. A gully is deeper and longer than a RILL. (See also ARROYO; CHANNEL CLASSIFICATION; FLUVIAL; WADI.)

**gully erosion** The pronounced erosion, by ephemeral streams, of soils and other poorly consolidated sediments, producing networks of steep-sided channels.

**Gumbel extreme value theory** A theory appropriate for the analysis of extreme values; it has been applied particularly to flood frequency analysis where the Gumbel extreme value 1 distribution may be used as the theoretical distribution to fit to the distribution of flood frequency values for a particular gauging station. KJG

**Reading**
Gumbel, E.V. 1958: *Statistics of extremes*. New York: Columbia University Press.

**gumbo** An area of clayey soil which turns to sticky mud when wet. Any damp, sticky clay.

**gustiness factor** An index of the variations in windspeed. It is calculated from the ratio of the total range of windspeed between gusts and lulls to the mean windspeed in the given period. Gustiness factors are highest in urban areas where the surface roughness is high, and lowest in coastal sites or exposed upland locations where surface friction is small. A gustiness factor may also be defined in terms of wind direction whereby the angular width in radians is the measure of lateral gustiness. For small values it is nearly equivalent to the speed ratio. PS

**guyot** A flat-topped mountain on the sea floor, especially in the Pacific, which does not reach the sea surface. A sea mount or drowned island, which is a truncated volcano, formed as a result of subsidence associated with sea-floor spreading. ASG

**Reading**
Watts, A.B. 1984: The origin and evolution of seamounts. *Journal of geophysical research* 89 (B13), pp. 1106–286.

**gypcrete** Gypsum crusts, as found in deserts, and comprising loose, powdery or cemented crystalline accumulations dominated by calcium sulphate dihydrate at or near the ground surface. ASG

**Reading**
Watson, A. 1983: Gypsum crusts. In Goudie, A.S. and Pye, K. eds, *Chemical sediments and geomorphology*. London: Academic Press. Pp. 133–61.

**gypsum** The evaporite mineral calcium sulphate dihydrate ($CaSO_4 \cdot 2H_2O$) formed as an evaporite deposit in enclosed basins and also widespread as a cementing agent in surface or near-surface crusts (such as GYPCRETE) in some arid and hyperarid environments. DJN

**gyre** A large-scale ocean circulation system generated by atmospheric circulation. Gyres play a fundamental role in general ocean circulation, and in maintaining the global heat balance. Each gyre comprises four major currents: a western and eastern boundary current, and two transverse currents. Prevailing winds drive surface waters east or west across the ocean basins, and the CORIOLIS FORCE deflects the resulting currents to an angle of about 45° relative to the wind direction. The continents bounding the basins also contribute to the deflection.

The largest systems are the five sub-tropical gyres that are each centred at about 30° north or south latitude. There are north and south pairs in the Atlantic and Pacific Oceans, and a southern gyre only in the Indian Ocean. The flow in these gyres mimics that of the sub-tropical anticyclones that sit over them. The western boundary currents include five of the six largest currents on earth, including the Gulf Stream and the Kuroshio. The western boundary currents are substantially narrower and faster than the eastern boundary currents. There are also smaller, sub-polar gyres in the north Atlantic and north Pacific, centred around 50°. These exhibit counterclockwise circulation similar to that of the semi-permanent, mid-latitude cyclones. Sub-polar gyres are absent in the southern hemisphere because there are no continents to block and redirect the circumpolar circulation. DJS

**Reading**
Pedlosky, J. 1990: The dynamics of the oceanic subtropical gyres. *Science* 248, pp. 316–322. · Stewart, R.W. 1969: The atmosphere and the ocean. *Scientific American* 221, pp. 76–105.

**gyttja** A nutrient-rich peat or organic mud which contains much plankton.

# H

habitat   The overall environment, but more often specifically the physical environment, in which organisms live. It may be examined in a range of scales which extend from the macroscale (continental, subcontinental), to the mesoscale (regional, local) and microscale, the latter being of particular significance to the numerous small animal species of submicroscopic and microscopic size. All organisms must be morphologically and genetically adapted to the habitats in which they reside for any length of time. Such adaptation usually begins with the acquisition of tolerance to the conditions therein.

For plants on land, the major influential factors of habitat which can limit growth may be grouped under four headings: climatic, topographic, edaphic and biotic. Of these *climatic* factors especially those related to cold tolerance and the provision of adequate amounts of heat and moisture in the growing season, are normally regarded as being most important, certainly on a continental, subcontinental or regional scale, but they are frequently of much less significance at a local and a microscale level. Climatic influences are those of light (energy), heat, moisture availability and wind. Conditions of light which affect plant growth and survival are its intensity, its wavelength quality, and the photoperiod (period of daylength), all of which can interact either individually or together, at one and the same time. The intensity of light is equivalent to solar energy income, and this can affect the rates of PHOTOSYNTHESIS, the rate of formation of auxins (growth-forming substances) and the vertical structure of vegetation communities which develops in response to their need to utilize radiant energy as efficiently as possible. Vegetation structures are most complex in the energy-rich wet tropics (see TROPICAL FOREST). Some green plants do very well on small amounts of light (sciophytes); but others (heliophytes, e.g. pioneers, weeds, palms) require strong light levels throughout their life cycle. Still others, among them many tree seedlings, demand a sciophytic, followed by a heliophytic phase, to achieve their best growth results. Wavelength quality variations are best exemplified on an altitudinal basis: above 2000 m, augmented levels of ultraviolet energy are received as compared to those of visible light, and many more UV-tolerant species con-

sequently are present at higher altitudes. In contrast, differences in photoperiod are most clearly displayed latitudinally: on a year-round basis, they are minimal in the tropics, and at their maximum at high latitudes. In themselves, they can prevent the successful transfer of, say, mid-latitudinal species into the tropics, and vice-versa. For other features of light-energy control of plants and vegetation, see BIOLOGICAL PRODUCTIVITY; NET PRIMARY PRODUCTIVITY and TROPHIC LEVELS.

Those temperature conditions which influence plant growth and physiology are in most cases a response to latitude, altitude and distance from the sea. Mean annual temperatures are higher at low latitudes, and at low as compared to high altitudes; and both mean diurnal and annual temperature ranges are augmented as one moves away from oceans. All plants adapt to external temperature conditions by adjusting their own internal temperatures to them as much as possible; and they will die if the differences between them become too great. Heat is absorbed by plants either directly by radiation, or by conduction from the layer of heat or water directly over the leaf or stem; and it is lost by conduction or convection from plant surfaces, by the evaporation and transpiration of water as latent heat, by respiration, and by reradiation of long wavelengths (Oke 1987). Most plants also have a range of further specific adaptations designed to combat possible physiological or morphological damage to them which might result from extremes of heat or cold. Thus a thick, cuticular tissue on leaf surfaces insulates all cell mechanisms from extremes of heat in hot deserts; and dwarf forms may be the only ones capable of surviving in areas of persistent cold, where they are protected by a snow cover for much of the year. In midlatitudes, the phenomenon of cold-hardiness, which develops each year as growth ceases, and peaks at the height of winter, reduces the danger of frost damage there (Melzack and Watts 1982). In general, most of the higher plants stop growth activity once external temperatures fall to 5 °C, so as to enable cold-hardiness to form. There are very few trees in places where mean annual external temperatures are lower than 10 °C; and little activity is exhibited by plants generally when the immediate temperature around them exceeds 45 °C.

Habitat moisture circumstances (the balance between incoming precipitation and outgoing evapotranspiration, plus the effect of the soil moisture reserve) also vary widely. On a world scale, the atmosphere holds a reserve at any one time of only about ten days' supply of precipitation so that, bearing in mind the large quantities of water required by most plants to survive (see NET PRIMARY PRODUCTIVITY), very efficient methods of gaining and recycling it are necessary in all but the wettest terrestrial biological systems (e.g. bog and marsh land). Within the overall evapotranspirational limits, which are set by energy availability (Penman 1963), the flow of water into plants generally rises with an increase in TRANSPIRATION, which in turn is usually augmented by the maximum opening of stomata under strong light conditions, high temperatures, and above-minimum windspeeds. Most plants can withstand some moisture stress, though if this becomes too great, or continues for too long, they may wilt and die. In arid areas, in which such stresses are severe and prolonged, xerophytes (i.e. drought-tolerant plants) have an inbuilt range of defences against them. Some may become dormant for long periods, e.g. aloes and sage brush (*Artemisia tridentata*); and others adopt an annual, rapid-growth cycle, springing up and maturing quickly after rain. Still others raise the efficiency with which they can extract water from a dry soil, by physiological means, while at the same time minimizing rates of transpiration through selection either of a small leaf form, leaves with thick cuticular surfaces and few stomata, or extremely narrow leaves. Some families, including Old World euphorbias and New World cacti, become succulents, storing water for relatively long periods in their cells. In contrast, hydrophytes prefer habitats in which their roots are permanently placed in water, or in wet soil; and they cannot exist elsewhere. But most plants are mesophytes, tolerant neither of prolonged excesses, nor deficiencies of water availability, but of a nice balance of both.

The wind factor in terrestrial habitats influences plant growth in four major respects. First, although of relatively small import in most land biological systems, in which existing vegetation acts as a protective agent itself against the possible adverse consequences to it of strong winds, physical damage may well result even so from the passage of hurricanes or severe storms. Secondly, a combination of strong winds and small particles (dust, sea salt) often produces a blasting effect which is capable of destroying plant cells at heights of 1 m and more above the ground surface; and a bevelling of vegetation occurs, notably along windward coasts. Thirdly,

similar effects develop at high altitudes and high latitudes through a wind-chill factor (wind and severe cold temperatures). Fourthly, more generally, but predominantly in mid-latitudes, a sudden rise in windspeed in spring may increase rates of plant transpiration to beyond levels at which it can respond, bearing in mind that it is still then in its sluggish, cold-hardiness phase; the condition is termed 'physiological drought', and through it cell damage is initiated through lack of water.

On a more local scale climatic influences of habitat may be overshadowed by factors of topography, and edaphic and biotic controls. Three conditions of *topography* are important. First, there is a direct altitudinal effect, arising from the normal decline in temperature with increase in height of $c.0.6\,°C\ 100\,m^{-1}$ which is prevalent in most parts of the world: and this causes a distinct and well-known zonation of plants and vegetation upslope on mountains, the essential features of which will vary according to latitude and precise location. Secondly, once slope angles have reached $c.15°$, patterns of vegetation will be further modified, often to include more xerophytes, since run-off will have become that much greater; and, at 35°, they are often so unstable as not to allow the development of vegetation at all. Thirdly, differences in slope aspect in relation to the angle of incident solar radiation can change temperature–water availability patterns (and, in consequence, plant growth patterns too) on alternate sides of the same valley; and the impact of this is greatest between latitudes 35° and 45° north and south of the equator (Holland and Steyn 1975). Among *edaphic controls* are those of soil, soil chemistry (see NET PRIMARY PRODUCTIVITY) and soil water: and any material deficiencies or excesses of soil nutrients or soil water as compared to the mean are likely to restrain plant growth. *Biotic controls* include especially the influence of grazing (see HERBIVORES), of fire, and of humans.

These external habitat factors may influence land-based animals too, both directly (temperature, water availability, etc.) and also indirectly, in that they determine to a large extent the nature of local and regional FOOD CHAINS. In consequence of this, the NICHE role of animals is a further important habitat constraint, as also is their precise relationship with other animals (see COMMENSALISM; COMPETITION; PARASITE; SYMBIOSIS). For water-based organisms the main habitat conditions to which they respond are differences in water chemistry, temperature, light penetration and the general state of the food-chain web within the water body.                                                    DW

**Reading and References**

Holland, P.G. and Steyn, D.G. 1975: Vegetational responses to latitudinal variations in slope angle and aspect. *Journal of biogeography* 2, pp. 179–84. · Melzack, R.N. and Watts, D. 1982: Cold hardiness in the yew (*Taxus baccata*, L.) in Britain. *Journal of biogeography* 9, pp. 231–41. · Oke, T.R. 1987: *Boundary-layer climates*. London: Routledge. · Penman, H.L. 1963: *Vegetation and hydrology*. Commonwealth Agricultural Bureau, Farnham Royal. · Prentice, T.C. 1992: A global biome model based on plant physiology and dominance, soil properties and climate. *Journal of biogeography* 19, pp. 117–34. · Watts, D. 1971: *Principles of biogeography*. Maidenhead and New York: McGraw-Hill. Ch. 4. · Woodward, F.T. 1987: *Climate and plant distribution*. Cambridge: Cambridge University Press.

**haboob** 'Derived from the Arabic word *habb*, to blow, refers to any duststorm raised by the action of the wind. More specifically, in a meteorological context, the term refers to a duststorm generated by the evaporative outflow of a parent cumulonimbus, an outflow that can exceed 80 km in extreme cases. Very turbulent conditions are experienced along the boundary of the cool, dense outflow as it undercuts hot stagnant air leading to vigorous dust raising activity' (Membery 1985, p. 217).                ASG

**Reference**

Membery, D.A. 1985: A gravity-wave haboob? *Weather* 40, pp. 214–21.

**hadal zone** Pertaining to the greatest depths (more than 6000 m) of the oceans.                ASG

**Hadley cell** The name often given to the large-scale thermally driven circulations existing in tropical latitudes and most prominent over the Atlantic and Pacific Oceans. There is one Hadley cell in each hemisphere, heated air rising near the equator in the INTERTROPICAL CONVERGENCE ZONE (ITCZ), flowing poleward aloft, descending at a latitude of 30–40°, especially in the eastern half of the very intense subtropical high pressure areas at these latitudes, and then flowing either poleward or equatorward near the earth's surface. Because of the earth's rotation the equatorward moving currents are deflected towards the west and become the north-east and south-east trade winds of the northern and southern hemispheres, respectively. The poleward moving currents are deflected towards the east and become the middle latitude westerlies. The upper poleward moving branch of the Hadley cell rapidly gains westerly momentum, which is concentrated in the subtropical JET STREAM at a height of 12–15 km above the tropical highs.

The intensity and position of the Hadley cells vary seasonally, the one in the winter hemisphere being stronger and latitudinally more extensive than the one in the summer hemisphere. The boundary between the two cells shifts from an average latitude of about 5 °S in February to 10 °N in August.

The Hadley cells are a very important source of energy for driving the circulation at higher latitudes. Although they transfer LATENT HEAT and SENSIBLE HEAT energy towards the equator, this is offset by an intense poleward transport of POTENTIAL ENERGY aloft. This transport is enhanced by heat added to the air by the condensation of the water vapour drawn into the ITCZ at low levels.                WDS

**Reading**

Palmén, E. and Newton, C.W. 1969: *Atmospheric circulation systems*. New York: Academic Press.

**haff** A coastal lagoon separated from the open seas by a sand pit formed by longshore drifting of sediments.

**hagg** A channel which separates hummocks in a peat bog.

**hail** Solid precipitation which falls in the form of ice particles from cumulonimbus clouds. The high concentration of liquid water in these clouds provides an environment which is favourable for the quick growth of ice particles by both coalescence and collision with supercooled water droplets. Hail is commonly spherical and the larger stones are composed of concentric shells of clear and opaque ice with a diameter as large as 10 cm in severe convective storms. Hailstones grow larger by being transported in the strong up- and downdraughts which characterize cumulonimbus clouds.                RR

**haldenhang** A degrading rock slope which underlies an accumulation of talus or scree.

**half-life** A parameter used to describe aspects of the radioactive decay process. It is the time required for one half of a given number of a radionuclide to decay to a protégé product. Alternatively this may be expressed in the quantum-mechanical view as the interval of time within which an unstable atomic nucleus has a probability of decay of 0.5. When large numbers of atoms are involved, the two definitions are in effect equivalent, but it is important to realize that a half-life is only exhibited because some atoms remain stable for much longer than others of the same radioactive element.

Each radioisotope has a fixed half-life which is typically determined by mass spectrometry measurements. If it is possible to assume that at the

| Element | Symbol and radiation type ($\alpha$ = alpha particle emission, $\beta-$ = electron emission) | Half-life (a = annum; Ga = giga annum ($10^9$ a) etc.) |
|---|---|---|
| Hydrogen (tritium) | $^3$H$(\beta-)$ | 12.43 a |
| Lead | $^{210}$Pb$(\beta-)$ | 22.3 a |
| Caesium | $^{137}$Cs$(\beta-)$ | 30.17 a |
| Carbon | $^{14}$C$(\beta-)$ | 5568 a |
| Chlorine | $^{36}$Cl$(\beta-)$ | 301 ka |
| Uranium | $^{238}$U$(\alpha)$ | 4.47 Ga |
| Potassium | $^{40}$K (electron capture) | 12.50 Ga |
| Thorium | $^{232}$Th$(\alpha)$ | 14.05 Ga |

*Source*: Geyh and Schleicher 1990.

time of formation of a deposit that no protégé isotopes were present, then measurement of the relative abundance of parent radioisotopes and protégé products, coupled with the knowledge of the half-life provide the basis for radiometric dating methods. The half-lives of some widely used radioisotopes are listed in the table (Geyh and Schleicher 1990).

Radiocarbon ($^{14}$C) and radioactive potassium ($^{40}$K) are both key RADIOISOTOPES for dating purposes. See also CARBON DATING, GEOCHRONOLOGY, POTASSIUM ARGON DATING, URANIUM SERIES DATING. SS / DLD

**Reading and Reference**
Faure, H. 1986: *Principles of isotope geology*. 2nd edn. Chichester: Wiley. · Geyh, M.A. and Schleicher, H. 1990: *Absolute age determination*. Trans. R.C. Newcomb. Berlin: Springer-Verlag.

**halocarbons** Carbon compounds containing fluorine, chlorine, bromine or iodine. The group includes halogenated fluorocarbons based on ethane and methane, such as CHLOROFLUOROCARBONS (CFCs and HCFCs), and halons containing bromine and perfluorocarbons (PFCs). Along with associated compounds such as sulphur hexafluoride, these halocarbons are effective greenhouse gases whose presence is solely due to anthropogenic activity. They have both a direct radiative effect and an impact through depletion of ozone.

Although the emission of CFCs and HCFCs has been controlled by the Montreal Protocol and atmospheric concentrations are expected to decline over the next century, PFCs and some of the halon group have residence times exceeding 1000 years and may have an influence on future climate. PSh

**haloclasty** The disintegration of rock as a result of the action of salts, which may result from salt crystallization, salt hydration, or the thermal expansion of salts. It is especially important in arid areas, but may also play a role in building decay in cities. (See also SALT WEATHERING.) ASG

**halons** Members of the halogenated fluorocarbon (HF) group of ethane- or methane-based compounds in which H$^+$ ions are partially or completely replaced by chlorine, fluorine and/or bromine. This group also includes chlorofluorocarbons. Halons are HFs which contain bromine, for example, halon 1211 (CF$_2$BrCl) and halon 1301 (CF$_3$Br). They are long-lived and have been implicated in stratospheric ozone depletion, where their damage potential has been estimated at 3–10 times that of equivalent CFC molecules. BJS

**haloturbation** The disturbance of soils and surficial sediments by the growth of salt crystals by a process akin to cryoturbation. Common in salt lakes and playas, haloturbation leads to the sorting of coarser particles to form patterned ground. PSh

**hamada, hammada** A desert region which does not have any surficial materials other than boulders and exposed bedrock.

**hamra** A red, sandy soil which also contains clay.

**hanging valley** A tributary valley whose floor is discordant with the floor of the main valley. Hanging valleys are a hallmark of glacial erosion in mountains. The discordance of the valley floors was one of the arguments used in the early nineteenth century to suggest that rivers did not cut valleys and the objection was only removed when the role of glacier activity was appreciated. The valley cross-sections were adjusted to the glaciers they held and it is likely that the glacier surfaces met concordantly (Penck 1905). DES

**Reference**

Penck, A. 1905: Glacial features in the surface of the Alps. *Journal of geology* 13, pp. 1–17.

**hardness** The resistance of a material to scratching. Moh's scale, which is not quantitative, depends upon the ability of one mineral in a series (usually 1 to 10, but also 1 to 8 or 1 to 12) to scratch others below it but not those above, e.g. quartz (7) will scratch feldspar (6) but not topaz (8). More quantitative measures, e.g. those of Brinell, Rockwell, Vickers, depend on the load applied to a standard indenter applied for a given time to produce a given pattern on the material under test. The SCHMIDT HAMMER test used in concrete research gives an approximate value of hardness via the 'rebound value' (in reality a measure of the coefficient of restitution of the material). This can be used to determine the compressive strength of the material.                          WBW

**hardpan** A compacted or cemented subsurface horizon within the soil zone, sometimes termed a duripan.

**harmattan** A dry, north-easterly wind of West Africa which blows in the winter months. Blowing out of the Sahara, it frequently carries much sand and dust.

**harmonic analysis** See FOURIER ANALYSIS.

**Hawaiian eruption** A type of volcanic eruption characterized by the emission of large quantities of highly fluid, basic lava issuing from vents and fissures. Explosive eruptions of material are rare.

**hazard** See NATURAL HAZARD.

**haze** A term used to describe a reduction in visibility due to the suspension of fine aerosols in the atmosphere. Caused by DUST particles of one micron ($10^{-6}$ m) or less entrained into the atmosphere by wind or human activity (commonly off-road vehicle driving) a haze results in a reduction in visibility to less than 2 km (but more than 1 km, otherwise it is termed a DUST STORM). A haze may last many hours or days owing to the long-term suspension of fine particles by the TURBULENT nature of the wind. Haze may also result from smoke from fires.     GFSW

**head** See SOLIFLUCTION.

**headcut** The upslope limit of a gully system, characterized by a steep wall which is cut back, migrating upslope, as further erosion occurs.

**headwall** A steep, arcuate slope around the head of a CIRQUE or LANDSLIDE. Gradients exceed the ANGLE OF REPOSE (ANGLE OF INITIAL YIELD) of granular materials (around 34°). Sometimes a headwall is divided into a backwall and two sidewalls.     IE

**headward erosion** The processes involved in the upslope migration of the head of a gully or source of a stream.

**headwater** A stream which forms the source and upper limit of a river, especially a large one.

**heat budget** Heat is a form of energy and it defines in a general way the aggregate internal energy of motion of the atoms and molecules of a body. It may be taken as being equivalent to the specific heat of a body multiplied by its absolute temperature in degrees Kelvin and by its mass, where the specific heat of a substance is the heat required to raise the temperature of a unit mass by one degree. Temperature is the condition which determines the flow of heat from one substance to another, the direction being from high to low temperatures. So long as only one object is considered its temperature changes represent proportional changes in heat content. The definition of heat content suggests that when a variety of masses and types of material are considered the equivalence of heat and temperature disappears. Often a small hot object will contain considerably less heat than a large cool one, and even if both have the same mass and temperature their heat contents can differ because of differing specific heats.

The transfer of heat to or from a substance is effected by one or more of the processes of conduction, CONVECTION or RADIATION. The common effect of such a transfer is to alter either the temperature or the state of the substance or both. A heated body may acquire a higher temperature (sensible heat) or change to a higher state (liquid to gas, or solid to liquid) and therefore acquire latent or hidden heat. Conduction is the process of heat transfer through matter by molecular impact from regions of high temperature to regions of low temperature without the transfer of the matter itself. It is the process by which heat passes through solids but its effects in fluids (liquids and gases) are usually negligible in comparison with those of convection. In contrast, convection is a mode of heat transfer in a fluid, involving the movement of substantial volumes of the substance concerned. Conduction is the main method of heat transfer in solid rocks and the soil, while the convection process frequently operates in the atmosphere and oceans.

A dry surface with no atmosphere will assume a very simple heat balance, such that:

$$R_N = R_T(1 - \alpha) - \varepsilon\sigma T^4 = H$$

where $R_N$ is the net radiation, $R_T$ is the global (solar plus sky) radiation, $\alpha$ is the albedo, $\varepsilon$ is the infrared emissivity, $\varepsilon\sigma T^4$ is the long-wave radiation loss from a surface at temperature $T$ K, and $H$ is the heat flux into/out of the soil.

Energy fluxes towards the soil surface are taken as being positive. If there is no heat source other than the global radiation, then over long time periods the net radiation $R_N$ will be zero, since the radiative energy gained must equal the radiative energy lost. The flow of heat $H$ into or out of the soil can produce a small imbalance in the short term, but in general $H$ is very small compared with the magnitude of the other energy fluxes. It follows therefore that in general the temperature $T$ will change in close accord with the daily march of incoming radiation, and that it will vary greatly between day and night.

The temperature of the surface for a given global radiation flux depends on its ALBEDO, on its infrared emissivity and on its thermal inertia. The higher the albedo, the more radiation is reflected and the lower the surface temperature. While the emissivity of many natural substances approaches unity, it is usually not exactly equal to unity but a few per cent less. It is the surface emissivity which controls the infrared radiation loss by the surface to space. Surfaces with a low emissivity will lose heat by radiation more slowly than surfaces with higher emissivities at the same temperature. Similarly, to lose the same amount of heat by radiation, a surface with a low emissivity will have to be at a higher temperature than a surface with a high emissivity.

The flux of sensible heat into the surface is controlled by the thermal properties of the surface and the subsurface. A natural parameter which expresses the thermal properties of a soil is the thermal inertia $(\rho c \lambda)^{1/2}$, where $\rho, c$ and $\lambda$ are the density, specific heat and thermal conductivity respectively. If the thermal inertia is large, the subsurface absorbs a large amount of heat during the day and then conducts a large amount of heat to the radiating surface during the night. Temperature variations of the surface will in consequence be moderate. In contrast, soils of poor thermal inertia conduct little heat to the subsurface, and attain high temperatures in the day-time and low temperatures at night. Obviously there are no dry surfaces of this type on earth but the lunar surface is of this nature.

The surface heat balance equation with the atmosphere present is the same as before, except that the atmosphere is capable of radiating energy $R_L$ and advecting heat $S$. Thus:

$$R_N = R_T(1 - \alpha) - \varepsilon\sigma T^4 + R_L = H + S.$$

The atmosphere will also modify the incoming solar beam by scattering and absorbing it, so not all the radiant energy will come from the direction of the sun. Since the atmosphere also absorbs and re-radiates infrared radiation there is an infrared flux towards the surface as well as away from it. Two general classes of dry land-surfaces can be found on earth. The first is fairly common and is found in the tropical deserts, where it is dominated by the global radiation flux. The second sometimes occurs over dry surfaces at high latitudes in winter and is dominated by the atmospheric heat flux $S$.

In the tropical deserts there is, because of the clear skies, a large solar radiation input, which leads to high surface temperatures. Since the winds are normally light, horizontal heat transfer by the atmosphere is small, and as a consequence the main heat loss from the surface is by infrared radiation. The infrared radiation losses can be large, leading to relatively small net radiation values despite the large solar radiation input. The surface temperature follows the variations of incoming solar radiation closely and is controlled by it. Great variations of temperature occur under these conditions and numerous stations in the Sahara have recorded maximum temperatures above 45 °C and minimum temperatures below 0 °C. Even higher maximum temperatures would be recorded but for the fact that the albedo of desert surfaces tends to be high and therefore a fair amount of incoming radiation is reflected.

During winter in high latitudes the incoming solar radiation $(S)$ is small, and under these conditions the heat transported by the atmosphere is considerably greater than the incoming dominated by the sensible heat flux(es). Under these conditions, with moderate winds, the surface temperature follows the air temperature which in turn depends on the prevailing synoptic conditions. While wet surfaces of this nature are common, extensive dry surfaces are rare.

Wet surfaces are the most common type of surface found in nature and the surface energy balance may be described by the equation in the previous section together with the addition of the energy lost during EVAPORATION $(LE_a)$:

$$R_N = R_T(1 - \alpha) - \varepsilon\sigma T^4 + R_L$$
$$= H + LE_a + S$$

where $L$ is the latent heat of vaporization, and $E_a$ is the actual evaporation. A wet surface in this context is one from which the evaporation is controlled solely by the prevailing radiational and meteorological conditions, that is to say the evaporation is independent of the rate of water supply. Such a surface could be one which is physically wet, such as an ocean or a landscape after rain, or it could be one in which evaporation is occurring freely from water moving through plants (transpiration).

The study of the heat budget of a natural surface falls into two stages. First, there is the study of the radiation balance, which leads to an estimation of the available net radiation. Secondly, the net radiation has to be divided among the sensible heat flows to the soil and the atmosphere, and the latent heat flow to the atmosphere, to produce the full heat balance. The ratio of the sensible heat flow to the atmosphere to the latent heat flow is known as the Bowen ratio, $\beta$, and can be written as:

$$\beta = \frac{\text{sensible heat loss to atmosphere}(S)}{\text{latent heat loss to atmosphere}(LE_t)}$$

In the absence of atmosphere advection, $\beta$ can vary between $+\propto$ for a dry surface with no evaporation to zero for an evaporating wet surface with no sensible heat loss. If there is atmospheric heat advection, $\beta$ may become negative, indicating a flow of heat from the atmosphere to the surface. JGL

**Reading**
Budyko, M.I. 1974: *Climate and life*. New York: Academic Press. · Houghton, J.T. 1977: *The physics of atmospheres*. Cambridge: Cambridge University Press. · Lockwood, J.G. 1979: *Causes of climate*. London: Edward Arnold. · Miller, D.H. 1981: *Energy at the surface of the earth*. International geophysical series 27. New York and London: Academic Press. · Sellers, W.D. 1965: *Physical climatology*. Chicago: University of Chicago Press.

**heat island**  See URBAN METEOROLOGY.

**heathlands**  These are mineral-based and usually distinguished by acidophilous vegetation. In Britain, heathlands are commonly associated with the sandstones of southern and eastern England. Heathlands exhibit a wide range of moisture conditions; the vegetation is usually dominated by heather (*Calluna*) communities.

Heathland in Britain mainly formed as a secondary habitat resulting from tree clearance in areas such as Breckland in East Anglia. This habitat is now relatively rare. For example, Breckland covered an area of 22,000 ha in 1880 and was reduced to just 7500 ha in 100 years as a result of agricultural improvement. ALH

**Reading**
Evans, D. 1992: *A history of nature conservation in Britain*. London: Routledge. · Rodwell, J.S. ed. 1991: *British plant communities, Vol. 2: mires and heaths*. Cambridge: Cambridge University Press.

**Heinrich events**  Layers of rock, debris-rich but FORAMINIFERA-poor sediment, identified in the deep sea core record from the North Atlantic were reported by Heinrich (1988). These have been attributed to deposition from the bottom of iceberg armadas resulting from the catastrophic break-up of the margins of the polar ice cap and adjacent ice sheets. These resulted from relatively rapid ice decay due to global warming events, and have been interpreted as important evidence of the link between oceans and the atmosphere. Heinrich events occurred irregularly during the last Glacial cycle with the five most recent dated to 14.5–13.5 ka, 22–19 ka, 27 ka, 35.5 ka and 52 ka. Heinrich events are probably linked to DANSGAARD–OESCHGER (D–O) EVENTS. See also BOND CYCLES. DSGT

**Reading and Reference**
Heinrich, H. 1988: Origin and consequences of cyclic ice-rafting in the northeast Atlantic Ocean during the past 130,000 years. *Quaternary research* 29, pp. 142–52. · Williams, M.A.J., Dunkerley, D., de Deckker, P., Kershaw, P. and Chappell, J. 1998: *Quarternary environments*. 2nd edn. London: Arnold.

**helical flow**  Spiral (roll) airflow vortices in the lower atmosphere, that develop in unstable thermal conditions. Parallel helical vortices were once thought to lead to the formation of linear sand dunes. DSGT

**helictite**  A small calcium carbonate SPELEOTHEM which grows in curved or spiralling forms in limestone caves. It grows by the slow accretion of calcium carbonate crystals from the top of the structure, fed by an internal canal. When the crystal-growth forces exceed that of the normal hydraulic force of the flowing water, the crystals can grow away from the vertical. Helping this erratic crystal growth, there may often be clay particle blockages in the feeder canal or even draughts within the cave passage. Helictites are quite common in limestone caves of all environments and can often produce intricate tangled structures. Famous ones can be found in the Kango Caves of South Africa. PAB

**helm wind**  A strong, cold, north-easterly wind that occasionally blows down the western slope of the Cross Fell range into the Vale of Eden in north-west England. It occurs when the horizontal component of airflow is virtually perpendicular to the hills, which restricts directions to the north-east, and when a stable layer of air lies about 600 m above the summit of the range. These conditions accord with the requirements for mountain and LEE WAVES in the atmosphere. The resultant airflow resembles that of water flowing over a weir. The classic study of the helm wind is by Manley (1945). BWA

**Reading and Reference**
Atkinson, B.W. 1981: *Meso-scale atmospheric circulations.* London and New York: Academic Press. Esp. pp. 25–108. · Manley, G. 1945: The helm wind of Cross Fell 1937–1938. *Quarterly journal of the Royal Meteorological Society.* 71, pp. 197–215.

**hemera** A zone or any other period of geological time as determined from an assemblage of fossils.

**herbivore** A plant-eating organism; an animal which feeds directly on photosynthetic species. Other than the plant itself, it is the organism which lies at the base of the food chain. Most herbivores are very small, e.g. the aphid species which live off cultivated plants; the largest are the grazing and browsing mammals of Africa (such as the giraffe or elephant). The herbivore group also includes man's domesticated animals. On land the smaller the size of the herbivore, the more likely it is to be linked with one plant species alone. Overall, in terrestrial environments, the ecological effect of herbivores under normal conditions is greatest within grasslands, where up to 35 per cent of production may be consumed; within forests, the equivalent figure is customarily 4–7 per cent (Hughes 1971; Whittaker 1975). Overstocking may result in the unbalanced and excessive removal of vegetative resources in all land habitats. In aquatic systems herbivores are relatively unimportant, most lighter plants being eaten by decomposers, moulds and bacteria.
DW

**References**
Hughes, M.K. 1971: Tree biocontent, net production and litter falls in a deciduous woodland. *Oikos* 22, pp. 62–73. · Whittaker, R.H. 1975: *Communities and ecosystems.* 2nd edn. New York: Macmillan.

**herbivore use intensity (HUI)** This measures the impact of grazing and browsing animals on the environment, reflecting the density of animals per unit area (Georgiardis 1987). This can show marked spatial variations in some managed environments, particularly when activities are focused on single water sources. (See also PIOSPHERE.) DSGT

**Reference**
Georgiardis, N.J. 1987: Responses of savanna grasslands to extreme use by pastoral livestock. Ph.D. thesis, Syracuse University. Unpublished.

**heterosphere** The outer portion of the earth's atmosphere beyond the hemosphere.

**heterotrophs** Organisms which depend upon organic foods in order to obtain energy. Most bacteria, fungi and animals fall into this category.

**hiatus** A gap in the stratigraphic record or in geological time which is not represented by any sediments.

**high energy window** The suggestion, first introduced by Neumann (1972), that in the mid-Holocene on tropical coasts there was a period when wave energy was higher than at present. This occurred during the phase when the present sea level was being first approached by the Flandrian (Holocene) transgression and prior to the protective development of coral reefs. The 'window' may have operated on a more local scale on individual reefs with waves breaking not on margins of an extensive reef flat as at the present time, but more extensively over a shallowly submerged reef top prior to the development of the reef flat (Hopley 1984).
ASG

**References**
Hopley, D. 1984: The Holocene 'high energy window' on the central Great Barrier Reef. In B.G. Thom, ed., *Coastal geomorphology in Australia.* Sydney: Academic Press. Pp. 135–50. · Neumann, A.C. 1972: Quaternary sea level history of Bermuda and the Bahamas. *American Quaternary Association Second National Conference Abstracts*, pp. 41–4.

**hillslope flow processes** The mechanisms and routes of flow followed by precipitation down hillsides to a stream channel.

Some precipitation is intercepted before reaching the ground, and if not evaporated will reach it directly or by stemflow, with some delay and with some spatial redistribution, particularly concentrating round tree-trunks, for example. If rainfall intensity exceeds the INFILTRATION capacity, or if the soil is saturated, surface depressions will fill and overflow to give OVERLAND FLOW which normally reaches the stream channels but may infiltrate into the soil downslope, especially if a rainstorm is very brief or very local, or if the soil thickens appreciably. Infiltration occurs most readily into soil crumbs or peds, but also overflows down soil cracks and other structural voids, thereby bypassing the surface soil layers and usually infiltrating into peds farther down but in some cases connecting directly to the groundwater table.

Within the soil flow is mainly vertical in the unsaturated zone, flowing under the predominant influence of hydraulic potential gradients. Soil porosity normally decreases with depth, though there are many exceptions from this

245

general rule. If downward flow is impeded a saturated zone may develop, either in contact with the groundwater body as a whole or perched above it. In such a layer hydraulic potential gradients are negligible and saturated subsurface flow or THROUGHFLOW predominates. As throughflow accumulates downslope the level of saturation in the soil rises towards the surface, or the SATURATION DEFICIT falls, and flow may re-emerge on the surface as return flow. The area of zero saturation deficit defines the SOURCE AREA for a rapid streamflow response to subsequent rain by overland flow production as described above.                                MJK

### Reading
Kirkby, M.J. ed. 1976: *Hillslope hydrology.* Chichester: Wiley.

### Hjulström curve
In 1935 the Swedish geomorphologist Filip Hjulström presented an empirical curve defining the threshold flow velocities required to initiate motion of grains of different sizes on a stream bed. By displaying this curve in conjunction with curves of the settling velocity or depositional velocity, the conditions required for transport, traction and deposition of sediment can be defined (see diagram). The threshold flow velocity is at a minimum for well-graded sand particles in the 0.2–0.5 mm size range, and higher velocities are necessary to entrain both finer and coarser sediments. Finer sediment includes cohesive silt and clay show whose entrainment is inhibited by particle aggregation and by submergence in the laminar sublayer (see BOUNDARY LAYER). Coarser sediments (gravel, pebbles and cobbles) require higher threshold velocities simply because they are heavier. However, there is no unique mean velocity at which grains of given size first move, because the bed velocity is a more critical control of entrainment by flowing water. For a particular threshold bed velocity the mean velocity is greater in deeper water.

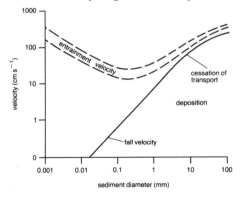

Hjulström curve.

Also, the particle size distribution affects the threshold velocity, since coarse grains in a poorly sorted sediment may shelter finer grains and prevent their entrainment. A similar shape of curve defines the fluid threshold in air (Bagnold 1941), but the different fluid densities result in different entrainment velocities in the two media.                                KSR

### References
Bagnold, R.A. 1941: *The physics of blown sand and desert dunes.* London: Chapman & Hall. · Hjulström, F. 1935: Studies of the morphological activities of rivers as illustrated by the River Fyris. *Bulletin of the Geological Institute, University of Uppsala* 25, pp. 221–527.

### hoar frost
The deposition of ice crystals usually by the direct sublimation of water molecules, but also by the freezing of dew, on a solid surface exposed to the atmosphere. The surface, if cooled at night by loss of radiation, in turn cools the water vapour immediately above it to its frost point and sublimation begins. The crystals are usually white.                                JET

### hodograph
A scheme for representing windspeeds and directions at different heights at a given time by plotting them on a polar coordinate diagram. Successive winds are plotted as vectors so that they all 'blow' from the respective directions towards the central point, which represents the station at which the observations were made. The hodograph is then constructed by drawing a series of vectors which join the successive end points of the plotted winds.

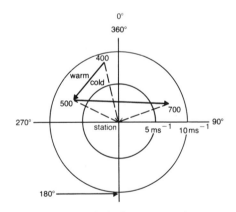

**Hodograph.** The blank diagram comprises compass directions and a series of concentric circles, the radii of which represent windspeed. The centre of all the circles represents the observing station. Three winds are shown (dashed lines) for the levels 400, 500 and 700 mb. The wind at 700 mb, for example, blows from a direction of about 70° with a speed of about 8 m s$^{-1}$. The solid arrows show the size and direction of the vector difference of the winds at the top and bottom of any layer. This vector difference is known as the thermal wind.

For example, the three winds plotted in the diagram for 700, 500 and 400 mb (broken lines) are shown with the hodograph construction (solid lines) which represents the shear vectors for the 700–500 mb and 500–400 mb layers. Assuming the observed winds to be geostrophic, the shear vectors can be considered to be THERMAL WINDS and therefore to provide information about the nature of temperature ADVECTION.

Between 700 and 500 mb the wind backs with height and within this layer the air is blowing from a relatively cold region (cold air lies to the left of thermal wind). So winds that back with height signify cold advection and those that veer (500–400 mb here) mean warm advection. The thermal winds can be measured directly from the scale on the polar diagram. RR

**Reading**
Atkinson, B.W. 1968: *The weather business*. London: Aldus Books.

**hogback** A long ridge of rock dipping steeply on both sides that is the exposure of a stratum or rock which has been tilted until the originally horizontal beds are almost vertical.

**Holocene** The second series of the Quaternary period, following the Pleistocene. Often called the postglacial, it has extended from *c.*11,000 years ago until the present and has been characterized by interglacial conditions. It has been marked by various climatic fluctuations (see ALTITHERMAL; BLYTT–SERNANDER MODEL; HYPSITHERMAL; LITTLE CLIMATIC OPTIMUM; NEOGLACIAL etc.) and by rapid sea-level rise (see FLANDRIAN TRANSGRESSION). The term was introduced during the 1860s, and means 'entirely modern'. ASG

**Reading**
Roberts, N. 1989: *The Holocene: an environmental history*. Oxford: Blackwell.

**holokarst** A term coined to describe the coastal Dinaric karst belt of the former Yugoslavia where the full suite of karst landforms is found and limestone solution processes dominate the landscape. Cvijic contrasted this landscape with the less well-developed karst landscape further inland which he termed merokarst. The term has been generally used to describe any limestone landscape with a fully developed range of karst features. HAV

**Reading**
Cvijic, J. 1893: Das Karstphänomen. *Geographische Abhandlung* 5, pp. 217–329. · — 1925: Types morphologiques des terrains calcaires. Le holokarst. *Comptes Rendus de l'Académie des Sciences* 180, pp. 592–4.

**homoclines** Locations which experience the same types of climatic regimes.

**homoiothermy** (homeothermy) The maintenance by an organism of a relatively constant body temperature, independent of, and usually somewhat above, that of the organism's immediate surroundings. Mammals and birds are homoiothermic or 'warmblooded'. PHA

**hoodoo** An unusually shaped pillar or outcrop of rock produced by erosion.

**horizon, soil** A recognizable layer within a soil, running essentially parallel to the soil surface, that has been differentiated from the materials above and below it by the operation of various weathering and soil-forming processes such as ELUVIATION. Horizons may be differentiated on the basis of properties like colour, texture, structure, stoniness, or accumulations or deficits of iron, manganese, carbonates or other materials. The principal horizon designations, O, A, B, C and R, are used to indicate characteristic kinds of modifications to the original parent material of the soil, and indicate something of the genetic linkages between the horizons found in the soil 'profile' (the characteristic series of layers from the surface downwards). 'O' horizons are accumulations of organic matter lying on top of the mineral soil; A horizons form the uppermost layer of the mineral soil, and display accumulations of humus and often a relative deficit of clay, iron and other materials, resulting in a sandier texture than the B horizon below, where the mobile materials have accumulated. 'C' horizons represent parent materials altered in limited ways, generally unaffected by biota or organic materials, and form a transition to the R horizon of underlying unweathered rock. These major horizon types are further subdivided both on the basis of depth and distinctive properties, e.g. into A1 where the major feature is accumulated organic matter, and A2 where loss of mobile materials has been most influential on the soil character. Suffixes are also used, e.g. Bca to indicate a B horizon with clear carbonate accumulation. Horizonation may reflect in part the operation of processes of hillslope transport, so that the notion of layers genetically linked only through the *vertical* flux of material cannot be applied in all situations. In the US *Soil Taxonomy*, an important diagnostic role is given to various additional kinds of horizons that are recognized in tandem with the O, A, B, C, R system. These include the *epipedon*, a horizon formed at the surface. If darkened by organic matter, for example, and the colouring affects the top of the B horizon as well as the A, then the epipedon includes part of both the A and the B horizons. Similar subsoil horizons include *argillic horizons* (where illuviated clays have accumulated), *natric* and *calcic* horizons of sodium and carbonate accumulation, *oxic* horizons and *duripans*, hardpans which are cemented by silica. DLD

### Reading
Brady, N.C. and Weil, R.R. 1996: *The nature and properties of soils*. 11th edn. New Jersey: Prentice-Hall. · United States Department of Agriculture, Soil Conservation Service. 1988: *Soil taxonomy*. Florida: Krieger Publishing Co.

**horn, glacial** A pyramidal peak with three or more distinct faces steepened by glacial undercutting. The classic situation occurs when cirque glaciers encroach on a mountain from all sides.

**horse latitudes** The latitude belts over the oceans at approximately 30–35° north and south where winds are predominantly calm or very light and the weather often hot and dry. These latitudes represent the normal axes of the subtropical high pressure belts. The origin of the name is uncertain but it is suggested that the crews of sailing ships carrying horses to the West Indies had occasionally when becalmed either to jettison their live cargo as fodder ran out or eat it. The term 'subtropical high pressure areas' is now frequently used in preference to 'horse latitude'. AHP

### Reading
Boucher, K. 1975: *Global climate*. London: English University Press.

**horst** An upstanding block of the earth's crust that is bounded by faults and has been uplifted by tectonic processes. The down-faulted areas which bound horsts are called graben.

**Horton overland flow model** Forecasts overland flow per unit area as rainfall intensity minus the INFILTRATION capacity, where this difference is positive. In Horton's original version of the model (1933), infiltration capacity was calculated as an exponential decay of the form:

$$f = f_c + (f_o - f_c)\exp(-ct)$$

where $f$ is the infiltration capacity at time $t$, and $f_c$, $f_o$ and c are constants, which depend on the soil and its moisture distribution at the start of infiltration. At the start of a rainstorm all rainfall is considered to infiltrate until the infiltration capacity falls to the current rainfall intensity. Subsequently the infiltration curve determines how much water enters the soil, and any excess is diverted as OVERLAND FLOW. Total overland flow discharge is obtained by routing this flow downslope, combined with the overland flow produced at each site. This overland flow is then considered to supply the sharply rising and falling peak of stream hydrograph response to rainstorms. The model was first developed for agricultural soils in an area of very intense rainfalls, and it works very well under these conditions. At the simplest level the model appears to imply a uniform rate of overland flow production so that flow discharge increases linearly downslope. Allowance for wetter soils downslope, however, forecasts increasing production downslope, so that Horton's model is able to perform well as a forecasting tool, even in humid forested environments where its physical basis is undermined by the rarity of observable overland flow, and where PARTIAL AREA MODELS are now preferred by many researchers.

Criticisms of the Horton model have been made at several levels. Attempts to improve the model may be made by improving the infiltration equation used, or by using an equation which takes into account the reduced rate at which infiltration capacity falls if rainfall intensity is at less than the maximum capacity. More fundamental criticisms note the existence of storm hydrograph peaks in the absence of observable overland flow, particularly in forested catchments, and have to take greater account of THROUGHFLOW within the soil as a direct source of stormflow and to forecast saturated SOURCE AREAS in which overland flow production is concentrated. MJK

### Reading and Reference
Horton, R.E. 1933: The role of infiltration in the hydrological cycle. *Transactions of the American Geophysical Union* 14, pp. 446–60. · Kirkby, M.J. ed. 1978: *Hillslope hydrology*. Chichester: Wiley.

**Horton's laws** Two laws of drainage composition were suggested by R.E. Horton (1945): the law of stream *numbers* whereby an inverse geometric series related number of streams of a particular ORDER *and order*, and the law of stream *lengths* which was based on a geometric series between mean lengths of streams of each order *and order*. These two laws were subsequently complemented by three others giving an inverse geometric series relation between mean slope of streams of a particular order *and order*, a geometric series relating mean basin area of streams of a particular order *to order*, and the law of contributing areas which was the logarithmic relation of drainage areas of each order and the total stream lengths which they contained and supported. Although modifications to these five laws were made with the advent of different systems of stream ordering and some useful comparisons were made between areas using the five relationships, it was realized by 1970 that the 'laws' were at least in part a consequence of the definition of stream ordering and therefore largely statistical relationships. Recent research has demonstrated that other parameters defined by MORPHOMETRY are more useful to relate to indices of drainage basin process and that the structure of drainage networks can be illuminated by topology of the NETWORK.

KJG

**Reading and Reference**
Gardiner, V. 1976: *Drainage basin morphometry*. British Geomorphological Research Group technical bulletin 14. Norwich: Geo Abstracts. · Horton, R.E. 1945: Erosional development of streams and their drainage basins; hydrophysical approach to quantitative morphology. *Bulletin of the Geological Society of America* 56, pp. 275–370.

**hot spot**  A small area of the earth's crust where an unusually high heat flow is associated with volcanic activity. Of approximately 125 hot spots thought to have been active over the past 10 million years most are located well away from plate boundaries (see PLATE TECTONICS). The major theory of hot spot formation involves the effects of a PLUME of hot MANTLE rising to the surface. Some researchers consider that hot spots are sufficiently stationary with respect to the mantle to provide a reference frame for determining plate motions and CONTINENTAL DRIFT.

MAS

**Reading**
Burke, K.C. and Wilson, J.T. 1976: Hot spots and the earth's surface. In J.T. Wilson ed., *Continents adrift and continents aground*. San Francisco: W.H. Freeman, Pp. 58–69. · Duncan, R.A. 1981: Hot spots in the southern oceans – an absolute frame of reference for motion of the Gondwana continents. *Tectonophysics* 74, pp. 29–42. · Morgan, W.J. 1972: *Plate motions and deep mantle convection*. Geological Society of America memoir 132, pp. 7–22. · Vink, G.E., Morgan, W.J. and Vogt, P.R. 1985: The earth's hot spots. *Scientific American* 252, pp. 32–9.

**hot spring**  An emission of hot, usually geothermally heated, water at the landsurface.

**hum**  A residual hill in limestone country formed through surface lowering of the surrounding landsurface.

**human impacts**  Many activities practised by humankind, whatever the level of social organization have a direct impact on the physical environment. Some impacts are of a scale, extent or frequency that no lasting change occurs in the environment or to a part of the earth system. Such activities are sustainable. Other activities leave a lasting impression on

natural systems, either because they are carried out on a large scale, or at a rate that exceeds the ability of natural systems to replace or repair what has been done. Such activities are unsustainable, i.e. their occurrence will alter or destroy permanently parts of the environment.

As levels of economic 'development' have increased, so humankind's ability to impact on the environment have grown in magnitude, diversity and scale. The development of agriculture, perhaps 10,000 years ago, represented a deliberate attempt to manage part of the environment in a positive way, to increase food production. The development of industrially based societies, founded on the extraction and use of STOCK RESOURCES, expanded both the diversity of direct, deliberate impacts, the necessary 'price' of development, and indirect, accidental impacts, in the form, for example, of pollution of waterways and the atmosphere. In turn, industrial developments fuelled more effective (and destructive) agricultural methods and, perhaps, led to developments that have contributed to rapid global human population growth. Thus, a growth in impacts, perhaps at an exponential rate in the last two centuries, has led to an increase in concern about the negative, often unknown or uncertain, effects of human activities.

Concern with human impacts on the environment is not, however, new. The literature of both Greek and Roman civilizations contain significant reference to negative human effects on the environment. George Perkins Marsh's book, *Man and nature*, was published in 1864 and stemmed from his observations of the rapid effects of European cultures on the North American environment, some of which he saw as highly negative, others as positive. Within Marsh's views are ideas that have subsequently developed into or contributed to debates about society's stewardship of the earth system, and sustainable use of the environment, that in turn contributed to the growth, in the mid and late twentieth century, of environmentalism.

Human impacts on the environment are not by any means all negative. Societies have the wherewithal to counter negative effects and carry out remedial actions. Perhaps the biggest challenges and difficulties are identifying the causes and scales of certain impacts, distinctions between natural and human-induced effects and the time lags between impact-causing activities and the full range of impacts being identified. Some human impacts are not always agreed upon (see, e.g. GREENHOUSE EFFECT). Overall, impacts are certainly diverse, affecting all systems and components of systems. Good exemplification of human impacts on vegetation, animals, the soil, water, landforms, climate and the atmosphere are contained within Andrew Goudie's book *The human impact*, first published in 1981. (See also, e.g., ENVIRONMENTAL IMPACT; CONSERVATION.)                                    DSGT

**References**
Goudie, A.S. 2000: *The human impact.* 5th edn. Oxford: Blackwell. · Marsh, G.P. 1864: *Man and nature.* New York: Scribner.

**humate** A collective term for the dark-brown to black gel-like humic substances formed as a result of the decomposition of organic matter in soils and sediments. It may be translocated down the profile by vadose water. Subsurface accumulations of this material occur most commonly in podzols. If the humate dries out it may harden sufficiently to create a humicrete.   ASG

**Reading**
Pye, K. 1982: Characteristics and significance of some humate-cemented sands (humicretes) at Cape Flattery, Queensland, Australia. *Geological magazine* 119, pp. 229–42.

**humic acid** Humic colloids may be chemically fractionated into fulvic acid, humic acid and humin, although the divisions are not clearcut. Humic colloids are built up of carbon, oxygen, hydrogen, nitrogen, sulphur and phosphorus, and contain polysaccharides, proteins and other organic materials including waxes and asphalts. Humic colloids are negatively charged and may be bound to clay through polyvalent cations (e.g. $Ca^{2+}$ and $Al^{3+}$) or to positively charged surfaces of iron or aluminium hydrated oxides.   ALH

**Reading**
Russell, E.W. 1973: *Soil conditions and plant growth.* Harlow: Longman.

**humidity** A term which relates to the water vapour content of the atmosphere and is expressed in a variety of ways.

Thus the relative humidity of an air sample is expressed as a percentage which is found by relating the observed VAPOUR PRESSURE at a given temperature to the saturation value of vapour pressure at that temperature. It is evaluated by using tables or a special slide rule in conjunction with dry-bulb and wet-bulb temperature readings. Relative humidity commonly displays a daily cycle which has a phase opposite to that of temperature so that the highest values are observed near dawn and the lowest during the afternoon.

Absolute humidity indicates the actual amount of water vapour present in a sample of air. For example, the humidity mixing ratio of an air parcel is defined to be the ratio of the mass of water vapour to the mass of dry air with which the water vapour is associated. At the surface it may range between near zero in frigid polar areas to $25\,g\,kg^{-1}$ in very humid tropical air. Absolute humidities can be deduced from wet-bulb temperature readings.   RR

**humus** Partially decomposed organic matter which accumulates on and within the soil zone.

**hunting and gathering** A human lifestyle, sometimes regarded as undeveloped, certainly primitive (in the broadest meaning) and certainly in harmony with nature in its purest form, whereby societies hunt animals and gather wild plants as the source of foodstuffs and clothing. It is certain, from archaeological evidence, that hunting and gathering was the basis of early, middle and late stone age societies, and therefore has been a fundamental component of human development and history. Today, few if any pure (i.e. not gaining resources from other means) hunter-gatherer societies exist, and where they do, numbers are small and the range (land area) for successful practice of this lifestyle is restricted and under threat from other social styles. The ENVIRONMENTAL IMPACTS of traditional hunting and gathering are small, and practices are sustainable.   DSGT

**hurricane** See TROPICAL CYCLONE.

**hydration** The process whereby an anhydrous mineral, one not containing water molecules within its crystal structure, takes up water to form a crystallographically distinct mineral. The partial decomposition of rocks by water.

**hydraulic conductivity** See CONDUCTIVITY, HYDRAULIC.

**hydraulic diffusivity** ($D$) The ratio of the hydraulic CONDUCTIVITY $K$ to the specific storage $S_s$ or transmissivity $T$ to storativity $S$, as follows:

$$D = K/S_s = T/S$$

The square root of hydraulic diffusivity is proportional to the velocity of HYDRAULIC HEAD transmission in the aquifer.

The specific storage is the volume of water that a unit volume of aquifer releases from storage per unit decline in hydraulic head, whereas the dimensionless parameter storativity is the volume of water it releases per unit surface area per unit decline in hydraulic head above the surface. Transmissivity (or transmissibility) is the product of the hydraulic conductivity and aquifer thickness.

The hydraulic diffusivity is therefore an aquifer parameter that combines the transmission characteristics and the storage properties. PWW

**Reading**
Freeze, R.A. and Cherry, J.A. 1979: *Groundwater*. Englewood Cliffs, NJ: Prentice-Hall. · Ward, R.C. and Robinson, M. 1990: *Principles of hydrology*. 3rd edn. Maidenhead: McGraw-Hill.

**hydraulic force** The water flowing in a channel reach exerts a hydraulic force on the bed and may move grains of unconsolidated bed material. This is usually known as the TRACTIVE FORCE, and is the effective component of the weight of water acting parallel to the bed in the direction of flow. It is usually expressed in relation to the bed area over which it acts, and is then called the unit tractive force, or the bed shear stress (force per unit area). The roughness of the channel bed imposes a frictional drag on the flow (see BOUNDARY LAYER), and in a reach where no flow acceleration occurs, this is equal and opposite to the hydraulic force of the flow on the bed. KSR

**hydraulic geometry** The hydraulic geometry of alluvial channels was defined by Leopold and Maddock (1953) as: (1) the at-a-station adjustment of flow characteristics such as discharge ($w$), depth ($d$) and velocity ($v$) within a section as discharge ($Q$) varies; and (2) the general downstream adjustment of these flow properties as discharge increases at a constant flow frequency. Both at-a-station and downstream,

these adjustments are commonly described by the power-functions:

$$w = aQ^b;\, d = cQ^f;\, v = kQ^m$$

and since $wdv = Q$, it follows that $ack = 1$ and $b + f + m = 1$. The overall flow geometry of a river system is represented by relationships for an upstream and a downstream section at both a high and a low frequency flow (see diagram). However, the downstream trends are only approximate because factors other than discharge, such as bed and bank sediment properties, control the changing shape of the channel within which the flow then adjusts at discharges which are not competent to change the section.

At-a-station flow geometry varies between rectangular sections with steep cohesive banks in humid areas, and parabolic sections in sandy sediments in semi-arid regions. In the former, $b = 0.05$, $f = 0.45$ and $m = 0.50$, and in the latter $b = f = m = 0.33$. Downstream the adjustment of width and depth accommodates more of the increase in discharge, and $b = 0.50$, $f = 0.40$ and $m = 0.1$ (Richards 1982, pp. 148–59). Downstream exponents vary according to the trends in ROUGHNESS and long-profile slope, but their general similarity reflects the tendency of all rivers to create forms which allow equal dissipation of energy and minimize total work.

At the local reach scale hydraulic autogeometry describes a further pattern in flow and channel form. KSR

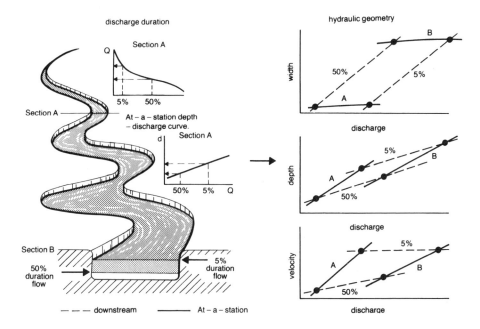

251

**Reading and References**

Ferguson, R.I. 1986: Hydraulics and hydraulic geometry. *Progress in physical geography* 10, pp. 1–31. · Leopold, L.B. and Maddock, T. 1953: The hydraulic geometry of stream channels and some physiographic implications. *United States Geological Survey professional paper 252.* · Richards, K.S. 1982: *Rivers: form and process in alluvial channels.* London and New York: Methuen.

**hydraulic gradient**  If piezometric tubes are inserted into flowing water (soil throughflow, groundwater flow, pipe flow or open channel flow) water will rise to the hydraulic grade line. The slope of this line is the hydraulic gradient. So long as conditions remain hydrostatic, which requires that no extreme flow curvature occurs and that slopes are less than 1 in 10, the hydraulic gradient is the water table or water surface slope, or the piezometric surface slope in the case of confined groundwater flow or pipe flow.

KSR

**hydraulic head**  (*h*) is the sum of the elevation head (*z*) and the pressure head (*hp*). Each of these parameters has the dimension of length and is measured in metres above a convenient datum, usually taken as sea level. The change of hydraulic head with distance (*l*) is known as the hydraulic gradient (*dh/dl*) and is an important term in the equation defining DARCY'S LAW.

PWW

**Reading**

Freeze, R.A. and Cherry, J.A. 1979: *Groundwater.* Englewood Cliffs, NJ: Prentice-Hall.

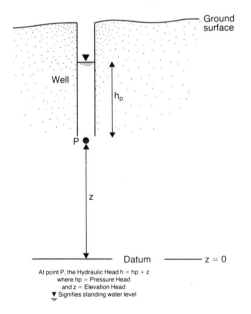

Ground surface

Well

$h_p$

P

z

Datum     z = 0

At point P, the Hydraulic Head h = hp + z
where hp = Pressure Head
and z = Elevation Head
▼ Signifies standing water level

**hydraulic jump**  A given discharge in an open channel of constant width can be conveyed either at a high velocity and shallow depth or at a low velocity in deep water; a sudden change between these states along a channel is a hydraulic jump. The FROUDE NUMBER $F_r$ passes from $F_r > 1$ (supercritical flow) to $F_r < 1$ (subscritical flow) during this transition. A hydraulic jump may be caused naturally by a sudden reduction of bed slope, or a large submerged obstacle (a boulder or bank slump), and is deliberately generated by hydraulic engineers to dissipate energy over weirs and dams, to ensure that upstream flow is critical in measuring installations such as flumes, and below sluices to ensure that they are not drowned (Sellin 1969, pp. 52–68).

KSR

**Reference**

Sellin, R.J.H. 1969: *Flow in channels.* London: Macmillan.

**hydraulic potential**  or fluid potential at any point in a porous substance is the product of the HYDRAULIC HEAD and the acceleration due to gravity. It is the mechanical energy of water (in this case) per unit mass. Since gravitational acceleration is almost constant near the earth's surface, hydraulic potential is very closely correlated with hydraulic head (Freeze and Cherry 1979). In accordance with DARCY'S LAW, subsurface water flows from points of high hydraulic potential towards points of low hydraulic potential. This occurs down the hydraulic gradient perpendicular to the EQUIPOTENTIALS.

PWW

**Reference**

Freeze, R.A. and Cherry, J.A. 1979: *Groundwater.* Englewood Cliffs, NJ: Prentice-Hall.

**hydraulic radius**  The hydraulic radius *R*, or the hydraulic mean depth, is the ratio of wetted perimeter (*P*) to the cross-sectional area of flow in a channel (*A*):

$$R = A/P.$$

It measures the efficiency of a section in conveying flow, as it is the area of flow per unit length of water–solid contact. In wide, shallow channels the mean depth is a convenient approximation for *R*; when the width–depth ratio of a rectangular channel is 50, this approximation results in a 4 per cent error, but when it is 10 the error increases to 17 per cent.

KSR

**hydraulics**  The science in which the principles of FLUID MECHANICS are applied to the behaviour of water flowing in pipes or open channels over rigid or loose boundaries. Basic hydraulic theories relating to water flow deal with the development and refinement of FLOW EQUATIONS which define the flow velocity in terms of depth, slope and boundary roughness, with assessment of the forces, momentum, and energy losses within the flow, and with analysis of velocity variations in two- and three-dimensional BOUNDARY LAYERS. Some of the most difficult problems in hydraulics occur in the area where it impinges most obviously on physical geography – namely, in the analysis of flow over loose boundaries, when the flow transports sediment at the bed, creates bedforms, and therefore affects its own flow resistance in a complex feedback process.

Applied hydraulics is concerned with the economic development of water supply, and therefore deals with the storage, conveyance and control of water. Storage problems include assessment of the ability of the physical supply within an area to cope with estimated and forecast demand, as well as the difficulties of maintenance of supply – for example, because of over-exploitation of groundwater, or reservoir sedimentation and loss of storage capacity. Hydraulic design is concerned with dams, spillways, and conveyance and diversion systems such as canals and irrigation ditches. There are close links between REGIME THEORY, which considers the design of stable sediment-bearing artificial channels, and the geomorphological study of natural river morphology. The design of control and measurement systems is a further important aspect of hydraulic analysis, in which spatially varied flow including free overfalls and HYDRAULIC JUMPS is deliberately manipulated to provide control conditions for the purpose of flow measurement in weirs and flumes.                                                          KSR

**Reading**
Chow, V.T. 1959: *Open-channel hydraulics*. Tokyo: McGraw-Hill. · Sellin, R.H.J. 1969: *Flow in open channels*. London: Macmillan.

**hydrodynamic levelling**  The transfer of survey datum levels by comparing mean sea level at two sites, and adjusting them to allow for gradients on the sea surface due to currents, water density, winds and atmospheric pressures.                                                          DTP

**hydrofracturing**  The rupture of rock caused when water enters cracks in rock which may be little wider than the combined diameters of a few molecules of water. Under cold conditions if ice seals off the end of such a crack, liquid water may be forced towards the tip of the crack and so extend it.                                         ASG

**hydrogeological map**  A cartographic representation of subterranean water resources information with associated surface water and geological data. Hydrogeological maps may be prepared from continental to local scales, depending on their purpose. Continental and regional scale maps commonly depict the major aquifers and identify water bearing lithologies and GROUNDWATER basins. Major SPRINGS and their discharges are also frequently depicted. Water balance information is sometimes contoured. Larger-scale local maps, in addition, usually show WATER TABLE (or piezometric) contours (or EQUIPOTENTIALS in the case of ARTESIAN aquifers) and may provide information on the direction of groundwater flow, aquifer thickness, borehole locations, recharge areas, and saltwater intrusions. Water quality data are also sometimes presented.                          PWW

**hydrographs**  Show discharge plotted against time for a point in a drainage basin. The discharge is usually monitored at a site on the stream channel network (stream hydrograph) but it may be measured at a section on a hillslope (hillslope hydrograph).

Stream hydrographs represent the pattern of total discharge through time and are made up of flow from different stores within the drainage basin (see HYDROLOGICAL CYCLE). During dry periods discharge gradually recedes and the hydrograph at a particular site may often be represented by a standard DEPLETION CURVE or base-flow recession curve. During rainfall events, the stream discharge responds in the form of a storm hydrograph (see diagram). The shape of the storm hydrograph is affected by both the spatial and temporal pattern of the rainfall, ANTECEDENT MOISTURE conditions, and by the drainage basin characteristics, including catchment geometry, rock and soil types, vegetation and land use pattern and drainage network structure.

Hydrograph analysis often begins with a separation of the hydrograph into components.

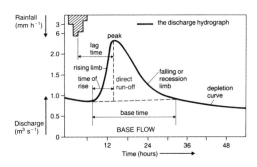

Storm hydrograph.

Traditional hydrograph separation subdivides the hydrograph into surface run-off or direct run-off, base flow and sometimes interflow. The subdivision is essentially empirical with straight lines joining either the beginning of the storm hydrograph rise or a point beneath the hydrograph peak to one or more points on the falling limb of the hydrograph. These points on the falling limb may be identified from the curvature of the falling limb, from the point of divergence of the falling limb from the depletion curve or by identifying breaks of slope on semi-logarithmic plots of the hydrograph (Barnes 1939; Gregory and Walling 1976). Hibbert and Cunningham (1967) avoided the generic terms for describing the components of the separated storm hydrograph by suggesting a separation into quick flow and delayed flow by a straight line from the point of hydrograph rise to the falling limb of the hydrograph with a slope of $0.551 \, s^{-1} \, km^{-2} \, h^{-1}$. Hydrograph separation based upon water quality variations (e.g. Pinder and Jones 1969; Pilgrim *et al.* 1979) show that straight line separations do not accurately reflect the way in which storm hydrographs are composed of water from different source areas.

Whichever separation technique is used, parameters of the storm hydrograph may be analysed to represent hydrograph size and shape and some widely used parameters include the time of rise, base time, run-off volume, peak discharge, percentage run-off and hydrograph peakedness. The unit hydrograph technique (Sherman 1932; see UNIT RESPONSE GRAPH) is often applied to separated storm hydrographs since it provides a simple means of summarizing and predicting the storm hydrograph at a site.

AMG

**Reading and References**
Barnes, B.S. 1939: The structure of discharge recession curves. *Transactions of the American Geophysical Union* 20, pp. 721–5. · Gregory, K.J. and Walling, D.E. 1976: *Drainage basin form and process*. London: Edward Arnold. · Hibbert, A.R. and Cunningham, G.B. 1967: Streamflow data processing opportunities and applications. In W.E. Sopper and H.W. Lull eds, *International symposium on forest hydrology*. Oxford and New York: Pergamon. · Linsley, R.K., Kohler, M.A. and Paulhus, J.L.H. 1988: *Hydrology for engineers* 3rd edn. New York: McGraw-Hill. · Pilgrim, D.H., Huff, D.D. and Steele, D.M. 1979: Use of specific conductance and contact time relations for separating flow components in storm runoff. *Water Resources Research* 15, pp. 329–39. · Pinder, G.F. and Jones, J.F. 1969: Determination of the groundwater component of peak discharge from the chemistry of total runoff. *Water resources research* 5, pp. 438–45. · Robson, A.J. and Neal, C. 1990: Hydrograph separation using chemical techniques: an application to catchments in Mid-Wales. *Journal of hydrology* 116, pp. 345–63. · Sherman, L.K. 1932: Stream flow from rainfall by the unit-graph method. *Engineering news record* 108, pp. 501–5.

**hydroisostasy** The reaction of the earth's crust to the application and removal of a mass of water, as when eustatic sea-level changes have affected the depth of water over the continental shelves causing the crust to be depressed at times of high sea level and to be elevated at times of low sea level. The same process can operate in lake basins like Lake Bonneville in response to climatically induced fluctuations in lake level, and may create warped shorelines.

ASG

**Reading**
Bloom, A.L. 1967: Pleistocene shorelines: a new test of isostasy. *Bulletin of the Geological Society of America* 78, pp. 1477–94.

**hydrolaccolith** A pingo, a dome of ice just beneath the surface in areas of permafrost. The product of the freezing of artesian spring water as it nears the surface.

**hydrological cycle** The hydrological cycle describes the continuous movement of all forms of water (vapour, liquid and solid) on, in and above the earth's surface and it is the central concept of HYDROLOGY.

The hydrological cycle includes the condensation and freezing of water vapour in the atmosphere to form liquid or solid precipitation, the movement of water from precipitation through one or more of a range of conceptual stores, including SURFACE STORAGE, soil moisture storage, groundwater storage, stream channels and the oceans until at some stage the water returns to the atmosphere as water vapour through the processes of evaporation and transpiration. Figure 1 schematically presents the drainage basin hydrological cycle or the possible routes that water may follow within a drainage basin to form part of stream discharge at the basin outlet or to be lost to the basin through evapotranspiration. The time taken for water to pass to and from a store, the capacity of the store and the volume of water contained in each store at a particular time will strongly influence the discharge response of the drainage basin to a rainfall event. Figure 2 summarizes the way in which the different contributions to run-off might be generated in a drainage basin in response to a storm.

Figure 2 illustrates, first, that more inaccessible stores will respond more slowly to rainfall, if they generate a measurable response. For example, water must infiltrate the soil and percolate to the water table before groundwater storage can be increased and so produce increased base flow to the stream. All these processes may take a long time and so introduce a long lag before the peak of base flow passes the

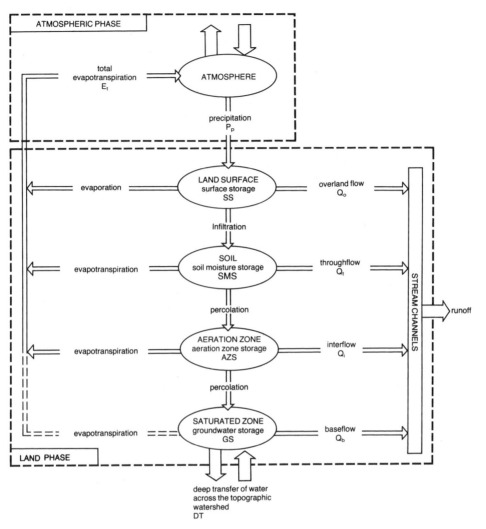

**Hydrological cycle: 1.** Drainage basin.

gauging station on the stream at the catchment outlet. Kirkby (1975), for example, suggested orders of magnitude for drainage times in hours to surface, soil and groundwater storage as 0, 5 and 500 h respectively, and drainage from these stores downslope to channel storage as 0.5, 30 and 200 h respectively. Once in the channel network, travel time to the gauging station may also be substantial in large drainage basins. Secondly, larger stores, containing more water, will produce a more attenuated response to a particular input of water. Thus, in figure 2, outflow from surface storage is less attenuated than that from soil moisture storage which in turn is less attenuated than that from groundwater storage, and this reflects the condition of the stores in the particular drainage basin con-

sidered in this example. Finally, if a store is full of water, it can only absorb as much as can drain from the store in a fixed time period and, as a result, excess water may build up in the preceding store in the cycle. An example of this which has important effects on the basin discharge characteristics is the effect of soil saturation on overland flow. ANTECEDENT conditions will control the extent to which areas of the soil in a drainage basin are saturated and so the dynamic area of the soil in a drainage basin will affect the area of the basin over which saturation overland flow will occur during a storm. This example also illustrates the fact that each of the conceptual lumped stores of figure 1 may be represented by a large number of distributed stores in a real drainage basin.

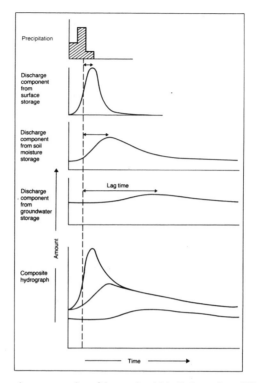

**Hydrological cycle: 2.** Schematic representation of the way in which discharge from different stores might combine to produce a composite storm hydrograph in a small drainage basin.

The hydrological cycle provides a conceptual framework for hydrological studies and it may be extended to include the movement of solutes and sediments, heat and biota as they are transported and stored within the water cycle. The cycle presents a logical structure for the measurement, analysis and modelling of hydrological processes and for the identification of the impact of man on the quality, quantity, routing and storage of water.          AMG

**Reading and Reference**
Kirkby, M.J. 1975: Hydrograph modelling strategies. In R. Peel, M. Chisholm and P. Haggett eds, *Processes in physical and human geography*. London: Heinemann. · Kirkby, M.J. ed. 1978: *Hillslope hydrology*. Chichester: Wiley. · Shaw, E.M. 1988: *Hydrology in practice*. 2nd edn. New York: Van Nostrand Reinhold.

**hydrology**  The science concerned with the study of the different forms of water as they exist in the natural environment. Its central focus is the circulation and distribution of water as it is expressed by the WATER BALANCE and HYDROLOGICAL CYCLE. Hydrology embraces not only the study of water quantity and movement but also the degree to which these are affected by man's activities, including deliberate management of water resources and the inad-

vertent effects of man on hydrological processes. It is often subdivided into physical and applied hydrology. Physical hydrology includes the detailed measurement and analysis of information on hydrological processes to improve understanding of the functioning of the hydrological system and also the refinement of statistical and mathematical methods of predicting and modelling these physical processes. Applied hydrology is concerned with the application of the understanding of hydrological processes to their modification and management. 'Water resources and pollution on the one hand and flooding and erosion on the other, are the chief concerns of the hydrologist' (Rodda *et al.* 1976).

The study of hydrology is at least as old as the ancient civilizations of Egypt, because the provision of a reliable water supply is essential to the survival of man. However, the development of plausible theories concerning the circulation of water in the hydrological cycle did not appear until the seventeenth century. These were largely based on observations of rainfall and river flow in the Seine basin by Pierre Perault and Edmé Mariotte and on the ideas of Edmund Halley, who simulated evaporation from the Mediterranean and concluded that this could account for all surface drainage.

Hydrology is not only a long-established science which has been given major impetus at the international level, it is also an interdisciplinary subject bringing together specialists from an enormous range of disciplines including biology, chemistry, civil engineering, environmental planning, forestry, geology, geomorphology, mathematics and physics. AMG

**Reading and References**
Dunne, T. and Leopold, L.B. 1978: *Water in environmental planning*. San Francisco: W.H. Freeman. · Gregory, K.J. and Walling, D.E. 1976: *Drainage basin form and process*. London: Edward Arnold. · Kirkby, M.J. 1978: *Hillslope hydrology*. Chichester: Wiley. · Linsley, R.K., Kohler, M.A. and Paulhus, J.L. 1982: *Hydrology for engineers*. 3rd edn. New York: McGraw-Hill. · —, Downing, R.A. and Law, F.M. 1976: *Systematic hydrology*. London: Butterworth. · Rodda, J.C. 1976: *Facets of hydrology*. Chichester: Wiley. · Shaw, E.M. 1988: *Hydrology in practice*. 2nd edn. New York: Van Nostrand Reinhold. · Ward, R.C. and Robinson, M. 1990: *Principles of hydrology*. 3rd edn. Maidenhead, Berks: McGraw-Hill.

**hydrolysis** *a*. The disintegration of organic compounds through their reaction with water.
*b*. The formation of both an acid and a base by a salt when it dissociates with water. This is an important mechanism of chemical weathering of rock.

**hydrometeorology** The application of meteorology to hydrological problems. Hydrometeorology is concerned particularly with the atmospheric part of the HYDROLOGICAL CYCLE. This involves the input of water to the surface through PRECIPITATION and outputs by EVAPORATION and TRANSPIRATION. Precipitation input to a catchment area depends upon atmospheric factors which can be examined at both short and long time-scales. Rainfall from an individual cloud will be determined by *microphysical processes* within the CLOUD as well as by the synoptic controls on cloud development. The extent to which suitable atmospheric conditions for cloud growth and precipitation formation prevail will be influenced by atmospheric circulation changes not only on a daily scale but also at the seasonal and annual time-scales. One of the main applications of hydrometeorology is for flood prediction. Statistical techniques are used to analyse existing precipitation data in order to calculate the return of flood events, i.e. how often a particular flood level is likely to occur, to estimate the PROBABLE MAXIMUM PRECIPITATION over an area, or even to simulate river-level variations. Some engineering design problems need similar information to provide cost-effective solutions when building reservoirs, bridges, sewers and planning irrigation schemes. PS

**Reading**
Shaw, E.M. 1988: *Hydrology in practice*. 2nd edn. London: Van Nostrand Reinhold.

**hydrometry** The measurement of water flow in channels. Accurate flow measurements are difficult to achieve, particularly when the cross-section of the flowing water is large or when the channel is shifting, irregular or subject to heavy weed growth. The techniques of flow or discharge measurement fall into three groups. Since discharge at a cross-section is equal to the product of the mean velocity of flow and the cross-sectional area of the flowing water, methods of gauging the flow are based upon either the direct measurement of discharge or on the product of an indicator of water cross-sectional area (usually through water depth or stage at a fixed site) and a means of estimating velocity of flow. In addition, measurements of the discharge may be just point measurements in time or they may be continuous. Measurements of stage, velocity or discharge are usually undertaken at fixed sites known as gauging stations. The table summarizes the main methods of flow measurements and recent developments of these techniques are discussed in IAHS (1982) (see DISCHARGE). AMG

**Reading and Reference**
Gregory, K.J. and Walling, D.E. 1976: *Drainage basin form and process*. London: Edward Arnold. · Herschy, R.W. 1976; New methods of river gauging. In J.C. Rodda ed., *Facets of hydrology*. Chichester: Wiley. · IAHS 1982: *Advances in hydrometry*. Proceedings of the Exeter symposium. International Association of Hydrological Sciences publication 134.

**hydromorphy** A process that occurs in soils which produces gleying and mottling as a result of the intermittent or permanent presence of excess water.

**hydrophobic soil** A soil that resists wetting by water. Hydrophobicity may be caused by fats (lipids) released by certain plants, and which coat particles of the soil. Hot fires may also distil volatile compounds, including various aliphatic hydrocarbons, from litter and these compounds may migrate into the upper few centimetres of the soil, there cooling and coming to rest. These distillation compounds may also trigger soil hydrophobicity. The outcome in either case is limited infiltration of water, and a greater risk of surface run-off and attendant soil erosion. DLD

Hydrometry: Methods for river flow measurement

| Variable | Point measurement | Continuous measurement | Comments |
|---|---|---|---|
| Stage | 1 Stage board<br>2 Float and counter-<br>weight<br>3 Tape and electrical<br>contact<br>4 Pressure bulb<br>5 Crest stage gauge<br>(for peak flow only) | 1 Float and counter-<br>weight or pressure<br>bulb attached<br>to continuous recorder<br>2 Bubble gauge and<br>recorder | Float and counterweight or<br>tape and electrical contact<br>both require a level water<br>surface and so are usually<br>used with a stilling well |
| Velocity | 1 Current meter<br>2 Floats<br>3 Tracer (e.g. NaCl or<br>$NaCr_2O_3$) velocity<br>4 Electromagnetic<br>flowmeter<br><br>5 Ultrasonic flowmeter | Electromagnetic flowmeter<br>and recorder (average<br>velocity in cross-section)<br>Ultrasonic flowmeter and<br>recorder (average velocity<br>at one or more depths<br>in the cross-section) | |
| Discharge | 1 Current meter (mean or<br>mid section method)<br><br>2 Tracer dilution (gulp or<br>continuous injection)<br><br>3 Electromagenetic<br>gauging<br><br>4 Ultrasonic gauging<br><br>5 Structures (weirs and<br>flumes)<br><br>6 Volumetric<br><br>7 Slope area method (e.g.<br>Manning equation) | Continuously monitored<br>stage with established<br>stage – discharge<br>relationship<br><br><br><br><br><br>Continuous measurement<br>of velocity and stage<br><br>Continuous measurement<br>of velocity and stage<br>Continuous stage<br>measurement coupled<br>with a stage – discharge<br>relationship | Choice of wading or cableway<br>or moving boat methods.<br>Smooth, straight section<br>required.<br>Suitable for turbulent sections<br>or where high velocities or<br>irregular bed make the use of<br>current meter difficult<br>Suitable for sections with<br>aquatic growth and unstable<br>bed<br>Suitable for very wide rivers<br>and estuaries<br>Suitable for small rivers<br><br><br><br>Suitable for measuring very<br>low flows<br>Suitable for estimating flows<br>at an ungauged site |

**Reading**
Debano, L.F., Mann, L.D. and Hamilton, D.A. 1970:
Translocation of hydrophobic substances into soil by
burning organic matter. *Soil Science Society of America
Proceedings* 34, pp. 130–3.

**hydrophyte** A herbaceous plant which has
parts beneath water which survive when the
parts above water die back.

**hydrosphere** The earth's water, which exists
in both fresh and saline form and may occur in a
liquid, solid or gaseous state. Land, sea and air
each contribute to the total volume of water,
which is conveyed between various locations
and transformed from one state to another

(HYDROLOGICAL CYCLE). The overall quantity of
water in the hydrosphere remains more or less
constant. An insignificant increment accrues
from volcanic water vapour, while dissociation
of water vapour in the upper atmosphere
(photodissociation) represents a minor loss.

About 70 per cent of the earth's surface is
occupied by water. Some 97.3 per cent of its
volume is currently in the oceans
($1350 \times 10^{15} \, m^3$), the maximum extent of
which is in the southern hemisphere. Of the
2.7 per cent of terrestrial water, most is polar
snow and ice ($29 \times 10^{15} \, m^3$). Groundwater (the
majority below soil level) accounts for
$8.4 \times 10^{15} \, m^3$, lakes (its main superficial terrestrial location) and rivers $0.2 \times 10^{15} \, m^3$, and

water in living organisms $0.0006 \times 10^{15}\,\mathrm{m}^3$. Water vapour is the most important variable constituent of the atmosphere, but only accounts for $0.013 \times 10^{15}\,\mathrm{m}^3$ of the total amount of water in the hydrosphere (Peixoto and Ali Kettani 1973).                    RLJ

**Reading and Reference**
Chorley, R.J. ed. 1969: *Water, earth and man: a synthesis of hydrology, geomorphology and socio-economic geography.* London: Methuen. · Cloud, P. 1968: Atmospheric and hydrospheric evolution on the primitive earth. *Science* 160, pp. 729–36. · Peixoto, J.P. and Ali Kettani, M. 1973: The control of the water cycle. In F. Press and R. Siever eds., *Planet earth (readings from Scientific American).* San Francisco: W.H. Freeman.

**hydrostatic equation**   This equation is as follows:

$$\frac{\partial p}{\partial z}r = \rho g \qquad (1)$$

where $p$ is pressure, $z$ is height, $\rho$ is air density, and $g$ is the acceleration of gravity. Alternatively it may be written as:

$$-\frac{1}{\rho}\frac{\partial p}{\partial z} = g \qquad (2)$$

Equation 2 reveals that the upward pressure gradient force (left-hand side) is balanced by the downward force of gravity (righthand side). Such a balance is known as hydrostatic equilibrium. The assumption that the atmosphere is a hydrostatic equilibrium has proved very useful in the analysis of large-scale motion, because at that scale, buoyancy forces may safely be ignored. At smaller scales, e.g. motion within clouds, we cannot validly assume that hydrostatic equilibrium exists.                    BWA

**Reading**
Hess, S.L. 1959: *Introduction to theoretical meteorology.* New York: Henry Holt.

**hydrostatic pressure**   The fluid pressure $p$ exerted by the underlying column of water in a body of water when at rest. It is given by Pascal's law as:

$$p = \rho g \psi + p_o$$

where $\rho$ is the mass density of the water, $g$ is the acceleration due to gravity, $\psi$ is the pressure head (or vertical depth of water) and $p_o$ is atmospheric pressure at the surface of the water. In groundwater hydrology practice, the latter is usually set equal to zero and $p$ is calculated in pressure above atmospheric and expressed in $\mathrm{N\,m}^{-2}$ or Pa.                    PWW

**Reading**
Freeze, R.A. and Cherry, J.A. 1979: *Groundwater.* Englewood Cliffs, NJ: Prentice-Hall.

**hydrothermal alteration**   Alteration of rocks due to earth pressures and temperatures, which are milder than those which produce metamorphism. It may give rise to chemical changes which are similar to weathering. Generally, weathering effects decrease from the surface downwards, while hydrothermal alteration increases downwards towards the fluids which give rise to the activity. Pneumatolysis is similar but relates to gases (including steam) at still higher temperatures and pressures.                    WBW

**hyetograph**   A self-recording instrument for measuring rainfall continuously. A chart showing the distribution of rainfall over a region.

**hygrograph**   A self-recording instrument for measuring atmospheric humidity.

**hygrometer**   A device for measuring the relative humidity of the atmosphere. Those which give a continuous record are called hygrographs.                    ASG

**hypabyssal rock**   Igneous rock which is intrusive but has consolidated in a zone above the base of the earth's crust and hence has distinct structural characteristics.

**hyper-arid**   These areas are the most arid DRYLANDS and cover c.7.5 per cent of the world's land surface. They represent the true DESERTS where biological productivity is lowest and rainfall most sporadic and variable. Hyper-arid areas have P/PET values of less than 0.05 (see DRYLANDS for methodology). The world's hyper-arid areas include much of the Sahara, Arabia, parts of central Asia and the coastal Namib and Atacama deserts in Namibia and Peru respectively. In most hyper-arid areas periods in excess of 12 months without any rainfall have been recorded.                    DSGT

**hyperconcentrated flow**   A river flow transporting an exceptionally large concentration of suspended particulates. Desert streams in flood commonly carry high concentrations of 5–$50\,\mathrm{g\,L}^{-1}$, but concentrations of hundreds of grams/L have been reported, and in rare cases more than half of the weight of a water sample has been composed of sediment. These extraordinary concentrations fall into the range termed hyperconcentrated, and clearly grade progressively into mud flows, one form of MASS MOVEMENT. Beverage and Culbertson (1964) suggested that any streamflow carrying in excess of 40 per cent by weight sediment be classed as hyperconcentrated. For ordinary sediment densities, this corresponds to flows composed of < 80 per cent water by volume.　　　DLD

**Reference**
Beverage, J.P. and Culbertson, J.K. 1964: Hyperconcentrations of suspended sediment. *Proceedings of the American Society of Civil Engineers, Journal of the Hydraulics Division*, 90, pp. 117–28.

**hypogene** Term applied to mineral and ore deposits formed by water rising towards the surface, in contrast to supergene.

**hypolimnion** The lowest layers of cold water which occur at the bottom of an ocean, sea or lake whereas the EPILIMNION consists of the warmer, less dense upper layers. The intermediate zone is called the thermocline.　　　ASG

**hypothesis** A set of beliefs or assumptions that we hold about the structure of the world or the way it behaves. Hypotheses are the basis of explanation in science, in that explanations are the logical consequence of the hypotheses we have accepted (Haines-Young and Petch 1986). For example, explanations of many contemporary global biogeographical patterns are based on accepted hypotheses about the movement of the continents over geological time.

But why do we accept one hypothesis and not another? The choice is based on the judgements we make about the possible truth of the hypothesis. Hypotheses can be mistaken. Thus we demand that hypotheses should be *testable* (Popper 1972). In principle we should be able to refute them by making some measurement or observation. We also demand that hypotheses should be consistent with the other hypotheses we have accepted. As scientists we cannot hold inconsistent ideas. Thus conflicts between hypotheses have to be resolved by eliminating those that are wrong, or by developing other testable hypotheses that account for the conflict. In the end, the hypotheses we accept and use are those that have withstood our best attempts to refute them.

Baker (1996) has recently described the role of hypothesis testing in the context of Quaternary geology and geomorphology, while Llambi (1998) has considered such ideas in an ecological context. Case studies that illustrate the role of hypothesis testing include Avery and Haines-Young (1990), Flather and Sauer (1996) and Carpinter *et al.* (1998).　　　RH-Y

**References**
Avery, M.I. and Haines-Young, R.H. 1990: Population estimates for the dunlin (*Calidris alpina*) derived from remotely sensed satellite imagery of the Flow Country of northern Scotland. *Nature* 344, pp. 860–2. · Baker, V.R. 1996: The pragmatic roots of American quaternary geology and geomorphology, *Geomorphology* 16.3, pp. 197–215. · Carpinter, S.R., Cole, J.T., Essington, T.E, Hodgson, J.R., Houser, J.N. Kitchell, J.F. and Pace, M.L. 1998: Evaluating alternative explanations in ecosystem experiments. *Ecosystems* 1.4, pp. 335–44. · Flather, C.H. and Sauer, J.R. 1996: Using landscape ecology to test hypotheses about large-scale abundance patterns in migratory birds. *Ecology* 77. 1, pp. 28–35. · Haines-Young, R.H. and Petch, J.R. 1986: *Physical geography: its nature and methods*. London: Paul Chapman. · Llambi, L.D. 1998: An epistemological debate in ecology: Popper and the hypothesis test. *Interciencia*, 23.5, pp. 286–94. · Popper, K.R. 1972: *The logic of scientific discovery*. London: Hutchinson.

**hypsithermal** A term introduced by Deevey and Flint (1957) for a warm HOLOCENE phase covering four of the traditional Blytt–Sernander pollen zones, embracing the Boreal through to the Sub-Boreal (8950–2550 BP). It is broadly equivalent in meaning to ALTITHERMAL or climatic optimum.　　　ASG

**Reference**
Deevey, F.S. and Flint, R.F. 1957: Post-glacial hypsithermal interval. *Science* 125, pp. 182–4.

**hypsographic curve** A generalized profile of the surface of the earth and the ocean floors. A curve or graph which represents the proportions of the area of the surface at various altitudes above or below a datum.

**hypsometric curve** See HYPSOGRAPHIC CURVE.

**hypsometry** The measurement of the elevation of the landsurface or sea floor above or below a given datum, usually mean sea level.　　　ASG

**Reading**
Cogley, J.G. 1985: Hypsometry of the continents. *Zeitschrift für Geomorphologie Supplementband* 53, pp. 1–48.

**hysteresis** A term borrowed from the study of magnetism and used to describe a bivariate plot, which evidences a looped form and therefore a different value of the dependent variable, according to whether the independent variable is increasing or decreasing. Examples include relationships between matric potential and soil moisture content, river discharge and water stage, and suspended sediment and solute concentrations and river discharge. In the latter case different concentrations are associated with a given level of discharge on the rising and falling limbs of a hydrograph, and this hysteresis effect may be ascribed either to a difference in timing or lag between the response of the two parameters or to asymmetry in either relative behaviour. Both clockwise and anticlockwise hysteresis loops have been reported for relationships between suspended sediment and solute concentrations and discharge. From a study of the River Rother in the UK, Wood (1977) suggested that the precise form of the hysteresis loop associated with the relationship between suspended sediment concentration and discharge for specific events could be related to the period between successive flow events and to the duration and intensity of each event.

DEW

### Reference
Wood, P.A. 1977: Controls of variation in suspended sediment concentration in the River Rother, West Sussex, England, *Sedimentology* 24, pp. 437–45.

# I

**ice** Normally the solid form of water formed by: (a) the freezing of water; (b) the condensation of atmospheric water vapour directly into ice crystals; (c) the sublimation of solid ice crystals directly from water vapour in the air; or (d) the compaction of snow. Each form of ice has important implications for the physical geography of the earth.

(a) Ice derived from the freezing of water involves SEA ICE and lake ice which contribute a characteristic morphology to water surfaces and shorelines in high and mid-latitudes, also GROUND ICE which forms a significant component of permafrost landscapes (see also ICE WEDGE; PINGO), and the freezing of water within a snow pack to form ice lenses or, in the case of a glacier, SUPERIMPOSED ICE. Repeated freezing and thawing is also an efficient means of WEATHERING and can cause rapid rock breakdown.

(b) Ice crystals in the atmosphere begin as minute ice particles which form round condensation nuclei. These nuclei may be dust particles with favourable molecular structure or even crystals of sea salt. They will grow so long as they exist in an atmosphere with an excess supply of water vapour and will form an ice cloud or, if conditions are suitable near the ground surface, an ice fog. If the ice crystals reach a critical size they begin to fall as snow. RIME can contribute to the growth of the falling ice crystal and also directly to the ground surface whenever supercooled water droplets freeze on impact with a solid object.

(c) The sublimation of ice crystals directly from water vapour in the air can take place within a snowpack in response to strong temperature gradients and causes the growth of fragile ice crystals known as *depth hoar*. The process also produces hoar frost, the ice equivalent of dew.

(d) The compaction of snow to form GLACIER ice involves a number of metamorphic processes whose overall effect is to increase the crystal size and eliminate air passage. Snow which has survived a summer melt season and begun this process of transformation is known as FIRN. When consolidation has proceeded sufficiently far to isolate the air into separate bubbles then the firn becomes glacier ice. Carbon dioxide ice occurs extra-terrestrially and is of geomorphological importance, e.g. on Mars (Lucchitta 1981).  DES

**Reading and Reference**
LaChapelle, E.R. 1973: *Field guide to snow crystals*. 3rd edn. Washington: University of Washington Press. · Lucchitta, B.K. 1981: Mars and earth: comparison of cold-climate features. *Icarus* 45, pp. 264–303. · Paterson, W.S.B. 1981: *The physics of glaciers*. Oxford and New York: Pergamon.

**ice age** A period in the earth's history when ice sheets were extensive in mid and high latitudes. Such conditions were often accompanied by the widespread occurrence of SEA ICE, PERMAFROST, mountain glaciers in all latitudes and sea level fluctuations.

Ice ages have affected the earth on many occasions. There are records of major glaciations in the Precambrian (on several occasions), the Eocambrian (650–700 million years ago), the Ordovician (c.450 million years ago), the Permo-Carboniferous (250–300 million years ago) and during the Cainozoic (the past 15 million years or so). On occasions the evidence of even the ancient ice ages is splendidly clear and ice directions can be inferred from striations and grooves, basal temperatures and basal water conditions from the type of TILL and sea level fluctuations from rhythmic marine sediments. The freshness of the striations cut by the Ordovician ice sheet in what is now the Sahara is quite stunning to the visitor.

Within an ice age there are sharp fluctuations of ice extent. During the Cainozoic, mid-latitude ice sheets built up on many occasions (see GLACIAL) only to disappear during interglacials. The fluctuations within the ice age can be related to cyclic variations in the amount of solar radiation received by the earth. Cycles of varying radiation receipt were postulated by Milankovitch (see MILANKOVITCH HYPOTHESIS) in 1924 and related to variation in the earth's orbital eccentricity, tilt and precession. Evidence from deep sea cores reveals a similar cyclic variability of ice build-up and decay (Hays *et al.* 1976) and current research is trying to understand the link between the two.

It is not clear why ice ages have occurred during part of the earth's history only. Perhaps an ice age needs a continent to be positioned in an approximately polar position so that ice can build up permanently and chill the earth as a whole.  DES

**Reading and Reference**
Hays, J.D., Imbrie, J. and Shackleton, N.J. 1976: Variations in the earth's orbit; pacemaker of the Ice Ages.

*Science* 194. 4270, pp. 1121–32. · Imbrie, J. and Imbrie, K.P. 1979: *Ice ages*. London: Macmillan. · John, B. ed. 1979: *The winters of the world*. Newton Abbot: David & Charles.

**ice blink**   A mariners' term for the white glare on the underneath of clouds indicating the presence of pack ice or glacier ice.

**ice cap**   A dome-shaped GLACIER with a generally outward and radial flow of ice. The difference between an ice cap and an ice sheet is normally taken to be one of scale with the former being less than 50,000 km$^2$ in area and the latter larger. The marginal regions of the ice cap may be drained by OUTLET GLACIERS which flow beyond the ice cap in rock-walled valleys.   DES

**ice contact slope**   A slope, formerly banked up against an ice mass, which has experienced slumping as a result of ice melting. Such slopes are commonly associated with KAME TERRACES and ESKERS.   DES

**ice core**   A column of ice extracted by drilling into an ice sheet. An ice core is typically of the order of 10 cm in diameter and up to 3000 metres long. Since the 1960s, many have been taken from parts of the Greenland and Antarctic ice sheets for the purpose of deriving information about past climates. Glacier ice accumulates in annual layers and so increasing depth down a core corresponds to increasing age. By measuring properties of ice of different ages throughout the core, it is possible to reconstruct changes over time-scales from annual/ decadal up to a complete glacial–interglacial cycle (i.e. 100,000 years). Analysis of the data obtained from ice cores has provided an uninterrupted record of change that has revolutionized our understanding of rates of climate change and the interactions between components of the climate system.

Air bubbles within the ice provide samples of the atmosphere at the time they were formed and so can be used for measurements of past concentrations of greenhouse gases such as carbon dioxide and methane. The relative abundance of small traces from atmospheric fallout such as wind-blown dust, sea salt, volcanic ash, and pollen can be used to provide evidence of particular events such as volcanic eruptions or aeolian activity, and can also inform about past wind directions and strength. The OXYGEN ISOTOPE ratio ($O_{18}/O_{16}$) of ice can be used as an estimate of global ice volumes and temperatures in a similar way to the methods used in ocean cores. Measured or observed annual ice increments can be used to count down from the surface to derive ages and information about relative changes in snow accumulation and ablation.   CDC

**Reading**
Lowe, J.J., and Walker, M.J.C. 1997: *Reconstructing quaternary environments*. Harlow: Longman. · Patterson, W.S.B. 1994: *The physics of glaciers*. Trowbridge: Pergamon.

**ice cored moraine**   A ridge or spread of glacial rock debris which contains a buried ice core. Protected from melting by the overlying debris, the ice core can survive for considerable periods of time.   DES

**ice dam**   A blockage of drainage caused by ice which leads to periodic and/or rapid fluctuations in meltwater drainage. Large ice dams occur subglacially and are particularly catastrophic. The drainage outbursts are usually triggered when the glacier internal drainage network developing during the ablation season taps the subglacial lake. An initially small outflow melts open a very large passage in a matter of hours. The Icelandic word JÖKULHLAUP is often used to describe the increased discharge associated with the breaching of a subglacial ice dam.

In High Arctic regions streamflow occurs in late spring when valleys are choked with snow. Channel development begins with a saturation of the valley snowpack and water movement within or through the snow. The ponding and subsequent release of water behind snow dams formed by drifts is a common occurrence.   HMF

**Reading**
Sugden, D.E. and John, B.S. 1976: *Glaciers and landscape*. London: Edward Arnold. Esp. pp. 295–7. · Woo, M-K. and Sauriol, J. 1980: Channel development in snow-filled valleys, Resolute, NWT, Canada. *Geografiska annaler* 62A, pp. 37–56.

**ice dome**   A term used to describe the main form of an ice sheet or ice cap and to distinguish it from streaming flow associated with OUTLET GLACIERS.   DES

**ice fall**   A heavily crevassed area of a glacier associated with flow down a steep rock slope. The zone is one of EXTENDING FLOW and is marked by arcuate rotational slumps.   DES

**ice field**   An approximately level area of ice which is distinguished from an ice cap because its surface does not achieve a dome-like shape and because flow is not radial outwards.   DES

**ice floe**   A piece of floating sea or lake ice which is not attached to the land. In the Arctic and Antarctic ice floes are commonly from tens

Immense ice floes photographed from the mast-head of the Imperial Trans-Antarctic Expedition, 1914–16.

of metres to several kilometres across and 2–3 m thick. Some ice floes form in one winter and melt the following summer. Where they survive from one year to the next they are known as multi-year ice and tend to be tougher and, in places where they have been rafted on top of one another or crushed together, thicker. (See also SEA ICE.) DES

**Reading**
Nansen, F. 1897: *Farthest north*. 2 vols. London: Constable.

**ice flow** The movement of ice by internal deformation or BASAL SLIDING. Most studies of ice flow concern glaciers and relatively little is understood, for example, about the flow of debris-rich ground ice in permafrost areas.

A glacier flows in response to shear stresses set up in the ice mass by the force of gravity. These vary according to the thickness of the glacier and its surface slope and can be calculated from the equation: $\tau = \rho gh \sin \alpha$ where $\tau$ is the shear stress, $\rho gh$ is the weight of overlying ice and $\alpha$ is the slope of the ice surface. Internal deformation of the glacier takes place mainly through the action of CREEP which is modelled for glaciers by GLEN's LAW. This shows that glacier flow is highly sensitive to an increase in shear stress and this is why most internal deformation occurs at the bottom of a glacier. Near the glacier bottom bedrock obstacles set up locally high stresses and enhanced basal creep is the mechanism by which they are passed. In

situations where the base of the glacier is at the pressure melting point and a film of water exists at much of the ice–rock interface, basal sliding contributes to glacier flow and rates may equal or exceed the rate of internal deformation for the glacier as a whole. Where glaciers overlie saturated soft sediments it has been discovered that deformation of the sediment may contribute to the forward movement of the glacier (Boulton 1979).

Most glaciers flow at the rate of 10–100 m per year but where rates of basal sliding are high, as in the case of SURGING GLACIERS, flow rates may exceed several kilometres per year. DES

**Reading and Reference**
Boulton, G.S. 1979: Processes of glacier erosion on different substrata. *Journal of glaciology* 23.89, pp. 15–38. · Paterson, W.S.B. 1981: *The physics of glaciers*. Oxford and New York: Pergamon.

**ice fog** A suspension of minute ice crystals in the air reducing visibility at the earth's surface. The optimum conditions for ice fog build-up are temperatures below $-30\,°C$ and a supply of water vapour. Such conditions are common in and around Arctic settlements in winter, where an inversion causes low air temperatures and vehicles and heating plants contribute more water vapour than can be absorbed without condensing. Such fogs are associated with severe pollution. DES

**Reading**
Benson, C.S. 1969: The role of air pollution in Arctic planning and development. *Polar record* 14.93, pp. 783–90.

**ice front** The vertical cliff forming the seaward edge of an ice shelf or floating glacier.

**ice jam** A blockage caused by the accumulation of pieces of river ice or sea ice in a narrow channel.

**ice rind** A stage in the growth of sea ice when the accumulating ice crystals coagulate to form a brittle skin.

**ice segregation** Sometimes called ice lensing, ice segregation is the process by which water freezes, thereby causing heave of the ground surface. Primary (i.e. capillary) and secondary heave can be distinguished. In primary heave, the critical conditions for the growth of segregated ice are:

$$P_i - P_w = \frac{2\sigma}{r_{iw}} < \frac{2\sigma}{r}$$

where $P_i$ is the pressure of ice, $P_w$ is the pressure of water, $\sigma$ is the surface tension of ice to water, $r_{iw}$ is the radius of the ice–water interface, and $r$

is the radius of the largest continuous pore openings.

Secondary heave is not clearly understood but may occur at temperatures below $0\,°C$ and at some distance behind the freezing front. Porewater expulsion from an advancing freezing front is another mechanism for ice segregation, especially massive ice bodies, provided that porewater pressures are adequate to replenish groundwater that is transformed into ice.    HMF

**Reading**
Miller, R.D. 1972: Freezing and heaving of saturated and unsaturated soils. In *Frost action in soils*. National Academy of Sciences–National Academy of Engineering highway research record 393. · Washburn, A.L. 1979: *Geocryology: a survey of periglacial processes and environments*. New York: Wiley. Esp. pp. 68–70. · Williams, P.J. 1977: General properties of freezing soil. In P.J. Williams and M. Fremond eds, *Soil freezing and highway construction*. Ottawa: Paterson Centre, Carleton University.

**ice sheet**    A large dome-shaped glacier (over $50,000\,km^2$ in area) with a generally outward and radial flow of ice. On a continental scale such ice sheets can exceed a thickness of $4\,km$, as they do in Antarctica. A simple model can be used to approximate the surface slope of an ice sheet, assuming the ice is perfectly plastic. The profile equation is:

$$h = 3.4(L - x)^{\frac{1}{2}}$$

where $L$ is the distance from ice sheet centre to edge and $h$ is the ice thickness at a distance $L - x$ from the edge and both quantities are in metres (Paterson 1981).    DES

**Reading and Reference**
Nansen, F. 1890: *The first crossing of Greenland*. London: Longman. · Paterson, W.S.B. 1994: *The physics of glaciers*. Trowbridge: Pergamon. P. 163. · Robin, G. de Q., Drewry, D.J. and Meldrum, D.T. 1977: International studies of ice sheet and bedrock. *Philosophical transactions of the Royal Society London* Series B. 279, pp. 185–96.

**ice shelf**    A floating sheet of ice attached to an embayment in the coast. It is nourished by snow falling onto its surface and by land-based glaciers discharging into it. The seaward edge is a sheer cliff rising some $30\,m$ above sea level. The shelf surface is virtually flat although it rises slightly inland with an increase in ice thickness in the same direction. Freed of basal friction associated with land-based glaciers, ice velocities are high and commonly $1-3\,km\ year^{-1}$. Periodically calving removes huge tabular ICEBERGS from the front. Thirty per cent of the Antarctic coastline is fringed by ice shelves. The Ross Ice Shelf extends $900\,km$ inland and is $800\,km$ across.    DES

**Reading**
Robin, G. de Q. 1975: Ice shelves and ice flow. *Nature* 253, pp. 168–72. · Thomas, R.H. 1974: Ice shelves: a review. *Journal of glaciology* 24.90, pp. 273–86.

**ice stream**    A relatively narrow zone of swiftly moving ice within an ice sheet or ice cap, often bordered by spectacular crevasses. The high velocities probably reflect high sliding velocities associated with a basal water film. Ice streams often form the heads of OUTLET GLACIERS.    DES

**Reading**
Drewry, D.J. 1983: Antarctic ice sheet: aspects of current configuration and flow. In R. Gardner and H. Scoging eds, *Megageomorphology*. Oxford: Oxford University Press.

**ice wedge**    A massive, generally wedge-shaped, ground ice body composed of foliated or layered, vertically oriented ice which extends below the permafrost table. Large ice wedges may be $1-2\,m$ wide near the top and extend downwards for $8-10\,m$. They form in cracks in polygonal patterns originating in winter by thermal contraction of the ground into which water from melting snow penetrates in the spring. Repeated annual contraction and subsequent cracking of the ice in the wedge, followed by freezing of water in the crack, lead to an increase in width and depth of the wedge. Ice wedges require PERMAFROST for their formation and existence. They often give a distinct polygonal micro-relief to the tundra surface.    HMF

**Reading**
Mackay, J.R. 1974: Ice-wedge cracks. Garry Island. NWT. *Canadian journal of earth sciences* 11, pp. 1336–83.

**iceberg**    A floating mass of ice which has been either calved or torn off from the snout of a glacier or the floating margin or ice shelf at the edge of an ice sheet. The shape and size of an iceberg is related to the dynamics and morphology of the glacier or ice sheet, particularly the distribution of crevasses. Heavily-crevassed surging glaciers typically produce small icebergs (termed 'brash ice' if $< 2\,m$ across or 'bergy bits' if $< 10\,m$ across) whilst large outlet glaciers and ice shelves produce large 'tabular icebergs' ($> 500\,m$ across). Exceptionally large icebergs are termed 'ice islands'.

Icebergs are most common in polar regions but, if either exceptionally large or carried by strong ocean currents, may drift as far south as Newfoundland or as far north as the Falkland Islands. Icebergs are important media for the transportation and deposition of glacially eroded debris from polar regions, with sediment deposited when icebergs melt, overturn or are grounded in shallow water. Grounded icebergs are also erosional agents, creating flat-bottomed troughs or furrows where they are dragged along lake or sea beds. Increased calving of icebergs from Arctic and Antarctic ice shelves in recent years is widely considered to be a direct result of global climatic change.                    DJN

**icing**   A mass of surface ice formed during the winter by successive freezing of sheets of water that may seep from the ground, from a river or from a spring. Icings are widespread in periglacial areas.                    DES

**igneous rock**   Rock formed when molten material, magma, solidifies, either within the earth's crust or at the surface.

**illuviation**   The precipitation and accumulation of material within the B horizon of a soil after the material has been leached from the surface or overlying soil horizons.

**imbrication**   A regular overlapping, or shingling, of non-spherical sedimentary particles as a result of their deposition by fluids. Imbrication, one form of an ANISOTROPHIC sedimentary FABRIC, is most commonly associated with pebble and cobble deposits, but may occur in sand and boulder deposits. The plane described by the long (A) and intermediate (B) axes of an imbricate clast tends to DIP toward the flow at angles less than 20°, and this affords potential palaeoflow evidence in the stratigraphic record. Imbricated deposits represent a relatively stable bed configuration, and may result in either bed ARMOURING or bed pavement.

Preferential orientations of the long axes may also be palaeoflow indicators. The A axis tends to be aligned perpendicular to the flow when the FROUDE NUMBER is low and transport is mainly as BEDLOAD. The A axis tends to assume a flow-parallel orientation with higher Froude numbers or when there is substantial transport via SALTATION. Imbrication is common on gravel beaches, on bars and outwash fans in BRAIDED RIVERS, and glacial TILLS, and is often a key to interpreting FACIES. The term imbrication is also used to describe overlapping tabular THRUST sheets.                    DJS

**Reading**
Leeder, M.R. 1982: *Sedimentology: process and product.* London: George Allen & Unwin. · Reineck, H.-E., and Singh, I.B. 1986: *Depositional sedimentary environments.* 2nd edn. Berlin: Springer-Verlag. · Rust, B.R. 1972: Pebble orientation in fluvial sediments. *Journal of sedimentary petrology* 42, pp. 384–8.

**impermeable**   Having a structure or texture which does not allow the movement of water

through a rock or soil material under the natural conditions in the groundwater zone.

**impervious**   Impermeable. Having a texture which does not allow the movement of water, oil or gas through a rock or soil material. Under certain conditions a rock may have an impervious texture though a stratum of the rock is permeable owing to joints and fractures.

**in and out channel**   The name given to a small, discontinuous channel produced by meltwater flow from a glacier onto the adjacent hillside.

**incised meander**   See MEANDERING.

**inconsequent stream**   A stream not apparently related to landsurface features or major geological controls, but following minor surface features without being developed into an organized pattern overall. The term was used by G.K. Gillbert (1877). It became largely superseded by the synonymous 'insequent' of W.M. Davis (1894) and is now used hardly at all.                    JL

**Reading and References**
Davis, W.M. 1894: Physical geography as a university study. *Journal of geology* 2, pp. 66–100. · Davis, W.M. 1899: The geographical cycle. *Geographical journal* 14, pp. 481–504. · Gilbert, G.K. 1877: *Report of the geology of the Henry Mountains.* Washington: US Geographical and Geological Survey of the Rocky Mountain Region.

**indeterminacy**   The situation in which the results of an investigation remain open to two or more conflicting interpretations. Indeterminacy is a common fact of life in many investigations of natural phenomena. It is not unique to physical geography, nor even to the earth sciences, but it is unusually common throughout the historical sciences in general (see Simpson 1963). Given the potential problem, the most useful method for proceeding seems to be to recognize this at the outset of any study and to take the steps outlined above so that the nature and extent of any indeterminate solution are clear to future workers.                    BAK

**Reading and Reference**
Chamberlin, T.C. 1890: The method of multiple working hypotheses. *Science* old series 15.92. Reprinted 1965, new series 148, pp. 754–9. · Chorley, R.J. and Kennedy, B.A. 1971: *Physical geography: a system approach.* London: Prentice-Hall International. · Simpson, G.G. 1963: Historical science. In C.C. Albritton ed., *The fabric of geology.* Stanford: W.H. Freeman. Pp. 24–48.

**inductive**   Science based on the principle that accumulated knowledge allows generalizations to be developed which provide theories or laws. In its purest form, data would be collected without the collector (scientist) prejudging what that data might demonstrate in relation to the phenomena being assessed. This may be believed to lead to clear objectivity in analysis, but in reality the distinction between observation and theory that this approach requires is not clear cut (Haynes-Young and Petch 1986). A DEDUCTIVE approach to scientific research is more widely employed in response to limitations with the inductive method (Popper 1972). DSGT

**Reading and References**
Haines-Young, R.H. and Petch, J.R. 1986: *Physical geography: its nature and methods*. London: Paul Chapman. · Popper, K.R. 1972: *The logic of scientific discovery*. London: Hutchinson.

**induration** The process of hardening through cementation, desiccation, pressure or other causes, applied particularly to sedimentary materials.

**infiltration** The process by which water percolates into the soil surface. Two main zones can be observed in the soil when infiltration is proceeding at its maximum rate, that is from a surface which is saturated with a thin layer of standing water. There is an upper transmission zone with an almost constant moisture content close to saturation. Below this is a sharp wetting front in which the moisture content declines rapidly towards its pre-infiltration value. Within the transmission zone, flow is driven mainly by gravitational forces. Across the wetting front there is a strong hydraulic or tension gradient, tending to push the water into the dryer soil in the way that water is drawn into a fine capillary tube. This hydraulic gradient advances the wetting front down into the soil and so allows additional water to infiltrate at the surface. The rate at which water can infiltrate under these ideal circumstances is called the infiltration capacity and it decreases as the wetting front advances deeper into the soil.

Infiltration capacity may be expressed either in terms of time since the process began, or in terms of current moisture storage. The main advantage of the storage expressions is that they may remain valid during infiltration at less than the capacity rate, as commonly occurs at the start of a rainstorm when infiltration capacity tends to be very high. One example of a widely used empirical infiltration equation is that used in the HORTON OVERLAND FLOW MODEL. Another equation with a better theoretical basis was put forward by Philip (1957/8):

$$f = A + Bt^{\frac{1}{2}}$$

where $f$ is the infiltration capacity at time $t$ and A and B are constants which depend on the soil and its initial moisture distribution.

In this expression the constant term A mainly represents the steady rate of infiltration under gravitational potential (i.e. the weight of the water) and the time-dependent term is due to the hydraulic potential gradient at the advancing wetting front. It may be seen that the infiltration capacity is initially very high, and decreases steadily towards a constant rate, which is usually achieved within one to two hours.

An example of a storage-based infiltration equation is the Green and Ampt (1911) equation:

$$f = A + C/S$$

where $f$ and A are as above, $S$ is a soil water storage value, and C is a constant of the soil and its initial moisture.

In the original formulation of this expression the storage $S$ was the total amount of water infiltrated since the start, but an alternative is to budget $S$ as representing a store of infiltrated water which leaks at steady rate A. This version has the advantage that if converted to a time-dependent form under conditions of surface ponding, it is exactly equivalent to the Philip equation above (with $B = (2C)^{\frac{1}{2}}$). During a rainstorm of constant or varying intensity, this kind of storage-based model allows estimation of the infiltration capacity at any time in terms of the water that has actually entered the soil previously, which may have been at any rate less than (or equal to) the current infiltration capacity.

Infiltration may be likened to the process of pouring water into a bottle: it may fail to get in either because it is being poured in too fast or because the bottle is already full. On a hillside, saturated THROUGHFLOW may increase downslope or in areas of flow concentration until the SATURATION DEFICIT is zero: in other words the bottle may be full or almost full. A second criterion for infiltration may therefore also be expressed in storage terms in that ponding will occur when soil water storage reaches a critical level, $S_c$.

In areas of high rainfall intensities and low infiltration capacity, infiltration capacity is commonly exceeded and the Horton overland flow model is generally applicable to the estimation of streamflow. This includes areas which are naturally or artificially clear of dense vegetation, that is to say semi-arid areas and (seasonally) cultivated fields, as will be seen below. In areas of low rainfall intensity and/or dense vegetation cover, including forested areas and much of the humid-temperate zone, infiltration capacities are seldom exceeded, and PARTIAL AREA MODELS, which estimate the areas of saturation and the volumes of THROUGHFLOW, are generally more appropriate for estimating streamflow volumes and HILLSLOPE FLOW PROCESSES.

Rates of infiltration are usually compared in terms of the steady long-term rate (A in the equations above). This rate responds to some extent to soil texture, typically ranging from 0–4 $mm^{-1}h^{-1}$ on clays through to 3–12 $mm^{-1}h^{-1}$ on sands where the soils are initially wet and unvegetated. Vegetation cover and protection by coarse particles shields the surface from raindrop impact which is otherwise liable to break down the top layer of soil peds and pack the resulting soil grains down into the next layer as a thin, impermeable crust. On a crusted surface infiltration capacity is commonly very low except in extremely intense storms which may break the crust: where crusting is prevented, capacities are much higher. Within the soil, structure has a greater influence on infiltration capacity than does texture, and vegetation and its associated organic soil have a strong influence on soil structure. Thus vegetation cover increases infiltration capacity in two ways, so that steady rates may be 50–100 $mm^{-1}h^{-1}$ under a good cover, compared to less than 10 mm on bare crusted soil.

Where soil structural voids are marked and soil textural pores fine, as for example for a cracked clay soil, soil water may bypass much of the soil mass. At the surface soil peds allow infiltration at maximum capacity, and excess rainfall overflows down the structural voids. In the largest voids water flows as a film down each wall and infiltrates into additional peds below the surface. The advance of water down each void is limited by this infiltration into the walls. In most cases rainfall enters the soil peds within a few tenths of a metre of the surface, but along a highly convoluted wetting front which follows the geometry of the largest voids. The resulting pattern may then be conceived either as a greater average depth of penetration of infiltrated water; or as shallow penetration to match the current rainfall intensity. In extreme cases water may bypass the soil as a whole and make direct contact with groundwater via bedrock fissures, etc. It should be pointed out that almost all real soils show some structural voids, but that local bypassing is only important where there is a very marked contrast between the textural and structural pore size distributions.

The importance of infiltration in physical geography lies in its role within the catchment and hillslopes hydrological cycle in partly determining the flow routes of precipitation to the streams and so determining the timing of streamflow response. Infiltration also plays an important part in separating groups of hillslope processes. Water which travels as overland flow takes little part in supplying soil moisture for plant growth and, because it comes into little and rather brief physical contact with mineral soil, picks up little solute load except from the litter layer. OVERLAND FLOW therefore tends to dilute stream solute concentrations, which therefore tend to be lower during intense storms when overland flow is greatest. Overland flow also erodes and transports all surface wash/soil erosion. The infiltrated water flows through the soil as throughflow which is able to carry negligible amounts of suspended material, but is in intimate contact with mineral soil grains from which it is effective in leaching solutes and promoting chemical weathering. The infiltrated water is also responsible for providing water for plant growth and for establishing patterns of hydraulic potential which have a powerful influence on slope stability in the context of mass movement. In other words the process of infiltration is a critical regulator of the landscape system in both the short, hydrological term and in longer erosional time spans.          MJK

**Reading and References**
Goudie, A. ed. 1990: *Geomorphological techniques*. 2nd edn. London: Unwin Hyman. · Green, W.H. and Ampt, G.A. 1911: Studies on soil physics I: the flow of air and water through soils. *Journal of agricultural science* 4.1, pp. 1–24. · Horton, R.E. 1945: Erosional development of streams and their drainage basins: hydrophysical approach to quantitative morphology. *Bulletin of the Geological Society of America* 56, pp. 275–370. · Knapp, B.J. 1978: Infiltration and storage of soil water. In M.J. Kirkby ed., *Hillslope hydrology*, Chichester: Wiley. Pp. 43–72. · Philip, J.R. 1957/8: The theory of infiltration. *Soil science* 83, pp. 345–57 and 435–48; 84, pp. 163–77 and 257–64; 85, pp. 278–86 and 333–7.

**inflorescence**   The flowering shoot of a plant.

**influent**   Either a tributary stream or river, or a term applied to a stream which supplies water to the groundwater zone.

**infrared imagery**      See THERMAL INFRARED LINESCANNER.

**infrared thermometer/thermometry**   Bodies reflect, emit and sometimes transmit radiation falling upon them. The efficiency of the surface in generating contributions are known, respectively, as reflectivity, emissivity and transmissivity. If the latter is zero (i.e. an opaque material) then reflectivity + emissivity = 1. The wavelengths near, but longer than, the visible spectrum are termed infrared and constitute most of the radiation from a 'hot' body. This radiation can be detected by an instrument which does not require contact with the body; when allowance is made for the emissivity of the surface its temperature can be measured. A THERMAL INFRARED LINESCANNER is a more complex instrument used in remote sensing.    WBW

**ingrown meander**   See MEANDERING.

**inheritance**   See DARWINISM; EVOLUTION.

**inlier**   An outcrop of rock which is completely surrounded by younger formations, frequently the result of erosion of the crest of an anticline.

**inselberg**   (German for island hill) A general class of large residual hill which usually surmounts an eroded plain. Small residual rock masses tend to be called TORS, large domed residuals tend to be called domed inselbergs or bornhardts, while large accumulations of boulders in the form of a hill are called koppies.

Inselbergs of resistance are those that are left as prominent landforms as a result of their superior resistance (brought about by the jointing density of the rock or its mineralogical composition), while inselbergs of position remain as prominent features because they are on divides farthest from lines of active erosion.

There has been considerable debate as to inselberg origin, and three main mechanisms have been proposed: that they are produced by scarp retreat across bedrock; that they are a result of scrap retreat across deeply weathered rocks; or that they result from differential weathering followed by stripping of the regolith.

Inselbergs occur in a wide range of rock types, but the most common lithologies appear to be sandstones and conglomerates (e.g. Ayers Rock in Australia or Meteora in Greece) or gneisses and granites, especially those that have widely spaced joints and a high potassium content.    ASG

**Reading**
Bremer, H. and Jennings, J. eds 1978: Inselbergs. *Zeitschrift für Geomorphologie Supplementband* 31. · King, L.C. 1975: Bornhardt landforms and what they teach. *Zeitschrift für Geomorphologie*, NF 19, pp. 299–318. · Pye, K., Goudie, A.S. and Thomas, D.S.G. 1984: A test of petrological control in the development of bornhardts and koppies on the Matopos Batholith, Zimbabwe. *Earth Surface processes and landforms* 9, pp. 455–67. · Thomas, M.F. 1974: *Tropical geomorphology*. London: Macmillan. · Twidale, C.R. 1982: The evolution of bornhardts. *American scientist* 70, pp. 268–76.

**insequent stream**   A drainage network that has developed as a result of factors which are not determinable.

**insolation**   a term used in two senses:

1   The intensity at a specified time, or the amount in a specified period, of direct solar radiation incident on unit area of a horizontal surface on or above the earth's surface.

A spectacular granite inselberg, the Spitzkoppje, rising from the granite plains of the Namib Desert in central Namibia.

2 The intensity at a specified time, or the amount in a specified period, of total (direct and diffuse) solar radiation incident on unit area of a specified surface of arbitrary slope and aspect.

In general, insolation depends on the solar constant, calendar date, latitude, slope and aspect of surface, and degree of transparency of the atmosphere.     JGL

**insolation weathering**   (or heating and cooling weathering) The disintegration of rock in response to temperature changes setting up stresses. Early travellers heard, or reported they heard, sounds like pistol shots as rocks cooled in the evening, and thus arose a classic process in desert geomorphology. Experimental work in the twentieth century and recognition that weathering appears to be far more effective in the presence of moisture have led to pure insolation weathering being related to a position of relatively lowly importance (Schattner 1961).     ASG

**Reference**
Schattner, I. 1961: Weathering phenomena in the crystalline of the Sinai, in the light of current notions. *Bulletin of the Research Council of Israel* 10G, pp. 247–65.

**instantaneous unit hydrograph**   (IUH) See UNIT HYDROGRAPH.

**intact strength**   The strength of a rock sample which is free of large-scale structural discontinuities such as joints, fissures or foliation partings. It is usually expressed as a measure of unconfined compressive strength.     MJS

**intensity of rainfall**   The rate at which rain falls in unit time, usually expressed in mm h$^{-1}$. It is measured using autographic (or recording) rain gauges and is important because of its impact on run-off characteristics. More intense rainfall will produce a more peaked run-off response. Rainfall intensity is inversely related to rainfall duration at a site and the intensity–duration–frequency analysis of rainfall records provides a good indication of the potential run-off characteristics of an area. (See also DEPTH–DURATION CURVE; RAIN GAUGE.)     AMG

**Reading**
Dunne, T. and Leopold, L.B. 1978: *Water in environmental planning*. San Francisco: W.H. Freeman.

**inter-annual variability**   This measures the likely probable departure from the annual mean value of rainfall in any one year, and is usually expressed as a percentage. Higher values are experienced in drier environments, where the likelihood of DROUGHTs is considerable.     DSGT

**interbasin transfers**   A method of water supply whereby the natural or regulated flow from one river system is transferred usually by pumping to another river system. This method of water supply is likely to increase in future decades and many schemes have already been evaluated such as the possibility of the southward diversion of flow from Siberian rivers such as the Lena and the Ob and the transfer of water from the Chiang Jiang (Yangtze) to the Huang He (Yellow River) basin in China. Developments have been approached cautiously because of the possible ENVIRONMENTAL IMPACTS of such interbasin transfers and it is known that in addition to changes of the water balance, especially in the evaporation term, there may also be substantial changes in the morphology and ecology of the river channels themselves because of the increase or decrease in streamflows.     KJG

**interception**   The process by which precipitation is trapped on vegetation and other surfaces before reaching the ground. Interception loss is the component of intercepted precipitation which is subsequently evaporated, although this is also frequently described as interception. The character of the intercepting surfaces has a major impact on the amount of precipitation that is intercepted and then lost through evaporation. (See also STEM FLOW and THROUGHFLOW.)     AMG

**Reading**
Courtney, F.M. 1981: Developments in forest hydrology. *Progress in physical geography* 5, pp. 217–41. · Crockford, R.H. and Richardson, D.P. 1990: Partitioning of rainfall in a eucalyptus forest and pine plantation in southeastern Australia (four papers). *Hydrological processes* 4, pp. 131–88. · Durocher, M.G. 1990: Monitoring spatial variability of forest interception. *Hydrological processes* 4, pp. 215–29.

**interception capacity**   The maximum volume or depth of water that can be retained on a plant canopy exposed to rain, generally under windless conditions. Interception capacity is determined in various ways, often by noting the weight gain on specimens that are uprooted for the purpose. It may also be estimated by measuring rainfall above the canopy, STEM FLOW, and THROUGHFALL beneath the canopy, and finding the intercepted amount from {rainfall – (stemflow + throughfall)}. For many plants, interception capacity amounts to a few mm of water depth over the projected area of the canopy. The value of interception capacity is influenced by leaf shape, leaf surface texture, leaf geometry, and the rigidity or flexibility of the foliage under the imposed weight of water. Interception capacity may be reached in large storms but not in small; overall, canopy interception, followed by evaporative return of the water to the atmosphere, consumes 10–20 per cent of the rainfall that falls over a dense forest. DLD

**Reading**
Aston, R.R. 1979: Rainfall interception by eight small trees. *Journal of hydrology* 42, pp. 383–96.

**interflow** A component of streamflow which responds to rainfall more slowly than surface run-off and more rapidly than BASE FLOW. In the HORTON OVERLAND FLOW MODEL streamflow was initially separated into overland flow and groundwater flow, and various procedures were used for partitioning the stream hydrograph into these components. Interflow was originally introduced as an intermediate hydrograph component which fell between the other two. It was considered to represent groundwater which re-emerged as overland flow; or else it was considered to be shallow groundwater flow. Some literature now uses the term interflow interchangeably with subsurface soil flow or THROUGHFLOW, but its original physical identification was rather tenuous. MJK

**Reading**
Linsley, R.K., Kohler, M.A. and Paulhus, J.L.H. 1949: *Applied hydrology*. New York: McGraw-Hill.

**interfluve** The area of high ground which separates two adjacent river valleys.

**interglacial** A phase of warmth between glacials when the great ice sheets retreated and decayed, and tundra conditions were replaced by forest over the now temperate lands of the northern hemisphere. The Holocene is an interglacial, but some of the Quaternary interglacials may have been slightly warmer than today. Iversen (1958) identified various stages in a glacial–interglacial cycle in north-western Europe. Development begins with a *protocratic* phase, characterized by rising temperature, raw, basic or neutral mineral soils, favourable light conditions, and a pioneer vegetation of small plants, with exacting requirements for both nutrients and light. In the following *mesocratic* phase, comprising the climax of the interglacial, there are maximum temperatures, brown forest soils, mull plants, dense climax forest, and a vegetation that while still demanding nutrients is tolerant of shade. The *oligocratic* phase arises as a result of soil development, and involves more acid soils and more open vegetation. The *telocratic* phase marks the end of interglacial forest development. Heaths expand, and bogs develop in response to climatic deterioration. The climatic deterioration culminates in a *cryocratic*

phase when cold conditions and soil instability are hostile to tree growth. ASG

**Reference**
Iversen, J. 1958: The bearing of glacial and interglacial epochs on the formation and extinction of plant taxa. In O. Hedberg ed., *Systematics of today. Acta Universitets Upsaliensis* 1958, pp. 210–15.

**intermittent spring** A natural outflow point of underground water, that sometimes dries up completely (see SPRINGS). Normally such a spring is intermittent because the WATER TABLE upstream of the spring has fallen to or below the elevation of the spring, with the result that the hydraulic gradient leading to the spring is zero. PWW

**intermittent stream** A stream is classified as intermittent if flow occurs only seasonally when the water table is at the maximum level. The drainage network is composed of ephemeral, intermittent, and perennial streams and the network expands during rainstorms and extends to limits affected by antecedent conditions especially antecedent moisture. Flow may occur along intermittent streams for several months each year but will seldom occur when the water table is lowered during the dry season. KJG

**interpluvial** A relatively dry phase interspersed with the wetter phases (pluvials) of the Pleistocene and Holocene. In many parts of the tropics the period at the end of the Late-Glacial Maximum (between *c.*18,000–13,000 years ago) was dry enough to cause lake levels to fall and dune fields to expand. ASG

**Reading**
Goudie, A.S. 1983: The arid earth? In R.A.M. Gardner and H. Scoging eds, *Mega-geomorphology*. Oxford: Oxford University Press.

**interrill flow** The overland flow that moves as a thin layer, perhaps with some deeper and faster flow filaments within it, but which is not yet organized into small channels (RILLS) that are excavated by the erosive power that results from channelization. The zones of a soil surface carrying interrill flow expand and contract dynamically with run-off rate, as rills too enlarge or contract. Thus, the division cannot be applied rigorously in the spatial sense, and is rather a distinction based on erosional processes. Interill areas are the primary sediment sources, where raindrop splash is the major detachment agency. Rills, in contrast, are principally zones of transport, and receive their sediment load from the influx of interrill flow. DLD

271

**Reading**
Sharma P.P. 1996: Interrill erosion. In M. Agassi ed. *Soil erosion, conservation, and rehabilitation*. New York: Marcel Dekker. Ch. 7, pp. 125–52.

**interstadial**   There is as yet no universally accepted definition which differentiates an interstadial from an interglacial. However, it may be defined as a relatively short-lived period of lesser glaciation and relatively greater warmth and thermal improvement during the course of a major glacial phase. During such phases conditions were not of sufficient magnitude and/or duration to permit the development of temperate deciduous forest of the full interglacial type. Information about interstadial environment conditions have been obtained and assessed using faunal and floral evidence, the timing of which have been obtained mainly from radiocarbon dating where this technique permits.   AP

**Reading**
Goudie, A.S. 1992: *Environmental change*. 3rd edn. Oxford: Clarendon Press. · Lowe, J.J. and Walker, M.J.C. 1984: *Reconstructing Quaternary environments*. London: Longman.

**interstices**   Voids such as pores and fissures that occur within a rock. They can be classified according to their origin, shape and size. Primary interstices are those formed when the rock was created, such as intergranular pores in a sandstone, whereas secondary interstices are the result of later tectonic activity or weathering, such as fault planes and voids left by the differential corrosion of minerals. Interstices are most often small and interconnected, although large isolated interstices termed *vugs* also sometimes occur. Both small primary and large secondary interstices can be present simultaneously in a rock. For example, well-jointed and bedded sedimentary rocks, especially if rich in carbonate, may have a geometrical lattice of large interstices produced by solution along the fissures system and a fine intergranular POROSITY within the main body of the rock.   PWW

**intertropical convergence zone**   (ITCZ) A band of nearly continuous low pressure, light and variable winds, high humidity, and intermittent heavy rain showers found near the equator. The name is derived from its location between the tropics of the two hemispheres and from the fact that it represents a narrow zone along which the trade winds of the two hemispheres converge. The ITCZ is often clearly visible on satellite photographs, especially over the oceans. It appears as a narrow, well-defined cloud band. Occasionally two ITCZs will be visible, most frequently in the eastern tropical Pacific Ocean shortly after the equinoxes. At that time the mean position of the ITCZ is near the equator. In the eastern Pacific cool surface water, produced by upwelling, splits the convergence zone into a northern and southern branch.

The ITCZ meanders with both longitude and season. The extreme positions occur in February and August when temperatures are highest in the respective summer hemisphere. In February its location ranges from 18°S over South Africa and Australia to 7°N in the eastern Pacific Ocean and in August from 3°N over India and South-East Asia. The MONSOON rains of the latter region and also of the Sahel region of Africa are triggered by the ITCZ.   WDS

**Reading**
Gedzelmen, S.D. 1980: *The science and wonders of the atmosphere*. New York: Wiley. · Riehl, H. 1979: *Climate and weather in the tropics*. New York: Academic Press.

**intrazonal soil**   A soil group comprising well-developed soils, the main characteristics of which can be attributed more to local factors such as relief, drainage or parent material than to climate factors.

**intrenched meander**   A meander or bend in a river channel that has become incised into the surrounding landscape as a result of local tectonic uplift (see MEANDERING).

**intrinsic permeability**   (or specific permeability) A measure of the capacity of a rock or soil under natural conditions to transmit fluids. It depends upon the physical properties of the porous medium, such as pore size, shape and distribution. It is measured in $m^2$ or darcy units (one darcy is approximately equal to $10^{-8} cm^2$). Intrinsic permeability $k$ is related to hydraulic CONDUCTIVITY $K$ (which takes into account the physical properties of the liquid as well as the rock) as follows:

$$K = \frac{\kappa \rho g}{\mu}$$

where $\rho$ mass density and $\mu$ dynamic viscosity are functions of the fluid alone, and $g$ is acceleration due to gravity (Freeze and Cherry 1979).   PWW

**Reference**
Freeze, R.A. and Cherry, J.A. 1979: *Groundwater*. Englewood Cliffs, NJ: Prentice-Hall.

**introductions, ecological**   Introductions, usually deliberate, of an organism into new regions lying outside the range of its natural occurrence, by which it is hoped to bring about some specific ecological condition or control in the receiving areas or habitats. Although any introduction of a species, accidental or otherwise, into a new environment is likely to have considerable ecological repercussions, the concept of ecological introductions is mostly used to describe carefully planned introductions made with specific ecological intentions in mind. For example, where aquatic weeds are an increasing problem, water authorities may decide to introduce the herbivorous grass carp (*Ctenopharyngdon idella*) to bring them under control. In many instances, the aim is to create or recreate some particularly effective food chain or ecological relationship.

A famous example of such an introduction is afforded by the deliberate spread of an Australian ladybird, *Novius (Vedalia) cardinalis*, to California. Around 1968, the fluted or cottony cushion scale insect (*Icerya purchasi*), another native of Australia, appeared in California to threaten the famous citrus groves of the area. Within a few years, this threat had been overcome by the numerous descendants of the 139 specimens of the ladybird introduced from Australia, where it was a natural enemy of the fluted scale insect and from where it had been introduced to California to control the alien pest. As Charles Elton (1958) has written, this was a 'miracle of ecological healing' in which 'Australia administered the poison, but it also supplied the antidote.' The success was repeated in many other countries, e.g. New Zealand and Egypt, where the fluted scale became a problem. It is a perfect example of an ecological introduction, in which one alien species is controlled by a natural predator from its own ecosystem, thus using and re-establishing a natural ecological chain. Another pest, this time from the plant kingdom, required the introduction of the cinnabar moth, which cleared vast areas of the prickly pear (*Opuntia* species) in Australia.

In most cases of predator control, the number of introductions need not be great, partly because of the speed of breeding, but also because of the fundamental principle of the ecological pyramid, in which predators are virtually always scarcer than their prey. Many ecological introductions, however, fail and some may even go badly wrong, leading to new and more serious problems including ECOLOGICAL EXPLOSIONS, the cure proving worse than the disease. Finally, there have been attempts to reintroduce formerly native animals and plants into their old habitats in order to try to establish the past ecological order, e.g. the wolf, a top carnivore, in certain forest areas in the west of Germany. (See also ALIENS; BIOLOGICAL CONTROL.)   PAS

**Reading and Reference**
Elton, C.S. 1958: *The ecology of invasions by animals and plants*. London: Methuen. · Emden, H.F. van 1974: *Pest control and its ecology*. London: Edward Arnold. · Sam-

ways, M.J. 1981: *Biological control of pests and weeds*. London: Edward Arnold. · Simmons, I.G. 1979: *Biogeography: natural and cultural*. London: Edward Arnold.

**intrusion**   A mass of igneous rock that has penetrated older rocks through cracks and faults before cooling. The process of emplacement of such a mass of rock.

**inversion of temperature**   An increase of temperature with height, the inverse of the normal decrease of temperature with height that occurs in the TROPOSPHERE. Temperature inversion layers are very stable and greatly restrict the vertical dispersion of atmospheric pollutants. They can form in several different ways. (1) Radiative cooling of the air near the ground at night. These inversions are very common on clear nights, but dissipate rapidly after sunrise. (2) Advective cooling of warm air passing over a cold surface. These persistent inversions may be accompanied by thick fog if the air is moist. (3) A cold air mass undercutting a warm air mass along a FRONT. These frontal inversions act as an invisible barrier between the two AIR MASSES. (4) Radiative heating in the upper atmosphere. The STRATOSPHERE and thermosphere are examples of this type of inversion. Thunderstorm clouds and pollutants rarely penetrate far into the stratosphere because of its great stability. (5) Descent and ADIABATIC heating of air from the upper troposphere. These subsidence inversions are most common near and to the east of ANTICYCLONES. They may be very intense and persist for days, trapping noxious pollution in a thin air layer near the ground.   WDS

**Reading**
Battan, L.J. 1979: *Fundamentals of meteorology*. Englewood Cliffs, NJ: Prentice-Hall.

**inverted relief**   The condition resulting from the erosion of areas of high relief, such as anticlines, to produce low-lying areas, such as valleys, which simultaneously results in the originally low-lying inclines becoming hills. Equally the deposition of resistant lag gravels, lava streams or duricrusts in river valleys may cause them to be left upstanding in a subsequent phase of erosion.

**involution**   The refolding of large nappes, producing complex structures of more recent nappes within old nappes. Also a term synonymous with cryoturbation.

**ion concentrations**   Studies of the DISSOLVED SOLIDS content of precipitation, run-off and water from other phases of the hydrological

273

cycle will frequently consider the concentrations of individual constituents. With the exception of dissolved silica and small quantities of dissolved organic matter, the dissolved material is largely dissociated into charged particles or ions, and water analyses are therefore generally expressed in terms of concentrations of individual ions ($mg\ l^{-1}$ or milliequivalents $l^{-1}$). The major cations (positively charged ions) in natural waters are $Ca^{2+}$, $Mg^{2+}$, $Na^+$, and $K^+$; and the major anions (negatively charged ions) are $HCO_3^{2-}$ and $CO_3^{2-}$, $SO_4^{2-}$, $Cl^-$, $F^-$, and $NO_3^-$.                                   DEW

**ionic wave technique**    See DILUTION; DISCHARGE.

**ionosphere**   Region above a height of about 50 km in which the gas density is so low and the temperature so high that positive and negative ions can move with some degree of independence. Electric currents so generated cause daily fluctuations in the earth's magnetic field, affect the propagation of radio waves and respond to solar flares. Other planets seem to have similar layers at similar values of pressure.      JSAG

**Reading**
Goody, R.M. and Walker, J.C.G. 1972: *Atmospheres*. Englewood Cliffs, NJ: Prentice-Hall.

**IPCC**   The Intergovernmental Panel on Climate Change (IPCC) was established in 1988 by the World Meteorological Organization (WMO) and the United Nations Environment Programme (UNEP) to assess the available scientific, technical, and socio-economic information in the field of climate change. The IPCC is organized into three interrelated working groups: Working Group I concentrates on the science of the climate system, Working Group II focuses on impacts of climate variability and response options, and Working Group III deals with economic and social dimensions of climate change. Approximately 2500 scientists from throughout the world are directly involved with the IPCC as editors, lead authors, contributing authors, and reviewers. The IPCC released its First Assessment Report in 1990 and its Second Assessment Report in 1995; IPCC continues to produce Technical Papers and develop methodologies (e.g., national greenhouse gas inventories) for use by parties to the Climate Change Convention. The Third Assessment Report will be completed around the year 2000.      RCB

**iridium layer**   Iridium (Ir) is an element (atomic number, 77; atomic weight, 192.2) that occurs in concentrations averaging about 0.5 parts per billion (ppb) by weight in the earth's crust, and about 2 ppb in the universe. In carbonaceous meteorites, the abundance averages 550 ppb. The iridium layer is a clay deposit found at the Cretaceous–Tertiary boundary, where concentrations are commonly about 50 ppb. There are two plausible sources for the excess iridium, either large-scale volcanic eruption, associated with the formation of the Deccan Traps, for example, or a massive extraterrestrial impact. It is widely assumed that the CHICXULUB IMPACT is the source of the iridium layer.      DJS

**Reading**
Alvarez, L.W., Alvarez, W., Asaro, F., and Michel, H.V. 1980: Extraterrestrial cause for the Cretaceous/Tertiary extinction. *Science* 208, pp. 1095–108.

**irrigation**   This is 'the practice of applying water to the soil to supplement the natural rainfall and provide moisture for plant growth' (Wiesner 1972: 23). The frequency and method of water delivery to the area to be irrigated can vary significantly. It may occur only once during a crop's growth cycle, for example to stimulate germination, or every day – more than once in some cases if, for example, the crop is being grown in glass houses. Barrow (1987) notes that there are three basic irrigation strategies:

1  complete irrigation, in low rainfall areas and most crop water requirements must be met;
2  supplementary irrigation, when rainfall is adequate but improved crop yields can be gained by irrigating;
3  protective irrigation, where there is a risk of crop quality being damaged by a rainfall shortage.

Irrigation has probably existed for thousands of years, and many qanats (underground tunnel transfer) systems in the Middle East may be very old. Many different irrigation methods exist. Water may be derived from perennial rivers and transported by channels and canals to areas of need. Groundwater may be pumped to the surface and used locally by an individual farmer or collectively. Water may also be harvested, either from slopes surrounding a farmed area by constructing bunds to catch and divert runoff, or by building dams on seasonal water courses. In many situations it is common for as little as 20 per cent of the water involved to reach the crop that it is intended for. To this end, subsurface and trickle irrigation systems (Agnew and Anderson 1992) may offer a number of advantages.

The most appropriate method of irrigation for a particular situation depends on a range of factors including water availability, crop type, soil type and the seasonality of climate. What-ever method is used, environmental damage can result from excessive watering. SALINIZATION is a particular effect in this respect, especially in DRYLAND areas or where centre pivot irrigation is employed. Water tables may be lowered too by excessive pumping of groundwater. Large-scale irrigation schemes may cause especial problems, including in the context of DESERTIF-ICATION. Especial concern on the direct and indirect consequences of irrigation have been associated with schemes in the Aral Sea basin and on the High Plains of the USA. It is estim-ated that all of Egypt's agriculture is irrigation dependent, 60 per cent of Japan's and 50 per cent of India's. ·            DSGT

### References and Reading

Agnew, C. and Anderson, E. 1992: *Water resources in the arid realm*. London: Routledge. · Barrow, C.J. 1987: *Water resources and agricultural development in the tropics*. London: Longman. · Wiesner, A. 1972: *The role of water in development: an analysis of principles of comprehensive plan-ning*. New York: McGraw-Hill.

## island arc

A chain of islands, mostly volcanic in origin, with a characteristic arcuate planform, rising from the deep ocean and associated with an ocean trench. Island arcs, such as those of the south-western Pacific, are generally located fairly near to continental masses and their cur-vature is typically convex towards the open ocean. Some island arcs comprise an inner arc of active volcanoes and an outer arc of non-volcanic origin formed from sediments thrust up from the ocean floor. According to the PLATE TECTONICS model, island arcs form as a result of the volcanism induced by subduction of oceanic LITHOSPHERE.       MAS

### Reading

Karig, D.E. 1974: Evolution of arc systems in the western Pacific. *Annual review of earth and planetary sciences* 2, pp. 51–75. · Sigimura, A. and Uyeda, S. 1973: *Island arcs: Japan and its environs*. Amsterdam: Elsevier.

## island biogeography

In general terms, the study of the distribution and evolution of organ-isms on islands; more narrowly, the examination of MacArthur and Wilson's equilibrium theory of island biogeography. Island floras and faunas have always fascinated biogeographers and bio-logists, and all the great nineteenth-century bio-logists were intrigued by the highly distinctive plants and animals and geographically isolated environments they found on these 'natural laboratories'. In a famous lecture to the British Association given in 1866, J.D. Hooker outlined the main characteristics and origins of island floras, while Darwin wrote of his famous Galápa-gos finches (the Geospizinae) and in 1880 A.R.

Wallace published his classic book entitled *Island life or the phenomena and causes of insular faunas and floras*.

In the main, island biota are more polar in character than their neighbouring mainland counterparts and there are usually fewer species than on a similar-sized continental area. More-over, the species mix tends to be disharmonic, being different from that on the mainland and often seeming out of harmony ecologically. For example, there are sometimes no top carnivores. The mix is usually an assemblage of taxa noted for their capacity to accomplish long-range dis-persal and migration. In many instances, the island progeny of these able dispersers have lost their dispersal ability, a phenomenon well exemplified by the ill-fated dodo (*Raphus cucul-latus*) of Mauritius, a large flightless bird related to the pigeon. The small populations on islands are subject to a range of distinctive ecological pressures and under these conditions evolution is accelerated, with new forms developing through ADAPTIVE RADIATION and hybridization.

Islands are therefore noted for their endemic organisms and for the large and abnormal per-centage of their floras and faunas which are endemic. For example, some 45 per cent of the birds of the Canaries are endemic and no less than 90 per cent of the flora of the Hawaiian Islands, the most isolated of all floristic regions. Islands are also the homes of relict organisms, which have survived there, but have become extinct elsewhere. This is especially the case on islands which were once part of a continental system, as with Crete, a former remnant of an old mountain system that connected the Balkans with southern Anatolia. Extinction, however, is also known in island biota and it is the balance between immigration and extinction which is at the heart of the now much discussed equilibrium theory of island biogeography.

This theory was first published by R.H. MacArthur and E.O. Wilson in 1963 as the equi-librium theory of island zoogeography, but was widened to include plants in their book of 1967, simply entitled *The theory of island biogeography*. The theory was mainly stimulated by thoughts on the distributions found across oceanic islands in the Pacific. The core of the theory is simple. It is argued that the number of species on an island is determined by a balance between immigration (which is a function of the distance of the island from the mainland) and extinction (which is a function of island area). The theory assumes a pool of species $P$, which can immigrate to the island and which is the number of species on the neighbouring landmass.

The theory can be presented in the form of an equation:

275

$$S_{t+1} = S_t + I + V - E,$$

where $S_t$ is the number of species at time $t$, $I$ is the number of immigrants by time $t + 1$, $V$ the number of new species evolving *in situ* on the island, and $E$ the number of extinctions.

This simple and seemingly logical theory has come under a wide range of criticism. First, it deals only with the number of species and not the number of individuals of the species on the island. In other words, it ignores population numbers. Secondly, it does not really deal with evolution, although in chapter 7 of the book the theory is tentatively extended to include this process. Thirdly, it appears to ignore historical factors which might, for example, mean that many organisms are relicts, subject only to extinction and with no potential for immigration. Fourthly, the theory lumps together all species and treats them as functioning in a similar manner. Fifthly, there are serious problems in defining both immigration and extinction and many argue that it is unacceptable to make immigration solely a function of distance. Finally, of course, it must not be assumed that all islands are in 'equilibrium', for in many this has yet to be reached, even in terms of the theory. Yet, despite all these criticisms, the theory has stimulated a new and invigorated interest in island biogeography.

It should also be noted that the theory has received a much wider application than its use on oceanic islands and that it has been related to the fauna and flora of biological 'islands' on continental areas, such as the remaining relicts of tropical rain forest. It is now being used to help determine the minimum size of viable conservation areas in which the local populations will be able to maintain themselves in some form of equilibrium. In general, it is true to say that the main tenets of island biogeography also apply to a wide range of such biological islands, including isolated mountain tops, ponds and lakes, and tracts of woodland surrounded by agriculture.

One thing is certain, namely, that the arrival of humans on many isolated islands has seriously disrupted and altered their ecosystems. Extinction rates have increased markedly with their presence, especially where they have created ecological disharmony by the introduction of alien species (see ALIENS). The Atlantic island of St Helena, for example, has seen the demise of its endemic St Helena ebony (*Trochetia melanoxylon*) which was destroyed by goats, first introduced in 1513, and by the deforestation of the island for fuel. All that is left is a barren landscape with a few relict fragments of the original biota persisting on cliffs and ridges. On the other hand, many species introduced by humans have themselves begun to change and form distinctive island races. It is believed that the special forms of the long-tailed field mouse (*Apodemus sylvaticus*) found on the Scottish and Scandinavian islands have developed from ancestors brought to these scattered locations by the Vikings. The house mouse (*Mus musculus*) on the island of Skokholm off Wales is some 30 per cent different in form from the mainland populations in Pembrokeshire. Yet it was probably only introduced to the island around 1900 by rabbit catchers and farmers. Thus, islands continue to be wonderful laboratories for the study of immigration, extinction and evolution and they will remain at the centre of biogeography for a long time to come. (See also ALIENS; ENDEMISM; EXTINCTION; REFUGIA; SPECIES–AREA CURVE.)                                    PAS

**Reading**
Brown, J.H. 1971: Mammals on mountain tops: non-equilibrium insular biogeography. *American naturalist* 105, pp. 467–78. · Carlquist, S. 1974: *Island biology.* New York: Columbia University Press. · Gilbert, F.S. 1980: The equilibrium theory of island biogeography: fact or fiction? *Journal of biogeography* 7, pp. 209–35. · Gorman, M. 1979: *Island ecology.* Outline studies in ecology. London: Chapman & Hall. · Lack, D. 1947: *Darwin's finches.* Cambridge: Cambridge University Press. · —1976: *Island biology illustrated by the land birds of Jamaica.* Oxford: Blackwell Scientific. · MacArthur, R.H. and Wilson, E.O. 1967: *The theory of island biogeography.* Princeton, NJ: Princeton University Press. · Stoddart, D.R. 1977, 1983: Biogeography. *Progress in physical geography* 1, pp. 537–43; 7, pp. 256–64. · Williamson, M. 1981: *Island populations.* Oxford: Oxford University Press.

**islands** Landsurfaces totally surrounded by water and smaller in size than the smallest continent (Australia). Oceanic islands are built up from the ocean floor and are part of the basal structure, not attached to continents, as in the example of the Hawaiian group. Continental islands are part of the neighbouring continental geological structure, as exemplified by the British Isles. The dispersal and colonization of plants and animals to islands is related to the distance from the species source. Hence oceanic islands tend to be occupied by a smaller number of species, highly adapted to the available HABITAT or NICHE and frequently endemic.                                    PAF

**Reading**
Gorman, M. 1979: *Island ecology.* London: Chapman & Hall. · MacArthur, R.H. and Wilson, E.O. 1967: *The theory of island biogeography.* Princeton, NJ: Princeton University Press. · Menard, H.W. 1986: *Islands.* New

York: Scientific American Books. · Nunn, P. 1993: *Oceanic islands*. Oxford: Blackwell.

**isochrones** Lines joining points on the earth's surface at which the time is the same. Lines joining points which experienced a seismic wave at the same time.

**isocline** A fold which is so pronounced that the strata forming the limbs of the fold dip in the same direction at the same angle.

**isolation, ecological** The ecological or habitat separation of one population from another so that interbreeding is normally prevented, even though the organisms involved may have overlapping geographical ranges (see SYMPATRY). Thus, although two closely related organisms live in the same region, interbreeding does not take place because they occupy different habitats. A classic example is afforded by two sympatric African species of *Anopheles* mosquito, the one, *A. melas*, confined to brackish water habitats, the other, *A. gambiae*, to freshwater. PAS

**Reading**
Ross, H.H. 1974: *Biological systematics*. Reading, Mass.: Addison-Wesley.

**isopleths** Lines drawn on maps connecting points which are assumed to be of equal value (e.g. contours on a topographical map). Among specific types of isopleth are those shown in the table.

**isostasy** The condition of hydrostatic equilibrium between sections of the LITHOSPHERE with respect to the underlying MANTLE. Units of the comparatively rigid outer layer of the earth in effect 'float' in the more mobile and denser material at greater depth. Isostatic adjustment was originally thought to occur by vertical movements of the crust with respect to the underlying MANTLE, but some isostatic models now assume that adjustment occurs through the movement of the lithosphere, which comprises not only the crust but an underlying zone of comparatively rigid mantle.

Two models of isostasy were proposed during the nineteenth century. G.B. Airy noted that the gravitational attraction of the Himalayan mountains was less than could be explained if the range was simply above a radially homogeneous crust and mantle. The gravity anomaly was explained by Airy as a result of crustal blocks of equal density but different thicknesses.

The thickest blocks form the highest topography and are supported by roots of light crust, which have displaced the denser underlying mantle. At a depth equal to, or greater than, the thickness of the crust, the pressure in the mantle is constant and hydrostatic equilibrium is attained. An alternative model proposed by J.H. Pratt attempted to explain isostasy by variations in the density rather than the thickness of crustal blocks. In this model the crust is assumed to be of equal thickness and areas of high elevation are associated with low density crust, which is more buoyant with respect to the underlying mantle than adjacent areas of denser crust. Although these models make unrealistic assumptions about the fluid nature

Some isopleth types

| Type | Connects up points of equal |
| --- | --- |
| Isobar | Barometric pressure |
| Isobase | Uplift or subsidence during a specified time period |
| Isobath | Distance beneath the surface of a water body |
| Isobathytherm | Temperature at a given depth below sea level |
| Isocheim | Mean winter temperature |
| Isoflor | Floral character |
| Isoglacihypse | Altitude of the firn line |
| Isohaline | Salinity in the oceans |
| Isohel | Recorded sunshine hours |
| Isohyet | Rainfall amount |
| Isomer | Mean monthly rainfall expressed as a percentage of the mean annual rainfall |
| Isoneph | Cloudiness |
| Isonif | Snow depth |
| Isopach | Rock-stratum thickness |
| Isoryme | Frost intensity |
| Isotach | Wind or sound velocity |
| Isothere | Mean summer temperature |
| Isotherm | Temperature |
| Isothermobath | Seawater temperature at a given depth |

of the mantle and the ability of crustal blocks to move independently, they describe adequately the gross variations in gravitational attraction over the earth's surface. The crustal thickness model is particularly applicable to most continental mountain systems, whereas the density model provides a more adequate explanation of the relief of the mid-ocean ridges.

While the lithosphere has a tendency to attain isostatic equilibrium, several factors may prevent this from occurring. For example, temperature and density variations associated with convection in the mantle can lead to marked gravity anomalies indicative of a lack of isostatic equilibrium. Another factor is the rigidity of the lithosphere, which means that variations in surface loading over a small area may not promote isostatic compensation, while compensations to a really extensive change in loading may produce vertical movements well beyond the zone of loading itself. This is especially well illustrated by glacial isostasy, the response of the lithosphere to the loading and unloading of the surface by ice. Rates of uplift estimated from raised shorelines and other features may exceed 20 mm year$^{-1}$. Much slower rates of isostatic compensation are associated with erosional unloading of the continents, although these rates are sustained over much longer periods and are consequently an important factor in continental uplift. (See GLACIOISOSTASY; RHEOLOGY.)                                   MAS

**Reading**
Andrews, J.T. ed. 1974: *Glacial isostasy*. Stroudsburg, Pa.: Dowden, Hutchinson & Ross. · Lyustikh, E.N. 1960: *Isostasy and isostatic hypotheses*. New York: American Geophysical Union Consultants Bureau. · Mörner, N-A. ed. 1980: *Earth rheology, isostasy and eustasy*. Chichester: Wiley. · Smith, D.E. and Dawson, A.G. 1983: *Shorelines and isostasy*. London: Academic Press.

**isotope** A form of an element which, while always having the same number of protons in the nucleus, has another form or forms with differing numbers of neutrons. There are three main types of isotope.

*Radiogenic or radioactive isotopes* (see also RADIOISOTOPES) are unstable and decay to a more stable isotopic form over predictable (typically very long) periods of time. These isotopes are utilized extensively in geological dating.

*Stable isotopes* are chemically stable and do not undergo transformation to other isotopic forms. Though chemically the same, their physical properties are slightly different. These differences result in heavier isotopes having lower mobility (which means that heavier molecules have lower diffusion velocities and a lower collision frequency with other molecules) and generally (although not always) having higher binding energies.

During natural cycles of evaporation, condensation, photosynthesis and diagenesis a natural fractionation takes place which is related to temperature and other factors. The value of the ratio $^{18}O/^{16}O$ for example is used in ice and marine core determinations of palaeotemperature (see also OXYGEN ISOTOPE). Hydrogen/deuterium (dD) variations are also used for a similar purpose.

COSMOGENIC ISOTOPES are produced by the interaction of extraterrestrial cosmic rays and other atoms.                                   SS

# J

**jet stream** A band of fast-moving ($> 30\,\mathrm{m\,s^{-1}}$) air usually found in middle latitudes in the upper TROPOSPHERE, and associated with strong horizontal gradients of density and temperature below (BAROCLINITY). Mean zonal cross-sections show a subtropical and polar-front jet. These probably define the equatorial and poleward limit of the excursions of a single jet distorted by successive weather systems.     JSAG

**Reading**
Ludlam, F.H. 1980: *Clouds and storms*. University Park: Pennsylvania State University.

**joint probability estimates** Estimates of probabilities of extreme sea levels and currents based on the probabilities of independent occurrence of the contributing tidal and surge events. By separating the individual statistics it is possible to calculate more reliable estimates from the available observations.     DTP

**jökulhlaup** An expressive Icelandic term for catastrophic drainage of a subglacial or ice-dammed lake. The lake may build up seasonally or over several years only to drain in a matter of hours when conditions are suitable for melt-water to open up tunnels in the glacier, mainly by frictional heating. In Iceland some jökul-hlaups may be triggered by volcanic activity.

DES

**Reading**
Björnsson, H. 1992: Jökulhlaups in Iceland: prediction, characteristics and simulation. *Annals of glaciology* 16, pp. 95–106. · Röthlisberger, H. 1972: Water pressure in intra- and subglacial channels. *Journal of glaciology* 11.62, pp. 177–203. · Thorarinsson, S. 1953: Some new aspects of the Grimsvötn problem. *Journal of glaciology* 2.14, pp. 267–74.

**juvenile water** Water that originates from the interior of the earth and has not previously existed as water in any state. Consequently, it has not previously participated in the hydro-logical cycle. The term was coined by Meinzer (1923), who contrasted juvenile with METEORIC WATER.     PWW

**Reference**
Meinzer, O.E. 1923: Outline of ground-water hydrology. *US Geological Survey water-supply paper* 494.

# K

**K- and r-selection**  See *r*- AND *K*-SELECTION.

**K-cycle**  The name given to a concept of landscape development involving the cyclic erosion of soils on upper hillslopes during unstable climatic phases and soil development during stable phases. The term is much used in Australia.

**kame**  An irregular mound of stratified sediment associated with GLACIOFLUVIAL activity during ice stagnation. It is a Scottish term for a landform much prized for the variety it adds to golf courses.                                         DES

**kame terrace**  A terrace formed between a hillside and a glacier by glaciofluvial activity. The landform is commonly associated with the former presence of stagnant ice down-wasting in valleys.

**kamenitza**  A generally small solutional basin developed on the surfaces of soluble rocks such as limestones. They are a type of LAPIÉ.   ASG

**kaolin**  A clay mineral, mainly hydrated aluminium silicate, or any rock or deposit composed predominantly of kaolinite. China clay or other material from which porcelain can be manufactured.

**karren**  (singular karre) A collective name describing small limestone ridges and pool structures which have developed as a result of the solution of rock by running or standing water. There are many types of karren, all differentiated by morphology (Bögli 1960). The commonest are rillenkarren (sharp ridges between rounded channels). In Britain the best examples are on Hutton Roof Crag, Kirkby Lonsdale (Lancs), while spectacular karren scenery can be found at Lluc in Mallorca. The term is German in origin; the French equivalent is *lapiés*.                                         PAB

**Reference**
Bögli, A. 1960: Kalklosung und Karrenbildung. *Zeitschrift für Geomorphologie Supplementband* 2, pp. 4–21.

One class of karren comprises the narrow vertical solution flutes called rillenkarren. These examples, which are 2 cm across, are from Mallorca.

**karst**  Generally, the term given to limestone areas which contain topographically distinct scenery, including CAVES, SPRINGS, BLIND VALLEYS, KARREN and DOLINES. Specifically, Karst is a region of limestone country between Carniola and the Adriatic coast, which is characterized by typical limestone topography.

Karst regions are typified by the dominant erosional process of solution, the lack of surface water and the development of stream sinks (dolines), cave systems and resurgences or springs. Indeed, the process of stream sinking is known as karstification. All the resultant landforms associated with karst scenery depend upon this phenomenon of stream sinking.

*Classification of solutional microforms developed on limestone*

| | Form | Typical dimensions | Comments |
|---|---|---|---|
| **Forms developed on bare limestone** — *Developed through areal wetting* | Rainpit | <30 mm across, <20mm deep | Produced by rain falling on bare rock. Occurs in fields on gentle rather than steep slopes.Can coalesce to give irregular, carious appearance |
| | Solution ripples | 20–30 mm high; may extend horizontally for >100 mm | Wave-like form transverse to downward water movement under gravity. Rhythmic form implies that periodic flows or chemical reactions are important in their development |
| | Solution flutes (rillenkarren) | 20–40 mm across, 10–20 mm deep | Develop due to channelled flow down steep slopes. Cross-sectional form ranges from semi-circular to V-shaped but is constant along flute |
| | Solution bevels | 0.2–1 m long, 30–50mm high | Flat, smooth elements usually found below flutes, Flow over them occurs as a thin sheet |
| | Solution runnels (rinnenkarren) | 400–500 mm across, 300–400 mm deep, 10–20m long | Down runnel increase in water flow leads to increase in cross-sectional area. May have meandering form. Ribs between runnels may be covered with solution flutes |
| *Developed through concentration of run-off* | Grikes (kluftkarren) | 500 mm across, up to several metres deep | Formed through the solutional widening of joints or, if bedding is nearly vertical, of bedding planes |
| | Clints (flackkarren) | Up to several metres across | Tabular blocks detached through the concentration of solution along near-surface bedding planes in horizontally bedded limestone |
| | Solution spikes (spitzkarren) | Up to several metres | Sharply pointed projections between grikes |
| **Forms developed on partly covered limestone** | Solution pans | 10–500 mm deep, 0.03–3m wide | Dish-shaped depressions usually floored by a thin layer of soil, vegetation or algal remains. CO2 contributed to water from organic decay enhances dissolution. |
| | Undercut solution runnels (hohlkarren) | 400–500 mm across, 300–400 mm deep 10–20 m long | Like runnels but become larger with depth. Recession at depth probably associated with accumulation of humus or soil which keeps sides at base constantly wet |
| | Solution notches (korrosionkehlen) | 1 m high and wide, 10 m long | Produced by active solution where soil abuts against projecting rock giving rise to curved incuts |
| **Forms developed on covered limestone** | Rounded solution runnels (rundkarren) | 400–500 mm across, 300–400 mm deep 10–20 m long | Runnels developed beneath a soil cover which become smoothed by the more active corrosion associated with acid soil waters |
| | Solution pipes | 1 m across, 2–5 m deep | Usually become narrower with depth. Found on soft limestones such as chalk as well as mechanically stronger and less permeable varieties |

*Note*: The commonly encountered German terms are given in parentheses.

*Source*: Summerfield, M.A. 1991: *Global geomorphology*. London and New York: Longman Scientific and Technical and Wiley. Table 6.6. Based largely on discussion in Jennings, J.N. 1985: *Karst geomorphology*. Oxford: Blackwell. Pp. 73–82.

The most abundant rock type which exhibits karst features is limestone and the best karst scenery can be found when that limestone is pure, very thick in areas of upstanding relief, in an environment which provides enough water for solutional processes. Other calcareous rocks such as chalk fit many but not all of the above prerequisites. Chalk is often too soft to give rise to distinctive karst scenery. PAB

**Reading**
Ford, D. and Williams, P. 1989: *Karst geomorphology and hydrology*. London: Unwin Hyman. · Ford, T.D. and

Cullingford, C.H.D. eds 1976: *The science of speleology.* London: Academic Press. · Jennings, J.N. 1985: *Karst geomorphology.* Oxford: Basil Blackwell.

**katabatic flows** Downslope winds, often coupled with, or induced by, the large-scale atmospheric circulation. These flows may reach surrounding lowlands as dry warm or cold winds, blowing at speeds in excess of 50 m s$^{-1}$ for several days. Examples of warm katabatic wind are the FÖHN on the north slopes of the Alps in Europe and the Chinook on the east slopes of the Rockies in the USA. These strong winds derive their warmth either from ADIABATIC compression during descent or from heat released by condensation on the windward slopes of the mountains or from both mechanisms together. This heat can increase the temperature of the air by 20 °C or more. Warm katabatic winds occur most frequently during the cooler months and when there is a rapid sea-level pressure from the highlands to the lowlands. Many people become depressed or irritable when these winds blow.

Cold katabatic winds occur when a large pool of cold air, forming perhaps over a mountain glacier or on ice caps, becomes so deep that it spills over into the highlands. Heat release by condensation is not involved here, so the air remains cold. The glacier winds of Greenland and Antarctica, the BORA along the Adriatic coast of Yugoslavia, and the MISTRAL along the French Mediterranean coast are good examples of cold katabatic winds. WDS

**Reading**
Atkinson, B.W. 1981: *Meso-scale atmospheric circulations.* New York and London: Academic Press. · Gedzelman, S.D. 1980: *The science and wonders of the atmosphere.* New York: Wiley.

**kata-front** Any front at which the warm air is subsiding relative to the cold air. As a result frontal activity is weak with only a belt of shallow stratiform cloud marking its presence. The change from an ana- to a kata-frontal structure can be seen sometimes on satellite images by their different cloud characteristics. Kata-fronts tend to develop at some distance from the cyclone centre where uplift in the warm air is less marked. PAS

**kavir** Iranian term for PLAYA.

**kegelkarst** Groups of residual conical-shaped limestone hills produced by limestone solution in adjoining DOLINES or SHAKEHOLES. The remnant limestone blocks are steep-sided and often heavily vegetated. They are also called cone-karst, COCKPIT KARST or MOGOTES. PAB

**Reading**
Sweeting, M.M. 1972: *Karst landforms.* London: Macmillan.

**kelvin wave** A long wave in the oceans whose characteristics are altered by the rotation of the earth. In the northern hemisphere the amplitude of the wave decreases from right to left along the crest, viewed in the direction of wave travel. DTP

**kettle, kettle hole** An enclosed depression resulting from the melting of buried ice. Kettle holes are characteristic features of STAGNANT ICE TOPOGRAPHY. DES

**khamsin** A hot, dry wind which blows from the desert to the south across the north African coast. Also called the sirocco and ghibli.

**kinematic wave** Consists of zones of high and low density which travel through a medium at a velocity which is generally different from that of the medium as a whole.

One set of solutions of the CONTINUITY EQUATION for material in a medium is generally dominant under certain circumstances. The theory was originally developed by Lighthill and Whitham (1955) and has been applied in a number of contexts within physical geography. The continuity equation in its differential form may be rewritten, using the same notation, as:

$$c\partial s/dx + \partial s/dt = a.$$

In this formulation, $c$ is defined as the kinematic wave velocity, equal to $[dQ/dS]$ evaluated at $x$, and not necessarily or usually a constant. For the simplest case where $c$ is constant and $a = 0$, the complete solution to the above equation is:

$$S = f(x - ct)$$

for an arbitrary function $f$. This represents a wave travelling without change of form at velocity $c$. In more complex cases different parts of the wave travel at different velocities, but the concept of a wave remains.

Kinematic waves have been applied to glacier response to climatic changes (Nye 1960), to the movement of riffles in stream beds (Langbein and Leopold 1968), to the movement of stream knick-points and valley-side terraces, and to the routing of river and overland and flow flood peaks. MJK

**Reading and References**
Freeze, R.A. 1978: Mathematical models of hillslope hydrology. In M.J. Kirkby ed., *Hillslope hydrology.* Chichester: Wiley. · Langbein, W.B. and Leopold, L.B. 1968: River channel bars and dunes – theory of kinematic waves. *US Geological Survey professional paper* 122L,

L1–209. · Lighthill, M.J. and Whitham, G.B. 1955: On kinematic waves I: flood movements in long rivers. *Proceedings of the Royal Society* Series A. 229, pp. 281–316. · Nye, J.F. 1960: The response of glaciers and ice sheets to seasonal and climatic changes. *Proceedings of the Royal Society* Series A. 256, pp. 559–84.

**kinematics**   The branch of mechanics dealing with the description of the motion of bodies without reference to the force producing the motion.                                                           BWA

**Reading**
Petterssen, S. 1956: *Weather analysis and forecasting.* New York: McGraw-Hill. Esp. chs 2 and 3.

**kinetic energy**   Energy due to the translational movement of a body. It is not so definitive as it looks. It depends on the frame of reference, e.g. an object lightly tossed from a rapidly moving railway train has potentially lethal energy for a bystander and vice versa and also upon scale, e.g. 'temperature' of a gas represents the kinetic energy of individual molecules of the gas.   JSAG

**kingdoms of animals and plants**   The simple classic division of all living things (except the non-cellular, problematical viruses) into two categories of plants and animals (*plantae* and *animalia*) is no longer found entirely satisfactory and has been abandoned by most life scientists. The principal problem with the traditional twofold classification is that it groups organisms that are very unlike one another under the same heading. A few organisms lie uneasily in either category.

No classification of kingdoms is entirely satisfactory; the following is in widespread use.

*Monera*
Not having a well-defined nucleus: bacteria and blue-green algae. Acellular organisms, i.e. those lacking clear division into cells.

*Protista*
Acellular organisms mostly lacking chlorophyll: flagellates (some of which do possess chlorophyll), amoebae, foraminifera, sporozoans, ciliates. Some of these form colonies, being incipiently multi-celled.

*Fungi*
All kinds of fungi, including slime moulds. This group has long been included with the plants, but its members have a long, quite distinct evolutionary history. Non-photosynthetic organisms, mostly with definite cell walls.

*Plantae*
Six phyla (major groups) of photosynthetic organisms with cell walls, ranging from acellular forms (some algae) to much more complex organisms with highly developed organs and systems of organs (ferns, flowering plants).

*Animalia*
Multi-celled, non-photosynthetic organisms without cell walls. Some classifications recognize over 300 phyla, varying from simple forms with few cells, the mesozoa, through sponges which have partly differentiated tissues, to a great diversity of complex metazoans with well-developed organs and organ systems, e.g. worms, insects, molluscs, vertebrates. (See also FAUNAL REALMS; FLORISTIC REALMS.)       PHA

**klippe**   An outcrop of rock that is separated from the rocks upon which it rests by a fault. It may represent an erosional remnant of a nappe or may have been emplaced by gravity sliding.

**knickpoint**   A break in profile, generally in the long profile of a river. This was especially thought of as the product of REJUVENATION, where a steeper-gradient lower reach is receding headward as a result of local or general lowering of base level, and this steeper profile intersects with a gentler upper one. Some knickpoints may be sharply defined, as in a waterfall, or they may be much less distinguished and only apparent after detailed field survey of stream profiles.

The term is also applied to any profile irregularity, as at tributary confluences or associated with lithological or structural controls, and not just those produced following rejuvenation. Furthermore, rejuvenation may itself be generated in alternative ways – extensively by eustatic sea level change, or by isostatic and tectonic movements, or by changes in river discharge or sediment load. The early geomorphologists W.M. Davis and W. Penck (writing of *Knikpunkte* in German) were among the foremost in developing studies of such phenomena, though they used the word differently and in more restricted senses than are now generally adhered to. Emphasis is now placed on the interaction of many factors in stream system development, so that several possibilities for breaks in long profile would need to be explored.       JL

**Reading**
Small, R.J. 1970: *The study of landforms.* Cambridge and New York: Cambridge University Press.

**knock-and-lochan topography**   A landscape of ice-moulded rock knobs with intervening lochans which have been eroded along lines of structural weakness. The type site is in the north-western highlands of Scotland (Linton 1963), but it is also characteristic of much of the Canadian and Scandinavian shields (Sugden 1978).       DES

### Reading and References
Gordon, J.E. 1981: Ice-scoured topography and its relationships to bedrock structure and ice movement in parts of northern Scotland and West Greenland. *Geografiska annaler* 63A. 1–2, pp. 55–65. · Linton, D.L. 1963: The forms of glacial erosion. *Transactions of the Institute of British Geographers* 33, pp. 1–28. · Sugden, D.E. 1978: Glacial erosion by the Laurentide ice sheet. *Journal of glaciology* 20.83, pp. 367–91.

**kolk** A form of macroturbulence, or large-scale vortex structure, that arises in the turbulent flow of water in rivers. The term was introduced by Matthes in 1947. A kolk involves a rapidly rotating cylinder of flow, or vortex, that is initially attached at its lower end to the bed surface, and rises away from it. The kolk subsequently breaks away and rises toward the water surface, where its presence is marked by a 'boil' on the water surface that dissipates after a short time (Jackson 1976). Kolks may entrain sediment particles from the bed and release them some distance higher, so fostering the downstream motion of the bed material.    DLD

### Reference
Jackson, R.G. 1976: Sedimentological and fluid-dynamic implications of the turbulent bursting phenomenon in geophysical flows. *Journal of Fluid Mechanics* 77, pp. 531–60.

**koniology** (coniology) The scientific study of atmospheric DUST together with its solid pollutants, such as soot, pollen, microbial spores, etc.

**kopje** A hillock or rock outcrop, applied especially in South Africa.

**Köppen's climatic classification** A system of climatic differentiation based upon TEMPERATURE and PRECIPITATION linked to vegetation zones. It is one of the most widely used methods of classification but has undergone numerous modifications since it was devised about 1900. Five major categories subdivide the earth's climates: tropical forest, dry, warm temperate rainy, cold forest and polar. Further subdivisions are made on the basis of the rainfall regime, temperature characteristics and any other special features. Each region can then be identified on the basis of a sequence of letters, e.g. Csb indicates a coastal Mediterranean climate with a mild winter and a dry but warm summer.    PS

### Reading
Barry, R.G. and Chorley, R.J. 1992: *Atmosphere, weather and climate.* 6th edn. London: Routledge.

**krotovina** Infilled animal burrows or filaments found in soils and sediments such as loess.

**krummholz** From the German, meaning crooked wood, refers to the stunted and gnarled woodlands characteristic of forest margins at high altitudes and high latitudes. The dwarfing, distortion and, in extreme conditions, the prostrate habit of trees is a result of the combined effects of wind and cold. Such features are common in the transition zone between the subalpine forest and alpine tundra in high latitudes, or in the elfin woods of high tropical elevations, bearing a heavy cover of epiphytes and a ground

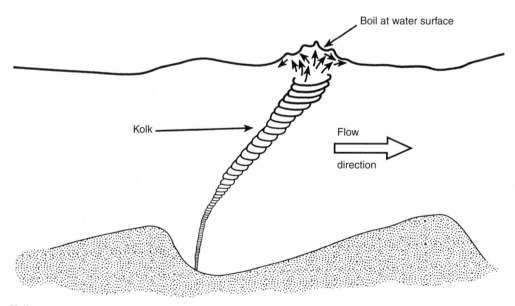

**Kolk.**

layer cushioned by mosses, herbaceous plants and grasses. The word is also used to describe dense, tangled thickets in tropical forest.      PAF

**kumatology** A neglected term developed by the great British amateur geographer, Vaughan Cornish, for the study of wave-like forms encountered in nature.      ASG

**kunkar** See CALCRETE.

**kurtosis** (particle size) A measure, as in statistics, of the peakedness of distribution; in the case of sediments it relates to both sorting (standard deviation) and differences from a normal distribution (where a normal distribution would have a kurtosis value of 1.0). A flat (platykurtic) distribution would be found in a poorly sorted sediment such as a till; a peaked (leptokurtic) distribution would be found in a well-sorted sediment such as a wind-blown sand. The usefulness of this measure has been questioned.      WBW

**Reading**
Briggs, D. 1977: *Sediments*. London: Butterworth. · Tucker, M.E. 1981: *Sedimentary petrology: an introduction*. Oxford: Blackwell Scientific.

# L

**laccolith** A mass of intrusive rock which though concordant with the host rocks has domed up the overlying strata. The base of the laccolith is either horizontal or convex downward.

**Reading**
Corry, C.E. 1988: Laccoliths: mechanisms of emplacement and growth. *Geological Society of America special paper* 220.

**lacustral** See PLUVIAL.

**lacustrine** Of or pertaining to lakes.

**lag gravel** An accumulation of coarse rock fragments at the landsurface which has been produced by the removal of finer particles, generally by deflation.

**lag time** The period elapsing between the occurrence of a causative phenomenon and its resulting effect, as in the time difference between peak storm rainfall and the later peak in stream discharge that results from it. This may be an important consideration in many physical processes. For example, BEDFORMS in rivers may be related to river flows, but such flows vary over time and it also takes a finite time for the bedforms themselves to develop. Thus there may be delayed response between development and a change in flow.      JL

**Reading**
Allen, J.R.L. 1974: Reaction, relaxation and lag in natural sedimentary systems: general principles, examples and lessons. *Earth science reviews* 10, pp. 263–342.

**lahar** A mass movement feature on the flank of a volcanic cone. The volcanic ash may move either as a mudslide when saturated with water or as a dry landslide as a result of earth tremors.

**lake** A lake is a body of water contained within continental boundaries. Worldwide, almost all natural lakes were formed by tectonic, volcanic or glacial activity, with the majority being glacial in origin. Others may occupy depressions formed by DEFLATION (see PAN; PLAYA). The relatively still waters of lakes have led to them being termed lentic environments. The size, depth and shape of lakes varies enormously from shallow, ephemeral systems, playa lakes, to huge deep lakes, such as the Siberian Lake

Baikal. This 25 million year old lake is over 1600 m in depth and contains about 20 per cent of the earth's unfrozen freshwater, more than the combined total of the five Great Lakes of North America. By far the largest lake in terms of area and volume is the vast saline Caspian Sea that covers an area of nearly 370,000 km$^2$ and contains over 78,000 km$^3$ of water. Most of the world's one million plus lakes are however very small freshwater bodies.

Once a lake basin has formed, physical, biological and chemical factors interact to produce discernible structures in the water. Thermal stratification is one of the most important physical events in a lake's annual cycle and it dominates most aspects of lake structure. If a lake is deep enough, heating of its surface waters by the sun will cause a warm, light layer to form, the epilimnion, above a colder dense layer, the hypolimnion. A third layer the metalimnion, forms a transitional zone between the two. The temperature gradient between the layers is termed the thermocline. During the summer the volume of the epilimnion increases at the expense of the hypolimnion. In autumn heat loss exceeds the input of solar heat and destratification will occur in all but the deepest lakes. If the waters become fully mixed the lake is termed holomictic. Meromictic lakes are deep water bodies, such as Lake Tanganyika, that remain stratified throughout the year. The majority of lakes are dimictic, with their waters mixing twice a year during the spring and autumn. Shallow lakes that are mixed throughout the year are termed polymictic. Not only is the heat transmitted by light responsible for various kinds of thermal stratification in water bodies, but it also regulates the rate of chemical reactions and biological processes. An increase in heat during the summer, for example, will increase metabolic activity and the rate of recycling of organic and mineral components.

Lakes can also be classified based on their primary biological productivity or trophic status. Those that are nutrient poor and experience low productivity are termed oligotrophic. Oligotrophic lakes are often deep with steep slopes and relatively small drainage areas. With few nutrients, oligotrophic lakes support relatively low levels of algae and consequently are characterized by clear blue waters which are highly transparent. Lakes that are nutrient rich and display high levels of primary productivity are termed eutrophic (see EUTROPHICATION). In many cases these are shallow water bodies, less than 10 m deep, with gentle sloping edges and a large drainage area-to-lake surface ratio. Characterized by high nutrient levels, the water of eutrophic lakes are characterized by low clarity and in extreme cases unsightly blooms of blue-green algae can form a crust on the surface of the lake. The sediments of eutrophic lakes become enriched with nutrients as organic matter accumulates. Initially this tends to increase the biomass of rooted macrophytes but phytoplankton growth can become so dense that it can shade out submerged plants. The trophic status of a lake will often change over time with waters usually becoming more nutrient rich as they get older. In recent decades, however, there has been growing concern about cultural eutrophication as a result of human activities. Cultural eutrophication is not a new phenomenon. For example, Lake Patzcuaro, in the highlands of central Mexico, became eutrophic following widespread deforestation and increased catchment erosion some 900 years ago (Metcalfe *et al.* 1989). Elsewhere, discharge of domestic waste into lakes and rivers has resulted in widespread eutrophication and in recent years problems have been associated with the intensification of agriculture and the increased use of organic fertilizer. Invariably some of the fertilizer applied is lost from the field and washes into lakes and rivers. Given the higher productivity of eutrophic lakes, eutrophication is not always undesirable. And there are many naturally eutrophic lakes where large quantities of fish are harvested for food and in some cases water bodies are managed to maximize productivity. (See also LIMNOLOGY.)

SLO

**Reading and Reference**
Hutchinson, G.E. 1957 and 1967: *A treatise on limnology.* Vols I and II. New York: Wiley. · Horne, A.J. and Goldman, C.R. 1994: *Limnology.* New York: McGraw-Hill International Editions. · Metcalfe, S.E., Street-Perrott, F.A, Brown, R.B. Hales, P.E., Perrott, R.A. and Steininger. F.M. 1989: Late Holocene human impact on lake basins in Central Mexico. *Geoarchaeology* 4, pp. 119–41.

**lake breeze**   See SEA/LAND BREEZE.

**laminar flow**   A type of flow in which the movement of each fluid element is along a specific path with uniform velocity, with no diffusion between adjacent 'layers' of fluid. Injected dye maintains a straight, coherent thread. The shear stress between adjacent layers increases from zero at the surface to a maximum at the fluid–solid contact (e.g. the river bed), and the flow velocity increases parabolically with height above the bed. Fluid motion is laminar if viscous forces are so strong relative to inertial force that the fluid viscosity significantly influences flow behaviour. The viscosities of air and water are so low that laminar flow is rare in these fluids. Laminar flow is characterized by a REYNOLDS NUMBER of below 500, e.g. in the very shallow water of overland flow on hillslopes, and then only when the water is undisturbed by raindrop impact.

KSR

**La Niña**   Trade winds usually return to normal following an ENSO (El Niño-Southern Oscillation) event. However, if the trade winds are unusually strong, cold surface water moves over the central and eastern Pacific, and warm water and rainy weather is confined mostly to the western tropical Pacific Ocean.

La Niña ('the girl child') is the cold counterpart of EL NIÑO. In La Niña conditions, the sea surface temperatures in the eastern equatorial Pacific Ocean fall below normal. La Niña episodes have been observed fairly often before or after (or both) an El Niño occurrence. Substantial cooling in the tropical troposphere during La Niña has been accompanied by significant changes in the atmospheric circulation in the middle range latitudes of both hemispheres. La Niña is often the time of extreme drought in Peru. La Niña is generally characterized by warm winters in the south-eastern United States, colder than normal winters in the Pacific north-west to the Great Lakes, and unsettled winters in the north-east and mid-Atlantic states. In six of the past ten La Niña winters, the northern Sierra had more rain than normal. The La Niña episode of 1988 has been tied to floods in Bangladesh and droughts in the mid-western United States.

MLH

**land breeze**   See SEA/LAND BREEZE.

**land capability**   A measure of the value of land for agricultural purposes. The capability unit is described as a group of soils that are nearly alike, based on an interpretation of soil data. The soils in each will: (1) produce similar kinds of cultivated crops and pasture plants with similar management practices; (2) require a similar conservation treatment and management under the same kind and condition of vegetative growth; and (3) have comparable potential productivity.

Subclasses are defined according to their limitations for agricultural use and hazards to which they are exposed. In the USA four general limitations are recognized: erosion hazard, wetness, rooting zone limitations and climate, of which all but the last are closely related to geomorphology. In the UK another subdivision – gradient or soil pattern – is also added.                                                                    SMP

**Reading**
Bibby, J.C. and Mackney, D. 1969: *Land use capability classification*. London: Soil Survey.

**land systems** Subdivisions of a region into areas having within them common physical attributes which are different from those of adjacent areas. Any one land system normally has a recurring pattern of topography, soils and vegetation, reflecting the underlying rocks (geology), erosional and depositional processes (geomorphology) and the climate under which these processes operate. A land unit, the detailed component of a land system, is particularly useful in evaluating land for agricultural and engineering purposes and in devising problem-orientated classifications. The resultant land systems maps are easily interpreted, and both rapid and economical to produce.                SMP

**Reading**
Veateh, J.O. 1983: *Agricultural land classification and land types of Michigan*. Michigan Agricultural and Experimental Station special bulletin 544.

**land use, hydrological effects of** Land use affects the hydrological cycle by altering the natural rates of infiltration, run-off, erosion, and sedimentation. Rural land-use practices alter natural vegetation diversity by limiting the number of species and forms, and agricultural activities change soil conditions and leave the surface barren for parts of the year in many cropping schemes. In urban areas impervious surfaces replace porous natural ones, and drainage networks are altered to include many artificial channels. These rural and urban changes have definable consequences for the hydrological cycle and for the products of its operation, especially the integrative measure of sediment yield.

The connections between land use, run-off, and sediment yield were demonstrated for rangelands by Noble (1965) who measured run-off and soil loss under three conditions of vegetation cover. Under good range conditions with diverse cover he found that only 2 per cent of the precipitation became run-off and that soil loss amounted to only 12 t km$^{-2}$. When land use of the area reduced the diversity and cover to fair conditions run-off increased to 14 per cent and soil loss to 122 t km$^{-2}$. Finally, when land use of the area resulted in the elimination of some species and the development of areas without vegetation cover, run-off increased to 73 per cent and soil loss to 1349 t km$^{-2}$.

Wolman (1967) has shown that when land use of a given humid area passes through a series of changes, sediment production also changes in a predictable fashion. Under conditions essentially unaltered by human activities sediment production of the example area was relatively low. The introduction of primitive farming resulted in a marked increase of sediment production. Mechanized farming without conservation practices resulted in further increases that raised the sediment production to a level several times the natural rate. Conservation practices reduced the sediment yield, and the abandonment of the fields which allowed the return of shrubs, dense grasses, and trees also reduced sediment yields to levels approaching their

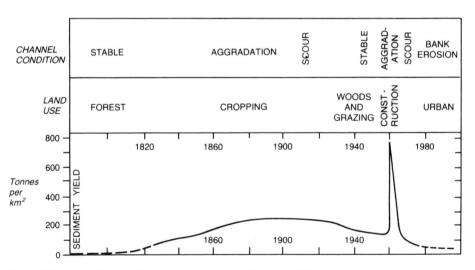

**Land use.** Effects on sediment yield as constructed for the Piedmont, USA.
*Source*: Wolman 1967.

natural values. When the area became urbanized construction activities increased sediment yields to levels 50 or more times the natural levels. This brief peak was followed by a precipitous decline as the impervious surfaces of the city reduced sediment yield to very low levels. Similar studies by Ursic and Dendy (1965) in the southern USA showed that sediment yield for abandoned fields was ten times the yield for nearby pine forests, while the yield for cultivated fields was more than 100 times that of the pine forests.

Conservation practices can reduce excessive soil loss by as much as 90 per cent (Baird 1984). Remedial measures include crop rotation to include species that do not leave large areas of the surface exposed to erosion for long periods. Strip planting can be used to alternate protective crops with those that are more damaging. By leaving some fields fallow for some periods of time the land manager can reduce erosion by increasing the amount of organic material in the soil. On sloping lands terraces may be useful in reducing the gradient of field surfaces, thereby reducing the amount and velocity of run-off and its accompanying erosion.

In urban areas the development of large areas of impervious surfaces and the installation of artificial drainways ensure that run-off increases in amount and collects more quickly than in unaltered environments. Leopold (1968) showed a systematic relationship among these variables in urban areas. The consequence is that downstream from urban areas flood peaks rise more rapidly and are higher than in natural circumstances, resulting in increased flood damage. The flashy discharges occasionally erode newly expanded floodplains that accumulate during the high sediment yield period of construction (Graf 1975).

Control of the urban run-off problem may take either structural or non-structural approaches. The structural alternatives include the construction of retention basins in upstream areas to retard the rate of accumulation in main streams, or the construction of channel improvements in the downstream areas to speed the water out of the area. Unfortunately, this latter strategy merely transfers the problem to other areas downstream. Non-structural alternatives include the setting aside of the undeveloped areas in the upstream sections to act as sinks for run-off, and the avoidance of flood hazard zones downstream through building restrictions and zoning. (See also URBAN HYDROLOGY.) WLG

**References**

Baird, R.W. 1984: Sediment yields from Blackland watersheds. *Transactions of the American Society of Agricultural Engineers* 7, pp. 454–6. · Graf, W.L. 1975: The impact of suburbanization on fluvial geomorphology. *Water resources research* 11, pp. 690–3. · Leopold, L.B. 1968: Hydrology for urban land planning: a guidebook on the hydrologic effects of urban land use. *US Geological Survey circular* 554. · Noble, E.L. 1965: Sediment reduction through watershed rehabilitation. *Proceedings of the Federal Interagency Sedimentation Conference*, US Department of Agriculture miscellaneous publication 970, pp. 114–23. · Ursic, S.J. and Dendy, F.E. 1965: Sediment yields from small watersheds under various land uses and forest covers. *Proceedings of the Federal Interagency Sedimentation Conference*, US Department of Agriculture miscellaneous publication 970, pp. 47–52. · Wolman, M.G. 1967: A circle of sedimentation and erosion in urban river channels. *Geografiska Annaler* 49A, pp. 385–95.

**land-bridge** An isthmus or other connection between two land masses across which animals and plants move to colonize a new environment.

**landfill or sanitary landfill** A method of disposing of refuse on land with the intention of creating the minimum nuisance or hazard by confining the refuse to the smallest practical area, reducing it (e.g. by compaction) to the smallest practical volume, and covering it with a capping. Landfill sites can include natural depressions and old man-made depressions, specially dug trenches or, in flat areas, artificial mounds. Sites need to be lined to prevent leaching of chemicals and organic wastes into groundwater. Problems also result from biochemical degradation which can lead to subsidence, the formation of disagreeable odours, and the formation of potentially explosive gases (e.g. methane). ASG

**Reading**
Costa, J.E. and Baker, V.R. 1981: *Surficial geology: building with the earth*. New York: Wiley.

**Landsat** See UNMANNED EARTH RESOURCES SATELLITES.

**landscape ecology** A term introduced by the German geographer Carl Troll who also later used the term geo-ecology. It has two components (Vink 1983): an approach to the study of the landscape which interprets it as supporting natural and cultural ecosystems; and the science which investigates the relationships between the biosphere and anthroposphere and either the earth's surface or the abiotic components. ASG

**Reference**
Vink, A.P.A. 1983: *Landscape ecology and land use*. London: Longman.

**landscape evaluation** The classification of rural landscapes so that appropriate planning action may be taken for their future management. The approach has developed since the early 1960s as the problem of defining the aesthetic quality of the landscape, so that it has real parity with social and economic factors, has been evident.

The evaluation operation involves more than the simple identification and mapping of land use changes, although many early subjective attempts used empirical landscape components, which were then scored and aggregated (Linton 1968) to identify spatial variations in landscape characteristics. It also incorporates the values that people attach to landscapes, which causes complications as these values are likely to be very subjective.

The development of objective techniques during the 1970s was a major advance in the field. These techniques attempt to weight factors according to their contribution to or detraction from landscape quality in a given area (Coventry Subregional Study 1971). Although involving computer analysis, many initial decisions concerning factor scores are still subjective.

There is a range of preference techniques which identifies people's reactions to landscape by asking them to rank photographs in order of landscape quality (Fines 1968). However, finding representative individuals to undertake the ranking, standardizing of photographic quality and translating of preferences into planning decisions cause many problems.

The social basis for evaluation has recently received attention. In this approach value is defined in terms of social indicators rather than landscape components (Penning Rowsell et al. 1979). Unlike most other methods these are not merely theoretically based, but aim to solve resource management problems.          SMP

**Reading and References**
Coventry, Solihull, Warwicks Subregional Study 1971: *Journal of the Town Planning Institute* 57, pp. 481–4. · Fines, K.D. 1968: Landscape evaluation: a research project in East Sussex. *Regional studies* 2, pp. 41–55. · Linton, D.L. 1968: The assessment of scenery as a natural resource. *Scottish geographical magazine* 84, pp. 219–38. · Penning Rowsell, E.C., Gullet, G.H., Seale, G.H. and Witham, S.A. 1979: *Public evaluation of landscape quality.* Planning research group report 13. Middlesex Polytechnic. · *Transactions of the Institute of British Geographers* 66 (old series), 1975. Special edition on the topic.

**landslide**   A landslip. The movement downslope under the influence of gravity of a mass of rock or earth. A mass of rock or earth that has moved in such a way. (See MASS MOVEMENT TYPES).

**lapié**   See KARREN.

**lapse rate**   The rate of *decrease* of a quantity with height, usually applied to temperature but sometimes also to the mixing ratio (see HUMIDITY) of water vapour to air. The typical rate of temperature change in the TROPOSPHERE is $6.5 \, \mathrm{K \, km^{-1}}$ whereas for the temperature of dry air displaced adiabatically it is $10 \, \mathrm{K \, km^{-1}}$ (dry adiabatic lapse rate). It follows that the troposphere is usually stable to dry adiabatic processes.          JSAG

**Reading**
Ludlam, F.H. 1980: *Clouds and storms.* University Park, Pa.: Pennsylvania State University Press.

**Late Glacial**   A term used for the span of time between the maximum of the Last (in Britain, the Devensian) Glacial (*c.*18,000 BP) and the beginning of the Holocene interglacial (*c.*11,000–10,000 BP). It was marked by various minor stadials and interstadials (e.g. see ALLERØD).          ASG

**latent heat**   That part of the thermal energy involved in a change of state, like the 2.4 $\times 10^6 \, \mathrm{J \, kg^{-1}}$ of energy released when water vapour condenses to liquid. This process makes rising cloudy air cool less rapidly than does the ambient air with height and hence the cloudy air becomes buoyant. EVAPORATION of liquid water at the ground 'saves up' solar energy in latent form until it can be released in cloudy convection. Typical English thunderstorms rain French water.          JSAG

**Reading**
Ludlam, F.H. 1980: *Clouds and storms*: University Park, Pa.: Pennsylvania State University Press.

**lateral accretion**   The process by which bed sediments accumulate at the side of a channel as it shifts laterally. The term applies notably to the sediment accumulating in POINT BAR DEPOSITS, but lateral accretion can also occur in BRAIDED RIVERS as channels shift and bars enlarge. Such deposits may contrast with the vertical accretion of sediments deposited from suspension which are usually finer in size and may accrete beyond the confines of channels, and they may be particularly important volumetrically among the near-surface sediments of many FLOODPLAINS. Such sediments may possess a distinctive type of cross-bedding, called epsilon cross-bedding by J.R.L. Allen, in which the dipping or sigmoid-shaped beds represent successive increments of accretion developed at right-angles to the general flow direction.          JL

**Reading**
Allen, J.R.L. 1970: *Physical processes of sedimentation.* London: Allen & Unwin. · Collinson, J.D. and Thompson, D.B. 1982: *Sedimentary structures.* London: Allen & Unwin. · Reading, H.G. ed. 1986: *Sedimentary environments and facies.* 2nd edn. Oxford: Blackwell Scientific.

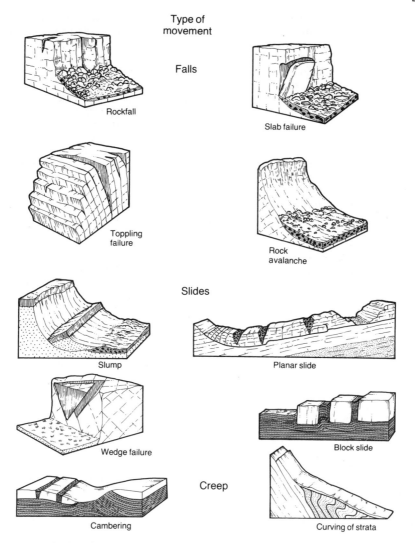

Type of
movement

Falls

Rockfall

Slab failure

Toppling
failure

Rock
avalanche

Slides

Slump

Planar slide

Wedge failure

Block slide

Creep

Cambering

Curving of strata

**Landslides.** A classification of landslides in rock.
*Source*: M.J. Selby 1982: *Hillslope materials and processes*. Oxford: Oxford University Press. Figure 7.18.

**lateral flow** Applied particularly to sub-surface near-horizontal or ground slope-aligned water movement, in contradistinction to vertical water movement. Water flow along permeable soil horizons may be an important mechanism in the transfer of water from soils and hillslopes into streams without such flow taking place over the surface of the ground or through deep percolation to groundwater. (See also INTER-FLOW; THROUGHFLOW.)                                    JL

**lateral migration** The movement of stream channels across valley floors through bank erosion and accompanying deposition on the opposite bank. This may proceed at varying rates on different rivers, on some amounting

to several metres a year and on others none at all.                                    JL

**Reading**
Osborn, G. and Du Toit, C. 1991: Lateral planation of rivers as a geomorphic agent. *Geomorphology* 4, pp. 249–60.

**lateral moraine** See MORAINE.

**laterite** A surface accumulation of the products of rigorous chemical selection, developing where conditions favour greater mobility of alkalis, alkali earths and silica than of iron and aluminium. Bauxite is a laterite rich in aluminium. Laterite was originally defined by its ability to harden rapidly and irreversibly on

291

exposure to the air, a property which led to its use as building bricks in southern India (Buchanan 1807). The term has been extended to include related materials (MacFarlane 1983) which were hard or contained hard parts, even though this induration may be an original result of iron segregation rather than of exposure.

Laterite profiles vary enormously in scale. The laterite may be a few centimetres to tens of metres thick, and below this the saprolite, leached or unleached of iron, varies from a few centimetres to over 100 m. Thick profiles develop on low angle slopes (Goudie 1973), and laterites can be divided into those that result from relative accumulation of iron and aluminium sesquioxides, and those that result from absolute accumulation.

Relative accumulations owe their concentration to the removal of more mobile components, and absolute accumulations to the physical addition of materials.                                      ASG

### References

Buchanan, F. 1807: *A journey from Madras through the countries of Mysore, Kanara and Malabar*. London: East India Company. · Goudie, A.S. 1973: *Duricrusts of tropical and subtropical landscapes*. Oxford: Clarendon Press. · MacFarlane, M. 1983: Laterites. In A.S. Goudie and K. Pye eds, *Chemical sediments and geomorphology*. London: Academic Press.

**latosol**  A deep, red, yellow and brown, highly weathered, residual iron-rich soil typically found in the humid tropics. The term is used to embrace a range of residual iron soils including lateritic soils, highly weathered ferralitic soils and moderately weathered ferruginous soils, and is broadly synonymous with the United States Department of Agriculture (USDA) term oxisol. In addition to being iron-rich, latosols tend to have a moderate to high clay content, dominated by kaolinite, goethite, haematite and gibbsite, and may merge at depth into saprolite and unweathered bedrock (see also LATERITE and SAPROLITE).          DJN

**Laurasia**  The northern part of Pangaea, a super-continent thought to have been broken up by continental drift. The southern continent, Gondwanaland, was separated from it by the Tethys Ocean.

**Laurentide ice sheet**  The ICE SHEET that covered much of North America during the last glaciation. At its maximum extent it covered most of Canada, ranging from the Queen Elizabeth Islands in the north to just south of the US border, and from beyond the current coastline in the east to where it met the Cordilleran Ice Sheet covering the Rocky Mountains in the

west. It is thought to be the largest ice sheet to grow and decay during glaciations of the QUATERNARY period, and at its maximum is estimated to have made up 35 per cent of the world's ice volume, containing sufficient water to lower global sea level by around 50 m. It occupied an area of around $11 \times 10^6 \, \text{km}^{-2}$ and reached a probable thickness of 2–3 km. The ice sheet comprised three main ice dispersal centres, called the Keewatin, Baffin and Labrador sectors. In its strict sense the name Laurentide ice sheet only applies to the ice mass during the last (Wisconsinan) glacial period, with its maximum extent at 21,000 years before present, although we know that similar ice sheets existed during earlier glacials.

The significance of this ice sheet lies in its interaction with the climate system. It was large enough to alter atmospheric circulation patterns, and during its decay, we now know that abrupt releases of meltwater and icebergs were of great enough magnitude to modify ocean circulation, thus forcing abrupt flips in the climate of the north Atlantic region.   CDC

### Reading

Benn, D.I., and Evans, D.J.A., 1998: *Glaciers and glaciation*. London: Arnold. · Fulton, R.J. and Prest, V.K. 1987. The Laurentide Ice Sheet and its significance. *Geographie Physique et Quaternaire* 41, pp. 181–6. · Clark, P.U., Licciardi, J.M., MacAyeal, D.R., and Jenson, J.W. 1996. Numerical reconstruction of a soft-bedded Laurentide Ice Sheet during the last glacial maximum. *Geology* 24, pp. 679–82.

**lava**  Molten rock material which is extruded from volcanoes and volcanic fissures.

**law of the wall**  A relationship describing the semi-logarithmic increase in fluid (normally wind) velocity at an increasing distance from a rough bed under conditions of TURBULENCE. Under normal conditions on flat, unvegetated surfaces and in the absence of thermal effects a turbulent velocity profile will plot as a straight line on a semi-logarithmic chart with velocity increasing away from the bed. The gradient of the semi-logarithmic profile is proportional to the shear velocity ($u*$) and is a result of the surface ROUGHNESS producing a drag on the overlying fluid flow. The relationships between the principal controlling parameters on the profile can be described by the Karman-Prandtl velocity distribution of:

$$\frac{u}{u*} = \frac{1}{\kappa} ln \frac{z}{z_0}$$

where: $u$ = fluid velocity; $u*$ = shear velocity; $k$ = von Karman's constant (0.4); $z$ = height; $z_0$ = ROUGHNESS LENGTH.

The law of the wall provides a particularly useful technique by which shear velocity can be calculated if fluid velocity is known at a number of heights from the bed. Such an approach is commonly used when determining sediment ENTRAINMENT and the CRITICAL EROSION VELOCITY in fluid flows because the shear velocity represented by the velocity profile is proportional to the SHEAR STRESS at the surface.

GFSW

**Reading**
Livingstone, I. and Warren, A. 1996. *Aeolian geomorphology: an introduction*. Longman: Harlow. · Wiggs, G.F.S. 1997: Sediment mobilisation by the wind. 2nd edn. In D.S.G. Thomas ed., *Arid zone geomorphology*. London: Wiley. Pp. 351–72.

**leachate** Liquid that passes through a soil, sediment or other permeable material, carrying away substances taken into solution (leached) from it. Applied to the water that passes through litter and the upper parts of the soil, which may be made more potent as a solutional agency by organic acids derived from the plant residues. Leachates also commonly require management at landfill and mine tailings sites, where percolating rainwater is capable of creating a contaminated leachate that must be prevented from escaping into the environment (Sophocleous *et al.* 1996).

DLD

**Reference**
Sophocleous, M., Stadnyk, N.G. and Stotts, M. 1996: Modelling impact of small Kansas landfills on underlying aquifers. *Journal of environmental engineering* 122, pp. 1067–77.

**leaching** The downward movement of water through the soil zone which results in the removal of water-soluble minerals from the upper horizons and their accumulation in the lower soil zone or groundwater.

**leaching requirement** The fraction of irrigation water that must be leached through the root zone to maintain the soluble salt level in the soil at an acceptable level in relation to the salt tolerance of the proposed crop.

**le Chatelier principle** Named after the French inorganic chemist, H.L. le Chatelier (1850–1936), it defines a condition of a system in STABLE EQUILIBRIUM in which a change in any one of the governing forces will cause the equilibrium to shift so that the original condition is restored. In other words, the initial change sets up an internal reaction which is equal and opposite to that change and there is no net alteration in the system.

As used by physical geographers this concept is generally conflated with ideas of homeostasis, negative feedback and self-regulation.

In its original form and related to the thermodynamics of (strictly, isolated) systems, it states: 'If the temperature of a system in equilibrium be raised, the equilibrium will shift in favour of the reaction in which heat is absorbed (endothermic reaction); the converse also applies.' This shows how weathering of rocks, more or less in equilibrium in the lithosphere, tends towards exothermic (heat evolving) reactions. Oxidation, hydration and carbonation are volume-increasing reactions in weathering typically of this kind.

BAK/WBW

**lee depression** Region of low pressure found downwind of a mountain chain, representing the large-scale part of the drag of the ground on the air. The cyclonic circulation is due to the interplay of the vorticity and divergence of the air to the lee of the obstacle. Occasionally such features appear to develop and move away, as in cyclogenesis in the lee of the European Alps, which results in a depression over northern Italy that forces water up the Adriatic sea. This occasionally results in the flooding of Venice.

JSAG

**Reading**
Harwood, R.S. 1981: Atmospheric vorticity and divergence. In B.W. Atkinson ed., *Dynamical meteorology: an introductory selection*. London and New York: Methuen. Pp. 35–54. · McIntosh, D.H. and Thom, A.S. 1969: *Essentials of meteorology*. London: Wykeham Publications.

**lee dune** When winds transporting sand encounter a topographic obstacle, air and sand flow may be accelerated around the obstacle to accumulate in a zone of slack (or separated) flow on the lee side. Aeolian accumulation may then extend down wind parallel to the direction of sand transport, so that lee dunes are usually linear forms. (See also TOPOGRAPHIC DUNE.)

DSGT

**lee eddy** A closed circulation (primarily in the vertical plane) often found on the downward side of steep obstacles. On the smaller scale, it defines a good place for scenic picnics; on the larger scale, it identifies places prone to recirculation of pollutants.

JSAG

**lee waves** Waves in the atmosphere of about 6 km wavelength (see GRAVITY) extending downwind of an obstacle in trains that may be up to 400 km long. Often made visible (remarkably so on satellite pictures) by alternating bands of clear and cloudy air (lenticular CLOUDS), they are conceptually important as identifying some aspects of the irreversibility of atmospheric processes.                    JSAG

**Reading**
Atkinson, B.W. 1981: *Meso-scale atmospheric circulations.* London and New York: Academic Press. · Ludlam, F.H. 1980: *Clouds and storms.* University Park, Pa.: Pennsylvania State University Press.

**lessivage**   Involves the translocation of silicate clays in colloidal suspension in a soil profile without any change in their chemical composition. It contrasts with podzolization in which the clay materials decompose and the hydrous oxides of iron and aluminium are mobilized.

**levée**   A broad, long-crested ridge running alongside a FLOODPLAIN stream or intertidal inlet, composed generally of coarse sand to silt grade suspended sediment deposited by floodwaters as they overtop channel banks. The levée may slope gently away from the river and consist of progressively finer sediment as distance from the channel increases. Rather different features of the same name may also be created on steeper slopes by debris flows: here they may comprise boulders or coarse material. Levées may also be artificially created or raised for flood protection.
JL

**Reading**
Reading, H.G. ed. 1986: *Sedimentary environments and facies.* 2nd edn. Oxford: Blackwell Scientific.

**lichenometry**   A technique for dating Holocene events that was developed in the 1950s. It is especially useful for dating glacial fluctuations over the past 5000 or so years. It is believed that most glacial deposits are largely free of lichens when they are formed, but that once they become stable, lichens colonize their surfaces. The lichens become progressively larger through time. By measurement of the largest lichen thallus of one or more common species, such as *Rhizocarpon geographicum*, an indication of the date when the deposit became stable can be obtained.                    ASG

**Reading**
Innes, J.L. 1985: Lichenometry. *Progress in physical geography* 9, pp. 187–254. · Worsley, P. 1990: Lichenometry. In A.S. Goudie ed., *Geomorphological techniques.* 2nd edn. London: Unwin Hyman. Pp. 422–8.

**life cycle analysis**   (alternatively environmental life cycle analysis or product life analysis). The evaluation of the environmental burdens associated with a product, process or activity. It involves the quantification of the amounts of energy and materials used and the wastes released to the environment during the entire life of product, process or activity, from extraction and processing of raw materials through to transport, manufacturing, maintenance, recyling and final disposal. The purpose of the analysis is to identify opportunities to implement improvements.                    ASG

**life form**   The body shape of an organism at maturity, most commonly used in reference to plants. RAUNKIAER's classification, for example, is based mainly on the nature of perennating buds and their position and protection. This distinguishes plants over or below 25 cm from the ground, at soil level, below ground or lying in mud or water. Other life-form features of plants include the length of shoots or the nature and density of the root system. Animal classifications also employ life-form attributes which result from morphological adaptations to the environment.                    PAF

**lightning**   A luminous discharge associated with a thunderstorm. Several types can be distinguished, principally:

1 cloud discharges (also called sheet lightning) occur between different parts of a thunderstorm, giving a diffuse illumination;
2 ground discharges (also called forked lightning) occur between cloud and ground along a tortuous path with side branches from a main channel;
3 air discharges occur between a part of the cloud and the adjacent air, but otherwise are similar in appearance to a ground discharge (see illustration on p. 295); and
4 ball lightning, reported to have the appearance of a moving, luminous globe-discharge about 20 cm in diameter, sometimes disappearing in a violent explosion.

In addition to being a significant natural hazard, lightning is an important agent in fire ecology and may also have miscellaneous geomorphological effects.                    KJW

**Reading**
Golde, R.H. ed. 1977: *Lightning.* Vol. 1. London and New York: Academic Press. · Norin, J. 1986: Geomorphological effects of lightning. *Zeitschrift für Geomorphologie* 30, pp. 141–50.

**limiting angles**   (of slopes) Describe the upper and lower angles at which distinct processes or forms may occur either in a given locality or under particular environmental conditions. The upper (maximum) limiting angles for continuous regolith and vegetation cover are commonly regarded as being in the range of

**Lightning** strikes the earth's surface an average 100,000 times each day. This spectacular example was photographed over the new volcanic island of Surtsey, Iceland.

40 to 45° in western Europe: in Papua New Guinea on mudstones under rain forest the limiting angle is in the range 70–80°. Lower limiting angles for the occurrence of landslides have been quoted for a few areas: in thin regolith under temperate climates this may be in the range of 18 to 40°, but in periglacial environments the angles are much lower and in the range of 1 to 8°. Further examples are given by Young (1972, p. 165). MJS

**Reading and Reference**
Young, A. 1972: *Slopes*. Edinburgh: Oliver & Boyd.

**limiting factors** Those factors in ecosystems which are in short supply and can thus inhibit efficient and productive ecological development. The concept of limiting factors was introduced in the 1840s by the German chemist, Justus von Liebig, who found that the yield of a crop could be increased only by supplying the plants with more of the nutrient which was present in the smallest quantities. ASG

**Reading**
Blackman, F. 1905: Optimal and limiting factors. *Annals of botany* 19, pp. 281–95. · Park, C.C. 1980: *Ecology and environmental management*. Folkestone: Dawson. Pp. 94–9.

**limnology** Derived from the Greek *limnos* (pool, lake, swamp), limnology is the study of the physical, chemical and biological processes of fresh or saline waters surrounded by land. The actual term limnology was first used by Forel in his 1892 book on the limnology of Lake Leman, although biological limnology began in 1674 when Leewenhoek provided the first microscopic description of the green ALGAE, *Spirogyra*. His pioneering study not only outlined the seasonal cycle of algae in lakes, but gave the first insights about food chain dynamics. By the late 1700s Saussure had devised a way of measuring temperature in deep water and discovered that lakes are thermally stratified. An enormous amount of information on lakes is now available and limnologists are not only concerned with investigating the physical, chemical and biological properties of lake systems but their preservation and enhancement. Researchers are not only interested in the present and future condition of lakes but the past, with the reconstruction of past lake environments being the domain of palaeolimnologists. (See also LAKES.) SLO

**limon** French term used for fine silty sediments laid down by the wind (or possibly water) and therefore equivalent to LOESS.

**line squall** See SQUALL LINE.

**lineament** A large-scale linear feature on the landsurface, such as a trough or ridge, that is the product of the structural geology of a region.

295

**linear dune** Linear dunes can be over 50 m high and may extend for tens of kilometres in length. They are the most common type of desert sand dune (Fryberger and Goudie 1981) and usually occur in extensive dunefields in which they are the dominant or only major dune type that is present, for example in the Kalahari, Australian continental dunefields, and the western Saharan ergs. Despite this, the mechanisms that lead to the formation of linear dunes have proved controversial: for example, one unsubstantiated theory was that they formed under the influence of parallel HELICAL FLOW vortices in the lower atmosphere. It is now known, and substantiated by field investigations by Tsoar (1978) and Livingstone (1986) that linear dunes form in acute bimodal wind regimes, with overall dune orientation parallel/subparallel to the resultant direction of sand transport. In most situations it is seasonal differences in wind regime that make up the components of the formative regime; if one of these results in greater net transport than the other, a slow lateral movement of the dune may be caused. Overall, however, linear dunes do not migrate, but extend slowly in the net transport direction.

The SLIP FACE on a linear dune may be relatively weakly developed, as is the case on a SAND RIDGE, or it may alternate seasonally from one side of the dune to the other. This may give rise to the dune developing a sinuous crestal profile, termed a SEIF DUNE. Overall, the lack of migratory behaviour, and occurrence of dune extension at the downwind end, means that linear dunes can be regarded as 'sand passing' forms. This type of activity can mean that the plinths of linear dunes are relatively stable, to the extent that they may support comparatively dense vegetation covers, with only upper slopes experiencing sand movement. Livingstone and Thomas (1993) have noted that the activity of linear dunes may be extremely episodic, making it difficult in some situations to distinguish between relict forms, inherited from drier climatic periods, and episodically active dunes with limited surface activity.          DSGT

**References**

Livingstone, I. 1986: Geomorphological significance of wind flow patterns over a Namib linear dune. In W.G. Nickling ed., *Aeolian geomorphology*. Boston: Allen and Unwin. Pp. 97–112. · —and Thomas, D.S.G. 1993: Modes of linear dune activity and their palaeoenvironmental significance: an evaluation with reference to southern African examples. In K. Pye ed., *The dynamics and context of aeolian sedimentary systems*. Geological Society of London special publication 72, pp. 91–101. · Tsoar, H. 1978: *The dynamics of longitudinal dunes: final technical report*, European Research Office, US Army.

**liquid limit** The maximum amount of water an unconsolidated sediment or material can hold before it becomes a turbid liquid.

**lithification** The process of the formation of a consolidated rock from originally unconsolidated sediments through cementation or other diagenetic processes.

**lithology** The macroscopic physical characteristics of a rock.

**lithosol** Surficial deposits which do not exhibit soil horizons.

**lithosphere** The earth's crust and a portion of the upper MANTLE, which together constitute a layer of strength, relative to the more easily deformable ASTHENOSPHERE below. On the basis of worldwide heat flow measurements, it has been estimated that the lithosphere varies in thickness from only a few kilometres along the crest of mid-ocean ridges where, according to the PLATE TECTONICS model, new lithosphere is being created, to over 300 km beneath some continental areas. Oceanic lithosphere capped by continental crust tends to be thinner but more dense than continental lithosphere, which is capped by continental crust.          MAS

**Reading**

Pollack, H.N. and Chapman, D.S. 1977: On the regional variations of heat flow, geotherms and lithospheric thickness. *Tectonophysics* 38, pp. 279–96. · Walcott, R.I. 1970: Flexural rigidity, thickness and viscosity of the lithosphere. *Journal of geophysical research* 75, pp. 3941–54. · Wilson, J.T. ed. 1976: *Continents adrift and continents aground*. San Francisco: W.H. Freeman.

**litter** The remains of dead vegetation material, especially tree leaves which are present on the ground surface. They are broken down into essential nutrients by a wide range of DECOMPOSER organisms (bacteria, fungi) and other associates, among which are earthworms, springtails, mites and millipedes. The rate of breakdown and, conversely, the degree of accumulation of litter varies with climate. The amount of litter present in tropical rain forests, where breakdown is fast, may be only $20 \, kg \, ha^{-1}$, or less than 1 per cent of the above-ground BIOMASS; whereas in the cold, dry climate of high-latitude boreal forests, litter accumulation is substantial due to slow rates of breakdown, reaching $300 \, kg \, ha^{-1}$, or *c.*30 per cent of the above-ground biomass (Rodin and Basilevich 1967).          DW

**Reference**

Rodin, L.E. and Basilevich, N.I. 1967: *Production and mineral cycling in terrestrial vegetation*. Edinburgh: Oliver and Boyd.

**litter dam** A small barrier made up of particles of plant litter, especially leaves, twigs, and flower parts. Litter dams are built up when these particles, floating on surface run-off, periodically come to rest and then trap additional particles delivered by the flow. In this way dams may grow to heights of 5 cm or more. The dams run approximately along the contour, and often occur in tiers. They are important, especially in small run-off events, because they retard or hold water and so promote infiltration. They also trap eroded soil, which settles in the impoundment. In dry climates especially, seeds also lodge within litter dams, and the water and nutrients available there may support germination.   DLD

Example of a litter dam.

**Little Climatic Optimum** A phase in early medieval times (*c.*AD 750–1200) when conditions were relatively clement in Europe and North America, allowing settlement in inhospitable parts of Greenland, reducing the problems of ice on the coast of Iceland, and allowing widespread cultivation of the vine in Britain.   ASG

**Reading**
Lamb, H.H. 1982: *Climate, history, and the modern world.* London: Methuen.

**Little Ice Age** See NEOGLACIAL.

**load, stream** The total mass of material transported by a stream. The units employed vary according to the time-base considered. Short-term loads may be expressed in $kg^{-1}$ or $t$ $day^{-1}$, whereas annual loads are expressed in $t$ $year^{-1}$. The total includes both organic and inorganic material and comprises three major components: first, material carried in solution, secondly, material transported in suspension and, thirdly, material moving on the bed of the stream as bedload. The magnitude of the load and the relative importance of the three load

components may vary markedly in both time and space. (See also BEDLOAD; DISSOLVED LOAD; SUSPENDED LOAD.)   DEW

**load structures** Irregular contortions found in fine-grained deposits where sands have been deposited on water-saturated hydroplastic silts or muds. Differences in density, compaction and pore-fluid pressures cause lobes of sand to sink into the underlying silts and/or tongues of mud to rise up into the sand horizons. The resulting load structures exhibit contorted and deformed laminae, often folded, festooned or detached.   JM

**Reading**
Rodin, L.E. and Basilevich, N.I. 1967: *Production and mineral cycling in terrestrial vegetation.* Edinburgh: Oliver & Boyd.

**local climate** See MESOCLIMATE.

**local winds** Those winds which differ from the general winds expected from the pressure pattern due to topographical or urban or other effects. Four main types of local wind have been identified: (1) those winds intensified by topographical features such as a narrow mountain gap or urban canyon; (2) winds that blow along the pressure gradient such as land and sea breezes, mountain and valley winds, and on larger scales föhn, chinook, bora and mistral winds; (3) winds associated with vertical instability such as those accompanying thunderstorms; (4) strong winds due to flow over a level surface with a strong pressure gradient, such as uninterrupted flow over a level surface with a strong pressure gradient, such as sirocco and blizzard.   JET

**lodgement till** See TILL.

**loess** The original German word *Löss* was simply a name for a particular form of loose, crumbly earth. In due course, definitions became more constricted, and that of Flint (1957, p. 181) has received wide currency: 'a sediment, commonly nonstratified and commonly unconsolidated, composed predominantly of silt-sized materials, ordinarily with accessory clay and sand, and deposited primarily by wind'. This definition involves a mechanism of formation that is not universally acceptable, though a hundred years ago Ferdinand von Richthofen, after visits to Tibet and Central Asia, had championed the aeolian cause. Some earlier workers, such as Lyell, had envisaged a fluvial origin, while later workers have proposed non-aeolian mechanisms of formation (e.g. the cosmic origin ideas of Keilhack; the *in situ* formation ideas of Berg; see Smalley 1975 for a selection of earlier papers on loess). The other prime argument about loess formation concerns the mechanism whereby silt-sized quartz material is generated. Some workers stress very strongly the importance of glacial grinding (e.g. Smalley and Vita-Finzi 1968), and the very widespread development of loess deposits around the borders of Pleistocene ice sheets lends some support to this view. These workers have tended to doubt the existence of a mechanism to produce material of appropriate grain size in quartz which is characteristic of the other environment from which loess might be derived – deserts. Processes such as salt weathering may produce silt-sized material in deserts, and the existence of frequent dust storms shows that silt-sized material is present in desert areas and available for wind transportation (Goudie *et al.* 1979).

Whatever their origin, loess deposits are undoubtedly of great importance, partly because of their great areal extent (Mississippi valley, Patagonia, New Zealand, Tunisia, Negev, Central Europe, Soviet Central Asia, China, Pakistan) but also because of their importance as a record of Pleistocene climatic fluctuations. The fossil soils and faunal and floral remains in thick, dated loess sections provide an environmental record that is equalled only by that preserved in the deep sea cores (Kukla 1977). Loess has also been an important influence on human settlement.                           ASG

**Reading and References**

Flint, R.F. 1957: *Glacial and Pleistocene geology.* New York: Wiley. · Goudie, A.S., Cooke, R.U. and Doornkamp, J.C. 1979: The formation of silt from quartz dune sand by salt-weathering processes in deserts. *Journal of arid environments* 2, pp. 105–12. · Kukla, G.J. 1977: Pleistocene land–sea correlations: I. Europe. *Earth science reviews* 13, pp. 307–74. · Pye, K. 1987: *Aeolian dust and dust deposits.* London: Academic Press. · Smalley, I.J. 1975: *Loess lithology and genesis.* Stroudsburg, Pa.: Dowden, Hutchinson & Ross. · —and Vita-Finzi, I.J. 1968: The formation of fine particles in sandy deserts and the nature of 'desert' loess. *Journal of sedimentary petrology* 38, pp. 766–74.

**log skew Laplace distribution** The particles within a sediment population do not always have a normal distribution, but are a mixture of several populations. The Folk and Ward (1957) statistics, widely used in the description of PARTICLE SIZE distribution of sediments, are underpinned by the assumption of a normal distribution. Where a distribution is not normal, but skewed, the Folk and Ward approach is in fact inappropriate (and erroneous). Log hyperbolic distributions were proposed by Bagnold

and Barndorff-Nielsen (1980) as more appropriate from mathematical and sedimentological perspectives, but the calculation of distributions can be complex. Log skew Laplace distributions are robust and relatively simple to calculate (Fieller *et al.* 1990). The adoption of the approach, and of log hyperbolic approaches, has however been very limited in the sedimentological literature.                           DSGT

**References**

Bagnold, R.A. and Barndorff-Nielsen, O., 1980: The pattern of natural size distributions. *Sedimentology* 27, pp. 199–207. · Fieller, N.R.J., Flenley, E.C., Gilbertson, D.D. and Thomas, D.S.G. 1990: Dumb-bells: a plotting convention for 'mixed' grain size populations. *Sedimentary geology* 69, pp. 7–12. · Folk, R.L. and Ward, W.C., 1957: Brazos River bar: a study in the significance of grain size parameters. *Journal of sedimentary petrology* 27, pp. 3–26.

**logan stone** Any large boulder that is so balanced that it readily rocks.

**long profile, river** The graph representing the relation between altitude and distance along the course of the river. The profile is usually concave upwards, is graded to a local or regional BASE LEVEL, and may be punctuated by KNICKPOINTS where the river cuts through former valley floors or river terraces. The profile may be plotted for the bed of the river channel or for the bankfull or channel capacity stage where analysis is to be related to contemporary processes, but it will be plotted for the floodplain or valley floor when related to valley development. The long profile of an entire river or valley may be approximated by one of several equations but in detail over short distances the river long profile is punctuated by pools and riffles.                           KJG

**Reading**

Richards, K.S. 1982: *Rivers: form and process in alluvial channels.* London and New York: Methuen. Pp. 222–51.

**longitudinal dune** Term sometimes used in place of LINEAR DUNE.

**longshore drift** Longshore drift is the transport of beach material along the coast. There are two main processes involved. Beach drifting is caused by the oblique upward transport of material by the swash of short, little refracted WAVES, and its return straight down the swash slope by the BACKWASH, thus moving it alongshore. The process takes place only in the swash zone. Longshore currents in the surfzone, generated by waves approaching the coast at an angle, also move material alongshore. The transport rate depends upon the wave energy and angle of approach mainly. Longshore transport with steep waves is usually at a maximum along the submarine bar crest. The process is of great importance in accounting for long-term coastal erosion and deposition.                    CAMK

**Reading**
Hardisty, J. 1990: *Beaches: form and process*. London: Unwin Hyman. · Komar, P.D. 1971: The mechanism of sand transport on beaches. *Journal of geophysical research* 76.3, pp. 713–21. · —and Inmam, D.L. 1970: Longshore sand transport on beaches. *Journal of geophysical research* 75, pp. 5914–27.

**lopolith**  An igneous intrusion similar to a laccolith but saucer-shaped on both its upper and lower surfaces.

**löss**  See LOESS.

**louderback**  A lava flow on the surface of the dip slope of a faulted block, the presence of which proves that the topography is the product of faulting rather than erosion.

**low flow analysis**  An analysis of the frequency, magnitude and persistence of low discharge and its relationship with climatic and catchment characteristics.                    AMG

**Reading**
Institute of Hydrology 1980: *Low flow studies*. Wallingford, England: Institute of Hydrology. · Task Committee on Low Flows 1981: Characteristics of low flows. *American Society of Civil Engineers: Journal of the Hydraulics Division* 106, pp. 717–32.

**luminescence dating methods**  These provide a means of directly determining the age of mineral grains in numerous sedimentary environments. They are based on estimating the time elapsed since a sediment was last exposed to daylight. They are radiation damage (or sometimes termed 'trapped charge') GEOCHRONOLOGY methods involving quantifying the accumulation of a radiation-related signal. Initial developments in the field, resulting in THERMOLUMINESCENCE (TL) dating, focused on the use of heat to stimulate a time-dependent signal from mineral grains which had been 'zeroed' by exposure to heat during formation or usage (e.g., pottery, burnt flint). In the early 1980s it was realized that exposure to daylight was also capable of 'zeroing' previously accumulated radiation damage and the method was applied to sediment. The date so established was the time elapsed since a sediment was last exposed to daylight (a depositional age). A significant drawback to TL methods was the difficulty in establishing the completeness of the resetting (bleaching) process at deposition. A significant advantage of the related development optical dating methods is the effectiveness of signal resetting at deposition.

Optical dating involves the generation of a signal from a sample by optical stimulation (optically stimulated luminescence [OSL] being produced). Convention now is to further classify the form of signal according to the wavelength of optical stimulation (for example green light (GLSL) and infrared (IRSL) stimulated luminescence are frequently described in the literature). Unlike TL, the eviction of charge from traps is observed only as a rapidly depleting signal when expressed as a function of light exposure. Minerals are variously sensitive to differing wavelengths of light; quartz for example is sensitive to green ($c.1 = 500\,\text{nm}$) and shorter wavelengths; feldspars exhibit a wide range of sensitivities to visible, red and near IR photons. The age of a sample is derived from the general age equation:

$$\text{Age(ka)} = \frac{\text{Equivalent dose(Gy)}}{\text{Dose rate(Gy.ka}^{-1})}$$

where the palaeodose (P) (sometimes referred to as equivalent dose [ED] or accumulated dose [AD]), is the accumulated radiation damage (measured in grays [Gy]); and the dose rate is the rate at which the sample absorbs energy from its immediate proximity; this comprises alpha, beta, and gamma radiation from $^{40}$K, $^{238}$U, $^{235}$U and $^{232}$Th and their protégé products, together with a typically small cosmic ray contribution (Aitken 1985, 1998).

There are a variety of methods which have been employed for evaluation of the palaeodose. Most utilize measurements of the response of single or multiple aliquots of refined samples of quartz, feldspar or mineral mixtures to laboratory radiation to quantify the radiation sensitivity of the mineral used, and all require some form of thermal pre-treatment to remove unstable luminescence components from those which are stable over geological time periods. Palaeodoses are typically calculated either for progressive temperature (TL) or exposure time (OSL) intervals as a means of testing the stability and reproducibility of the estimate. Dose rate evaluation may be achieved by a wide variety of field or laboratory-based spectroscopic, nuclear and chemical methods (Aitken 1985).

The maximum age range of TL and OSL methods is controlled both by the quantity of radiation damage which a given mineral species can accommodate, and the rate at which the damage accumulates. At typical environmental dose rates quartz saturates at between 100 and 150 ka, although under certain circumstances ages of over 700 ka may be obtainable. Feldspars demonstrate a considerably greater capacity to accumulate charge, and correspondingly exhibit a maximum age range up to 800 ka though many feldspars exhibit anomalies which may result in age underestimation. Fine-grained polymineral mixtures from loess have also been dated up to 800 ka.

The minimum age range of the methods is determined by both the completeness of resetting of previously accumulated luminescence at deposition (including the degree to which the solar spectrum is filtered in subaqueous depositional environments), samples sensitivity to ionizing radiation, and subtle charge reorganization effects which may occur during deposition, or which are the result of laboratory pre-treatment procedures.                                    ss

**Reading and Reference**
Aitken, M.J. (1985): *Thermoluminescence dating*. London: Academic Press. · — (1998): *Optical dating*. Oxford: Oxford University Press.

**lunette dune**  A crescent-shaped sand dune found on the downwind margin of some PAN or PLAYA basins (Hills 1940) and widely identified in Australia, southern Africa, Tunisia and Texas. Lunettes are in fact a form of TRANSVERSE DUNE, and while not present on the margins of all pans they are widespread in environments whether the pan depression occurs in a sandy substrate, or where processes on the pan floor lead to the release of sand-size sediment particles. Lunette dunes may form directly by the deflation of sediments from dry pan surfaces, or from deflation from beach sediments that result from wave transport, so that their development could occur during dry or wet phases. While this may influence the palaeoenvironmental significance of lunettes dated to the late Quaternary period, those containing CLAY PELLETS (also called CLAY DUNES) are almost certainly a result of deflation directly from dry floors.                        DSGT

**Reading and Reference**
Bowler, J.M. 1986: Spatial variability and hydrological evolution of Australian lake basins: analogue for Pleistocene hydrological change and evaporite formation. *Palaeogeography, palaeoclimatology, palaeoecology* 54, pp. 21–41. · Hills, E.S. 1940: The lunette: a new landform of aeolian origin. *Australian Geographer* 3, pp. 1–7. · Lancaster, I.N. 1978: Composition and formation of southern Kalahari pan margin dunes. *Zeitschrift fur Geomorphologie* NF 22, pp. 148–69.

**lynchet**  A terrace on a hillside, generally held to be human-made and produced by ploughing. Lynchets are widespread in southern England and northern France.

**lysimeter**  An instrument for assessing evapotranspiration losses from a vegetated soil column. Lysimeters may be used to assess either actual or potential evapotranspiration losses and the estimates are derived using a WATER BALANCE approach. A column of soil and vegetation is placed in a container and replaced in the soil so that the vegetation and soil conditions are as similar as possible to their surroundings. The container should be as large as possible to allow free growth of the vegetation and to reduce the significance of boundary effects. There are two main types of lysimeter, the drainage type and the weighing type, although some weighing lysimeters also allow drainage. Input of water to the lysimeter is assessed using rain gauges, output is measured as drainage from the base of the container enclosing the soil column and changes in soil moisture storage are estimated by repeatedly weighing the soil column. These measurements allow the estimation of losses of water through evapotranspiration. If estimates of potential evapotranspiration are required the lysimeter and a surrounding area are irrigated to ensure that the soil moisture is maintained at field capacity.

There is no standard size for a lysimeter. Larger instruments are less influenced by boundary effects but present problems if soil moisture changes are to be determined accurately by changes in weight of the soil column. Occasionally, special environmental circumstances allow the construction of very large drainage lysimeters. For example, large lysimeters with a Sitka spruce cover were employed by both Law (1957) and Calder (1976) where an impermeable subsoil allowed the construction of lysimeters by isolating a slope plot with an impermeable wall penetrating the soil and collecting the drainage from the soil above the impermeable layer. (See also POTENTIAL EVAPORATION.)                        AMG

**Reading and References**
Calder, I.R. 1976: The measurement of water losses from a forested area using a 'natural' lysimeter. *Journal of hydrology* 30, pp. 311–25. · Kovacs, G. 1976: The use of lysimeters in the hydrological investigation of the unsaturated zone. *Hydrological sciences bulletin* 21, pp. 499–516. · Law, F. 1957: Measurement of rainfall, interception and evaporation losses in a plantation of Sitka spruce trees. *Publications of the International Association of Scientific Hydrology* 44, pp. 397–411. · Reyenga, W., Dunin, F.X., Bautovich, B.C., Rath, C.R. and Hulse, L.B. 1988: A weighing lysimeter in a regenerating eucalyptus forest: design, construction and performance. *Hydrological processes* 2, pp. 301–14.

# M

**maar**  An old volcanic crater. A pond or lake formed in such a depression.

**machair**  A term commonly applied to the landform/vegetation systems of many dune pasture areas of parts of the highlands and islands of Scotland. The essentials of machair have been summarized by Ritchie (1976).

**Reference**
Ritchie, W. 1976: The meaning and definition of Machair. *Transactions of the Botanical Society of Edinburgh* 42, 431–40.

**macrofossils**  Animal or plant fossil remains visible with the naked eye but which usually require microscopic examination for identification. The commonest macrofossils are those of the genus *Sphagnum,* which occur in and may comprise the bulk of many PEAT deposits; other common macrofossils include seeds and fruits (often abundant in lake sediments), insect remains and molluscs. Many can be identified down to species level. The study of all three groups has been well developed in Quaternary palaeoecology to demonstrate local vegetational changes and hydroseral development, and to investigate wider phenomena such as climatic change (insects, molluscs and peat macrofossils) and marine transgressions (molluscs).    KEB

**Reading**
Barber, K.E. 1981: *Peat stratigraphy and climatic change.* Rotterdam: Balkema. · Birks, H.J.B. and Birks, H.H. 1980: *Quaternary palaeoecology.* London: Edward Arnold. · Godwin, H. 1975: *History of the British flora.* 2nd edn. Cambridge: Cambridge University Press.

**macrometeorology**  The study of weather systems of large scale, up to and including the scale of the earth itself. It studies the largest in the classification of micro-, meso-, and macro-scales of atmospheric motion. The lower limit of the macroscale is variously defined in the range a hundred to a few thousand kilometres. If the lower value is used hurricanes, cyclones and anti-cyclones are included in macroscale, but these are often classed separately as synoptic scale systems. Macroscale systems of larger scale include waves in the westerlies (ROSSBY WAVES), MONSOONS, the southern oscillation, the quasi-biennial oscillation and, the largest motion system of all, the mean global flow pattern or GENERAL CIRCULATION.

Waves in the westerlies are vast meanders of the basically westerly flow in the upper troposphere and lower stratosphere, usually numbering about four around a latitude circle.

Monsoonal circulations, on the scale of the continents, develop in response to the differing behaviour of ocean and land in the annual variation of solar input. The processes involved in monsoons are similar to those associated with land and sea breeze circulations, but are on much larger time and space scales. In summer, air over a continent is warmed much more than that over an ocean, so that a thermally induced low pressure area over the land leads to a convergent, cyclonically rotating flow at low levels. In winter the situation is reversed, giving a high over the cold continent and winds in the reverse direction. The principal monsoon circulation is that associated with the land mass of Asia.

The southern oscillation is a fluctuation of the intertropical general circulation, and in particular that part in the Indian and Pacific Ocean regions, also called a Walker circulation. In this there is an exchange of air between the southeast Pacific subtropical high and the Indonesian equatorial low. The circulation is driven by temperature differences between the two areas – the relatively cool south-east Pacific and the warm western Pacific/Indian Ocean region. The complex climatological relationships between these two areas (and others) was first introduced by Sir Gilbert Walker who found that when pressure is high over the Pacific Ocean, it tends to be low in the Indian Ocean; rainfall amount varies in the opposite direction to pressure. The southern oscillation is by no means regular in time, and in this regard the word 'oscillation' is somewhat misleading.

The quasi-biennial oscillation (QBO) is a major reversal of wind direction in the stratosphere with a period of between 22 and 29 months, with westerly winds for roughly one year and easterly winds for the following year. The oscillation has its largest amplitude near the 25 km level, but disappears at the tropopause. The phase of the oscillation varies with height, with the wind direction changes first appearing at about 30 km and propagating downward at a rate of about 1 km per month. The oscillation has its largest amplitude above the equator, decreasing poleward and becoming very small at about 30° latitude. There is a small temperature oscillation associated with the oscillation in wind, of amplitude of about $2\,°C$ near 25 km height above the equator. Observations and theory provide evidence that the energy for the QBO is provided by vertically propagating tropospheric waves.     KJW

**Reading**
Atkinson, B.W. 1981: *Meso-scale atmospheric circulation.* London and New York: Academic Press. · Holton, J.R. 1972: *An introduction to dynamic meteorology.* New York and London: Academic Press. · Lockwood, J.G. 1979: *Causes of climate.* London: Edward Arnold. · Riehl, H. 1979: *Climate and weather in the tropics.* London and New York: Academic Press.

**macrophyte** Although in the strict sense of the word, this is a plant large enough to be seen without the aid of a microscope, in practice it refers to the larger photosynthesizing plants in a community. More typically used in relation to aquatic or marine habitats (see AQUATIC MACROPHYTE), in the context of terrestrial environments the term most likely refers to plants of more substantial physical stature, for example trees and shrubs as opposed to grasses and herbs. In the terrestrial sense, then, the term is relative and would not be used to describe all those plants large enough to be seen by the naked eye. Thus, in a temperate deciduous forest, the imposing oak and beech trees, reaching perhaps 30 m in height, would be described as macrophytes, whereas the smaller herbaceous flowering plants in the understorey would not.     MEM

**maelstrom** A powerful tidal current or whirlpool.

**magma** Fused, molten rock material found beneath the earth's crust from which igneous rocks are formed. Magma may contain gases and some solid mineral particles.

**magnetic anomaly** A fluctuation in the strength of the magnetic field mapped over large areas of the ocean floor. Magnetic anomalies were so named when they had been mapped but not yet explained. Their origin lies in seafloor spreading, which results when new basaltic rock wells up along the mid-oceanic fracture system, and, cooling as it does so, is magnetized in the direction of the earth's own magnetic field. The geomagnetic field periodically reverses, however, with the southern magnetic pole exchanging locations with the northern magnetic pole. New basalts intruded into the sea floor during a time of REVERSED POLARITY then have a magnetic orientation opposite to that of the present-day field. Thus, when the apparent strength of the field in this vicinity is measured, it appears anomalously weak, the field in the basalts opposing the earth's field. Likewise, above rocks magnetized normally, the field appears anomalously strong since the field of the rocks reinforces the earth's field.     DLD

**Reading**
McElhinny M.W. 1973: *Palaeomagnetism and plate tectonics.* Cambridge: Cambridge University Press.

**magnetic declination** The angle separating true north (the direction to the northern geographic pole of rotation) from magnetic north, the direction to the northern magnetic pole. Since it is related to fluid processes in the outer core, the geomagnetic field is not completely stable in its orientation, and indeed drifts westward, lagging the earth's rotation.     DLD

**Reading**
Merrill, R.T., McElhinny, M.W. and McFadden, P.L. 1996: *The magnetic field of the earth: paleomagnetism, the core, and the deep mantle.* San Diego: Academic Press.

**magnetic storm** A high level of magnetic disturbance produced by particles of solar origin, causing rapid field variations over the earth. Such storms disturb the ionosphere causing anomalous radio propagation and adversely affecting cable telegraphy. During storms, aurora, arcs and rays of coloured light appearing in the sky are visible much further towards the equator than their usual position.     KJW

**Reading**
*Scientific American* 1979: *The physics of everyday phenomena.* San Francisco: W.H. Freeman.

**magnetic susceptibility** Fe-bearing minerals are widespread in the natural environment and their presence may be diagnostic of geology, soil processes, sediment pathways, pollution and biological conditions. Large differences in the way Fe is organized at the atomic level allow these minerals to be studied and classified by their magnetic properties. The most simple and fundamental property is the ease by which a mineral is magnetized, which is termed *magnetic susceptibility*. Measurements of environmental samples provide first order information about the type of Fe-bearing minerals and their concentrations. Routine measurements are rapid, non-destructive and may be made in the laboratory or field using portable probes and sensors. In the natural environment, metals, Fe-bearing minerals and materials can be divided according to their magnetic properties: ferromagnetic (e.g. iron, nickel); ferrimagnetic (e.g. magnetite, maghemite, titanomagnetite, greigite); imperfect antiferromagnetic (e.g. goethite, haematite) and paramagnetic (e.g. ferrihydrite, lepidocrocite, ilmenite, Fe-silicates). These all show a positive susceptibility but it is the strongly magnetic ferrimagnetic group which dominates a bulk magnetic susceptibility value in the majority of samples. The susceptibility of magnetite is about 1000 times higher than a typical paramagnetic mineral. Materials which are generally considered as non-magnetic (e.g. quartz, water, organic matter, $CaCO_3$) are termed diamagnetic and show a very weak and negative susceptibility. Within the ferrimagnetic group especially, minerals may also be divided according to the type of internal organization into magnetic domains controlled by crystal dimensions. Thus for magnetite there are multidomain, pseudo-single domain, stable single domain ($<0.2$ mim) and superparamagnetic ($< 0.03 \, \mu$m) domain states. Identification of the domain state provides important information about the origins and formation of ferrimagnetic minerals. Primary minerals derived from igneous rock are predominantly multidomain pseudo-single domain and stable single domain, and secondary minerals produced biogeochemically in soil and sediments, or through fire are mainly stable single domain and superparamagnetic. (See also ENVIRONMENTAL MAGNETISM.)

*Volume susceptibility* ($\kappa$) is defined as the ratio of magnetization (M) to an applied magnetic field (H), thus $\kappa =$M/H. *Mass specific susceptibility* ($\chi$) is defined as the volume susceptibility divided by sample density (p), thus $\chi = \kappa$/p. Preferred measurement units are SI where magnetization and field are expressed in $A \, m^{-1}$, and density is expressed as $kg \, m^{-3}$. Thus $\kappa$ is dimensionless, and $\chi$ is expressed in units of $m^3 \, kg^{-1}$. For magnetite, $\kappa$ is $\sim 3.1$ SI and $\chi$ is $\sim 600 \times 10^{-6} m^3 \, kg^{-1}$. Most routine measurements are made in low magnetic fields (typically $80 \, A \, m^{-1}$) at room temperature, but there are three additional sets of measurements: (i) in high magnetic fields (typically $> 10^6 \, A \, m^{-1}$); (ii) in low fields using two AC frequencies set a decade apart, and (iii) at low or high temperatures. *High field susceptibility* is controlled by non-ferrimagnetic behaviour. In practice, it is often a good estimate of the concentration of total Fe contained in the paramagnetic iron oxides, iron hydroxides and Fe-bearing silicates, which in terms of mass contribution, may outweigh the ferrimagnetic minerals by orders of magnitude.

*Frequency-dependent susceptibility* ($\chi_{FD}$) detects very fine-grained ferrimagnetic minerals, lying within the superparamagnetic domain range. Expressions of $\chi_{FD}$ are in mass specific ($\chi_{LF} - \chi_{HF}$) or percentage terms ($[\chi_{LF} - \chi_{HF}/\chi_{LF}] \times 100$) where *LF* and *HF* denote low and high frequency (typically 470 and 4700 Hz). Percentage $\chi_{FD}$ values range from zero in samples containing no superparamagnetic grains to $\sim 15\%$ in samples where superparamagnetic grains completely dominate the magnetic mineral assemblage. The measurements are most useful in studies of modern and old soils where superparamagnetic grains are an important component of secondary ferrimagnetic mineral formation.

*Thermal susceptibility measurements* ($\chi$-T) from $-196 \, °C$ (liquid nitrogen) to room temperature, and from room temperature to $\sim 700 \, °C$, may provide valuable information about mineral and domain type. Two types of diagnostic behaviour are common. Thermal disordering causes the susceptibility of paramagnetic minerals to decline with increasing temperature (Curie–Weiss law) and in other minerals and domains occurs at specific temperatures known as Curie points, Néel points or blocking temperatures. For instance, the ordered ferrimagnetic behaviour of magnetite becomes disordered and paramagnetic at $\sim 580 \, °C$ with an almost total loss of susceptibility. Transitions in magnetocrystalline properties cause changes in susceptibility at specific temperatures, notably the Verwey transition in multidomain magnetite ($-155 \, °C$) and Morin transition in multidomain haematite ($-10 \, °C$). Measurements made at low temperature are essentially non-destructive with reversible $\chi$-T changes, while those at high temperature often lead to irreversible changes, mineral destruction and the formation of new minerals. Magnetic susceptibility measurements are based on the singular property of in-field magnetization: further room temperature and thermal measurements, especially of remanent magnetization in a zero field, are needed in many studies in order to identify mineral types, domains and their concentrations using their magnetic properties alone.                    JAD

**Reading**

Dearing, J.A. 1999: *Environmental magnetic susceptibility.* Kenilworth: Chi Publishing. · Dunlop, D.J. and Özdemir, Ö. 1998: *Rock magnetism: fundamentals and frontiers.* Cambridge: Cambridge University Press. · O'Reilly, W. 1984: *Rock and mineral magnetism.* Glasgow: Blackie. · Thompson, R. and Oldfield, F. 1986: *Environmental magnetism.* London: George Allen and Unwin.

**magnitude and frequency effects** Discrete events in the natural world can be characterized by their magnitude and frequency of occurrence. Generally speaking, large events of the same process occur seldom and small events occur often. When the magnitude of the event is plotted against its frequency the resulting curve usually shows an exponential decline, as in the case of earthquakes, storms and floods. For example, large damaging earthquakes are relatively rare for any given place on the surface of the earth, while small imperceptible ones are an everyday occurrence. There are some exceptions, however. Small rockfalls, for example, occur with modest frequency while intermediate-sized ones occur most often (Gardner 1977).

In the most common applications the systematic decline of frequency with increasing magnitude lends itself well to statistical modelling. Research on the subject is especially common in FLOOD FREQUENCY analysis, where four types of probability distributions have seen application: lognormal, Gumbel Type I, Gumbel Type III extreme value (a logarithmic transformation of the Gumbel Type I), and Pearson Type III (see Chow 1964). Special graph paper is available for use with some of the distributions that allows straight-line plotting of untransformed data (Craver 1980). The choice of which of these extremal frequency functions to use in a particular application has little theoretical support and usually depends on convenience or goodness of fit between the data and the selected distribution.

The declining exponential function describing the relationship between the magnitude and frequency of floods shows some important identifiable values, especially if the annual flood series (a series consisting of the largest flood of each year of record) is considered. On a Gumbel Type I distribution, for example, the arithmetic mean is at 2.33 years, so that the discharge of the mean annual flood is the magnitude that corresponds to the 2.33 year return interval. Similarly, the median annual flood is the one with a magnitude corresponding to the 2.00 year return interval. The most probable annual flood is the one with a return interval of 1.58 years, but when records shorter than one year are included in the analysis the most probable annual flood has the expected value for a return interval of 1.00 year.

Magnitude and frequency are sometimes considered on the basis of the concept of recurrence interval or return interval which is defined as the amount of time that an event of a given magnitude is expected to be equalled or exceeded. Therefore a flood with a 100 year recurrence interval is one that has a magnitude that is expected to be equalled or exceeded only once in a century. The recurrence interval ($T$) of a given event is defined as:

$$T = (n+1)/m$$

where $n$ is the number of years of record and $m$ is the rank of the event magnitude, with the largest event having $m = 1$. A plot of recurrence intervals versus associated magnitudes usually produces a group of points that approximates a straight line on semi-logarithmic paper. A predictive function may be fitted to the data numerically, but in many cases a simple graphic line fit is adequate on account of one or two outlying points which unduly affect the numerical fit. The definition of recurrence interval is as sensitive to length of record as it is to the rank of a given event, so it is sometimes difficult to interpret in those cases where the record is short.

Recurrence interval is directly related to the probability of occurrence or exceedence in any given year as follows:

$$q = 1 - (1 - (1/T))\exp n$$

where $q$ is the annual probability, $T$ is the recurrence interval of an event of a given magnitude, and $n$ is the number of years considered. Thus in one year's time ($n = 1$) the probability of experiencing the 100-year flood ($T = 100$) is 0.01 or 1 per cent.

The effects of hydrological events such as floods are different for events with different magnitudes and frequencies. For example, in humid regions in stream channels with no entrenchment the channel cross-section is usually adjusted so that the flood with the recurrence interval of 1.50 in the annual flood series fills the channel but does not overflow.

An important consideration in geomorphic studies is the understanding of which events accomplish the most work. If work is defined as sediment being transported out of the basin, in humid regions 90 per cent of the work is performed by processes operating at least once every five to ten years (Wolman and Miller 1960). This implies that although large, rare events appear to be impressive in their impacts it is the more frequent events that are the most important in the long run because they do intermediate amounts of work but they do that work often. This generalization is not consistent from one climatic region to another: Neff (1967) found that in arid regions only 40 per cent of the work (defined as sediment transport) was performed by the events occurring with less than a ten-year recurrence interval. In arid regions, therefore, the less common events are responsible for most of the work.

Different processes may have different relationships between magnitude, frequency, and amount of work. In his study of rockfalls Gardner (1977) found that the most work was performed by the largest event during his observation period, even though the event occurred only once. Landslides demonstrate still other relationships (Wolman and Gerson 1978).

In a larger scale of analysis the relative importance of large, infrequent events and smaller, more frequent ones has been a long-standing debate in the science of geomorphology (Pitty 1982). Two hundred years ago landscape scientists thought that many of the features of the surface of the earth were the product of catastrophic events such as the biblical flood. The concept of UNIFORMITARIANISM provided an alternative view that specified frequent operation of small-scale events over a long time as an explanation of the modern landscape. These two opposing views have merged to a certain degree so that the landscape is now viewed as the product of a long series of events of relatively low magnitude with some evidence of catastrophic events superimposed. Processes in the past are considered to have operated in the same fashion as we observe now, but at some times with different intensities (Pitty 1982).   WLG

**References**

Chow, V.T. 1964: Frequency analysis. In V.T. Chow ed., *Handbook of applied hydrology*. New York: McGraw-Hill. · Craver, J.S. 1980: *Graph paper from your copier*. Tucson, Arizona: HP Books. · Dunne, T. and Leopold, L.B. 1978: *Water in environmental planning*. San Francisco: W.H. Freeman. · Gardner, J.S. 1977: *Physical geography*. London: Harper's College Press. · Neff, E.L. 1967: Discharge frequency compared to long-term sediment yields. *Publications of the International Association of Scientific Hydrology* 75, pp. 236–42. · Pitty, A.F. 1982: *The nature of geomorphology*. London and New York: Methuen. · Wolman, M.G. and Gerson, R. 1978: Relative scales of time and effectiveness of climate in watershed geomorphology. *Earth surface processes* 3, pp. 189–208. · — and Miller, J.P. 1960: Magnitude and frequency of forces in geomorphic processes. *Journal of geology* 68, pp. 54–74.

**mallee**   Eucalyptus scrub characteristic of the semi-arid areas of Australia.

**mammilated surface**   The surface of a rock outcrop that has been smoothed and rounded by erosional processes.

**mangrove**   Plant communities dominated by the genera *Rhizophora*, *Bruguieria* and *Avicennia* which colonize tidal flats, estuaries and muddy coasts in tropical and subtropical areas. Communities of mangroves, termed *mangals*, play an important role on many tropical coasts. They are highly productive ecosystems which are capable of exporting energy and materials to adjacent communities, they support a diverse heterotrophic food chain, act as nurseries in the life cycles of some organisms, and offer some protection against coast erosion and storm surge attack. At the present time, like many types of wetland, they are under severe anthropogenic pressures.   ASG

**Reading**

Tomlinson, P.B. 1986: *The botany of mangroves*. Cambridge: Cambridge University Press.

**manned earth resources satellites**   Satellites carrying a human crew plus photographic and other REMOTE SENSING devices for the production of images of the earth's surface. The US manned satellites of Mercury, Gemini and Apollo are known for their participation in the 'space race' while the later US manned satellites of Skylab and Space Shuttle were experimental space stations. All were designed and operated by the National Aeronautics and Space Administration (NASA). A complementary series of manned earth resources satellites were launched by the CIS but as their image products are not readily available to physical geographers they will not be discussed here.

The first successful photographs of the earth from space were taken in 1961 from a Mercury satellite on its fourth mission. The interest provoked by these photographs led to colour photographs being taken of areas of exposed geology during the satellite's later missions. Following the photographic success of the Mercury missions the two-man crews of the Gemini series of satellites were asked to obtain photographs of 'interesting' areas on the earth's surface. On the third Gemini mission in 1965 photographs were taken of the south-west United States. The fourth Gemini mission was much longer and enabled over 100 photographs to be taken. The fifth Gemini mission suffered from a lack of power that prevented optimal alignment for ground photography. Nevertheless 1275 photographs were taken of many parts of the world. The sixth and seventh Gemini missions returned 310 photographs between them. During 1966 many more colour photographs were taken as part of a further five Gemini missions, increasing the total number of photographs taken from a Gemini satellite to over 2400.

By the time that the Apollo series of satellites was launched an experiment had been formalized. The aims were first to obtain automatic colour photography from Apollo 6 in 1967, second to obtain handheld colour photography from Apollos 7 and 9 for areas of the earth for which aerial photography and ground data were also available, and third to obtain multiband photography using a multi-camera array from Apollo 9. The first two objectives were fulfilled and over 600 colour photographs were taken.

The third objective was also successful and provided the stimulus for NASA to push ahead with the development of the satellite Landsat.

The space station Skylab was first launched by NASA in 1973 into a near-circular orbit 435 km above the earth's surface. It circled the earth every 93 min and passed over each point on its surface between 50° north and 50° south once every five days. Skylab was designed for a number of experiments, one of which was concerned with remote sensing and was named the Earth Resources Experiment Package (EREP). This comprised three types of aerial camera, a 13-channel multispectral scanner, and a number of non-imaging remote sensors. Of greatest interest to physical geographers was the S190A multi-camera array aerial camera which provided good quality photographs with a spatial resolution of 30–80 m in six wavebands and the S190B earth terrain aerial camera which provided good quality photographs with a spatial resolution of 20 m in one or three wavebands.

The Space Shuttle is a series of NASA spacecraft designed to shuttle backwards and forwards between earth and space. They started flying in April 1981 and are likely to continue well into the twenty-first century. Each Space Shuttle contains two rocket boosters, one liquid propellant tank and an orbiter vehicle. The orbiter vehicle carries astronauts who take photographs of the earth's surface on a regular basis and two remote sensing packages named OSTA and Spacelab.

The OSTA package is a collection of remote sensing experiments and is named after its NASA designers, the former Office of Space and Terrestrial Applications. The package has included three environmentally useful sensors, a SIDEWAYS LOOKING AIRBORNE RADAR (SAR type), a pair of television cameras and an optical imager. Spacelab, which is built and operated by the European Space Agency, holds a number of sensors, including an aerial camera and a sideways looking airborne radar (SAR type).     PJC

**Reading**
Barrett, E.C. and Curtis, L.F. 1992: *Introduction to environmental remote sensing*. 3rd edn. London and New York: Chapman & Hall. · Lowman, P.D. 1980: The evolution of geological space photography. In B.S. Siegal and A.R. Gillespie eds, *Remote sensing in geology*. New York: Wiley. Pp. 91–115.

**Manning equation**   In 1891 the Irish engineer Robert Manning summarized uniform flow data in the popular, widely used FLOW EQUATION:

$$\nu = (kR^{\frac{2}{3}}s^{\frac{1}{2}})/n$$

Here $\nu$ is velocity (m$^3$ s$^{-1}$), $R$ is the HYDRAULIC RADIUS (mean depth (m) in wide channels), and $s$ is the slope (m m$^{-1}$). The Manning ROUGHNESS coefficient $n$ is a general measure of channel resistance which is numerically constant regardless of the measurement units being used. The coefficient $k$ therefore accommodates the variation of measurement system, and is 1 for SI units and 1.49 for Imperial units. In straight natural streams, $n$ ranges from 0.03 for smooth sections to 0.10 for rocky or heavily vegetated sections (Chow 1959, pp. 98–123; Barnes 1967). It is related to the coefficient $C$ in the CHÉZY EQUATION by:

$$C = R^{\frac{1}{6}}/n.$$     KSR

**Reading and References**
Barnes, H.H. 1967: Roughness characteristics of natural channels. *US Geological Survey water supply paper* 1849. · Chow, V.T. 1959: *Open channel hydraulics*. London: McGraw-Hill. · Manning, R. 1891: On the flow of water in open channels and pipes. *Transactions of the Institution of Civil Engineers of Ireland* 20, pp. 161–207.

Values of Manning's roughness coefficient for various types of natural channel

| Channel type | Normal value | Range |
| --- | --- | --- |
| Small channels (width < 30 m) | | |
| Low-gradient streams | | |
|    Unvegetated straight channels at bankfull stage | 0.030 | 0.025–0.033 |
|    Unvegetated winding channels with some pools and shallows | 0.040 | 0.033–0.045 |
|    Winding vegetated channels with stones on bed | 0.050 | 0.045–0.060 |
|    Sluggish vegetated channels with deep pools | 0.070 | 0.050–0.080 |
|    Heavily vegetated channels with deep pools | 0.100 | 0.075–0.150 |
| Mountain streams (with steep unvegetated banks) | | |
|    Few boulders on channel bed | 0.040 | 0.030–0.050 |
|    Abundant cobbles and large boulders on channel bed | 0.050 | 0.040–0.070 |
| Large channels (width > 30 m) | | |
|    Regular channel lacking boulders or vegetation | – | 0.025–0.060 |
|    Irregular channel | – | 0.035–0.100 |

*Source*: Based on data in Chow, V.T. ed. 1964: *Handbook of applied hydrology*. New York: McGraw-Hill.

**mantle** The zone within the earth's interior extending from 25 to 70 km below the surface to a depth of 2900 km and lying between the partially molten core and the thin surface crust. The uppermost rigid section of the mantle forms the lower part of the LITHOSPHERE. Extending a few hundred kilometres below this region is the ASTHENOSPHERE, a layer in which the magnesium and iron-rich silicate rocks of the mantle are probably partially molten. Most of the convection within the mantle, viewed by some as a crucial mechanism of PLATE TEC-TONICS, is generally thought to occur in this zone.                                    MAS

**Reading**
Davies, P.A. and Runcorn, S.K. eds 1980: *Mechanisms of continental drift and plate tectonics*. London: Academic Press. · Ringwood, A.E. 1975: *Composition and petrology of the earth's mantle*. New York: McGraw-Hill. · Wyllie, P.J. 1976: The earth's mantle. In J.T. Wilson, *Continents adrift and continents aground*. San Francisco: W.H. Freeman. Pp. 46–57.

**mantle plume** A convectional flow of hot rock that rises through the MANTLE to the base of the LITHOSPHERE and which gives rise to a HOT SPOT on the surface. In the oceans, mantle plumes create lines of large volcanoes (such as the Hawaiian chain) if the overlying plate is moving with respect to the plume. On the continents, mantle plumes have probably been responsible for generating voluminous and extensive accumulations of BASALT flows, such as those of the Karoo (South Africa) and the Deccan (India), many of which are located at PASSIVE MARGINS and were thus originally close to sites of continental breakup. They are also probably capable of causing surface uplift over areas in excess of 1000 km across, both by the direct effects of heating of the lithosphere and by the associated thickening of the crust through the addition of enormous quantities of igneous material.                              MAS

**Reading**
Summerfield, M.A. 1991: *Global geomorphology*. London and New York: Longman Scientific and Technical and Wiley. · White, R.S. and McKenzie, D. 1989: Magmatism at rift zones: The generation of volcanic continental margins and flood basalts. *Journal of geophysical research* 94, pp. 7685–729.

**maquis** Scrub vegetation of evergreen shrubs characteristic of the western Mediterranean, and broadly equivalent to chaparral. Many areas of maquis probably represent human-induced degradation of the natural forest vegetation.

**margalitic** Soil A horizons which are dark coloured with a high base status, Ca and Mg being the predominant exchangeable cations.

**marginal channel** A meltwater stream flowing at the margin of a glacier.

**marine pollution** The pollution of the oceans and seas. At first sight, as Jickells *et al.* (1991, p. 313) point out, two contradictory thoughts may cross our minds on this issue:

The first is the observation of ocean explorers, such as Thor Heyerdahl, of lumps of tar, flotsam and jetsam, and other products of human society thousands of kilometres from inhabited land. An alternative, vaguer feeling is that given the vastness of the oceans (more than 1,000 billion billion litres of water!), how can man have significantly polluted them?

What is the answer to this conundrum? Jickells *et al.* (p. 330) draw a clear distinction between the open oceans and regional seas and in part come up with an answer:

The physical and chemical environment of the open oceans has not been greatly affected by events over the past 300 years, principally because of their large diluting capacity... Material that floats and is therefore not diluted, such as tar balls and litter, can be shown to have increased in amount and to have changed character over the past 300 years.

In contrast to the open oceans, regional seas in close proximity to large concentrations of population show evidence of increasing concentrations of various substances that are almost certainly linked to human activities. Thus the partially enclosed North Sea and Baltic show increases in phosphate concentrations as a result of discharges from sewage and agriculture.

Likewise, it is clear that pollution in the open ocean is, as yet, of limited biological significance. GESAMP (1990), an authoritative review of the state of the marine environment for the United Nations Environment Programme, reported (p. 1):

The open sea is still relatively clean. Low levels of lead, synthetic organic compounds and artificial radionuclides, though widely detectable, are biologically insignificant. Oil slicks and litter are common along sea lanes, but are, at present, of minor consequence to communities of organisms living in open-ocean waters.

On coastal waters it reported (p. 1):

The rate of introduction of nutrients, chiefly nitrates but sometimes also phosphates, is increasing, and areas of entrophication are expanding, along with enhanced frequency and scale of unusual plankton blooms and excessive seaweed growth. The two major sources of nutrients to coastal waters are sewage disposal and agricultural run-off from fertilizer-treated fields and from intensive stock raising.

Attention is also drawn to the presence of synthetic organic compounds – chlorinated hydrocarbons, which build up in the fatty tissue of top

307

predators such as seals which dwell in coastal waters. Levels of contamination are decreasing in northern temperate areas but rising in tropical and subtropical areas due to continued use of chlorinated pesticides there.                    ASG

**References**
GESAMP    (IMO.FAO/UNESCO/WMO/WHO/IAEA/ UN/ENEP Joint Group of Experts on the Scientific Aspects of Marine Pollution) 1990: The state of the marine environment. *UNEP regional seas reports studies* 115, Nairobi. · Jickells, T.D., Carpenter, R. and Liss, P.S. 1991: Marine environment. In B.L. Turner, W.C. Clark, R.W. Kates, J.F. Richards, J.T. Mathews and W.B. Meyer eds, *The earth as transformed by human action*. Cambridge: Cambridge University Press. Pp. 313–34.

**Markov process**   A statistical process in which the probability of an event, in a sequence of random events, is influenced by the outcome of the preceding event. Any system in which an outcome depends in part on the previous outcome may be described as exhibiting a Markov property. Any system in which successive events are partially dependent is said to have a 'memory' of the preceding event. The methodology is formalized using a transition matrix in which entries represent the probabilities of transition from one state to another. When the matrix is powered the probabilities represent transitions from one state to another in two steps. Repeated powering will develop a matrix in which all rows are the same. States may also be defined which are absorptive, when the realization of the process terminates, or reflexive, when a process enters a state and then returns to the previous state. The utility of the process depends on the length of the record on which the probabilities are established, whether the series exhibits the Markov property, is stationary and whether the transition probabilities are invariant with time.
                                                   DB

**Reading**
Harbaugh, J.W. and Bonham-Carter, G. 1970: *Computer simulation in geology*. New York: Wiley. · Scheidegger, A.E. 1966: Effect of map scale on stream orders. *Bulletin of the International Association of Scientific Hydrology* 11, pp. 56–61. · Thornes, J.B. and Brunsden, D. 1977: *Geomorphology and time*. London: Methuen.

**marl**   An argillaceous, non-indurated calcium carbonate sediment, frequently grey to blue-grey in colour, formed in freshwater environments.                                               PSh

**mass balance**   Changes in the mass of a sedimentary landform over time. The concept is fundamental to the understanding of a glacier where a zero mass balance describes a situation where accumulation balances ABLATION, a positive mass balance describes an increase in glacier mass and a negative mass balance a reduction in glacier mass. The concept can also be applied to other sedimentary landforms, for example a beach, talus or sand dune.                    DES

**mass movement types**   The modes of hillslope failure. Various methods have been used to classify mass movements including reference to parent material, the causes and mechanics, slope and landslide geometry, shape of failure surface, post-failure debris distribution, soil moisture and physical properties. The most generally accepted method is the complex variable classification of Varnes (1958). The most common general groupings are linked to the processes of fall, slide, flow and creep.

Falls of rock or soil are the free movement downslope of slope-forming materials. Failures may be in circular, plane, wedge, toppling or settling modes. Slides of rock, mud and soil take place by movement on a sliding surface which may be circular, non-circular or planar in form. Circular features may be single, multiple or successive in character. Non-circular failures are usually controlled by lithology or rock structures such as bedding planes which impart a translational component to the slide. A graben form at the head or retrogressive failure may also be evident. Planar failures are usually shallow and translational including debris slides and mudslides, but may be of immense size such as the major rock slides along discontinuities in mountain regions.

All slides may disintegrate upon failure to develop into an 'avalanche' or 'flow'. These involve complex processes such as fluidization, liquefaction, cohesionless grain flow, remoulding and possibly air lubrication. Flows represent a transitional set of processes lying between streamflow or mass transport and mass movement. The term debris flow is used for the mass movement of a wet mixture of granular slides, clay, water and air under the influence of gravity with intergranular shear distributed evenly through the mass. They usually occur in three modes, as simple flows on hillsides, as valley-confined flows debouching onto fans or as catastrophic flows which effectively destroy everything in their path and override topography. This is a convenient division based on scale but the processes are probably similar. Creep includes the continuous, gravity or mass creep of hillslopes affected by the force of gravity in which low-continued stresses deform the slope materials at a fairly constant rate but to a considerable depth. The forms which result are gravity slides or *sackung*, cambering and valley bulging (also produced by periglacial conditions

TYPE OF
MOVEMENT

MATERIAL

RESISTANT
ROCK

NONRESISTANT
SOILS & ROCKS

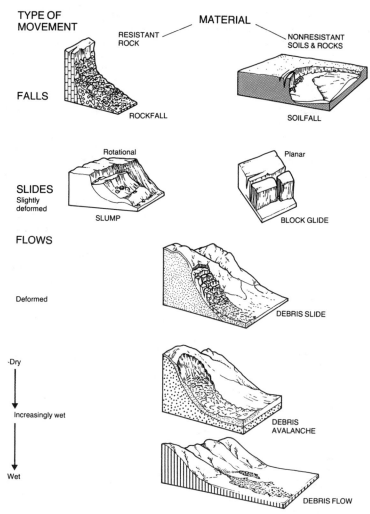

FALLS

ROCKFALL

SOILFALL

SLIDES
Slightly
deformed

Rotational

Planar

SLUMP

BLOCK GLIDE

FLOWS

Deformed

DEBRIS SLIDE

·Dry

Increasingly wet

DEBRIS
AVALANCHE

Wet

DEBRIS FLOW

Mass movement types.
*Source*: M.J. Selby 1982: *Hillslope materials and processes*. Oxford: Oxford University Press. Figure 6.1.

which accelerate the process), outcrop curvature and botanic deformation. Creep is more commonly regarded as a seasonal process in which surface materials are continuously affected by the expansion and contraction of the soil and by soil heave caused by changes of temperature, moisture, freezing, crystal growth or chemical changes. Varieties occur as soils, scree or talus and intense activity develops into faster solifluction processes under periglacial conditions (gelifluction congelifluction, congeliturbation).   DB

**Reading and References**
Brunsden, D. 1979: Mass movements. In C.E. Embleton and J.B. Thomes, eds, *Process in geomorphology*. London: Edward Arnold. · Sharpe, C.F.S. 1938: *Landslides and related phenomena*. New York: Columbia University Press. · Terzaghi, K. 1960: Mechanism of landslides. *Bulletin of the Geological Society of America Berkey volume,*

pp. 83–122. · Varnes, D.J. 1958: Landslide types and processes. In E.B. Eckel ed., *Landslides and engineering practice*. Highway Research Board special report 29, pp. 20–47. · Zaruba, Q. and Mench, V. 1969: *Landslides and their control*. Prague: Academic and Elsevier.

**mass strength**   (of rock). A measure of the resistance to erosion and instability of an entire rock mass inclusive of its discontinuities, contained water and weathering products. For geomorphic purposes mass strength has been assessed quantitatively by taking into account the following parameters: (1) strength of intact rock; (2) state of weathering of the rock; (3) spacing of joints, bedding planes, foliations or other discontinuities within the rock mass; (4) orientation of discontinuities with respect to a cut slope; (5) width of the discontinuities; (6)

309

lateral or vertical continuity of the discontinuities; (7) infill of the discontinuities; and (8) movement of water within or out of the rock mass.                                                   MJS

**Reading**
Dackombe, R.V. and Gardiner, V. 1983: *Geomorphological field manual*. London: Allen & Unwin. · Selby, M.J. 1980: A rock mass strength classification for geomorphic purposes: with tests from Antarctica and New Zealand. *Zeitschrift für Geomorphologie NF* 24, pp. 31–51.

**matorral** The Chilean equivalent of the evergreen, xerophilous, woody plants of Mediterranean environments. (See also CHAPARRAL and MAQUIS).

**Reading**
Miller, P.C. 1981: *Resource use by chaparral and matorral*. Berlin: Springer Verlag.

**matrix flow** The flow of water through the ordinary microscopic pore spaces that make up perhaps 50 per cent of the volume of many soils. There is significant attraction between the water molecules and the walls of the micropores ($<$ 10 $\mu$m diameter) and mesopores (10–1000 $\mu$m diameter), and in addition the pathway that must be followed by water in order to traverse a path through the soil is very tortuous, and therefore longer than the actual straight-line length would suggest. Both these factors, together with others like obstruction of some pores and dead-end passages, retard water traversing the micropores. The flow travels slowly and in the laminar flow state, at speeds of mm–cm per day, in accordance with DARCY'S LAW. Much higher speeds are achieved if there are larger conduits such as faunal burrows within the soil (see BYPASS FLOW) that can permit turbulent flow. Speeds may then reach mm–cm per second.                                              DLD

**Reading**
Dingman, S.L. 1994: *Physical hydrology*. New Jersey: Prentice-Hall.

**maturity** The stage of development of a river system or landscape at which processes are most efficient and vigorous and are tending towards maximum development. In landscape evolution, the period of maximum diversity of form.

**Maunder minimum** A proposed period from about 1650 to 1700 when there was a relative absence of sunspots, and which may be associated with a period of especially inclement weather during the Little Ice Age.            ASG

**Reading**
Eddy, J.A. 1976: The Maunder minimum. *Science* 192, pp. 1189–202.

**mean sea level** An average elevation of the sea surface based upon elevation data that are referenced to an established datum. Mean SEA LEVEL may be estimated using temporal data for a specific location, or from temporal and spatial data to represent the sea surface. These estimates are used to establish baseline conditions for the measurement of sea level changes. Global mean sea level closely approximates the GEOID, with maximum differences of the order of 1 metre, as a result of, for example, ocean currents, atmospheric pressure systems, and different water densities (Pugh 1987). Compared to the earth's ellipsoid, there is more than 150 metres of relief across the time-averaged sea surface. For example, there is a depression averaging about $-100$ metres off the coast of India, and a bulge of about 70 metres near New Guinea (Tooley 1993).

Contemporary data are usually obtained from tide gauges or satellite altimetry. The latter has been especially valuable in providing open ocean data. Time and space averages of mean sea level are useful for the assessment of seasonal changes in water surface elevation, and changes caused by other atmospheric phenomena, such as EL NIÑO. The 1997–8 El Niño was characterized by sea levels that were, relative to the annual mean, higher than normal in the eastern Pacific, and lower than normal in the western Pacific.

Carefully obtained estimates of mean sea level are also valuable in the detection and quantification of sea-level trends. Recent estimates of the rate of global sea-level rise (Intergovernmental Panel on Climate Change 1996) are between 1.0 and 2.5 mm/yr. The range in these estimates derives largely from uncertainty over the quality of the tide gauge records (especially vis-à-vis the vertical stability of the gauge location) used to obtain local mean sea-level data.            DJS

**References**
Intergovernmental Panel on Climate Change 1996: *Climate change 1995: the science of climate change*. Cambridge: Cambridge University Press. · Pugh, D.T. 1987: *Tides, surges and mean sea level*. Chichester: Wiley. · Tooley, M.J. 1993: Long term changes in eustatic sea level. In R.A. Warrick, E.M. Barrow and T.M.L. Wigley eds, *Climate and sea level change: observations, projections and implications*. Cambridge: Cambridge University Press. Pp. 81–107.

**meandering** The sinuous winding of a river, as in the River Menderes in south-west Turkey. Various more restricted definitions have been applied, e.g. that a river may be arbitrarily considered as meandering if it flows in a single channel that is more than one and a half times the direct downvalley length of the reach in question. Other definitions have suggested that meandering must involve the development of regular bends of repeated geometry; these bends may have characteristic amplitude, wavelength and curvature, and they may be approximated by geometrical curves such as circular arcs or so-called sine-generated curves, or recently by migration-rate models. Irregular and compound bend patterns are also possible.

The term is also applied to analogous flow patterns (as in the Gulf Stream or atmospheric jet streams) and to the dynamic processes by which such planforms may develop. For rivers, these include outer-bank erosion and inner-bank deposition on channel bends, and these may be related to the hydraulics of flow and sediment flux in curving channels.

Different kinds of meandering may be distinguished: these include free (developing in alluvial materials without hindrance), forced or confined (where barriers such as bedrock valley sides affect meander generation) and valley (rather than just channel) meandering. Meanders in bedrock may be incised, either as they develop (ingrown) or vertically intrenched. (See also CHANNEL CLASSIFICATION.) JL

**Reading**
Collinson, J.D. and Lewin, J. eds 1983: *Modern and ancient fluvial systems*. International Association of Sedimentologists special publication 6. Oxford: Blackwell Scientific. · Gregory, K.J. ed. 1977: *River channel changes*. Chichester: Wiley. · Richards, K. 1982: *Rivers*. London and New York: Methuen.

## Mediterranean climate

Characterized by hot, dry, sunny summers and a winter rainy season. In KÖPPEN'S CLIMATIC CLASSIFICATION this is the Cs climate, while in Thornthwaite's (1948) it is designated as a subhumid mesothermal climate. Circulation in summer is dominated by expansions of the subtropical high-pressure cells and in winter by travelling mid-latitude depressions (Perry 1981).

Typical monthly mean temperatures are between 21–27 °C in summer and 4–13 °C in winter with annual rainfall totals between 38 and 76 cm. A number of distinctive LOCAL WINDS are associated with Mediterranean climates, e.g. mistral, bora and Santa Ana (Meteorological Office 1964).

This climatic type is found around the Mediterranean Sea, in southern California, the central Chilean coast, the South African coast near Cape Town and in western Australia. AHP

**References**
Meteorological Office 1964: *Weather in the Mediterranean.* 2 vols. London: HMSO. · Perry, A. 1981: Mediterranean climate – a synoptic reappraisal. *Progress in physical geography* 5, pp. 105–13. · Thornthwaite, C.W. 1948:

An approach to a rational classification of climate. *Geographical review* 38, pp. 55–94.

## mega dune

(or DRAA) The largest aeolian bedform in Wilson's (1972) hierarchy, with a wavelength of up to 5500 m and a height of 20–450 m. Given their size, mega dunes are in many cases compound or complex DUNE forms. DSGT

**Reading**
Wilson, I.G. 1972: Aeolian bedforms: their development and origins. *Sedimentology* 19, pp. 173–210.

## mega-scale glacial lineations

Elongated streamlined ridges of sediment produced beneath an ICE SHEET, similar to FLUTES, but much larger in all dimensions. Typical lengths are 6 to 70 km, widths of 200 to 1300 m and spacings of 300 m to 5 km. Their large scale requires satellite images to observe them, as they often appear too fragmented at the scale of aerial photographs. This great length requires fast ice flows for their formation and when found they may indicate the location of former ice streams. They have been found in Canada, Ireland, Scandinavia and adjacent to the west Antarctic ice sheet. CDC

**Reading**
Clark, C.D. 1993. Mega-scale glacial lineations and cross-cutting ice flow landforms. *Earth surface processes and landforms* 18, pp. 1–29.

## megashear

A term introduced by S.W. Carey (1976) to describe a strike-slip fault with a large lateral displacement greatly exceeding the thickness of the crust. In a strike-slip fault a section of the crust moves horizontally with respect to an adjoining section, as on the San Andreas fault system. Megashears form a component of Carey's model of global TECTONICS based on the assumption of an expanding earth. MAS

**Reading and Reference**
Carey, S.W. 1976: *The expanding earth*. Amsterdam: Elsevier. · Neev, D., Hall, J.K. and Saul, J.M. 1982: The Plesium megashear system, across Africa and associated lineament swarms. *Journal of geophysical research* 87, pp. 1015–30.

## megathermal climate

A climate in which no month has a mean temperature below 18 °C. Such conditions are found in the humid tropics and subtropics.

## mekgacha

Dry or fossil valley systems from the Kalahari of southern Africa. They snake through a largely featureless terrain where at present surface run-off is of very limited occurrence. The origins of mekgacha have been variously ascribed to fluvial activity during wetter episodes, ephemeral flow during high rainfall events, river capture, and groundwater sapping along ancient lineaments. They may be of considerable antiquity. ASG

**Reading**
Shaw, P.A., Thomas, D.S.G. and Nash, D.J. 1992: Late Quaternary fluvial activity in the dry valleys (mekgacha) of the Middle and Southern Kalahari, southern Africa. *Journal of Quaternary science* 7, pp. 273–81.

**meltout till** See TILL.

**meltwater** Water produced by the melting of snow or ice. In polar periglacial areas the melting of the winter snow cover and ice in the river channels occurs in the spring and early summer. The resulting flood may last from a few days to a few weeks and account for up to 90 per cent of the total discharge. Sediment loads of such floods are commonly suppressed by the still frozen state of the ground and river channel.

Glacier meltwater flow peaks somewhat later in the season and is less sudden in its build up and decline, but sediment loads are high. Meltwater in a glacier is derived mainly from summer melting of the glacier surface, but a small though significant amount is derived from bottom melting of the glacier by geothermal heat and by frictional heating associated with glacier flow (so long as the glacier base is at the pressure melting point).

On glaciers which are below the pressure melting point the meltwater is normally restricted to surface flow. Where the ice is at the pressure melting point, meltwater penetrates the ice via MOULINS until it flows at the glacier bed. It can be distinguished from basally derived meltwater by its lower solute content (Collins 1979). Two hydrological systems exist at the glacier bed, that of the conduits and that of the intervening areas where a film of water exists between glacier ice and bedrock. In certain bedrock depressions water can be trapped to form subglacial lakes up to many kilometres across (Oswald and Robin 1973). DES

**Reading and References**
Arnborg, L., Walker, H.J. and Peippo, J. 1966: Water discharge in the Colville River, Alaska, 1962. *Geografiska annaler* 48A, pp. 195–210. · Collins, D.N. 1979: Quantitative determination of the subglacial hydrology of two alpine glaciers. *Journal of glaciology* 23.89, pp. 347–62. · Lliboutry, L. 1983: Modifications to the theory of intraglacial water-ways for the case of subglacial ones. *Journal of glaciology* 29.102, pp. 216–26. · Oswald, G.K.A. and Robin, G. de Q. 1973: Lakes beneath the Antarctic ice sheet. *Nature* 245.5423, pp. 251–4. · Shreve, R.L. 1972: Movement of water in glaciers. *Journal of glaciology* 11.62,

pp. 205–14. · Vivian, R. and Zumstein, J. 1973: Hydrologie sousglaciaire au glacier d'Argentière (Mont Blanc, France). *Symposium on the hydrology of glaciers, Cambridge, 9–13 September 1969.* International Association of Scientific Hydrology 95, pp. 53–64.

**memory capacity (dune)** The ability of a sand dune to retain a form resulting from development in one wind regime after environmental conditions have changed. 'Dune memory' was coined by Warren and Kay (1987). All other things being equal, a dune containing a low volume of sand has a short memory, perhaps of only a few hours or days, while a large dune has a long memory that may extend for thousands of years. Dune memory is related to the principle of BEDFORM RECONSTITUTION, but can be applied equally to dune forms that do or do not migrate. DSGT

**Reference**
Warren, A. and Kay, S. 1987: The dynamics of dune networks. In L.E Frostick, and I. Reid, eds, *Desert sediments ancient and modern.* Geological society of London, special publication 35, Oxford: Blackwell. Pp. 205–12.

**Mercalli scale** A scale between 1 and 12 for measuring the intensity of earthquakes based on the amount of structural damage they cause. (See EARTHQUAKE.)

**mere** A lake, especially in Cheshire and East Anglia where meres are developed in glacial outwash deposits.

**meridional circulation** Usually defined as the average, over all longitudes, of the flow in the meridional plane. Though very much weaker than the zonal (i.e. along latitudes) component, the convergence-divergence of the meridional component defines the main climatic zones. Spatial unrepresentivity of observations makes the arithmetic unreliable compared to the values of wind speeds of $0.1$ to $1\,\mathrm{m\,s^{-1}}$ observed. Annual averages confirm the picture of a direct (upward moving air warmer than downward moving air) tropical cell, bounded on the equatorial side by the INTERTROPICAL CONVERGENCE ZONE (ITCZ) and by an indirect cell (upward moving air cooler than downward moving air) between $30°$ and $60°$ latitude. There is some suggestion of a second direct cell between $60°$ and $90°$. Downward velocities bringing dry air towards the surface at $30°$ (and possibly $90°$) coincide with the desert belts at these latitudes. Fluxes of sensible, gravitational potential and latent energy in the meridional circulation are each much greater than their sum. Fluxes of ANGULAR MOMENTUM at high and low levels are each large compared to their sum. Thus the circulation acts largely to convert one type of thermal energy into another and to adjust the vertical distribution of zonal wind. To some extent the circulation is an arithmetic artefact, in that the actual flow takes place in specific geographical areas (such as Indonesia) often on occasions of intense tropical cumulonimbus (see CLOUDS) rather than being spread uniformly in space and time. JSAG

### Reading
Lorenz, E.N. 1967: *The nature and theory of the general circulation of the atmosphere.* Geneva: World Meteorological Organization.

**mesa** A steep-sided plateau of rock, often in horizontally bedded rocks, surrounded by a plain.

**mesoclimate** The intermediate scale of climatic investigation between the MICROCLIMATE and the macroclimate. Although there is not a clear demarcation between these classes, the mesoscale is often taken to include the horizontal scale between 100 m and 200 km (Oke 1987). Some features of the atmosphere, such as sea breezes, are the product of MESOMETEOROLOGY at this scale, though in many cases, they are insufficiently frequent to form part of the mesoclimate of an area. At the lower end of the scale, the term, local climate, is often used. The distinguishing features of a local climate are usually taken to include topographic features as, being static, they provide a more permanent foundation for the generation of surface climate. An extensive, south-facing slope would be an example of a favourable local climate in the northern hemisphere as more solar energy would be received than on a horizontal surface and so it is likely to be warmer. A similar north-facing slope would be colder because of less radiation input.

The processes operating at the microscale are distinctive in generating a particular surface climate. At the mesoscale the same processes will be operating near the ground but because of the larger scale of interest they will affect larger areas and the individual surface contributions may be modified. An example of this scale problem would be a city. Each individual element of the urban surface – roads, buildings and green space – could generate their own microclimate. As cities may occupy a considerable area, these microclimates become aggregated into a mosaic of an urban climate that is different from its rural surroundings and possesses considerable intra-urban differences. If the city is sufficiently large, it can be said to have its own mesoscale urban climate with distinctive temperatures, air-flow and even precipitation distribution as a result of the surface modifications. The zone of greater warmth over the city has been termed the urban heat island as it is a frequent feature, particularly at night. It is strongest where building density is highest and under favourable circumstances of clear skies and light winds. Maximum temperature differences between city centre and rural surrounds of up to 12 °C have been reported by Oke (1987) in North American cities with high ratios of height to width in their central areas. Under these

**A mesa** created by erosion in the Monument Valley Tribal Park, USA. Such features are a good example of one form of slope evolution: parallel retreat.

conditions, the thermal anomaly would be expected to generate an urban breeze as a meso-climatic feature, though observations to confirm its existence are limited.

In many cases, the surface processes are modified by topographic features as well as by the varied nature of the surface itself. The earth's surface is rarely flat. Slopes generated by mountain building, erosion and deposition will modify the amount of radiation received from the sun; they may produce a screening or shadowing onto nearby surfaces. In all cases, the amount of energy received will be changed. At night, slopes can allow cold, denser air to drain towards lower topographic basins to produce pools of cold air known as frost hollows. If the areas affected are sufficiently large, the modification could be described as a local or mesoscale climate. Where large-scale or macroclimatic winds are light, then these mesoscale thermal differences can produce distinctive wind systems. In mountainous areas, we can find MOUNTAIN/VALLEY WINDS, and near the coast there are SEA/LAND BREEZES. In some parts of the world where mountains are close to the coast, such winds have been found to interact to produce a complex interplay of winds. Mesoscale winds can be important in assisting the dispersion of air pollution as they do blow when the prevailing winds are light. Unfortunately they do not always produce a clearance of the air. In Los Angeles, pollution carried inland during the daytime sea-breeze can be returned across the city during the night as the land breeze reverses the flow without being strong or unstable enough to mix clean with polluted air.

As computational technology and meteorological understanding have improved, so it is now possible to produce realistic mesoscale models to predict the patterns of climatic elements which should be experienced. Static energy balance based models have been developed at the microscale that can be scaled upwards if the nature of the ground surface is known. The area of greatest progress has been on mesoscale dynamical models of boundary layer flow over complex terrain. With three-dimensional versions, the effects of deflection around mountains, valley and slope wind effects and leeslope winds can be modelled. Such mesoscale models are now incorporated into GENERAL CIRCULATION MODELS (GCMs) to provide regional detail within the large-scale model. PS

**Reading and Reference**
Atkinson, B.W. 1981: *Meso-scale atmospheric circulations.* London: Academic Press. · Barry, R.G. 1992: *Mountain weather and climate.* London: Routledge. · Oke, T.R. 1987: *Boundary layer climates.* London: Routledge. ·

Yoshino, M.M. 1975: *Climate in a small area.* Tokyo: Tokyo University Press.

**mesometeorology** In the analysis of the environment, in whole or in part, an appreciation of scale is vital. In meteorology this appreciation is primarily manifest in a recognition of entities of air motion that have significantly different characteristic horizontal sizes. Thus we may visualize the general, global atmospheric circulation to comprise other smaller circulations of different sizes, all interrelated in various ways. It is not unreasonable to use the analogy of a car engine, the whole being the 'general circulation' and all the various components being the 'smaller circulations'. Analysis of the energy content of the atmosphere suggests a crude three-tier hierarchy of circulations based upon their characteristic horizontal dimensions (see table). The middle tier of this hierarchy is called the mesoscale, and includes circulations with horizontal sizes of the order of 20 km to 250 km and lifetimes of a few to a few tens of hours. Mesometeorology is primarily concerned with understanding these circulations.

Mesoscale atmospheric circulations may be conveniently categorized into those that are topographically induced and those that are intrinsically products of the free atmosphere. Each of these categories may be divided into two parts. Thus topographically induced systems may result from mechanical and thermal processes and the free atmosphere systems may result from convective and non-convective processes.

Mechanically forced circulations include LEE WAVES, downslope winds (see BORA; FÖHN) and circulations within wakes. Lee waves are just that – waves in the airflow to the lee of the hill that triggers them. Their typical wavelength is about 5–15 km and their crest-to-trough amplitude is about 0.5 km. They are ubiquitous, occurring to the lee of hills as small as the Chilterns and as large as the Rockies. They result primarily from vertical oscillations induced in a stable airstream by the hill. A wind with strong vertical shear and a direction almost perpendicular to the main axis of the hill is also

Mesometeorology: three scales of atmospheric motion

| | Macro | Meso | Local |
|---|---|---|---|
| Characteristic dimension (km) | > 483 | 16–160 | < 8 |
| | Synoptic | Meso | Micro |
| Period (h) | > 48 | 1–48 | < 1 |
| Wavelength (km) | > 500 | 20–500 | < 20 |

*Source*: Atkinson 1981.

favourable. Strong winds down the leeward slope of the hills frequently accompany lee waves. Circulations in wakes usually take the form of vortices, frequently in trains known as vortex streets. These streets develop behind an obstacle, usually an island, when the flow in the wake region does not mix with that in the surrounding region.

Thermally induced circulations include SEA/LAND BREEZES and slope and VALLEY WINDS. Both are the result of a thermally direct, diurnally varying, vertical circulation. In the case of the sea/land breeze, the land–sea thermal contrast generates low-level flow from the sea by day and low-level flow from the land by night. In hilly areas the daytime mechanism generates upslope and up-valley airflow near the surface, overlain by a 'return-current' aloft that flows from the hills to the plains. At night the reverse occurs, resulting in katabatic flows downslope and down-valley.

Moving gravity waves are non-convective, free-atmosphere systems. Little is currently known about this type of wave but we do know that they have long wavelengths, up to 500 km, and that they are frequently associated with strong vertical shear of the horizontal wind. Convective, free-atmosphere circulations comprise severe local storms (see THUNDERSTORM), shallow cellular circulations and circulations in CYCLONES. Severe local storms have been studied for decades yet still remain a challenge. It now appears that in addition to ATMOSPHERIC INSTABILITY a strong vertical shear of horizontal wind is vital for long-lived (several hours), large (up to 100 km across) storms to exist.

Shallow cellular circulations were first fully documented only in the satellite era. The cells are usually hexagonal, may be open (i.e. cloud-free area surrounded by cloud) or closed (i.e. cloud area surrounded by a hexagonal ribbon of clear air), have diameters of about 30 km and depths of about 2 km. Their full explanation is as yet unknown. Circulations within cyclones are also recently observed features. Radar and autographic rain gauge records have revealed systems about 50 km across, with life times of a few hours, that form, move and dissipate within synoptic-scale frontal areas. At present they are thought to result from CONVECTION which is released in layers of potentially unstable air over the frontal surfaces.

Much remains to be discovered and explained within the realms of mesometeorology. The sub-discipline is taking an increasingly important role within meteorology as a whole. The next decade holds exciting prospects for both observational and theoretical studies of mesoscale atmospheric circulations.                    BWA

**Reading**
Atkinson, B.W. 1981: *Meso-scale atmospheric circulations.* London and New York: Academic Press.

**mesophyte**  A plant which flourishes under conditions which are neither very wet nor very dry.

**mesoscale cellular convection**  Over millions of square kilometres of oceans' heat transfer to the atmosphere takes the form of mesoscale cellular convection (MCC). The convection is visible in the particular cloud forms produced and satellite imagery first noted this form of convection about 1970. The convection is shallow (c. 2 km deep) and forms distinctive horizontal patterns. In planform the convection takes two main forms: parallel lines of clouds, known as cloud streets; and hexagonal cells. The former is essentially two-dimensional, the latter three-dimensional convection.

The cloud streets may extend for a few hundred kilometres and they lie a few kilometres apart. The hexagonal cells are typically 30 km across and are of two types: closed cells and open cells. The former have uplift in their centre (hence are cloudy and 'closed') and compensatory subsidence on their sides, giving clear air. The open cells have a reversed configuration of vertical motion: uplift on the 'walls' and subsidence in the middle. These different forms are related to vertical stability and vertical shear of horizontal winds, particularly the latter. The details of these relationships are not yet fully clear.

Both types of MCC tend to occur over warm ocean surfaces adjacent to cold land surfaces, such as the Sea of Japan and the Atlantic east of Greenland. In such areas very cold air frequently flows over warm ocean surfaces thus creating the instability required to drive this kind of convection.                    BWA

**mesosphere**  A portion of the earth's atmosphere. Either that between the stratosphere (at about 40 km above the surface) and the thermosphere (80 km) or that between the ionosphere (400 km) and the exosphere (1000 km).

**mesothermal climate**  Characterized by moderate temperatures, with the mean temperature of the coldest month between $-3\,°C$ and $+18\,°C$. Climates of this type are found mainly between latitudes $30°$ and $45°$ but may extend up to latitude $60°$ on the windward side of continents.

**mesotrophic**  See NUTRIENT STATUS.

315

**Messinian salinity crisis** An event that took place around 6.5 to 5.0 million years ago, at the end of the Miocene. The growth of the Antarctic Ice Sheet at that time contributed to a marked sea-level reduction so that the western TETHYS OCEAN lost its connection with the Atlantic. Tectonic activity may have been a contributory factor. Isolation of this ancient Mediterranean from the world ocean caused it to become highly saline, so that enormous gypsum and halite deposits were precipitated. Complete evaporation of this brine is thought to have occurred between 5.0 and 5.5 million years ago. The drying up of so large a body of water over a brief geological period would have had major consequences, facilitating, for example, the incision of rivers and the movements of fauna and flora.                                    ASG

**Reading**
Adams, C.G., Benson, R.H., Kidd, R.B., Ryan, W.B.F. and Wright, R.C. 1977: The Messinian salinity crisis and evidence of late Miocene eustatic changes in the world ocean. *Nature* 269, pp. 383–6

**metamorphism** The processes by which the composition, structure and texture of consolidated rocks are significantly altered through the action of heat and pressure greater than that produced normally by burial, or by the introduction of additional minerals as a result of a marked change in the thermodynamic environment, e.g. limestone may be converted to marble, mudstone to slate, and granite to gneiss.

The intensity of metamorphism is referred to as its grade, and high grade metamorphic rocks include gneiss, granulite, blueschist, amphibolites and eclogites.

There are three classes of metamorphism: regional metamorphism which occurs in orogenic belts; contact metamorphism around the boundaries of igneous intrusions: and dislocation metamorphism resulting from friction along fault or thrust planes.                      ASG

**metapedogenesis** The modification of soil properties by human agency (Yaalon and Yaron 1966).                                    ASG

**Reference**
Yaalon, D.H. and Yaron, B. 1966: Framework for man-made soil changes – an outline of metapedogenesis. *Soil science* 102, pp. 272–7.

**metasomatism** The processes by which a rock is physically or chemically altered, partially or wholly, as a result of the introduction of new materials. Contact metamorphism is the alteration of rocks adjacent to an igneous intrusion by metasomatism and high temperatures.

**meteoric water/groundwater** Water that is derived by precipitation in any form from the atmosphere. Originally defined by Meinzer (1923) and contrasted with JUVENILE WATER, meteoric groundwater is water with an atmospheric or surface origin rather than with a source in the interior of the earth. (See also CONNATE WATER.)                      PWW

**Reference**
Meinzer, O.E. 1923: Outline of groundwater hydrology. *US Geological Survey water supply paper* 494.

**meteorological satellites** Artificial, earth-orbiting devices designed solely or largely for meteorological observation, supplemented in some cases by meteorological data collection and / or relay functions. They should be distinguished from spacecraft (deep space probes) designed to investigate other bodies in the solar system, particularly other planets and their moons although, in several cases, these have investigated atmospheric conditions by similar means to those deployed on meteorological satellites themselves. Observations from meteorological satellites have commonly been made by sensor systems designed to provide cloud imagery, radiation budget data, and vertical profiles through the atmosphere. Although American Vanguard and Explorer satellites orbited in 1959 and 1960 were equipped for earth imaging, this facet of their missions was secondary, and the cloud pictures returned were low in resolution, and therefore indistinct. The first purpose-built meteorological satellite was the American Tiros-I, launched on 1 April 1960. This, a 'television and infrared observation satellite', carried a vidicon camera for visible (reflected) light imaging, and an infrared radiometer to provide target temperature data in the invisible thermal region of the electromagnetic spectrum. The data were transmitted to earth by radio, a procedure followed by all subsequent meteorological satellites. As time passed the value of such satellites was confirmed; in particular it was recognized that they provided more uniform and more complete views of global weather than were available from conventional ('*in situ*') sources. Additionally, their views of weather were new, and supplementary to those obtained from within the atmosphere or from the interface with its parent planet.

Many new satellite configurations, systems and sensors were developed during the 1960s and early 1970s. Russian, European, Japanese and Indian satellites have joined those of the USA to provide an almost overwhelming flood of data useful not only in day-to-day meteorological operations (especially weather forecasting), but also meteorological and climatological research. The last is expanding rapidly now that satellite data sets have begun to achieve climatologically suitable lengths. Key activities in the meteorological satellite arena, organized for the most part chronologically, have included the following:

1 Early testing and development of low-altitude, polar-orbiting satellites by the USA (Tiros series) and the former USSR (some of the Cosmos series), early to mid 1960s.
2 Inauguration of the first fully operational global weather satellite system (Essa series) by the USA in 1966. This consisted of two polar-orbiting satellites, one giving direct read-out data for local use, the other tape-recording data for global archiving. The USSR effectively followed suit with its Meteor series, inaugurated in 1969.
3 Very active testing and research of new sensors and modes of operation, impressively effected through the US Nimbus satellite series from 1964 to the mid 1980s.
4 Implementation of improved operational polar-orbiting satellite systems for civilian use by the American National Oceanic and Atmosphere Administration (NOAA) series inaugurated in 1970, and subsequently by an advanced NOAA series in 1979, capitalizing on experience with earlier civilian satellites, and military satellites of the US Defense Meteorological Satellite Program (DMSP series).
5 Early testing, in the late 1960s, of US 'geostationary' satellites of the Applications Technology Satellite (ATS) series (deployed at 35,400 km above the equator, where they appear to be stationary above a preselected point on the surface of the earth).
6 Inauguration of the first operational geostationary satellites by the USA in 1975, the Geostationary Operational Environmental Satellite (GOES) series, leading to a complete global encirclement by geostationary satellites in low and middle latitudes for the First GARP Global Experiment (FGGE) year from 1 December 1978 to 30 November 1979. The FGGE coverage necessitated five geostationary orbiters, three American (GOES), one European (Meteosat-1) and one Japanese (Geostationary Meteorological Satellite: GMS (or 'Himawari')-1).
7 The implementation in the early to mid 1980s of geostationary satellites for operational coverage of the European/African, and Indian/Indian Ocean sectors of the globe by Eumetsat (on behalf of a consortium of European meteorological offices) and India respectively.

The continued internationalization of meteorological satellite activities indicated by (7) above confirms the key role which such satellites are now adjudged to play in meteorology and climatology both globally and regionally. A wide range of satellite products and derived products are now available for weather and climate research. Increasingly *in situ* observing networks are being planned and operated in full conjunction with meteorological satellites, which are now recognized as indispensable parts of any effective broad-scale atmospheric monitoring system.                                        ECB

**Reading**
American Society of Photogrammetry 1983: Meteorology satellites. In *Manual of remote sensing.* 2nd edn. Pp. 651–78. · Anderson, R.K. and Veltischev, N.F. 1973: *The use of satellite pictures in weather analysis and forecasting.* WMO technical note 124. Geneva: World Meteorological Organisation. · Brimacombe, C.A. 1981: *Atlas of Meteosat imagery.* ESA SP-1030. Paris: European Space Agency. · National Space Science Data Center periodically: *Reports on active and planned spacecraft and experiments.* NSSDC/WDC-A-R & S. Greenbelt, Md: Goddard Space Flight Center. · NASA 1982: *Meteorological satellites: past, present and future.* NASA conference publication 2227. Washington DC: NASA.

**meteorology** The science concerned with understanding atmospheres. In recent years since satellites have probed the atmospheres of planets other than earth, it has become acceptable to use the term 'atmospheric science' and to apply the term to any atmosphere, not just that on earth.                                   BWA

**microclimate** The climate near the ground surface in which plants and animals live. There is no distinct and internationally agreed separation between the study of climate at the smallest or microscale and that of larger scales. This is not surprising as the processes operating at the ground surface form part of a continuum that is not readily separated into discrete subdivisions. As a result, the term is used frequently without a precise meaning. Yoshino (1975) suggests a horizontal scale of 0.01 m to 100 m with anything larger belonging to the local, meso- or macroscales of climate, whilst Oke (1987) takes 1000 m as his upper limit of the microscale. Averaging of temporal scales also produces variety. Yoshino suggests 10 seconds for the microscale in contrast to $10^6$ (approximately 1 day) for the macroscale. Both spatial and temporal definitions indicate a highly variable environment in contrast to that of the macroscale.

Definitions of climate scales

| Climate | Horizontal distribution | Climatic example | Life-time of phenomenon |
|---------|------------------------|------------------|-------------------------|
| Microclimate | $10^{-2}$ to $10^2$ m | Small-field climate | $10^{-1}$ to $10^1$ seconds |
| Local climate | $10^2$ to $10^4$ m | Slope climate | $10^1$ to $10^4$ seconds |
| Mesoclimate | $10^3$ to $2 \times 10^5$ m | Urban climate | $10^4$ to $10^5$ seconds |
| Macroclimate | $2 \times 10^5$ to $5 \times 10^7$ m | Regional climate | $10^5$ to $10^6$ seconds |

Microclimate is inextricably linked to the processes taking place in the immediate earth–air interface. Hence the nature of the surface is extremely important in determining the type of microclimate likely to be experienced. The prime factor is the treatment of solar energy input. Surfaces will respond differently to absorption through their different ALBEDOS, or reflective capacities. How this energy is converted into heat to warm up the surface will depend upon the nature of the vegetation canopy and the degree of moisture availability which in turn influences how much energy is used for LATENT HEAT and SENSIBLE HEAT. Soil factors may be significant as dry soils conduct heat away from the surface more slowly than wet soils and the colour of the soil will affect how much solar energy will be absorbed. Aspect and shadowing may affect the amount of energy a surface receives; in the northern hemisphere, south-facing slopes receive more energy and therefore are potentially warmer, than north-facing slopes. At night, cold air may drain into topographic basins to produce a TEMPERATURE INVERSION. The sky-view factor (or proportion of the sky visible for long-wave radiational cooling) can be important in determining the degree of cooling at night. If the sky view is unobstructed by buildings, trees or slopes, the loss of long-wave radiation will be much greater than if the site is very sheltered as in a deep, narrow valley. As a result temperatures may fall to lower values. In winter the consequence of these processes may be seen from the distribution of surface frost.

Of equal importance is the influence of wind speed and turbulence in mixing the air. If wind speeds are light, then the surface properties mentioned above can dominate. If winds are stronger, then the surface-generated thermal microclimates can become mixed to give a relatively uniform temperature where microscale differences are limited, Normally wind speeds do decrease as the ground surface is approached, as a result of friction and shelter. This gives a logarithmic wind speed profile above the surface.

As a result of this potential shelter, the thermal climate near the ground can become much more extreme than that observed at a standard climatological site about 1.5 m above the ground with instruments cased in a wooden screen. As the ground surface is the main area of absorption and emission of energy, it tends to become warmer by day and cooler by night than standard observation sites. Only if the atmosphere is cloudy with a strong wind is the effect nullified. For example, in a desert environment, the daily range of temperature in the air a few centimetres above the surface may be 50 °C greater than the range measured at a standard height. In practice, the majority of this difference occurs within the lowest 50 centimetres. Grass minimum temperatures are the only routine observations at climate stations that give any indications of conditions near the ground surface and these are only read once per day. Under clear skies and light winds, minimum temperatures recorded here can be up to 10 °C cooler than in a Stevenson screen.

The greater range of temperatures is important for the growth and survival of plants and animals living in this environment. Seedlings emerging from the soil have to withstand greater daytime heating and night-time cooling than might be expected from normal climatological values. Ground frost can cause serious damage to newly emerging plants. Animals (including humans) can select appropriate microclimates for comfort and survival, and several useful schemes have been developed to relate possible microclimates to known physiological responses, thereby establishing the microclimate space in which an animal can exist.          PS

**Reading and References**
Geiger, R., Aron, R.H. and Todhunter, P. 1995: *Climate near the ground*. Wiesbaden: Vieweg. · Oke, T.R. 1987: *Boundary layer climates*. London: Routledge. · Yoshino, M.M. 1975: *Climate in a small area*. Tokyo: Tokyo University Press.

**microcracks** Can occur in rocks of all kinds. They are generally short (a few centimetres long) and narrow (less than one millimetre) but they are important because they provide one of the main means by which rocks break down. Both physical and chemical weathering can take place within them as they provide a much enlarged surface area for chemical action to take place and the crack tip is an important stress concentrator in brittle fracture. A general classification of such cracks has been put forward by Farran and Thenoz (1965).

microfissures: less than 1 $\mu$m in width and about the length of a crystal;
microfractures: about 0.1 mm or less wide;
macrofractures: greater than 0.1 mm wide and may be several metres long.　　　　　　　WBW

### Reading and Reference
Farran, J. and Thenoz, B. 1965: L'alterabilité des roches, ses fractures, sa prevision. *Ann. Inst. Tech. Batiments Travaux Publiques* no. 25. · Whalley, W.B., Douglas, G.R. and McGreevy, J.P. 1982: Crack propagation and associated weathering in igneous rocks. *Zeitschrift für Geomorphologie* 26, pp. 33–54.

**micro-erosion meter** An instrument used to measure accurately the erosion or weathering of a rock from direct measurement on the rock surface. It comprises a spring-loaded probe connected to an engineer's dial gauge. In order to measure exactly the same area each time, the instrument is mounted on three legs which are placed onto studs drilled in the rock surface. Measurements have been achieved to the nearest 0.00001 mm.　　　　　　　　　　PAB

**microfossils** These are normally less than 2.00 mm in size and include the mineralized shell remains of animals (e.g. COCCOLITHS; DIATOMS; FORAMINIFERA; OSTRACODS, etc.) and the organic-walled remains of plants (POLLEN). Because of similar processing and study methods, several microfossil groups are studied under the umbrella term 'micropalaeontology'. Their small size means that they can only be studied with the aid of microscopes and in the case of extremely small microfossils scanning electron microscopes.　　　　　　　　　SLO

**micrometeorology** The science concerned with the study of physical phenomena taking place on a microscale in the atmosphere close to the surface. The development of micrometeorology has involved not only the examination and analysis of highly accurate observations made in the air layers adjacent to surfaces but also detailed study of processes at surfaces which determine the properties of air in such layers. A principal feature of air near the ground is its turbulence, the complex mixing which is produced by eddies with sizes ranging from microscopic to tens of metres. The turbulence is generated by friction between moving air and the surface, and may be modified by temperature differences between air and ground. Understanding of the structure of turbulence has been a major theme of micrometeorology, of funda-

mental importance and with practical applications such as the dispersal of spores, and the diffusion of air pollution. Exchange between the surface and the atmosphere leads to vertical gradients of temperature, windspeed and gas concentrations. Much progress in micrometeorology has been made by the detailed analysis of mean gradients (over periods of about 30 min) at sites where there is extensive uniform horizontal ground cover. Based on such analysis, rates of exchange at the surface can be deduced from measurements made entirely in the atmosphere. For example, rates of evaporation or of carbon dioxide uptake by growing crops can be studied in relation to diurnal changes in the weather. Gradient methods in micrometeorology have a sound physical basis but must rely on empirical corrections to allow for effects of temperature gradients on turbulence. Recent developments in instrumentation and computers suggest that, in future, methods of measuring turbulent transfer more directly by sensing the motion and composition of individual eddies – an eddy correlation approach – will become more readily available for applied research.

The surface processes most directly affecting the structure of the lower atmosphere begin with the absorption of radiant energy at the ground. Micrometeorology is concerned with the interception and fate of that energy, with the way it is used to heat vegetation, soil and water, and with the conversion of some of the energy to latent heat when water is evaporated (see EVAPORATION). Heat stored in the soil may be of great importance as an energy source at night, and the understanding of factors determining the temperature structure of the lower atmosphere at night is a challenging problem for micrometeorologists.

Although this topic can be pursued as a fundamental aspect of meteorological physics, much of micrometeorology has practical applications. Examples are particularly common in AGROMETEOROLOGY, including the study of forest meteorology. Particular progress has been made in understanding atmospheric and surface controls of evaporation, diffusion of pollution from point and area sources, uptake of air pollutants by plants, and the spread of animal and plant disease.　　　　　　　　　　MHU

### Reading
Arya, S.P. 1988: *Introduction to micrometeorology*. San Diego: Academic Press. · Munn, R.E. 1966: *Descriptive micrometeorology*. New York and London: Academic Press. · Oke, T.R. 1978: *Boundary layer climates*. London: Routledge. · de Vries, D.A. and Afgan, N.H. eds 1975: *Heat and mass transfer in the biosphere*. Washington: Scripta Press.

**microtopography** A term used to describe small-scale irregularities of the ground surface. Traditionally, geomorphologists have regarded microtopography as recognizable elements superimposed on the overall hillslope form, which occur in a variety of settings and environments. More recent interest in the application of FRACTALS has led to speculation that hillslope forms contain a spectrum of features at all scales. On slopes, microtopography is most frequently encountered in environments where soils are generally thin and bedrock is at, or close to, the surface. In KARST environments, water flow paths across rock surfaces generate a variety of microtopographic features, generally referred to as KARREN. In semi-arid environments, the swelling and heaving of clays, associated with wetting and drying cycles, can lead to the sorting of stone layers and the formation of stepped hillslope microtopography, termed desert GILGAI. The banded pattern of treads and risers is often reinforced by vegetation (Dunkerley and Brown 1995). Similar sorting of fine and coarse particles is observed in periglacial environments due to FROST HEAVING and SOLIFLUCTION processes. The most conspicuous form of microtopography on humid temperate hillslopes is the TERRACETTE, a stepped sequence of treads and risers. Their origin remains unclear, but they are thought to be relict periglacial features accentuated by the action of trampling animals (Vincent and Clarke 1976). Fluvial microtopography includes river BEDFORMS and clusters. A precise definition of where microtopography begins and ceases is difficult. Reviewing previous studies of microtopography (or microrelief), Parsons (1988) notes its application to features with elevation differences between 2.5 cm and 3 m. More recent innovations in measuring techniques, such as *in situ* laser reflection, permit irregularities to be characterized much more precisely. The statistical description of surfaces is fundamental to tribology, the science of wear.            DH

**References**
Dunkerley, D.L. and Brown, K.J. 1995: Runoff and runon areas in patterned chenopod shrubland, arid western New South Wales, Australia. *Journal of arid environments* 30, pp. 41–55. · Parsons, A.J. 1988: *Hillslope form*. London: Routledge. · Vincent, P.J. and Clarke, J.V. 1976: The terracette enigma. *Biuletyn peryglacjalny* 25, pp. 65–77.

**mid-ocean ridge** Large linear arches on the sea floor which mark the lines of volcanic activity along which basaltic rocks are added to the sea floor as it separates. Volcanic activity along the ridges may produce sea-mounts, guyots and islands.

**migration** In a wide sense, any movement of animals or plants from one region or habitat to another (see DISPERSAL); more specifically, the seasonal or cyclical movement of animals between two or more areas, a phenomenon common in mammals, birds, fishes and insects. The classic example is probably that of the Arctic tern, which migrates each year from its summer breeding grounds in the Arctic circle right across the world to winter on the coast of Antarctica.

Exactly how and why animals carry out these great feats of migration still remain largely unanswered questions, although it is clear that no one theory is universally applicable throughout the animal kingdom. Navigation appears to be possible through the use of the sun as a compass, a similar use of the stars, the use of the earth's magnetic field and the Coriolis effect, the recognition of the lie of the land and distinctive landmarks, and the use of smell.            PAS

**Reading**
Baker, R.R. 1978: *The evolutionary ecology of animal migration*. London: Hodder & Stoughton. · Dunbar, R. 1984: How animals know which way to go. *New scientist* 101, 12 January, pp. 26–30.

**Milankovitch hypothesis** One of the most significant models of earth history, which is seen by some (e.g. Imbrie and Imbrie 1979) as the key to understanding the causes of the climatic fluctuations of the PLEISTOCENE. The hypothesis is based on the fact that the position and configuration of the earth as a planet in relation to the sun is prone to change, thereby affecting the receipt of insolation at the earth's surface. There are three such changes which have been identified, all three occurring in a cyclic manner: changes in the eccentricity of the earth's orbit (a 96,000 year cycle), the precession of the equinoxes (with a periodicity of 21,000 years), and changes in the obliquity of the ecliptic (the angle between the plane of the earth's orbit and the plane of its rotational equator). This last has a periodicity of about 40,000 years.

The earth's orbit around the sun is not a perfect circle but an ellipse. If the orbit were a perfect circle then the summer and winter parts of the year would be equal in their length. With greater eccentricity there will be a greater difference in the length of the seasons. Over a period of about 96,000 years the earth's orbit can stretch by departing much further from a circle and then reverting to almost true circularity.

The precession of the equinoxes simply means that the time of year at which the earth is nearest the sun (perihelion) varies. The reason is that the earth wobbles like a top and swivels

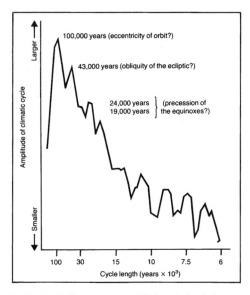

Amplitude of climate cycles and links to orbital changes.

its axis round. At the moment the perihelion comes in January. In 10,500 years it will occur in July.

The third cyclic perturbation, change in the obliquity of the ecliptic, involves the variability of the tilt of the axis about which the earth rotates. The values vary from 21°39′ to 24°36′. This movement has been likened to the roll of a ship. The greater the tilt, the more pronounced is the difference between winter and summer (Calder 1974).

Appreciation of the possible significance of these three astronomical fluctuations of the earth goes back to at least 1842, when J.F. Adhemar made the suggestion that climate might be affected by them. His views were developed by Croll in the 1860s and by Milankovitch in the 1920s (Mitchell 1965).

The major attraction of these ideas is that while the extent of temperature change caused by them may well only be of the order of 1 or 2 °C, the periodicity of these fluctuations seems to be largely comparable with the periodicity of the ice advances and retreats of the Pleistocene. Recent isotopic dating has shown that the record of sea-level changes preserved in the coral terraces of Barbados and elsewhere, and the record of heating and cooling in deep-sea cores, correlates well with theoretical insolation curves based on those of Milankovitch (Mesollela et al. 1969).

The variations in the earth's orbit have recently been seen as 'the pacemaker of the ice ages' (Imbrie and Imbrie 1979) for detailed statistical analysis of ocean cores shows that they possess statistically significant wavelike fluctuations with amplitudes of the order of around 100,000 years, 43,000 years and 19–24,000 years (see diagram). The most important of these cycles is the longest one, corresponding to variations in eccentricity (Hays et al. 1976). This applies back to 900,000 years ago, but probably not further (Pisias and Moore 1981).

Thus there is some substantial evidence to suggest that astronomical theories may be valid as an explanation of the longer scale of environmental changes (Goreau 1980).

The Croll–Milankovitch model produces a near cyclical series of events which is too long to be relevant to most postglacial fluctuations of climate and too short to throw much light on the spacing of the major ice ages. In addition, the model advocates the development of glaciation in high latitudes by insolation variations, whereas from a glacial mass budget viewpoint, an increase in precipitation over the present minimal levels now received in polar areas may be more important (Andrews 1975). Finally, the computed variations in insolation resulting from this model are never more than a very few per cent so that it is likely that even if this mechanism can initiate change, other factors would be necessary to intensify it.

They may also help to explain the very marked expansion of lakes that took place in tropical and subtropical areas around 9000 BP. Recent analyses of theoretical insolation levels by Kutzbach (1981) indicate that radiation receipts in July at that time were larger than now (by about 7 per cent) and that this led to an intensification of monsoonal circulation and associated precipitation.                    ASG

**Reading and References**
Andrews, J.T. 1975: *Glacial systems: an approach to glaciers and their environments*. North Scituate, Mass.: Duxbury. · Berger, A., Imbrie, J., Hays, J., Kukla, G. and Saltzman, B. eds 1984: *Milankovitch and climate*. 2 vols. Dordrecht: Reidel. · Calder, N. 1974: *The weather machine*. London: BBC. · Goreau, T. 1980: Frequency sensitivity of the deep-sea climatic record. *Nature* 287, pp. 620–2. · Hays, J.D., Imbrie, J. and Shackleton, N.J. 1976: Variations in the earth's orbit: pacemaker of the Ice Ages. *Science* 194, pp. 1121–32. · Imbrie, J. and Imbrie, K.P. 1979: *Ice ages: solving the mystery*. London: Macmillan. · Kutzbach, J.E. 1981: Monsoon climate of the early Holocene: climate experiment with the earth's orbital parameters for 9000 years ago. *Science* 214, pp. 59–61. · Mesollela, K.J., Matthews, R.K., Broecker, W.S. and Thurber, D.L. 1969: The astronomical theory of climatic change: Barbados data. *Journal of geology* 77, pp. 250–74. · Mitchell, J.M. 1965: Theoretical paleoclimatology. In H.E. Wright and D.G. Frey eds, *The Quaternary of the USA*. Princeton, NJ: Princeton University Press. Pp. 881–901. · Pisias, N.G. and Moore, T.C. 1981: The evolution of Pleistocene climate: a time-series approach. *Earth and planetary science letters* 52, pp. 450–8.

**minimum acceptable flow** This concept dates from the Water Resources Act 1963 which requires the UK River Authorities to determine and to keep under review minimum acceptable flows for all the main rivers in their areas. Such a flow is very difficult to determine because it necessarily depends upon local circumstances but it should take account of the character of the inland water and its surroundings and should be 'not less than the minimum which in the opinion of the river authority is needed for safeguarding the public health and for meeting (in respect of both quantity and quality of water) the requirements of existing lawful uses of the inland water, whether for agriculture, industry, water supply and other purposes, and the requirements of land drainage, navigation and fisheries' (Water Resources Act 1963, section 19). AMG

**mirage** An optical illusion produced by refraction in the lower atmosphere as a result of differential heating of the air. Objects beyond the horizon may become visible.

**mire** A peat-accumulating wetland. This is a general term applied to peat-producing ecosystems that develop in sites of abundant water supply. Mires are made up of microtopes, mesotopes and macrotopes. Mire microtopes are small-scale topographical features such as a regular arrangement of ridges or hummocks and hollows associated with the mire surface. Mesotopes are mire systems developed from one original centre of peat formation, they may join to form a mire macrotope or mire complex. (See also RAISED MIRE; PEAT.) ALH

**Reading**
Heathwaite, A.L. and Gottlich, Kh. 1993: *Mires: process, exploitation and conservation*. Wiley: Chichester.

**misfit** See UNDERFIT STREAM.

**mist** A suspension in the air of very small water drops or wet hygroscopic particles, causing a reduction in the horizontal visibility at the earth's surface. In meteorology the term is used when visibility is so reduced but is still 1000 m or more. Water contents of mists are typically a very small fraction of $1 \, \text{g m}^{-3}$ and average drop radii are typically less than $1 \, \mu\text{m}$. Owing to relatively high concentrations of solute within the drops, mists may persist with relative humidities as low as 80 per cent. KJW

**mistral** A cold dry north or north-west wind affecting the Rhone valley, particularly to the south of Valence. It is typically strong and squally and most violent in winter and spring.

A depression over the Tyrrhenian Sea or Gulf of Genoa with high pressure advancing from the west towards Spain provides the necessary synoptic gradient. Market gardens and orchards require protection from the mistral by windbreaks. Topographic channelling causes local strengthening of windspeeds in the Rhone valley (Barsch 1965), while marked diurnal variations are common. In the area of maximum frequency of mistral in the Rhone delta an average of over 100 days per year are recorded. AHP

**Reading and References**
Barsch, D. 1965: Les arbres et le vent dans la vallée meridionale du Rhone. *Revue de géographie de Lyons* 40, pp. 35–45. · Boyer, F., Orieux, A. and Powger, E. 1970: *Le Mistral en Provence occidentale*. Monographs de la météorologie 79. Paris: Nationale.

**mixing corrosion** The increased degree of solutional corrosion which occurs when two saturated karst waters of different composition mix. (See also TROMBE'S CURVES.)

**mixing models** Models used to explain or predict temporal variations in the solute concentrations of streamflow by taking account of the mixing of water from different sources or the mixing of water within a store. Gregory and Walling (1976) report several studies where simple mass balance models, based on the mixing of individual run-off components, have been developed. Johnson *et al.* (1969) also describe a model involving the mixing of incoming precipitation with water stored within the basin which in turn supplies the streamflow output. DEW

**References**
Gregory, K.J. and Walling, D.E. 1976: *Drainage basin form and process*. London: Edward Arnold. Pp. 222–3. · Johnson, N.M., Likens, G.E., Bormann, F.H., Fisher, D.W. and Pierce, R.S. 1969: A working model for the variation in stream water chemistry at the Hubbard Brook Experimental Forest, New Hampshire. *Water resources research* 5, pp. 1353–63.

**mobile belt** A linear crustal zone characterized by tectonic activity. The term is applied to both contemporary zones of OROGENY, where mountain ranges are actively being formed, and to ancient zones of intense crustal activity indicated by the effects of metamorphism, granite emplacement and faulting. In the latter sense mobile belts are contrasted with the stable crustal regions of continents or CRATONS. MAS

**Reading**
Spencer, E.W. 1977: *Introduction to the structure of the earth*. 2nd edn. New York: McGraw-Hill. · Windley, B.F. 1984: *The evolving continents* 2nd edn. London and New York: Wiley.

**moder**  One of the three main forms of organic matter in soils. See also MOR and MULL. Between the neutral or slightly acid conditions in which mull develops and the very acid conditions of mor, there is an intergrade known as moder. Although much of the organic fraction is well decomposed and incorporated into the soil's mineral profile, the binding between the two remains weak. There is therefore no strong structural development, and a thin layer of litter and fermented material accumulates at the surface.

**modulus of elasticity**  A spring is a mechanical model for behaviour of an elastic material under stress. Such a material responds immediately to an applied load and its change of dimensions is directly proportional to the applied stress. When the load is removed the entire strain (i.e. deformation) is recoverable. The magnitude of the stress ($\sigma$) required to produce a given strain ($\varepsilon$) is a characteristic of the material and is known as Young's modulus ($E$) of elasticity (a modulus is a constant factor):

$$E = \sigma/\varepsilon$$

This relationship between stress and strain is termed Hooke's Law and a perfectly elastic substance is said to be Hookean. The greater the value of $E$ for a material, the less will be the deformation produced by a given value of stress, and the stronger will be the material. Strong rocks are near to being Hookean solids and have values of $E$ in the range 50–100 GPa.  MJS

**Reading**
Selby, M.J. 1993: *Hillslope materials and processes*. 2nd edn. Oxford: Oxford University Press. Ch. 4.

**mogote**  (haystack hill) A large residual limestone hill which is a remnant of limestone solution and erosional processes. It is roughly circular in shape, with steep sides which terminate abruptly in a flat alluvial plain. Mogotes are usually found in large numbers and are common in south China (where they inspired many classical landscape pictures), North Vietnam, France (Massif Central), the Philippines, Yugoslavia and Java. In each area they have different local names. (See also COCKPIT KARST; KEGELKARST.)  PAB

**Reading**
Klimaszewski, M. 1964: The karst relief of the Kuelin area (South China). *Geographia Polonica* 1, pp. 187–212.

**Mohorovičić discontinuity**  Sometimes referred to as the 'moho', lies between the crust of the earth and the underlying MANTLE. Occurring at around 30–40 km below the continents and at about 10 km beneath the oceans,

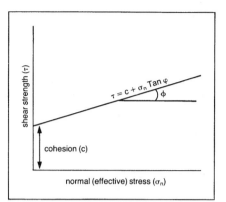

it is a zone where seismic waves are significantly modified.  ASG

**Mohr–coulomb equation**  A widely used relationship to describe the strength of soils in terms of COHESION and FRICTION of materials:

$$s = c + \sigma \tan \phi$$

where $s$ is the shear strength at failure (sometimes denoted by $T$), $c$ is the cohesion component and $\phi$ the friction of the soil; $\sigma$ (sometimes $\sigma_n$) is the normal stress. Either cohesive or frictional components may be absent but a plot of normal stress and shear stress for a soil with both is of the form shown above. The strength parameters $c$ and $\phi$ are usually obtained by SHEAR BOX or TRIAXIAL APPARATUS tests. The equation is modified if pore water pressures are taken into account to give an EFFECTIVE STRESS analysis.  WBW

**Reading**
Mitchell, J.K. 1976: *Fundamentals of soil behaviour*. New York: Wiley. · Statham, I. 1977: *Earth surface sediment transport*. Oxford: Clarendon Press. · Whalley, W.B. 1976: *Properties of materials and geomorphological explanation*. Oxford: Oxford University Press.

**moisture index**  Relates water loss through potential evapotranspiration ($PE$) to water gain through precipitation ($P$), as a basis for climatic classification:

$$MI = \frac{100(P - PE)}{PE}$$

The moisture index ($MI$) falls to $-100$ when there is no precipitation, and exceeds $+100$ when precipitation greatly exceeds potential evapotranspirational losses.  ASG

**molard**  A term from the French Alps for 'conical mounds of broken slide rock deposited along the typically lobate margins of rock avalanche spoil debris – a debris cone' (Cassie *et al.* 1988). They may be up to 20 m high.  ASG

**Reference**
Cassie, J.W., Van Gassen, W. and Coudess, D.M. 1988: Laboratory analogue of the formation of molards, cones of rock-avalanche debris. *Geology* 16, pp. 735–78.

**molasse** A term applied to any thick succession of continental deposits consisting in part of sandstones and conglomerates which were formed as a result of mountain building. The molasse facies is the main diagnostic feature of orogeny (mountain building).

**momentum budget** Except for the small effects of tidal friction, the absolute ANGULAR MOMENTUM of the earth and atmosphere combined must remain constant in time; and because the annual average rotation rate of the earth is observed to remain almost constant, the atmosphere alone must also on average conserve its angular momentum. Thus it is possible to consider the fluxes of angular momentum in an angular momentum budget. Such a momentum budget is applied to the time-averaged flow of the atmosphere, that is the GENERAL CIRCULATION.

Angular momentum per unit mass of a westerly current is the product of angular velocity with the distance about the axis of rotation. In the tropics, where surface winds are easterly, the torque, or twisting moment, about the earth's axis exerted by the earth on the atmosphere is such that the atmosphere gains angular momentum from the earth; while in the middle latitudes, where surface winds are westerly, the atmosphere gives up angular momentum to the earth. Therefore there must be a net poleward flux of angular momentum in the atmosphere between the tropics and middle latitudes, otherwise surface friction would decelerate both the easterlies and westerlies. This poleward flux of angular momentum must increase with latitude in the regions of the easterlies and decrease with latitude in the westerlies. In the northern hemisphere it is observed to reach a maximum at about 30° latitude. Poleward of the mid-latitude westerlies, the surface area of the earth is relatively small, so that angular momentum exchange is relatively unimportant. If the total angular momentum of the atmosphere is to remain constant the exchanges between atmosphere and earth in the easterlies and westerlies must be equal but of opposite sense.

In low latitudes the poleward momentum transport is effected partly by the mean meridional flow and partly by eddies. In middle latitudes the mean meridional flow is much too weak to effect a significant flux so that the poleward transport is accomplished mainly by eddies. For eddies to effect a meridional transport of angular momentum they must be asymmetric, with the axes of troughs and ridges tilting from south-west to north-east. In such a trough–ridge system zonal flow is larger in those portions where the meridional flow is poleward, so that a poleward angular momentum flux results. Observations show that eddies account for most of the poleward transport except within about 10° latitude of the equator.

The torque exerted by the earth on the atmosphere is due partly to surface friction and partly to pressure differences across mountains. For instance, in the mid-latitude westerlies, observations show that surface pressures on the western slopes of mountains tend to exceed those on the eastern slopes at the same heights. In mid-latitudes this mountain pressure torque is estimated to be about as large as the torque due to surface friction.                           KJW

**Reading**
Holton, J.R. 1992: *An introduction to dynamic meteorology.* New York and London: Academic Press.

**monadnock** Any residual hill or mountain which is isolated on a flat plain produced by erosion. It is a product of the late stages of the Davisian cycle of erosion, and rises above a peneplain.

**monoclimax** A theory of vegetation requiring that all the SERES (community sequences) in an area converge on a uniform and stable plant community, the composition of which depends solely on regional climate. Given sufficient time, it is argued, the processes of SUCCESSION overcome any major effects on vegetation of differences in other environmental factors, such as topography and soils.

It is now generally replaced by other theories of CLIMAX VEGETATION. Many vegetational terms had their origin in monoclimax theory. (See also POLYCLIMAX.)                           JAM

**Reading**
Clements, F.E. 1916: *Plant succession.* Publication 242. Washington: Carnegie Institute. · Matthews, J.A. 1979: Refutation of convergence in a vegetation succession. *Naturwissenschaften* 66, pp. 47–9. · Walker, D. 1970: Direction and rate in some British postglacial hydroseres. In D. Walker and R.G. West eds, *Studies in the vegetation history of the British Isles.* Cambridge: Cambridge University Press. Pp. 117–39.

**monocline** A zone of steeply dipping strata in an area of horizontally bedded rocks. A zone of rocks which dip steeply to great depth.

**monophylesis** The origin of a taxonomic group at one point in space and time, by EVOLUTION.

**monsoon** Derived from the Arabic word *mausim*, meaning season, the term originally referred to the winds of the Arabian Sea, which blow for about six months from the north-east and for six months from the south-west. The term is now used for other markedly seasonal winds, e.g. the stratospheric monsoon. The characteristics of the monsoon climate are to be found mainly in the Indian subcontinent, where over much of the region annual changes may conveniently be divided as follows:

1  The season of the north-east monsoon: (*a*) January and February, winter season; (*b*) March to May, hot weather season.
2  The season of the south-west monsoon: (*a*) June to September, season of general rains; (*b*) October to December, post-monsoonal season.

Over much of India the bulk of the rainfall comes in the south-west monsoon season.  JGL

**Reading**
Chang, C.P. 1987: *Monsoon meteorology*. Oxford: Oxford University Press. · Ramage, C.S. 1971: *Monsoon meteorology*. London and New York: Academic Press.

**monsoon forest** A lowland and montane division of TROPICAL FOREST characterized by seasonal climates with pressure and wind reversal. Monsoon forests occur when the total rainfall during the monsoon is sufficient to maintain forest, but where there is also a marked dry season. The resulting trees may be predominantly evergreen, depending upon the intensity of the wet season. They are, however, more characteristically semi-deciduous, particularly in the upper canopy, as a result of the seasonal water deficiency. The structure is more open than lowland evergreen forest, with greater light penetration to the ground and denser undergrowth, with locally abundant bamboos and lianas.  PAF

**Reading**
Golley, F.B., Lieth, H. and Werger, M.J.A. eds 1982: *Tropical rain forest ecosystems*. Amsterdam: Elsevier. · Longman, K.A. and Jenik, N. 1987: *Tropical forest and its environment* 2nd edn. London: Longman.

**monumented sections** River channel cross-sections which are precisely surveyed in relation to at least two fixed points and which can be resurveyed on a number of occasions to indicate the amount of channel change. Such sections were advocated as part of the Vigil Network of REPRESENTATIVE AND EXPERIMENTAL BASINS.

KJG

**mor** Raw humus which is not admixed with mineral material.

**moraine** A distinct landform fashioned by the direct action of a glacier. In the past the term was also used to describe glacial sediments, e.g. ground moraine, but now it is accepted that *moraine* should refer to the landforms and the word *till* to the sediment.

One group of moraines exists on the surfaces of glaciers and includes *lateral* moraines, which form through the accumulation of valley-side material on either side of the glaciers, and *medial* moraines which form from the junction of lateral moraines as two glaciers meet. In the ABLATION areas of glaciers in particular such moraines can form prominent upstanding ridges where the debris has protected the underlying ice from melting. The material in lateral and

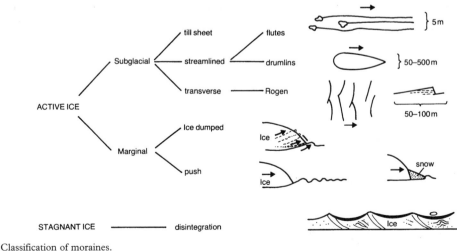

Classification of moraines.
*Source*: Chorley *et al.* 1984. Figure 17.23.

Classification of the major types of moraine

| Parallel to ice flow | Transverse to ice flow | Lacking consistent orientation |
|---|---|---|
| Subglacial forms with streamlining<br>Fluted and drumlinized ground moraine<br>Drumlins and drumlinoid ridges<br>Crag-and-tail ridges | Subglacial forms<br>Rogen or ribbed moraine<br>De Geer or washboard moraine<br>Subglacial thrust moraine<br>Sublacustrine moraine | Subglacial forms<br>Low-relief ground moraine<br>Hummocky ground<br>moraine |
| Ice-pressed forms<br>Longitudinal squeezed ridges | Ice-pressed forms<br>Minor transverse squeezed<br>ridges | Ice-pressed forms<br>Random or rectilinear<br>squeezed ridges |
| Ice marginal forms<br>Lateral and medial moraines<br>Some interlobate and kame moraines | Ice front forms<br>End moraines<br>Push moraines<br>Ice thrust/shear moraines<br>Some kame and delta moraines | Ice surface forms<br>Disintegration moraines |

*Source*: Modified from Sugden, D.E. and John, B.S. 1976: *Glaciers and landscape*. London: Edward Arnold. Table 12.1, p. 236. After Prest, V.K. 1968: *Geological survey papers Canada* 67–57.

medial moraines is characteristically angular rockfall debris and undergoes minimum modification during transport.

A second group of moraines occurs at the edge of existing glaciers or in areas formerly covered by glaciers. One classification scheme, based on glacier dynamics and position with regard to a glacier is shown in the diagram. Subglacial forms constructed beneath moving glacier ice include uniform till sheets, as for example occur widely in Iowa and Illinois (Kemmis 1981), streamlined and transverse features. Glacial flutes are small-scale examples of ridges streamlined parallel to the direction of ice flow and are generally a few tens or hundreds of metres long and up to 2 m high. They build up in the lee of boulders which become lodged on the bed and thereby create a cavity or low pressure zone in their lee. Drumlins are larger forms streamlined parallel to the direction of ice flow (see DRUMLIN). ROGEN MORAINE forms transverse to ice flow.

Active glaciers which end on land build up moraines at the ice margin. The size of the moraine depends on the period that the margin lies in the same location and also on the amount of rock debris transported to the edge by the glacier. The common 'ice-dumped' moraine is typically a complex landform reflecting many different processes of debris accumulation, including slumping or flow off the glacier surface, and lodgement or deformation of basal till. Another common moraine is associated with deformation of sediments by pushing. Glaciers tend to advance in winter when ablation rates at the snout are low and this advance may physically push sediments into a ridge up to 2 m high. Sometimes it may push a frontal snow bank or

sheet of lake ice which itself deforms sediments (Birnie 1977). If a glacier is in overall retreat these annual advances are marked by a succession of small ridges whose spacing can be directly related to the amount of annual ablation. At a larger scale push moraines may involve much larger areas of sediments in front of a glacier, as for example, can occur in the case of a SURGING GLACIER.

Disintegration moraine is the complex remains of a process whereby debris-bearing ice stagnates and melts *in situ*. An irregular landscape of hummocks and kettles, with differing combinations of subglacial and slumped surface debris is the end product. Such moraines form best where compressing flow brings large quantities of debris up to the ice surface.

Moraines have been widely used to delimit the former extent of glaciers. When they are dated unambiguously a relatively sophisticated reconstruction of past climate can be made. Problems have arisen in recent years, however, because many moraines are found to have been built up by successive glacier advances over long time spans.                                    DES

**Reading and References**
Birnie, R.V. 1977: A snow-bank push mechanism for the formation of some 'annual' moraine ridges. *Journal of glaciology* 18.78, pp. 77–85. · Chorley, R.J., Schumm, S.A. and Sugden, D.E. 1984: *Geomorphology*. London: Methuen. · Kemmis, T.J. 1981: Importance of the regelation process to certain properties of basal tills deposited by the Laurentide ice sheet in Iowa and Illinois. *Annals of glaciology* 2, pp. 147–52. · Moran, S.R., Clayton, L., Hooke, R. Le B., Fenton, M.M. and Andriashek, L.D. 1981: Glacier-bed landforms of the prairie region of North America. *Journal of glaciology* 25.93, pp. 457–76.

**morphogenetic regions** Those regions in which it is claimed that certain geomorphological processes result from a particular set of climatic conditions, thereby giving distinctive regional landscapes.

**morphological mapping** A means of mapping landforms. In the strict sense morphological maps display only the shape of the ground with breaks of slope and gradients indicated. Such maps have been found useful by, for example, South African engineers. In a wider sense morphological mapping has been used to map landforms in terms of their origin. This form of morphogenetic mapping, popular in central Europe, has found wide applications in terms of engineering geomorphology, resource surveys and pure research.                                    DES

**Reading**
Brunsden, D., Doornkamp, J.C. and Jones, D.K.C. 1979: The Bahrain surface materials resources survey. *Geographical journal* 145, pp. 1–35. · Crofts, R.S. 1974: Detailed geomorphological mapping and land evaluation in Highland Scotland. *Institute of British Geographers special publication* 7, pp. 231–51.

**morphometry** The quantitative description of forms; in physical geography it refers to the earth's surface (strictly geomorphometry) but in other sciences it is an approach which can relate to fossils or crystals, for example.

Evans (1972, 1990) has distinguished *general geomorphometry* which is based upon an analysis of the entire landsurface as a continuous, rough surface described by the attributes at a sample of points or from arbitrary areas, and *specific geomorphometry* which relates to specific landforms and to the measurement of their size, shape and relationships. In both approaches definitions should be made to allow relationships to process indices. Morphometry is necessary to characterize areas and landforms quantitatively, to allow areas studied by different scientists to be compared easily, to demonstrate how aspects of the landsurface are inter-related and to provide parameters which can relate to processes in relationships from which processes may be estimated where only morphometric parameters are available.

In general geomorphometry, a specific part of the landsurface could be described by an equation but this would have so many terms and be so complex that it has not been used except for specific areas of landforms. Therefore altitude at a point and the derivatives of slope and curvature have often been used as the basis for general geomorphometry and Evans (1990) and others have identified five fundamental attributes which are altitude, gradient, aspect, profile convexity and plan convexity. Profile convexity is the rate of change along a line of maximum gradient and plan convexity is the rate of change of aspect along a contour. These five attributes relate to a point or small area and systems of general geomorphometry which have been put forward are based upon analysis of spatial patterns of some or all of the attributes. Variations between methods which have been suggested depend upon the relative significance accorded to the five attributes but several schemes have been devised because of the relevance to trafficability, drainage, suitability for different types of land use, and susceptibility to erosion hazard.

Specific geomorphometry has been used as a more realistic way of simplifying the task of earth surface description and morphometric methods have been devised for the description of coral atolls, karst depressions, glacial cirques, sand dunes, lake basins and many other landforms. Because the morphometry of drainage basins has attracted much attention morphometry has sometimes been associated mainly with drainage networks and drainage basin morphometry. The earliest developments in drainage basin morphometry were based upon stream ORDER and HORTON'S LAWS and these could provide the basis for comparisons between areas but have been less valuable because of the statistical nature of these so-called 'laws'. The morphometry of drainage basins has focused (Gregory and Walling 1976) upon the area, length, shape and relief attributes and these have parallels at the level of the drainage NETWORK when DRAINAGE DENSITY is an important measure of relative length. A wide range of morphometric measures has been devised for drainage basins and these have usually been defined as either ratio measures such as ratio of maximum width to breadth giving an index of drainage basin shape, or as measures which depend upon comparison with an ideal shape and drainage basins have been compared with a circle or with a lemniscate (pear-shaped) loop for example. (See also CIRQUES; DRAINAGE DENSITY; DUNES; ORDER, STREAM.)        KJG

**Reading and References**
Chorley, R.J. ed. 1972: *Spatial analysis in geomorphology*. London: Methuen. · —ed. and Kennedy, B.A. 1971: *Physical geography: a systems approach*. London: Prentice-Hall. · Evans, I.S. 1972: General geomorphometry, derivatives of altitude and description statistics. In R.J. Chorley ed., *Spatial analysis in geomorphology*. London and New York: Methuen. Pp. 17–90. · —1990: General geomorphometry. In A. Goudie ed., *Geomorphological techniques*. London: Unwin Hyman. Pp. 31–7. · Goudie, A. ed. 1990: *Geomorphological techniques*. London: Unwin Hyman. · Gregory, K.J. and Walling, D.E. 1976: *Drainage basin form and process*. London: Edward Arnold. Pp. 37–60.

**morphotectonics** (or tectonic geomorphology) The study of the interaction of tectonics and geomorphology.                      ASG

**Reading**
Ollier, C.D. 1991: Morphotectonic and structural geomorphology. *Zeitschrift für Geomorphologie Supplementband* 82, pp. 1–161.

**mosaic-cycle concept** An ecological concept associated with forest growth. Natural forests exhibit a cyclic alternation between an optimal phase consisting of trees roughly the same height and age (usually of a single species) which then deteriorates into a phase of decay. This is succeeded by a phase of rejuvenation which in time becomes an optimal phase. Stages with other tree species (some of them pioneer species) are often interpolated between the phases of decay and rejuvenation. A primary forest is thus a mosaic of areas in which the same cyclic succession of growth and decay is going on but in which the cycles are out of step.                      ASG

**Reading**
Remmert, H. ed. 1991: *The mosaic-cycle concept of ecosystems.* Berlin: Springer Verlag.

**mottled zone** The portion of a soil zone or weathering profile immediately beneath a ferricrete or silcrete horizon, in which bleached kaolinitic material occurs with patches of iron staining.

**moulin** A vertical cylindrical shaft by which surface meltwater flows into a glacier. Moulins tend to form at lines of structural weakness in the glacier and are usually 0.5 to 1.0 m in diameter and up to 25 to 30 m deep.                      DES

**Reading**
Stenborg, T. 1969: Studies of the internal drainage of glaciers. *Geografiska annaler* 51A. 1–2, pp. 13–41.

**mountain meteorology** Mountains exert an influence on the atmosphere both *mechanically*: by blocking the airflow, deflecting it over and around the barrier, and through frictional drag; and *thermodynamically*: by acting as a direct, elevated heat source, as an indirect heat source through latent heat release in clouds formed over the mountains, and as a moisture sink through precipitation. The scales of mountain effects on the atmospheric circulation include: the planetary wave scale, with upper-air low-pressure troughs located over eastern North America and eastern Asia related, respectively, to the Rocky Mountains and the Tibetan Plateau upwind; the regional-synoptic scale, with the modification of frontal systems as they move across major mountain ranges and the formation of lee cyclones; the mesoscale of mountain-induced lee wave clouds and fall winds (föhn, bora); and the local scale of mountain/valley and slope winds systems resulting from topoclimatic contrasts in diurnal heating patterns. The characteristics of weather and climate in mountain areas are most closely related to the last three categories of meteorological phenomena.

A mountain climate can be considered to exist whenever the relief creates an altitudinal zonation of climatic elements (temperature, precipitation) sufficient to change the local vegetation characteristics. Exceptions to this criterion may occur, however, where vegetation is absent on hyperarid subtropical or polar mountains. The effect of altitude causes a temperature decrease (environmental lapse rate) of about 5–6 $°C\,km^{-1}$, on average, although a temperature increase with height often occurs in mountain valleys and basins, with wintertime and/or nocturnal temperature inversions. There is also a general altitudinal decrease of water vapour content, a decrease of pressure (approximately 100 mb $km^{-1}$ in the lower troposphere), and an increase of incoming solar radiation (about 5–15 per cent $km^{-1}$ under cloudless skies). Orography redistributes and in many cases augments the precipitation that would otherwise have occurred through cyclonic or convective processes. The altitudinal enhancement mostly occurs as a result of increased amounts rather than greater frequency of precipitation. On windward mountain slopes, the zone of maximum precipitation, in a climatic sense, typically occurs at low elevations in equatorial zones, about 700–1200 m in the tropical (trade wind) zones, and at higher levels (up to 3000 m and above) in mid-latitudes. In the lee of many mountain ranges, with respect to the prevailing wind direction, there is a reduction in average precipitation giving rise to a so-called 'rain-shadow'.                      RGB

**Reading**
Barry, R.G. 1992: *Mountain weather and climate.* 2nd edn. London: Routledge. · Browning, K.A. and Hill, F.F. 1981: Orographic rain. *Weather* 36, pp. 326–9. · Smith, R.B. 1979: The influence of mountains on the atmosphere. *Advances in geophysics* 21, pp. 87–230. · Yoshino, M.M. 1975: *Climate in a small area: an introduction to local meteorology.* Tokyo: University of Tokyo Press.

**mountain/valley wind** A local wind system produced in mountainous regions as a result of temperature differences. The circulation is best developed in summer, when the skies are clear and large-scale motions are weak, and in deep, straight valleys with a north–south axis. During the day the air above the slopes and floor of the valley is heated to a temperature well above that over the centre of the valley. Shallow upslope (ANABATIC) flow results, and is compensated for by air sinking in the valley centre. If the ascending air is moist enough convective clouds may form along the valley ridges. Superimposed on this cross-valley flow is a VALLEY WIND blowing up-valley at low levels from the adjacent plains.

At night the valley surface and the overlying air cool by the emission of infrared radiation, causing the air to flow downslope under the influence of gravity. The convergence of these slope winds near the valley centre produces both a weak ascending motion and a low-level down-valley or mountain wind which flows out of the mountains onto the adjacent plains. At higher elevations a counter flow occurs from the plains to the valleys.                                    WDS

**Reading**
Atkinson, B.W. 1981: *Meso-scale atmospheric circulations*. London: Academic Press. · Oke, T.R. 1987: *Boundary layer climates*. 2nd edn. London: Routledge.

**mountains**  Substantial elevations of the earth's crust above sea level, which result in localized disruptions to climate, drainage, soils, plants and animals. Increases in altitude tend to repeat the bioclimatic patterns associated with a move towards higher latitudes, although cloudiness, day length and seasonal variations differ from the latitudinal progression. Temperature drops at a rate of approximately $0.5\,°C$ $100\,m^{-1}$ and rainfall is often heaviest where moisture-laden winds are forced to rise (over 1500 m in the tropics). Vertical ZONATION of plants and animals is most clearly illustrated where mountains rise from tropical forest to tundra environments.                          PAF

**Reading**
Gerrard, A.J. 1990: *Mountain environments*. London: Belhaven Press. · Moore, D.M. ed. 1983: *Green planet: the story of plant life on earth*. Cambridge and New York: Cambridge University Press. · Price, L.W. 1981: *Mountains and man: a study of process and environment*. Berkeley: University of California Press.

**mud lumps**  Small-scale landforms found from the Mississippi delta region of the USA. Rapid forward growth of distributary channels deposits deltaic sand, mud and organic sediment on top of unstable prodelta clay. This causes loading which in turn causes diapiric intrusions of plastic clays through the overlying sands. Updoming or extrusion occurs, producing the mud lumps.                          ASG

**Reading**
Morgan, J.P., Coleman, J.M. and Gagliano, S.M. 1968: Mudlumps: diapiric structures in Mississippi delta sediments. *Memoir of the American Association of Petroleum Geologists* 8, pp. 145–61.

**mud volcano**  A mount built up of mud carried to the surface by geysers or gap eruptions in volcanically active regions.

**mull**  Humus admixed with mineral material in the surface horizons of the soil zone.

**multiple working hypotheses**  These represent a method used to test alternative explanations against each other (Haines-Young and Petch 1986). When trying to explain aspects of the world we can often think of several plausible HYPOTHESES. Using this method we can test the logical consequences of each hypothesis and attempt to eliminate those ideas that are wrong. The availability of competing hypotheses helps us focus on just what observations or experiments are needed to examine ideas critically.

The method was initially discussed by Chamberlin (1890) and Gilbert (1896). Baker (1996) describes their ideas in a wider context. Examples of its use include Batterbee *et al.* (1986) and Turner (1997).                          RH-Y

**References**
Baker, V.R. 1996: The pragmatic roots of American Quaternary geology and geomorphology. *Geomorphology* 16.3, pp. 197–215. · Batterbee, R.W, Flower, R.J., Stevenson, J. and Rippy, B. 1985: Lake acidification in Galloway: a palaeological test of competing hypotheses. *Nature* 314, pp. 350–52. · Chamberlin, T.C. 1890: The method of multiple working hypotheses. Initially printed in *Science*. Reprinted in *Journal of geology* 1995, 103.3, pp. 349–54. · Gilbert, G.K. 1896: The origin of hypotheses, illustrated by the discussion of a topographical problem. *Science* 3, pp. 1–3. · Haines-Young, R.H. and Petch, J.R. 1986: *Physical geography: its nature and methods*. London: Paul Chapman. · Turner, R.E. 1997: Wetland loss in the northern Gulf of Mexico: multiple working hypotheses. *Estuaries* 20.1, pp. 1–13.

**multispectral scanner**  An optical remote sensor which is used to derive images of the earth's surface. It possesses three advantages over AERIAL PHOTOGRAPHY. First, it has a very fine radiometric resolution in narrow and simultaneously recorded wavebands. Secondly, these wavebands span a relatively large range of ELECTROMAGNETIC RADIATION from ultraviolet wavelengths to thermal infrared wavelengths and, thirdly, the data can be stored in digital form for DIGITAL IMAGE PROCESSING.

A multispectral scanner measures the radiance of the earth's surface along a scan line perpendicular to the line of aircraft flight. As the aircraft moves forward repeated measurement of radiance enables a two-dimensional image to be built up. These scanners comprise a collecting section, a detecting section and a recording section. A telescope directs radiation onto the rotating mirror. The rotating mirror reflects the radiation into the optics. The optics focus the radiation into a narrow beam. The dichroic (doubly refracting) grid splits the radiation into its reflected and emitted components. The reflected radiation is further divided and the emitted radiation goes to the thermal infrared detectors. A prism is placed in the path of the reflected radiation, to divide it into its spectral components. The radiation detectors sense the reflected and emitted radiation. The reflected radiation is usually detected by silicon photodiodes that are placed in their correct geometric position behind the prism. The emitted (thermal infrared) radiation is usually detected by photon detectors, which are held in a cooling flask. After detection the signal is amplified by the preamplifier and is passed in electronic form to the control box. The electronic control console has three components: a signal processor to format the data as required for the recorders, an amplifier to boost the signal level even further and a power distribution unit to balance the signal strength in each waveband. The type of recorder depends upon the make and model of the multispectral scanner. The majority of multispectral scanners have a monitor to enable the operator to observe the data as they are recorded. Simple multispectral scanners also tend to have analogue recorders; either a film recorder, where the electrical impulses are recorded directly onto film, or an analogue tape recorder, where the electrical impulses are stored on magnetic tape. The analogue tape recorder records the electronic signal in the aircraft. The signals can either be fed into a film recorder to produce images, or can be digitized for later digital image processing. Current scanners use analogue to digital converters to produce a digital output which is recorded by a digital tape recorder.

The two most popular applications for these images are the mapping of vegetation and soil, often as a prerequisite to studies of water quality. PJC

**Reading**
Curran, P.J. 1985: *Principles of remote sensing.* Harlow: Longman Scientific and Technical. · Lowe, D.S. 1976: Nonphotographic optical sensors. In J. Lintz and D.S. Simonett eds, *Remote sensing of environment.* Reading, Mass. and London: Addison-Wesley.

**murram**  Laterite or ferricrete in East Africa.

**muskeg**  A Canadian-Indian term for waterlogged depressions in the subarctic zone of Canada and Alaska. There are some 500,000 square miles of such marsh in Canada alone. These depressions are largely filled with peat and characterized by *Sphagnum* moss. It is a region of marshy depressions with scattered lakes, stagnant mosquito-infested pools and slow meandering streams. AP

**Reading**
Radforth, N.W. 1969: Environmental and structural differentials in peatland development. In E.C. Dapples and M.E. Hophins, eds, *Environments of coal deposition.* Geological Society of America special paper 114, pp. 87–104. · — and Branner, C.O. eds 1977: *Muskeg and the northern environment in Canada.* Toronto: University of Toronto Press.

**Muskingum method**  A method of FLOOD ROUTING which assumes that along any channel reach the difference between the inflow hydrograph ($I$) to the reach and the outflow hydrograph ($O$) from the reach is equal to the stored or depleted water in a specified time interval. Two simultaneous equations can be solved, namely the water balance equation which expresses change in storage:

$$\Delta S = I - O$$

and an equation for storage:

$$S = K[xI + (1 - x)O]$$

where x is a dimensionless constant for the channel reach and K is a storage constant which is obtained from hydrographs of $I$ and $O$ at each end of the reach. KJG

**Reading**
Lawler, E.A. 1964: Flood routing. In V.T. Chow, *Handbook of applied hydrology.* New York: McGraw-Hill. Sect. 25–II.

**mutation**  A change in the structure or amount of DNA in the chromosomes in the cells of an organism, or the resulting change in the organism's characteristics. If a mutation occurs in the gametes (reproductive cells) it is inherited; if it occurs elsewhere (in somatic or non-reproductive cells) it is not. Inherited mutations caused by a change in the structure of the DNA molecule are known as gene mutations; those produced by a change in the amount of DNA are known as chromosomal mutations. These errors in the coding of inherited information occur at a low frequency, apparently spontaneously.

Most mutations are deleterious because of the long period of testing (by natural selection) the genome (the package of genes within the gamete or germ-cell) has undergone. They will be eliminated. The extremely rare beneficial mutation will be incorporated into the genome by the process of natural selection. PHA

**mutualism**  An interaction that benefits the species involved. The most widespread mutualisms are between plants and animals, to use the animals to improve the efficiency of plants' reproduction and to provide food for the animals in return. For example, birds are attracted to fruits and eat them. The fruits pass through the gut and a portion is defecated shortly afterward. Meanwhile the bird has flown from the site of the parent plant and the seed is left in a supply of fertilizer, ready to germinate.                                                      ASG

**mycorrhizal fungi**  These infect the roots of host plants, and exist in a mutualistic relationship that is beneficial to both symbionts. The term 'mycorrhiza' was coined by Frank in 1885 from the Greek meaning 'fungus-root' (*myko – rhiza*). Mycorrhizae are very widespread, ranging from Arctic to desert to equatorial plant communities. They also occur through a considerable range of soil depths, spanning near-surface locations to depths of some metres. The fungi produce filaments, or hyphae (about 3 $\mu$m in diameter) that individually may extend up to a few centimetres into the soil surrounding the root, but which may reach total lengths of 50 m/gram of soil. They reach further into the soil than do root hairs, which may only be 1 mm in length. The hyphae provide additional area exposed to the soil solution, and provide an important source of nutrients like P, N, and K, especially when these are only available at low levels in the soil. Mycorrhizal fungi are capable of transporting nitrogen in the form of ammonium ($NH_4^+$) as well as NITRATE, so further enhancing the supply of this nutrient. Fungal remains also contribute organic matter to the soil in substantial amounts, and may enhance the flow of water to the host plant. The hyphae bind soil particles together physically, and also by producing polysaccharides that act as binding agents.                                                      DLD

**Reading**
Allen, M.F. 1991: *The ecology of mycorrhizae*. Cambridge: Cambridge University Press. · Jungk, A. and Claassen, N. 1997: Ion diffusion in the soil-root system. *Advances in agronomy* 61, pp. 53–110.

# N

**nanism** (or microsomia) The condition of being dwarfed, often implying stunted, and refers to both plants and animals. Small size is implicit in the expression nanophyllous (small-leaved), or nanoplankton (the smallest plankton), or nanophanerophytes (shrubs under 2 m in height) although in SI terminology the prefix nano- strictly signifies a unit $\times 10^{-9}$. Artificial breeding of dwarf animals and plants is sometimes referred to as nanization.          PAF

**nappe** A mass of rock which is thrust over other rocks by thrust faulting or a recumbent fold or both.

**natural disaster** A NATURAL HAZARD that actually happens. An event that has a dramatically negative effect on humans but which is due to the occurrence of a natural event or process. For example, a famine may be due to the occurrence of a to-be-expected DROUGHT event in a DRYLAND region, or large-scale death and destruction may result from the passage of a TORNADO through a populated area. If blame has to be ascribed in the context of the disaster, it lies not with the natural event but with the location of human activities or settlements in the locality or pathway of that event.          DSGT

**Reading**
Alexander, D. 1993: *Natural disasters*. London: University College London Press.

**natural hazard** Any aspect of the physical environment's natural functions that may adversely affect human society to cause social disruption, material damage and/or loss of life, in which case the impact is referred to as a natural disaster or catastrophe. Physical geographers may classify natural hazards according to their sphere of occurrence into geological (e.g. earthquake), hydrological (e.g. flood), atmospheric (e.g. snowstorm) and biological (e.g. disease). Another approach divides hazards into rapid-onset, intensive events (short, sharp shocks such as tornadoes) and slow-onset pervasive events which often affect larger areas over longer periods of time (such as droughts).

Although such classifications provide useful summaries of hazard types, many hazards can have a variety of effects. Hence, an earthquake may cause a tsunami wave at sea, landslides or avalanches on slopes, building damage and fires

in urban areas, and flooding due to the failure of dams, as well as ground shaking and displacement along faults. Similarly, many natural disasters cause disruption to public hygiene and consequently result in heightened risks of disease transmission.

In practice, it is often difficult to distinguish between purely 'natural' events and human-induced events. In one sense, all 'natural' disasters can be thought of as human-induced since it is the presence of people which defines whether a hazard creates a disaster or not, and many of the natural physical processes that cause disasters can also be triggered or exacerbated by human action. The composite nature of many disasters also blurs the distinction: the major cause of death due to earthquakes is usually crushing beneath buildings, so is the disaster natural or human-induced? Some researchers believe that the difficulty of making this distinction has rendered the division pointless, and prefer to talk of 'environmental' hazards which refer to a spectrum with purely natural events at one end and distinctly human-induced events at the other (e.g. Smith 1996).          NJM

**Reading and Reference**
Alexander, D. 1993: *Natural disasters*. London: University College London Press. · Blong, R. 1997: A geography of natural perils. *Australian geographer* 28, pp. 7–28. · Smith, K. 1996: *Environmental hazards: assessing risk and reducing disaster*. 2nd edn. London: Routledge.

**natural resources** Components of the natural environment that have a utility to humankind, and following Zimmerman (1933), it is sometimes said that resources 'are not, they become'. This indicates that the natural environment contains components that are both useful and not useful to people, the utility being derived from societal needs and the technical ability to extract usage, rather than the mere existence of the elements in the environment. To this end, elements without utility were termed 'neutral stuff' by Zimmerman. The natural resource/neutral stuff dichotomy is not a static one, since over time different societies have found use for elements that were previously unused, and vice versa. Thus at the end of the twentieth century, flints, so vital for stone age cultures, had very limited practical usage (except perhaps as a luxury building material), while stone age people had no use for uranium ore; indeed they did not know what it was.

It is clear, from the perspective of geography, that natural resources can be considered from both their physical and human dimensions. In the latter case, demand, supply, differences in resource perceptions based on cultural development, wealth and so on are all relevant issues for consideration. From a physical perspective, the natural distribution of resources in the biosphere and lithosphere, resource types, and the impacts of resource use on the wider environment are all currently topics of interest. In practice, however, the physical and human dimensions coexist and impinge upon each other, as is well illustrated in the works of Ian Simmons (for example, 1974 and 1991).

Natural resources can be classified in many ways. A useful division can be made between STOCK RESOURCES and FLOW RESOURCES, with division between the two based on the time it takes for a resource to form relative to human lifespans. For stock resources, availability is a function of the natural abundance of the resource in the earth system, knowledge of the distribution of reserves, having the means to extract reserves of that resource (which may be defined by technological developments; for example, sinking an oil well on land is generally easier and cheaper than tapping oil fields that lie beneath the North Sea), and the rate of usage, itself affected by demand. Over time, the status of a particular component of the base can change, for example from being hypothetical, to conditional upon technology allowing exploitation, to proven and awaiting use, to used. The future availability of a resource may be revised on the basis of factors including new reserves being identified and exploited, and changes in demand that may relate, for example, to substitutes being developed or recycling rates increasing.

Many environmental issues and problems relate to the use of natural resources. Overuse of soil, excessive cultivation and the expansion of agriculture to marginal lands may lead to SOIL EROSION and in drylands to DESERTIFICATION, turning a flow resource into one that is being used unsustainably. The growth in use of fossil fuels in the twentieth century has been seen as a key cause of the CARBON DIOXIDE PROBLEM and GLOBAL WARMING. Damage to the OZONE layer may result from the release of certain chemicals into the atmosphere. Spatial inequalities and mismatches between the occurrence of particular natural resources and human demands contribute to the transport of resources over long distances, adding to various forms of POLLUTION and the consumption of energy

resources. Overall therefore, natural resources are a fundamental component of human use and occupation of the earth, their use and temporal changes in demand relate to cultural developments and advancements, as well as to the rise of environmental issues and concerns.

DSGT

**Reading and References**
Rees, J. 1990: *Natural resources: allocation, economics and policy.* 2nd edn. London: Routledge. · Simmons, I.G. 1974: *The ecology of natural resources.* London: Edward Arnold. · —1991: *Earth, air and water: resources and environment in the late 20th century.* London: Edward Arnold. · Zimmerman, E.W. 1933: *World resources and industries.* New York: Harper.

**natural selection**   See DARWINISM; EVOLUTION.

**natural vegetation**   A general term for the total sum of plants in an area, grouped by communities but not as part of a taxonomic system. 'Natural' signifies the sum total of inheritance or genotype, but the term natural vegetation is also associated with environmental factors which encourage or constrain plant growth after an equilibrium between plants and their surroundings has been established. The larger groupings of plants illustrate a ZONATION which has a combined biological and environmental basis. Perhaps the most common usage is to denote the plant cover of any area prior to its modification by humans.   PAF

**Reading**
Walter, H. 1973: *The vegetation of the earth.* London: English Universities Press.

**neap tide**   See TIDES.

**nebkha**   (also nabkha) A small sand dune, from *c.*10 cm to *c.*2–3 m high, formed when windblown sand is trapped within or accumulates around a plant. Nebkhas are discrete forms, that may occur individually or in nebkha fields, that develop in dryland areas where the vegetation cover is discontinuous.   DSGT

**neck**   A narrow isthmus or channel. A mass of lava which has solidified in the pipe or vent of a volcano.

**neck cut-off**   The process in which a tight meander loop on an alluvial river is abandoned by the incision of a new linking channel that bypasses the loop. The other common mode of abandonment is by chute cut-off, when the flow exploits the low-lying part of an old meander scroll system to bypass part of a meander loop. (See also CUT-OFF.)   DLD

**needle ice** A small-scale heave phenomenon produced by freezing and associated ice segregation at or just beneath the ground surface. Cooling at the ground surface results in ice crystals which grow upwards in the direction of heat loss. The needles, which can range in length from a few mm to several cm, may lift small pebbles or soil particles. The growth of needle ice is usually associated with diurnal freezing and thawing. It is widespread and particularly common in alpine locations in mid-latitudes where the frequency of freeze–thaw cycles is at its greatest. Wet, silty, frost-susceptible soils are the sites of most intense needle ice activity. Needle ice frequently occurs in orientated stripes, and both wind direction and sun have been suggested as explanations for the pattern; it is not clear whether orientated needle ice patterns are primarily a shadow effect developed by thawing or a freezing effect.

Thawing and collapse of needle ice is thought significant for frost sorting, frost creep, the differential downslope movement of fine and coarse material, and the origin of certain micro-patterned ground forms. The importance of needle ice as a disruptive agent has probably been underestimated, especially in exposing soil to wind and water in periglacial regions. In other areas it may be responsible for damage to plant materials when freezing causes vertical mechanical stress within the root zone. HMF

**Reading**
Lawler, D.M. 1988: Environmental limits of needle ice: a global survey. *Arctic and alpine research* 20, pp. 137–59. · Mackay, J.R. and Mathews, W.H. 1974: Needle ice striped ground. *Arctic and alpine research* 6, pp. 79–84. · Washburn, A.L. 1979: *Geocryology: a survey of periglacial processes and environments*. New York: Wiley. Esp. pp. 91–3.

**negative feedback** See SYSTEMS.

**nehrung** A sand or shingle spit which separates a haff from the open sea. A bar which isolates an estuary or lagoon from the sea.

**nekton** Free-swimming aquatic plants and animals, especially strong swimmers such as fish.

**neocatastrophism** A term introduced into geoscience in the mid-twentieth century by palaeontologists concerned with sudden and massive extinctions of life forms, such as that which afflicted the great mammals at the end of the Pleistocene. It has been extended into geomorphology by those dealing with rapid outputs from, and rapid inputs to, interfluve systems (Dury 1980). Much modern geomorphology is concerned with events of great magnitude and low frequency, and problems have been encountered with accommodating such an approach within the context of UNIFORMITARIANISM. ASG

**Reference**
Dury, G.H. 1980: Neocatastrophism? A further look. *Progress in physical geography* 4, pp. 391–413.

**neo-Darwinism** An evolutionary theory which combines DARWINISM with modern genetics. It regards the gene pool of a population as the fundamental unit in evolution, and takes into account larger mutations as well as the small heritable variations of Darwin. ASG

**Reading**
Berry, R.J. 1982: *Neo-Darwinism*. London: Edward Arnold.

**neoglacial** A small-scale glacial advance that occurred in the Holocene, after the time of maximum HYPSITHERMAL glacier shrinkage (Denton and Porter 1970). Fluctuations appear to have been frequent and to have shown sparse temporal correlation between different areas (Grove 1979), though the latest advance, the so-called Little Ice Age, was widespread between *c.*AD 1550 and 1850. ASG

**Reading and References**
Denton, G.H. and Porter, S.C. 1970: Neo-glaciation. *Scientific American* 222, pp. 101–10. · Grove, J.M. 1979: The glacial history of the Holocene. *Progress in physical geography* 3, pp. 1–54. · —1988: *The Little Ice Age*. London: Routledge.

**neotectonics** The study of the processes and effects of movements of the earth's crust that have occurred during the Late Cainozoic (Neogene). Some investigators use the term in a more restricted temporal sense to refer to post-Miocene or even just Quaternary movements, while others regard neotectonics as involving all tectonic activity which has been instrumental in forming present-day topography. Neotectonic activity which has been directly monitored by geodetic relevelling or other measurement techniques during the present century is commonly referred to as recent crustal movements.

Several lines of evidence have been employed to establish the nature of neotectonic activity, depending on the size of area and time period being considered. Regional subsidence and uplift over several millions of years are largely investigated by the usual methods of structural geology. As the temporal and spatial scale contracts, geomorphological and sedimentological data become more important. For instance, coastal movements can be monitored by raised or down-warped shorelines, while inland the mapping of fluvial features and erosion surfaces in conjunction with detailed study (and especially dating) of associated deposits can provide valuable information. Rapid uplift in mountainous areas may be indicated by geomorphological evidence of the onset of glaciation, while both horizontal and vertical movements along faults can in many cases be related, respectively, to offset drainage and knickpoints.

Maximum rates of uplift estimated from geomorphological and other evidence, or measured directly by relevelling, vary from several orders of magnitude over the earth's surface. Average rates of uplift over several millions of years rarely exceed $300\,\text{mm}\,\text{ka}^{-1}$. However, rates of postglacial isostatic uplift (see ISOSTASY) may exceed $20\,\text{m}\,\text{ka}^{-1}$, while contemporary crustal movements in currently highly active tectonic zones average up to $10\,\text{m}\,\text{ka}^{-1}$, or more. Such high short-term rates are clearly not sustained for more than a very limited period in geological terms. MAS

**Reading**
Fairbridge, R.W. ed. 1981: Neotectonics. *Zeitschrift für Geomorphologie supplementband* 40. · Vita-Finzi, C. 1986: *Recent earth movements – an introduction to neotectonics.* London: Academic Press. · Vyskocil, P., Green, R. and Maelzer, H. eds 1981: *Recent crustal movements, 1979.* Amsterdam: Elsevier. · — Wasset, A.M. and Green, R. eds 1983: *Recent crustal movements, 1982.* Amsterdam: Elsevier.

**nephanalysis** The term used to cover the analysis and interpretation of spatially organized cloud data. Coined in presatellite days, the term originally applied to 'the study of synoptic charts on which only clouds and weather are plotted' (Berry *et al.* 1945). The observations plotted were of cloud type, cloud amount, precipitation, weather, cloud ceilings and cloud-top heights. With the advent of METEOROLOGICAL SATELLITES it soon became clear that the contents of the satellite imagery were so rich and complex that many users would prefer to be provided with simpler cloud charts instead. Such charts are known as *satellite nephanalyses*. The earliest type was designed for use in the US Weather Bureau (see Godshall 1968). Similar manual schemes were implemented in the late 1960s and early 1970s in other major meteorological centres. Proposals for improved satellite nephanalysis, designed in part to standardize results from different analysts, were made for visible imagery by Harris and Barrett (1975), and for infrared imagery by Barrett and Harris (1977). Detailed objective (computer-based) schemes have been proposed by Shenk and Holub (1972), but the problem of automatic cloud identification is not straightforward. Therefore today's quasi-operational objective satellite nepahanalyses are the '3-D nephanalysis' of the US Weather Bureau (mapping broad categories of cloud top heights: see Decotiis and Coulan 1971), and a bispectral 'cloud type' procedure applied to Meteosat image data by the European Space Operations Centre. The USAF purports to combine satellite data with conventional data in a more comprehensive nephanalysis, but its procedures have been subject to variation and change. ECB

**Reading and References**
Barrett, E.C. and Harris, R. 1977: Satellite infra-red nephanalyses. *Meteorological magazine* 106, pp. 11–26. · Berry, F.A., Bollay, E. and Beers, N.R. 1945: *Handbook of meteorology.* New York: McGraw-Hill. · Decotiis, A.G. and Coulan, E.F. 1971: Cloud formation in three spatial dimensions using infra-red thermal imagery and vertical temperature profile data. In *Proceedings of the Seventh International Symposium of Remote Sensing of the Environment.* Ann Arbor, Michigan. · European Space Agency 1977: *Meteosat meteorological users' handbook.* Meteorological Information Extraction Centre, Darmstadt: European Space Agency. · Godshall, F.A. 1968: Intertropical convergence zone and mean cloud amount in the tropical Pacific Ocean. *Monthly weather review* 96, pp. 172–5. · Harris, R. and Barrett, E.C. 1975: An improved satellite nephanalysis. *Meteorological magazine* 104, pp. 9–16. · Shenk, W.E. and Holub, R.J. 1972: A multispectral cloud type identification method using Nimbus 3 HRIR measurements. In *Proceedings of the Conference on Atmospheric Radiation, Fort Collins, Colorado.* Boston: American Meteorological Society. Pp. 152–4.

**nephoscope** An instrument for measuring the height, direction of movement and velocity of clouds from a point on the ground.

**neptunism** The belief that a large proportion of the earth's rocks are precipitates laid down in some chaotic fluid, a theory that was devised and popularized by the German geologist. A.G. Werner in the late eighteenth century, and imported into Britain in the early nineteenth century by R. Jameson. It contrasts with plutonism. ASG

**Reading**
Davies, G.L. 1969: *The earth in decay.* London: Macdonald.

**neritic** Pertaining to the part of the seas and oceans above the continental shelf.

**ness** A promontory or headland, especially in Scotland, but also in eastern and southern England.

**net primary productivity** (NPP) The net augmentation of green plant material per unit area per unit time on land, and of blue-green and other algae, phytoplankton, and higher plants in water bodies; or, the amount of stored cell energy produced by PHOTOSYNTHESIS. It may be expressed by the equation:

NPP = gross production − respiration

(see BIOLOGICAL PRODUCTIVITY), and is measured in dry weight $g\,m^{-2}day^{-1}$ or year, dry weight $t\,ha^{-1}year^{-1}$, or in assimilated carbon equivalents, or energy equivalents. Although photosynthetic energy fixation is its main determinant, NPP may also be constrained by limiting factors which restrict growth, in particular cold and/or drought, and nutrient availability inadequacies.

On land, NPP may be estimated by means of the harvest method, in which all parts of living plants (including roots if possible) are cut and weighed at the end of a set period of time, and this is especially useful for crops, grasslands and forest plantings. For natural forests and woodlands, some form of forest dimension analysis is normally adopted. The NPP of plankton communities is more often ascertained by the technique of measuring uptake rates of $CO_2$ labelled with radiocarbon, $^{14}C$, which provide accurate values over one or two days.

For land plant communities adequate quantities of energy, water, $CO_2$ and soil nutrients are required for optimum rates of NPP. Most land plants need very large amounts of water to survive and prosper: e.g. a beech forest in southern England may take in 25,000 to 30,000 kg of water $ha^{-1}day^{-1}$ in summer. The majority of this goes in transpiration, so as to keep open the stomata through which $CO_2$ is received: but some is utilized in a range of metabolic reactions including photosynthesis, and a good deal is settled in cell structures themselves. In dry climates there is an almost linear relationship between water availability and net primary productivity, but this breaks down in more humid areas, for there is a point beyond which an increase in water availability has no effect at all on NPP. Values of NPP are also augmented substantially in areas of high temperatures, especially in the energy-rich tropics. $CO_2$ is normally present at a level of $c.0.03$ per cent of atmospheric gases: and slight variations in this are known to modify rates of NPP fairly quickly. Most plants demand a large range of nutrients, and shortages in any one may also influence the NPP: in Australia, for example, deficiencies in molybdenum may lead to the establishment of a heath vegetation, instead of the more productive mallee, or eucalyptus forest. In oceanic communities, the three main restrictions to NPP are, first, the inability of incident solar energy to penetrate very far into the water, so effectively limiting the major productive layer to depths of 30–120 m; secondly, the tendency for a large number of plankton to sink beneath this zone; and, thirdly, a marked deficiency in some nutrients, especially nitrogen and phosphorus.

As might be expected, bearing in mind the low BIOMASS of autotropic organisms in oceans compared to that on land, the mean NPP of the former is low ($147\,g\,m^{-2}\,year^{-1}$ cf. $780\,g\,m^{-2}\,year^{-1}$ on land: Whittaker 1975). In more detail, NPP in oceans can range from $c.2\,g\,m^{-2}\,year^{-1}$ in Arctic waters under the ice cap, to almost 5000 in some coral reefs, mangrove swamps, and tropical estuaries. Most open oceans, which are poor in nutrients, have values which are equivalent to semi-desert on land, namely $40–200\,g\,m^{-2}\,year^{-1}$. Greater net primary productivities are found in regions of upwelling, and on continental shelves, both of which are nutrient-rich ($400–600\,g\,m^{-2}\,year^{-1}$), and when these spatially coincide they can reach $1000\,g\,m^{-2}\,year^{-1}$. Similar rates occur in many temperate-latitude estuaries. A good deal of variation is also found in freshwater systems: OLIGOTROPHIC streams normally have very low NPPs, while those of EUTROPHIC cattail marshes in Minnesota can reach $2500\,g\,m^{-2}\,year^{-1}$, with $5600\,g\,m^{-2}\,year^{-1}$ attained in swamps artificially enriched with sewage in California (Woodwell 1970).

On land the lowest NPPs are located in deserts, semi-deserts and tundras, at $0–250\,g\,m^{-2}\,year^{-1}$. Systems whose growth is restricted by cold and/or drought (boreal forest, semi-desert shrublands, tropical savannas, steppe grasslands) give values of $250–1000\,g\,m^{-2}\,year^{-1}$; temperate-latitude forests, in which NPP is restrained by seasonal cold, lie within $1000–2000\,g\,m^{-2}\,year^{-1}$; and tropical rain forest and some of the most nutrient-rich marshland may attain $2000–3000\,g\,m^{-2}\,year^{-1}$ and occasionally as high as $5000\,g\,m^{-2}\,year^{-1}$ in the case of the tropical rain forest. This latter system, though it covers only $c.40$ per cent of the earth's surface, may account for 25 per cent of the new land biomass each year. Ranges of NPP in agricultural systems also vary widely, from $c.250–500\,g\,m^{-2}\,year^{-1}$ in the case of many tropical subsistence crops, and for most non-intensive temperate-latitude farmland, to $750–1500\,g\,m^{-2}\,year^{-1}$ under modern intensive, and frequently irrigated agriculture (see Leith and Aselmann 1983). Sugar cane is perhaps worthy of special mention, since it is one of the most productive of crops: its mean NPP is $c.1725\,g\,m^{-2}\,year^{-1}$, but under favourable circumstances this can be augmented to $6700\,g\,m^{-2}\,year^{-1}$, and occasionally even higher.                    DW

**Reading and References**
Leith, H. and Aselmann, I. 1983: Comparing the primary productivity of natural and managed vegetation: an example from Germany. In W. Holzner, M.J.A. Werger and I. Ikusima eds, *Man's impact on vegetation*. The Hague, Boston and London: W. Junk. Pp. 25–40. · Whittaker, R.H. 1975: *Communities and ecosystems*. 2nd edn. New York: Macmillan. · Woodwell, G.M. 1970: The energy cycle of the biosphere. *Scientific American* 223, pp. 64–74.

**net radiation** The resultant flux of the solar and terrestrial radiation through a horizontal surface. The downwards (positive) flux of radiation consists of shortwave solar radiation plus infrared atmospheric counter-radiation. The upward (negative) flux consists of reflected shortwave radiation and infrared radiation from the ground surface. The net radiation is considered positive if the flux downwards exceeds that upwards, and in this case will add energy to the surface. Net radiation is also known as radiation balance. Typically, it is positive during the day and negative at night. JGL

**network** The structure composed of links and nodes which are the junctions of at least three links. Networks can be identified for any linear form or process, and Haggett and Chorley (1969) recognized two fundamental types of networks, open and closed. Most emphasis has been placed upon drainage networks where network delimitation and density are important considerations (see DRAINAGE DENSITY; DRAINAGE NETWORK). KJG

**Reading and References**
Abrahams, A.D. and Flint, J.J. 1983: Geological controls on the topological properties of some trellis channel networks. *Bulletin of the Geological Society of America* 94, pp. 80–91. · Haggett, P. and Chorley, R.J. 1969: *Network analysis in geography*. London: Edward Arnold.

**neutron probe** An instrument for determining soil moisture content. It consists of a radioactive source of fast (or high energy) neutrons, a slow neutron detector and a counter unit. The method is based on the principle that fast neutrons emitted into the soil collide with the nuclei of atoms in the soil, notably the hydrogen nuclei of soil water, and as a result lose energy and slow down. A proportion of the resulting cloud of slow neutrons is scattered towards the probe where it is sensed by the slow neutron detector and translated into an estimate of soil moisture content using the mean count rate displayed on the counter unit and a soil moisture calibration curve. The neutron probe is introduced into the soil using a permanently sited access tube and so repeated measurements of soil moisture content can be made at the same site without destruction of the site. AMG

**Reading**
Schmugge, T.J., Jackson, T.J. and McKim, H.L. 1980: Survey of methods for soil moisture determination. *Water resources research* 16, pp. 961–79.

**neve** Another, less widely used, word for FIRN.

**niche** This refers to the precise way a species relates to its environment and to the other species with which it interacts. While the HABITAT of a species describes its physical environmental circumstances, the word niche deals in addition with the functioning of the species in the community as a whole. In this sense, habitat is the species' 'address', while the niche is its 'profession'. Clearly, the occurrence or otherwise of a particular species depends on the availability of suitable habitat, but is also determined by the ability of that species to occupy its own niche within that space. No two species occupy precisely the same niche, an idea referred to as Gause's competitive exclusion principle, although possibly even a relatively large proportion of their habitat and resource requirements may overlap. Different species manage to coexist by partitioning their resources, perhaps the most well known example of this being provided by the herbivorous megafauna of the African savannas. The grasslands of the Serengeti plains of Tanzania, for example, support numerous large herbivores that appear to occupy identical niches (i.e. they are all grazers and all live within the same habitat). However, on closer inspection, it can be observed that due to migration and other smaller spatial rearrangements of populations, different species rarely come into direct competition. Even when occupying the same locality, different species may be selectively grazing distinctive grass species, or different parts of the same species. G. Evelyn Hutchinson further defined the niche as a multidimensional volume that must take into consideration a species' resource and habitat requirements, as well as its physical environment tolerances. Generalist species have correspondingly wide niches, while those with more exacting requirements by definition occupy narrower niche spaces. MEM

**Reading**
Ricklefs, R.E. 1990: *Ecology*. 3rd edn. New York: Freeman. Pp. 728–47.

**nick point** A break in the long profile of a river.

**nimbostratus** See CLOUDS.

**nimbus** See CLOUDS.

**nitrate** An oxidized form of nitrogen, in combination with three atoms of oxygen. The chemical symbol for the nitrate ion is $NO_3^-$. Commonly produced in the soil by the bacterial conversion of ammonia ($NH_4^+$), one of the breakdown products produced by the decay of organic compounds. This process is known as nitrification. The nitrate ion is also a common anion in aqueous (water) solutions. It is delivered in precipitation, having resulted from processes involving free nitrogen gas ($N_2$) in the atmosphere, and is also derived from natural and artificial agricultural fertilizer. Nitrates are also generated as a by-product when fossil fuels are burned, and when in the atmosphere contribute to the acidification of rain. Consequently, $NO_x$ compounds (a mixture of nitric oxide, NO, and nitrogen dioxide, $NO_2$) are increasingly removed from flue gases by various DENITRIFICATION processes.          DLD

**nitrification and denitrification** Two essential components of the nitrogen cycle. The former involves the capture of nitrogen from the air, a process which is carried out by bacteria which live symbiotically in association with leguminous plants. They fix nitrogen in the form of highly soluble nitrates, which are taken up by plants, which convert them into useful organic compounds such as proteins. These in turn may be consumed by animals. The process of denitrification takes place when bacteria and fungi decompose plant or animal waste and convert it back again into nitrogen gases.          ASG

**nitrogen cycle** Nitrogen (N), together with phosphorus (P) are the rate-limiting nutrients in freshwaters, thus they control the productivity of freshwater ecosystems and are implicated in the accelerated EUTROPHICATION of many rivers and lakes, particularly in the developed

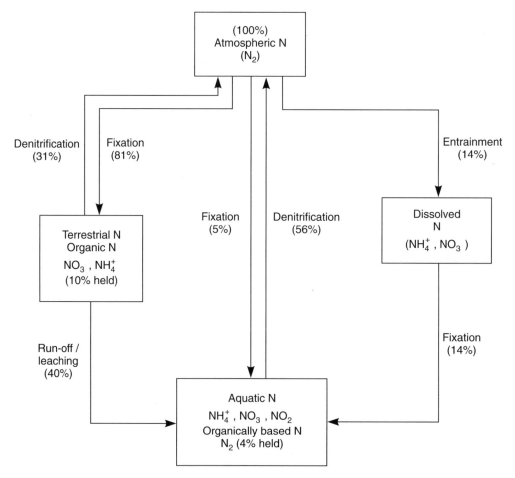

**The nitrogen cycle.**
*Source*: T.P. Burt, A.L. Heathwaite and S.T. Trudgill 1993: *Nitrate: process, pattern and management*. Chichester: Wiley.

world. Nitrogen in freshwaters takes several forms. Excluding molecular $N_2$, the dominant combined N fractions in freshwaters are, dissolved inorganic N ($NH_4^+$, $NO_2^-$, $NO_3^-$), dissolved organic N, and particulate N, which is usually organic but may contain inorganic N. Organic N usually exists as an integral part of protein molecules or in the partial breakdown of these molecules (e.g. peptides, urea, amino acids). Ammoniacal N is usually present in freshwaters as a result of the biological decomposition of organic N.

Terrestrial N is an important source for the aquatic nitrogen system. Up to 40 per cent of total N flux reaches the aquatic system through direct surface run-off or subsurface flow from the catchment. As a consequence, patterns of use, particularly land use, in the terrestrial system determine the magnitude and form of N inputs to the aquatic system.

There are five main reactions in nitrogen cycling. *Fixation* converts molecular N to ammonia through bacterial mediation. Some cyanobacteria can fix N directly and often form dense planktonic mats at the surface of lake waters. Nitrogen fixation is primarily important in eutrophic lakes with large populations of cyanobacteria. Photosynthetic bacteria may fix nitrogen in the anoxic (no oxygen) zone. *Assimilation* of nitrate to organic N is mediated primarily by plankton in freshwaters. The rate of assimilation varies with nitrate concentration and ammonia may be assimilated if available. *Mineralization* (ammonification) may occur both in the water column of lakes and rivers and in their sediments but is relatively more important in lake sediments. Mineralization may be rapid when plankton biomass dominates lake waters because this creates a low carbon: nitrogen ratio. Ammonification converts organic N in sediments to ammonium via microbial decomposition processes. The process is oxygen-demanding and regenerates N in a form that is available for re-assimilation by primary producers. The rate of release of N from decomposing organic matter can be an important factor in determining nutrient limitation in freshwaters.

*Nitrification* is a two-stage oxidation process mediated by the chemoautotrophic bacteria, nitrosomonas (ammonia to nitrite) and nitrobacter (nitrite to nitrate). The oxidation of ammonium to nitrite by nitrosomonas is usually rate-limiting, so nitrite is rarely present in appreciable concentrations in freshwaters. Nitrate, the end-product, is highly oxidized, soluble and biologically available. A high rate of nitrification is essential for efficient N cycling in freshwaters, particularly as nitrate is an important substrate for denitrification. *Denitrification* is a loss process for nitrate from rivers and lakes. Loss may also occur via dissimilatory nitrate reduction but this process is less important than denitrification in freshwaters. Denitrification is controlled by the oxygen supply and available energy provided by organic matter. To function, this N cycling pathway requires anaerobic conditions and a fixed bacterial carbon supply. ALH

**Reading**
Burt, T.P., Heathwaite, A.L. and Trudgill, S.T. 1993: *Nitrate: process, patterns and management.* Chichester: Wiley.

**nivation** The localized erosion of a hillside by frost action, mass wasting and the sheet flow or rill work of meltwater at the edges of, and beneath, lingering snow patches. The term was introduced by Matthes (1900). The main effect of nivation is to produce nivation hollows, which, as they grow in depth, trap more snow and thereby enhance the process of deepening. Given adequate time and suitable conditions a nivation hollow may evolve into a cirque. Topographic and climatic controls strongly influence the distribution and orientation of nivation hollows. The most favoured locations are on hillsides protected from the sun and with an ample supply of drifted snow. In mid-latitudes these factors favour a north-eastern orientation in the northern hemisphere and south-eastern orientation in the southern hemisphere. DES

**Reading and Reference**
Embleton, C. and King, C.A.M. 1975: *Periglacial geomorphology.* London: Edward Arnold. · Matthes, F.E. 1900: Glacial sculpture of the Bighorn Mountains, Wyoming. *US Geological Survey 21st annual report* 2, pp. 173–90. · Thorn, C.E. 1988: Nivation: a geomorphic chimera. In M.J. Clark ed., *Advances in periglacial geomorphology.* Chichester: Wiley. Pp. 3–31. · Washburn, A.L. 1979: *Geocryology.* London: Edward Arnold.

**nivometric coefficient** An index of snowfall efficacy, being the ratio of snowfall (water equivalent) to total annual precipitation. A coefficient of 1 implies precipitation entirely of snow.

**Reading**
Tricart, J. 1969: *Geomorphology of cold environments.* London: Macmillan.

**noctilucent clouds** Are found near a height of 80 km where there is a minimum in the temperature. They are so extremely tenuous that they can be seen only against the light scattered by air molecules on summer nights, between latitudes of 50° and 70°, at least one hour after the sun has set at the ground. While the cloud particles are likely to be ice, its origin, and the mechanism for the formation of the clouds is under considerable debate.                    JSAG

**Reading**
Ludlum, F.H. 1980: *Clouds and storms*. University Park, Pa.: Pennsylvania State Press.

**nonconformity**  An angular unconformity or other discontinuity between strata where the older rocks are of plutonic origin.

**non-linear system**  In nature, most systems are non-linear; they are viewed for simplicity as being linear in operation over a restricted range of action e.g. extension of a spiral spring (Hooke's law). More complicated systems can be modelled by approximating or simplifying to the linear case. True non-linear systems, however, do not have an easily derived solution by formula, although they can be computed. Non-linear dynamic systems may give rise to responses which are described as chaotic; where the system does not settle down to a fixed equilibrium condition or value. The earliest 'natural' chaotic systems studied were simple meteorological (Edward Lorenz) and the logistic equation in ecology (Robert May). For a steady state condition a 'control parameter' gives a single value (e.g. population in the logistic equation) but at a further value of the control parameter two solutions may be given and at still higher values four, eight etc. and so increasingly rapidly up to the chaotic regime. This is related to the Feigenbaum number.          WBW

**Reading**
Gleick, J. 1987: *Chaos*. Harmondsworth: Sphere/Cardinal/Penguin.

**non-renewable  resource** See  STOCK RESOURCE.

**normal cycle**  See CYCLE OF EROSION.

**normal fault**  A fault with the fault-plane inclined towards the side which has been down-thrown.

**normal stress**  The stress or load applied to the surface of an object (as either compression or tension). It is perpendicular to the SHEAR STRESSES which act parallel to the surface. A normal stress produces a strain or deformation in the material.                    WBW

**notch**  Landform that develops at the base of a cliff, platform or reef flat, especially in limestone and on tropical coasts. Deep narrow notches are characteristic of areas with a low tidal range.

Their positions are related to lithological and structural controls, tidal characteristics and sea-level history. In general, the higher the amplitude of the waves and the higher the tidal range, the greater is the difference in elevation between the notch roof and the floor. Notches in the humid tropics may be 1–5 m in depth. Although mechanical action of waves may contribute to their development, most investigators now believe that chemical or biochemical corrosion, or biological boring and grazing activities, are important.                    ASG

**Reading**
Woodroffe, C.D., Stoddart, D.R., Harmon, R.S. and Spencer, T. 1983: Coastal morphology and Late Quaternary history, Cayman Islands, West Indies. *Quaternary research* 19, pp. 64–84.

**nubbins**  Small lumps of earth produced by heaving owing to the growth of needle ice.

**nuclear waste**  (radwaste) Waste produced during the operation of nuclear facilities and as a result of decommissioning. Much comes from nuclear power stations, but other sources include hospitals and research institutions. Radwaste can be classified according to its volume, level of activity (high, medium, low) and its form (liquid, solid or gas).

*High-volume low-activity solid wastes* result from mining and uranium ore processing, from reactor operations, from final plant dismantling (decommissioning) and from soiled clothing, etc. Generally speaking, low-activity wastes are characterized by radionuclides with short half-lives. Commonly, this type of waste is buried in designated shallow trenches, but in the past sea disposal has been used to remove much of this waste.

The disposal of *low-activity liquid waste* from nuclear power plants and fuel reprocessing factories depends upon the siting of the works. Those with a coastal location, or on a large river or lake, remove sufficient radionuclides from liquid streams by distillation or floc precipitation to produce effluents of 'acceptable' purity prior to discharge into the adjacent water body.

*Medium-volume medium-activity wastes* are produced by both reactor operation and fuel reprocessing, e.g. ion exchange resins, sludges and precipitates, and may include some plutonium-contaminated material. *Solid low-volume high-activity waste* comprises mainly fuel element cladding and solidified material from reprocessing. *High-activity liquid waste* is produced entirely in fuel reprocessing operations.

Disposal of high-activity wastes currently presents problems because of their potential as biological hazards. Not only are they highly active, but they also contain some very long-lived activity. The optional sequence of events in the management of high-level waste is:

(a) Storage of fuel elements in ponds for months to several years.
(b) Storage of highly active liquor produced in reprocessing fuel for not more than two decades.
(c) Solidification into borosilicate glass, after which the glass blocks will be artificially cooled, for between ten and twenty years.
(c) The encapsulation of the blocks and their emplacement in a final deep geological repository.                                    ASG

**Reading**
Mounfield, P. 1991: *World nuclear power.* London: Routledge.

**nuclear winter**   A severe deterioration of climate that might take place as a result of multiple nuclear explosions. They might generate so much fire and wind that large quantities of smoke and dust would be emitted into the atmosphere, thereby causing darkness and great cold, the latter resulting from backscattering of incoming solar radiation caused by the reflectance of the fine soot. If an exchange of several thousand megatons took place, ambient land temperatures might be reduced to between – 15 and – 25 °C (Turco *et al.* 1983).         ASG

**Reference**
Turco, R.P., Toon, O.B., Ackerman, T.P., Pollack, J.B. and Sagan, C. 1983: Nuclear winter: global consequences of multiple nuclear explosions. *Science* 222, pp. 1283–92.

**nudation**   See SUCCESSION.

**nuée ardente**   (glowing cloud) A cloud of super-heated gas-charged ash produced by certain acidic volcanic eruptions (e.g. the eruption of Mt Pelée on Martinique in 1902 and 1903). The deposits produced by nuées ardentes are termed ash-flow tuffs, welded tuffs or ignimbrites.                                             ASG

**numerical modelling**   A method for obtaining particular solutions or deductions for a model which is expressed in mathematical or logical form, and for which general mathematical solutions are not appropriate and/or not available.

Numerical models are commonly although not necessarily implemented on digital computers, although many methods in use predate their development. A numerical solution is often the only one available for any but the simplest model, but has the disadvantage that it lacks the generality of an analytical solution. In numerical modelling all parameters must be given definite values, and all variables assigned initial values. A model run is then a single realization of the model constrained by these particular values. A large number of trials is therefore needed satisfactorily to explore all the possibilities inherent in any model.

Models vary considerably in style and complexity. In 'black box' models either the whole of a system or parts of it are considered solely in terms of empirical relationships between the input and output of the system. Most models in use in physical geography contain at least elements of this type. The commonest source of material for this type of model is a regression equation based on field observations. As understanding advances the black-box components within the total system become less significant as more components are based on established scientific principles. The level of empirical relationships in a useful forecasting model is partly constrained by our state of knowledge and partly by the level of detail which it is appropriate to represent. Such practical models are commonly developed drawing on the methods of systems analysis.

A stochastic element is often found in numerical models. For example, rainfall inputs to a hydrological model may be drawn at random from a known distribution in order to generate a probability distribution of high and low flows for hydraulic engineering design. Stochastic elements are usually included either to represent a model input of which a direct model is not needed; or to cover variability at scales below the level of resolution of the model. In the example above, a stochastic rainfall model may well be more appropriate and economic as an input than an independent model of the general atmospheric circulation! As another example, microtopography might be included within a model of long-term hillslope evolution as a random variation in process rate. The microtopography is below the general level of resolution of the model but could cause some of the observed variability in the forms of neighbouring hillslope profiles apparently subject to identical processes.

Numerical models require an underlying formulation in mathematical terms, and the range of models reflects the range of mathematical possibilities, which is immense. Perhaps the most common type of numerical model in physical geography is a solution to one or a family of differential equations, which are ultimately based on the CONTINUITY EQUATION with the substitution of suitable EQUATIONS OF MOTION or other expressions for the rates of the relevant

341

processes of material or energy transport. In some cases particular classes of solution are sought, e.g. EQUILIBRIUM or KINEMATIC WAVE solutions. In all cases the numerical model can only be run when boundary and initial conditions are specified. Initial conditions must specify the relationships which inputs and outputs must satisfy where they enter/leave the system of interest.

An example of differential equation type is the model of hillslope evolution based on continuity of downslope sediment transport. Ignoring wind-blown dust, the continuity equation is:

$$\partial Q/\mathrm{d}x + \partial z/\mathrm{d}t = 0$$

where $Q$ is the rate of downslope sediment transport at distance $x$ from the divide, and $z$ is the elevation at $x$ and time $t$.

For hillslope processes such as soil creep or soil EROSION which are largely transport-limited removal, the sediment transport may be expressed in the form $Q = f(x) \times s^n$ for some function of slope distance on slope gradient $s$. Where $f(x)$ is constant an analytical solution is available (Culling 1963) but otherwise numerical modelling is the best method. In this case any initial conditions may be used to describe the original slope profile form. The simplest boundary conditions are a fixed divide ($Q = 0$ at $x = 0$) and basal removal at a fixed base level ($z = 0$ at $x = x_1$), although others may be used. Runs of this model may be carried out on small micro-computers to follow long-term profile development.                                    MJK

### Reading and Reference
Carson, M.A. and Kirkby, M.J. 1972: *Hillslope form and process.* Cambridge and New York: Cambridge University Press. · Culling, W.E.H. 1963: Soil creep and the development of hillside slopes. *Journal of geology* 71, pp. 127–61. · Thomas, R.W. and Huggett, R.J. 1980: *Modelling in geography: a mathematical approach.* London: Harper & Row.

**nunatak** An Inuit-derived word describing a mountain completely surrounded by glacier ice, normally an ICE CAP or ICE SHEET. The nunatak hypothesis is the idea that plant and animal (but especially the former) communities have been isolated as refugia on nunataks. Sometimes these refugia may be as mountains, as with the normal usage of nunatak, but the nunatak hypothesis may also relate to much larger areas of isolated, ice-free land.                    WBW

### Reference
Gjaerveroll, O. 1963: Survival of plants on nunataks in Norway during the Pleistocene glaciation. In A. Löve and D. Löve eds, *North Atlantic biota and their history.* Oxford: Oxford University Press. Pp. 261–83.

**nutrient** A biologically essential chemical element and one that is required for the maintenance of life processes. In practice, the nutrients are classified into macronutrients, which are required in proportionally large quantities, and micronutrients, only trace concentrations of which are needed to sustain life but which are, nevertheless, essential. Macronutrients include: nitrogen, phosphorus, sulphur, potassium, calcium, magnesium and iron, whereas the micronutrients, of which there are approximately 30, include manganese, zinc, copper, iodine, molybdenum and sodium. Because of adaptational differences, not all organisms require the same amounts of nutrients, nor even the same nutrients, for example, silicon is an essential nutrient for diatoms but is of no consequence to the higher plants. Humans and other vertebrates require relatively large quantities of calcium and phosphorus for the construction and maintenance of bone and need iodine in the manufacture of the thyroid hormone that controls the metabolic rate. Nutrients have uneven distributions within the environment; for example, carbon and oxygen elements are obtained directly from the atmosphere via PHOTOSYNTHESIS of green plants and can be made available to other organisms thereafter. Other nutrient elements enter the FOOD CHAIN via the soil or sediments. Complex linkages between the atmosphere, lithosphere, hydrosphere and biosphere exist such that nutrients are constantly circulated in BIOGEOCHEMICAL CYCLES.      MEM

**nutrient status** A collective term usually applied to soils, peatlands and lakes whose nutrient status may be EUTROPHIC (rich in nutrients), *oligotrophic* (poor) or *mesotrophic* (transitional). Mires are also referred to as OMBROTROPHIC (rain-feeding) or *rheotrophic* (flow-feeding). In this context the use of the suffix 'trophic' is not to be confused with the trophic levels of animal communities feeding off each other and off the autotrophs or primary producers, the plants.

The nutrient status of soils is clearly discussed by Trudgill (1988) who considers in turn the inputs and outputs of the soil–vegetation system and illustrates the general principles with a number of diagrammatic models of soils of low and moderate nutrient status. Soil nutrient status is basically determined by the rock weathering input and the atmospheric input, on the one hand, and the leaching output on the other hand, moderated by biological recycling and storage of nutrients. In a hot, wet climate the potential for vegetative cycling and for weathering will be high but so will the potential for leaching – hence the concern over wholesale clearing of tropical rain forest. The opposite extreme, of a dry cold climate, will limit cycling and leaching, and therefore reduce chemical weathering to a large degree although physical fragmentation of rock may be important. Nutrient status of the soil is therefore low. As Trudgill (1988) goes on to note, there is a real need for integrated studies of the factors leading to low or high nutrient status, for although many studies of important cations, such as calcium, have been made, they have been at different scales, for different purposes and under different climatic regimes, making comparisons difficult. In a humid temperate context one is considering different orders of magnitude in $CaCO_3$ content (e.g. in river water, values of 1–20 ppm for non-limestone areas; 150–200 ppm for chalky areas) and similarly distinct differences in soil biochemistry and vegetative cycling. There are many complicating factors, especially with regard to the relative uptake, storage and release of calcium and other major mineral nutrients (magnesium, potassium, iron, aluminium, sodium, phosphorus and silica).

The nutrient status of lakes, and the way in which conditions have changed over time, have been extensively studied and are a topic of growing importance and scientific effort (Oldfield 1977; Birks and Birks 1980). At the beginning of the Holocene, 10,000 years ago, many lakes, especially those in glaciated lowlands, were eutrophic by virtue of mineral soil erosion inputs. Rising organic productivity and soil stabilization during the early to mid-Holocene changed this situation to one of mesotrophic/oligotrophic status, until man began affecting the soils within the catchment leading to greater nutrient inputs, culminating in some cases in gross 'cultural eutrophication' through sewage and other domestic wastes. The question of whether a lake, in its natural state, has eutrophic or oligotrophic status and what changes will occur in the absence of man, is a complex one. Some lakes in areas rich in nutrients appear to have remained eutrophic throughout more or less the whole Holocene, such as the meres of the Shropshire-Cheshire till plain described by Reynolds (1979), whereas some lakes in the English Lake District and north-west Scotland have been shown (by diatom stratigraphy of their sediments) to have been oligotrophic throughout almost the whole Holocene (Battarbee 1984). The usefulness of lake sediments, with their record of diatoms, chemicals, magnetism, pollen and other indicators, to indicate change within the water body and the catchment, has been amply demonstrated by a number of studies (Birks and Birks 1980; Battarbee 1984).

In the final stages of the terrestrialization of a lake basin *Sphagnum* bog mosses often invade to change the course of the hydrosere from the woodland climax postulated by Tansley and others earlier this century to what we now know as the almost inevitable endpoint of the hydrosere, at least in the British Isles: raised bog (Walker 1970). The nutrient status of these communities is discussed under PEAT. The actual levels of pH and calcium ions used to differentiate the various grades of peatland have been reviewed by Ratcliffe (1964).      KEB

## References

Battarbee, R.W. 1984: Diatom analysis and the acidification of lakes. *Philosophical transactions of the Royal Society* B, 305, pp. 451–77. · Birks, H.J.B. and Birks, H.H. 1980: *Quaternary palaeoecology*. London: Edward Arnold. · Oldfield, F. 1977: Lakes and their drainage basins as units of sediment-based ecological study. *Progress in physical geography*, 1, pp. 460–504. · Ratcliffe, D.A. 1964: Mires and bogs. In J.H. Burnett, ed., *The vegetation of Scotland*. Edinburgh: Oliver and Boyd. · Reynolds, C.S. 1979: The limnology of the eutrophic meres of the Shropshire-Cheshire plain. *Field studies* 5, pp. 93–173. · Trudgill, S.T. 1988: *Soil and vegetation systems*. 2nd edn. Oxford: Oxford University Press. · Walker, D. 1970: Direction and rate in some British post-glacial hydroseres. In D. Walker and R.G. West, eds, *Studies in the vegetational history of the British Isles*. Cambridge: Cambridge University Press.

# O

**oasis** An area within a desert region where there is sufficient water to sustain animal and plant life throughout the year.

**obsequent stream** A stream or river which is the tributary of a subsequent stream and flows in a direction opposite to the regional dip of the landsurface.

**obsidian hydration dating** (OHD) Used by archaeologists and geologists to date events ranging in age from a few hundred years to several million years. The principle is based on the dating of obsidian (volcanic glass), which when a fresh surface is formed and exposed to the atmosphere, the diffusion of ambient water will proceed to the formation of a hydration layer. The hydration layers are firmly adherent to the parent glass and resistant to chemical dissolution. The thickness of the hydration layer, which varies from 1 $\mu$m to more than 50 $\mu$m, depends on the time of exposure. OHD requires that a measurement of hydration thickness or the depth of penetration of water into obsidian be measured and a rate of hydration be known. The measurement can be made optically or by using particle accelerators. AP

**Reading**
Trembour, F. and Friedman, I. 1984: The present status of obsidian hydration dating. In W.C. Mahaney, ed., *Quaternary dating methods: developments in palaeontology and stratigraphy 7*. Amsterdam: Elsevier. Pp. 141–51.

**obstacle dune** See TOPOGRAPHIC DUNE.

**occlusion** A complex frontal zone associated with the later stages of the life cycle of an extratropical cyclone. The name is derived from the associated occluding or uplifting of the warm sector air from the earth's surface. Cold fronts tend to travel more quickly than warm fronts, so the area of warm air between narrows. Eventually the warm air is entirely aloft with the occluded front marking the surface juxtaposition of the two cool or cold air masses. The detailed structure of the occlusion will be determined by the temperature difference of these two air masses. Where the cold air behind the cold front is cooler (warmer) than that ahead of the original warm front it is known as a cold (warm) occlusion (see figure). In practice the

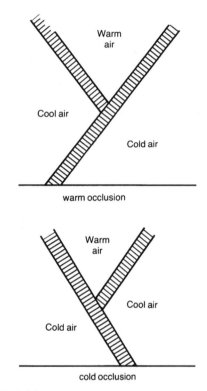

warm occlusion

cold occlusion

**Occluded** front types.

temperature difference may be quite small and the occlusion difficult to classify. PS

**occult deposition** The wet deposition of acidic pollutants, particularly $SO_2$ and $NO_x$, onto surfaces by the impaction of fog and cloud droplets. Patterns of deposition are influenced by climatic factors which encourage fog and mist and by local variations in wind direction and intensity. Concentration of pollutants in occult deposition can be considerably higher (up to $\times$ 20) than those in wet deposition by rainfall. The term is particularly used in connection with acid deposition in urban environments where it precipitates and concentrates pollutants on surfaces protected from rainwash. The process is important in, for example, coastal areas of low rainfall but high relative humidity and fog frequency and deposition can be increased by land-use changes, for example, afforestation, which increase surface area. BJS

**Reference**
Building Effects Review Group 1989: *The effects of acid deposition on buildings and building materials in the United Kingdom.* London: HMSO.

**ocean** The general name for large bodies of saltwater making up around 70 per cent of the earth's surface. Open oceans or oceanic zones are those parts deeper than 200 m whereas shallow coastal waters or neritic zones lie over continental shelves and are usually less than 200 m deep. There is only one ocean basin; the geographical subdivisions are made for convenience, because they are all interconnected. Shallow waters are more affected by changes in temperature, salinity, sedimentation and water movements. They are reached by sunlight and are richer in nutrients, and hence in plant and animal life than the deeper more constant oceans. In general, the nearer the land the higher the NET PRIMARY PRODUCTIVITY.         PAF

**Reading**
Hedgpeth, J.W. 1957: *Classification of marine environments.* Geological Society of America memoir 671. Pp. 17–28.

**Ocean Drilling Programme** or Program (ODP) An international co-operative science venture focused on fundamental research into the history of the ocean basins and the nature of the crust beneath the ocean floor. The programme is funded principally by the US National Science Foundation with substantial contributions from several international partners. Deep-ocean drilling began initially with the Deep Sea Drilling Project (DSDP), headquartered at Scripps Institution of Oceanography. In 1964, the Joint Oceanographic Institutions for Deep Earth Sampling (JOIDES) was formed in the United States and in 1968 drilling operations began under their leadership with DSDP and its drilling vessel, *Glomar Challenger*. In 1983, Texas A&M University was designated as the principal science operator and the name Ocean Drilling Program was adopted. The drilling vessel currently in use by ODP is *JOIDES Resolution,* and it is this vessel which is responsible for the major ocean sediment coring activities of the programme.         MEM

**ocean sediments** These have accumulated over millennia, particularly in deep ocean locations. On CONTINENTAL SHELVES ocean sediments may preserve materials derived from terrestrial erosion that have been transported to the oceans by fluvial and aeolian processes. In ocean deeps, erosion or removal of sediments may be virtually absent such that long records of accumulation are preserved. Sediments may predominantly comprise the detritus derived from the accumulated remains of ocean-dwelling organisms, though in high latitudes ice-rafted glacial sediments may be a significant component of the total sediment. From carbonate sediments, long records of OXYGEN ISOTOPE changes may be determined, reflecting global environmental changes and glacial–interglacial cycles. Long deep sea core records, provided by the OCEAN DRILLING PROGRAMME, have yielded records of environmental change that have revolutionized understanding of Quaternary and late Tertiary period environments and climates.         DSGT

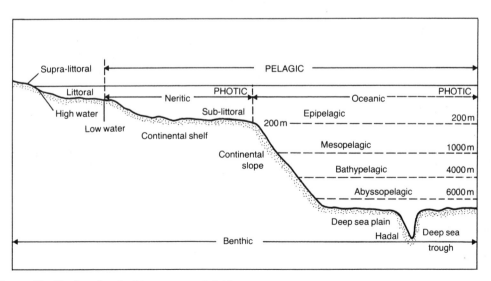

**Ocean.** Classification of marine biome-zones and divisions.
*Source:* P. Furley and W. Newey 1984: *The geography of the biosphere.* London: Butterworth. Figure 13.1.

**oceanic crust** A portion of the earth's crust underlying most ocean basins. There are two major crustal types, continental and oceanic. Oceanic crust comprises about 59% of the crust. The basaltic oceanic crust is sometimes called SIMA because of the abundance of silicon (50%) and magnesium (8%). It is denser ($3000\,\mathrm{kg\,m^{-3}}$) than the granitic continental crust ($2700\,\mathrm{kg\,m^{-3}}$), and thinner ($c.5\,\mathrm{km}$ in thickness vs. $c.30\,\mathrm{km}$). Both oceanic and continental crusts float on the even denser mantle ($4500\,\mathrm{kg\,m^{-3}}$). The crust and mantle are separated by the MOHOROVIČIĆ DISCONTINUITY.

Studies of oceanic crust provided several key pieces of evidence supporting the hypothesis of PLATE TECTONICS. Oceanic crust is not more than about 200 million years old, remarkably young compared to some continental crust that is several billion years old. This suggests that some mechanism of destruction and generation of oceanic crust must exist, and this is presumed to be sea floor spreading. When collisions occur between continental and oceanic plates, the former tend to ride up over the latter because continental crust is less dense than oceanic crust. Sea floor spreading forces oceanic crust beneath the continents in SUBDUCTION ZONES, where the rock is remelted. In the Pacific basin, the length of time required for crust to spread from the ridge to a subduction zone is about 200 million years, thus setting the maximum ages for oceanic crust.

As a result of sea floor spreading, distinct, paired bands of palaeomagnetic signatures are found symmetrical about mid-ocean ridges. When new crust is formed, the axes of magnetite crystals within the molten rock align with the earth's magnetic field. The rapid cooling of the sea floor basalts freezes those crystals into place. A record of periodic reversals of the earth's polarity is thus preserved in the basalt.      DJS

**Reading**
Allegre, C. 1988: *The behavior of the earth*. Cambridge, Mass.: Harvard University Press. · Condie, K.C. 1989: *Plate tectonics and crustal evolution*. Oxford: Pergamon.

**oceanography** The study or description of the oceans encompassing the sea floor, the physics and chemistry of the seas and all aspects of marine biology.

**Reading**
Pickard, G.L. and Emery, W.J. 1990: *Descriptive physical oceanography: an introduction*. 5th edn. Oxford: Pergamon.

**ogive** (also known as Forbes bands) Alternating bands of light and dark ice that extend across the surface of some glaciers below ice falls. They are arcuate in response to the normal pattern of ice flow across a glacier. The combined width of a dark and light band corresponds to the distance the glacier moves in a year. The dark ice corresponds to the ice that traverses the ice fall in summer and is thus exposed to melting, while the light ice reflects the incorporation of snow as the ice traverses the ice falls in winter.      DES

**Reading**
Paterson, W.S.B. 1981: *The physics of glaciers*. Oxford and New York: Pergamon.

**oil-shale** A shale which contains sufficient quantities of hydrocarbons to yield oil or petroleum gas when distilled.

**okta** Measurement of the amount of CLOUD cover is one of the two standard scales used by surface meteorological observers worldwide (the other scale being tenths). The observer reports the number of eighths or oktas of the celestial dome which is covered by clouds. Total cloud cover and layer cloud amount are reported in this fashion. Care must be taken to give equal weight to all areas of the sky especially in the case of *cumuliform* clouds when cloud sides as well as cloud bases may be viewed.      AH-S

**oligotrophic** See NUTRIENT STATUS.

**ombrotrophic** Description of plants or plant communities which are associated with a rain-fed substrate which is poor in nutrients.

**omnivore** See DIVERSIVORE.

**onion-weathering** Exfoliation. The destruction of a rock or outcrop through the peeling off of the surface layers.

**ontogeny** (ontogenesis) The sequence of development during the whole life history of an organism. The term is also applied to the life history of lakes and other systems.      ASG

**oolite** A sedimentary rock, usually calcareous but sometimes dolomitic or siliceous, which is composed of concentrically layered spheres – ooliths – which have formed by accretion on the surface of a grain. The Jurassic rocks of southern England contain the Greater and the Inferior Oolite, and originally obtained their name because of the supposed resemblance of their fabric to fish roe.

**ooze** Fine-grained organic-rich sediments on the floors of lakes and oceans.

**open channel flow** See FLOW REGIMES.

**open system** See SYSTEMS.

**opisometer** An instrument for measuring distances on a map.

**order, stream** Identification of the links in a stream NETWORK is by the process of stream ordering. At least eleven methods of stream ordering have been devised. The first method to be adopted was proposed by R.E. Horton and led to the development of HORTON'S LAWS. In Horton ordering, all unbranched streams were designated first order, two first combined to make a second order and so on, and then the highest order streams were projected to the headwaters. Subsequently Strahler (1952) suggested that the second stage should not be effected and Strahler ordering became widely used. Later developments have endeavoured to propose ordering methods which are mathematically consistent and which allow all components of the network to be differentiated according to their position. Thus in the Horton and Strahler systems of ordering the order of a stream of order $n$ is unaffected by the entry of a tributary of order less than $n$. Later ordering systems such as those proposed by Shreve (1967) and by Scheidegger (1965) have overcome this fundamental objection. The objective of all ordering systems is to be able to describe a link in a drainage network anywhere in the world in an unambiguous manner, and also to provide an ordering system that can readily provide an indication of the discharge from a network.
KJG

**Reading and References**
Gardiner, V. 1976: *Drainage basin morphometry*. British Geomorphological Research Group technical bulletin 14. Norwich: Geo Abstracts. · Gregory, K.J. and Walling, D.E. 1976: *Drainage basin form and process*. London: Edward Arnold. · Horton, R.E. 1945: Erosional development of streams and their drainage basins: hydrophysical approach to quantitative morphology. *Bulletin of the Geological Society of America* 56, pp. 275–370. · Scheidegger, A.E. 1965: The algebra of stream-order numbers. *US Geological Society professional paper* 525B, pp. 187–9. · Shreve, R.L. 1967: Infinite topologically random channel networks. *Journal of geology* 75, pp. 178–86. · Strahler,

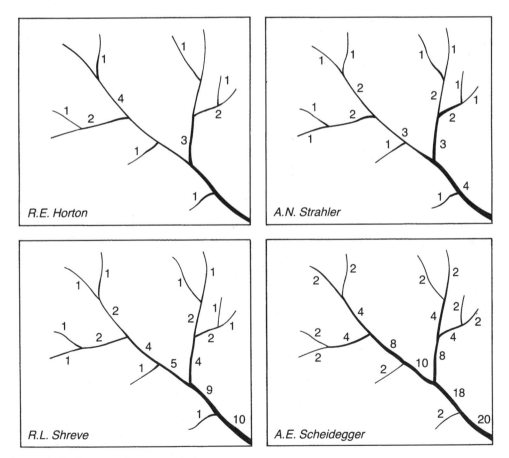

Four methods of stream and segment ordering.

A.N. 1952: Hypsometric (area-attitude) analysis of erosional topography. *Bulletin of the Geological Society of America* 63, pp. 923–38.

**organic weathering** The disintegration or destruction of a rock by living organisms or organic processes. It is a much neglected cause of weathering.

**orocline** A structural arc which has formed by horizontal displacement subsequent to the development of the main structural features of an area.

**orogens** Total masses of rock deformed during an orogeny (mountain-building episode).

**orogeny** The event, or mechanism, of the construction of characteristically linear or arcuate mountain chains formed on continents. Such mountain chains, of which the Andes and the Himalayas are contemporary examples, are known as orogenic belts or simply orogens. Investigation of the results of recent orogeny shows it to produce crustal thickening, deformation and associated volcanic activity, although in some cases, such as the Himalayas, the latter is relatively less important. Many earlier theories of orogeny emphasized the accumulation of a thick wedge of sediments forming a geosyncline before uplift, but these ideas have now been largely subsumed within the more comprehensive model of PLATE TECTONICS. Within this scheme orogeny is of two fundamental types. Where a continent margin, such as that of western South America, is under-ridden by oceanic LITHOSPHERE being reabsorbed or subducted into the earth's interior, largely thermal effects generate uplift and volcanic activity. In contrast, mainly mechanical effects are responsible for the Himalayan orogeny, which is associated with the collision of the northward moving Indian continent and the Eurasian continent.                MAS

**Reading**
Dennis, J.G. ed. 1982: *Orogeny*. Stroudsburg, Pa.: Hutchinson Ross. · Hsü, K.J. ed. 1982: *Mountain building processes*. London: Academic Press. · Miyashiro, A., Aki, K. and Cecal Sengor, A.M. 1982: *Orogeny*. Chichester: Wiley.

**orographic precipitation** Precipitation caused by the forced ascent of air over high

A view to the Kali Gandaki gorge in Nepal. This is one of the steepest places on earth and is created by a combination of mountain building (orogeny) and erosion.

ground. Uplift of air leads to cooling which, if the air is moist, may lead to condensation and eventually precipitation. The warm sector of an intense extra-tropical cyclone is the synoptic situation which demonstrates the orographic effect most clearly. Even where rain of convectional or cyclonic origins is falling, the orographic influence can still be seen in larger and sometimes longer precipitation events over the hills. The extra uplift will ensure that the precipitation processes in the clouds operate more effectively. PS

**osage-type underfits** See UNDERFIT STREAM.

**osmosis** Three forces act on water in soils: (1) gravity, (2) suction and (3) osmosis. Water moves by osmosis from regions of low solute concentration to regions of higher concentration separated by a semi-permeable membrane. Osmosis is rarely involved in large-scale water movement in soils owing to the lack of semipermeable membranes which means that difference in salt concentration are quickly balanced by diffusion of dissolved salts. The exception is dry soils where water moves as vapour and the air–water interface acts as a membrane allowing the passage of water but not solutes. At the small-scale, osmosis results in localized movement of water into interlayer spaces of clay particles resulting in swelling. Osmosis is important in plants where the uptake of water traverses plant membranes. ALH

**ostracods** Small crustaceans 0.3 to 30 mm in length. Their body and limbs are surrounded by two valves which are hinged dorsally to form the shell. When the animal dies the shell may be preserved as a fossil and it is possible to identify species based on the shell structure and morphology. Ostracods can be found in almost any global environment where water is present and a wide range of factors including water temperature, salinity and the nature of the substrate governs their distribution. The analysis of ostracod shells, particularly their geochemistry, is proving a valuable tool in reconstructing past climates (e.g. Holmes 1992). SLO

**Reference**
Holmes, J.A. 1992: Non-marine ostracods as Quaternary palaeoenvironmental indicators. *Progress in physical geography* 16, pp. 405–431.

**outlet glacier** A type of glacier which radiates out from an ICE DOME and often occupies significant depressions. Within an ice dome they can frequently be distinguished by a zone of high-velocity ice termed an ICE STREAM.

**outlier** An isolated hill lying beyond the scarp slope of a cuesta. A rock outcrop that is surrounded by rocks that are of an older age.

**outwash** Comprises stratified GLACIOFLUVIAL sands and gravels deposited at or beyond the ice margin. Outwash usually forms fan, valley bottom (valley train) or plain (SANDUR) deposits, often hundreds of metres thick, built up by aggrading braided or anastomosing meltwater channels which migrate laterally across the outwash surface. Periodic high energy flood events are marked by high rates of transport of sediment reworked from older outwash and till. Where outwash has accumulated on the glacier margin itself, differential ice melt may produce pitted outwash. Close to the glacier outwash is often steeply graded, comprising coarsegrained, imbricated, non-cohesive sediments deposited rapidly during unidirectional, high flow regime conditions, both within channels and on longitudinal bars. Bar deposits exhibit crude horizontal bedding truncated by erosional contacts and scour-and-fill structures representing successive discrete flood events. The bar surfaces are often characterized by SAND LENSES, SILT drapes, TRANSVERSE RIBS and coarse gravel lags. CLASTS are typically poorly sorted, subangular to subrounded, and comprise heterogeneous lithologies. Farther from the glacier outwash is more gently graded, with finergrained, more cohesive, sandy facies types forming transverse and linguoid bars characterized by dune and ripple forms and by bar avalanche face sediments. These distal sediments often exhibit planar and trough cross-bedding, and ripple and ripple drift CROSS-LAMINATION; LOAD STRUCTURES may also be present. Clasts tend to be better sorted and more rounded, while fewer, more resistant lithologies are represented in both the clast and heavy mineral populations. Coarsening or fining upward sequences may reflect local flood events or periods of channel scour and infill, or longer-term periods of ice advance or recession. JM

**Reading**
Boothroyd, J.C. and Ashly, G.M. 1975: Process, bar morphology and sedimentary structures on braided outwash fans, northeastern gulf of Alaska. In A.V. Jopling and B.C. McDonald eds. *Glaciofluvial and glaciolacustrine sedimentation*. Society of Economic Paleontologists and Mineralogists special publication 23. Pp. 193–222. · Church, M. 1972: Baffin Island sandurs: a study of Arctic fluvial processes. *Geological Survey of Canada Bulletin* 216. · —and Gilbert, R. 1975: Proglacial fluvial and lacustrine environments. In A.V. Jopling and B.C. McDonald eds, *Glaciofluvial and glaciolacustrine sedimentation*. Society of Economic Paleontologists and Mineralogists special publication 23. Pp. 22–100. · Rust, B.R. 1978: Depositional models for braided alluvium. In A.D. Miall

ed., *Fluvial sedimentology*. Canadian Society of Petroleum Geologists memoir 5. Pp. 605–25.

**outwash terrace** An outwash deposit that has been incised by meltwater to form a terrace. Incision of the outwash deposit is a response by meltwater streams to an increase in channel slope and/or an increase in the discharge/sediment load balance such that a period of proglacial aggradation is followed by a period of increased meltwater flow capacity and stream degradation. This change is likely to occur during glacial retreat when an overall decrease in meltwater and sediment supply occurs. A stepped sequence of outwash terraces may relate to successive periods of ice advance and retreat, or possibly to a more complex response to a single glacial event. Many outwash terraces can be traced up-valley to an associated terminal moraine; the terrace surfaces often exhibit traces of former braided channel networks and are pitted with kettle holes. JM

**overbank deposit** Flood sediment laid down by a river beyond its normal-flow channel, generally fine materials deposited from suspension in floodwaters, but it is possible for sheets of coarser bed material to be deposited overbank as well. Finer material may be deposited considerable distances from the channel. JL

**overbank flow** River discharge that has escaped from a channel when its banks were overtopped. Overbank flow inundates adjoining land, often the surface of a floodplain. Overbank flow is a normal occurrence that is expected every few years, because the geometry and capacity of channels is adjusted to conditions existing during the much more common lower flows. Overbank flows are sometimes erosive, but are frequently shallower and therefore slower-moving than the stream itself. They also exert a drag on the channel flow, as momentum is exchanged between the relatively fast and slower-moving flows. Deleterious consequences have sometimes followed from the control of overbank flows with artificial levée banks, since floodplain ecosystems that had evolved in the presence of periodic inundation no longer experienced this, or experienced only much less frequent flooding. DLD

**overdeepening, glacial** Often regarded as a prime characteristic of glacial erosion, overdeepening refers to the long profile of glacial troughs which tend to have a 'down-at-heel' profile with a steep gradient near or at their heads and a gentler slope, sometimes a reverse slope, towards their mouths (Linton 1963). They are 'overdeepened' only when compared to river-long profiles and indeed from a glacial viewpoint river valleys can be regarded as underdeepened! One explanation of the 'overdeepened' glacial long profile is that, unlike most river valleys, a glacier discharge is greatest at the midway EQUILIBRIUM line and decreases towards the snout. One might expect most erosion at the point of maximum discharge. Overdeepening at a different scale involves the excavation of rock basins by glaciers. This is feasible where rock conditions are favourable because glaciers can flow up a bed slope so long as the overall ice surface gradient is in the contrary direction. DES

**Reference**
Linton, D.L. 1963: The forms of glacial erosion. *Transactions of the Institute of British Geographers* 33, pp. 1–28.

**overflow channel** A channel incised into the landscape by a lake overflow. In the glacial landscape overflows were considered a form of meltwater channel, where water from an englacial lake cut across low cols, through rock or surficial sediment, to release water. By the 1950s they had been identified in large numbers throughout the formerly glaciated areas of Britain, though little consideration had been given to the generation of the hydraulic energy necessary to incise through rock in a short period of time, or the lack of shoreline features associated with them. Subsequent research has established that many overflows, for example those associated with the deglaciation of the Canadian prairies, have been cut by sudden and catastrophic subglacial flow. Overflow channels have also been identified in lake basins subject to past changes in level in tropical and arid regions, such as east Africa and the south-west USA, where overflow has taken place at the lowest point on the catchment divide. PSh

**overland flow** A visible flow of water over the ground surface, however produced.

Surface flow may be generated through a number of HILLSLOPE FLOW PROCESSES, including excess of rainfall intensity over infiltration capacity; excess of rainfall amount over soil storage capacity and the seepage of return flow. Where storms are brief or local, flow may run overland for some distance and then infiltrate. Flow velocities are typically of $3$–$150\,\mathrm{mm\,s^{-1}}$, and flow depths normally of a few millimetres. Flow rarely consists of a uniform sheet of water, and commonly concentrates in threads of higher velocity. Overland flow is very important for the removal and transport of debris in SOIL EROSION by water. MJK

**Reading**
Emmett, W.W. 1978: Overland flow. In M.J. Kirkby ed., *Hillslope hydrology.* Chichester: Wiley. Pp. 145–76.

**overthrust** A thrust fault with a faultplane dipping at a low angle and considerable horizontal displacement.

**overtopping** The process by which coastal barrier crests are built up as swash flows, of insufficient magnitude to reach across the crest and cause overwash, terminate on top of the crest. Sediment carried in overtopping flows is deposited at the swash limit and hence increments the barrier crest height. There should be a relationship between barrier crest height and the return period of storms generating the swash flows. Any increase in storm return period would be associated with a reduction in barrier crest height as increasing wave height would generate flows capable of forming overwash down the back of the barrier eroding material from the crest in the process. Therefore, overtopping is part of a continuum of cross-barrier flow types, merging into overwash as the magnitude of the flow increases (Orford and Carter 1982). The balance between overtopping and OVERWASHING controls the instability and migration of gravel barriers, whereas on sand barriers aeolian deposition is a more effective control on barrier crestal elevation than wave-generated overtopping. JO

**Reference**
Orford, J.D. and Carter, R.W.G. 1982: The structure and origins of recent sandy gravel overtopping and overwashing features at Carnsore Point, southeast Ireland. *Journal of sedimentary petrology* 52, pp. 265–78.

**overwashing** The process by which storm-generated swash flows transport beach face sediment over the top of the beach ridge and deposit it on the back slope. The definition of a coastal barrier is that it should exhibit some form of landward-dipping back slope; therefore overwash should be recognized as a major contributor to the creation and renewal of the back slope of a barrier. Overwash is regarded as a dominant process by which sand barrier islands are generated (with aeolian deposition as a co-dominant process; Leatherman 1976). Transport is often constrained sufficiently to cut through the crest via an overwash throat. Back slope deposition often occurs in the form of discrete washover fans, but washover can occur along a broad front parallel to the barrier crest. This has been described as the result of sluicing overwash (Orford and Carter 1982) and leaves no obvious sign of overwash throats. Note that 'overwash' is the process and 'washover' the sedimentary

result. Fan volume generally correlates with the height of the breaking wave (controlled for sediment size). Some washover fans may exhibit a longshore periodicity. Spatially periodic overwash has been related to the presence of transverse or edge waves (prestorm or during the storm) influencing the morphology of the beach (Orford and Carter 1984). In particular, high-level periodic cusps on the beach (formed by edge wave interaction) channel and spatially constrain overwash, hence imparting the same periodicity to washover fans. The reflective nature of gravel-dominated barriers tends to increase the likelihood of periodic fans, whereas the dissipative nature of low angle sand barriers militates against periodic overwash flows. Continual overwashing erodes a barrier's seaward side and extends the landward side. Over time, a barrier appears to be rolling onshore as sediment on the backslope is buried by subsequent overwash and eventually exhumed on the seaward barrier slope as the form of the barrier gradually retreats over its washover foundation. Overwash works towards a reduction of barrier height. The counter process which builds up the barrier crest is known as barrier OVERTOPPING. JO

**References**
Leatherman, S.P. 1976: Barrier island dynamics: overwash processes and eolian transport. *Proceedings of the 15th conference of Coastal Engineers, American Society of Civil Engineers* 3, pp. 1958–74. · Orford, J.D. and Carter, R.W.G. 1982: The structure and origins of recent sandy gravel overtopping and overwashing features at Carnsore Point, southeast Ireland. *Journal of sedimentary petrology* 52, pp. 265–78. · — 1984: Mechanisms to account for the longshore spacing of overwash throats on a coarse clastic barrier in southeast Ireland. *Marine geology* 56, pp. 207–26.

**oxbow** A lake, usually curved in plan, occupying a CUT-OFF channel reach that has been abandoned. The term may be applied also to an extremely curved active channel meander with only a narrow neck between adjacent reaches or even to the land within such a reach. The term derives from the U-shaped piece of wood fitted around the neck of a harnessed ox. Lakes of this type may become plugged with sediment where they adjoin the channel and then progressively fill in. JL

**oxidation** In general terms the loss of electrons from an atom but specifically the loss of oxygen from, or addition of hydrogen to, a substance. It is an important weathering mechanism, and the oxidation of iron compounds in rocks and soil can cause reddening to occur.

**oxygen isotope** Oxygen is the most abundant element on earth. It occurs in three stable isotopic forms (see also ISOTOPES): $^{16}O$(99.759%), $^{17}O$(0.0374%), and $^{18}O$ (0.2039%). It is common to evaluate the ratio of these isotopes and they are found to exhibit systematic variations which reflect physical and chemical processes. As $^{17}O$ is the least abundant it is generally not used or described and the abundance of the more easily measured $^{18}O$ is usually compared to $^{16}O$. Measurements are made in a mass spectrometer. This ratio is normally expressed by the $\delta^{18}O$ value (the Greek letter $\delta$ stands for 'difference') which is a measure of how different a measured ratio is from a standard (modern seawater (SMOW) or Pee Dee Belemnite (PDB)). These differences are extremely small so are expressed as parts per thousand (per mil) rather than parts per hundred (per cent). Owing to fractionation processes the ratio of the oxygen isotopes varies. As a result, observing the ratio of oxygen isotopes in a variety of deposits is the most widely employed proxy record of past climates. The principal deposits which have been used in oxygen isotope research are marine shells and ice preserved in high latitude and high altitude ice cores.

In the case of oxygen isotopic records preserved in ice, water ($H_2O$) molecules consisting of the most abundant $^{16}O$ isotope are slightly lighter than water formed from either $^{17}O$ or $^{18}O$. This lighter water will preferentially undergo evaporation from oceans into the atmosphere and is more likely to remain there during condensation in comparison to water molecules which include the heavier ($^{18}O$) oxygen isotope. The latter process is controlled by temperature so that the oxygen isotope ratio of past rainfall or snow events can be used to retrospectively estimate palaeotemperatures.

Similar isotopic fractionation effects occur when calcareous marine organisms incorporate oxygen from seawater to form their calcium carbonate ($CaCO_3$) exoskeletons. The most important group of organisms used for this are the bottom (benthonic) and surface (planktonic) dwelling micro-organisms called *foraminifera*. Measuring the isotopic ratio of the oxygen preserved in fossils was initially thought to provide a direct record of past ocean temperatures. It was then realized that the fractionation of oxygen isotopes from the seawater into the skeletal structure was also dependent on the isotopic composition of the water. The composition of the water varied dramatically during the changes in global climate between glacial and interglacial periods. In glacial periods the continents store vast amounts of water in the form of ice which is enriched in $^{16}O$. As a result, global sea level falls by around 100 m and the water becomes enriched in $^{18}O$. The systematic

shift in the oxygen isotopic ratios of shells collected from deep ocean sediments has been used as a means of constructing a named marine $\delta^{18}O$ sequence of stages which is now the most widely used basis to subdivide portions of time during the past 2 million years.                                    SS

**ozone**   A form of oxygen, but whereas the molecules of ordinary oxygen each contain two atoms, the ozone molecule has three. The ozone layer is a relatively high ozone concentration zone which occurs at a height of 16–18 km in polar latitudes and at about 25 km over the equator. This layer helps to control the temperature gradient of the atmosphere and also through absorption controls the amount of ultraviolet (UV-B) radiation reaching the ground.

Ozone is constantly created and destroyed through natural chemical reactions. However, human actions are increasing the concentrations of certain substances that may accelerate the rate of ozone destruction in the stratosphere: oxides of nitrogen, hydrogen, bromide and chlorine. The oxides of nitrogen might be generated by high-flying supersonic aircraft emissions, while the offending chlorine comes from such sources as the chlorofluorocarbons (CFCs) and carbon tetrachloride ($CCl_4$). In recent years it has become apparent that the concentrations of ozone in the stratosphere have declined, most notably over the Antarctic, where an 'OZONE HOLE' has been detected by a combination of ground monitoring and satellite observations. The reasons for the development of this zone of depletion in the south polar area include the presence of a well-established polar vortex and the extreme cold which generate ice particles that play a role in the crucial chemical reactions. However, stratospheric ozone depletion is increasingly being recognized in other geographical regions. The most obvious cause for concern is that this depletion will reduce the effectiveness of the ozone layer as a filter for incoming UV-B radiation.

At lower levels in the atmosphere ozone levels may, paradoxically, increase as a result of anthropogenic pollutants. Photochemical reactions can produce ozone. High ozone concentrations can adversely affect human health and damage vegetation.                          ASG

**Reading**
Elsom, D. 1992: *Atmospheric pollution: a global problem.* 2nd edn. Oxford: Blackwell. · Wayne, R.P. 1991: *Chemistry of atmospheres.* Oxford: Clarendon Press.

**ozone hole**   A region in the stratosphere 19–48 km above the earth's surface has a high concentration of ozone ($O_3$). This layer of ozone absorbs most of the ultraviolet radiation from the sun preventing the damaging radiation from reaching the earth's surface. In 1985 researchers reported finding a hole in the ozone layer occurring during several spring seasons since the late 1970s over Antarctica. The cause of the ozone hole was attributed to CHLORO-FLUOROCARBONS (CFCs). Normally chlorine (Cl) atoms are largely locked into 'safe' compounds that cannot harm ozone. The stratosphere above Antarctica has one of the highest ozone concentrations. During the polar night (winter) temperatures drop below −80 °C (−112 °F). This frigid air allows the formation of polar stratospheric clouds. These clouds activate inert chlorine into chlorine that can be broken down by sunlight. This chlorine builds during winter and early spring. During late September and October (spring), a belt of stratospheric winds called the polar vortex encircles Antarctica. Once the sun reaches the chlorine it activates it, destroying the ozone molecules. As the vortex weakens during the summer, ozone depleted air spills out into the mid-latitudes. The same process occurs in the Arctic, but to a lesser extent. The Arctic polar vortex is not as strong and the temperatures are warmer than the Antarctic. Depletion of ozone over the Arctic exceeds 35 per cent. As a consequence of ozone depletion, the 1987 Montreal Protocol established a timetable for diminishing CFC emissions. The treaty was revised in 1989, 1990, 1992, and 1995 and called for a complete phase-out of CFCs and other ozone destroying chlorine compounds by the year 2000, though full compliance has not been achieved.     JAS

# P

**pacific type coast** A longitudinal coast where folded belts trend parallel to the coastline.

**pack ice** See SEA ICE.

**padang** A vegetation type, developed on poor soils in south-east Asia, consisting of grass and shrubs.

**palaeobotany** The study of ancient or fossil plants and plant communities.

**palaeochannel** A river or stream channel (or more generally, river valleys) which no longer conveys discharge and are no longer part of the contemporary fluvial system. Palaeochannels may be preserved as abandoned surface channels on river floodplains or may have been infilled by fluvial or other sediment and are exposed as isolated sediment sections. It is rare that complete channel systems are preserved unless there has either been rapid aggradation in the system or a relatively abrupt cessation of flow. The extent of many larger buried palaeochannel networks, for example in Australia and northern Africa, have only been identified through the use of remote-sensing techniques. Many palaeochannels are abandoned as a result of changes in the fluvial regime, the most common of which is reduction in flow due to climate change. Other causes of abandonment include STREAM CAPTURE which may divert flow away from a section of the system and RIVER METAMORPHOSIS which may lead to the development of a new river system. Data derived from well-preserved and unmodified palaeochannels, such as the channel dimensions, cross-sectional shape, planform, sedimentary infill and terrace materials can be used to estimate palaeodischarge values and can be compared with contemporary channels (see PALAEOHYDROLOGY).

DJN

**Reading**
Maizels, J. 1990: Raised channel systems as indicators of palaeohydrologic change: a case study from Oman. *Palaeogeography, palaeoclimatology, palaeoecology* 76, pp. 241–77. · Van De Graaff W.E.J. 1977: Relict early Cainozoic drainage in arid Western Australia. *Zeitschrift für Geomorphologie* 21, pp. 379–400.

**palaeoclimatology** The study of climate prior to the period of instrumental measure-

ments, few of which pre-date the nineteenth century. Indeed, instrumental records only span a tiny fraction (less than $10^{-7}$) of the earth's climatic history and so provide a record that is both an inadequate perspective on climatic variation and a very limited view of the evolution of climates. The foundation of palaeoclimatology is the use of climate-dependent proxy data. The principal sources of proxy data are ice cores, ocean cores, various types of terrestrial sediment (e.g. glacial deposits, periglacial features, loess, relict sand dunes, speleothems, tufas etc.), biological evidence (e.g. tree rings, pollen, plant microfossils, diatoms, insect fossils) and historical records (e.g. writings, paintings, tax returns, phenological records).

ASG

**Reading**
Bradley, R.S. 1985: *Quaternary paleoclimatology: methods of palaeoclimatic reconstruction.* Boston: Allen & Unwin.

**palaeoecology** The term widely used for the research field that involves the study of fossils in order to infer past ecological processes, past biological environments and hence past biogeographical patterns. Increasingly these inferences are becoming quantified as we learn more about present plant-environment and animal-environment relationships and fossils become more accurate indicators of past ecological conditions. Fossil plant remains commonly studied are pollen and spores, diatoms, phytoliths and the so-called 'MACROFOSSIL' remains of plants, seeds, fruits, leaves and wood. Fossil animal remains commonly studied are cladocera, molluscs, coleoptera (beetles), and skeletal fragments. Using the language of biology and geology, palaeoecologists describe the changing fossil composition of the sedimentary sequence. From these descriptions they infer changing biotic compositions and hence changing environments, and deduce changes in environmental variables such as climate, soil and human influence.

RHS

**Reading**
Birks, H.J.B. and Birks, H.H. 1980: *Quaternary palaeoecology.* London: Edward Arnold. · — and West, R.G. eds 1973: *Quaternary plant ecology.* Oxford: Blackwell Scientific.

**palaeoenvironment** A past environment, with the term usually applied to environments beyond the time of historical records or at the scale of $10^4$ years or longer. Palaeoenvironments and the occurrence of environmental changes at a range of spatial scales are investigated through the analysis of a diverse range of PALAEOENVIRONMENTAL INDICATORS, with the timing of palaeoenvironmental changes established through the use of GEOCHRONOLOGY methods. The QUATERNARY period is especially investigated within physical geography and allied disciplines.                                                    DSGT

**palaeoenvironmental indicators** Lines of PROXY evidence for past environmental conditions. Palaeoenvironmental indicators may be derived through the analysis of sedimentary deposits or RELICT landforms, and include for example fossil pollen (see POLLEN ANALYSIS), preserved bone or shell remains and deep ocean cores whose isotopic content may be analysed, and RELICT landforms. The diversity of palaeoenvironmental indicators used in the study of QUATERNARY environmental conditions is immense, with their utility enhanced through the application of one of many GEOCHRONOLOGY techniques that are available. Since the availability of palaeoenvironmental indicators is dependent upon their preservation and lack of subsequent disturbance or alteration, the availability and utility may decrease with age, as may the existence of a suitable dating method. The interpretation of palaeoenvironmental indicators may be somewhat imprecise or subject to revision in the light of interpretative and chronometric developments.                    DSGT

**Palaeogene** A subdivision of the TERTIARY period of the CAINOZOIC era, encompassing the Palaeocene, Eocene and Oligocene epochs. It is followed by the Neogene, containing the Miocene and Pliocene epochs.           PSh

**palaeogeography** The geography of a former time, especially a specific geological epoch.

**palaeohydrology** The science of the waters of the earth, their composition, distribution and movement on ancient landscapes from the occurrence of the first rainfall to the beginning of continuous hydrological records (Gregory 1983). The advance of palaeohydrology has been possible with greater understanding of contemporary HYDROLOGY so that morphological evidence, including that from PALAEOCHANNELS, sedimentological evidence including characteristics of PALAEOMAGNETISM and information from organic deposits especially by palynology and diatom analysis, and knowledge of the mechanisms of the hydrological cycle, can be employed to make quantitative estimates of hydrological conditions in the past. Most emphasis has been placed upon Quaternary palaeohydrology but it has also been possible to indicate the palaeohydrological features of geological periods (Schumm 1977). The advance of palaeohydrological investigation depends upon refinement of the relationships between parameters such as precipitation, run-off and temperature and also upon a closer liaison between palaeohydrology and PALAEOCLIMATOLOGY and PALAEOECOLOGY.                     KJG

**Reading and References**
Gregory, K.J. ed. 1983: *Background to palaeohydrology.* Chichester: Wiley. · Schumm, S.A. 1977: *The fluvial system.* Chichester: Wiley. · Starkel, L. and Thornes, J.B. 1981: *Palaeohydrology of river basins.* British Geomorphological Research Group technical bulletin 28. · Starkel, L., Gregory, K.J. and Thornes, J.B. 1991: *Temperate palaeohydrology: fluvial processes in the temperate zone during the last 15,000 years.* Chichester: Wiley.

**palaeomagnetism** The intensity, direction and polarity of the earth's magnetic field throughout geological time. Palaeomagnetic studies are possible because certain iron-rich rocks containing magnetic minerals become more or less permanently magnetized at the time they are formed. This can occur when a rock cools (thermal remanent magnetism), when an iron mineral is chemically altered to another form (chemical remanent magnetism) or when magnetic particles are deposited in calm water (detrital remanent magnetism). Systematic deviations in the magnetic orientation of rocks of different ages have enabled the previous latitudes of continents to be determined, while periodic polarity reversals of the magnetic field now provide both a record of the creation of new oceanic LITHOSPHERE and a basis for dating suitable deposits.                          MAS

**Reading**
Cox, A. ed. 1973: *Plate tectonics and geomagnetic reversals.* San Francisco: W.H. Freeman. · Glen, W. 1982: *The road to Jaramillo.* Stanford, Cal.: Stanford University Press. · Kennett, J.P. ed. 1980: *Magnetic stratigraphy of sediments.* Stroudsburg, Pa.: Dowden, Hutchinson and Ross.

**palaeosol** A soil of an environment of the past, formed either by burial under later geological materials or because of a change in the climate or topographical conditions of soil formation. They are identifiable by any evidence that indicates the presence of a former landsurface that has undergone some form of alteration in response to *in situ* surface processes. Palaeosols represent periods of geomorphic stability.                      ASG

**Reading**
Fenwick, I. 1985: Palaeosols – problems of recognition and interpretation. In J. Boardman ed., *Soils and Quaternary landscape evolution*. Chichester: Wiley. Pp. 3–21.

**palaeovelocity** The velocity of a past flood event that was not directly observed, but for which an estimate is made using various traces of the flow that remain in the channel or the landscape. One method involves determining the slope and roughness of the former channel bed, and from these estimating the palaeovelocity from one of many standard flow equations. Various relations can be used to estimate the roughness from values of $D_{50}$ or $D_{84}$, grain size percentiles for which the nominated fraction (50 per cent or 84 per cent) of the bed sediment is finer. These measures are used to indicate the size of roughness elements that would have been present on the bed, but neglect other sources of roughness such as the channel banks, vegetation, etc. Deriving estimates of $D_{84}$ or other grainsize percentiles from field measurement can be very difficult, and many workers have used approximations to $D_{84}$ by measuring only the largest 10–20 stones carried by the former flows. The final piece of information required to estimate the palaeovelocity is the former depth of the flow, and this again must be estimated from limited evidence available in the field.

DLD

**Reading**
Church, M., Wolcott, J. and Maizels, J. 1990: Palaeovelocity: a parsimonious proposal. *Earth surface processes and landforms* 15, pp. 475–80.

**pali ridge** A sharply pointed ridge between two stream valleys on deeply dissected volcanic domes, especially in Hawaii.

**Palmer Drought Severity Index** (PDSI) A single numerical value that takes into account PRECIPITATION, potential EVAPOTRANSPIRATION, RUN-OFF, and SOIL MOISTURE DEFICIT to depict extended periods (months, years) of abnormal dryness or wetness. The Palmer Index value is 'standardized' so that comparisons can be made between geographic locations and over months. The index uses an arbitrarily selected scale ranging from below −4.0 (extreme drought) to above 4.0 (extreme moist spell). The index has been used to delineate drought and wetness disaster areas, for monitoring the status of aquifer, stream and reservoir water supplies, and to indicate the potential intensity of forest fires. The index generally does not perform as well over short time periods (days, few weeks) which are used for monitoring the impact on crops and agricultural operations.

LN

**palsa** A peat mound associated with the development of an ice lens. Genetically the feature is similar to a PINGO, but is restricted to peat bogs.

DES

**Reading**
Seppälä, M. 1972: The term 'palsa'. *Zeitschrift für Geomorphologie* NF 16, p. 463.

**paludal sediments** Deposits of marshes and swamps formed in areas of low and irregular topography, as along the banks of lakes, river floodplains, deltas, etc.

**paludification** The expansion of a bog caused by the gradual rising of the water table as accumulation of peat impedes water drainage.

ASG

**palynology** See POLLEN ANALYSIS.

**pan** A closed (i.e. usually with no inlet or outlet) dryland depression, that may hold an ephemeral shallow water body or which may have been occupied by a lake under past positive water balance conditions. The term is widely used in southern Africa and is equivalent to PLAYA and deflation basin in other localities, though pans are usually no more that $10^1$–$10^4$ m in diameter. Although aeolian deflation, from a susceptible surface, has often been regarded as the principal mechanism of formation, with the presence of a fringing LUNETTE or CLAY DUNE taken as evidence for this, other factors, notably groundwater fluctuations and the concentration of salts, are now regarded as vital prerequisites for their development. The term pan is also applied to very small closed depressions that result from the drinking and wallowing activities of large mammals.

DSGT

**Reading**
Goudie, A.S. and Wells, G.L. 1995: The nature, distribution and formation of pans in arid zones. *Earth science reviews* 38, pp. 1–69. · Shaw, P.A. and Thomas, D.S.G. 1997: Pans, playas and salt lakes. In D.S.G. Thomas, ed., *Arid zone geomorphology: process, form and change in drylands*. 2nd edn, Chichester: Wiley. Pp. 293–317.

**panbiogeography** See VICARIANCE BIOGEOGRAPHY.

**panfan** The surface produced when a hill or mountain is completely eroded so that the peripheral fans coalesce, as in the end stage of landscape evolution in an arid region.

**Pangaea** The name given to a postulated continental landmass which split up to produce most of the present northern hemisphere continents. Also the name applied to the former landmass that comprised all the present continental landmasses.

**panplain** A flat or almost flat landscape that has been produced by lateral erosion by rivers and lowering of divides and interfluves.

**pantanal** A type of savanna area along the sides of some Brazilian rivers. The land is seasonally flooded by river water but is very dry for most of the year.

**parabolic dunes** (also called hairpin dunes) Crescentic sand accumulations in which the horns point away from the direction of dune movement (the opposite of BARCHANS). They may occur as rake-like clusters and can develop from blowouts in transverse ridges. Parabolic dunes occur in areas where vegetation provides some impedence to sand flow, as in some coastal dune fields and on desert margins.        ASG

Reading
McKee, E.D. 1979: A study of global sand seas. *US Geological Survey professional paper* 1052.

**parallel retreat** The phenomenon of denudation of a landscape by lateral erosion of scarp slopes and hills which maintain their slope angle as erosion progresses. It is one of three classic models of slope evolution, the other two being slope downwearing (as in the Davisian cycle) and slope replacement (as in the model by W. Penck). L.C. King explained the great escarpments and inselbergs of Africa through this process.

**parameterization** A process undertaken during development of numerical simulation models in which unknown values describing the strength (and structure) of relationships are assigned to particular model components. This is undertaken in three main ways: (1) direct derivation of parameter values from specially designed laboratory or field experiments; (2) indirect derivation through calibration, where parameter values within a numerical model of a system are changed until optimal agreement with previously measured observations of the real system are obtained (optimization); and (3) sensitivity analysis, where effects of parameter changes upon model response are quantified to define probable parameter values. An example of this is the HORTON OVERLAND FLOW MODEL which requires parameters which depend upon soil and soil moisture characteristics, and which may be determined directly by field experiments (i.e. 1); indirectly by changing these parameters with respect to measured over-land flow (i.e. 2); or quantifying the effects of different parameter combinations upon model predictions (i.e. 3). Model parameterization raises difficult issues (Beven 1989): (a) parameters may have poor physical meaning or are not readily measured such that direct derivation from experimentation is not possible; (b) if a large number of parameter values need to be specified, there may be many different possible parameter combinations for testing against real world data; (c) different combinations of parameter values may produce similar results, such that the right results may be obtained for the wrong reasons; and (d) both (b) and (c) assume that high quality test data are available. If they are not available, or are of poor quality, model parameterization may result in wrong results for the wrong reasons! Thus, modellers now give explicit attention to the effects of parameter uncertainty upon model predictions by undertaking many model simulations with different combinations of parameter values to assign uncertainty bounds to those predictions.        KSR

Reading and Reference
Beven, K.J., 1989. Changing ideas in hydrology: the case of physically-based models. *Journal of hydrology* 105, pp. 157–72. · Howes, S. and Anderson, M.G., 1988: Computer simulation in Geomorphology. In M.G. Anderson, ed., *Modelling geomorphological systems*. Chichester: Wiley. pp. 421–40.

**parasite** An organism which exists in a somewhat imbalanced SYMBIOSIS with another, in which, although it is able to extract energy (food) from its host for its own preservation it is unlikely to develop this ability to the point where the host is killed. Some human parasites are, however, exceptions to this general rule: and there are others (e.g. the larvae of some Diptera and Hymenoptera). Since parasites do not require their own energy-production systems, they are often structurally simple: for example, mistletoe, which lives through tapping the phloem tissue of certain deciduous trees, often has no chlorophyll. Parasites are usually much smaller than their hosts; and they may co-exist with them externally or internally.        DW

**parna** A word coined by Butler (1956) for aeolian clay deposits found in Australia. Parna deposits occur either as discrete dunes or as thin, discontinuous, widespread sheets. They contain some 'companion sand', in addition to the clay pellets which make up the greater portion of the material. They may be derived from the deflation of material from unvegetated, saline lake floors, or from other soil or alluvial surfaces. Parna is in effect a loessic clay. (See also CLAY DUNE; LUNETTE DUNE.)

**Reading and Reference**
Butler, B.E. 1956: Parna – an aeolian clay. *Australian journal of science* 18, pp. 145–51. · — 1974: A contribution to the better specification of parna and other aeolian clays in Australia. *Zeitschrift für Geomorphologie* 20, pp. 106–16.

**partial area model**  Forecasts the saturated area around streams and channel heads as a basis for flood hydrograph prediction. This type of model is in direct contrast to the HORTON OVERLAND FLOW MODEL. THROUGHFLOW is generally routed downslope to estimate the areas where SATURATION DEFICIT is zero. This DYNAMIC SOURCE AREA is then assumed to generate OVERLAND FLOW which provides the most rapidly responding part of the stream hydrograph.  MJK

**Reading**
Kirkby, M.J. ed. 1978: *Hillslope hydrology*. Chichester: Wiley. Esp. chs 6–9.

**partial duration series**  A series of events analysed in flood frequency analysis and consisting of all flood peak discharges above a specified threshold discharge. The series is particularly useful when it is needed to know the frequency with which a particular flood discharge is exceeded. (See also FLOOD FREQUENCY.)  KJG

**particle form**  Has been used as a composite term to cover several properties of the morphology of sedimentary particles. Thus, in a functional expression:

$$F = f(Sh, A, R, T, F, Sp)$$

where $Sh$ denotes the shape of the particle, $A$ its angularity and $R$ its roundness, $T$ is the surface texture and $Sp$ is the sphericity. Thus a complete description of the form would entail observations of these component parts. However, the ease with which this may be done depends upon factors such as the size of the particles to be examined. Particle size itself does not come into the expression but it is relatively easy to measure axial ratio lengths on gravel size particles, much less so with sands. Both shape and sphericity of particles can be determined by measurement of $a$, $b$ and $c$ axes. Shape has been defined as 'a measure of the relation between the three axial dimensions of an object'; this is related to, but not the same as, sphericity, 'a measure of the approach of a particle to the shape of a sphere'. Angularity and roundness are at different ends of a continuum and are usually measured by comparing the particle with a 'standard' chart. This method has problems with operator variance and it is likely that computer-based methods using Fourier analysis will supersede it. Analysis of surface texture has been a much-used technique on its own with the advent of the scanning electron microscope, but it too is largely based on qualitative assessments and comparisons.  WBW

**Reading**
Bull, P.A. 1981: Environmental reconstruction by electron microscopy. *Progress in physical geography* 5, pp. 368–97. · Clark, M.W. 1981: Quantitative shape analysis: a review. *Mathematical geology* 13, pp. 303–20. · Goudie, A.S. ed. 1990: *Geomorphological techniques*. 2nd edn. London: Unwin Hyman. · Whalley, W.B. 1972: The description and measurement of sedimentary particles and the concept of form. *Journal of sedimentary petrology* 42, pp. 961–5.

**particle size**  Can refer either to an individual particle or to a mass of particles. It is useful for characterizing or describing one aspect of a particle's morphology, form being another important component. Various names have been given to size ranges, e.g. clay, silt, sand, etc. but these are useful as rough, ordinal scale descriptors only and the actual size range covered by each name varies from scale to scale. For particles that can be measured by picking up an individual and measuring lengths with a rule or calipers, three orthogonal axes give a good measure of size. For smaller particles, especially those which can only be seen easily with a microscope or hand lens, it is more usual to measure a single (long) axis. Because the size of particles may range from microscopic aerosols (less than $1\,\mu m$) through clay-size (less than $2\,\mu m$), silt, sand and cobbles to even larger blocks, no single measurement technique can be applied. This means that the definition of size varies according to the technique used. This restriction applies even to the normal range of particles seen in a soil.

For instance, if the long $a$ axes of cobbles (above 60 mm) are measured with calipers but smaller particles are measured by sieving, the sizes obtained by these methods are not strictly comparable. This difficulty of making measurements on a continuous scale also occurs when sieving becomes difficult and precipitation methods are employed. Sieves give 'nominal diameters' which roughly correspond to the intermediate or $b$ axis of a grain. When, as is usual, a sample is taken containing perhaps many thousands of grains, size fractions are determined by sieving through progressively finer sieves. The weights held on each sieve are recorded and expressed as a 'percentage passing' or 'percentage finer' than a given mesh. Such data may be graphed as histograms or as a cumulative size distribution. Although millimetre and micrometre ($\mu m$) sizes can be recorded, it is often convenient to transform the sizes into the $\phi$ scale by:

$$\phi = -\log_2 d$$

Major class intervals used in description of sediment sizes

| mm | | Size classes | |
|---|---|---|---|
| | | Boulder | |
| 256 | −8 | | Gravel |
| 64 | −6 | Cobble | |
| 32 | −5 | | |
| 16 | −4 | Pebble | |
| 4 | −2 | | |
| 2.83 | −1.5 | Granule | |
| 2.00 | −1.0 | | |
| 1.41 | −0.5 | Very coarse sand | |
| 1.00 | 0.0 | | |
| 0.71 | −0.5 | Coarse sand | |
| 0.50 | 1.0 | | Sand |
| 0.35 | 1.5 | Medium sand | |
| 0.25 | 2.0 | | |
| 0.177 | 2.5 | Fine sand | |
| 0.125 | 3.0 | | |
| 0.088 | 3.5 | Very fine sand | |
| 0.0625 | 4.0 | | |
| 0.031 | 5.0 | Coarse silt | |
| 0.0156 | 6.0 | | Silt |
| 0.0078 | 7.0 | Fine silt | |
| 0.0039 | 8.0 | | |
| | | Clay | Clay |

Source: Based on Pettijohn, F.J., Potter, P.E. and Siever, R. 1972: *Sand and sandstone*. New York: Springer-Verlag. Table 3–2, p. 71.

where *d* is the particle size in millimetres. This transform takes into account the generally log-normal size distribution of most sediments. The usual statistical treatments of distributions can be applied to the data obtained by sieving, for example. Thus mean, SKEWNESS, SORTING and KURTOSIS of a size range can be derived by various methods (most often graphically). Such size parameters can be used to compare sediments, to determine their possible origin, or to use in some other way, e.g. to suggest their geotechnical behaviour.                            WBW

**Reading**
Briggs, D. 1977: *Sediments*. London: Butterworth. · Goudie, A.S. ed. 1990: *Geomorphological techniques*. 2nd edn. London: Unwin Hyman. · Syvitski, J.P.M. ed. 1991: *Principles, methods and applications of particle size analysis*. Cambridge: Cambridge University Press.

**passive margin**   The margin of a continent formed as a result of the breakup of a large continental mass, or supercontinent. Passive margins are so called since, in comparison with the active continental margins which coincide with zones of plate convergence, they are characterized by low levels of tectonic activity once continental rupture has been completed. The fragmentation of the supercontinent of Pangaea, which began about 180 Ma ago, created the many new passive continental margins which form a substantial proportion of the perimeters of the present-day continents. There are two

types of passive margin; rifted margins are formed by divergent plate motion, whereas sheared margins are produced where the movement between two adjacent continental blocks has been essentially transform.              MAS

**Reading**
Summerfield, M.A. 1991: *Global geomorphology*. London and New York: Longman Scientific and Technical and Wiley.

**pastoralism**   System of land use based on the rearing of livestock. Pastoralism may be nomadic, whereby the herders move their animals from place to place, either seasonally or more frequently to reflect grazing opportunities and availability, or sedentary, as in ranch-based systems. Nomadic pastoralism is particularly well-suited to areas with highly seasonal and / or low rainfall levels, for example in semi-arid areas or in SAVANNA ecosystems, or in areas where marked seasonal temperature regimes make higher-altitude areas usable only in summer months. In African savanna areas, nomadic pastoralism has often, and wrongly, been blamed as a major cause of DESERTIFICATION, contributing to its demise and replacement by sedentary ranch systems that are largely unsuited to highly variable environments and which may be susceptible to DROUGHT impacts.            DSGT

**Reading**
Livingstone, I., 1991: Livestock management and 'overgrazing' among pastoralists. *Ambio* 20, pp. 80–5. ·

359

Mace, R. 1991: Overgrazing overstated. *Nature* 349, pp. 280–1.

**paternoster lake** One of a series of lakes occupying basins in a glacial trough. The name is taken from the term for rosary prayer-beads.

**patterned ground** A general term for the more or less symmetrical forms, such as circles, polygons, nets and stripes, that are characteristic of, but not necessarily confined to, soils subject to intense frost action. Description is usually based upon geometric form and the degree of uniformity in grain size (i.e. sorting). The most common type of patterned ground is the non-sorted circle, sometimes termed earth hummock or mud-boil. The processes responsible for the formation of patterned ground are not clear; some forms may be polygenetic and others may be combination products in a continuous system having different processes as end members. In some instances thermal or desiccation cracking is clearly essential. In others cryoturbation, the lateral and vertical displacement of particles associated with repeated freezing and thawing, combined with site-specific factors such as moisture availability, lithology, and slope appear essential.                    HMF

**Reading**
Mackay, J.R. 1980: The origin of hummocks, western Arctic coast, Canada. *Canadian journal of earth sciences* 17, pp. 996–1006. · Shilts, W.W. 1978: Nature and genesis of mudboils, central Keewatin, Canada. *Canadian journal of earth sciences* 15, pp. 1053–68. · Washburn, A.L. 1979: *Geocryology: a survey of periglacial processes and environments.* New York: Wiley. Esp. pp. 119–70.

**pavement** See STONE PAVEMENT.

**peak discharge** The highest discharge achieved at a stream gauging site within a specific time period, usually during the passage of a storm hydrograph (see HYDROGRAPHS). Peak discharge is a widely used parameter of the flood characteristics of a site and is the usual parameter to be analysed in FLOOD FREQUENCY analysis. (See also DISCHARGE.)                    AMG

**Pearson type III distribution** A theoretical probability distribution which may be used to fit the flood frequency distribution in FLOOD FREQUENCY analysis. There are several distributions available and the log-Pearson type III is a further form which may be used.                    KJG

**peat** A deposit of partially decomposed or undecomposed organic material derived from plants. Peat deposits form where the rate of accumulation of dead plant remains exceeds the rate at which they are decomposed. This imbalance

results where microbial decomposition is retarded, rather than because primary productivity is high. Waterlogging is one of the major causes of peat formation although any factor that reduces the respiratory activity of aerobic microorganisms may promote peat accumulation. Such factors include, low pH, mineral and nutrient deficiency and low temperatures.

Peatland is a generic term used to refer to all peat-accumulating wetlands from bogs to fens. According to Maltby *et al.* (1992), the term peatland is not synonymous with MIRE because it includes areas that may no longer carry peat-forming communities, but where peat soils occur, for example, agricultural land. Peatlands cover around 3% of global land mass or 385–410 million hectares. Most peatlands form in high latitudes, although about 20% of the total peatland area is recorded in warmer climates. On a global scale, around 28 million hectares of peatland have been exploited for agriculture and forestry (*c.*7% of the total peatland area). The UK Joint Nature Conservation Committee estimate that prior to disturbance, peatlands covered around 1,640,000 hectares in the UK. Around 14% of this total formed in lowland areas. The former peatland area in Ireland is estimated to have covered 1,180,000 hectares (Maltby *et al.* 1992).

The main peat bog types are:

*Blanket mires*: rain-fed, terrain-covering peatlands, generally 1–3 m deep with a global extent of over 10 million hectares. Approximately 13% of the total global blanket mire area is found in the UK. Generally develop in cool climates with small seasonal temperature fluctuations and over 1000 mm of rainfall and over 160 rain days each year.
*Raised mires*: rain-fed, potentially deep peatlands usually forming in complexes and having a domed microtopography. Primarily recorded in lowland areas across much of Northern Europe, including the UK. Also recorded in the former USSR, North America and parts of the southern hemisphere.
*String mires*: flat or concave peatlands usually showing a string-like pattern of hummocks. Primarily found in the cold northern forest (boreal) zone of Scandinavia but also recorded in the western former USSR and North America. A few examples exist in northern Britain.
*Tundra mires*: describe peatlands with a shallow peat layer, *c.*50 cm thick, dominated by sedges and grasses. They cover around 11–16 million hectares in Alaska, Canada, and the former USSR. This mire type forms in areas where the ground is continuously frozen (permafrost zone).

*Palsa mires*: describe a peat microtopography with characteristic high peat mound with a permanently frozen core with wet depressions between the mounds. They develop where the ground surface is only frozen for part of the year, and are common in the former USSR, Canada and parts of Scandinavia.

*Peat swamps*: are forested peatlands and include both rain- and groundwater-fed types. They are commonly recorded in tropical regions with high rainfall. This peat type covers around 35 million hectares, primarily in south-east Asia.

A wider definition of peatlands may include wet heath, marshes and meadows but these are more strictly WETLANDS because they contain significant proportions of mineral material and are not necessarily accumulating peat.                              ALH

**Reading and Reference**
Hughes, J.M.R. and Heathwaite, A.L. 1995: *Hydrology and hydrochemistry of British wetlands*. Chichester: Wiley. · Finlayson, M. and Moser, M. 1991: *Wetlands: international*. Oxford: Waterfowl and wetlands research bureau. Pp. 224. · Maltby, E., Immirzi, C.P. and McLaren, D.P. 1992: *Do not disturb! Peatbogs and the greenhouse effect*. London: Friends of the Earth.

**pedalfer**   A well-drained soil which has had most of its soluble minerals leached from it. They occur in more humid situations than PEDOCALS.

**pediment**   A term applied by G.K. Gilbert (1880) to alluvial fans flanking the mountains near Lake Bonneville in Utah. Since then it has changed its meaning and is now defined (Adams 1975) as follows: a smooth planoconcave upward erosion surface, typically sloping down from the foot of a highland area and graded to either a local or more general base level. It is an element of a piedmont belt, which may include depositional elements such as fans and playas. The pediment, as defined, excludes such depositional components, although an alluvial cover is frequently present. It is broadly synonymous with the French term, *glacis*. Coalescing pediments create pediplains.

Pediments were initially recognized as being of wide extent in the American south-west. McGee (1897) proposed that they were planed off by sheetfloods, whereas others have invoked the role of lateral planation by rivers or the backwearing of mountain fronts (Cooke and Warren 1973, p. 188 ff). Most pediments have low angle surfaces (generally less than 10°) and the junction between them and the hill mass behind is often marked by an abrupt change of angle. For this to happen a sharp contrast is needed between the nature of sound rock and the weathering debris produced. In particular it is necessary for the products of weathering to be relatively fine-grained and to have a limited size range, so that they can be transported across the low angle pediment slope, thereby permitting development or maintenance of an abrupt break of slope. ASG

**References**
Adams, G. ed. 1975: *Planation surfaces*. Stroudsburg, Pa.: Dowden, Hutchinson & Ross. · Cooke, R.U. and Warren, A. 1973: *Geomorphology in deserts*. London: Batsford. ·

**A rock-cut pediment** in granitic desert terrain near Springbok in Namaqualand, South Africa.

Gilbert, G.K. 1880: Contributions to the history of Lake Bonneville. *US Geological Survey annual report* 2, pp. 167–200. · McGee, W.J. 1897: Sheetflood erosion. *Bulletin of the Geological Society of America* 8, pp. 87–112.

**pedocal** A poorly drained soil which has no soluble salts leached from it. Free calcium occurs in the profile, in the form of concretions, veins, nodules, or layers.

**pedon** A term used in soil science, and defined as the smallest volume that can be called soil. Its lateral area ranges from 1 to $10\,m^2$ and is large enough to permit the study of the nature of any horizon present, for a horizon may be variable in thickness and even discontinuous.

**peds** Soil aggregates: their strength, size and shape give the soil its structure. (See also SOIL STRUCTURE.)

**pelagic** Refers to that component of an aquatic ecosystem which excludes its margins and substrate. The pelagic environment corresponds to that of the main part of the body of water (Barnes 1980).

The pelagic environment varies in depth. In deep water (e.g. of oceans), light is able to penetrate only a limited distance below the water surface. The lighted or euphotic zone is the site of abundant organic productivity. Here, phytoplanktonic primary producers are consumed by zooplankton (see PLANKTON) and small fish. Beneath the euphotic zone are the mesopelagic and bathypelagic zones, mainly populated by predatory nekton (see NEKTON). These zones benefit from downward-moving biological detritus and are visited by migratory organisms. In the bathypelagic zone, organisms are relatively infrequent. Numerous species emit their own light in order to obtain their prey (Isaacs 1977). RLJ

**Reading and References**
Barnes, R.S.K. 1980: The unity and diversity of aquatic systems. In R.S.K. Barnes and K.H. Mann eds, *Fundamentals of aquatic ecosystems*. Oxford: Blackwell Scientific. Pp. 5–23. · Isaacs, J.D. 1977: The nature of oceanic life. In H.W. Menard ed., *Ocean science (Readings from Scientific American)*. San Francisco: W.H. Freeman. Pp. 189–201. · Moss, B. 1980: *Ecology of fresh waters*. Oxford: Blackwell Scientific.

**Penck and Brückner model** Developed during the first decade of the twentieth century by A. Penck and E. Brückner, working primarily in Bavaria, to provide a framework for understanding the PLEISTOCENE history of the Alps. Four main glacial phases were initially identified – Günz, Mindel, Riss and Würm – together with various interglacials, including the GREAT INTER-

GLACIAL. A more complex Pleistocene history has now been determined on the basis of deep sea core studies, but the Penck and Brückner model has been immensely influential and extended uncritically worldwide. ASG

**Reference**
Penck, A. and Brückner, E. 1909: *Die Alpen im Eiszeitalter*. Leipzig: C.H. Tauchnitz.

**peneplain** A term adopted by W.M. Davis, to describe 'an almost featureless plain, showing little sympathy with structure' (Davis 1899b), and considered as 'the penultimate form developed in a cycle of erosion' (Davis 1902). It, therefore, invokes a cycle of uplift and lowering by DENUDATION of mountain ranges to form plains of low relief, gentle slope and extensive weathering cover close to *base level*. Davis developed his concept from the pioneer work on the Colorado River by Powell and Dutton, and on the Henry Mountains by Gilbert (Davis 1902), especially by reference to the geological concept of *unconformity*, but it was applied by Davis (1899a) to the evolution of sub-aerial relief within his 'cycle of erosion'.

The apparent absence of erosional plains, cut across structure, from many regional landscapes led to early criticism that peneplains were abstract forms nowhere observed intact and not seen close to present-day base levels. The central problem for Davis' theory was not his ideas about slope evolution, as often implied by later critics, but the nature and frequency of crustal movements and the time needed to effect peneplanation. Tarr discussed these limitations as early as 1898. Penck (1924) constructed an alternative model, relating planation to uplift rates, but this was widely neglected or misunderstood by anglophone readers and interest waned in the topic after the 1950s. Modern plate tectonic theory and its links to mantle convection and gravity tectonics, together with a better understanding of the time-scale involved in planation (Ahnert 1970), suggest that extensive peneplains must date from the Mesozoic or before. Many authors use the term 'peneplain' in a generic rather than a genetic sense, to describe landscapes of low relief, without determining formational process or history. Such landscapes on the old Gondwana continents, date from before the modern era (Mesozoic-Cainozoic) of continental movement and Alpine tectonics. Key papers by Davis were re-published in convenient form in 1954; reviews of planation surfaces are contained in Adams (1975) and in Melhorn and Flemel (1975), while Summerfield (1991) offers a modern comment on these issues. MFT

**Reading and References**
Adams, G. ed. 1975: *Planation surfaces*. Benchmark Papers in Geology 22. Pennsylvania: Dowden, Hutchinson & Ross. · Ahnert, F. 1970: Functional relationships between denudation relief and uplift in large, mid-latitude drainage basins. *American journal of science* 268, pp. 243–63. · Davis, W.M. (1899a): The geographical cycle. *Geographical journal* 14, pp. 481–504. · — (1899b): The peneplain. *American geologist* 22b, pp. 207–239. · — (1902): Base-level, grade and peneplain. *Journal of Geology* 10, pp. 77–111. · — 1954: *Geographical essays*. ed. D.W. Johnson. New York: Dover Publications. · Melhorn, W.N. and Flemel, R.C. eds 1975: *Theories of landform development*. London: George Allen and Unwin. Pp. 45–68. · Penck, W. 1924: Die Morphologische Analyse. *Geographische Abhandlungen*. Vol. 2. · Summerfield, M.A. 1991: *Global geomorphology*. Harlow: Longman. · Tarr, R.S. 1898: The peneplain. *American geologist* 21, pp. 351–70.

**penetrometer**   A device used to measure soil strength. Hand-held penetrometers consist of a steel rod housed within a metal casing with either a line engraved 6 mm from the tip of the rod or a standard-sized metal cone attached to the end. The rod or cone is pressed into the soil until the engraved line or the base of the cone is reached. The rod compresses a calibrated spring and a pointer attached to the rod allows the soil strength in $kgf\,cm^{-2}$ ($1\,kgf\,cm^{-2}$ is equal to $98\,kPa$) to be measured. Laboratory penetrometers are similar in principle but rely on gravity, with a standard cone positioned at the surface of a soil sample allowed to penetrate into the soil for 5 seconds. The depth of penetration is measured and converted to a value in $kgf\,cm^{-2}$.

DJN

**peninsula**   A headland or promontory surrounded by water but connected to the mainland by a neck or isthmus.

**Penman formula**   See EVAPOTRANSPIRATION.

**perched block**   A boulder or block of rock that is balanced on another rock or outcrop having been deposited there by ice.

**Reading**
Patterson, E.A. 1984: A mathematic model for perched block formation. *Journal of glaciology* 30, pp. 296–301.

**perched groundwater**   An isolated body of unconfined GROUNDWATER suspended by a discontinuous relatively impervious layer above the main saturated zone and separated from it by unsaturated rock. Groundwater that is perched also has a perched WATER TABLE.   PWW

**percolation**   The process of essentially vertical water movement downwards through soil or rock in the unsaturated (or vadose) zone. Percolation water is the water that has passed through the soil or rock by this process. It may be measured by a PERCOLATION GAUGE.   PWW

**percolation gauge**   An instrument for measuring the quantity of water that passes vertically through soil or rock by the process of PERCOLATION. Such gauges may be established beneath lysimeters to catch the excess water after losses due to evapotranspiration and sometimes they are installed in caves to measure the response at percolation input points to rainfall at the surface.   PWW

**percoline**   A path or seepage line along which moisture flow becomes concentrated, is particularly well developed where soils are relatively deep, and usually presents a dendritic pattern tributary to surface stream channels. After water from precipitation infiltrates it may flow throughout the soil profile as matrix flow, but once it becomes more concentrated – but before a definite stream channel is produced – there will be a percoline which may be indicated by a broad linear depression on the surface and this may be inter-related with SUFFOSION and the feature at the head of seasonally occupied channels such as DAMBOS in tropical landscapes.

KJG

**Reading**
Bunting, B.T. 1961: The role of seepage moisture in soil formation, slope development and stream initiation. *American journal of science* 259, pp. 503–18.

**pereletok**   A Russian term for a layer of ground between the active layer and the permafrost below, which remains frozen for one or several years and then thaws. Use of this term is not recommended since it presupposes that pereletok is not permafrost, although the definition assigns a sufficient duration of time for it to be considered as such.   HMF

**Reading**
Brown, R.J.E. and Kupsch, W.O. 1974: *Permafrost terminology*. Ottawa: National Research Council of Canada publication 14274.

**perennial stream**   A stream which flows all year. A dynamic drainage network also includes INTERMITTENT STREAMS and EPHEMERAL STREAMS but there should always be flow in a perennial stream channel. For much of the time this flow may be in the form of BASE FLOW or DELAYED FLOW except when QUICKFLOW occurs after rainstorms.   KJG

**pericline**   A dome produced by folding. An ANTICLINE which pitches at both ends.

**perigee** The point in its orbit at which the moon is closest to the earth.

**periglacial** A term first used by Walery von Lozinski in 1909 to describe frost weathering conditions in the Carpathian Mountains of Central Europe. The concept of a 'periglacial zone' subsequently developed referring to the climatic and geomorphic conditions of areas peripheral to Pleistocene ice sheets and glaciers. Theoretically, it was a tundra region extending as far south as the treeline. Modern usage refers to a wide range of cold non-glacial conditions regardless of their proximity to glaciers, either in time or space. Periglacial environments exist not only in high latitudes and tundra regions but also below the treeline and in high altitude (alpine) regions of temperate latitudes.

Approximately 20 per cent of the earth's land-surface currently experiences periglacial conditions in the form of either intense frost action or the presence of permafrost, or both. There are all gradations between environments in which frost processes dominate, and where a whole or major part of the landscape is the result of such processes, and those in which frost action processes are subservient to others. Complicating factors are that certain lithologies are more prone to frost action than others, and no perfect correlation exists between areas of intense frost action and areas underlain by permafrost.

Classification of periglacial climates

| Polar lowlands | Mean temperature of coldest month $< -3\,°C$. Zone is characterized by ice caps, bare rock surfaces and tundra vegetation. |
|---|---|
| Subpolar lowlands | Mean temperature of coldest month $< -3\,°C$ and of warmest month $> 10\,°C$. Taiga type of vegetation. The $10\,°C$ isotherm for warmest month roughly coincides with treeline in northern hemisphere. |
| Mid-latitude lowlands | Mean temperature of coldest month is $< -3\,°C$ but mean temperature $> 10\,°C$ for at least four months per year. |
| Highlands | Climate influenced by altitude as well as latitude. Considerable variability over short distance depending on aspect. Diurnal temperature ranges tend to be large. |

*Source*: Based on the classification presented by Washburn, A.L. 1979: *Geocryology*. London: Edward Arnold. Pp. 7–8.

Unique periglacial processes are the formation of PERMAFROST, the development of thermal contraction cracks, the thawing of permafrost (THERMOKARST), and the formation of wedge and injection ice. Other processes, not necessarily restricted to periglacial regions, are important on account of their high magnitude or frequency. These include ICE SEGREGATION, seasonal frost action, frost (i.e. cryogenic) weathering, and rapid mass movements.

The most distinctive periglacial landforms are those associated with permafrost. The most widespread are tundra polygons, formed by thermal contraction cracking. They divide the ground surface into polygonal nets 20–30 m in dimension. Ice-cored hills, or pingos, are a less widespread but equally classic periglacial landform; they form when water moves to the freezing plane under hydraulic or hydrostatic pressure. Other aggradational landforms, such as palsas and peat plateaux, are associated with ice segregation. Ground ice slumps, thaw lakes, and irregular hummocky topography with enclosed depressions (thermokarst) result from the melt of ice-rich permafrost.

Many periglacial phenomena form by frost wedging and the cryogenic weathering of exposed bedrock. Frost wedging is associated with the freezing and expansion of water which penetrates joints and bedding planes. The details of cryogenic weathering are poorly understood. Coarse angular rock debris (blockfields), upthrust bedrock blocks, talus (scree) slopes, and certain types of patterned ground are usually attributed to frost action. Angular bedrock masses (tors) may stand out above the debris-covered surfaces, reflecting more resistant bedrock. Flat erosional surfaces, termed cryoplanation terraces, are sometimes associated with tors but, equally, can occur quite independently.

The overall flattening of landscape and smoothing of slopes, thought typical of many periglacial regions, is generally attributed to mass wasting. Agents of transport include frost creep and solifluction. NIVATION, a combination of frost wedging, solifluction and sheetwash operating beneath and downslope of snowbanks, is often regarded as important. In areas dominated by extreme nival regimes and underlain by unconsolidated sediments, fluvial activity can be a significant landscape modifier.

HMF

**Reading**
Clark, M.J. 1988: *Advances in periglacial geomorphology*. Chichester: Wiley. · French, H.M. 1976: *The periglacial environment*. London and New York: Longman. · Lozinski, W. von 1909: Uber die mechanische Verwitterung der Sandsteine im gemässigten Klima. *Bulletin*

*International de l'Academie des Sciences et des Lettres de Cracovie, Classe des Sciences Mathématiques et Naturelles* 1, pp. 1–25. · Washburn, A.L. 1979: *Geocryology: a survey of periglacial processes and environments*. New York: Wiley.

**perihelion**  The point in its orbit about the sun that a planet or comet is closest to the sun.

**permafrost**  The thermal condition in soil and rock where temperatures below 0 °C persist over at least two consecutive winters and the intervening summer. Permafrost is defined purely as a thermal condition: moisture, in the form of water and/or ice may or may not be present. The term was first introduced by S.W. Muller in 1945, as a shortened form of permanently frozen ground. Permafrost is not necessarily synonymous with 'frozen ground', however, since earth materials may be below 0 °C in temperature but essentially unfrozen on account of depressed freezing points due to mineralized groundwaters or other causes. One solution is to differentiate between 'cryotic' (i.e. below 0 °C) and 'non cryotic' (i.e. above 0 °C) ground, and to subdivide the former into 'unfrozen', 'partially frozen' and 'frozen', depending upon the amount of unfrozen water present. Equally, permafrost is not 'permanently frozen ground', since changes in climate and terrain may cause it to degrade.

The upper boundary of permafrost is known as the permafrost table, and the near-surface layer which is subject to seasonal thaw is called the active layer. The depth at which annual temperature fluctuations are minimized is termed the depth of zero annual amplitude; this usually varies between 10 and 20 m depending

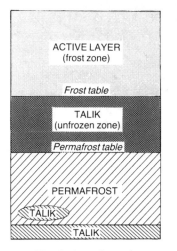

**Permafrost: 1.** Terminology of some features associated with permafrost.
*Source*: R.U. Cooke and J.C. Doornkamp 1974: *Geomorphology in environmental management*. Oxford: Clarendon Press. Figure 9.1.

upon climate and terrain factors such as amplitude of annual surface temperature variation, snow cover, and effective thermal diffusivity of the soil and rock. In polar regions, permafrost temperatures may be as low as −15 °C at the depth of zero annual amplitude.

Permafrost underlies approximately 20–25 per cent of the earth's landsurface. It is widespread in Siberia, northern Canada, Alaska and China (see diagram on p. 366). Zones of continuous, discontinuous and sporadic (isolated) permafrost are generally recognized, together with alpine, intermontane and subsea (offshore) permafrost. Thicknesses range from a few metres at the southern limits to approximately 500 m in parts of northern Canada and Siberia where it may be relic.                    HMF

**Reading**
Brown, R.J.E. and Kupsch, W.O. 1974: *Permafrost terminology*. National Research Council of Canada publication 14274. · French, H.M. 1982: *The Roger J.E. Brown memorial volume. Proceedings of the Fourth Canadian Permafrost Conference*. National Research Council of Canada. · Muller, S.W. 1947: *Permafrost or permanently frozen ground and related engineering problems*. Ann Arbor, Mich.: J.W. Edwards. · National Academy of Sciences 1973: North American contribution. *Permafrost. Second international conference*. Yakutsk, USSR, 13–28 July 1973. · — 1983 *Permafrost. Fourth international conference*. Vol. I. Washington DC, National Academy Press.

**permeability**  A measure of the capacity of a rock or soil to transmit fluids. It is often termed INTRINSIC PERMEABILITY or specific permeability to distinguish it from hydraulic CONDUCTIVITY which is sometimes (but now less frequently) called the coefficient of permeability. Permeability (or PERVIOUSNESS) depends upon the physical characteristics of the rock, whereas hydraulic conductivity also takes account of the physical characteristics of the fluid.                    PWW

**Reading**
Freeze, R.A. and Cherry, J.A. 1979: *Groundwater*. Englewood Cliffs. NJ: Prentice-Hall.

**permeameter**  An instrument for measuring the saturated hydraulic CONDUCTIVITY or $K$ of soils or other sediments. Two types of apparatus are in common use. One is a constant head permeameter, the other a falling head permeameter. In the constant head case:

$$K = \frac{QL}{Ah}$$

and in the falling head case:

$$K = \frac{aL}{aT}\ln\frac{h_0}{h_1}$$                    PWW

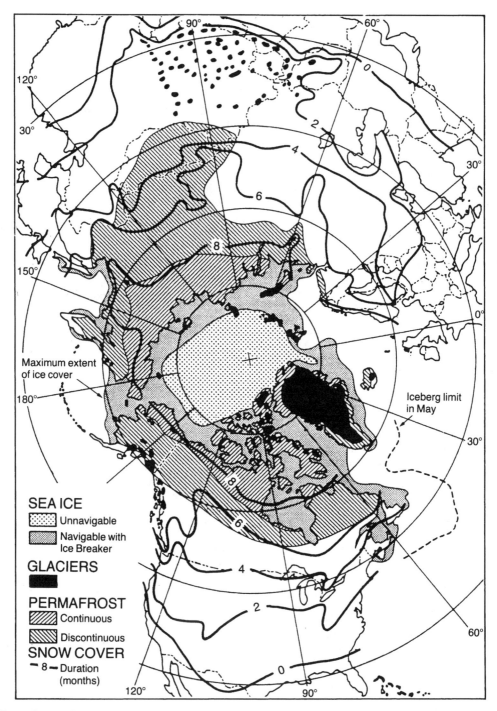

**Permafrost: 2**. Snow, ice and permafrost in the northern hemisphere.
*Source*: H.J. Walker 1983: *Mega-geomorphology*. Oxford: Clarendon Press. Figure 3.2.

### Reading and Reference

Freeze, R.A. and Cherry, J.A. 1979: *Groundwater*. Englewood Cliffs, NJ: Prentice-Hall. · McGreal, W.A. 1981: Permeability and infiltration capacity. In A. Goudie ed., *Geomorphological techniques*. London: Allen & Unwin. Pp. 94–6. · Todd, D.K. 1980: *Groundwater hydrology*. 2nd edn. New York and Chichester: Wiley.

**persistence** The continued existence of an ecosystem without significant fluctuations in overall species composition or in the relative number of individuals in different species. Persistence represents one concept of STABILITY and has also been referred to as constancy or 'no-oscillation' stability. This aspect of stability does not take into account the ability of a system to adjust to disturbance.

The linkage of the term persistence to the notion of constancy of ecosystem species populations is unfortunate. Although major fluctuations in population size are thought to increase the risk of extinction during episodes of low population, there is considerable evidence that oscillations of population numbers may enhance the survival of many species by producing episodes of saturation and scarcity with respect to predation.                                              ARH

**Reading**
Dunbar, M.J. 1973: Stability and fragility in Arctic ecosystems. *Arctic* 26, pp. 179–85.

**perturbation** Variation, usually of small amplitude, about some well-defined, usually uninteresting, basic state. It is the conceptual equivalent of gently shaking an unknown package before embarking on more irreversible probing. Perturbation analysis allows all possible states similar to the initial one to be followed at least for a short time. In contrast, numerical integration (see NUMERICAL MODELLING) allows a few possible states to be followed for a longer time. The term is also used to denote various types of weather disturbance such as the waves of the trades and equatorial easterlies.     JSAG

**Reading**
Haurwitz, B. 1941: *Dynamic meteorology*. New York and London: McGraw-Hill.

**perviousness** A property of soils and rocks indicating its capacity for transmitting fluids. It is equivalent to PERMEABILITY. Impervious rocks behave hydrologically like an AQUIFUGE or aquiclude.     PWW

**pesticides** Poisonous chemicals used by man to regulate or eliminate plant and animal pests. The term biocide is sometimes used in the same context but, as Ware (1983) points out, is without a precise scientific meaning and is often used in a rather emotive manner to describe the lethal effects of chemicals on living organisms.

Pesticides are mainly synthetic organic chemicals which possess varying levels of toxicity. They may be selective or non-selective and are aimed at 'target' organisms which are economically or socially undesirable (Cremlyn 1978).

Among elementary forms of life, viruses and bacteria are responsible for numerous plant and animal diseases. Of lower plants, algae which bloom in bodies of water and fungi parasitic on living tissue are pests. Fungi are also the cause of many diseases of flora and fauna. The 30,000 or so higher plants which are pests because they shade crops and/or diminish their supply of moisture and nutrients are termed weeds. Animal pests include some Nematoda (roundworms), Arthropoda (insects), Arachnida (spiders, mites, ticks), Mollusca (slugs and snails), birds and rodents.

There are three main categories of pesticide – herbicides, insecticides and fungicides – in the production of which four major groups of chemicals are employed. Herbicides may operate on either a systemic (penetrative) or non-systemic (contact) basis. Phenoxyaliphatic acids (for example, 2,4-D and 2,4,5-T) are important herbicides, the collapse, wilting and death of target weeds under their application being due to amended phosphate and nucleic acid metabolism and cell division.

Among insecticides, the best known, DDT, is an organochlorine (or chlorinated hydrocarbon) which causes death by upsetting sodium and potassium equilibrium in the nervous system, thereby altering its normal functions. Organophosphates (for example, parathion) have largely supplanted organochlorines as insecticides. Additionally, these are very toxic to vertebrates as they also inhibit normal functioning of the nervous system.

Fungicides need to kill the parasite but not its host. Their usual function is to halt the germination of fungal spores. Inorganic compounds of sulphur, copper and mercury are long-established fungicides, but there is an increasing number of organic fungicides such as dithocarbamates. After metabolism these influence amino-acid activity in disease cells and cause their elimination.

During the first millennium BC, sulphur was used to fumigate Greek houses. Around 150 years ago, sulphur had been joined by arsenic and phosphorus as pesticides. The impetus for much pesticide use came during the Second World War. Some pesticides (for example, DDT) were then used in the control of human disease carriers such as mosquitoes. At this time, research into nerve gas led to the discovery of the insecticidal property of organophosphates. Subsequent use of pesticides was often enthusiastic and without regard to their possible effects upon non-target organisms and their environment (see BIOLOGICAL MAGNIFICATION).

Pests can also become resistant to pesticides. This occurs because a pesticide disrupts a single genetically controlled process in the metabolism of the pest (Ware 1983). This resistance is often a sudden occurrence, and takes place either by the natural selection of hardy individuals which form the basis of subsequent populations, or by mutation to form resistant genotypes.      RLJ

### References
Cremlyn, R. 1978: *Pesticides*. Chichester and New York: Wiley. · Ware, G. 1983: *Pesticides: theory and application*. 2nd edn. New York and Oxford: W.H. Freeman.

**pF**   The $\log_{10}$ of the negative head of water (in centimetres) indicating the strength of soil moisture SUCTION at a site. Soil moisture suction is often termed the matric or capillary potential and it results from the attraction of solid surfaces (the soil matrix) for water and also of the water molecules for each other, so that water is held in the soil against gravity by adsorptive and CAPILLARY FORCES. These forces lower the potential energy of the soil water and the degree of this reduction in potential energy in comparison with free water at the same elevation and under the same air pressure may be indicated by a negative head of water. The negative head is measured by TENSIOMETERS and because the negative head may increase rapidly with quite small changes in soil moisture content, it is usually expressed in relation to a $\log_{10}$ base as a pF value. The graph of pF plotted against soil moisture content (by volume) for a soil sample that has been progressively drained of water is called the soil moisture characteristic curve or pF curve.      AMG

### Reading
Schofield, R.K. 1935: The pF of water in the soil. *Transactions of the Third International Congress on Soil Science* 2, pp. 37–48. · Smedema, L.K. and Rycroft, D.W. 1983: *Land drainage*. London: Batsford.

**p-form**   Smoothed, apparently plastically sculptured forms caused by the action of glaciers. They comprise *grooves* which may be slightly sinuous with soft flowing outlines and sometimes display an overhanging lip, crescentic-shaped depressions or SICHELWANNEN 1–10 m in length and 5–6 m wide, with horns pointing down glacier, and POT-HOLES and a variety of shallower forms, ranging in size from depressions a few centimetres across to giant potholes 15–20 m deep and 16 m in diameter. Good descriptions are given by Dahl (1965).

Although Gjessing (1965) argued in favour of erosion by saturated till, it may be that the grooved features can be explained by normal glacial abrasion and the remaining forms by high velocity subglacial meltwater action.   DES

### References
Dahl, R. 1965: Plastically sculptured detail forms on rock surfaces in northern Nordland, Norway. *Geografisker annaler* 47, pp. 83–140. · Gjessing, J. 1965: One 'plastic scouring' and 'subglacial erosion'. *Norsk Geografisk Tiddskrift* 20, pp. 1–37.

**pH**   The measure of the acidity or alkalinity of a substance measured as the number of hydrogen ions present in one litre of the substance and reported as a figure on a scale the centre of which is 7, representing neutrality. Acid substances have a pH of less than 7 and alkaline substances have a higher pH.

**phacolith**   A lens-shaped igneous intrusion usually situated beneath an anticlinal fold or in the base of a syncline.

**phenology**   The scientific study of the timing of recurring natural phenomena in the life cycle of plants and animals in nature. While all natural phenomena may be included (e.g. the timing of animal migrations, flowering and fruiting of plants, harvest, ripening and seed-time), phenological observations are often restricted to the time at which certain plants come into leaf and flower each year, and to the date of the first and last appearance of birds and animals. Phenological observations have been used as a source of proxy climate change, in order to describe temperature patterns during pre-instrumental times, based on long-term observations of the growth and maturity of cultivated plants, for example wine harvest records (Pfister 1992).      AP

### Reading and Reference
Jefree, E.P. 1960: Some long-term meanings from the Phenological Reports (1891–1948) of the Royal Meteorological Society. *Quarterly Journal of the Royal Meteorological Society* 86, pp. 95–103. · Pfister, C. 1992: Monthly temperature and precipitation in central Europe 1525–1979: quantifying documentary evidence on weather and its effects. In R.S. Bradley and P.D. Jones eds, *Climate since AD 1500*. London: Routledge. Pp. 118–43.

**phoresy**   The transport by one animal of another of a different species to a new feeding site.

**phosphate rock**   An indurated sedimentary deposit which is rich in apatite, a calcium phosphate. Much of it results from the interaction of guano (seabird excrement) and the calcium carbonate of reef sands and elevated limestones.      ASG

### Reading
Stoddart, D. and Scoffin, T. 1983: Phosphate rock on coral reef islands. In A.S. Goudie and K. Pye eds, *Chemical sediments and geomorphology*. London: Academic Press. Ch. 12.

**photic zone**   The surface zone of a lake, sea or ocean above the maximum depth to which sunlight penetrates.

**photochemical smog**   See SMOG.

**photogrammetry**   The science and technology of obtaining reliable measurements by means of photography. The measurement of aerial photographs to provide details of area, distance and height are skills readily practised by physical geographers. The task of the photogrammetrist is usually considered to be the measurement of aerial photographs both with sufficient accuracy in two dimensions to enable features to be plotted with respect to a national or international grid coordinate system, and with sufficient accuracy in three dimensions to enable them to be located in relation to their height above sea level.                                    PJC

**Reading**
American Society of Photogrammetry 1981: *Manual of photogrammetry*. 4th edn. Falls Church, Virginia: American Society of Photogrammetry. · Paine, D.P. 1981: *Aerial photography and image interpretation for resource management*. New York: Wiley.

**photosynthesis**   A chemical process, which takes place in the cellular structures of green plants, blue-green algae, phytoplankton and certain other organisms, and which transforms received solar energy into chemically stored foodstuffs, through the conversion of carbon dioxide and water into carbohydrates with the simultaneous release of oxygen, as indicated in the following simplified formula:

$$6CO_2 + 6H_2O \rightarrow C_6H_{12}O_6 + 6O_2 + 673\text{kcal}$$

the carbohydrate in this case being glucose. The radiant energy reacts initially with the green pigment chlorophyll, with the result that one electron of this complex molecule is raised above its normal energy level for some $10^{-7}$ to $10^{-8}$ s, so triggering other chemical changes (Hutchinson 1970). The 673 kcal noted above represents a fairly large store of energy, which is then used for plant growth and respiration. Photosynthesis is solely a daytime phenomenon. It is thus the basis of the flow of energy through the TROPHIC LEVELS of ECOSYSTEMS and indeed the fundamental process that enables life itself to occur on this planet, with the exception of a few species (mainly bacteria) which obtain their energy from chemicals.                                    DW

**Reading and Reference**
Gregory, R.P.F. 1989: *Photosynthesis*. Glasgow: Blackie. · Hutchinson, G.E. 1970: The biosphere. *Scientific American* 223, pp. 44–53.

**phreatic**   GROUNDWATER may be divided into zones of aeration (interstices partially occupied by water and partially by air) and saturation (all interstices filled with water under hydrostatic pressure). Usually, a single zone of aeration overlies a single zone of saturation and extends up to the ground surface. Overlying impermeable strata usually bound the upper saturated zone, which extends down to underlying impermeable strata such as clay beds or bedrock. Where overlying impermeable strata are absent, the upper surface of the zone of saturation is the water table or phreatic surface, and is defined as the surface of atmospheric pressure. It would be revealed by the level at which water stands in a well penetrating the aquifer. Phreatic is derived from the Greek *phrear, -atos*, meaning a well.                                    ALH

**phreatic divide**   An underground watershed. The WATER TABLE or upper surface of the permanently saturated zone usually has an undulating topography that depends upon variations in hydraulic CONDUCTIVITY in the bedrock and on the location of zones of RECHARGE and discharge. Water flows down the hydraulic gradient (see DARCY'S LAW) away from high points on the water table, which are therefore phreatic divides, i.e. the watersheds of GROUNDWATER basins. But the pattern of groundwater divides need not mirror the surface pattern of topographic watersheds. Unlike the watersheds of surface catchment areas which are permanent, phreatic divides can shift as the water table rises, lowers or changes its configuration because of localized recharge. (See also GROUNDWATER.)
                                    PWW

**phreatophytes**   Plants that have developed root systems with the capability to penetrate great depths in order to draw water directly from groundwater reservoirs. The name is derived from Greek meaning 'well plant'. In arid and semi-arid regions they make up almost all the plants in riparian habitats, where they affect the geomorphological processes of the nearby stream by causing increased hydraulic roughness and concomitant sedimentation. Phreatophytes have important economic impacts on the hydrological cycle because they transpire moisture from the groundwater table which might otherwise be used for pumped irrigation water (Horton and Campbell 1974).
                                    WLG

**Reference**
Horton, J.S. and Campbell, C.J. 1974: *Management of phreatophyte and riparian vegetation for maximum multiple use values*. US Department of Agriculture Forest Service research paper RM-117.

**phylogenesis** The emergence of new taxa through the splitting of evolutionary lineages, the word is derived from the Greek *phylon* (tribe) and *genesis* (origin). Phylogeny is an approach to the classification of organisms based on the evolutionary history of a species or group of related species and describes, in effect, its 'family history'. The phylogeny is usually depicted in the form of a dendrogram, known as a phylogenetic tree, which illustrates the evolutionary relationship. Phylogenesis describes the process giving rise to the characteristic branching pattern of the evolutionary tree. The taxonomic group in question is said to be monophyletic if a single ancestor species gave rise to all the species placed within that taxon. MEM

**physical geography** Mary Somerville's *Physical geography* (1848) was one of the first and most influential of textbooks in physical geography and gave a clear definition of the field: 'Physical geography is a description of the earth, the sea, and the air, with their inhabitants animal and vegetable, of the distribution of these organized beings and the causes of that distribution... man himself is viewed but as a fellow-inhabitant of the globe with other created things, yet influencing them to a certain extent by his actions and influenced in return. The effects of his intellectual superiority on the inferior animals, and even on his own condition, by the subjection of some of the most powerful agents in nature to his will, together with the other causes which have had the greatest influence on his physical and moral state, are among the most important subjects of this science' (quotation from 4th edn, 1858).

Somerville's view of physical geography had certain similarities with that of Arnold Guyot (1850) who thought that physical geography should be more than 'mere description'. 'It should not only describe, it should compare, it should interpret, it should rise to the *how* and the *wherefore* of the phenomena which it describes'. Guyot regarded physical geography as 'the science of the general phenomena of the present life of the globe, in reference to their connection and their mutual dependence'. As part of his concern with 'connection' and 'mutual dependence' he included a consideration of racial differences and performance in relation to environmental controls, a theme that was to be developed by the North American school of environmental determinists over the next six to seven decades.

Thus both Guyot and Somerville saw a human dimension in physical geography and believed that it was more than an incoherent catalogue. This was a view shared by Thomas Huxley, who used the term 'physiography' in preference to 'physical geography'. Similarly, in France, Emmanuel de Martonne (1909), in addition to covering what he regarded as the four main components of physical geography – climatology, hydrography, geomorphology and biogeography – also included a consideration of environmental influences.

In the twentieth century, physical geography was dominated for much of the time by geomorphology and as Stoddart (1987) remarked about the British situation during the inter-war years, 'in research, if not teaching, "physical geography" meant geomorphology: for while some attention was given to meteorology, climatology and to some extent pedology and biogeography, it was on the level of elementary service courses for students rather than as a contribution to new knowledge'.

In contrast to the works of Guyot, Somerville and de Martonne, physical geography tomes tended increasingly to ignore human and environmental influences. So, for example, Pierre Birot (1966) saw physical geography as the study of 'the visible surface of natural landscapes as they would appear to the naked eye of an observer travelling over the globe before the interaction of mankind'.

Physical geography possibly reached its nadir within geography in the late 1960s and early 1970s when spatial modellers, particularly of urban systems, saw little room for it in the discipline, and even some physical geographers doubted the role of the subdiscipline in a world where regional differences were seen to be declining and where many people were thought to be becoming progressively divorced from the reality of their immediate physical surroundings (Chorley 1971).

In recent decades physical geography has become more concerned with the integration of its various elements (Goudie 1994) and has re-established its concern with human issues (Gregory 1985). Physical geography texts now tend to include the word 'environment' in their titles. The human impact (Goudie 2000) has become a major concern and many of the great issues facing the world today have proved susceptible to geographical treatment, including acid deposition (Battarbee *et al.* 1988), forest decline (Innes 1992), desertification (Thomas and Middleton 1994), salinity problems (Goudie and Viles 1997) and climate change (Williams *et al.* 1998).

Another major theme of modern physical geography is natural environmental change. It is remarkable, but partly uncoincidental, that at the same time that scientists have been concerned with anthropogenic impacts on the environment, they have also become increasingly aware of the frequency, magnitude and consequences of natural environmental changes at a whole range of time-scales from relatively short-lived events like ENSO phenomena, to events at the decade and century scales (e.g. Grove 1988 on the Little Ice Age), to the major fluctuations of the HOLOCENE (Roberts 1998) and the Younger Dryas (Anderson 1997), to the cyclic events of the PLEISTOCENE and to the longer-term causes of the Cainozoic climate decline. Much of the reason for this concern arises from the development in the last four decades of new technologies for dating and environmental reconstruction, including the coring of ocean floors, lakes and ice sheets (Lowe and Walker 1997).

There has also been an increasing concern with the application of physical geography to societal needs (Jones 1980; Cooke and Doornkamp 1992). Likewise, environmental management has become a major field in many branches of physical geography (O'Riordan 1995), including management of water resources (Beaumont 1988), water pollution (Burt et al. 1993) and coasts (Viles and Spencer 1995). In recent years, physical geographers have made many contributions to the study of hazards (e.g. Jones 1993; Smith 1992) and disasters (Alexander 1993) and in some cases this has also brought them to consider societal issues (see, for example, Chester 1993) at the same time as they consider geomorphological, hydrological or climatic events.

Perhaps a major reason for a tendency towards integration of the elements of physical geography has been the resurgence of biogeography, which for too long, with climatology, was one of the less vibrant parts of the discipline. This is in part reflected in the success of a journal, The journal of biogeography (started in 1974), but is also seen in broad ranging texts that attempt to give an integrated view of landscape types as diverse as ocean islands (Nunn 1994), caves (Gillieson 1996) and rain forests (Milington et al. 1995), and also of whole continents (e.g. Adams et al. 1996). Exciting new developments are taking place in our understanding of environments like savannas through an increasing concern with forces such as fire (natural and human), herbivores, soil nutrient status and soil hydrology, and a long history of human land use practices (see, for example, Fairhead and Leach 1997).

Building upon some of these tendencies, Slaymaker and Spencer (1998, p. 7) have sought to define physical geography so that it is redirected from an emphasis on 'the pot-pourri of information about the Earth and its atmosphere to a coherent integrating theme of global environmental change'. They believe that to achieve that, physical geography should be re-defined as that branch of geography concerned with:

1 identifying, describing and analysing the distribution of biogeochemical elements in the environment;
2 interpreting environmental systems at all scales, both spatial and temporal, at the interface between atmosphere, biosphere, hydrosphere, lithosphere and society; and
3 determining the resilience of such systems in response to perturbations, including human activities.

They also argue that a commitment to the understanding of human–environmental linkages is crucial to the sustainability of our planet and that this should be the mandate for physical geography in the twenty-first century.      ASG

### References

Adams, W., Goudie, A.S. and Orme, A.R. 1996: The physical geography of Africa. Oxford: Oxford University Press. · Alexander, D. 1993: Natural disasters London: UCL Press. · Anderson, D. 1997: Younger Dryas research and its implications for understanding abrupt climatic change. Progress in physical geography 21, pp. 230–49. · Anderson, M.G. and Brooks, S.M. eds, 1996: Advances in hillslope processes. Chichester: Wiley. · Battarbee, R. and 15 collaborators 1988: Lake acidification in the United Kingdom 1800–1986. London: Ensis. · Beaumont, P. 1988: Environmental management and development in drylands. London: Routledge. · Birot, P. 1966: General physical geography. London: Harrap. · Burt, T.P., Heathwaite, A.L. and Trudgill S.T. eds 1993: Nitrates: processes, patterns and control. Chichester: Wiley. · Chester, D. 1993: Volcanoes and society. London: Arnold. · Chorley, R.J. 1971: The role and relations of physical geography. Progress in geography 3, pp. 87–109. · Cooke, R.U. and Doornkamp, J.C. 1992: Geomorphology in environmental management. 2nd edn. Oxford: Oxford University Press. · de Martonne, E. 1909: Géographie physique. Paris: Colin. · Fairhead, J. and Leach, M. 1997: Misreading the African landscape. Cambridge: Cambridge University Press. · Gillieson, D. 1996: Caves. Oxford: Blackwell. · Gregory, K.J. 1985: The nature of physical geography. London: Arnold. · Goudie, A.S. 1994 The nature of physical geography: a view from the drylands. Geography 79, pp. 194–209. · — 2000: The human impact on the natural environment. 5th edn. Oxford: Blackwell. · —and Viles, H. 1997: Salt weathering hazards. Chichester: Wiley. · Grove, J.M. 1988: The little Ice Age. London: Routledge. · Guyot, A. 1850: The earth and man: lectures on comparative physical geography in its relation to the history of mankind. New York: Scribners. · Innes, J. 1992: Forest decline. Progress in physical geography 16, pp. 1–64. · Jones, D.K.C. 1980: British applied geomorphology: an appraisal, Zeitschrift für Geomorphologie NF, Supplementband 36, pp. 48–73. · — ed. 1993: Environmental hazards: the challenge of change. Geography 78, pp. 161–98. · Lowe, J. and Walker, M. 1997: Reconstructing

*Quaternary environments*. 2nd edn. Harlow: Longman. · Milington, A.C., Thompson, R.D. and Reading, A.J. 1995: *Humid tropical environments*. Oxford: Blackwell. · Nunn, P.D. 1994: *Oceanic islands*. Oxford: Blackwell. · O'Riordan, T. ed. 1995: *Environmental science for environmental management*. Harlow: Longman Scientific and Technical. · Roberts, N. 1998: *The Holocene*. 2nd edn. Oxford: Blackwell. · Slaymaker, O. and Spencer, T. 1998: *Physical geography and global environmental change*. Harlow: Longman. · Smith, K. 1992: *Environmental hazards: assessing risk and reducing disaster*, London: Routledge. · Somerville, M. 1848: *Physical geography*. London: Murray. · Stoddart, D.R. 1987: Geographers and geomorphology in Britain between the wars. In R.W. Steel ed. *British geography 1918–1945*. London: Institute of British geographers. Pp. 156–76. · Thomas, D.S.G. and Middleton, N.J. 1994: *Desertification: exploding the myth*, Chichester: Wiley. · Viles, H.A. and Spencer, T. 1995: *Coastal problems*. London: Arnold. · Williams, M.A.J., Dunkerley, D., de Deckker, P., Kershaw, P. and Chappell, J. 1998: *Quaternary environments*. 2nd edn. London: Arnold.

**physical meteorology** That part of meteorology or the atmospheric sciences which deals with the physical properties of the atmosphere and the processes occurring therein. Usually included are atmospheric chemistry, electricity, radiation, thermodynamics, optics and acoustics, cloud and precipitation physics, AEROSOL physics, and physical climatology. Because all of these fields of study interact in one way or another with atmospheric motions, it is somewhat artificial to distinguish between physical meteorology and DYNAMICAL METEOROLOGY. This is especially true in the present day when so much research activity in the atmospheric sciences deals with the numerical modelling of physical processes and their interaction with the atmospheric circulation.

Of the various sub-branches of physical meteorology atmospheric thermodynamics is perhaps the most basic and the most closely related to dynamical meteorology. Usually included under this heading are the gas laws (see EQUATION OF STATE), the HYDROSTATIC EQUATION, the first and second laws of thermodynamics, latent heats and water vapour in the air, ADIABATIC processes, static stability (see VERTICAL STABILITY/INSTABILITY), and entropy. Cloud and precipitation physics (see CLOUD MICROPHYSICS) includes not only studies of the formation and growth of cloud droplets, raindrops and ice crystals but also studies involving the processes whereby CLOUDS may be modified to increase or decrease PRECIPITATION, to suppress HAIL, LIGHTNING, and HURRICANE winds, and to dissipate FOG.

The sub-branches of physical meteorology are not autonomous. For example, a study of the effect of AEROSOLS on the ALBEDO of clouds would involve aerosol physics, radiative transfer, cloud physics, and possibly atmospheric chemistry. Modern research in physical meteorology may also use REMOTE SENSING, either from SATELLITES or from the ground using radiometers, radar and lidar.                    WDS

**Reading**
Wallace, J.M. and Hobbs, P.V. 1977: *Atmospheric science*. New York: Academic Press.

**physiography** A word that has obscure origins, although in common currency in eighteenth-century Scandinavia, and in regular usage in the English-speaking world in the nineteenth century (Stoddart 1975). It was defined by Dana in 1863:

Physiography, which begins where geology ends – that is, with the adult or finished earth – and treats (1) of the earth's final surface arrangements (as to its features, climates, magnetism, life, etc.), and (2) its systems of physical movements and changes (as atmospheric and oceanic currents, and other secular variations in heat, moisture, magnetism, etc.).

One of the most notable exponents of physiography was T.H. Huxley, who published a highly successful text, *Physiography*, in 1877. In the USA W.M. Davis preferred the term to GEOMORPHOLOGY, but he used it without the catholicity of meaning that it had for Huxley.      ASG

**References**
Dana, J.D. 1863: *Manual of geology: treating on the principles of the science*. Philadelphia: Bliss. · Huxley, T.H. 1877: *Physiography: an introduction to the study of nature*. London: Macmillan. · Stoddart, D.R. 1975: 'That Victorian science': Huxley's *Physiography* and its impact on geography. *Transactions of the Institute of British Geographers*, 66, pp. 17–40.

**phytogeography** The scientific study of the distribution of plants on the earth. Although traceable as a theme to the 'father of botany', Theophrastus (*c.*370–287 BC), the first comprehensive studies in plant geography were the *Essai sur la géographie des plantes* of Humboldt and Bonpland (1805, German edition 1807) and the *Géographie botanique raisonnée* of Alphonse de Candolle (1855). Used narrowly, the term tends to be confined to the study of geographical or spatial distribution of plant species, genera and families over the surface of the earth; more generally, it is often used, particularly by geographers, to include many aspects of plant ecology and plant biology as well.      PAS

**Reading**
Moore, D.M. ed. 1982: *Green planet: the story of plant life on earth*. Cambridge: Cambridge University Press. · Stott, P.A. 1981: *Historical plant geography*. London: Allen & Unwin.

**phytogeomorphology** A concept that 'reflects those sensitive landform–vegetation relationships that are visibly dominant on the landscape' (Howard and Mitchell 1985, p. 5). A major component of its study is the use of REMOTE SENSING as most satellite images of the land surface reflect terrain and vegetation types which have become the basis for the interpretation of many less clearly visible phenomena, including soils. ASG

**Reference**
Howard, J.A. and Mitchell, C.W. 1985: *Phytogeomorphology*. New York: Wiley-Interscience.

**phytokarst** Features produced by the weathering and erosive action of plants and animals on limestone rocks. It is also called phytokarren or biokarst. The identification of erosive (boring and digestive) action or chemical weathering (chelation) on rocks is very problematic. While phytokarst can form spectacular features (Bull and Laverty 1982), it usually produces random spongework forms (Folk *et al.* 1973). The term BIOKARST is preferred to phytokarst (Viles 1984). PAB

**References**
Bull, P.A. and Laverty, M. 1982: Observations on phytokarst. *Zeitschrift für Geomorphologie* 26, pp. 437–57. · Folk, R.L., Roberts, H.H. and Moore, C.H. 1973: Black phytokarst from Hell, Cayman Islands. *Bulletin of the Geological Society of America* 84, pp. 2351–60. · Viles, H. 1984: A review of biokarst. *Progress in physical geography* 8, pp. 523–43.

**phytosociology** In its widest sense, the study of plants as social or gregarious organisms and thus the study of plant communities; more specifically, the floristic description, classification and naming of community types and the study of their distribution and chief ecological characteristics. The most fully developed approach to phytosociology is that of the Zürich–Montpellier school developed by Braun-Blanquet around 1913, now much refined and widely applied throughout the world. The basic unit of classification in this approach is the ASSOCIATION, which is a recurrent grouping of plant species recorded by means of sampling units termed *relevés*. In recent years, the application of quantitative methods and of computers has revolutionized the study of phytosociology. PAS

**Reading**
Chapman, S.B. ed. 1976: *Methods in plant ecology*. Oxford: Blackwell Scientific. · Harrison, C.M. 1971: Recent approaches to the description and analysis of vegetation. *Transactions of the Institute of British Geographers* 52, pp. 113–27. · Randall, R.E. 1978: *Theories and techniques in vegetation analysis*. Oxford: Oxford University Press. · Shimwell, D.W. 1971: *The description and classification of vegetation*. London: Sidgwick & Jackson.

**piedmont** An area of relatively gentle slopes and low relief flanking an upland area. The term is commonly applied to assemblages of planar alluvial surfaces flanking an area of mountains or rocky desert uplands. Desert piedmont surfaces may be underlain by sandy and gravelly ALLUVIUM tens of metres in thickness and of Quaternary and pre-Quaternary age, or may carry only an alluvial veneer 1–2 m thick. Alluvial veneers may overlie rock-cut surfaces. Commonly, piedmont surfaces are dissected by drainage systems, and the incisions made by these reveal that the piedmont surfaces often reflect a long history of AGGRADATION, pedogenesis, and incision, the patterns of which may be quite complex and variable in different areas of the piedmont (Bourne and Twidale 1998).

The development of desert piedmont surfaces by multiple episodes of aggradation and incision has been hypothesized to relate to many potential causes of changes in the run-off regime and associated erosional processes active in the uplands and upon the piedmont surfaces themselves. Triggering processes for such changes include: (1) CLIMATIC CHANGE (and associated vegetation change); (2) regional warping or tilting of the terrain; (3) deposition of dust following AEOLIAN transport – dust deposition may be involved in the development of silty surfaces that alter the infiltration properties of the surface of deposition, and so the nature of runoff generated from it; (4) pedogenesis, including the slow accumulation of carbonate materials in the sub-surface materials of the piedmont, so building up carbonate hardpans which again may alter the ability of the sediments to take in and transmit water.

Remnant surfaces of different ages within a desert piedmont often display differences in surficial rock weathering, subsoil carbonate accumulation, soil texture, and extent of stone pavement and desert varnish development as a result of differences in the age of the surfaces. DLD

**Reading**
Bourne, J.A. and Twidale, C.R. 1998: Pediments and alluvial fans: genesis and relationships in the western piedmont of the Flinders Ranges, South Australia. *Australian journal of earth sciences* 45, pp. 123–35.

**piedmont glacier** A glacier which spreads out into a piedmont lobe as it debouches onto a lowland.

**piezometer** An instrument for measuring pressure head at a point within a saturated porous medium. A piezometer consists of a tube of greater than capillary cross-section, placed in the soil or rock so that water may only enter the tube at a fixed level (usually at the base of the tube through a porous pot). Water enters the tube and the water level rises until the head of water in the tube balances the water pressure at the entry point. The depth of water in the piezometer is known as the pressure or piezometric head.                                    AMG

**Reading**
Curtis, L.F. and Trudgill, S. 1974: The measurement of soil moisture. *British Geomorphological Research Group technical bulletin* 13, Norwich: Geo Abstracts.

**piezometric surface**  An imaginary surface to which water levels rise in wells tapping confined aquifers. Confined aquifers (rock formations which contain and can yield significant quantities of water) occur where the aquifer contains GROUNDWATER which is confined at a pressure greater than air pressure by overlying, relatively impermeable rock formations. A well which penetrates a confined aquifer will experience water levels higher than the junction between the aquifer and the overlying rock formations and the piezometric surface is the surface which passes through these well water levels. If the piezometric surface rises above the ground surface a well located at that point will produce flowing water at the surface and is described as an artesian well.                                AMG

**Reading**
Walton, W.C. 1970: *Groundwater resource evaluation.* New York: McGraw-Hill Kogakusha International.

**pingo**  An ice-cored hill which is typically conical in shape and can only grow and persist in permafrost. The term is of Inuit origin but has been widely adopted elsewhere.

Pingos form through the freezing of water which moves under a pressure gradient to the site of the pingo. If water moves from a distant elevated source, the pingo is hydraulic (i.e. open system) in nature. If water moves under pressure arising from local permafrost aggradation and associated pore water expulsion, the pingo is hydrostatic in nature (i.e. closed system). The greatest concentration, about 1450, and some of the largest in the world, occur in the Mackenzie Delta region of Canada where they commonly form in drained lake basins.                          HMF

**Reading**
Mackay, J.R. 1979: Pingos of the Tuktoyaktuk Peninsula area, North-west Territories. *Géographie physique et quaternaire* 23, pp. 3–61.

**piosphere**  (or piosphere effect) This was used by Andrew (1988) to describe the spatially variable impact on the environment of animals using a point (usually well or borehole) water source. The term is applicable to situations in DRYLANDS and SAVANNAS where the development of ranching has occurred in environments with no surface water to support livestock. By tapping groundwater, and given that animals need to drink and therefore have to focus their movements on the water source, a radial pattern of pressure on the environment arises, which is greatest close to the water source, and diminishes with distance away from it. The word is derived from *pios*, Greek for 'drink'. (See also HERBIVORE USE INTENSITY.)          DSGT

**Reading and Reference**
Andrew, M.H. 1988: Grazing impact in relation to livestock watering points. *Trends in ecology and evolution* 3, pp. 336–9. · Perkins, J.S. and Thomas, D.S.G. 1993: Environmental responses and sensitivity to permanent cattle ranching, semi-arid western central Botswana. In D.S.G. Thomas, and R.J. Allison eds, *Landscape sensitivity.* Chichester: Wiley. Pp. 273–86.

**pipes**  Subsurface channels up to several metres in diameter caused by deflocculation of clay particles in fine-grained, highly permeable soils. Pipes are commonly found in arid or semi-arid regions, less commonly elsewhere. They are usually formed where the soils contain significant amounts of swelling clays such as montmorillonite, illite or bentonite, and where there is a low water table with a steep hydraulic gradient in the near-surface environment. Locally, pipes are usually found in steep slopes and on gully and arroyo sides (Heede 1971). The pipes carry sediment as well as water, and if erosion continues for a long enough period the conduit may enlarge so much that the roof collapses, forming a gully. Pipes are economically important because they are a sign of deteriorating soil conditions and represent accelerated erosion.
                                         WLG

**Reading and Reference**
Heede, B. 1971: *Characteristics and processes of soil piping in gullies.* US Department of Agriculture Forest Service research paper RM-68. · Jones, J.A.A. 1981: *The nature of soil piping: a review of research.* British Geomorphological Research Group monograph 3. Norwich: Geo Books.

**pipkrakes**  See NEEDLE ICE.

**pisoliths**  Spherical rock particles of around 5–6 mm in diameter which are formed by the gradual accretion of material around a nucleus. Laterites and calcretes often display pisolithic textures.                                      ASG

**pitometer**  (or Pitot tube) A means of estimating flow velocity by measuring the pressure of the fluid as it hits an immersed rounded body. It was invented by Henri Pitot in 1732.     AMG

**pixel**  The area corresponding to one number in a digital image. Pixel is a contraction of the word PIcture ELement and is the basic unit for storing and displaying a digital data set, such as a satellite image, digital elevation model, or computer-graphics image. Such images are composed of a 2-dimensional grid, which typically defines thousands to millions of pixels. Each pixel will have a single number assigned to it that represents the parameter of interest for that area, such as reflectance in the case of a satellite image, or height above sea level for a digital elevation model. Individual pixels are not normally visible to the human eye unless the image has been considerably enlarged.

Note that the pixel is simply a convenient data structure and is not usually the framework within which the actual measurements were made. A height of 100 m above sea level for a pixel within a digital elevation model does not mean that the whole area of the pixel is exactly at this height, but that this is a fair approximation.

Pixels should be described in areal units (e.g. $m^{-2}$), although it is far more common for the term pixel size (or resolution) to be taken as its X or Y dimensions. The pixel size of a satellite image from Landsat THEMATIC MAPPER, for example, is commonly given as 30 m, which means that each pixel is 30 by 30 m in dimensions giving a ground area of $900\,m^2$.  CDC

**planation surface**  See EROSION SURFACE.

**plane bed**  Applied to the deformable bed of a fluid flow on which there are no organized BED-FORMS and where the relief is that of individual grains. Such quasi-flat beds can occur when sediment is hardly moving under close-to-threshold conditions of fluid shear (lower stage), and under high shear and high transport rate conditions (upper stage). The latter may be a transitional stage which develops when the Froude number approaches unity and dune forms are destroyed before upper regime bed-forms develop.  JL

**Reading**
Allen, J.R.L. 1970: *Physical processes of sedimentation*. London: Allen & Unwin.

**planimeter**  An instrument for the measurement of areas on maps and plans, which is less time consuming and more accurate than counting squares or estimation but which is less efficient than digitizing methods associated with a mainframe or microcomputer.  KJG

**plankton**  Small freshwater and marine organisms, a substantial number of which are microscopic. While some plankton possess limited mobility, many are inactive. The movement of plankton mainly depends upon the motion of the water in which they are suspended.

There are three major planktonic categories: phytoplankton, zooplankton and bacterioplankton. Phytoplankton (algae) account for the bulk of primary production in aquatic ecosystems, and their BIOLOGICAL PRODUCTIVITY is conventionally high (Barnes 1980). Zooplankton, which include mature and/or larval representatives of numerous important animal groups (e.g. Protozoa, Crustacea and Mollusca), may be herbivores, carnivores or omnivores, either filtering or seizing living planktonic or detrital organic matter (Parsons *et al.* 1977; Parsons 1980). Bacterioplankton (for instance, *Bacillus* and *Nitrosomonas*) are mainly decomposers. Some are able to perform photosynthesis and chemosynthesis, thereby contributing to primary production (Fogg 1980).

The distribution and productivity of plankton vary in both space and time. This is due to a combination of environmental factors, among which nutrient availability and climate are of considerable importance.  RLJ

**Reading and References**
The three following references appear in R.S.K. Barnes and K.H. Mann eds 1980: *Fundamentals of aquatic systems*. Oxford: Blackwell Scientific: · Barnes, R.S.K. 1980: The unity and diversity of aquatic systems. Pp. 5–23. · Fogg, G.E. 1980: Phytoplanktonic primary production. Pp. 24–45. · Parsons, T.R. 1980: Zooplanktonic production. Pp. 26–66. · Moss, B. 1980: *Ecology of fresh waters*. Oxford: Blackwell Scientific. · Parsons, T.R., Takahashi, M. and Hargrave, B. 1977: *Biological oceanographic processes*. 2nd edn. Oxford and New York: Pergamon Press.

**plastic limit**  The water content of an unconsolidated material when it is at the point of transition from a plastic solid to a rigid mass.

**plasticity**  The behaviour under stress of weak materials such as moist clays and weak rocks. Such weak materials do not deform under very low magnitudes of stress, but above a critical magnitude, called a yield stress, they deform at a continuous rate if the level of stress is constant: materials exhibiting such behaviours are said to be plastic substances.  MJS

**Reading**
Selby, M.J. 1993: *Hillslope materials and processes*. 2nd edn. Oxford: Oxford University Press. Ch. 4.

**plate tectonics**  A theory of global TECTONICS that holds that the LITHOSPHERE forming the earth's surface is divided into eight major and several minor internally rigid plates which are in motion with respect to each other and the

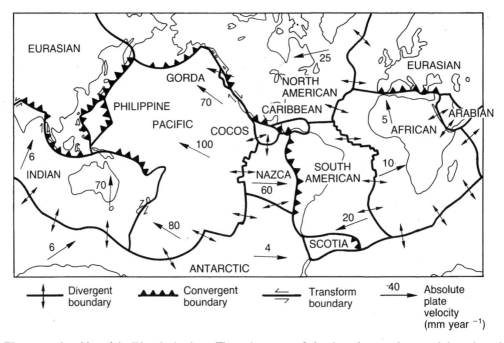

**Plate tectonics.** Map of the lithospheric plates. The various types of plate boundary are shown and the estimated current rates (mm year $^{-1}$) and directions of plate movements are indicated.
*Source*: M.A. Summerfield 1991: *Global geomorphology*. London and New York: Longman Scientific and Technical and Wiley.

underlying ASTHENOSPHERE. CONTINENTAL DRIFT is a consequence of plate motion, and earthquakes, volcanoes and mountain building are concentrated in the vicinity of, although are not entirely confined to, plate boundaries.

There are three main types of plate boundary. Divergent boundaries, which are mostly located along the extensive ridge system of the ocean basins, represent the sites of sea-floor spreading where new lithosphere is created and two plates move away from each other. Convergent plate boundaries occur when two plates move towards each other. This leads to the subduction (reabsorption into the sublithospheric mantle) of one of the plates as it plunges down under the leading edge of the other. Sites of subduction are marked by deep ocean trenches and are associated with intense seismicity and volcanic activity. Plate subduction involving only oceanic lithosphere leads to ISLAND ARC formation, but where subduction occurs along the margin of a continent a mountain belt develops (such as the Andes). If two converging plates are capped by continental crust the two continental masses will eventually collide and subduction will be halted. A complex and extensive zone of crustal deformation results, as exemplified by the Himalayas and the Tibetan Plateau which have been created as a result of the collision of India and Eurasia. The third major category of plate inter-

action occurs along a transform boundary where two plates slip horizontally past each other, such as along the San Andreas Fault System in California. This type of boundary is characterized by numerous earthquakes but low levels of volcanic activity.

Although convection currents in the mantle are clearly involved in plate motion, it does not seem that they are the main driving force. The most important mechanism is probably the pull exerted on the rest of a plate by those parts being actively subducted. Although the plate tectonics model has revolutionized our understanding of the oceans, research over the past decade has emphasized that it provides only a generalized understanding of the morphology and structure of the continents.                                                    MAS

**Reading**
Cox, A. and Hart, R.B. 1986: *Plate tectonics: how it works*. Palo Alto: Blackwell Scientific. · Kearey, P. and Vine, F.J. 1990: *Global tectonics*. Oxford: Blackwell Scientific. · Molnar, P. 1988: Continental tectonics in the aftermath of plate tectonics. *Nature* 335, pp. 131–7. · Summerfield, M.A. 1991: *Global geomorphology*. London and New York: Longman Scientific and Technical and Wiley.

**plateau**   An elevated area of relatively smooth terrain, frequently separated from adjacent areas by steep slopes. Plateaux are often composed of horizontally bedded rocks, and vary in size from sub-continental features such as the Deccan Plateau of India to small mesas, as in the American south-west. PSh

**plateau basalt** An extensive flow or flows of basalt rock which, owing to erosion of the surrounding less-resistant rocks, forms an upstanding plateau (e.g. the Deccan of India).

**playa** A closed depression in a DRYLAND area that is periodically inundated by surface water. The term is also used to refer solely to the salt flat that may occupy such a depression. The term is derived from the Spanish for beach and is thus incorrectly used by English-speaking geomorphologists. Terms such as pan, chott and kavir are used for the same features in some parts of the world.

Playas are highly variable in form and in terms of the sediments that they accumulate, but several general characteristics can be determined (Shaw and Thomas 1997):

1 occupy regional or local topographic lows;
2 lack surface outflows;
3 ephemerally, not permanently, occupied by surface waters;
4 usually have very flat surfaces;
5 have a hydrological budget in which evaporation greatly exceeds inputs;
6 are usually vegetation free, or if vegetation occurs, have distinct vegetation assemblages. DSGT

**Reference**
Shaw, P.A. and Thomas, D.S.G. 1997: Pans, playas and salt lakes. In D.S.G. Thomas ed., *Arid zone geomorphology: process, form and change in drylands*. Chichester: Wiley. pp. 293–317.

**Playfair's law** In a book of 1802 J. Playfair suggested that every river will flow in a valley proportional to the size of the river and that where rivers join their levels will be accordant. This law of accordant tributary junctions came to be known as Playfair's law. KJG

**Reading**
Kennedy, B. 1984: On Playfair's law of accordant tributary junctions. *Earth surface processes and landforms* 9, pp. 153–73. · Playfair, J. 1802; *Illustrations of the Huttonian theory of the earth*. Edinburgh: William Creech.

**Pleistocene** The Pleistocene is the first epoch of the Quaternary, preceded by the Pliocene and succeeded by the Holocene. The Pleistocene was composed of alternations of great cold (glacials, stadials) with stages of relatively greater warmth (interglacials, interstadials), during which worldwide sea levels fluctuated in response to the formation and melting of ice sheets. In glaciated regions, these eustatic changes were accompanied by isostatic depression under the weight of ice cover during glaciations and recovery during the interglacial phases.

The classic interpretation of the history of the Pleistocene, especially in the northern hemisphere, has been based on the study of the extent and character of these alternations of glacial and interglacial deposits on land. However, there is a marked degree of controversy over the number of glaciations, stadials, interglacials and interstadials. This is due to the problem of definition of these events, and also a lack of agreement with regard to correlations of events between different areas.

Until recent times, Pleistocene events were recorded in a chronology based on the location of evidence in geological or archaeological strata. Evidence was dated by correlation with known successions based on the typology, stratigraphy or prehistoric cultures. Increase in the use of new dating techniques and in deep sea core evidence have transformed Pleistocene studies. The traditional view from terrestrial studies indicated that four, five or possibly six glacials existed during the Pleistocene. Indications from ocean cores are that there have been no less than seventeen glacial cycles in the past 1.6 million years.

However, stratigraphic terminology is still to a great extent understandized, and often no clear distinction is made as to whether the classification system is: lithostratigraphic (based on rock or sediment classification), biostratigraphic (based on the occurrence of fossil fauna and/or flora), chronostratigraphic or a combination of these. A major point of contention in attempts to construct a framework for dating subdivisions of the Pleistocene is the location of the Pleistocene / Pliocene boundary and, as a corollary, the duration of the Pleistocene itself. Some authors suggest a short time-scale (600,000 years); some favour a medium time-scale, e.g. on faunal grounds the Pleistocene / Pliocene boundary has been placed at about 1.6 million years and thus conveniently coincides with a major geomagnetic reversal (the top of the Olduvai event). Others accept the long time-scale (up to 3 million years) on the basis of some major climatic deterioration, namely the marked appearance of mid-latitude, as opposed to polar, glaciers. The variations appear in the main to be due to differing forms of evidence, their interpretation and geographical location, especially marine versus terrestrial evidence. The boundary between the Pleistocene and the Holocene (oxygen stage 1) is arbitrary but is generally regarded as having occurred near 10,000 BP. AP

**Reading**
Goudie, A.S. 1992: *Environmental change*. 3rd edn. Oxford: Clarendon Press.

**plinian eruption**  An explosive volcanic eruption which is frequently so violent that the volcanic cone is destroyed.

**plots, erosion / run-off**  Plots are small areas of the landscape isolated from their surroundings by low walls that are embedded in the soil. The goal of this isolation is to be able to study the hydrologic and erosional response of the enclosed area and to apply the understanding gained from this to the larger landscape beyond. Plots receive no run-off or eroded sediment from areas upslope, so that they cannot truly encapsulate all of what would ordinarily take place over the area that they cover. On the other hand, many INTERRILL processes of the kind studied through the use of plots take place to a degree in isolation from upslope contributing areas, since the landscape is broken up into smaller sub-catchments by microrelief. A rill, for example, may be a response to run-off from a local contributing area and not strongly linked to areas more than a few metres distant. Therefore, the behaviour of part of this contributing area can appropriately be studied by plot methods.

Small plots (having sizes of a few square metres) have the advantage that their hydrological response involves little delay between the generation of run-off on the plot and its arrival at a measuring or sampling structure located on the downslope edge of the plot. Likewise, there is limited scope for sediment eroded on the upper parts of the plot to be redeposited on the lower margins. Progressively larger or longer plots (up to hundreds of square metres) increasingly involve the possibility of significant delays in run-off response or of sediment deposition before the plot border is reached. This means that derived data may need to be processed using a flow routing procedure. It also means that the plot ceases to exhibit the very simple behaviour that can be revealed on small plots.

Plot methods are widely employed in agricultural research, where they are appropriate in the investigation of the effects on soil erosion of factors such as varying styles of tillage, mulching of the soil, or fertilizer application (Andreu *et al.* 1998; Ghidey and Alberts 1998). They are also used in fundamental research, in work designed to understand the effects of changing gradient, plant canopy cover, soil type, and other factors. The advantage that small plots possess in this kind of work is that a plot can exhibit quite uniform characteristics of soil and gradient,

whereas a larger area (such as a small experimental drainage basin) would certainly not do so. This simplifies and facilitates the interpretation of derived data, and because of this, very large amounts of data derived from experimental plots have been used in the development and calibration of tools such as the USLE (universal soil loss equation). Plots are frequently applied to the assessment of hydrologic parameters such as infiltration rate. In this application, they have the advantage of being larger and potentially more representative than small areas tested by other techniques such as cylinder infiltrometry. Moreover, plots can be operated under simulated rain, so that soil crusting and sealing phenomena can be more realistically maintained than in infiltration trails using shallow ponds.

DLD

**Reading and References**
Andreu, V. Rubio, J.L. and Cerni, R. 1998: Effects of Mediterranean shrub cover on water erosion (Valencia, Spain). *Journal of soil and water conservation* 53, pp. 112–20. · Ghidey, F. and Alberts, E.E. 1998: Runoff and soil losses as affected by corn and soybean tillage systems. *Journal of soil and water conservation* 53, pp. 64–70. · Lal, R. 1994: *Soil erosion research methods*. 2nd edn. Delray Beach, Florida: St Lucie Press.

**plucking**  A process of glacial erosion describing the removal of discrete blocks of bedrock. It is commonly contrasted with the other main form of glacial erosion, ABRASION, which describes the process of rock wear. Plucking results from failure of the rock along joint planes and reflects two processes. The first is wedging by the pressure of over-riding rock particles. The second is the freezing of blocks to over-riding glacier ice in response to temperature fluctuations at the ice–rock interface as a result of pressure variations.  DES

**Reading**
Addison, K. 1981: The contribution of discontinuous rock-mass failure to glacier erosion. *Annals of glaciology* 2, pp. 3–10. · Röthlisberger, H. and Iken, A. 1981: Plucking as an effect of water pressure variations at the glacier bed. *Annals of glaciology* 2, pp. 57–62.

**plume**  See MANTLE PLUME.

**pluton**  A mass of rock which has solidified underground from intrusions of magma. Plutons have variable shapes, sizes and relationships with the country rock (the invaded rock) surrounding them. Batholiths, dykes, laccoliths, lopoliths, sills and stocks are the main forms.

**plutonic**  Refers to rock material that has formed at depth (e.g. igneous rocks such as granite) where cooling and crystallization have occurred slowly.

**plutonism** A term, at first used in derision, to describe the ideas of James Hutton in the late eighteenth century. In 1785 Hutton discovered that in Glen Tilt, Perthshire, granite veins were breaking and displacing local rocks and he postulated that granites were formed by the solidification of molten material intruded into the crust from the earth's hot interior. He invoked a similar igneous origin for basalt, and claimed that all sills, dykes and mineral veins had been filled with molten material rising from deep inside the earth. In contrast to the Neptunists he regarded many igneous rocks as younger than the surrounding strata. He relied heavily upon the power of subterranean heat to form rocks and to raise continents.                               ASG

Reading
Davies, G.L. 1969: *The earth in decay*. London: Macdonald.

**pluvial** Time of greater moisture availability, caused by increased precipitation and/or reduced evapotranspiration levels. Pluvials caused many lake levels in the arid and seasonally humid tropics to be high at various times in the Pleistocene and early Holocene (hence pluvials may also be called lacustrals), helped to recharge groundwater, and caused river systems to be integrated. Pluvials used to be equated in a simple temporal manner with glacials, but this point of view is no longer acceptable.            ASG

Reading
Goudie, A.S. 1992: *Environmental change*. 3rd edn. Oxford: Clarendon Press, Ch. 3. · Street, F.A. 1981: Tropical palaeoenvironments. *Progress in physical geography* 5, pp. 157–85.

**pluviometric coefficient** The ratio between the mean rainfall total of a particular month and the hypothetical amount equivalent to each month's rainfall were the total to be equally distributed throughout the year.

**pneumatolysis** See HYDROTHERMAL ALTERATION.

**podzol** In the traditional 'Great soil groups' system of taxonomy, a soil which has a distinctive and strongly leached upper HORIZON underlain by a genetically linked horizon strongly enriched in the materials carried downwards. The term is derived from Russian *pod* meaning 'under' and *zola*, 'ash', and refers to the distinctively pale, leached A horizon found in podzols. The translocated materials that accumulate in

the B horizon may include organic matter, carbonates, iron and aluminium. Since they require intense mobilization of materials, podzols are characteristic of soils where acid leachates can dissolve and carry substances downward. The leachates are organic acids derived from a surface litter layer, and are most active in permeable, sandy or loamy parent materials. Podzols are included in the Spodosol Order of the US system of soil taxonomy (from the Greek *spodos*, 'wood ash'), where the distinctive *spodic* horizon is one where amorphous organic matter and aluminium, with or without iron, have accumulated. (See LEACHING.)            DLD

Reading
United States Department of Agriculture, Soil Conservation Service. 1988: *Soil taxonomy*. Florida: Krieger Publishing Co.

**point bar deposits** Sediments laid down on the inside of a meander bend or 'point' largely by LATERAL ACCRETION. Individual attached BARS commonly form low arcuate ridges or scrolls. Units of accretion are added as the meander loop develops, eventually making up a complex of ridges separated by depressions or swales. The layout of the individual scroll bars may reveal the growth pattern of the point bar as a whole. In some environments, however, this pattern is either irregular or not even apparent at all on the inside of meander bends. The extent of point bar development depends on the amount of blanketing OVERBANK DEPOSIT and on the variable nature of the point bar accretion in conditions with contrasting sediments and river regimes.            JL

Reading
Reading, H.G. ed. 1986: *Sedimentary environments and facies* 2nd edn. Oxford: Blackwell Scientific.

**polar front** The front separating air of polar origin from that originating within the subtropics. In winter it can often be traced as a band of cloud over thousands of kilometres between 40 and 50° latitude especially over the oceans. Extra-tropical cyclones may be initiated along the strong thermal gradient of the front. Bjerknes (1921) based his theory of frontal evolution upon the presence of this front. In summer the front is more variable in its location and the temperature gradient is weaker.            PS

Reference
Bjerknes, V. 1921: On the dynamics of the circular vortex with applications to the atmosphere and atmospheric vortex and wave motions. *Geofysiske Publikationer* 2.4.

**polar wander** The progressive apparent shifting of the location of the earth's magnetic poles. Polar positions at past epochs are estimated from the dip angle of magnetic minerals in igneous rocks of known age, low dip angles indicating cooling at low latitudes and high dip angles, high latitudes. The polar wander that results when the position of the pole at a series of past times is plotted is largely an *apparent* wander, rather than an actual motion of the magnetic pole. The explanation is rather that the rocks recording former locations of the pole have themselves been moved by plate tectonic displacements. The magnetic poles do move by small amounts, but since the geomagnetic field has its origin is processes linked to the rotation of the planet, the magnetic and geographic poles do not diverge as widely as apparent polar wander curves suggest.          DLD

**Reading**
McElhinny, M.W. 1973: *Palaeomagnetism and plate tectonics*. Cambridge: Cambridge University Press. · Tarling, D.H. 1971: *Principles and applications of palaeomagnetism*. London: Chapman & Hall.

**polder**  A low-lying area of land that has been reclaimed from the sea or a lake by artificial means and is kept free of water by pumping.

**polje**  An extensive depression feature in karst, closed on all sides, mostly with an even floor, a steep border in places and a clear angle between the polje bottom and the slope. It has underground drainage and can be dry all year round, have ephemeral streams within it, or be inundated continually. The very diverse nature of poljes prevents a definition based on genesis; they are truly polygenetic features.          PAB

**Reading**
Gams, I. 1977: Towards a terminology of the polje. *Proceedings of the Seventh International Congress of Speleology*, Sheffield. Pp. 201–2.

**pollen analysis**  The scientific analysis of microscopic pollen grains and spores preserved in sediments, pollen analysis, or palynology, is utilized principally in the reconstruction of former vegetation types and, by extension, palaeoenvironments. The technique was first used in Sweden in 1916 by Lennart von Post, who recognized that relatively large numbers of pollen grains were preserved in Quaternary lake and mire sediments as microfossils. Subsequent statistical analysis of the inventory of various pollen types recorded with the aid of a microscope then facilitated a reconstruction of the relative proportions of the various plant taxa that produced the pollen types over the period of deposition. Today, pollen analysis is widely applied to Quaternary sediments in particular, but is utilized in geologically older sediments, as

well as in applications which deal with, for example, the pollen content of the contemporary atmosphere in allergy studies and even in forensic science. Pollen analysis has proved especially useful in reconstructing the increasing level of human impact on vegetation communities and in elucidating the changes in plant distribution consequent upon climate change over the late Quaternary.

Pollen analysis is based on the following principles:

1 all seed-plants (ANGIOSPERMS and gymnosperms) produce pollen as the male gametophyte.
2 many plant species produce large quantities of pollen and only small proportions are used in reproduction so that much is released into the environment where it may accumulate in sediments;
3 pollen grains, because of the chemistry of sporopollenin (the substance from which the pollen grain wall, or exine, is constructed), are extremely resistant to decay (especially, but not exclusively, in anaerobic conditions) and may be found in deposits in which other microfossils have been destroyed;
4 because of their small size (5 to 200 $\mu$m), pollen grains are widely and relatively evenly dispersed;
5 the pollen grain is an indicator of the plant which produced it because its size, structure, morphology and sculpturing is taxonspecific;
6 quantities of pollen become preserved in sediments accumulating over time so that analysis of grains from particular layers of sediment yields information about the nature of the vegetation types in the surrounding area at the time of deposition;
7 the quantities of pollen accumulating in sediments is such that a statistical approach to data analysis is possible; pollen frequency data may be compiled into pollen diagrams and subject to various multivariate statistical analyses to reveal additional information about trends over time.

These principles have led to the wide application of pollen analysis in the reconstruction of palaeoenvironments and the technique is arguably the most important in QUATERNARY ECOLOGY. There are, nevertheless, several methodological problems that need to be considered. Among these is the fact that pollen is produced in varying quantities by different taxa, for example, wind-pollinated (anemophilous) plants produce pollen in greater quantities and disperse it

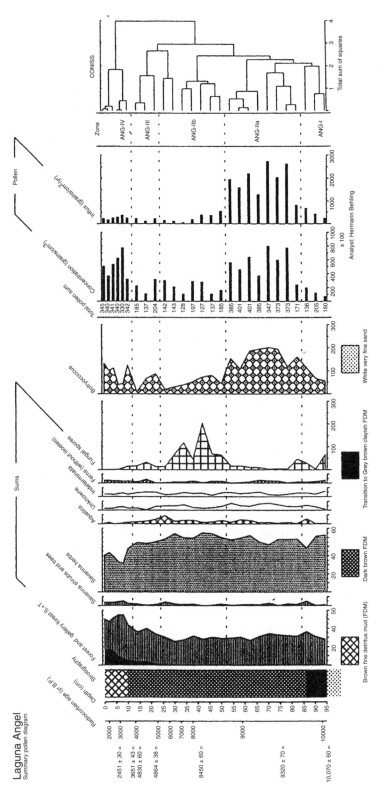

**Pollen analysis.** Pollen diagram for Laguna Angel, Colombia. This is a summary pollen diagram (i.e. simplified to show changes in ecological groups rather than individual taxa) and displays many of the features typical of this kind of representation of pollen data. For example, the x axis shows the frequency of the various pollen taxa as percentages, the y axis represents depth and age, stratigraphy is shown in stylized form on the left of the diagram, statistical manipulation producing the various pollen assemblage zones is shown on the right. *Source:* H. Behling and H. Hooghiemstra 1998: Late Quaternary palaeoecology and palaeoclimatology from pollen records of the savannas of the Llanos Orientales, Colombia. *Palaeogeography, palaeoclimatology, palaeoecology* 139, pp. 251–67.

over much greater distances than do those pollinated by insects or birds (zoophilous). Moreover, some pollen grains are more resilient than others, so that pollen diagrams require careful interpretation in the light of information, if possible, about contemporary pollen productivity and preservation. Although pollen is taxonomically specific, it is not always feasible, especially with fossil grains, to determine the precise taxonomic group beyond genus level, hence reducing the degree of resolution possible in the vegetation reconstruction.

Ultimately, the pollen diagram (see figure) is the fundamental interpretive tool in pollen analysis. Pollen data for the various stratigraphic layers are arranged in chronological sequence and plotted either as percentages or, in the case of so-called absolute pollen data, as pollen concentrations. Changing pollen frequencies over time are identified as distinctive pollen assemblage zones and these are then used as the basis of the reconstruction of the vegetation history of the site in question. Key to the analysis is the development of a precise chronology; in the case of late Quaternary sediments (the most common application of pollen analysis), RADIOCARBON DATING is frequently applied to resolve the chronology.

Although initially applied only to localities subject to cool temperate climates favouring high moisture contents and the development of anaerobic conditions within the sediments, pollen analysis is now successfully employed in a wide range of other types of environment. For example, pollen grains are preserved in lake, pan and fluvial sediments in arid environments, in marine sediments at all latitudes, in cave sediments and in COPROLITES. Even in the tropics, where prolific plant species diversity makes accurate pollen taxonomy difficult and where warm, humid conditions were thought to promote pollen decay, the technique has been applied to great effect.                    MEM

**Reading**
Faegri, K. and Iversen, J. 1989: *Textbook of pollen analysis*. London: Wiley. · Lowe, J.J. and Walker, M.J.C. 1997: *Reconstructing Quaternary environments*. 2nd edn. London: Longman. Pp. 163–75. · Moore, P.D., Webb, J.A. and Collinson, M.E. 1991: *Pollen analysis*. 2nd edn. Oxford: Blackwell Scientific Publications.

**pollution** A condition which ensues when environmental attributes become inimical to the normal existence of living organisms. A contaminant is a substance foreign to an environment and capable of pollution within it. A contaminant has a source from which it is dispersed, usually by means of an atmospheric or aquatic pathway. During this process it may be rendered harmless by transformation or dilution. If this does not occur, the contaminant becomes a pollutant which has a target (Holdgate 1979; Newson 1992). As Mellanby (1972) states, while there are numerous instances of natural pollution (volcanic emission which becomes toxic and inhibits the development of vegetation in the vicinity, for example), that resulting from human activity is the more significant.

Mature organisms are better able to cope with harmful effects than are young ones. However, as Bailey *et al.* (1978) point out, a substance may only need to reach a concentration of 1 ppm to become a pollutant, the presence of which could ultimately lead to the death of an organism. As pollutants are earth materials, they comprise part of a finite quantity. Thus their components may be changed from one state or position to another but not obliterated (Jørgensen and Johnsen 1981). When change is possible, dilution in air or water (of pesticides, heavy metals and toxic gases, for example), or degradation on land (of garbage and sewage, for instance) is normally involved. Some pollutants, though (certain nuclear wastes and lethal chemicals, for example), are so hazardous and/or of low degradability that they must be sealed and interred rather than released to the environment.

Pollution can occur at a variety of scales and in numerous circumstances. Atmospheric pollutants, for example, may give rise to serious local conditions, and can also be circulated widely. Smog is a localized atmospheric condition formed by the combination of pollutants (such as carbon monoxide and sulphur compounds) and fog. More widespread effects on the atmosphere can be brought about by $CO_2$, together with sulphur and nitrogen oxides (see ACID PRECIPITATION; GREENHOUSE EFFECT). In a similar vein, an increase in particulate concentration (by dusts, for example) lessens atmospheric transparency and affects the reflectivity of solar radiation.

Aquatic pollution can result from the addition of harmful substances such as acids or hydrocarbons. However, the gradual build-up of essential elements in freshwater subsequent to their application as terrestrial agricultural fertilizers (EUTROPHICATION), may also pollute. In the terrestrial environment, the major pollutant by volume is urban-industrial refuse which, if treated, is either stored, or reduced – usually by BIODEGRADATION and burning.

Pollution is a significant and developing environmental problem. None the less, specific instances involving potentially harmful substances and circumstances often lead to disagreement. As Barbour (1983) notes, fact is frequently obscured by conjecture, while there are sometimes political and socioeconomic undertones to particular cases. Some such factors may be exemplified with reference to noise pollution, much of which (up to the 120 decibel limit when human pain is felt) is a rather subjective experience.

Pollution control strategies vary. Control at source has been preferred in the United States and on mainland Europe, while in the United Kingdom it has been customary to manage the pathways of pollutants (Newson 1992). The law of England and Wales sets out few exact criteria pertaining to pollution control across the entire territory. Examples of legislation are the Control of Pollution Act 1974 which governs terrestrial waste disposal, and the Clean Air Acts (1956 and 1968) relating to air pollution. Enforcement of the controls has been largely entrusted to bodies (County Councils and Water Authorities for example) with local and regional responsibility for the environment (Macrory 1990). RLJ

**Reading and References**
Bailey, R.A., Clark, H.M., Ferris, J.P., Krause, S. and Strong, R.L. 1978: *Chemistry of the environment.* New York and London: Academic Press. · Barbour, A.K. 1983: The control of industrial pollution. In R.M. Harrison ed., *Pollution: causes, effects and control.* London: Royal Society of Chemistry. Pp. 1–18. · Holdgate, M.W. 1979: *A perspective of environmental pollution.* Cambridge: Cambridge University Press. · Jørgensen, S.E. and Johnsen, I. 1981: *Principles of environmental science and technology.* Amsterdam and New York: Elsevier. · Macrory, R. 1990: The legal control of pollution. In R.M. Harrison ed., *Pollution: causes, effects and control.* 2nd edn. Cambridge: Royal Society of Chemistry. Pp. 277–96. · Mellanby, K. 1972: *The biology of pollution.* London: Edward Arnold. · Newson, M. 1992: The geography of pollution. In M. Newson ed., *Managing the human impact on the natural environment: patterns and processes.* London and New York: Belhaven Press. Pp. 14–36.

**polyclimax** A theory of vegetation that allows the coexistence of a number of stable plant communities in an area. According to polyclimax theory, all the SERES (community sequences) in an area do not converge to identify in a MONOCLIMAX, but SUCCESSION produces a partial convergence to a mosaic of different stable communities in different HABITATS. In polyclimax theory, all climax types are of equal rank rather than subordinate to the climatic climax as is required by monoclimax theory. (See also CLIMAX VEGETATION.) JAM

**polygonal karst** A limestone landscape entirely pitted with depressions which are smooth rimmed and soil covered, producing a crude polygonal network when viewed from the air. The term was introduced by Williams (1971) when reporting the features from New Guinea. PAB

**Reference**
Williams, P.W. 1971: Illustrating morphometric analysis of karst with examples from New Guinea. *Zeitschrift für Geomorphologie* 15, pp. 40–61.

**polymorphism** Existing in more than one physical form. Applied to minerals which have the same chemical composition but different physical characteristics, to animals that undergo major physical alterations during individual life cycles and to animal and plant species which, while interbreeding, have markedly different physical forms.

**polynya** A pool of open water within pack ice or an ice floe.

**polypedon** A collection of small columns which run through the soil zone.

**polytopy** Loosely used, a term to describe the occurrence of any organism in two or more completely separate geographical areas; more specifically, a term referring to the process by which an organism may evolve two or more times, quite independently, in differing geographical localities. If the polytopic populations have developed at different times they are also termed polychronic in origin. The term is primarily employed by plant geographers and, with our present understanding of genetics and evolution, such an explanation for a disjunct distribution would only be accepted in very exceptional circumstances. A classic example of polytopy is afforded by the separate, but closely related, sand dune ecotypes of *Hieracium umbellatum* found along the coasts of Sweden. (See also DISJUNCT DISTRIBUTION.) PAS

**Reading**
Stott, P. 1981: *Historical plant geography.* London: Allen & Unwin.

**pool and riffle** The pool and riffle sequence is a large-scale bedform characteristic of streams with gravelly, heterogeneous bed material. Pools are closed hollows scoured in the bed and commonly floored by relatively fine gravel and sand, while riffles are topographical highs representing accumulations of coarser pebbles and cobbles. These features are created by the pattern of scour and deposition at bankfull discharge, when bed velocity is higher in the pool than in the adjacent riffle, and the coarse sediment then in motion is removed from the pools and deposited on the riffles. At low discharges the flow adjusts to the bed topography, and the water surface slope is flat over the deep, sluggish pools and steep over riffles where flow is shallow and rapid. Fine sediment mobile at this flow stage is removed from the riffles and deposited in pools. The pool–riffle sequence repeats with a mean wavelength of five to seven times the mean channel width, suggesting initial control by a large-scale turbulent eddy scaled to the channel size. Once formed, riffles and pools are fairly stable morphological features although individual sediment grains move through the sequence from riffle to riffle. In regular meander patterns, riffles tend to occur at inflection points and pools at bends, and the meander wavelength is twice the pool–riffle wavelength. This suggests that the pool–riffle feature is a fundamental bedform common to 'straight' and meandering rivers.                            KSR

**population dynamics**  The study of changes in population size. This involves a consideration of those factors which might give rise to population growth, and those which might lead to its decline. Growth may be achieved by an increase in the rates of natality over those of mortality, and / or by immigration; and decline normally results from an excess of mortality over natality, and / or emigration. The structure of a population may also influence its dynamics: for example, a population with a relatively high percentage of females of reproductive age is clearly in a favourable position for growth, whereas one without such a representation may not be.

The earliest theoretical studies of population dynamics date back to Malthus (1798), and the first mathematical representation of population growth, characterized as it is by a sigmoid curve, to P.F. Verhulst in Paris in 1838. The essentials of the latter are as follows. A population will establish itself slowly in a new environment and then, once it is adjusted ecologically and competitively to it, will grow rapidly within its determined NICHE. This explosive phase of growth has been noted widely among both plants and animals, and is associated with *r*-selection (see *r*- AND *K*-SELECTION). Subsequently, as it nears the BIOTIC POTENTIAL for that niche and area, environmental resistance will flatten the growth curve until it reaches an equilibrium, which is equivalent to the CARRYING CAPACITY of the area for that species. The process may be expressed by:

$$\frac{\Delta N}{\Delta t} = \gamma N \frac{(K - N)}{K},$$

in which $N$ refers to the numbers of a given population, $t$ to a given period of time, $\gamma$ to the rate of increase (as determined by the difference between specific birth and death rates for the population), and K is a constant which relates to the upper limit of population growth (in other words, the carrying capacity). Close to the top of the growth curve, most species will be associated with *K*-selection.

Once at equilibrium level populations may adopt several strategies in order to maintain themselves. In laboratories most species will keep as close to equilibrium as possible, but under field conditions many of the larger organisms will display slight cyclical oscillations around it. These are DENSITY-DEPENDENT, in the sense that when population densities become low enough for a real and permanent decline to become a possibility, in-built compensatory mechanisms (usually in the form of increased birth rates) set in to restore the balance; similar though reverse responses occur when densities become too high. The amplitude of such oscillations is normally greatest when organisms are small (e.g. reaching $\times 40,000$ the minimum in the case of locusts but only $c. \times 2$ for most birds); and, for larger animals, the periodicity customarily fits into the framework of a 4–5 year (mice, voles, foxes, lemming species on both sides of the Atlantic) or a 9–10 year (lynx, snowshoe hare) cycle. Should environmental circumstances change and organisms fail to adapt sufficiently, populations may begin to decline in numbers in a density-independent manner, seriously enough for EXTINCTION to threaten, though if the change is only temporary (e.g. an exceptionally cold winter, which may affect birds which eat freshwater organisms; or a single chemical or heat pollution event in water), such populations may still recover.   DW

**Reading and Reference**
Malthus, T.R. 1798: *An essay of the principle of population as it affects the future improvement of society.* London. (Various modern editions.)

**pore ice**  A type of ground ice occurring in the pores of soils and rocks. It is sometimes referred to as cement ice. On melting, pore ice does not yield water in excess of the pore volume, in contrast to SEGREGATED ICE. In terms of the total global ground-ice volume, pore ice probably constitutes the most important ground-ice type, primarily because of its ubiquitous distribution.                                    HMF

**Reading**
Mackay, J.R. 1972: The world of underground ice. *Annals of the Association of American Geographers* 62, pp. 1–22.

**pore water pressure**  The pressure exerted by water in the pores of a soil or other sediment.

Pressure is positive when below the WATER TABLE and negative when above it. Negative pore water pressure in soil is referred to as soil moisture tension (or suction). Pore water pressure can be measured by a tensiometer connected to a mercury manometer, vacuum gauge or pressure transducer (Burt 1978).     PWW

**Reading**
Burt, T.P. 1978: An automatic fluid-scanning switch tensiometer system. *British Geomorphological Research Group technical bulletin* 21.

**porosity**  A property of a rock or soil concerned with the extent to which it contains voids or INTERSTICES. It is usually defined as a ratio of the aggregate volume of voids to the total volume of the rock or soil, and is expressed as a percentage. A distinction is sometimes made between primary porosity, arising from intergranular interstices at the time of deposition of the rock, and secondary porosity, arising from later jointing or corrosion. All interstices whether primary or secondary are included in the estimation of the aggregate volume of voids for the purpose of measuring porosity.     PWW

**Reading**
Davis, S.N. 1969: Porosity and permeability of natural materials. In R.J.M. De Wiest ed., *Flow through porous media.* New York: Academic Press. Pp. 54–89.

**positive feedback**  See SYSTEMS.

**postglacial**  See HOLOCENE.

**potamology**  The scientific study of rivers.

**potassium argon (K/Ar) dating**  An isotopic dating technique which utilizes unaltered potassium-rich minerals of volcanic origin in basalts, obsidians, and the like. It is particularly useful for materials more than 50,000 years old.     ASG

**Reading**
Miller, G.H. 1990: Miscellaneous dating methods. In A.S. Goudie ed., *Geomorphological techniques.* 2nd edn. London: Unwin Hyman. Pp. 405–7.

**potential climax**  The mature vegetation community which develops following the removal of disturbance factors or a change in climate. The concept forms part of the elaborate, and now little-used terminology that F.E. Clements established to account for SUCCESSION towards CLIMAX VEGETATION. The potential climax vegetation is seen to replace the existing community when conditions alter due to, for example, the removal of human-induced disturbance processes or fluctuations in climate. Following a perturbation, 'secondary succession'

processes in effect produce directional and, in the Clementsian sense, predictable and progressive changes in the composition and structure of the vegetation until a new mature community forms in equilibrium with the novel environmental conditions. Potential climax, then, is a theoretical plant community in an area, taking into account possible changes in response to modifications in human interference or, possibly, climatic changes. The concept was regarded as necessary because of the universality of change and, given the ubiquity of human influences and the likelihood of climatic perturbation, a large proportion of observed plant communities have been substituted for the original vegetation, directly or indirectly, by human activities. Modern ecological thought would hold that special terminology of this sort is unnecessary, even unhelpful, in the light of the perceived dynamic nature of plant communities in time and space.     MEM

**potential energy**  The energy change when a system is reduced to some standard state, usually applied to gravitational potential energy, with mean sea level defining the reference state. The concept has very wide applicability in physics and, more particularly, in the physics of the natural environment. It has been used very frequently within meteorology for a greater understanding of the energetics and dynamics of atmospheric motion. For example, using:

$$v^2 = 2gh$$

where $v$ is air velocity, $g$ is acceleration due to gravity and $h$ is the height above a specified datum, a parcel of air in the upper TROPOSPHERE has potential energy to a speed of $450\,\mathrm{m\,s^{-1}}$. Natural parcels do not acquire such speeds because they are unable to fall freely, needing to push other air out of the way. This notion eventually defines the available potential energy and gives:

$$v^2 \simeq 2gh\delta T/T$$

where $T$ is the temperature difference, as they pass, of two parcels being exchanged between two levels in the atmosphere. Hence a more realistic magnitude of air speed is $30\,\mathrm{m\,s^{-1}}$.

JSAG

**potential evaporation**  The rate of water loss from a surface when water supply to the surface is sufficient to meet the evaporative demand. Since evaporation from a water surface will always be at the potential rate, potential evaporation is often called potential evapotranspiration and is the rate of water loss from a surface other than a water surface, through evaporation and transpiration processes and when these processes are not limited by a water deficiency. In order to ensure comparability of estimates from different areas, Penman (1956) defined potential water loss from a vegetated surface (which he called potential transpiration) as evaporation from an extended surface of 'fresh green crop of about the same colour as green, completely shading the ground, of fairly uniform height and never short of water'.　AMG

**Reading and Reference**
Allen, R.G. 1986: A Penman for all seasons. *American Society of Civil Engineers, Journal of the Irrigation and Drainage Division* 112, pp. 348–68. · Morton, F.I. 1983: Operational estimates of areal evapotranspiration and their significance to the science and practice of hydrology. *Journal of hydrology* 66, pp. 1–76. · Penman, H.L. 1956: Estimating evaporation. *Transactions of the American Geophysical Union* 37, pp. 43–6. · Ward, R.C. and Robinson, M. 1989: *Principles of hydrology*. 3rd edn. London: McGraw-Hill.

**potential evapotranspiration**　See EVAPO-TRANSPIRATION.

**potential sand transport**　See DRIFT POTENTIAL.

**potential temperature**　(θ) The temperature an air parcel would possess if it were moved from its level to a level with a pressure of 1000 mb dry adiabatically (a rate of $9.8\,K\,km^{-1}$). If ascending or descending air parcels are subject only to these ADIABATIC changes their potential temperature will remain constant. In fact, the motion of air within the atmosphere is often close to adiabatic and the θ value of a given sample of air is conserved and acts as a kind of 'label'. Thus when an unsaturated parcel of air ascends or descends it will do so dry adiabatically and will move up or down an imaginary surface of constant potential temperature. If it ascends it will cool at $9.8\,K\,km^{-1}$ and will warm at this rate in descent – so long as no CONDENSATION occurs.

Because it is a more conservative property than dry-bulb temperature it is often used to highlight frontal changes in vertical cross-section of the atmosphere. Lines of constant θ appear on THERMODYNAMIC DIAGRAMS.　RR

**pot-hole**　A deep, circular hole in the rocky bed of a river which has formed by abrasion by pebbles caught in eddies. Any vertical shaft in limestone.

**prairie**　North American term for mid-latitude grasslands, equivilent to the Eurasian STEPPES.

**Pre-Boreal**　See BLYTT–SERNANDER MODEL.

**precession of the equinoxes**　An orbital mechanism caused by the gravitational forces of the sun, moon, and other planets on the earth's equatorial bulge. Precession has two components. The first component (axial precession) causes the earth's axis of rotation to 'wobble' like a spinning top. As a result, the axis of rotation describes a circle in space. For example, the axis of rotation is currently oriented towards the North Star. In ∼13,000 years time the axis of rotation will point in the direction of the star Vega. A complete cycle takes ∼26,000 years. The second component (precession of the ellipse) changes the elliptical orbit of the earth around the sun about one focus. A complete cycle takes ∼22,000 years. The combined effect of the two precessional forces changes the calendar dates for aphelion, perihelion, the equinoxes, and the solstices. For example, today aphelion occurs near 4 July while perihelion is near 3 January. The positions of aphelion and perihelion will be reversed in ∼11,000 years. The climatic result of precession can cause a change in the distribution of seasonal insolation between the northern and southern hemispheres. The current orientation of precession enables the southern hemisphere to receive more radiation at the top of the atmosphere than the northern hemisphere. Also, the southern hemisphere currently receives as much as 10 per cent more insolation than in 11,000 years when the opposite conditions will exist.

JAS

**precipitation**　The deposition of water from the atmosphere in solid or liquid form. It covers a wide range of particle sizes and shapes such as RAIN, SNOW, HAIL and DEW. In most parts of the world rain is the only significant contribution to annual precipitation totals and the terms are frequently used synonymously. In polar regions and at high altitudes snow will be the dominant type of precipitation. The processes whereby water vapour is converted into precipitation are explained in CLOUD MICROPHYSICS.　PS

**Reading**
Sumner, G. 1988: *Precipitation: process and analysis*. Chichester: Wiley.

**predation**　The killing of one free-living animal by another for food. Technically this may involve the total removal of a species, or several species, from an environment by a predator, though in mature and/or complex communities, and in natural circumstances, it is unlikely that this would ever happen, for the predator would then have eliminated a potential food resource; moreover, most such communities possess a large number of prey species for each predator, so that the demands on any one are never too heavy. Also most predators prefer a range of different animals in their diet.

However, there is little doubt that continued predation modifies the patterns of COMPETITION in an area and often, in consequence, the local distribution of species. Through reducing their population densities predators tend to lower the competition pressure from prey species in similar NICHES; and this may result in two competitors surviving where, without predation, only one would. Further, it is possible that balanced predation may actually increase species diversities in many communities. On the Pacific Coast of North America, Paine (1966) has noted, in experiments on rock shores, that the removal of a major predator, the starfish *Pisaster ochraceus*, caused a diminution both in the number of species present in those communities (from 15 to 8), and in their functional variety (from acorn barnacles, limpets, chitons, dog whelks and mussels to predominantly barnacles and mussels, the population of the latter growing explosively). If the idea of HERBIVORES as plant predators is accepted similar consequences can be seen to attend the cessation of balanced grazing: in southern England, meadows closely grazed by sheep may have *c.*20 component species, while those which are taken out of grazing quickly establish a dominance structure in which several of the ground plants are shaded out, and not replaced.

It is considered by some authorities (e.g. MacArthur 1972) that both the number of predators per unit area, and predation pressure generally, increase towards the tropics, and in conditions of reduced physiological stress, both on the major landmasses and in the oceans. But this is not the case in tropical oceanic islands, many of which lack locally evolved predators; and in these the introduction of alien predator species has, unlike the patterns of 'natural' predation, frequently given rise to immense ecological disruption. In the West Indies the planned arrival of the mongoose, which was designed to reduce the number of snakes, also resulted in the total removal of many native birds; and the European dog, cat and rat, together and separately, have wrought similar ecological havoc in Pacific island systems. Recently, human beings themselves have directly augmented rates of predation everywhere, often in an unbalanced way.                                    DW

**References**

MacArthur, R.H. 1972: *Geographical ecology: patterns in the distribution of species.* New York: Harper & Row. · Paine, R.T. 1966: Food web complexity and species diversity. *American naturalist* 100, pp. 65–75.

**predator–prey relationships** The population and energy balances between predators and prey. An intimate relationship exists between these two groups of organisms, for while the former can easily reduce the population of the latter, they are themselves vulnerable to decline and possible EXTINCTION through starvation, should prey become too few. Accordingly, balanced predator–prey interactions depend in large measure on the effective control of the population size of, and by, both sets of participants. They are also important in the EVOLUTION of new species forms by natural selection, selection favouring the efficient predator and the elusive prey: in the latter case the development of a wide variety of cryptic and mimetic coloration in many species.

Animals which may be listed as predators include both 'true predators' and insect parasitoids (often incorrectly termed insect parasites). The latter are extremely numerous, and account for about 10 per cent of the approximately one million known insect species. Most belong to the Diptera (flies) and Hymenoptera (ants, bees, wasps) families, e.g. there are huge numbers of different species of parasitoid wasps, ranging in size from free-living forms to microscopic egg parasitoids. Most, too, are host-specific, i.e. they seek out one host species alone. Unlike true predators only the females of insect parasitoids look for hosts, and then usually only to lay eggs in or on them. The larvae which subsequently emerge feed from the host either internally or externally, but an effective energy balance between them is maintained until the larvae approach maturity, at which point the host's vital organs are eaten, and the host is killed. In this way insect parasitoids also differ from PARASITES, which tend to ensure that the host's life is secured. Relationships between the populations of both host and insect parasitoid are fairly simple in the sense that host mortality depends solely on the ability of females of the prey species, at one particular stage alone in their life cycle, to search out a host; and the reproductive rate of insect parasitoids may be seen to relate clearly to the number of hosts which are colonized. Although some insect parasitoids control the size of their host populations quite closely, this is not always the case.

In contrast, most true predators have more diffuse interactions with their prey, and this is especially the case for vertebrate predators, many of which are not prey-specific, having a wide range of preferred foods. Males, females, and often their young as well, all must search for prey throughout the year, unless they hibernate for part of it; and this is undertaken with different degrees of efficiency. Moreover, the reproductive rates of some true predators are more finely determined by the demographic characteristics of their own populations rather than by the numbers of prey, though this is not an invariable situation.

The population dynamics of the predator–prey dependency are by now fairly well known, and have been reviewed by Hassall (1976). For simple relationships the basic form is expressed by:

$$N_{t+1} = \lambda N_t f(N_t, P_t) \text{ and}$$

$$P_{t+1} = N_t[1 - f(N_t, P_t)]$$

where $N_t$, $N_{t+1}$ and $P_t$, $P_{t+1}$, are respectively the prey and predator populations in successive generations, and $\lambda$ is the finite net rate of increase of the prey (birth rate minus death rate). The predation rate at time $t$ is an unspecified function of prey and predator sizes. Nicholson (1933) considered that three further assumptions could be made about predator–prey interactions: first, that predators would search randomly for their prey without being influenced by the distribution and density of either the prey or of other potential predators; secondly, that predators' food requirements would be unlimited; and thirdly, that the area in which the search for prey is conducted (a in the equation below) is likely to be constant for a given predator population. With this in mind, and referring back to the first equation:

$$f(N_t, P_t) \text{ then} = \exp(-aP_t)$$

and the model changes to:

$$N_{t+1} = \lambda N_t \exp(-aP_t) \text{ and}$$

$$P_{t+1} = N_t[1 - \exp(-aP_t)]$$

This suggests that, first, for each predator–prey system an equilibrium level of population will exist; and, secondly, that this will be inherently unstable, for any movement away from the overall balance will lead quickly to oscillations in both populations, with that of the predator lagging behind that of the prey. In theory such oscillations may increase in size until one of the populations becomes extinct; or, they may subsequently stabilize. Stabilization can ensue, in a predetermined way, from the initial relative numerical balance between predators and prey, in which case it may be termed 'neutral stability'; or it may be derived more actively ('actively induced stability') from a density-dependent factor (see POPULATION DYNAMICS) which operates especially on the prey, from resource limitations on the prey (May 1972), or from particular homeostatic predator responses to increases in prey populations, which may be functionally or numerically based.

Of these several possibilities, unstable predator–prey oscillations have been observed in laboratory experiments involving small animals such as protozoa and mites, in simple relationships which become self-annihilating. When laboratory conditions were made more complex stable oscillations resulted and, at least in some cases, the amplitude of these declined over time: they were convergent. This latter event seems to arise from evolutionary response in both predators and prey, in which the prey became more resistant to PREDATION, and the predators in turn began to seek out the prey less assiduously. In the instance of the house fly (*Musca domestica*) and wasp parasite (*Nasonia vitripennis*) populations, these responses are known to be capable of developing in the laboratory in as few as twenty generations (Pimentel *et al.* 1963).

Although it is likely that much of the observed stability in natural biological systems is due to a long-term but similar pattern of co-evolution between predators and prey, there is very little direct evidence of this thus far. Nor has anyone yet discovered a flawless example of a maintained predator–prey oscillatory cycle in field populations. For a long time it was thought that the lynx (*Lynx canadensis*)–snowshoe hare (*Lepus americanus*) predator–prey relationship in Canada, which is centred around a 9–10 year cycle of population growth and decline (see POPULATION DYNAMICS), provided one, but this view was negated by the realization that, in those parts of Canada which were not inhabited by the lynx (especially Anticosti Island, in the Gulf of St Lawrence), snowshoe hares retained that cycle of their own accord (Keith 1963). However, some good field evidence of the general effect of predators on field populations of prey is available. Where predators have been removed entirely from large areas of land, e.g. through shooting, the numbers of prey have frequently been substantially augmented, even to a point where they could no longer be supported by the resources of the environment; the prey population then collapsed. A classic instance of this type of instability has been recorded for the Kaibab plateau of Arizona, in which deer populations soared then crashed after their main predators (wolves and cougar) had been eliminated. Conversely, the presence of predators often encourages both stability and variety among prey populations (see PREDATION): and some of the stability at least may be achieved by the periodic numerical and functional changes in predator feeding habits which are known to occur. Thus, if prey populations become too large, predators may consume more of them; and if they decline too much, predators may switch to an alternative source of food until they recover. Such switching may eventuate most frequently between prey which are almost equally preferred in a diet, and it may be most common in animals that feed in flocks (Murdoch and Oaten 1975) but it has been noted too in relatively solitary predators, such as the English tawny owl (Southern 1970). Curiously, controlled field experiments which have sought to clarify predator–prey population interactions have not yet produced standardized results; and clearly these can be influenced by many environmental factors which are difficult to quantify. But it is likely that the most stable natural predator–prey systems are those in which several species are present, particularly in respect of the prey; those in which safe refuge areas for prey may be found; and those in which the predators select a large number of prey individuals which are no longer of reproductive age.

DW

**Reading and References**
Hassall, M.P. 1976: Arthropod predator–prey systems. In R.M. May ed., *Theoretical ecology, principles and applications*. Oxford: Blackwell Scientific. · Keith, L.B. 1963: *Wildlife ten-year cycle*. Madison, Wisconsin: University of Wisconsin Press. · May, R.M. 1972: Limit cycles in predator–prey communities. *Science* 177, pp. 900–4. · Murdoch, W.W. and Oaten, A. 1975: Predation and population stability. *Advances in ecological research* 9, pp. 1–131. · Nicholson, A.J. 1933: The balance of animal populations. *Journal of animal ecology* 2, pp. 131–78. · Pimentel, D., Nagel, W.P. and Madden, J.L. 1963: Space–time structure of the environment and the survival of parasite–host systems. *American naturalist* 97, pp. 141–67. · Southern, H.N. 1970: The natural control of a population of tawny owls (*Strix aluco*). *London journal of zoology* 162, pp. 197–285.

**pressure, air** The force per unit horizontal area exerted at any given level in the atmosphere by the weight of the air above that level. At sea level the average air pressure is $14.7 \, \text{lb in}^{-2}$, $760 \, \text{mm Hg}$, $29.92 \, \text{in Hg}$, or $1013.25 \, \text{mbar}$ (or hectopascals). The air pressure decreases most rapidly with height near sea level where the air is most dense. It decreases by about 50 per cent for every 5 km of ascent. WDS

**pressure melting point** The temperature at which a liquid becomes solid at a particular pressure. The concept is fundamental to glacial geomorphology since glacier ice may exist at the pressure melting point. Since the temperature at which water freezes diminishes under additional pressure, by a rate of $1 \, °C$ for every $1400 \, \text{kPa}$ ($140 \, \text{bar}$), the melting point at depth will be below $0 \, °C$. For example, water was discovered beneath $2164 \, \text{m}$ of ice at Byrd Station in West Antarctica at a pressure melting point of $-1.6 \, °C$. DES

**Reading**
Paterson, W.S.B. 1981: *The physics of glaciers*. London and New York: Pergamon.

**pressure release** The process whereby large sheets of rock become detached from a rock mass owing to the continuing relaxation of the pressure within the mass which built up before it was exhumed. For example, the erosion of overlying sedimentary strata from above a granite intrusion may cause pressure release to open joints in the granite when it is exposed.

**prevailing wind** This term is really an abbreviation for 'prevailing wind direction' and means the wind direction most frequently observed during a given period. The periods most frequently used are days, months, seasons and years. BWA

**primarrumpf** An upwarped dome which though still undergoing uplift is being eroded at an equal rate.

**prisere** A primary successional sequence of plant communities. It is a series of communities that results from the processes of SUCCESSION on newly formed landsurfaces, e.g. emergent land at the coast, lava flows and recently deglaciated terrain. It is distinguished from a SUBSERE by the initial conditions, namely the absence of remnants of previous communities or soil.

Priseres, together with subseres, are commonly classified according to initial environmental conditions. Thus dry sites give rise to xeroseres, which include lithoseres (on rock) and psammoseres (on sand), whereas hydroseres are characteristic of wet sites and include haloseres (saline or alkaline conditions) and oxyseres (acidic conditions). (See also SERE.)

JAM

**Reading**
Matthews, J.A. 1979: The vegetation of the Storbrean gletschervorfeld, Jotunheimen, Norway. I: Approaches involving ordination and general conclusions. *Journal of biogeography* 6, pp. 133–67. · Olson, J.S. 1958: Rates of succession and soil changes on southern Lake Michigan sand dunes. *Botanical gazette* 119, pp. 125–70.

**probable maximum precipitation** The rainfall depth, for a particular size of catchment, that approaches the upper limit that the present climate can produce.

**process–response system** See SYSTEMS.

**productivity** A general term referring to the rate of BIOLOGICAL PRODUCTIVITY and accounting for both primary and secondary productivity in an ecosystem. Productivity refers to the accumulation of organic matter resulting from energy assimilation and transfer processes, including its initial entry into the system via PHOTOSYNTHESIS and its subsequent dissemination through the various TROPHIC LEVELS. There are several identifiable components to productivity, such as NET PRIMARY PRODUCTIVITY and gross primary productivity among the lower trophic levels, and secondary productivity at higher levels.                                    MEM

**profile**  See SOIL PROFILE.

**proglacial**  See ENGLACIAL.

**proglacial lake**  A lake impounded in a depression in front of a glacier. During glacial periods such lakes were well developed in front of the southern margins of the Laurentide and Eurasian ice sheets. The proglacial Lake Agassiz in the area north-west of Lake Superior had an area larger than that of the Great Lakes today and was over 1000 km long. Similar lakes were impounded in the lower valleys of the northward-flowing Asian rivers such as the Ob'.

DES

**progradation**  The extension of a shore-line into the sea through sedimentation.

**protalus rampart**  A narrow ridge, usually a metre or so high and tens of metres long in front of a mountain rock face, composed of rock fragments. It may look like a small moraine, with which they have sometimes been confused, but whereas the former are the result of glacier action a protalus rampart is generally considered to be formed by rock debris sliding over a snowpatch. They have also been called 'winter talus ridges' and 'nival ridges' but protalus rampart (Bryan 1934) is now the accepted name.  WBW

**Reading and Reference**
Ballantyne, C.K. and Kirkbride, M.P. 1986: Characteristics and significance of some Lateglacial protalus ramparts in upland Britain. *Earth surface processes and landforms* 11, pp. 659–71. · Bryan, K. 1934: Geomorphic processes at high altitudes. *Geographical review* 24, pp. 655–6. · Butler, D.R. 1986: Winter-talus ridges, nivation ridges and pro-talus ramparts. *Journal of glaciology* 32, p. 543.

**proximal trough**  A depression around a steep rock body caused by the increased velocity that can occur when a moving element (e.g. ice, water or wind) flows round an obstruction.

ASG

**Reading**
Lassila, M. 1986: Proximal troughs and ice movements in Gotland, southern Sweden. *Zeitschrift für Geomorphologie* 30, pp. 129–40.

**proxy**  A substitute or stand-in, the term proxy is used in the study of past environments to refer to preserved sedimentary or geomorphological evidence that can be used to give estimates of past climatic parameters or environmental conditions.                                    DSGT

**psammosere**  See PRISERE.

**pseudokarst**  Landforms produced in non-carbonate rocks which are morphologically similar to those normally associated with karst rocks. These non-calcareous rocks can produce features by solutional processes similar to those types of reaction found in limestone (karren features) or by processes entirely different from these (lava caves).                                    PAB

**Reading**
Warwick, G.T. 1976: Geomorphology and caves. In T.D. Ford and C.H.D. Cullingford eds, *The science of speleology*. London: Academic Press. Pp. 61–126.

**pumice**  A highly porous fine-grained volcanic rock produced when numerous gas bubbles are trapped within the lava when it solidifies.

**puna**  A cold desert, especially one at high altitude as in the Andes.

**punctuated aggradational cycles**  A stratigraphic model which states that most stratigraphic accumulation occurs episodically as thin (1–5 m thick) shallowing upward cycles separated by sharply defined non-depositional surfaces. They are created by geologically instantaneous basin-wide relative base-level rises (punctuation events), with deposition occurring during the intervening periods of base-level stability. Glacial eustatic changes driven by orbital perturbations (see MILANKOVITCH HYPOTHESIS) may be a preferred mechanism.

ASG

**Reading**
Goodwin, P.W. and Anderson, E.J. 1985: Punctuated aggradational cycles: a general hypothesis of episodic stratigraphic accumulation. *Journal of geology* 93, pp. 515–33.

**push moraine**  See MORAINE.

**pyroclastic**  Refers to fragmental rock products (e.g. ash, volcanic bombs, ignimbrite, etc.) ejected by volcanic explosions.

# Q

---

**quartz** An extremely abundant crystalline silica mineral, $SiO_2$ that is very resistant to breakdown. It is often the most common constituent of SAND and SILT deposits.

**quasi-equilibrium** The 'apparent' balance between opposing forces and resistances. In geomorphology it is applied often to the 'concept' of an apparent balance between the rate of supply, temporary storage and removal of material from a dynamic depositional landform such as a scree slope, alluvial fan, beach or dune. The geometry of the landform or depositional store is dependent on the nature of the balance, the length of time over which it is maintained and the overall volumes involved. In many cases the volume of material passing through the system may be much greater than that in the store at any one time.

The idea may be applied to a surface of transportation such as a hillslope or pediment where there is an apparently close relationship between the geometry of the debris mantle and the form of the surface over which it moves. This implies that there is a negative feedback between the balance of input–storage–output processes of the mantle and the adjustment of the bedrock surface. Thus a decrease in storage thickness may increase weathering rates on the bedrock to restore the mantle to its previous condition. In this way the balance of equilibrium conditions of the depositional layer may partially control overall landform development.

A major weakness of the concept is that the relevant time-scales, over which budgetary or geometrical relationships are achieved or maintained, are unknown. No evaluations have been made of the relaxation times involved and it is not possible, at present, to understand fully the effects of environmental change or the status of landscape relicts.                                    DB

**Quaternary** The Cainozoic era of geological time is divided into the Tertiary period, about 65–2 million years before present (Ma BP) and the Quaternary period, from about 2 Ma BP up to and including the present day. The major subdivisions of geological time have by tradition been named in accordance with fossil evidence of the stage reached in biological evolution; hence Cainozoic = 'recent life', and the epochs within it (Palaeocene, Eocene, Oligocene, Miocene, Pliocene, Pleistocene and Holocene) are derived from terms expressing the proportions of modern marine shells occurring in fossil form (e.g. 'Meion' less, 'Pleion' more, 'Pleiston' most and 'Holos' complete or whole). The Quaternary period, consisting of the PLEISTOCENE and HOLOCENE epochs, is today delimited on different criteria, and, depending upon the criterion in favour from time to time, the date of its commencement varies somewhat (see below). The Quaternary is the time during which modern human beings evolved from primate ancestors, becoming makers of primitive stone tools, developing agriculture, and finally, in the last half of the Holocene, modern mining, wheeled transport, agriculture and industry. It is also one of the intervals of geological time when extensive GLACIATION affected parts of the earth. The environmental stresses of this period may have been part of the driving force underlying human cultural development.

Following slow environmental deterioration in the Tertiary, glaciers began to form in high northern latitudes toward the end of the Tertiary and the beginning of the Quaternary. Reasons for this glacial onset remain unclear, but tectonic uplift of areas like the Himalayas, Sierra Nevada and Rocky Mountains may have provided areas suitable for the steady accumulation of snow and ice, whose reflectivity may have lessened the amount of solar heating that was absorbed. In bottom sediments of the north Atlantic, increasing amounts of siliceous rock debris are recorded at this time, reflecting transportation into the open ocean on floating ice calved from glaciers reaching down to sea level. Elsewhere, the commencement of major Quaternary environmental change was indicated in other ways. In China, rapid LOESS accumulation began, reflecting a strengthening of winds from the desert interior. On the basis of dating transitions of this kind, the start of the Quaternary has been placed at 2.58 Ma BP (the date of the Gauss–Matuyama magnetic reversal) (Suc *et al.* 1997).

During the Quaternary, great environmental changes affected most of the earth. With a somewhat changing periodicity that appears to be controlled astronomically (see MILANKOVITCH HYPOTHESIS), glaciers repeatedly extended over large areas of Europe, Canada, and North America, with smaller areas in South America and elsewhere, and again retreated. Sea ice also became periodically more extensive. Sufficient water was stored in the continental ice sheets, which reached thicknesses of more than 3 km, to lower sea level by 130 m or so, creating large areas of new land on the continental shelves as well as opening land bridges across former seaways that were traversed by animals, plants and people. The weight of ice depressed the crust, and together with the loss of ocean volume, created major perturbations in the relative levels of land and sea. The last ice melted by about 10 thousand years ago (10 ka BP), and the return of warm conditions marks the start of the Holocene epoch of the Quaternary. In areas heavily loaded down at the time of the last peak in ice extent (at about 18 ka BP in radio carbon years, termed the 'last glacial maximum'), such as Scandinavia and central Canada, the now unloaded crust is still in the process of rising to its former level.

On land, climatic changes, involving swings in rainfall, temperature, river runoff, windiness, continentality, and so on, led to parallel adjustments in flora, fauna, and landscape processes like erosion and soil formation. Global average temperatures periodically fell by perhaps 5 degrees, but cooling was much more marked, in excess of 15 degrees cooler than the present day, in the interiors of the continents. During the cool periods, termed GLACIALS, treelines were lowered, and vegetation cover changed structurally and floristically. Over large areas, forest retreated and was replaced by much more open communities of grasses. The repeated advance and retreat of forests, in successive glacial episodes, modified the carbon balance and so the abundance of carbon dioxide, a major greenhouse gas, in the atmosphere. Similar changes were triggered by altered biological productivity in the oceans, and by the increased solubility of the gas itself in colder sea water. Carbon dioxide concentrations fell by 50–100 ppmv, which can be compared to the pre-industrial level of about 280 ppmv.

Changes to the hydrological cycle were also marked. Cold oceans release less moisture to the atmosphere, so that as conditions descended toward glacial cold, rainfall declined in many areas, though the pattern of response is quite variable. In some locations, cold also appears to have curtailed evaporative losses from lakes, so that lake levels fluctuated dramatically, some closed basins periodically overflowing and later drying out completely. Some Quaternary lakes of the 'high lake' phases, like the large Willandra Lakes system of inland south-east Australia, which are now dry, periodically provided sustenance for early people, and were among the sites where early human cultural development took place.

In the last 750,000 years, the kinds of environmental changes noted here came and went with a periodicity of about 100,000 years. Palaeoenvironmental indicators of various kinds (marine microfossils, terrestrial pollen, isotopes trapped in deeply buried layers of ice in Greenland, Antarctica and other remaining ice caps) demonstrate that through most of this period, conditions have been colder than they are presently. The earth is now in a Quaternary warm phase (INTERGLACIAL); the great ice sheets have long since retreated, and sea level is at about its highest level with respect to the land. However, these interglacial periods are typically brief, lasting perhaps 10,000 years. Given that the interglacial that we now enjoy has already persisted for this long, we must anticipate renewed cooling in the coming millennia. Indeed, predictions based on what is known of the orbital influence on terrestrial climates suggest that the next full glaciation will be reached in about 60,000 years. Human activity in the meantime may of course upset these ancient rhythms.                                      DLD

**Reading and Reference**
Lowe, J.J. and Walker, M.J.C. 1997: *Reconstructing Quaternary environments*. 2nd edn. London: Addison-Wesley. · Suc, J.-P., Bertini, A., Leroy, S.A.G. and Suballyova, D. 1997: Towards lowering of the Pliocene/Pleistocene boundary to the Gauss–Matuyama reversal. *Quaternary international* 40: pp. 37–42. · Williams, M.A.J., Dunkerley, D.L., De Deckker, P., Kershaw, A.P. and Chappell, J.M.A. 1998: *Quaternary environments*. 2nd edn. London: Edward Arnold.

**quick clay** Water-saturated clay which has insufficient cohesion to prevent heavy objects from sinking into its surface.

**quickflow** The part of the stream hydrograph which lies above an arbitrary cut-off line drawn on the hydrograph, representing the most rapidly responding hydrological processes and parts of the catchment. The division between quickflow and DELAYED FLOW is usually made by a line which rises from the start of the hydrograph rise at a gradient of $0.551 \, \text{s}^{-1} \text{km}^{-2} \text{h}^{-1}$ until the line meets the falling limb of the hydrograph.      MJK

**Reading**
Hewlett, J.D. 1961: *Soil moisture as a source of base flow from steep mountain watersheds*. South-eastern Forest Experimental Station paper 132. US Department of Agriculture.

**quicksand** Water-saturated sand which is semi-liquid and cannot bear the weight of heavy objects.

# R

***r*- and *K*-selection** The influence of the physical environment on life history traits as expressed through two different kinds of pattern of population increase over time. Different kinds of habitat tend to favour species with different life history responses, especially with regard to the rate of reproduction. For example, in relatively stable habitats, where a given range and abundance of resources is consistently available for exploitation by resident species, populations of many of these species over time will tend to remain at or near the CARRYING CAPACITY of that habitat. By contrast, instability or uncertainty of conditions within a habitat will tend to favour those opportunistic species able to rapidly exploit the resources, which may be available only on a temporary basis. MacArthur and Wilson (1967) labelled the opportunists as *r*-selected, given their tendency to increase numbers rapidly from low initial populations, whereas the species found at population numbers approaching the carrying capacity were said to be *K*-selected.

*K*-selected populations, then, occupy equable habitats and as a result, the populations of residents tend to be both relatively high and reasonably constant. Intense competition between fertile adult members of the population determines the survival rate, while the young must compete equally intensively and only proportionally few reach maturity and are themselves able to reproduce. The characteristics of individuals belonging to the *K*-selected populations are as follows: larger body size, deferred and extended reproductive maturity, a smaller allocation of time and energy to reproduction, larger offspring with a higher degree of parental care. *r*-selected populations occupy unpredictable or temporary habitats and, as a result, the population may undergo periods of rapid growth during favourable periods followed by periods of equally rapid population decline when conditions become unfavourable – in essence a kind of 'boom and bust' behaviour. The characteristics of individuals belonging to *r*-selected populations are as follows: smaller body size, earlier age at and shorter period of maturity, a high proportion of time and energy allocated to reproduction, smaller and greater numbers of offspring.

As with other generalizations, this dichotomy is over-simplistic, but a useful means of appreciating that there are different ways for species to respond to habitat constraints. Actually, the *r*- and *K*-selected life-histories described are really special cases at opposing ends of what is probably a continuum. Certainly it is a useful way to identify differences among taxa. Trees in the *K*-selected habitat of, say, a mature oak woodland in north-west Europe, do indeed exhibit longevity, delayed and extended maturity, large size and produce large seeds. On the other hand, the plants found commonly in arid and semi-arid regions tend to follow the trends exhibited by *r*-selected individuals; many are EPHEMERAL PLANTS with very short life-spans that exhibit mass flowering and seed-set to take advantage of the transient nature of the soil moisture. In other temporary or disturbed habitats, for example road or railway verges, there is a predominance of *r*-selected ruderal plants typically thought of as WEEDS.

It must be emphasized that attempts to classify organisms with respect to their life history strategies can be misleading if interpreted too rigidly. Species make a trade-off in the allocation of resources to maintenance, growth and reproduction and may display life-cycle characteristics containing elements of both *r*- and *K*-categories. For example, Solbrig and Simpson (1974) demonstrated that a dandelion population in a highly disturbed grass lawn behaved as *r*-strategist with a high reproductive allocation, whereas the same species in a nearby undisturbed site set fewer seeds and allocated more resources to growth and maintenance. Different populations of the same species can therefore occupy different positions on the *r*–*K* continuum. In some cases, populations of species occupying variable habitats display dynamic life history responses and are able to shift between relative *r*- and *K*-positions over time.                    MEM

References
MacArthur, R.H. and Wilson, E.O. 1967: *The theory of island biogeography.* Princeton, New Jersey: Princeton University Press, 203 pp. · Solbrig, O.T. and Simpson, B.B. 1974: Components of regulation of a population of dandelions in Michigan. *Journal of ecology* 62, pp. 473–86.

**radiation** Any object not at a temperature of absolute zero ($-273\,^{\circ}$C) transmits energy to its surroundings by radiation, that is, by energy in the form of electromagnetic waves travelling with the speed of light and requiring no

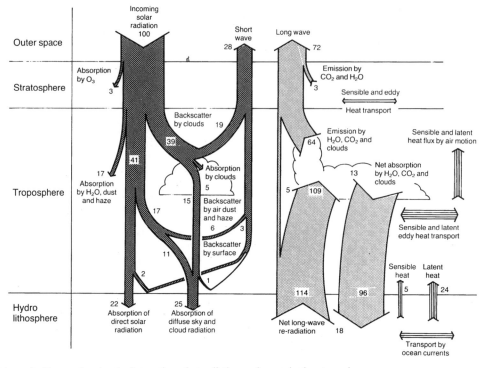

Schematic diagram showing the interactions that radiation undergoes in the atmosphere.
*Source*: T.G. Lockwood 1979: *Causes of climate*. London: Edward Arnold.

inter-vening medium. This radiation is charac-terized by its wavelength, of which there is a wide spectrum extending from the very short X-rays through the ultraviolet and visible to infrared, microwaves and radio waves.

A valuable theoretical concept in radiation studies is that of the black body, which is one that absorbs all the radiation falling on it and which emits, at any temperature, the maximum amount of radiant energy. The term arises from the relation between darkness of colour and the proportion of visible light absorbed, since a body which appears white scatters most of the visible light falling on it. For a perfect all-wave black body, the intensity of radiation emitted and the wavelength distribution depend only on the absolute temperature, and in this case several simple laws apply. The Stefan–Boltz-mann law states that the amount of energy ($F$) emitted in unit time from a unit area of a black body is proportional to the fourth power of its absolute temperature ($T$), i.e.:

$$F = \sigma T^4$$

where $\sigma$ is Stefan's constant (5.6697 $\times 10^{-12} \, W \, cm^{-2} \, K^{-4}$). The higher the temp-erature of an object the more radiation it will emit.

A black body does not radiate the same amount of energy at all wavelengths for any temperature. At a given temperature, the energy radiation reaches a maximum at some particular wavelength and then decreases for longer or shorter wavelengths. The Wien displacement law states that this wavelength of maximum energy ($\lambda_{max}$) is inversely proportional to the absolute temperature, i.e.:

$$\lambda_{max} = \frac{\propto}{T}$$

where $\propto$ is a constant (0.2897 cm K if $\lambda$ is in centimetres). As the temperature of an object increases, the wavelength of maximum energy decreases, passing from the infrared for objects at room temperature to the visible wavelengths for extremely hot objects.

If the sun is assumed to be a black body, an estimate of its effective radiating temperature may be obtained from the Stefan–Boltzmann law, which suggests an effective surface tem-perature of 5750 K. For the sun, the wavelength of maximum emission is near 0.5 $\mu m$ (1 $\mu m = 10^{-6}$ m), which is in the visible portion of the electromagnetic spectrum, and almost 99 per cent of the sun's radiation is contained in so-called short wavelengths from 0.15 to 4.0 $\mu m$. Observations show that 9 per cent of this short-wave radiation is in the ultraviolet (less than 0.4 μm), 45 per cent in the visible (0.4–0.6 μm), and 46 per cent in the infrared (greater than 0.74 μm).

The surface of the earth, when heated by the absorption of solar radiation, becomes a source of long-wave radiation. The average temperature of the earth's surface is about 285 K (12 °C), and therefore most of the radiation is emitted in the infrared spectral range from 4 to 50 μm, with a peak near 10 μm, as indicated by the Wien displacement law. This radiation may be referred to as long-wave, infrared, terrestrial or thermal radiation. JGL

**Reading**
Budyko, M.I. 1974: *Climate and life*. New York: Academic Press. · Houghton, J.T. 1977: *The physics of atmospheres*. Cambridge: Cambridge University Press. · Lockwood, J.G. 1979: *Causes of climate*. London: Edward Arnold. · Sellers, W.D. 1965: *Physical climatology*. Chicago: University of Chicago Press.

**radiative forcing** A change or perturbation which is imposed upon the climate system and modifies the radiative balance. The causes of such a change include changes in solar radiation input, cloud cover and character, ice, greenhouse gases, volcanic activity etc. Quantification of the effects of radiative forcing is one of the main goals of many climatic models. ASG

**radiocarbon dating** See CARBON DATING.

**radiocarbon years** The measure of time used in ages derived from CARBON DATING. Radiocarbon dates are usually expressed in the form $x$ years $\pm y$ before present (BP), where $y$ represents a standard deviation, and the present is defined as AD 1950. Initially Libby, the pioneer of radiocarbon dating, had assumed that levels of atmospheric carbon had remained stable over the time range used and that a radiocarbon year would be constant. However, there have been significant variations in $^{14}C$ production, arising from flux in incoming cosmic radiation and from long-term changes in the carbon cycle through the late Quaternary, leading to $^{14}C$ ages which are significantly younger than true calendar ages. For ages in the Holocene it has been possible to correct the radiocarbon dates using a calibration curve, based on radiocarbon assays of wood of known age, usually from dendrochronology studies. Bard *et al.* (1990) have extended calibration to 30,000 years using a comparison with uranium series ages from corals (see URANIUM SERIES DATING). As the limitations of radiocarbon dating have become recognized it has become common

practice to quote a corrected (calendar) age alongside the radiocarbon date. PSh

**Reading and Reference**
Bard, E., Hamelin, B., Fairbanks, R.G. and Zindler, A. 1990: Calibration of the $^{14}C$ timescale over the past 30,000 years using mass spectrometric |U-Th ages from Barbados corals. *Nature* 345, pp. 405–10. · Smart, P.L. and Frances, P.D. 1991: *Quaternary dating methods: a users guide*. Quaternary Research Association Technical Guide 4, University of Bristol, UK.

**radioisotope** The isotopes of some elements which undergo radioactive decay and alter either to the same or a different element. The decay process is a spontaneous process which involves the emission of alpha, beta or gamma radiation. The rate of decay of a radioisotope is proportional to number of atoms remaining. Mathematically this is referred to as the decay constant (*1*), this is specific to each radioisotope and is related to the HALF-LIFE ($T_{1/2}$) by the equation:

$$T_{1/2} = \frac{ln2}{\lambda}$$

SS

**Reading**
Faure, H. 1986: *Principles of isotope geology*. 2nd edn. Chichester: Wiley.

**radiolaria** Single-celled amoebic protozoans that secrete a delicate siliceous framework to support soft body parts. They are usually circular in shape and are between 100 and 2000 μms in diameter. They are present throughout the world's oceans and are able to live in surface waters down to depths of more than 4 km. Radiolaria can live in and are tolerant of a range of salinity and temperature conditions. They are particularly useful in biostratigraphic studies where calcareous microfossil components (e.g. foraminifera, ostracods etc.) have been removed due to post-burial chemical processes. SLO

**radiosonde** An instrument for measuring the pressure, temperature and humidity of the air at heights of greater than 10 m above the ground. The instrument is attached to a balloon which rises through the TROPOSPHERE, usually bursting near the TROPOPAUSE. The meteorological data are transmitted to a ground station as the balloon ascends. Upper air winds are established by tracking the whole package by radar. BWA

**radon gas** A colourless, odourless gas (radon-222) about eight times denser than air. It is derived largely from uranium, which is present in rocks such as granite. It poses an environmental problem when it accumulates in houses, for it may lead to cancer formation. Some areas of Britain are regarded as being at high risk (especially portions of Devon and Cornwall), but the threat can be reduced by appropriate building techniques and ventilation procedures. ASG

**Reading**
Clarke, R. and O'Riordan, M. 1990: Rumours of radon. *Science and public affairs* 5, pp. 23–36.

**rain** Precipitation in the form of liquid water drops. Drop sizes vary up to a maximum of about 0.5 cm in diameter. The smallest diameter of a raindrop is sometimes defined rather arbitrarily as being 0.02 cm. Drops in the range of 0.02–0.05 cm are classed as drizzle and are fairly common on windward locations in temperate latitudes. PS

**rain day** In Britain, a climatological period of 24 hours from 09.00 UT within which at least 0.2 mm of precipitation is recorded. The average number of rain days ranges from about 160 in south-eastern England to over 250 in the highlands of Scotland and the Outer Hebrides, where more than 300 rain days may be recorded in wet years. In some countries a different base is taken as the definition of a rain day, e.g. 0.01 inches in the USA. PS

**rain factor** A measure of the relationship between temperature and precipitation designed to provide an indication of the climatic aridity of a region. The formula used is:

$$\text{rainfall factor} = \frac{\text{mean annual precipitation in mm}}{\text{mean annual temperature in } °C}$$

It is clearly inappropriate for polar deserts where mean annual temperatures are below freezing. Elsewhere it gives an impression of the dryness of an area, although the seasonal distribution of rainfall will affect what plants can be grown. For example, Hobart in Tasmania and Kaduna in northern Nigeria have a similar rain factor but the former site has rain each month and the latter only for six months. PS

**rain gauge** An instrument used to measure rainfall amounts. Gauge characteristics differ in detail between countries, but basically it consists of a funnel and storage system shielded from the free air to prevent evaporation. In the UK the standard gauge made of copper has a collection funnel of 12.7 cm diameter placed on a cylinder 30.5 cm above the ground surface. The water is stored in a glass bottle held within the copper cylinder. Any overflow from the bottle is collected in another container to prevent loss during heavy storms. The rainfall is normally read daily (at 0.900 GMT) by emptying the contents of the collecting vessel into a cylindrical flask graduated to allow a direct reading of the rainfall total. Some gauges are read only weekly or even monthly in remote areas.

Rain gauges do not catch all the rain falling upon the surface because they present an obstruction to the airflow. In strong winds the actual catch may be depleted by up to 50 per cent of the true catch or even more during snowfall. The height of the rain gauge funnel above the ground will influence catch. A taller cylinder will generate more turbulence and reduce the catch compared with an identical gauge close to the surface. Other errors may be caused by splashing, evaporation or even observer mistakes. The total obtained by a rain gauge is therefore only an approximation to the true fall. Where a large proportion of the annual precipitation falls as snow, as in Canada and parts of the former USSR, separate snow gauges are used as standard rain gauges are unsuitable. Because of the different types and sizes of gauge used, comparison of rainfall totals across international frontiers is difficult.

Some gauges make a continuous record of precipitation and its time of occurrence, allowing the calculation of RAINFALL INTENSITY. This is achieved by collecting the water in a tipping bucket, on a weighing system or by a float which measures water level. Each type offers some advantages and disadvantages. PS

**Reading**
Sumner, G. 1988: *Precipitation: process and analysis*. Chichester: Wiley.

**rain gauge, tilting siphon** One type of autographic or recording rain gauge where the pen records on a rotating chart and is connected to a float in a precipitation collecting chamber which is emptied when full by a tilting siphon mechanism. The collecting chamber is precisely balanced against a counter weight, so that when the chamber is full it is sufficiently heavy to overcome the weight of the counter weight and so tilts causing the contained water to siphon away. The pen records on the rotating chart the rate at which the collecting chamber fills with water and this indicates rainfall intensity as well as amount. The collecting chamber empties rapidly so that the minimum of information on rainfall characteristics is lost. AMG

**Reading**
Meteorological Office 1969: *Observer's handbook*. London: HMSO.

**rain shadow** An area experiencing relatively low rainfall because of its position on the leeward side of a hill or mountain range. Uplift and precipitation over the hills decreases the water content of the air, and this, coupled with the descent of air down the leeward slope, reduces the capacity of the air to produce rain. A good example in the UK is the Moray Firth area to the lee of the Scottish Highlands, where rainfall totals are relatively low.                    PS

**rain splash**   The process in which the impact of raindrops ejects water droplets and soil particles from a soil surface. Ejected splash droplets may be composed largely of sediment-free water, or they may contain sediment particles derived from the soil surface. Rain splash, which has also been termed *airsplash* to emphasize that the particles follow a trajectory above the soil surface, is especially effective for silt and fine sand-sized particles. Particles carried in splash droplets may be carried more than 0.5 m in a single splash event. The splash process is driven by the energy possessed by raindrops, which commonly have masses of 5–30 mg and fall at 5–10 m s$^{-1}$. Even in a small storm, there may be millions of raindrop impacts on every square metre of exposed soil. The actual energy expended at the surface varies with rainfall intensity, because the distribution of drop sizes in rain shifts toward a larger median diameter at higher intensities, but is typically about 20 J m$^2$ per millimetre of rain.

Detailed investigations of the splash process have shown that many of the droplets thrown out from the point of impact of a raindrop are derived from the splash corona, which is a sheet of water forced upwards as the arriving droplet passes into any water film covering the surface. The ejected corona is rapidly drawn into a series of droplets by surface tension forces, and these continue on an outward trajectory. Droplets striking a dry surface, or one carrying only a very thin water film, cannot throw up a splash corona, and so they are less capable of rain splash. Splash is consequently accentuated when the surface is covered by a thin water film, but becomes relatively ineffective when the water layer exceeds about three drop diameters in thickness. At this point, the soil surface is sheltered from the impact-related forces by the water lying above it.

Net transport by rain splash is possible on a sloping surface, or when wind drives the rain. Splash processes are also of great importance in dislodging particles that may then be eroded by flowing water.                    DLD

**Reading**
Proffitt, A.P.B., Rose, C.W. and Hairsine, P.B. 1991: Rainfall detachment and deposition: experiments with low slopes and significant water depths. *Soil Science Society of America journal* 55, pp. 325–32. · Selby, M.J. 1993: *Hillslope materials and processes*. 2nd edn. Oxford: Oxford University Press. · Sharma, P.P., Gupta, S.C. and Rawls, W.J. 1991: Soil detachment by single raindrops of varying kinetic energy. *Soil Science Society of America journal* 55, pp. 301–7.

**rainbeat crust**   A thin layer at the soil surface displaying structural changes in the arrangement of soil particles that are caused by raindrop impact. Such crusts may be only a fraction of a millimetre in thickness. Their upper surface is composed of aligned clay particles that are packed tightly together and only weakly permeable to water. Below this, there may be a layer of clay particles 1–3 mm in thickness formed by the washing-in of clay particles which partially obstruct the soil pores. Rainbeat crusts permit very little infiltration of rainwater, and therefore promote ponding or surface run-off. Pioneering studies of surface crusts were made by McIntyre (1958).                    DLD

**Reading and Reference**
Mualem, Y., Assouline, S. and Rohdneburg, H. 1990: Rainfall induced soil seal (A): a critical review of observations and models. *Catena* 17, pp. 185–203. · McIntyre, D.S. 1958: Soil splash and the formation of surface crusts by raindrop impact. *Soil science* 85, pp. 261–6.

**rainbow**   An optical effect consisting of an arc of the spectral colours. It is formed by the passage of sunlight through raindrops. As the beam of light enters the raindrop it is first refracted, then internally reflected from the far side before being refracted again as it leaves the drop. Some of the light may be reflected twice to produce a double rainbow effect. White light from the sun is composed of the colours of the spectrum which are separated in the refraction process because of their slightly different wavelengths. The degree of coloration of the rainbow depends upon the size of the drop and the intensity of the sunlight.                    PS

**raindrop impact erosion**   Erosion that is attributable to, or accentuated by, physical forces that arise from the impact of raindrops at the soil surface. Raindrops are held quite strongly in an approximately spherical form by the surface tension of water, and consequently are not readily disrupted when they strike water ponded on or flowing over the surface. Rather, raindrops embed themselves into surface water, forcing a crater to be evacuated at the site of collision. Drops are progressively deformed as they strike the floor of the crater, and the water is thrown outward, merging with water ejected as the crater develops. Modelling studies (e.g. Harlow and Shannon 1967) have shown that high-speed flows are created in this way, whose outward velocity may exceed that of the arriving raindrop by up to five times. These high-speed flows, which may last for only 0.1–0.2 seconds, are thought to result in shear forces at the soil surface that are capable of rupturing soil aggregates. In this way, the physical forces following droplet impact can disaggregate soil materials so that the resulting particles are more readily moved by RAIN SPLASH or surface run-off. Once the outward flow ceases, the flow direction reverses and the crater collapses. The inward flow of water towards the impact point once again creates shearing forces at the surface that may further contribute to particle breakdown. Repeated raindrop impacts continually move disaggregated particles and lift them away from the surface, to which they may rapidly re-settle. However, these brief periods provide opportunities for lifted grains to be carried along in incremental steps in surface run-off, and this provides a second mechanism through which raindrop impact can increase the rate of erosion.     DLD

**Reading and Reference**
Al-Durrah, M.M. and Bradford, J.M. 1982: The mechanism of raindrop splash on soil surfaces. *Soil Science Society of America journal* 46, pp. 1086–90. · Bradford, J.M. and Nearing, C. 1996: Splash and detachment by waterdrops. In M. Agassi ed. *Soil erosion, conservation, and rehabilitation.* New York: Marcel Dekker. Pp. 61–76. · Harlow, F.H. and Shannon, J.P. 1967: The splash of a liquid drop. *Journal of applied physics* 38, pp. 3855–66.

**rainfall** The total water equivalent of all forms of precipitation or condensation from the atmosphere received and measured in a RAIN GAUGE. Values are normally expressed as daily, monthly or annual totals. On a world scale, mean annual rainfall amounts are very variable from almost zero in the driest deserts to above 10,000 mm in a few parts of India and Hawaii.     PS

**rainfall intensity** The rate at which rainfall passes a horizontal surface. Usually this will be the ground where a recording RAIN GAUGE can indicate the value, but it is possible to determine the rates beneath cloud-base by raindrop impaction onto specially designed aircraft equipment or by radar. Rates are normally expressed in $mm\,h^{-1}$ averaged over the hour so are not necessarily representative of short period bursts when much higher rates are possible. For the UK the annual mean rainfall intensity is only $1.25\,mm\,h^{-1}$. In the tropics much higher values are recorded with annual means above $15\,mm\,h^{-1}$.     PS

**rainfall run-off** That part of the hydrological cycle which connects rainfall and channel flow.

When rainfall exceeds the infiltration capacity of the surface materials water begins to collect in small surface depressions. Eventually these surface depressions fill and the water begins to move downslope. The entire overland process is termed Hortonian flow after R.E. Horton who first outlined the process. (See also OVERLAND FLOW; RATIONAL FORMULA.)     WLG

**rainfall simulator** A research tool designed to generate water droplets whose sizes correspond to those of natural rain, deliver them to the soil surface at speeds comparable to those reached by raindrops that have fallen from the cloud base, and operate at a range of simulated rainfall intensities. Rainfall simulators have been widely adopted in studies of erosion and flow mechanics, because of the many advantages they confer, such as freeing the investigator from the inefficiency of having to wait for natural storms to pass over a study site, and the ability to control rain intensity and duration. In many EXPERIMENTAL DESIGNS, another advantage is the ability to replicate as many identical rain events as are required while other factors, like the soil type exposed to rain, are changed. Rainfall simulators of many designs have been used, some for laboratory use and others for field operation. Among the means used to produce drops, the two principal techniques are *drop-formers*, often banks of syringes from whose tips drips are released, and *spray nozzles*, which are specially designed to yield known drop sizes when run at nominated water pressures (Bowyer-Bower and Burt 1989). It is difficult to generate high intensities using drop-formers, and hard to produce low intensities with spray nozzles. Indeed, no simulator is capable of generating rain with drop size, intensity, and fall speed or drop kinetic energy all correctly matching natural rain. Nevertheless, the advantages offered by rainfall simulation see it in continued use as a research tool.     DLD

**Reference**
Bowyer-Bower, T.A.S. and Burt, T.P. 1989: Rainfall simulators for investigating soil response to rainfall. *Soil technology* 2, pp. 1–16.

**raised beach** An emerged shoreline represented by stranded beach deposits, marine shell beds, and wave cut platforms backed by former sea cliffs. During the first decades of the twentieth century it was believed that many raised beaches were the result of eustatic changes brought about by changes in the volume of water stored in ice caps during glaciations and deglaciations, though it was also appreciated that in areas where the earth's crust had been weighed down by the presence

Spray nozzle rainfall simulator in use in the field.

of an ice mass that ISOSTASY would have been an important process. In reality, though both these processes are important, there is a wide range of factors that can cause sea-level changes. Raised beaches, which can occur at heights of as much as tens or hundreds of metres above current sea levels, may be warped, so that care needs to be exercised in correlating on the basis of height alone.                                                            ASG

**Reading**
Guilcher, A. 1969: Pleistocene and Holocene sea-level changes. *Earth science reviews* 5, pp. 69–97. · Rose, J. 1990: Raised shorelines. In A.S. Goudie ed., *Geomorphological techniques*. 2nd edn. London: Unwin Hyman.

**raised channel**   A channel whose form is preserved in the environment in negative relief long after it was active. Maizels (1987) described raised Plio-Pleistocene channels in Oman, where cementation of the original channel sediments had rendered them more resistant to erosion than the surrounding landscape. This had subsequently been lowered by erosion, leaving the raised channels as positive features in the landscape.                                                            DSGT

**Reference**
Maizels, J.K. 1987: Plio-Pleistocene raised channel systems in the western Sharqiya (Wahiba), Oman. In L. Frostick and I. Reid eds, *Desert sediments, ancient and*

*modern*. Geological Society of London special publication 35, pp. 31–50.

**raised mire**   An acid peatland dominated by Sphagnum mosses and supplied by precipitation solely from atmospheric sources (rain, snow, fog, etc.). Raised mires form characteristic shallow domes of peat where the topography is typically convex, with a gently sloping rand away from its centre towards the surrounding moat-like drainage channel or lagg (Swedish terms describing margin of raised bog, typically with a stream and/or minerotrophic poor-fen or fen woodland) surrounding the bog. This mire type is primarily a lowland system, and mainly occurs in broad, flat(ish) valleys or basins.

Raised mires have a wide distribution in Britain but predominate in the cooler, wetter north and west. Remnants of raised mires are recorded in Lincolnshire, the Cambridgeshire Fens, Somerset Levels and Amberley Wild Brooks in Sussex, but reclamation and ancient peat winning has removed most traces of ombrotrophic peat from these regions. In lowland Britain, raised mires are recorded in basins, floodplains and at the heads of estuaries, for example, Thorne Moors National Nature Reserve which forms part of the Humberhead Peatlands. Upland examples of raised mires exist at Tarn Moss,

Malham, North Yorkshire and Dun Moss in Perthshire. ALH

**randkluft** The gap between the back wall of a cirque and the glacial ice that fills the cirque.

**random-walk networks** Drainage networks can be simulated by a random-walk STOCHASTIC MODEL in which the network evolves on a regular grid, some cells being randomly selected to contain a stream source. The direction of exit from the current cell is then chosen by reference to random number tables. The model may be purely random with equal probabilities ($P = 0.25$) of a move in each cardinal direction (Leopold and Langbein 1962), or biased with, for example, $P = 0.25$ for left and right moves, $P = 0.5$ 'downhill' and $P = 0.0$ 'uphill', to simulate a regional slope. Spatial variation of bias can be allowed to simulate different topographical influences. Constraints to disallow triple junctions, reversals of direction and closed loops are necessary. These simulated networks commonly obey the laws of drainage network composition. KSR

Reading
Leopold, L.B. and Langbein, W.B. 1962: The concept of entropy in landscape evolution. *US Geological Survey professional paper 500–A.*

**ranker** The name given to a soil which has undergone limited development and has started to have some organic accumulation. Such soils occur on young geomorphological surfaces (e.g. recently deposited alluvium or dune sand).

**rapids** A steep section in a river channel where the velocity of flow increases and there is extreme turbulence.

**raster** One of the two main types of GEOGRAPHIC INFORMATION SYSTEM (GIS) for storing and analysing spatial data. A grid of square cells (PIXELS) is laid across the area of interest, and values of one geographical variable (e.g. landuse, temperature, population count) recorded for each pixel. One set of pixels forms a layer, with multiple variables being stored in further layers. Raster GIS are particularly good at handling information about phenomena which vary continuously across space (see VECTOR) and are widely used in the environmental sciences. SMW

**rated section** A method of obtaining a continuous record of discharge for a river section by continuously recording water stage or level and establishing a relationship between stage and discharge to rate the section. The discharge measurements required to establish this rela-

tionship are commonly made using the VELOCITY AREA METHOD. It is important to obtain a stable rating relationship and a section with a bedrock control or artificially stabilized bed and banks is frequently employed. The records obtained are generally less accurate than those provided by WEIRS and FLUMES, particularly for low flows. (See also DISCHARGE.) DEW

**rating curve** A term frequently used to describe a relationship between discharge and water stage or between suspended sediment and solute transport and water discharge, which can be used to estimate values of the former variable from measurements of the latter. In the case of suspended sediment and solute transport, values of either concentration (mg $l^{-1}$) or material discharge (kg s$^{-1}$) may be plotted against water discharge, and logarithmic axes are commonly employed. The characteristics of these plots, including slope, intercept, and degree of scatter, have frequently been used to characterize the sediment and solute response of a drainage basin. DEW

Reading
Gregory, K.J. and Walling, D.E. 1976: *Drainage basin form and process.* London: Edward Arnold. Pp. 215–25.

**rational formula** A simple and long used formula to estimate run-off ($Q$) in terms of a coefficient ($C$) depending upon the character of the surface of the drainage area, a measure of precipitation ($I$), and an index of the basin area ($A$) in the form: KJG

$$Q = CIA.$$

**Raunkiaer's life forms** The life form of a plant is its gross structure. The best-known classification of life forms is that of C. Raunkiaer (1934), who arranged the life forms of plant species into a series primarily based on the position of the perennating buds. He suggested that this reflected an adaptation to climate, arguing that plants' environments were formerly more uniformly hot and moist than at present, that the most primitive life form is the one dominating tropical environments, and that the more highly evolved forms are found particularly in areas with colder climates. The main categories he distinguished are as follows:
*Phanerophytes*, with the perennating buds on aerial shoots – the most primitive form, in Raunkiaer's view. Evergreen and deciduous phanerophytes are distinguished, and each of these categories can be subdivided:

*Nanophanerophytes*, 2 m in height (shrubs)
*Microphanerophytes*, 2–8 m

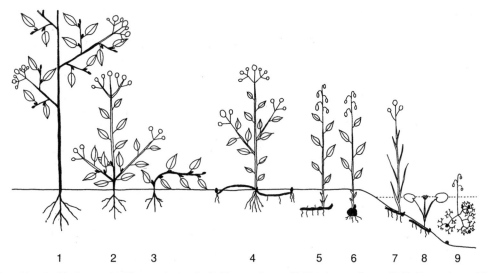

**Raunkiaer's life forms.** (1) Phanerophytes; (2–3) Chamaephytes; (4) Hemicryptophytes; (5–9) Cryptophytes. The parts of the plant that die during the unfavourable season are unshaded, the persistent portions and perennating buds are black.

*Mesophanerophytes*, 8–30 m
*Megaphanerophytes*, over 30 m (the larger forest trees).

*Chamaephytes*, with the perennating buds close to ground level. This category includes forms in which aerial shoots die away as the unfavourable season (winter or the dry period) approaches, plants in which the vegetative shoot grows along the ground, and cushion plants.
*Hemicryptophytes*, where the perennating buds are at ground level, almost all the above-ground material dying with the advance of unfavourable conditions. The group includes plants with stolons and rosette plants.
*Cryptophytes*, where the perennating buds are below the ground surface or submerged in water.                                    PHA

**Reference**
Raunkiaer, C. 1934: *The life forms of plants and statistical plant geography*. Translated by H. Gilbert-Carter, A. Fausboll and A.G. Tansley. Oxford: Clarendon Press.

**reach**  A length of channel, as applied to a coastal inlet, the arm of a lake and river channels. The term may be used in more specialized senses, as for a relatively straight section of a navigation waterway, or alternatively for a short length of channel for which discharge or other hydraulic conditions are approximately uniform.                                                          JL

**reaction time**  The time which elapses between the application of a change to, or a constraint upon, an earth system and the beginning

of adjustment of the system. Reaction time is therefore the period of time after the modification and before RELAXATION TIME begins.    KJG

**Reading**
Graf, W.L. 1977: The rate law in fluvial geomorphology. *American journal of science* 277, pp. 178–91.

**recession limb of hydrograph, recession curve**  The recession limb of the stream discharge hydrograph is the portion after the peak discharge and represents the time after precipitation has ceased when discharge gradually falls and is unaffected by rainfall (for diagram see HYDROGRAPHS). By analysing several hydrographs for the same gauging station it is possible to derive an average recession curve which may be expressed by a mathematical function such as the exponential form:

$$Q_t = Q_o e^{-at}$$

where $Q_t$ is discharge at time $t$, $Q_o$ is the initial discharge, a is a constant, $t$ is the time interval and $e$ is the base of natural logarithms.    KJG

**recharge**  The process by which precipitation is absorbed through the soil or regolith, transmitted, and ultimately added to the saturated or phreatic zone within the underlying soil or bedrock, thereby replenishing GROUNDWATER reserves. The term can also be used to describe mechanisms by which soil moisture is replenished after it has been depleted due to percolation loss to underlying horizons or by evapotranspiration. Recharge zones are those

**401**

areas in a landscape which are particularly important for the replenishment of groundwater. The location of such zones varies according to climate, soil permeability and thickness, permeability of underlying bedrock and the distribution of geomorphological surface depressions such as rivers and lakes. In arid areas the main recharge areas are often the beds of large ephemeral rivers whilst recharge may be more diffuse in humid temperate environments.  DJN

**recovery time**  A term sometimes used in the explanation of SYSTEMS to describe the length of time it takes a system to recover its original state after being disturbed by an external force or process. *Elasticity* is sometimes used as an alternative term.  DSGT

**recumbent fold**  A fold that is so overturned that the strata lie horizontally.

**recurrence interval or return period**  $(T_r)$ is the expected frequency of occurrence in years of a discharge of a particular magnitude. For a series of flood discharges, which may be obtained as ANNUAL SERIES, annual exceedance series, PARTIAL DURATION SERIES, ranked from the largest to the smallest the formula usually used to calculate recurrence interval is:

$$T_r = \frac{N+1}{m}$$

where N is the total number of items in the series and $m$ is the rank in the array numbering from the largest (1) downwards.  KJG

**red beds**  Sediments, soils or sedimentary rocks which possess red coloration because of the presence of finely divided ferric oxides, chiefly haematite. Most represent iron-stained or cemented clastic sediments. In the continental literature equivalent terms for red beds are *couche rouge*, *rotschicht* and *capa roja* (Pye 1983). A distinction may be drawn between *in situ red beds* (formed in place, without transportation, by processes of direct chemical precipitation or by weathering, soil formation and diagenesis), and *detrital red beds* (formed by erosion, transportation and redeposition of existing red soils or sediments). Only sediments and soils having hues redder than 5YR on the Munsell soil colour chart should be described as red beds. Recent work suggests that if a suitable source of iron and oxidizing conditions exist, reddening can occur very rapidly after deposition of a sediment.  ASG

**Reading and Reference**
Pye, K. 1983: Red beds. In A.S. Goudie and K. Pye eds, *Chemical sediments and geomorphology*. London: Academic

Press. · Turner, P. 1980: *Continental red beds*. Amsterdam: Elsevier.

**redox cycle**  When soils, whether mineral or organic, are inundated with water, anaerobic conditions usually result. As water fills the soil pore spaces, the rate at which oxygen can diffuse through the soil is around 10,000 times slower than diffusion through a porous medium such as a drained soil. Such low rates of diffusion lead to anaerobic or reduced conditions.

The redox cycle describes the response of a soil to inundation. Oxidation occurs during uptake of oxygen and/or when hydrogen is removed (e.g. $H_2S \Rightarrow S^{2-} \Rightarrow 2H^+$) or when a chemical gives up an electron (e.g. $Fe^{2+} \Rightarrow Fe^{3+} + e^-$). Reduction is the opposite and is the process of giving up oxygen and gaining hydrogen, or gaining an electron.

The redox potential is shorthand for the *oxygen-reduction potential*. It may be measured on a hydrogen scale (similar to pH measurement) and is referred to as $E_H$. It represents the concentration of oxidants {ox} and reductants {red} in a redox reaction and is defined by the Nernst equation:

$$E_H = E^\circ + 2.3[RT/nF]\log[\{ox\}/\{red\}]$$

where:

$E^\circ$ = potential or reference, mv
R = gas constant = 81.987 cal deg$^{-1}$mole$^{-1}$
T = temperature, °K
$n$ = number of moles of electrons transferred
F = Faraday constant = 23,061 cal/mole-volt
ALH

**Reading**
Mitsch, W.J. and Gosselink, J.G. 1993: *Wetlands*. New York: Van Nostrand Reinhold.

**redox potential**  Reduction is the gain of an electron in a chemical reaction; the opposite is known as oxidation. Oxidation is most common in weathering processes. As some elements exist in several oxidation states, e.g. ferrous, Fe(II) is oxidized to the ferric state, Fe(III). The stability of any state depends upon the ease with which oxidation or reduction can occur, i.e. electron transfer. This is often dependent on pH. Redox potential is a measure of this ease of transfer and is symbolized by $E_H$. Redox potential may be measured using a redox probe which measures electric potential in millivolts (mv) relative to a hydrogen electrode or a calomel reference electrode. A plot of redox potential (measured in volts) versus pH is a useful tool in determining the stability fields of products of reactions, especially for geochemical environments.

The presence of free dissolved oxygen in solution gives a fairly stable redox potential around +400 to +700 mv. The redox potential of wetland soils is generally in the range +400 (most oxidized) to −400 mv (reduced).     WBW ALH

**Reading**
Yatsu, E. 1988: *The nature of weathering*. Tokyo: Sozosha.

**reef** A rocky construction found at or near sea level, formed mainly from biogenically produced carbonates. The term bioherm is also used. Several forms of marine organisms are capable of precipitating calcium and other carbonates in skeletal or non-skeletal forms and it is the accumulation of this material that gives rise to the characteristic form of reefs. Coral reefs, whose form is usually dominated by coral skeletons, are the most common form of reef, but other organisms may also form or contribute to the development of reefs, e.g. algae, vermetids and serpulids. Reefs vary greatly in size from individual clusters of organisms (called patch reefs or micro-atolls) to large-scale rock masses. Large coral reefs exhibit a range of facies controlled by the positions of the corals relative to exposure. Four major forms of large-scale coral reefs have been identified, i.e. fringing reefs, barrier reefs, atolls and table reefs (Stoddart 1969).     HV

**Reference**
Stoddart, D.R. 1969: Ecology and morphology of recent coral reefs. *Biology reviews* 44, pp. 433–98.

**reflective beach** A BEACH that reflects a substantial fraction of the wave energy impinging upon it. The beach profile is steep relative to the steepness of the WAVES that it reflects. The steep profile is usually the result of accretion during swell wave conditions. The reflective condition represents one end of the spectrum of morphodynamic beach states described by Wright *et al.* (1979). A dissipative beach, where almost all wave energy is dissipated in the nearshore, is at the other end of the continuum, and intermediate beaches separate the extremes. Reflective beach profiles are morphologically simple. They are convex upward in shape, and are without nearshore BARS. They may be more complex in plan, as beach CUSPS are commonly present. Reflection is a necessary condition for the generation of EDGE WAVES on beaches.     DJS

**Reading and Reference**
Wright, L.D., Chappell, J., Thom, B.G., Bradshaw, M.P. and Cowell, P. 1979: Morphodynamics of reflective and dissipative beach and inshore systems, south-east Australia. *Marine geology* 32, pp. 104–40. · —and Short, A.D. 1984: Morphodynamic variability of surf zones and beaches: a synthesis. *Marine geology* 56, pp. 93–118.

**refraction, wave** Wave refraction is the bending of the wave front as water depth changes when it is less than half the wave length. The wave velocity decelerates as water depth decreases. The wave front, therefore, bends to become more nearly parallel to the bottom contours. Orthogonals, which are lines normal to the wave crests and between which energy is constant in deep water, converge on headlands and diverge in bays. Wave energy is concentrated on the headlands and dissipated in bays. Wave refraction diagrams can be drawn manually or by computer where the offshore relief, wave approach direction and wave period are known. These show zones of wave energy concentration and dissipation by constructing the orthogonals or wave rays.     CAMK

**Reading**
Authur, R.S., Munk, W.H. and Isaacs, J.D. 1952: The direct construction of wave rays. *American Geophysical Union transactions* 33, pp. 955–65. · Harrison, W. and Wilson, W.W. 1964: *Development of a method for numerical calculation of wave refraction*. Technical memoir 6. Washington, DC: Coastal Engineering Research Centre.

**refugia** Localities or habitats in which formerly widespread organisms survive in small restricted populations which are either disjunct or endemic in their distribution. Such localities usually possess some distinctive microclimatic, geomorphological, ecological or historic characteristic which accounts for the survival and persistence of the relict organism at that site. A classic example of such a refugium is the Upper Teesdale area of Yorkshire and Co. Durham in north-east England, which harbours

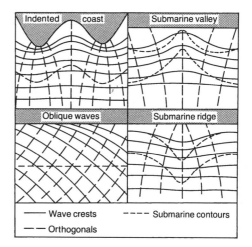

Examples of wave refraction showing wave crests and orthogonals relative to submarine contours.
*Source*: C.A.M. King 1980: *Physical geography*. Oxford: Basil Blackwell.

The refraction of waves in a bay at Rhossili, Gower, South Wales. Wave refraction is one of the most important controls of bay beach geometry.

survivals from the open-habitat flora that existed just after the last Ice Age (e.g. the arctic-alpine species, *betula nana*).                    PAS

**Reading**
Holmquist, C. 1962: The relict concept. *Oikos* 13, pp. 262–92. · Pennington, W.A. 1974: *The history of British vegetation*. 2nd edn. London: English Universities Press.

**reg**   A stony desert or any desert region where the surface consists of sheets of gravel.

**regelation**   The refreezing of meltwater. This commonly occurs at a glacier surface to form superimposed ice and at the base of a sliding glacier to form regelation ice. In the latter case ice at the base of a warm-based glacier melts on the upstream side of an obstacle where pressure is locally high and refreezes on the downstream side where pressure is lower.                    DES

**Reading**
Paterson, W.S.B. 1981: *The physics of glaciers*. Oxford and New York: Pergamon.

**regeneration complex**   A term coined by Scandinavian botanists in the early twentieth century to describe the small-scale mosaic pattern of vegetation on raised-bog plateaux surfaces which was thought to be the mechanism whereby such OMBROTROPHIC bogs grew or regenerated themselves in an autogenic cyclic

fashion. The mosaic of species – *Sphagna* (bog mosses) *Calluna* (heather), *Eriophorum* (cotton-grass) etc. – also reflects the microtopography of the mire surface, a pattern of hummocks and hollows, which may contain open water. Each species, because of its tolerance to waterlogging and acidity etc., has a micro-habitat range within this topography; *Calluna* and *Eriophorum vaginatum* generally grow on hummocks with *Sphagnum capillifolium* (formerly *S. rubellum*); the hollows and pools are inhabited by *S. cuspidatum* and other aquatic sphagna, and the hummock sides and flattish 'lawn' areas by *S. papillosum*, *S. magellanicum* and *Eriophorum angustifolium* etc. Of all the plant associations of an untouched raised bog it is this regeneration complex which is the most vigorous in growth and peat accumulation. Osvald, the Swedish bog ecologist who popularized the term in the 1920s and 1930s after his work on the Komosse bogs, influenced the British ecologists Tansley and Godwin, and the idea of the regeneration complex as a cyclic succession emerged during this period. The basic idea was that as a hummock became higher and drier than the surrounding hollows, its growth would slow down and it would become moribund. The hollows, accumulating peat rapidly, would then grow up to overtop and drown the former hummock which would then in turn become a hollow or pool. This hummock–

hollow cycle would be repeated all over the growing bog surface and was seen as the normal growth process over centuries and millennia. Lack of stratigraphic evidence, as well as doubts over climatic influences, did not prevent this theory becoming widely accepted in the ecological literature. Specific testing of the theory, from peat sections of the past two millennia, was carried out by Barber (1981) who rejected the theory in favour of close climatic control of the hydrology, and therefore of the floral composition and peat stratigraphy, of raised bogs.   KEB

**Reading and Reference**

Barber, K.E. 1981: *Peat stratigraphy and climatic change: a palaeoecological test of the theory of cyclic peat bog regeneration*. Rotterdam and Salem, New Hampshire: Balkema. · Godwin, H. 1981: *The archives of the peat bogs*, Cambridge: Cambridge University Press.

**regime**   The average annual variations of climatic or hydrological variables. Seasonal fluctuations of river discharge, represented by the mean monthly flows, are river regimes which vary regionally in relation to precipitation, temperature, evapotranspiration and drainage basin characteristics (Beckinsale 1969). Using the analogy of average climatic statistics, equilibrium channel morphology is also referred to as the regime state, which is constant over a period of years, is adjusted to prevailing hydrological and sedimentological influences, and is predicted using regime theory (Blench 1969). The term is also applied to flow regimes, which are defined by the Froude and Reynolds numbers and associated bed-form types.   KSR

**References**

Beckinsale, R.P. 1969: River regimes. In R.J. Chorley ed., *Water, earth and man*. London and New York: Methuen. · Blench, T. 1969: *Mobile bed fluviology*. Edmonton: University of Alberta Press.

**regime theory**   An approach to the design of river channel shape based upon theoretical and empirical assessment of the best shape for transporting a given DISCHARGE and sediment supply. It has its origins in engineering in early work on the design of canals through mobile bed sediments, notably for overseas irrigation. One of the most basic supporting concepts behind regime theory is the idea of HYDRAULIC GEOMETRY. This argued that width, depth and velocity in a channel all change in response to increases in discharge through the channel, something reflected in both: (1) trends in cross-section morphology from headwaters to the sea, as river discharge increases; and (2) changes in cross-section morphology at a fixed cross-section in response to discharge fluctuation through time.

Regime theory requires an assumption of EQUILIBRIUM, with the river being adjusted to channel discharge and the sediment supplied to it. The main support for this followed from MAGNITUDE AND FREQUENCY EFFECTS, in which Wolman and Miller (1960) argued that intermediate events in rivers systems (e.g. the bankfull discharge) do most work. This leads directly into the concept of DOMINANT DISCHARGE, which describes the average discharge of a river to which its average form is most directly related. Thus, regime theory can be based upon identification of associations between the dominant channel-forming discharge combined with relationships identified in a hydraulic geometry analysis, to design rivers whose shape is capable of transporting the discharge supplied to them.

Recent progress has recognized the problems of regime theory for a number of reasons. First, most river channel adjustments are essentially multivariate. In addition to changes in width, depth and velocity, it is also possible to change bed slope and boundary resistance. As a result, most river channel adjustments involve mutually interacting processes, where many variables may change simultaneously. As a result, rivers may display many different responses and hence morphologies, according to the interaction of feedbacks during the adjustment process (e.g. Miller 1984).

Second, many river channels will only show equilibrium characteristics over certain timescales. As discharge and sediment supply change, so the morphology required to achieve transport will change. Cross-section morphology rarely adjusts at a sufficient rate to a given change in discharge or sediment transport for the morphology to be always capable of transporting the water and sediment supplied to it: it is for this reason that rivers flood! Thus, even if it is possible to identify a stable morphology capable of transporting a particular discharge, this doesn't mean to say that the morphology will be stable at all discharges, particularly extreme ones.

Third, fluvial geomorphologists now recognize that river morphology does not just adjust to the discharge and sediment supplied to it, but can actively influence the processes operating within it. In a benchmark paper, Schumm and Lichty (1965) argued that variables changed their status according to the scale of interest. For instance, discharge is a product of catchment characteristics (e.g. basin relief, geology, hydrology) and is therefore a dependent variable. However, for smaller-scale processes, it is an independent variable. For instance, sediment transport may be explained by discharge variation. At these smallest scales, Schumm and

Lichty note that rather than river channel morphology being a product of variables like discharge, morphology could be an active influence upon the interactions between flow and sediment transport processes. This 'bottom-up' view of river channel change emphasizes the dynamics of that change in response to imposed external factors (e.g. sediment supply, discharge variation) and is one of the reasons for more intensive investigation of particular case-study rivers. It implies that morphology influences process, as well as process causing changes in morphology.  SNL

**Reading and References**
Blench, T. 1969: *Mobile-bed fluviology.* Edmonton: University of Alberta Press. · Hey, R.D. 1997: Stable river morphology. In C.R. Thorne, R.D. Hey and M.D. Newson eds, *Applied fluvial geomorphology for river engineering and management.* Chichester: Wiley. pp. 223–36. · Miller, T.K., 1984: A system model of stream-channel shape and size. *Bulletin of the Geological Society of America* 95. · Schumm, S.A. and Lichty, R.W., 1965: Time, space and causality in geomorphology. *American journal of science* 263, pp. 110–19. · Wolman, M.G. and Miller, J.P., 1960: Magnitude and frequency of forces in geomorphic processes. *Journal of geology* 68, pp. 54–74.

**regional circulation modelling**  A regional circulation model is a subset of one of the primary tools of investigation now used by climate researchers, GENERAL CIRCULATION MODELLING. These mathematical computer models describe the primary controls (such as radiative fluxes), energy transfers, circulations and feedbacks existing in the earth-ocean-climate system. Regional circulation models have their roots in the numerical weather models developed by the US National Weather Service to provide higher resolution of an important area's air flows and patterns. For example, many computer models now use the concept of 'nesting' in which higher resolution models are imbedded or 'nested' within lower resolution global models such that the coarse conditions of the global model provide the initial boundary conditions for the higher resolution model.

Other types of regional circulation models exist as stand-alone numerical models. For example, hurricane models built on the pioneering work of R. Anthes define and allow tropical cyclones to develop within a simulated geographic domain of a few hundred square kilometres. The simulation of other localized or regional phenomena such as the South Pacific's ENSO circulations have also been attempted using regional circulation models. The fundamental difference between such regional scale models and their large global counterparts is the definition of boundary conditions. For regional circulation models, continued input of data at the domain boundaries is required. For this reason, regional circulation models are more limited in their uses than GCMs.  RSC

**regolith**  A term coined by Merrill in 1897 (pp. 299–300) to describe the 'superficial and unconsolidated portion of the earth's crust...the entire mantle of unconsolidated material, whatever its nature or origin'. Some later workers have tended to use it as a synonym for weathering products, but such a narrow usage is incorrect (Gale 1992).  ASG

**References**
Gale, S.J. 1992: Regolith: the mantle of unconsolidated material at the Earth's surface. *Quaternary research* 37, pp. 261–2. · Merrill, G.P. 1897: *A treatise on rocks, rock weathering and soils.* New York: Macmillan.

**regosol**  Any weakly developed soil.

**regur**  Black, clayey soils which develop in tropical regions of high rainfall and temperatures, notably the north-west Deccan of India.

**rejuvenation**  The renewal of former activity, as of a fault which is reactivated by new tectonic movement or especially a stream whose erosional activity is redeveloped by uplift, a base-level fall or possibly a change in stream sediment load or discharge. The term may be applied to whole landscapes whose reduced relief may be redeveloped as a result of such changes, giving valley-in-valley forms.

The term especially derives from the interpretation of landforms developing in a cyclical fashion from youth to old age, with rejuvenation representing the redevelopment of 'young' surface forms in landscapes otherwise well advanced in a cycle of development. Some of the landforms that were so interpreted more than half a century ago might now be explained in alternative ways, with emphasis placed on equilibrium adjustments between forms, environments and processes.  JL

**relative age**  The age of an event or feature expressed not in terms of time units, such as years, but in relation to other phenomena (see ABSOLUTE AGE).

**relative dating**  A very simple form of GEOCHRONOLOGY whereby a sediment, landform or other feature is ascribed an age relative to other sediments, landforms or features. Relative dating may have two forms: the *relative position* of a feature, and the *relative change* of a feature. Relative position may be exemplified by the position of glacial end moraines in a formerly glaciated part of a valley, whereby the innermost (closest to source area) moraine is believed to be younger than another end moraine further away from the source area. If the period of time that lapsed between the two features is not known, and if the numerical age of one of the features is not known, then they are only differentiated by the position in the landscape. An example of relative change is when one soil is deemed to be older than another on the basis of mineral weathering or organic content. In this case, it is assumed that change is a function of time and that the sediments in which the two soils developed were initially the same. DSGT

## relative humidity See HUMIDITY.

## relaxation time
A term used in the study of SYSTEMS to describe the length of time it takes a system to establish a new quasi-stable state after an external disturbance has taken place. DSGT

## relic / relict
These words are variously used by different authors to describe landforms or features that are no longer experiencing the principal process that was responsible for their origin. In the *Oxford English Dictionary* relic is described as a dead body and relict as a geological or other object that has survived in a primitive form. The terms are widely used in palaeoenvironmental studies. DSGT

## remanié
A glacier which is not directly connected to a snow field but receives ice from avalanches.

## remote sensing
At its broadest the term refers to the observation and measurement of an object without touching it, and involves the use of force fields, acoustic energy and ELECTROMAGNETIC RADIATION. Within physical geography it involves the use of electromagnetic radiation sensors to record images of the environment which can be used to yield useful information. The four most important sensors are the AERIAL CAMERA, MULTISPECTRAL SCANNER, SIDEWAYS LOOKING AIRBORNE RADAR and THERMAL INFRARED LINE-SCANNER. The two most important platforms are the aircraft and satellite.

The 1960s were formative years for remote sensing during which visual interpretation of black and white aerial photography paralleled research into the use of data from the new aircraft and satellite borne sensors. Remote sensing, especially of the non-photographic type, grew rapidly after the launch of the Landsat 1 satellite in 1972. Today remote sensing is regarded as a powerful and useful technique in physical geography. PJC

### Reading
Barrett, E.C. and Curtis, L.F. 1992: *Introduction to environmental remote sensing*. 3rd edn. London and New York: Chapman & Hall. · Cracknell, A.P. and Hayes, L.W.B. 1991: *Introduction to remote sensing*. London: Taylor & Francis. · Curran, P.J. 1987: Remote sensing methodologies and geography. *International journal of remote sensing* 8, pp. 1255–75. · Lillesand, R.M. and Kiefer, R.W. 1987: *Remote sensing and image interpretation*. 2nd edn. New York: Wiley.

## rendzina
Dark, organic-rich soil horizons developed upon unconsolidated calcareous materials in areas of chalk and limestone bedrock.

## renewable resources See FLOW RESOURCES.

## representative and experimental basins
Small drainage basins where the representative basins are chosen to be representative of hydrological regions and should experience minimum change during monitoring and where the experimental basins are subjected to deliberate modification so that the impact of the modification on drainage basin dynamics may be identified. The establishment of a large number of representative and experimental basins was one of the major achievements of the International Hydrological Decade (IHD, see HYDROLOGY).

Studies in representative basins involve monitoring hydrological processes to improve understanding of these processes and their inter-relationships. If plentiful data can be built up from catchments which are representative of a particular hydrological region these data should help to improve estimation of the magnitude of hydrological processes, particularly discharge characteristics, in ungauged catchments in the same region. Benchmark catchments (Toebes and Ouryvaev 1970) have been identified as a special type of representative basin which are in their natural state and, therefore, permit the observation of hydrological processes without the effects of humans. The Vigil Network also forms a network of representative basins which was originally conceived within the USA (Leopold 1962; Slaymaker and Chorley 1964). In the case of the Vigil Network, the basins are not protected from artificial change but the aim is to monitor a large number of hydrological and geomorphological variables using simple and inexpensive techniques so that landscape and process change may be identified over long periods of time.

Work in experimental basins involves not only observation of hydrological processes but also deliberate modification of the basin so that the impact of cultural changes on hydrological

407

processes may be measured. Experimental basins are often subjected to a calibration period for several years before modification. This allows the magnitude and variability of processes before and after the change to be observed and compared. If more than one basin is available, where all the basins are virtually identical in size, physical character and vegetation cover, either a paired or a multiple watershed experiment can be carried out. In this type of experiment a long calibration period is desirable but not essential. One or more control basins can be monitored without receiving modification and the remaining basin or basins can be subjected to one or more experimental treatments, preferably using replicate basins for each treatment. In all the basins the same processes are observed so that differences between the control and the treated basins may be identified.                          AMG

### Reading and References
IAHS 1980: *The influence of man on the hydrological regime with special reference to representative and experimental basins*. Helsinki Symposium. International Association of Hydrological Sciences publication 130. · IASH 1965: *Representative and experimental areas*. Budapest Symposium. International Association of Scientific Hydrology publication 66, vols 1 and 2. · Leopold, L.B. 1962: The vigil network. *International Association of Scientific Hydrology bulletin* 7, pp. 5–9. · Slaymaker, H.O. and Chorley, R.J. 1964: The Vigil Network system. *Journal of hydrology* 2, pp. 19–24. · Swank, W.T. and Crossley, D.A. 1987: *Forest hydrology and ecology at Coweeta: ecological studies*, Vol. 66. Berlin: Springer-Verlag. · Toebes, C. and Ouryvaev, V. eds 1970: *Representative and experimental basins: an international guide for research and practice*. UNESCO studies and reports in hydrology 4. Paris: UNESCO.

**reptation**  Transitional between the creep and saltation of sand grains in aeolian transport, reptation (Anderson 1987) occurs when the high-velocity impact of a saltating grain sets several other grains moving through a low hopping process.                          DSGT

### Reference
Anderson, R.S. 1987: Eolian sediment transport as a stochastic process: the effects of a fluctuating wind on particle trajectories. *Journal of geology* 95, pp. 497–512.

**resequent stream**  A stream following the original direction of drainage, but developed at a later stage. This might apply to streams on the back slope of a cuesta of *resistant* rock which did not outcrop on the originally exposed landsurface which guided initial stream development. The term is not now commonly used.                          JL

**reservoir rocks**  See GROUNDWATER.

**reservoir, storage effects of**  Reservoirs may be constructed to regulate river flows for supply purposes downstream (regulating reservoirs), to provide water supply directly from direct supply reservoirs or to control floods downstream in which case the reservoir is maintained at less than capacity so that flood discharges may accumulate in the reservoir (flood control reservoirs). Immediately downstream of the dam, scour of the channel bed and banks may occur because the water released is comparatively free of sediment and further downstream there can be major changes of CHANNEL CAPACITY or of channel planform as a response to the changes in FLOOD FREQUENCY and to the altered sediment transport. There will also be changes in the river ecology as a response to changes in flow and to the alterations of aquatic habitats.                          KJG

### Reading
Eschner, T., Hadley, R.F. and Crowley, K.D. 1983: Hydrologic and morphologic changes in channels of the Platte River Basin in Colorado, Wyoming, and Nebraska: a historical perspective. *US Geological Survey professional paper* 1277-A. · Petts, G.E. 1979: Complex response of river channel morphology to reservoir construction. *Progress in physical geography* 3, pp. 329–62. · Williams, G.P. and Wolman, M.G. 1984: Downstream effects of dams on alluvial rivers. *US Geological Survey professional paper* 1286.

**residence time**  A concept employed in studies of chemical weathering and solute generation in drainage basins to describe the length of time between the input of water as precipitation and its output as run-off. The residence time is seen as a major control on the magnitude of solute uptake by water moving through the drainage basin system and therefore on solute concentrations in streamflow, since a significant period of time may be required for the water to reach chemical equilibrium with the soil and rock.                          DEW

**residual strength**  (or residual shear strength) The strength of a soil (usually a clay soil) which is below the peak SHEAR STRENGTH. Large shear strains (deformations) produce a reorientation of the clay mineral particles in the FABRIC and allow a shear plane to develop. The shear strength along this plane is less than the peak strength. The residual strength can be determined in a SHEAR BOX (or a modified version of this called a ring shear) apparatus for soil testing.                          WBW

**resistance**  *a.* Any property that opposes acceleration. Acceleration or movement of rock, soil, snow or ice is opposed by compressive strength, tensile strength and especially SHEAR STRENGTH. MASS STRENGTH is more important than INTACT STRENGTH. Components of strength include plane friction, interlocking and COHESION. More broadly, resistance is more than strength: it includes surface ROUGHNESS, vegetation, and permeability. (See also FORCE; FACTOR OF SAFETY.)

b. Term sometimes used in the study of SYSTEMS to explain the ability of a system to withstand disturbance by an external force or change.    IE

**Reading**
Carson, M.A. and Kirkby, M.J. 1972: *Hillslope form and process*. Cambridge: Cambridge University Press. Ch. 4.

**respiration**  For land plants the chemically reverse process of PHOTOSYNTHESIS or, in other words, the release of energy from food molecules in cells, which is then used in metabolism. Plant 'waste' products of carbon dioxide and water are also transferred back into the atmosphere. The reactions may be represented, in simplified form, by:

$$2(C_3H_6O_3) + 6O_2 \rightarrow 6CO_2 + 6H_2O + heat$$

The heat goes to form part of the long-wave energy component of the earth's atmosphere, and is unavailable thereafter for use in biological systems. The amount of the plant's available energy store which is used in respiration varies widely, but it is often in excess of 50 per cent of that used in growth, and may be very much more than this (Odum 1971; see also NET PRIMARY PRODUCTIVITY). Unlike photosynthesis, respiration is a day-long not just a day-time phenomenon. For higher land animals respiration is a broadly similar process after atmospheric oxygen has been taken in through breathing.    DW

**Reference**
Odum, E.P. 1971: *Fundamentals of ecology*. 3rd edn. Philadelphia: W.B. Saunders.

**resultant wind**  The vectorial average of all wind directions and speeds for a given level at a given place for a specified period, such as one month. It is obtained by resolving each wind observation into components from north and east, summing over the given period, obtaining averages and reconverting the average components into a single vector.    BWA

**resurgence**  An emergence point of underground water. The term is often used to describe a karst SPRING, the headwaters of which are initially on the surface but are lost underground by disappearing down a stream-sink (or SWALLOW HOLE). A distinction is sometimes made between this kind of spring and that with no known surface headwaters, which is called an exsurgence.    PWW

**Reading**
Bögli, A. 1980: *Karst hydrology and physical speleology*. Berlin: Springer-Verlag. · Smith, D.I., Atkinson, T.C. and Drew, D.P. 1976: The hydrology of limestone terrains. In T.D. Ford and C.H.D. Cullingford eds, *The science of speleology*. London: Academic Press. Pp. 179–212.

**retention curve**  Or soil moisture retention curve is that obtained by plotting soil moisture content against soil moisture suction. The curve will be of a different form according to whether the soil is drying or wetting. This is due to the phenomenon of capillary hysteresis.    PWW

**Reading**
Childs, E.C. 1969: *An introduction to the physical basis of soil water phenomena*. New York: Wiley. · Ward, R.C. and Robinson, M. 1990: *Principles of hydrology*. 3rd edn. Maidenhead: McGraw-Hill.

**retention forces**  Are those responsible for holding water in the pores of a rock or soil against the force of gravity. The forces involved are capillarity (surface tension), adsorption and osmosis.    PWW

**Reading**
Ward, R.C. and Robinson, M. 1990: *Principles of hydrology*. 3rd edn. Maidenhead: McGraw-Hill.

**retrogradation**  The destruction of a beach profile by large breakers resulting in retreat of the shoreline.

**return period**  The average time between events such as the flooding of a particular level. This information may also be expressed as the level which has a particular return period of flooding, for example, a hundred years. The inverse of the return period is the statistical probability of an event occurring in any individual year.    DTP

**reversed fault**  A fault whose fault-plane is inclined towards the upthrown side.

**reversed polarity**  The condition occurring when the earth's magnetic poles change their polarities, the positive becoming the negative and vice versa. Generally applied when the positive magnetic pole lies in the southern hemisphere.

**reversing dune**  A dune that tends to grow upwards but migrate only a limited distance because seasonal shifts in direction of the dominant wind cause it to move alternately in nearly opposite directions. See DUNE.

**Reynolds number**  ($R_e$) A dimensionless ratio defining the state of fluid motion as laminar or turbulent according to the relative magnitude of inertial and viscous forces. In general:

$$R_e = \rho f v L / \mu = v L / \nu$$

where $\rho f$ is fluid density, $v$ is velocity, $L$ is a characteristic length, $\mu$ is the dynamic viscosity and $\nu$ is the kinematic viscosity ($\nu = \mu / \rho f$), which is a measure of molecular interference between adjacent fluid layers. The Reynolds number for channel flow is defined by using the hydraulic radius $R$ or the mean depth as the characteristic length, and the mean flow velocity. Flow is turbulent if $R_e > 750$; laminar flow ($R_e < 500$) is rare in rivers and only occurs in shallow overland flow depths. Grain Reynolds numbers can be defined using grain diameter as the length, and grain fall velocity in still water as the velocity. If the grain Reynolds number is $< 0.1$, flow around the falling grain is streamlined and laminar, and the fall velocity varies as the square of the grain diameter (Stokes law). This is the case for silt and clay particle sizes; grain Reynolds numbers are larger for sand particles, and turbulence and flow separation occur around the falling grain as inertial forces dominate. For such grains the fall velocity varies as the square root of grain diameter (Richards 1982, pp. 76–9).                KSR

**Reference**
Richards, K.S. 1982: *Rivers: form and process in alluvial channels*. London and New York: Methuen.

**rheidity**  The capacity of some solid materials to flow under certain conditions.

**rheology**  The study of the deformation and flow of matter. In particular, rheology is concerned with whether a substance behaves as an elastic solid or effectively as a 'viscous' solid or in a manner intermediate between these two. The type of deformation experienced by a particular material is influenced by the duration of the applied stress. This is expressed by the property of rheidity which depends on the relationship between the resistance to viscous flow (viscosity) and the resistance to elastic deformation (rigidity) of a substance and is expressed in units of time. For ice a deforming stress must be applied for only a few weeks before it begins to flow but for subcrustal material the time required is several thousand years.                MAS

**Reading**
Hager, B.H. and O'Connell, R.J. 1980: Rheology, plate motions and mantle convection. In P.A. Davies and S.K. Runcorn eds, *Mechanisms of continental drift and plate tectonics*. London: Academic Press. · Mörner, N.-A. ed. 1980: *Earth rheology, isostasy and eustasy* Chichester: Wiley. · Ramberg, H. 1981: *Gravity, deformation and the earth's crust*. 2nd edn. London: Academic Press.

**rheotrophic**  See NUTRIENT STATUS.

**rhexistasy**  See BIOSTASY.

**rhizome**  The underground stem of some plants. The rootstock.

**rhizosphere**  The portion of the soil zone which immediately surrounds the root systems of plants.

**rhodoliths**  Free-living massive and branching spheroidal growths of calcareous red algae which occur in two different settings: (a) shallow lagoonal, reef-flat and back-reef environments, commonly in tidal channels and sea-grass beds, and (b) moderately deep water on fore-reef terraces or shelf edges.                ASG

**Reading**
Scoffin, T.P., Stoddart, D.R., Tudhope, A.W. and Woodroffe, C. 1985: Rhodoliths and coralliths of Muri Lagoon, Rarotonga, Cook Islands. *Coral reefs* 4, pp. 71–80.

**rhourd**  A pyramid-shaped dune akin to a STAR DUNE.

**ria**  An inlet of the sea formed by the flooding of river valleys, either by the rising of the sea during the FLANDRIAN TRANSGRESSION or as a consequence of sinking of the land. They contrast with fjords which are drowned glacial valleys. They are a feature of Pembrokeshire, the south-west peninsula of England, Brittany and Galicia (Spain).                ASG

**ribbed moraine**  See ROGEN MORAINE.

**Richardson number**  In the presence of a density gradient in a liquid, whether or not the liquid is stratified in a stable manner, stability can only be judged with respect to the level of turbulence, and hence the shear within the liquid. Thus, the Richardson number is a parameter that is used to classify a liquid in terms of its propensity to be stable, and is expressed as the ratio of the stabilizing forces that exist in a fluid due to its density to the destabilizing forces that arise from turbulent shear.                SNL

**Richter denudation slope**  A straight rock-slope unit with an angle of inclination which is at the maximum angle for stability (usually 32–36°) of its thin talus cover. Such slopes are relatively common in the Transantarctic Mountains, and have been recognized in the Cape Mountains of southern Africa and the European Alps. They form as rock fragments fall from the cliff at the crest of the Richter slope. If the newly fallen material just covers a little of the base of the cliff, the next fall will be over the new talus and hence the base of the cliff will then be higher, so the cliff will extend upwards by a series of minute steps at the angle of the talus. Eventually the free face will be eliminated and a smooth rock slope of uniform inclination will underlie a talus sheet. The talus may subsequently be removed.                                    MJS

**Reading**
Selby, M.J. 1993: *Hillslope materials and processes.* 2nd edn. Oxford: Oxford University Press.

**Richter scale**   The Richter scale is a measure of earthquake magnitude. The scale was introduced in 1935 by Charles Richter based on the amplitude of the largest wave form as recorded on a seismogram. Richter used a local seismogram network in southern California and the largest wave form was the S-wave. Subsequent work showed that for large and distant earthquakes the surface wave had the greatest amplitude and for deep focus quakes the body wave was the more appropriate measurement. To accommodate the wide range of values of wave form amplitude the scale is logarithmic. Each unit on the Richter scale represents a 10-fold increase in wave amplitude and around a 30-fold increase in the energy released. The largest recorded earthquakes have a magnitude of 8.6 and earthquakes with a magnitude of more than 5.5 typically cause structural damage. The measured values are corrected for depth and distance from the epicentre. The magnitude of an earthquake is a representation of the energy released and this is reflected not only by the wave form amplitude but also by the duration of ground shaking. Richter magnitude is represented by the expression:

$$M = \log X / T + Y$$

Where $M$ = Richter magnitude, $X$ = wave form amplitude, $T$ = frequency (measurement of duration) and $Y$ = difference in arrival time between P and S waves (a measure of the distance of the reading from epicentre). To achieve the best estimate for $M$, readings should be averaged from a number of different stations.                                    AD

**riedel shears**   A special form of shear fracture which forms roughly perpendicular to the main direction of shear strain or movement. Although they are mainly found in brittle materials they can occur in muds below the plastic limit.                                    WBW

**riegel**   A step in the rock floor of a glacial valley.

**riffle**   See POOL AND RIFFLE.

**rift valley**   A valley or linear trough formed by subsidence or downthrusting in areas of continental crust in plate interiors where tensional stresses predominate in the lithosphere. The classic interpretation of the morphology is that of a graben, with the rift floor being seen as a downthrown block bounded by normal faults, which create steep bounding escarpments. Seismic data suggest, however, that in many cases this is an oversimplification. The structure of many rifts appears to be asymmetric, with much of the downthrow occurring along a major boundary listric fault on one side only. These major faults tend to be discontinuous and may alternate along the rift, separated by transfer faults. Such rifts have a half-graben structure (see diagram).                    ASG

**Reading**
Summerfield, M.A. 1991: *Global geomorphology.* Harlow: Longman.

**rifting**   The tearing apart of the LITHOSPHERE along a roughly linear zone. This is due to tectonic stresses from either thermal doming or plate motions. If doming from upwelling ASTHENOSPHERE or MANTLE PLUMES comes first, the rifting is *active*, volcanism is abundant and uplift may extend for hundreds of kilometres. If stresses come from moving lithospheric plates or drag at the base of the lithosphere, the rifting is *passive*: uplift is confined to the shoulders of the rift zone and volcanism comes later as the crust and lithosphere are thinned. Production of a RIFT VALLEY by faulting leads to abundant clastic sedimentation as the steep margins are eroded.

In continental lithosphere, passive rifts may form at high angles to collision OROGENS, for example the Rhine GRABEN in relation to the Alps, and even the Baikal rift and Shanxi graben system (north-east Asia) in relation to the distant Himalayas. Down-faulting in Baikal is 5 or 6 km. Other passive rifts relate to transform faults, as in southern California and western Turkey, and the pull-apart basin of the Dead Sea. Rifts such as the Red Sea may be active and develop into sea-floor spreading ridges, that is, new divergent plate boundaries (see PLATE TECTONICS). In the Red Sea, continental crust is stretched, thinned and injected by basalt DYKES. Points of high heat flow about 50 km apart become the initial sites of oceanic crust (mid-ocean ridge basalt) formation. Magnetic anomalies develop as this process propagates along the rift axis and, after a few million years of rifting, the segments join to form a continuous sea-floor rift valley at about 1500 m depth (Bonatti 1985).                    IE

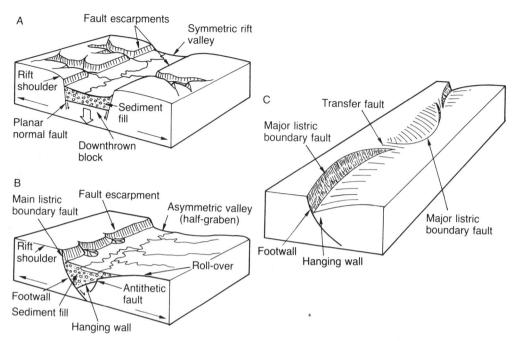

A

Fault escarpments

Symmetric rift valley

Rift shoulder

Planar normal fault

Sediment fill

Downthrown block

C

Transfer fault

Major listric boundary fault

Major listric boundary fault

Footwall

Hanging wall

B

Main listric boundary fault

Fault escarpment

Asymmetric valley (half-graben)

Rift shoulder

Roll-over

Footwall

Antithetic fault

Sediment fill

Hanging wall

**Rift valley.** Schematic representation of contrasting rift structures: (A) classic symmetric graben structure with a downthrown block bounded by normal faults. (B) asymmetric, half-graben structure. In both cases the number of faults in real rifts and the complexity of their structure are much greater than indicated here. (C) Highly schematic illustration of alternating half-graben along a rift valley.
*Source*: Summerfield 1991. Figures 4.9 and 4.10.

### Reference
Bonatti, E. 1985: Punctiform initiation of sea-floor spreading in the Red Sea during transition from a continental to an oceanic rift. *Nature* 316, pp. 33–7.

**rill**  Rills are small ephemeral channels that often form in sub-parallel sets on sloping agricultural land in response to intense run-off events. They are also common on steep and unprotected surfaces like road and other earthen embankments. The term 'rill' is also applied to small but longer-lived channels that carry run-off from hillslopes and convey it to larger stream channels (see INTERRILL FLOW). Rills are primarily found on hillslopes, though thalweg rills do occur (e.g. Slattery *et al.* 1994). On hillslopes, larger and deeper rills may capture the flow from adjoining ones, so that there is some integration of the rill network in the downslope direction. Rill spacing tends to diminish on steeper hillslope gradients. Rill dimensions lie in the range cm–dm for both width and depth, but rill length may reach hundreds of metres. Being a form of channelized flow, rills carry relatively deep and fast-flowing water, and are able to scour their beds efficiently and convey pebble-sized materials as bedload. Rills formed on tilled soils are ordinarily obliterated by ploughing,

**Rills** in an unprotected slope.

but can evolve again rapidly during subsequent run-off. Laboratory studies have shown that rill cutting can begin in shallow, laminar and subcritical flow, when a threshold of STREAM POWER is crossed that may be as low as 0.05 W/m2. The retreat of rill headcuts and the onset of active sediment transport along a rill appear to require power levels about 10 times higher than this.                                                    DLD

### Reading and Reference
Bryan, R.B. ed. 1987: Rill erosion: processes and significance. Cremlingen: *Catena* supplement 8. · — 1990: Knickpoint evolution in rillwash. In R.B. Bryan ed., Soil erosion – experiments and models. Cremlingen: *Catena* supplement 17, pp. 111–32. · Slattery, M.C., Burt, T.P. and Boardman, J. 1994: Rill erosion along the thalweg of a hillslope hollow: a case study from the Cotswold Hills, central England. *Earth surface processes and landforms* 19, pp. 377–85.

**rime ice**   Ice formed by the accumulation of supercooled water droplets when they strike a cold object and freeze on impact. Rime ice builds up most rapidly in cool humid conditions on surfaces exposed to the wind. On mountain peaks it can accumulate as large cauliflower-shaped excrescences.                              DES

### Reading
Koerner, R.M. 1961: Glaciological observations in Trinity Peninsula, Graham Land, Antarctica. *Journal of glaciology* 3.30, pp. 1063–74.

**ring complex**   'A petrologically variable but structurally distinctive group of hypabyssal or subvolcanic igneous intrusions that include ring dykes, partial ring dykes and cone sheets. Outcrop patterns are arcuate, annular, polygonal and elliptical with varying diameters ranging from less than 1 to 30 km or greater. The majority of ring complexes represent the eroded roots of volcanoes and their calderas' (Bowden 1985, p. 17).                              ASG

### Reference
Bowden, P. 1985: The geochemistry and mineralization of alkaline ring complexes in Africa (a review). *Journal of African earth sciences* 3, pp. 17–39.

**ring-dyke**   A funnel-shaped or cylindrical intrusion of igneous rock usually surrounded by an older intrusive mass. The dyke appears as a ring of rocks when viewed from the air.

**rip current**   A narrow, fast current flowing seaward through the breaker zone. The term, first introduced by F.P. Shepard in 1936, replaces undertow. Rip currents are fed by longshore currents in the surfzone. Their velocity ranges from about $1\,m\,s^{-1}$ to over $5\,m\,s^{-1}$ in severe storms. Channels 1–3 m deep can be scoured in the breaker zone by rip currents. Their spacing depends on variation of wave height alongshore, which may arise through wave refraction or may be related to the width of the surfzone. They may be associated with cuspate shoreline features and edge waves.         CAMK

### Reading
Bowen, A.J. 1969: Rip currents. *Journal of geophysical research* 74, pp. 5467–78. · Dalrymple, R.A. 1975: A mechanism for rip current generation on an open coast. *Journal of geophysical research.* 80, pp. 3485–7. · Gruszczyński, M., Rudowski, S., Semil J., Slomiński J., and Zrobek, J. 1983: Rip currents as a geological tool. *Sedimentology* 40, pp. 217–36.

**riparian**   A feature occurring in the region of a river bank. Thus, it is conventional to refer to riparian vegetation, riparian BUFFER STRIPS, conservation of the riparian zone, etc. The condition of the riparian zone is of great importance to the stability of the river banks, to the quality of the stream habitat, and provides a focus for management in catchments where extensive land use has changed the hydrologic and erosional behaviour of the broader landscape. In such cases, the riparian zone forms a link between the catchment and the stream. A number of specific values are associated with the riparian zone. These include the maintenance of water quality, especially in surface run-off and groundwater that pass over and through the soils of the riparian zone. Grass and forest can provide very efficient removal of excess nutrients such as N, perhaps sourced from fertilizers, that would otherwise enter the stream system. Riparian environments may also provide valuable remnant ecosystems, a corridor for the movement of animals, and an environment of considerable recreational and scenic amenity value. The condition of in-stream habitats may also be related to events in the riparian zone, which may release important organic materials that nourish the aquatic ecosystem, and also moderate water temperatures where shading of the stream occurs. Riparian vegetation may assist in the stabilization of stream banks and also contribute large fragments of organic debris (e.g. branches) which, singly or in debris jams, may temporarily obstruct flow in the stream, and contribute to the diversity of habitats there.                              DLD

### Reading
Large, A.R.G. and Petts, G.E. 1994: Rehabilitation of river margins. In P. Calow and G.E. Petts eds, *The rivers handbook*. Vol. 2. Oxford: Blackwell. Ch. 21, pp. 401–18.

**ripple**   See CURRENT RIPPLES.

**rising limb** The portion of the stream hydrograph for a storm event between the beginning of the rise in discharge and the peak of the hydrograph. (For diagram see HYDROGRAPH.)

**river basin planning** (integrated basin planning) The process of managing water resources within the drainage basin in a manner which optimizes water use throughout the basin and minimizes deleterious effects for water, river channels and land use. Although the idea of an integrated approach to river basin development is long established, experience of piecemeal development which has induced subsequent feedback effects in other parts of the drainage basin has fostered the movement towards a more integrated view. This movement has also been encouraged both by the need to design multiple purpose water resource projects that will, for example, supply water and power and reduce floods, and by the use of the drainage basin as the most appropriate unit for analysis in relation to planning. The United Nations first advocated integrated river basin development in 1958 and its report was reissued in 1970.    KJG

**Reading and Reference**
Saha, S.K. and Barrow, C.J. 1981: *River basin planning: theory and practice.* Chichester: Wiley. · United Nations 1970: *Integrated river basin development: report of panel of experts.* New York: Department of Economic and Social Affairs.

**river classification** See CHANNEL CLASSIFICATION.

**river discharge** See DISCHARGE.

**river metamorphosis** Refers to the change of river channel morphology that can occur when changes of discharge and sediment exceed a THRESHOLD condition. Channel changes which occur can be from a multi-thread to a single thread channel and S.A. Schumm (1969) introduced the term river metamorphosis to signify the range of changes that may arise. Metamorphosis may involve changes of size, shape or composition of aspects of river channel morphology as shown in the table. It is important to be able to predict the degrees of freedom which a river system possesses: width, depth, slope, velocity and plan shape. To predict the character of river metamorphosis it is necessary to suggest what changes will occur, where they will obtain, and when the changes will begin and end.    KJG

**Reading and References**
Gregory, K.J. 1981: River channels. In K.J. Gregory and D.E. Walling eds, *Man and environmental processes.* London: Butterworth. Pp. 123–43. · Schumm, S.A. 1969: River metamorphosis. *Proceedings of the American Society of Civil Engineers, Hydraulic Division* 95, pp. 251–73. · — 1977: *The fluvial system.* Chichester. Wiley.

**river regime** See REGIME.

River metamorphosis: potential channel adjustments

| | Fluvial landform | | |
|---|---|---|---|
| Potential adjustments of | River channel cross-section | River channel pattern | Drainage network |
| Size | INCREASE OR DECREASE OF RIVER CHANNEL CAPACITY Erosion of bed and banks can produce a larger channel which maintains the same shape. Sedimentation can produce a smaller channel which maintains the same shape. | INCREASE OR DECREASE OF SIZE OF PATTERN Increase or decrease of meander wavelength while preserving the same planform shape. | INCREASE OR DECREASE OF NETWORK EXTENT AND DENSITY Extension of channels or shrinkage of perennial, intermittent and ephemeral streams. |
| Shape | ADJUSTMENT OF SHAPE Width/depth ratio may be increased or decreased. | ALTERATION OF SHAPE OF PATTERN A change from regular to irregular meanders. | DRAINAGE PATTERN CHANGE IN SHAPE Inclusion of new stream channels after deforestation. |
| Composition | CHANGE IN CHANNEL SEDIMENTS Alteration of grain size of sediments in bed and banks possibly accompanied by development of berms and bars. | PLANFORM METAMORPHOSIS Change from single to multi-thread channel or converse. | NETWORK COMPOSITION CHANGE The replacement of channels with no definite stream channel (dambos in West Africa) by a clearly defined channel. |

General changes are indicated in capitals and examples given in lower case letters.

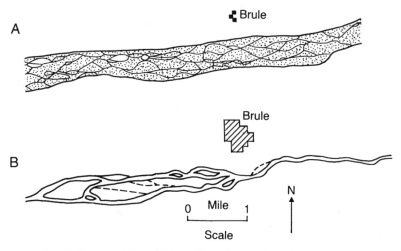

**River metamorphosis.** The South Platte River at Brule, Nebraska. The channel according to surveys made in 1897 is shown in A and the same channel in 1959 in B is based on an aerial photograph.
*Source*: S.A. Schumm 1971: Channel adjustment and river metamorphosis. In H.W. Shen ed., *River mechanics*. Vol. 1. Fort Collins: Water Resources Publications. Pp. 5.1–5.22.

**river terrace** A nearly flat area formed at river level but now above the river, separated at least by an eroded slope. Such a landform implies that production of a *valley floor* or alluvial plain, by a phase of deposition or lateral erosion by the river, was followed by a phase of *down-cutting* (incision). Terraces may be *rock-floored* (though a veneer of river deposits may remain) or *alluvial* (accumulation terraces). The simplest alluvial terraces consist of deposits from one phase, but often there are patches of older

deposits: just as a terrace may be cut into rock, it may be cut into any previous deposit, fluvial or otherwise. Thus the age of the terrace materials must not be confused with the age of the land-form.

Initially a river terrace should slope down-valley rather than toward the river, but later modification by compaction, erosion or deposition (slopewash or aeolian) commonly produces a riverward slope. There are other types of terrace or bench that might, after further erosion,

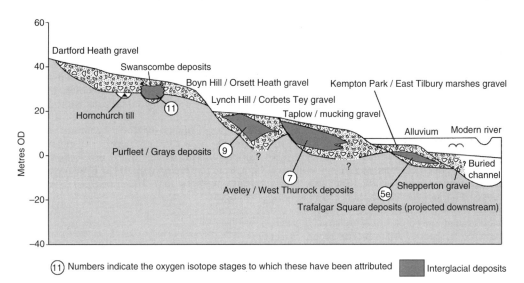

Idealized transverse sequence through the terraces of the lower Thames, England.
*Source*: Reproduced from D.R. Bridgland, P. Allen and B.A. Haggart eds 1995: *The Quaternary of the lower reaches of the Thames: field guide*. Durham: Quaternary Research Association.

be confused with river terraces. Colluvial terraces are formed of slope deposits and have greater slopes toward the river: they are common in British uplands, in soliflucted till. PEDIMENTS are erosional forms at the foot of highlands: they are usually attributed to slope retreat but lateral planation by rivers has also been invoked. Pediment slopes are usually steeper than those of rock-floored terraces. Structural benches relate to differential erosion of weaker rocks and are floored by more resistant rocks. In flat-lying rocks (e.g. most of the Grand Canyon) they may resemble river terraces, but their upslope margins are more gradual (not former river bluffs) and if traced down-valley they maintain their relation to a single bed of rock. Agricultural terraces are artificial, which may not be obvious long after their abandonment when even the remains of terrace-front walls may become obscure. KAME TERRACES are a special type of river terrace, whose formation was influenced by the presence of a glacier.

Traced downstream, a terrace provides the former gradient (as modified by tectonism), and its surface deposits show the competence of the former river, and perhaps the channel type. Later lateral erosion may completely destroy a terrace, but in a valley alluvial terrace remnants may be *rock-defended* as the river encounters the sloping surface of underlying bedrock. Terraces on either side of a river will be *paired* at the same relative height if downcutting is rapid, but *unpaired* (at varied heights) if it is steady, especially by a meandering stream. A series of river terraces provides valuable information on landscape development.

The alternation of alluviation or lateral erosion with downcutting may be produced in various ways, which must be worked out for each region. Regional uplift is the most obvious possibility, producing for example the mainly rock-floored terraces along the middle Rhine (Ahnert 1998), though the existence of a staircase of twelve terraces probably relates to climatic variation rather than interruptions of uplift. Changes in BASE LEVEL commonly cause downcutting or alluviation. A fall in sea level (see EUSTASY) has been assumed to cause river downcutting, working headward from the newly exposed course: the assumption that the new course has a steeper gradient than the old lower course may not apply to smaller rivers where CONTINENTAL SHELVES are broad. River diversion or capture may produce a single terrace set.

River aggradation or downcutting is controlled by the balance between the river's capacity to erode and transport, and the supply of material (mainly BEDLOAD, since suspended silts and clays are unlikely to settle out in flowing water). Both capacity and supply are changing continuously, giving ample opportunity for phases of aggradation and downcutting to alternate; in fact, this may occur in a single flood, with metres of scour on the rising stage followed by deposition as discharge wanes. Average discharge is less important than the frequency of major FLOODS. Delivery of material from slopes and tributaries may be on a different time-scale to their onward transport by the main stream, which may leave such material as terraces after a major flood. Minor floods may cause aggradation, but major ones take material farther down-valley leaving terraces upstream. Major phases of aggradation may come from glaciation, periglaciation, climatic changes or (anthropogenic) soil erosion. Downcutting is likely to follow the reduction of such high rates of sediment supply. When rivers are blocked by landslides, they initially aggrade but soon cut down as the slide material is eroded (see figure).

There have been debates, for example concerning the Thames, England, on whether terraces were built up in cold phases of the Quaternary, because of sediment supply; or in warm phases, because of high sea levels. The work of Bridgland (1994) shows that the Thames terraces are more complex. Although each terrace suite seems to relate to a major warm/cold cycle, it contains both fossiliferous fine-grained interglacial deposits and gravels which Bridgland relates to the beginning and end of cold phases. The common assumption that each terrace level relates to one change in base level or climate is challenged also by the concepts of *episodic* erosion and sediment supply, and of *complex response* in the relation of sediment transport to supply. Schumm and others (1987, pp. 123–6 and 211) have shown by experimentation that deposition and downcutting may overshoot sustainable levels, producing a damped oscillation: thus two or more terraces may relate to a single change. They give examples from Colorado and California of flights of four or more terraces formed in a century, over a 10 m range of altitude. The THRESHOLDS involved may be intrinsic to development of the river system, and some terraces may be unrelated to external changes. IE

**References**

Ahnert, F. 1998: *Introduction to geomorphology.* London: Arnold Ch. 15, pp. 183–190. · Bridgland, D.R. 1994: *Quaternary of the Thames.* Geological Conservation Review Series, 7. London: Chapman & Hall. · Richards, K. 1982: *Rivers: form and process in alluvial channels.* London: Methuen. Pp. 272–5. · Schumm, S.A., Mosley, M.P. and Weaver, W.E. 1987: *Experimental fluvial geomorphology.* New York: Wiley.

**roche moutonnée** An asymmetric rock bump with one side ice-moulded and the other side steepened and often cliffed, generally recognized as the hallmark of glacial erosion. The term was introduced by H.B. de Saussure in 1787 in recognition of the similarity of the rocks to the rippled appearance of wavy wigs styled *moutonnées* at the time. The morphology of roches moutonnées seems to reflect the contrast between ABRASION on the smoothed up-ice side and PLUCKING on the lee side.    DES

**Reading**
Chorley, R.J., Schumm, S.A. and Sugden, D.E. 1984: *Geomorphology*. London: Methuen. Ch. 17. · Sugden, D.E., Glasser, N. and Clapperton, C.M. 1992: Evolution of large roches moutonnées. *Geografiska Annaler* 74A, pp. 253–64.

**rock coatings** These are among the earliest phenomena that we know humans studied. Prehistoric artists used rock coatings as a backdrop for engravings. After rock coatings were mentioned in the Old Testament (Leviticus 13.14), Nabateans used visual differences between rock coatings and the underlying sandstone to highlight monumental architecture in Petra at about the same time as Nasca cultures juxtaposed dark and light rock coatings to create giant ground figures in the Peruvian Desert. The phrase 'bare or naked rock' is almost always a misnomer, for in every 'rock exposure' students of the earth are almost always studying rock coatings. Some 14 different classes of rock coatings have been identified thus far. (See also DURICRUSTS.)    RID

**Reading**
Dorn, R.I. 1998: *Rock coatings*. Amsterdam: Elsevier. · Krumbein, W.E. and Dyer, B.D. 1985: This planet is alive: weathering and biology, a multi-faceted problem. In J.E. Drever ed., *The chemistry of weathering*. Dordrecht: D. Reidel Publishing Co. Pp. 143–60.

**rock drumlin** A rock hill streamlined by the passage of over-running glacier ice.

**rock flour** The fine debris created by ABRASION beneath a glacier. The material, generally less than 0.2 mm in diameter, may be flushed away to give the characteristic brown and blue-green colour of glacial meltwater streams and lakes. (See GLACIER MILK.)    DES

**Reading**
Haldorsen, S. 1981: Grain-size distribution of subglacial till and its relation to glacial crushing and abrasion. *Boreas* 87, pp. 1003–15.

**rock glacier** A term which has been used to describe a feature, comparatively common in many alpine areas, which looks like a glacier (with apparent flow structures, etc.) but is composed of rock debris. Some workers use the term to imply that such a feature has a specific mode of origin. There are two main theories: that the rock debris has ice mixed in the spaces between the rock (the interstitial ice model) or that the debris is a thick covering on a thin, probably decaying true glacier (the glacier ice model). There is much disagreement as to which model is correct. One major feature is that rock glaciers usually exhibit slow movement, often less than a metre per year, i.e. at least an order of magnitude less than most true ice glaciers.    WBW

**Reading**
Giardino, J.R., Shroder, J.F. and Vitek, J.D. 1987: *Rock glaciers*. Boston: Allen & Unwin. · Martin, H.E. and Whalley, W.B. 1987: Rock glaciers a review: Part I. *Progress in physical geography* 11, pp. 260–86. · Whalley, W.B. and Martin, H.E. 1992: Rock glaciers: Part II. Models and mechanisms. *Progress in physical geography* 16, pp. 127–86.

**rock mass strength** An important concept in geomorphology, developed by Selby (1980) and others in an attempt to gain a quantitative measure of the resistance of a rock mass to erosion. It involves giving a rank of importance to a range of different rock parameters and then summing them to come up with a total rating of strength.    ASG

**Reference**
Selby, M.J. 1980: A rock mass strength classification for geomorphic purposes: with tests from Antarctica and New Zealand. *Zeitschrift für Geomorphologie* 24, pp. 31–51.

**rock quality indices** Measures used to relate the numerical intensity of fractures in a rock to the quality of the unweathered rock. One measure is the relationship between the compressional wave velocity measured *in situ* in massive rock and on a core of the intact rock (see INTACT STRENGTH). The fewer the joints, the nearer is the ratio to unity. Rock quality designation is the relationship between intact cored rock length to the total length drilled and the 'fracture index' is the frequency of fractures occurring within a rock unit.    WBW

**Reading**
Bell, F.G. 1992: *Engineering properties of soils and rocks*. 3rd edn. Oxford: Butterworth Heinemann.

**rock step** See RIEGEL.

**rock varnish** A type of a ROCK COATING that is dark in colour and is characterized by clay minerals (~40–60 per cent) cemented to rock surfaces by oxides and hydroxides of manganese (birnessite) and iron (goethite and haematite) that typically comprise 20–40 per cent. The

Nomenclature of rock coatings, in alphabetical order

| Term | Summary description | Related terms |
|------|---------------------|---------------|
| Carbonate skin | Coating composed primarily of carbonate, usually calcium carbonate, but could be combined with magnesium or other cations | Caliche, calcrete, patina, travertine, carbonate skin, dolocrete, dolomite |
| Case hardening agents | Addition of cementing agent to rock matrix material; the agent may be manganese, sulphate, carbonate, silica, iron, oxalate, organisms, or anthropogenic | Sometimes called a particular type of rock coating |
| Dust film | Light powder of clay- and silt-sized particles attached to rough surfaces and in rock fractures | Gesetz der Wüstenbildung; clay skins; clay films; soiling |
| Heavy metal skins | Coatings of iron, manganese, copper, zinc, nickel, mercury, lead and other heavy metals on rocks in natural and human-altered settings | Described by chemical composition of the film |
| Iron film | Composed primarily of iron oxides or oxyhydroxides; unlike orange rock varnish because it does not have clay as a major constitutent | Ground patina, ferric oxide coating, red staining, ferric hydroxides, iron staining, iron-rich rock varnish, red-brown coating |
| Lithobiontic coatings | Organic remains form the rock coating for example lichens, moss, fungi, cyanobacteria, algae | Organic mat, biofilms |
| Nitrate crust | Potassium and calcium nitrate coatings on rocks, often in caves and rock shelters in limestone areas | Saltpetre, nitre, icing |
| Oxalate crust | Mostly calcium oxalate and silica with variable concentrations of magnesium, aluminium, potassium, phosphorus, sulphur, barium, and manganese. Often found forming near or with lichens. Usually dark in colour, but can be as light as ivory | Oxalate patina, lichen-produced crusts, patina, scialbatura |
| Phosphate skin | Various phosphate minerals (e.g. iron phosphates or apatite) that are mixed with clays and sometimes manganese | Organophosphate film; epilithic biofilm |
| Pigment | Human-manufactured material placed on rock surfaces by people | Pictograph, paint, sometimes described by the nature of the material |
| Rock varnish | Clay minerals. Mn and Fe oxides, and minor and trace elements; colour ranges from orange to black in colour produced by variable concentrations of different manganese and iron oxides | Desert varnish, desert lacquer, patina, manteau protecteur, Wüstenlack, Schutzrinden, cataract films |
| Salt crust | The precipitation of sodium chloride on rock surfaces | Halite crust, efflorescence, salcrete |
| Silica glaze | Usually clear white to orange shiny lustre, but can be darker in appearance, composed primarily of amorphous silica and aluminium, but often with iron | Desert glaze, turtle-skin patina, siliceous crusts, silica-alumina coating, silica skins |
| Sulphate crust | Composed of the superposition of sulphates (e.g. barite, gypsum) on rocks; not gypsum crusts that are sedimentary deposits | Gypsum crusts; sulphate skin |

Classification of rock mass strength

| Variable | Weighting (%) | Very strong | Strong | Moderate | Weak | Very weak |
|---|---|---|---|---|---|---|
| Intact rock strength (Schmidt hammer rebound value) | 20 | 100–60 $r = 20$ | 60–50 $r = 18$ | 50–40 $r = 14$ | 40–35 $r = 10$ | 35–10 $r = 5$ |
| Weathering | 10 | Unweathered $r = 10$ | Slightly weathered $r = 9$ | Moderately weathered $r = 7$ | Highly weathered $r = 5$ | Completely weathered $r = 3$ |
| Joint spacing | 30 | > 3 m $r = 30$ | 3–1 m $r = 28$ | 1–0.3 m $r = 21$ | 300–50 mm $r = 15$ | < 50 mm $r = 8$ |
| Joint orientations | 20 | Very favourable. Steep dips into slope, cross joints interlock $r = 20$ | Favourable Moderate dips into slope $r = 18$ | Fair. Horizontal dips or nearly vertical dips (hard rocks only) $r = 14$ | Unfavourable. Moderate dips out of slope $r = 9$ | Very unfavourable. Steep dips out of slope $r = 5$ |
| Joint width | 7 | < 0.1 mm $r = 7$ | 0.1–1 mm $r = 6$ | 1–5 mm $r = 5$ | 5–20 mm $r = 4$ | > 20 mm $r = 2$ |
| Joint continuity and infill | 7 | None, continuous $r = 7$ | Few, continuous $r = 6$ | Continuous, no infill $r = 554$ | Continuous, thin infill $r = 4$ | Continuous, thick infill $r = 1$ |
| Groundwater outflow | 6 | None $r = 6$ | Trace $r = 5$ | Slight < 40 ml $s^{-1} m^{-2}$ $r = 4$ | Moderate 40–200 ml $s^{-1} m^{-2}$ $r = 3$ | Great > 200 ml $s^{-1} m^{-2}$ $r = 1$ |
| Total rating | | 100–91 | 90–71 | 70–51 | 50–26 | < 26 |

*Source*: Modified from Selby 1980. Table 6, pp. 44–5.

oxides are nanometres in thickness and they are a by-product of the weathering of bacterial casts. Clay minerals are ubiquitous, but they will not form rock varnish by themselves. Rock varnish grows only where and when the nanometre-scale remnants of bacteria manoeuvre in between the broken and decayed fragments of clay minerals weathered at the nanometre scale. Most varnishes seen at the surface today actually start in the subsurface in fissures. It forms in all terrestrial environments, but the term 'desert varnish' is common because varnishes are most geochemically stable in the subaerial environment in deserts. Typical thicknesses are less than 100 $\mu$m. Usually dull in lustre, its occasional sheen comes from a smooth surface micromorphology in combination with manganese enrichment at the very surface of the varnish. The constituents in varnish accrete on the host rock. The most volumetrically significant postdepositional modification is the leaching of cations from rock varnish. Varnishes are also eroded by lithobionts, but in places where varnishes are intact black (manganese-rich) and orange (iron-rich) microlaminae may indicate past changes in climate. Rock varnishes

sometimes act as case-hardening agents for the underlying rock. Many varnishes also encapsulate organic matter; unfortunately, radiocarbon dating of these organics is not useful because the entombed organics do not form an open system of older and younger organics.                    RID

**Reading**
Dorn, R.I. 1998. *Rock coatings*. Amsterdam: Elsevier. · Drake, N.A., Heydeman, M.T. and White, K.H. 1993. Distribution and formation of rock varnish in southern Tunisia. *Earth surface processes and landforms* 18, pp. 31–41.

**roddon** A sinuous, silty ridge that snakes about above the general level of the peat fenland of East Anglia (England). Roddons represent ancient river systems that may initially have flowed between levées above the general level of the surrounding land or which have subsequently become relatively elevated as a consequence of peat wastage. They are favoured sites for settlement.                    ASG

**Reading**
Fowler, G. 1932: Old river beds in the Fenlands. *Geographical journal* 79, pp. 210–12. · Godwin, H. 1938: The origin of roddons. *Geographical journal* 91, pp. 241–50.

**rogen moraine** Landform assemblage of numerous, parallel, closely spaced ridges consisting of glacial drift, usually TILL. The ridges are formed transverse to ice flow in a subglacial position and are usually found in the central portions of former ice sheets. Individual ridges are typically 10–30 m high, 300–1200 m long and 150–300 m wide, have a straight to arcuate planform, concave in the down-ice direction. Their name derives from Lake Rogen in Sweden. *Ribbed moraine* is the North American, and perhaps, the preferred term. Their formation is uncertain but is often closely linked to that of DRUMLINS. (See SUBGLACIAL BEDFORMS.)                               CDC

**Reading**
Hättestrand, C. and Kleman, J. 1999: Ribbed moraine formation. *Quaternary science reviews* 18.1, pp. 43–61.

**rollability** A property related to the angle of a slope down which a given sedimentary particle will roll. The concept was introduced by Winkelmolen to account for the ease with which particles (usually of sand size) can be rolled in unidirectional fluid flow. It is indexed by the time it takes for grains to travel down the inside of a revolving cylinder inclined at a slight angle. The time taken is a measure of the rollability potential. Usually, a few grammes of the sediment is used in a test. The measure is said to correlate with the influence of grain form on the particle settling velocity. Rollability can be considered as amalgamating several aspects of PARTICLE FORM.                               WBW

**Reading**
Goudie, A.S. ed. 1990: *Geomorphological techniques*, 2nd edn. London: Unwin Hyman. · Winkelmolen, A.M. 1971: Rollability, a functional shape property of sand grains. *Journal of sedimentary petrology* 41, pp. 703–14.

**Rossby waves** Wave motions in the atmosphere of planetary scale. They take the form of vast meanders of airflow around a hemisphere and are most clearly identifiable at upper levels. They owe their existence to the variation with the latitude ($\phi$) of Coriolis parameter ($2\Omega \sin \phi$), where $\Omega$ is the angular velocity of rotation of the earth).

In non-divergent, large-scale motion absolute VORTICITY, $\zeta_a$, can be considered constant, given by:

$$\frac{d}{dt}(\zeta_a) = \frac{d}{dt}(\zeta_r + 2\Omega \sin \phi) = 0$$

where $\zeta_r$ is the relative vorticity. In the northern hemisphere if a uniform westerly current with zero initial relative vorticity is displaced poleward, then, as latitude increases, the relative vorticity must become negative (i.e. anti-cyclo-nic) so that the air turns southward. After moving towards the equator of its original latitude its relative vorticity becomes positive (cyclonic) so that it turns northward. The current thus oscillates about its original latitude giving a series of waves called long waves or Rossby waves (named after Carl-Gustav Rossby, a Swedish-American meteorologist) usually numbering between two and six around a latitude circle.

Rossby wave theory predicts that in a basic westerly current and for a particular number of troughs and ridges around a latitude circle, there is a critical flow speed for which the waves are steady and stationary. If the flow is less than this critical speed the waves drift westward, while if it is greater they drift eastward.

Rossby waves are forced by three principal mechanisms: by orographic forcing resulting from a basic westerly flow impinging on a mountain range (especially the Rockies or the Andes); by thermal forcing due to differential heating of the oceans and continents; and by interaction with smaller scale disturbances such as extra-tropical cyclones.

Rossby waves have also been identified in the oceans.                               KJW

**Reading**
Gill, A.E. 1982: *Atmosphere–ocean dynamics*. London: Academic Press. · Houghton, J. 1986: *The physics of atmospheres*. Cambridge: Cambridge University Press.

**rotational failure** The name given to failure, usually in clays although also in weak rocks, where the shape of the slip surface (akin to a shear plane), which forms the boundary between the stable ground and the mass which has moved, is curved. At the top it frequently forms a step, where tension cracks develop in the early stages of failure. At the base there is usually a bulge. If the mass is rather mobile, as is frequently the case with QUICK CLAY, then there may be spreading at the foot, perhaps to give a secondary earthflow. In some cases the production of one rotational failure may give rise to a second failure.                               WBW

**Reading**
Selby, M.J. 1993: *Hillslope materials and processes*. 2nd edn. Oxford and New York: Oxford University Press.

**rotor streaming** A condition of unsteady flow over a mountain in which lee eddies are generated, then blow away. It probably involves a mechanism similar to that influencing the vortex streets noted by and named after von Karman.                               JSAG

**Reading**
Atkinson, B.W. 1981: *Meso scale atmospheric circulations*. London and New York: Academic Press.

**roughness** As a fluid moves over a surface (e.g. water over a river bed or wind over the earth's surface) it is retarded by friction at its base. The amount of friction at the interface between surface and fluid is dependent upon the roughness of the surface and the resultant retardation develops the fluid velocity profile in the BOUNDARY LAYER. Very close to the surface the drag imposed by the roughness results in a fluid velocity of zero and the depth of this zero-velocity layer is called the ROUGHNESS LENGTH. Surface roughness is a very important parameter in sediment transport modelling in both wind and water, because, by initiating a roughness length, it partly controls the gradient of the velocity profile and hence the shear velocity of the fluid.

The friction resulting from surface roughness can be effected by a number of parameters across a wide range of scales. In both wind (Bagnold 1941) and fluvial environments (Bathurst 1993) the roughness of the surface has been shown to increase with the mean or median particle size of the surface material. This is commonly termed *grain roughness*. The size, shape and arrangement of vegetation may also significantly contribute to the roughness evident in both wind and fluvial environments. At a larger scale, bedforms (such as ripples or dunes) may provide a *form roughness*. Similarly, *free surface resistance* may stem from surface waves and hydraulic jumps in rivers (Knighton 1998). At the largest scale, bends and islands in rivers may contribute to the total fluvial roughness and in meteorological studies roughness may be imparted by forests, cities or mountain fronts.

In fluvial studies roughness is commonly measured by friction co-efficients such as MANNING's $n$, Chézy $C$ or Darcy–Weisbach $f$, whilst in aeolian studies the aerodynamic roughness length $z_0$ is measured. A particular problem with the assessment of roughness is the great spatial and temporal variability which the property exhibits. For example, the form roughness provided by bedforms may alter substantially with increasing or decreasing flow velocities as the bedforms themselves grow or diminish. The relationships between surface roughness, roughness length, flow velocities and sediment transport are still not completely understood. GFSW

**Reading and References**
Bagnold, R.A. 1941: *The physics of blown sand and desert dunes*. London: Methuen. · Bathurst, J.C. 1993: Flow resistance through the channel network. In K. Beven and M.J. Kirkby eds, *Channel network hydrology*. Chichester: Wiley. Pp. 69–98. · Knighton, D. 1998: *Fluvial forms and processes: a new perspective*. London: Arnold. · Wiggs, G.F.S. 1997: Sediment mobilisation by the wind. In D.S.G. Thomas ed., *Arid zone geomorphology*. 2nd edn. London: Wiley: Pp. 351–72.

**roughness height** See ROUGHNESS LENGTH.

**roughness length** Fluid velocity over a surface (e.g. water over a river bed or wind over the earth's surface) is retarded by the surface ROUGHNESS and very close to that surface the velocity of the fluid becomes zero. The depth of this zero-velocity layer (termed $z_0$) is called the roughness length and is controlled by the surface roughness which develops the fluid velocity profile. The roughness length is a particularly important parameter in sediment transport modelling in both fluvial and aeolian studies as it partly controls the gradient of the velocity profile and hence the shear velocity of the fluid. The relationships between fluid velocity, roughness length and shear velocity are described by the LAW OF THE WALL.

More commonly measured in aeolian studies (in fluvial studies relative resistance coefficients such as MANNING's $n$ are usually identified) the aerodynamic roughness length has been shown to vary widely both temporally and spatially. Typical values for $z_0$ may be 0.0007 m for stationary sand surfaces or 0.2 m and above for vegetated or semi-vegetated desert surfaces. Bagnold (1941) identified a relationship between $z_0$ and surface grain diameter ($d$) where $z_0 = d/30$. More recent experiments have shown that $z_0$ is also a function of roughness element (e.g. large stones) spacing where $z_0$ may reach a maximum value of $d/8$ for widely spaced elements before returning to $d/30$ as element spacing increased further (Greeley and Iversen 1985). Roughness length can also be determined dynamically from measured velocity profiles and using the law of the wall to calculate the height intercept at zero velocity. Such an approach works well for flat and simple terrain but there is still no agreed method for determining $z_0$ on surfaces which are complex in character and may involve a changing mixture of sediment sizes and/or topography (Blumberg and Greeley 1993). GFSW

**Reading and References**
Bagnold, R.A. 1941: *The physics of blown sand and desert dunes*. London: Methuen. · Blumberg, D.G. and Greeley, R. 1993: Field studies of aerodynamic roughness length. *Journal of arid environments* 25, pp. 39–48. · Greeley, R. and Iversen, J.D. 1985. *Wind as a geological process*. Cambridge: Cambridge University Press. · Wiggs, G.F.S. 1997: Sediment mobilisation by the wind. In D.S.G. Thomas, ed., *Arid zone geomorphology*. 2nd edn. Chichester: Wiley. Pp. 351–72.

**roundness** Tending towards rounded edges, describing the degree of abrasion of a clastic fragment as shown by the sharpness of its edges and corners, independent of shape. Spherical particles are perfectly rounded, but well-rounded objects (such as an egg), need not be spherical (Waddell 1933).                ASG

**Reference**
Waddell, H. 1933: Sphericity and roundness of quartz particles. *Journal of geology* 41, pp. 310–31.

**routing, flood**   See FLOOD ROUTING.

**r-selection**   See r- AND K-SELECTION.

**ruderal vegetation**   Plants which grow in waste land, on or among rubbish and debris. The term was used originally in the sense of stone waste but has been extended to include roadsides, verges and old fields. The plant succession is usually rapid, with profusely seeded and rapidly growing annuals being replaced by hardy perennials, particularly grasses. Naturalized plants are often found in the early seral stages because of the relative absence of competition, but in time they tend to be eliminated in the struggle with native plants.        PAF

**run-off**   Water that drains from an area of the land is referred to as run-off. The process may be considered at various scales, spanning continental run-off, the run-off from a river basin or other catchment area, or the run-off from a small area such as a roof, a field, or an experimental run-off plot (see PLOTS, EROSION / RUN-OFF). In all cases, the run-off forms a link in the global hydrologic cycle, in which about 40,000 km$^3$ of water drain from the land areas of the earth annually.

Run-off follows both above-ground and sub-surface pathways, and may then be referred to explicitly as surface run-off, sub-surface run-off or groundwater run-off. The partitioning of water delivered as precipitation among these various possible flow pathways is influenced partly by the nature of the precipitation itself, and also by a range of land surface and soil characteristics (Brammer and McDonnell 1996; Newman *et al.* 1998). The intensity of rainfall exerts an important control on the flow pathway that run-off follows. Intense rain may exceed the ability of the soil to take in water, and in this case surface run-off is promoted. This may take the form of infiltration-excess or Hortonian surface run-off (see HORTON OVER-LAND FLOW MODEL), a form that is particularly common in desert areas where soils often display low infiltration rates, but is also characteristic of the extensive impervious surfaces of urban areas. However, surface run-off may also

arise in rain of lower intensity, and can be promoted where soils are partly saturated, such as hollows or low-lying areas flanking streams, or where soils are thin so that their ability to hold water is low. Surface run-off generated in such locations is referred to as saturation overland flow. Soils that are very permeable, such as the humus-rich soils of some forested regions, may have the capacity to take in almost all arriving rain water, so that surface run-off may be rare or absent, and the run-off dominantly follows the sub-surface pathways. Groundwater run-off in particular is an enormously important pathway, and this slow flow provides the continuous flow that sustains many perennial streams even through long periods of dry weather.

Storm characteristics are also important in run-off production. Small events can lose a significant fraction of the water they deliver to canopy interception, for example, and so their potential to generate run-off is diminished. Intense storm events in contrast can trigger widespread soil saturation and so generate large amounts of run-off.

The efficiency with which an area produces run-off can be expressed in terms of the run-off ratio that it exhibits. This is the ratio of the volume of run-off released to the volume of precipitation delivered over the area. Impervious urban areas can exhibit run-off ratios as high as 0.95. In contrast, a forested area with permeable soils may have a run-off ratio of only 0.1–0.2. The balance of the water is commonly lost to evaporation and transpired by vegetation.                DLD

**Reading and References**
Brammer, D.D. and McDonnell, J.J. 1996: An evolving perceptual model of hillslope flow at the Maimai catchment. In M.G. Anderson and S.M. Brooks eds, *Advances in hillslope processes*. Chichester: Wiley. Vol. 1, pp. 35–60. · Dingman, S.L. 1994: *Physical hydrology*. New Jersey: Prentice-Hall. · Newman, B.D., Campbell, A.R. and Wilcox, B.P. 1998: Lateral subsurface flow pathways in a semiarid ponderosa pine hillslope. *Water resources research* 34, pp. 3485–96.

**run-off plot**   An area of the landscape generally enclosed by leak-proof barriers so that all water and sediment leaving it is known to have come from within the borders of the plot. Run-off plots are used for such purposes as the comparison of the effects of land management treatments on run-off and erosion, the investigation of the erosional behaviour of differing soil types, and monitoring the effects of various experimental treatments such as vegetation removal by clipping or by fire. Dimensions range from 0.5 m$^2$ to hundreds of m$^2$. Run-off water and sediment shed from a plot are either measured continuously by appropriate apparatus, or directed into buried storage drums whose contents are later examined and subsampled. Plots may be exposed to natural or simulated rain.                                    DLD

**Reading**
Mohamoud, Y.M., Ewing, L.K. and Boast, C.W. 1990: Small plot hydrology. Vol. 1. Rainfall infiltration and depression storage determination. *Transactions of the American Society of Agricultural Engineers* 33, pp. 1121–31.

**ruware**  A low, dome-shaped exposure of bedrock projecting from a cover of alluvium or weathered bedrock. It is either an incipient or a relict INSELBERG.

# S

**sabkha** (also sabkhah) *a.* a closed depression, often with a saline surface (akin to a PAN or PLAYA), in an arid environment.
*b.* A saline flat in arid areas that is above the mean high tide level but subject to periodic inundation. Both terms are commonly used in Arabic countries.                                    DSGT

**salars** Basins of inland drainage in a desert which are only occasionally inundated with saline water. They are also known as PLAYAS, *sabkhas* or *chotts*.

**salcrete** A light-coloured surface crust of halite-cemented beach sand caused by the concentration by evaporation of swash or spray blown onshore by breaking waves (Yasso 1966).
                                                                ASG

**Reference**
Yasso, W.E. 1966: Heavy mineral concentration and sastrugi-like deflation furrows in a beach salcrete at Rockaway Point, NY. *Journal of sedimentary petrology* 36, pp. 836–8.

**salinization** The process of accumulation of soluble salts in upper soil horizons, thereby compromising plant growth. Though there are many plants that can grow in salt-rich soils, many plants grown as crops cannot do so effectively. In DRYLANDS especially, attempts to increase crop production by irrigation can, in a matter of a few years, result in salinization, either because water application exceeds the amount which plants can use, and drainage is poor, or/and because high EVAPOTRANSPIRATION rates cause salts to be precipitated in the soil. Salinization is in some areas, for example the San Joaquin basin of California, and parts of Pakistan and India, a major form of DESERTIFICATION (Thomas and Middleton 1994). Though remedial measures may be attempted, including efforts to flush excess salts out of the soil, salinization may simply be a facet of attempting to increase crop yields in environments generally unsuited to agriculture. It is not a new problem, and may have occurred in conjunction with early Mesopotamian agriculture (Jacobsen and Adams 1958).                    DSGT

**References**
Jacobsen, T. and Adams, R.M. 1958: Salt and silt in ancient Mesopotamian agriculture. *Science* 128, pp. 1251–8. · Thomas, D.S.G. and Middleton, N.J.

1994: *Desertification: exploding the myth*. Chichester: Wiley.

**salt marsh** Salt marshes are vegetated mudflats in the high intertidal zone found commonly on many low-lying coasts in a wide range of temperate environments. On tropical coastlines salt marshes tend to be replaced by MANGROVE swamps, although sometimes they are found together. Salt marshes, which support a range of halophytic vegetation, grade seawards into mud- or sand-flats. Salt marsh plants themselves play an important role in trapping sediment and in building up the marsh surface. In turn, the development of the marsh encourages a succession of plants from early colonizers such as *Salicornia* spp. and *Spartina* spp. to plants that are less tolerant of frequent inundation by seawater. Salt marshes vary greatly throughout the world in both ecology and geomorphology, but are often characterized by intricate creek systems and salt pan development.          HV

**Reading**
Adam, P. 1990: *Saltmarsh ecology*. Cambridge: Cambridge University Press. · Allen, J.R.L. and Pye, K. eds 1992: *Saltmarshes: morphodynamics, conservation and engineering significance*. Cambridge: Cambridge University Press.

**salt tectonics** Deformation of the earth's crust by the flow of salts from deep-seated evaporite deposits to form salt domes, salt pillows and associated structures. The intrusion into, and disturbance of existing strata is also known as diapirism. Salt diapirs are economically important as oil reservoirs and as sources of sulphur.                                            PSh

**salt weathering** The breakdown of rock by HALOCLASTY; it is caused primarily by physical changes produced by salt crystallization, salt hydration, or the thermal expansion of salts. Among the most effective salts are sodium sulphate, sodium carbonate and magnesium sulphate. Salt weathering has been recognized as an important process in desert, coastal, polar and urban areas, and is a serious hazard to concrete structures in saline regions.          ASG

**Reading**
Cooke, R.U. 1981: Salt weathering in deserts. *Proceedings of the Geologists' Association of London* 92, pp. 1–16. · Goudie, A.S. 1985: Salt weathering. *Research papers series, School of Geography, University of Oxford* 32.

**saltation** Has two main meanings in physical geography. The first refers to the hopping motion of sand grains transported by a fluid (water or air). The grains are ejected from the bed in a near-vertical trajectory by lift forces, accelerate in the flow direction when affected by fluid drag, then fall to the bed again on a path inclined at 6°–12° which is the result of gravitational and drag forces. Sand is only two to three times more dense than water, so the inertia of the rising grain only carries it to a height of about three grain diameters. The viscosity of water allows the grain to settle gently back to the bed. Saltation in air, however, involves trajectories up to 2–3 m high, especially after bouncing impacts on rock or pebble surfaces. The return impact also splashes other grains into the air, allowing a mechanical chain reaction of accelerated transport once motion has been initiated by fluid forces.

The second usage occurs in biogeography and refers to a theory which postulates the origin of major new taxonomic groups by the occurrence of single massive mutations. KSR

**salt-dome** A rounded hill produced by the upward doming of rock strata as a result of the diapiric movement of a halite bed or other evaporite deposit. ASG

**Reading**
Goudie, A.S. 1989: Salt tectonics and geomorphology. *Progress in physical geography* 13, pp. 597–605.

**salt-flat** A near horizontal stretch of salt crust representing the bed of a former salt lake.

**saltwater intrusion** See GHYBEN–HERZBERG PRINCIPLE.

**sand** Both a type of SOIL TEXTURE and a PARTICLE SIZE class, ranging from 63 to 2000 microns. Since sand is a particle size that is readily transported by both water and wind, it is commonly found in fluvial and aeolian deposits. DSGT

**sand banks** Form significant depositional features in many coastal regions and on continental shelves, and are often fed and capped by sand waves. Two main types have been identified (Dyer 1986). Linear sand banks or ridges occur in shallow tidal seas where sand is present and current velocities exceed about 0.5 m s$^{-1}$. They can be up to 80 km long, and typically 1–3 km wide and tens of metres high. Headland (or banner) banks develop in association with promontories, again where current strengths exceed about 0.5 m s$^{-1}$. They are only a few

kilometres in length and have an elongated pear-shaped form, the broad end being directed towards the top of the headland. ASG

**Reference**
Dyer, K.R. 1986: *Coastal and estuarine sediment dynamics.* Chichester: Wiley.

**sand lens** A discontinuous layer of sand in a sedimentary sequence representing the remnant of a former channel infill or overbank deposit comprising dunes, ripples, plane-bedded sands or sand sheets, or channel margin sediments.

JM

**sand ramp** An accumulation of sediment, either upwind or downwind of a topographical obstacle and along a sand transport pathway, that comprises interdigitated aeolian and slope or fluvial deposits. Palaeosols may also be present. Sand ramps may exceed 100 m in thickness and contain within their sediments a record of environmental changes, since sediments accumulate under fluctuations in conditions that favour aeolian transport and those favouring slope or fluvial processes. DSGT

**Reading**
Zimbleman, J.R., Williams, S.H. and Tchakerian, V.P. 1995. Sand transport paths in the Mojave Desert, southwestern United States. In V.P. Tchakerian ed., *Desert aeolian processes.* London: Chapman and Hall. Pp. 101–29.

**sand ridge** A type of low LINEAR DUNE that may have a pronounced vegetation cover on its lower slopes and only limited slip face development.

**sand rivers** See SANDBED CHANNELS.

**sand rose** A circular histogram that depicts the amount of sand that can potentially be moved by winds from various compass directions. A sand rose can be produced from surface wind velocity data collected at meteorological stations or from data collected in, for example, a desert dune field using a portable weather station. A sand rose differs from a WIND ROSE in that only wind speeds above the threshold velocity for sand transport ($c$.5 m s$^{-1}$ for sand with a mean particle diameter of 2 mm) are used in the calculation of the potential sand transport (usually expressed as vector units) from each direction for which wind data are recorded. A method for their construction is described by Fryberger (1979). Above the threshold velocity, wind data are usually grouped into several velocity classes, the median value of each class being multiplied by a weighting factor representing the rate of sand transport for each class and the percentage time the wind blew from that direction and in that velocity class. The arms of a sand rose are proportional to the potential sand drift from each direction, which total potential drift and the overall resultant vector, and its strength, can also be computed. Sand roses can be calculated for daily, weekly, monthly or annual data and are a useful tool for evaluating the aeolian environments in which sand drift is a problem or in which different dune types occur (e.g. Bullard *et al.* 1996). Since sand roses can be calculated for any environmental setting for which suitable data are available, and since their construction averages out many key variables that control sediment entrainment by the wind, ignoring surface factors such as soil moisture and vegetation cover, they only represent the generalized *potential* for sand transport, rather than actual rates of sand drift. DSGT

**Reading and References**
Bullard, J.E., Thomas, D.S.G., Livingstone, I. and Wiggs, G.F. 1996. Wind energy variations in the southwestern Kalahari desert and implications for linear dunefield activity. *Earth surface processes and landforms* 21, pp. 263–78. · Fryberger, S. 1979. Dune forms and wind regime. In E.D. McKee ed., *A study of global sand seas.* US Geological Survey Professional Paper 1052, pp. 137–69.

**sand sea** The largest unit of aeolian deposition; an extensive ($10^2$–$10^6$ km$^2$) area of aeolian sand that may comprise sand dunes and/or sand sheets. In excess of 99 per cent of dryland aeolian sand deposits are found in sand seas (also called ergs). Various definitions exist in the literature, for example Fryberger and Ahlbrandt (1979) cite a minimum size of 125 km$^2$, with at least a 20 per cent aeolian sand cover. Sand seas may be currently active with regard to aeolian processes or relict or fixed in terms of present day climates. For a sand sea to develop, it is necessary for there to be sufficient sand supply and winds strong enough to transport sand to a location where net accumulation occurs. Wilson (1971) noted that, at the regional scale, sand sea development is favoured in locations where sand flow is convergent and winds are decelerating (or where the actual net sand transport is decreasing). In the Sahara, for example, many sand seas occur in intermontane basins. Fryberger (1979) classified sand seas into low, intermediate and high energy environments, with many high energy sand seas occurring in the trade wind belt and low energy sand seas near the centre of semi-permanent high and low pressure cells. DSGT

**Reading and References**
Fryberger, S. 1979. Dune forms and wind regime. In E.D. McKee, ed., *A study of global sand seas.* US Geological Survey Professional Paper 1052, pp. 137–69. · —and Ahlbrandt, T.S. 1979. Mechanisms for the formation of aeolian sand seas. *Zeitschrift fur Geomorphologie* NF 23, pp. 440–60. · Thomas, D.S.G. 1997. Sand seas and aeolian bedforms. In D.S.G. Thomas ed., *Arid zone geomorphology.* Chichester: Wiley. Pp. 371–412. · Wilson, I.G. 1971. Desert sand flow basins and a model for the development of ergs. *Geographical journal* 137, pp. 180–97.

**sand sheet** A sand sheet is an accumulation of aeolian sand that is without significant morphological expression. While sand dunes are generally absent, sand sheets may possess aeolian ripples or even low ZIBAR dunes. Sand sheets are in fact very common in dryland environments and range in size from small features of a few km$^2$ to the vast 100,000 km$^2$ Selima sand sheet in the eastern Sahara. Five major controls on sand sheet development were recognized by Kocurek and Nielson (1986): vegetation, particularly grasses, which encourages low-angle laminations to develop; coarse sand, which does not readily form dunes other than zibar; a water table close to the surface (see STOKES' SURFACE) that may act as a base level to wind scour or an encouragement to 'sticky surface' deposition; periodic or seasonal flooding, that inhibits dune formation; and surface crusts such as SALCRETE, that inhibit deflation. DSGT

**Reference**
Kocurek, G. and Nielson, J. 1986. Conditions favourable for the formation of warm-climate eolian sand sheets. *Sedimentology* 33, pp. 795–816.

**sand trap** A device to measure the rate of sand transport (or flux) by the wind by efficiently removing airborne sediment over a known period of time whilst simultaneously allowing the free passage of airflow.

Horizontal traps consist of a partitioned rectangular box dug into the sand flush with the surface. Sand in SALTATION is trapped in the box, which is then retrieved and weighed. Typically the rate is calculated in kg.width$^{-1}$min$^{-1}$. Horizontal traps do not disturb the upstream airflow but sand can jump over the box if care is not taken in selecting its length.

Many vertical sand traps are based on a simple design (commonly referred to as the Aarhus type) whereby particles are collected through a vertical slot about 0.01 m wide, 0.5 m high with lengths up to 0.3 m. Commonly the traps are partitioned horizontally so that vertical flux can also be established. To limit back pressure within the trap, air is bled from the back through a thin gauze. A popular version is described by Leatherman (1978). A wedge-shape design (widening downwind), minimizes back-pressure through the Bernoulli effect by 'sucking' air through the trap (Nickling and McKenna Neuman 1997). Traps can also be mounted on rotating masts so that they continuously face into the oncoming sand stream. Sophisticated designs are now in use which include electronic load cells at their base giving a continuous high frequency recording of the mass of sand in the trap. Trap efficiencies (the proportion of moving sand collected) vary from as low as 20 per cent to 70 per cent depending upon design and field operation.                                        GFSW

**References**
Leatherman, S.P. 1978. A new aeolian sand trap design. *Sedimentology* 25, pp. 303–6. · Nickling, W.G. and McKenna Neuman, C.K. 1997. Wind tunnel evaluation of a wedge-shaped aeolian sediment trap. *Geomorphology* 18, pp. 333–45.

**sand volcano**   A small mount of sand with a smaller conical depression at the apex. Dimensions range from less than 2.5 cm in diameter up to 5 cm, and heights rarely exceed 3 cm. They are a surface expression of sediment compaction and dewatering. As the sediment settles, interstitial water may be expelled as small springs, at the mouth of which sand and mud particles may be deposited in a cone.                          ASG

**Reading**
Picard, M.D. and High, L.R. 1973: *Sedimentary structures of ephemeral streams.* Amsterdam: Elsevier Scientific. Pp. 139–41.

**sand wedge**   A relict ICE WEDGE, where the cast of the ice wedge has been infilled by wind blown or washed in sediment, usually sand. They can be found in environments that have experienced periglaciation and permafrost, for example, in parts of southern England where they are inherited from ice wedges formed during the last glacial cycle.                     DSGT

**sandbed channels**   Are common in semi-arid environments. They have bed material in the 0.0625–2 mm sand size-range which is mobile even during low sub-bankfull flows, and is transported by SALTATION, or in suspension at high flow stages. During transport the sand travels in bedforms (ripples, dunes, antidunes) which change systematically with the FLOW REGIME, strongly influencing the bed ROUGHNESS, and which may be preserved as sedimentary structures in the sand river deposits. Sandbed rivers have characteristically high rates of bedload transport, are migratory on their floodplains, and are often morphologically unstable with multiple, changing, braided channels and sand bars.                                        KSR

**sandstone**   An indurated sedimentary rock composed of cemented particles of sand, with a range of grain sizes between 0.0625 mm and 2 mm. They can be subdivided into various types on the basis of grain size and mineral composition. Types with a limited matrix content (less than 15 per cent) are called *arenites*, whereas those with a greater matrix content are termed *wackes*. Within the arenites, distinction must be made between *quartz* (less than 5 per cent feldspar or rock fragments), *lithic* (more than 25 per cent rock fragments, excluding feldspar), *arkose* (more than 25 per cent feldspar) and *volcanic* (more than 50 per cent volcanic fragments) subtypes. Sandstones cover very approximately the same area of the continents as do granites and carbonates, but have been much less the scene for the development of any particular geomorphological approach than the other two rock types. None the less, sandstone landscapes do have some distinctive geomorphological features that have been the subject of a review by Young and Young (1992).   ASG

**Reference**
Young, R. and Young, A. 1992: *Sandstone landforms.* Berlin: Springer-Verlag.

**sandstorm**   An atmospheric phenomenon occurring when strong winds entrain particles of dust and sand and transport them in the atmosphere.

**sandur**   (pl. sandar; Icelandic) An extensive plain of glaciofluvial sands and gravels deposited in front of an ice margin by a system of braided or anastomosing meltwater streams which migrate across the sandur surface. The whole sandur is rarely flooded except during jökulhlaup events. A valley sandur is confined between valley walls. Pitted sandur forms on an ice margin that melts out to produce kettle holes. For morphological and sedimentary characteristics, see OUTWASH.                              JM

**Reading**
Bluck, B.J. 1974: Structure and directional properties of some valley sandur deposits in Southern Iceland. *Sedimentology* 21, pp. 533–54. · Hjulström, F. 1952: The geomorphology of the alluvial outwash plains (sandurs) of Iceland, and the mechanics of braided rivers. *International Geographical Union, seventeenth congress proceedings.* Washington DC. Pp. 337–42. · Krigström, A. 1962: Geomorphological studies of sandur plains and their braided rivers in Iceland. *Geografiska annaler* 44A, pp. 328–45. · Price, R.J. 1969: Moraines, sandar, kames and eskers near Breidamerkurjökull, Iceland. *Transactions of the Institute of British Geographers* 46, pp. 17–43.

**sapping**   The undermining of the base of a cliff, rock or sediment face, usually associated with the subsequent failure of the face. It can occur in a number of ways, including undercutting by wave action, lateral stream erosion and erosion by emerging (exfiltrating) groundwater flow.

Erosion by emerging GROUNDWATER is often referred to as 'spring sapping' but is actually made up of two sets of processes, tunnel scour and seepage erosion. The relative operation of these processes depends upon whether groundwater emergence is focused upon one point or is along a more diffuse seepage zone, the latter most commonly occurring where permeable soil, sediment or rock overlies less permeable material. Tunnel scour operates when stress is applied to the walls of a pre-existing macropore by near-surface groundwater flow, commonly within a partially or fully consolidated material, leading to its widening and eventual collapse. Seepage erosion occurs when sufficient drag force is generated as water seeps through and exfiltrates from a porous material to entrain particles, cause failure or liquefy the material. It may also take place by the operation of biological, chemical and resultant physical weathering processes in the zone of groundwater emergence (e.g. algal growth or the precipitation of salts in pore spaces) which weaken the material and lead to mass wasting. Spring sapping and seepage erosion have been held responsible for the development of the steep heads of some DRY VALLEYS, scarp faces, ALCOVES and the amphitheatre-shaped heads of certain canyons (Howard *et al.* 1988). DJN

**Reading and Reference**
Higgins, C.G. and Coates, D.R. eds 1990: *Groundwater geomorphology; the role of subsurface water in earth-surface processes and landforms*. Special Paper 252, Boulder, Colorado: Geological Society of America. · Howard, A.D., Kochel, R.C. and Holt, H.E. 1988: *Sapping features of the Colorado Plateau: a comparative planetary geology fieldguide*. NASA Publication SP-491. · Nash, D.J. 1997: Groundwater as a geomorphological agent in drylands. In D.S.G. Thomas ed., *Arid zone geomorphology: process, form and change in drylands*. Chichester: Wiley. Pp. 319–48.

**saprolite** Weathered or partially weathered bedrock which is *in situ*.

**sapropel** Amorphous organic compounds which collect in various water basins: lakes, lagoons, shallow marine basins and estuaries are termed sapropels. The sapropel is formed by predominantly anaerobically decomposing remains of phyto- and zoo-plankton. These are richer in fatty and protein substances than is peat. The decomposition and putrefaction of the organic content leads to the formation of various hydrocarbons, which are believed to be the basis of the origin of petroleum and natural gas compounds which form after compression under accumulated sediment. The progressive accumulation of sapropel is governed largely by rapid multiplication of the organisms responsible for it. AP

**Reading**
Pettijohn, F.J. 1984: *Sedimentary rock*. 3rd edn. Delhi: CBS Publishers. · Rossignol-Strick, M. 1985: Mediterranean Quaternary sapropels, an immediate response of the African monsoon to variation of insolation. *Palaeogeography, palaeoclimatology, palaeoecology* 49, pp. 237–63.

**saprophyte** An organism, usually a plant, which obtains nutrients from dead or dying organisms. Most fungi are saprophytes.

**sarsen** A block of silica-cemented sandstone, breccia or conglomerate found in many parts of southern England, notably on the margins of the London and Hampshire basins. Sometimes called 'grey wethers' or 'pudding stone', sarsens are thought to be the result of weathering processes in a surface or near-surface environment under conditions of Tertiary warmth. Humans have often employed such blocks to make monuments, such as Avebury stone circle or Windsor Castle. Sarsens are in effect a fossil silcrete duricrust, and comparable deposits in the Paris Basin are called *meulières*. ASG

**Reading**
Summerfield, M.A. and Goudie, A.S. 1980: The sarsens of southern England: their palaeoenvironmental interpretation with reference to other silcretes. In D.K.C. Jones ed., *The shaping of southern England*. Institute of British Geographers special publication 11, pp. 71–100.

**sastrugi** Furrows and ridges in the surface of ice and snow accumulations through the action of wind.

**satellite meteorology** Meteorology which depends largely, or completely, on data generated by METEOROLOGICAL SATELLITES. As such it is a relatively young branch of the parent science, but one which grew very rapidly after the launching of the first specialized meteorological satellite in 1960. With meteorological satellite platforms and sensor systems still undergoing active development as part of the present day rise of environmental remote sensing, satellite meteorology continues to increase in both its scope and sophistication. Each meterological satellite or satellite system is supported by ground facilities which receive satellite data for dissemination to the user community. There, initial preprocessing (e.g. removal of extraneous data contents) and processing (e.g. geographical rectification) are carried out, and ranges of satellite products are prepared to meet the needs of the primary user community. These include:

1 Manual and basic products, including unenhanced and enhanced visible and infrared imagery; facsimile maps (e.g. nephanalyses, and snow and ice charts); and alphanumeric outputs (e.g. satellite weather bulletins, moisture analyses and plume winds).
2 Man–machine combined products, including cloud motion vectors and precipitation fields.
3 Computer-derived image products, including period minimum brightness composites, and cloud field analyses (cloud types, cloud top heights, etc.).
4 Computer-derived digital products, e.g. vertical temperature and moisture profiles, and sea surface temperature data.
5 Archival products, e.g. magnetic tapes of raw and/or processed data, and photographic images, for use in atmospheric research.

These products form a very important part of the complete data pool for use in weather forecasting. Some data (e.g. cloud images) may be used *qualitatively* in the analysis of synoptic situations, and forecasting for synoptic, subsynoptic, or mesoscale regions, either on the ground, or even aloft. Others are used *quantitatively*, e.g. satellite-derived vertical profiles, and satellite winds; such data may be added to the available conventional (*in situ*) observations to provide improved data arrays for numerical (computer-based) forecasting procedures. The satellite inputs are particularly important for tropical and polar regions, and some of the more remote continental areas in middle latitudes, i.e. those regions traditionally least well observed by surface weather stations. Satellite-derived quantitative data are also important in vertical profiling of the atmosphere in these regions because of the sparseness of their upper-air weather observatories – but satellite soundings are of value almost everywhere because they penetrate the atmosphere downwards, whereas radio-sondes and similar conventional devices inspect it from the bottom upwards, usually to the tropopause, and rarely far beyond it.

Equally important and varied uses are made of satellite data in *meteorological research*. On the global scale evaluations of earth/atmosphere radiation, and related energy budgets have benefited greatly from satellite radiation data. Since satellite sensor systems orbit above the top of the earth's atmosphere a number of budget components which were previously amenable only to estimation can now be measured. These include the earth/atmosphere albedo, longwave radiative fluxes towards space, and the net radiation budget for the globe, or any selected area, and their attendant columns of the atmosphere. Studies of spatial and temporal changes of these and other quantities are yielding vital insights into the behaviour of the atmosphere, involving both long-distance interactions ('teleconnections') and vertical inter-relationships among its different layers. Satellite imagery has revolutionized our knowledge and understanding of many regions, and types of weather systems, ranging in the first case from tropical latitudes to polar regions, and in the second from synoptic down to meso-scale features. Virtually no region or type of weather system has not been elucidated in some way. The First GARP Global Experiment (FGGE) in 1979 depended heavily on satellite data, and is likely to be the forerunner of other large and complex meteorological research projects in the future.

As the runs of data from meteorological satellites have lengthened, so increasing attention has been paid to *climatological analysis* of these types of data sets. Particular advantages have accrued in studies of the climatologies of cloud distributions, synoptic weather systems (e.g. the intertropical cloud band and associated features, hurricanes, jet streams, and mid-latitude depressions), and the structure and behaviour of the upper atmosphere. As the data sets lengthen further, so they may be expected to support an increasing number and range of studies of climatic change. It is also likely that microwave data will feature increasingly in both meteorological and climatological satellite applications.

ECB

**Reading**
Anderson, R.K. and Veltischev, N.F. 1973: *The use of satellite pictures in weather analysis and forecasting.* WMO technical note 124. Geneva: World Meteorological Organization. · Barrett, E.C. and Curtis, L.F. 1992: *Introduction to environmental remote sensing.* 3rd edn. London: Chapman & Hall. · —and Martin, D.W. 1981: *The use of satellites in rainfall monitoring.* London: Academic Press. · Cracknell, A.P. 1981: *Remote sensing in meteorology, oceanography and hydrology.* Chichester: Ellis Horwood; New York: Wiley. · Kondratiev, Ya. 1983: *Satellite climatology.* Leningrad: Hydrometeozdat. · NASA 1982: *Meterological satellites: past, present and future.* NASA conference publication 2227. Washington DC: NASA. · Tanczer, T., Gotz, G. and Major, G. 1981: First FGGE results from satellites. *Advances in space research* 1.4. Oxford: Pergamon.

**saturated wedge and zone** Consists of a layer of saturated soil at the base of a hillslope,

429

thickening downslope and in some areas reaching the soil surface to define an area of surface saturation.

Saturation develops in response to THROUGH-FLOW collecting from the hillside above. Its total volume generally increases downslope, whereas its rate of flow may decrease if the slope profile is concave or if water is backed up from the stream. The magnitude of this effect may be seen for conditions of steady net rainfall at intensity $I$, falling in a collecting area of a per unit contour width. The total outflow along 1 m of contour is then $Q = Ia$. This flow may also be expressed in terms of the depth h of saturated flow in the soil and the slope (or strictly the total potential) gradient $S$, in the form $Q = Sf(h)$ where $f$ is a function which increases with h and with soil permeability. Equating these expressions for the flow, the saturated wedge has thickness h given by

$$f(h) = Ia/S$$

The effect of slope profile concavity (decreasing $S$) and of a large collecting area or converging flow (large $a$) may readily be seen. It is also evident that saturated zones/wedges only persist seasonally in areas where net rainfall (rainfall minus evapotranspiration) remains positive seasonally. Saturated wedges rise to the soil surface where h is greater than the soil water storage, giving rise to a DYNAMIC SOURCE AREA of surface saturation.                                    MJK

**Reading**
Kirkby, M.J. 1978: *Hillslope hydrology*, Chichester: Wiley. Esp. chs 7 and 9.

**saturation coefficient**   or degree of saturation (S) is the water content of a rock after free saturation, expressed as a percentage of the maximum water content. The free saturation is the water which can be absorbed into the pore spaces of a sample when it is just immersed in water, while the maximum water content is the amount of water absorbed when it is forced in under vacuum. The saturation coefficient is important in the testing of the mechanical breakdown of rocks.                          WBW

**saturation deficit**   The depth of water required to bring saturation up to the soil surface and initiate overland flow. Since THROUGHFLOW is not necessarily connected to the groundwater table at depth, a zero saturation deficit need not imply complete saturation of the soil profile. The saturation deficit is important in establishing areas of current overland flow production, and gives directly the amount of storm rainfall needed to initiate overland flow. The deficit is also important as one of the

controls on evaporation, particularly from unvegetated soil surfaces.                          MJK

**saturation overland flow**   See OVERLAND FLOW.

**savanna**   (savannah) A grassland region of the tropics and subtropics. The word is probably of American Indian origin, signifying treeless grasslands. Savannas today are taken to mean plant formations dominated by grasses and grass-like species (graminoids) with herbaceous non-grass species (forbs), often possessing a light to dense scattering of trees. Woody savannas prevail where grasses are at least co-dominant, grading to savanna woodland where trees assume dominance. Seasonal climates impose water stresses during the dry period, which inhibit tree growth and encourage drought-resistant plants. Burning, both natural and human-induced, also favours a herbaceous formation with xeromorphic characteristics.           PAF

**Reading**
Bourlière, F. ed. 1983: *Tropical savannas*. Amsterdam: Elsevier. · Huntley, B.J. and Walker, B.H. 1982: *Ecology of tropical savanna*. Berlin: Springer-Verlag. · Werner, P.A. 1991: *Savanna ecology and management: Australian perspectives and intercontinental comparisons*. Oxford: Blackwell Scientific.

**scabland**   An area where, as a result of erosion, there is no soil cover and the landsurface consists of rock and rock fragments.

**scald**   Localized destruction of vegetation, in particular ground cover, caused by increased soil salinity. Scalding is a common, if ephemeral, feature of semi-arid regions.           PSh

**scar**   A cliff or very steep slope or a rocky outcrop.

**Schmidt hammer**   Originally, this device was invented to test the curing of concrete and as a measure of its 'strength'. It has been used in geomorphology to provide a measure of weathering of rock and stones as well as durability of building materials and as a surrogate measure of the age of deposits. It consists of a barrel which holds a steel rod which is made to impact upon the target area by a calibrated coil spring. The rod's rebound gives a value on a scale which can be read off. This is the R value (a measure of the coefficient of restitution between rod and target). For a calibrated rod and spring with (usually normal) direction of impact, the R value can be used as a measure of the compressive strength of the near surface material. Crystal type, grain size and distribution as well as small scale irregularities provide a scatter in the results, thus requiring careful sampling and statistical analysis. It is best used for comparative measures at a site. It has been used, where rock type is 'constant', to show a progression of weathering over time; moraine chronologies have been developed by this method.          WBW

**Reading**
Goudie, A.S. ed. 1991: *Geomorphological techniques.* 2nd edn. London: Unwin Hyman. Pp. 174–5, 218–19.

**sciophyte** A plant that lives in a well-shaded environment.

**sclerophyllous** Refers to species of evergreens that, like olive and cork oak, have adapted to the lengthy seasonal drought experienced in regions with a 'Mediterranean type' of climate, by producing tough, leathery leaves to cut down moisture loss caused by transpiration.

**scoria** A volcanic rock or slag consisting of angular rock fragments and numerous voids which were originally filled with volcanic gases.

**scour and fill** The processes of cutting and subsequent filling of fluvial channels.

**scree** An accumulation of primarily angular clasts which lies at an angle of around 36° beneath an exposed free face or cliff. The prime cause of deposition is rock fall, but other processes, such as debris flows, may contribute to their development. The largest clasts occur at the base of the scree.          ASG

**Reading**
Statham, I. 1973: Scree slope development under conditions of surface particle movement. *Transactions of the Institute of British Geographers* 59, pp. 41–53.

**S-curve** A method of extending a unit hydrograph. It represents the surface run-off hydrograph caused by an effective rainfall of intensity $I/T$ mm h$^{-1}$ applied indefinitely where $T$ is the duration of the effective rainfall (see UNIT HYDROGRAPH).

**sea cliffs** Steep slopes that border ocean coasts. They are ubiquitous and occur along approximately 80 per cent of the ocean coasts of the earth (Emery and Kuhn 1982). They can be classified into three main types according to their stage of development: *active cliffs*, where bedrock is exposed by continuous retreat under the influence of both marine and subaerial forces; *inactive*, where their bases are mantled by talus, and where there is some vegetation cover; and *former*, where the influences of

marine erosion have disappeared, so that subaerial erosion rounds the crests and provides material for stream deposition beyond their bases. Among the important factors in their development are the nature of the landward topography, the structure and lithology of the materials in which they have been eroded, the nature of marine erosion, and the power of subaerial processes (rockfalls, mass movements, frost weathering, etc.). Many cliff profiles display complex forms brought about by long continued changes in sea level, climate, and the balance between subaerial and marine processes (Steers 1962).          ASG

**References**
Emery, K.O. and Kuhn, G.G. 1982: Sea cliffs: processes, profiles, classification. *Bulletin of the Geological Society of America* 93, pp. 644–53. · Steers, J.A. 1962: Coastal cliffs: report of a symposium. *Geographical journal* 128, pp. 303–20.

**sea ice** Ice which forms on the sea surface when water temperatures fall to around $-1.9\,°C$. Such conditions apply to considerable areas of the Arctic and Antarctic. In polar latitudes the equilibrium thickness of sea ice is around 3 m and any surface melting is equalled by bottom freezing. When the sea ice is attached to the land it is known as *fast* ice and when floating free under the influence of currents and wind it is called *pack* ice. The latter consists of ICE FLOES centred on relatively resistant ice, leads of open water which may freeze over rapidly in winter, and irregular fractured and crumpled ice-forming features known as *pressure ridges* or *keels* (see illustration).          DES

**Reading and Reference**
Lewis, E.L. and Weeks, W.F. 1971: Sea ice: some polar contrasts. In G. Deacon ed., *Symposium on Antarctic ice and water masses, Tokyo, September 1970.* Cambridge: Scientific Committee for Antarctic Research. Pp. 23–31. · Nansen, F. 1897: *Farthest north.* 2 vols. London: Constable. · Sugden, D.E. 1982: *Arctic and Antarctic.* Oxford: Basil Blackwell; Totowa, NJ: Barnes & Noble. · Zwally, H.J. and Gloersen, P. 1977: Passive microwave images of the Polar Regions and research applications. *Polar record* 18, 116, pp. 431–50.

**sea/land breeze** The sea and land breezes form part of a diurnally varying, vertical circulation of air that is ultimately induced by temperature differences between land and sea. Given a morning of near calm conditions and clear skies, a land area heats up more rapidly than does the adjacent water body. Consequently, as a result of turbulent transfer and convection of heat, the air over the land becomes warmer than that at the same height over the sea. In turn, this means that the land air is less dense than the sea air and thus, by the HYDROSTATIC EQUATION, the vertical

431

**Sea ice** photographed on the Imperial Trans-Antarctic expedition, 1914–16.

gradient of pressure is less over land than over sea. Assuming that pressure was initially uniform over land and sea, the difference in vertical gradients must result in a higher pressure over the land at a height H (say) than at the same height over the sea. Consequently, at height H, air moves from land to sea. In turn, this disturbs the hydrostatic equilibrium of the air columns over both land and sea and also generates a horizontal pressure gradient from sea to land near the surface. The air motion resulting from this gradient force is the sea breeze. At the landward and seaward extremities of this breeze the air rises and sinks respectively. Consequently a vertical circulation is completed.

The land breeze results from a reversal of these processes when the land is cooler than the sea.                                    BWA

**Reading**
Atkinson, B.W. 1981: *Meso-scale atmospheric circulations.* London and New York: Academic Press. Esp. ch. 5.

**sea level**  The elevation of the ocean surface relative to a vertical datum. The term is often used synonymously with MEAN SEA LEVEL, although the latter is derived from a series of observations of the former. Relative sea level changes at a location can involve changes in either the absolute elevation of the ocean, or the land, or both. Instantaneous measurements of sea level exhibit cyclical and secular changes, and the averaging duration will vary depending upon a particular purpose. Monthly averages may be used to remove short-term cycles, such as those associated with WAVES and TIDES, and to allow seasonal comparisons of sea level. Averages comprising a year (or more) of data are appropriate for revealing seasonal cycles and anomalies, and for the compilation of data series for the analysis of long-term sea-level change.

Several geophysical processes affect sea level. Waves and tides cause regular, short-term fluctuations. There are also local fluctuations associated with atmospheric low pressure systems (the inverted barometer effect), STORM SURGES, and flood discharges from rivers. These transient phenomena can cause local sea level rises exceeding 5 metres, and may have catastrophic human impacts (e.g., Pugh 1987). There are also seasonal changes, on the order of 0.1 metre, caused by variations in the location and intensity of semi-permanent high and low pressure systems, and changes in ocean circulation. At return intervals of several years, atmospheric phenomena such as EL NIÑO are capable of changing regional sea levels by as much as 0.5 metres. In order to establish a stable estimate of mean sea level for purposes of tracking long-term global changes, averaging periods must be long enough to minimize the impacts of such local/regional effects.

The study of long-term sea-level changes involves several academic disciplines, including climatology, engineering, geography, geology, geophysics, and oceanography. The focus of such research may extend back through geological time for several billion years, and into the future for several centuries. In the former case, the motivation is to understand the evolution of the earth as a planet. In the latter case it is to understand earth as the home of humanity. Sea level studies for earth history are based largely on estimations of CONTINENTAL FREEBOARD and are concerned, necessarily, with EUSTATIC, ISO-STATIC and EPEIROGENIC changes. Eustatic changes are caused by changes in the volume of the ocean basins, or in the volume of the water in the oceans. Changes in ocean basin volume occur over time-scales of $10^4$ to $10^8$ years, and are associated with sea level changes of the order of one to 100 metres (Geophysics Study Committee 1990). Epeirogenic changes, vertical movements of the crust, occur at similar scales. Isostatic changes result from crustal deformation caused by the redistribution of mass over the earth's surface, and occur at time-scales of $10^2$ to $10^8$ years. The shorter time-frame is associated with redistribution of ice and water, while the latter are caused by sedimentation.

Changes in ocean volume can occur over substantially shorter time-frames usually caused by shifts in the volume of water (including ice) stored on the land, and by steric, or thermohaline changes. There are also longer-term changes in the total mass of water on the earth. According to Meier (1990), there are about $30 \times 10^6 \, \text{km}^3$ of water currently stored as ice and it is estimated that the volume was two to three times larger during the last glacial maximum, about 20,000 years ago (e.g., Pirazzoli 1996). The change in volume was caused by melting of the PLEISTOCENE ice sheets, and was accompanied by sea-level rise of more than 100 metres.

The density of seawater varies with temperature and salinity. Large-scale salinity changes will occur slowly, but ocean temperature responds relatively quickly to global atmospheric temperature changes. If the upper 500 metres of the ocean were warmed uniformly by 1 °C, thermal expansion of the water would raise sea level by about 0.06 m in high latitudes, and about 0.15 m in the tropics (Wigley and Raper 1993). Current rates of sea level rise, estimated to be about 0.1–0.2 mm per year (Pirazzoli 1996), are attributed largely to thermal expansion of seawater, and to a lesser degree, continued melting of alpine glaciers. Melting of Pleistocene ice also caused substantial isostatic adjustment, with land elevation exceeding 800 metres in some regions (Pirazzoli 1996).

Local changes in sea level may also be caused by SUBSIDENCE, FAULTING and FOLDING. Many deltaic coastlines, e.g., in Louisiana, Bangladesh, Venice, or Bangkok, are experiencing rapid sea-level rise caused by compression of sediments and crustal downwarping. Vertical movement along faults may cause almost instantaneous, metre-scale rise or fall of local sea level. Folding also causes vertical movements that may translate to a localized change in sea level.

Rates of sea level rise are projected to increase, perhaps doubling in the next century, caused by GREENHOUSE EFFECT warming and the resulting warming of the oceans and additional ice melting. The potential inundation of populous, low-lying coastlines has become an issue of international concern.                    DJS

**References**
Geophysics Study Committee 1990: Overview and recommendations. In National Research Council ed., *Sea-level change*. Washington, DC: National Academy Press. Pp. 3–34. · Meier, M.F. 1990: Role of land ice in present and future sea-level change. In National Research Council ed., *Sea-level change*. Washington, DC: National Academy Press. Pp. 171–84. · Pirazzoli, P.A. 1996: *Sea-level changes: the last 20,000 years*. Chichester: Wiley. · Pugh, D.T. 1987: *Tides, surges and mean sea level*. Chichester: Wiley. · Wigley, T.M.L. and Raper, S.C.B. 1993: Future changes in global mean temperature and sea level. In R.A. Warrick, E.M. Barrow, and T.M.L. Wigley eds, *Climate and sea level change: observations, projections and implications*. Cambridge: Cambridge University Press. Pp. 111–33.

**sea mount**    A mountain or other area of high relief on the sea floor which does not reach the surface. Flat-topped sea mounts are called GUYOTS.

**secondary depression**    A region of low pressure which forms within the circulation of a pre-existing depression. It sometimes occurs at the 'triple point' where the occlusion and warm and cold fronts meet, or along the individual fronts as warm or cold front waves. The development of the secondary is often preceded by an increase in the spacing of the isobars locally and once formed, the secondary can even absorb the parent low.

Secondary depressions also form occasionally in the unstable airstreams as polar lows.    RR

**secondary flows**    Those currents in moving fluids which have a velocity component transverse to the local axis of the primary, or main

433

flow direction. In river channels they are associated with the longitudinal vortex known as helical flow. Secondary flows in straight channels are probably caused by non-uniform distribution of boundary shear stress and by turbulence generated at the base of the channel banks. Two circulatory cells occur side by side, with flow alternately *converging* at the surface then downwelling, and *diverging* at the surface where upwelling occurs in mid-channel, at locations spaced apart at five to seven times the channel width. Downwelling intensifies the local bed shear stress exerted by the flow and encourages scour, while upwelling results in a low shear stress and deposition. This secondary circulation flow pattern thus relates to the POOL AND RIFFLE sequence. In curved channels, secondary flows result from skewing of the main flow towards one bank, and the consequent transverse HYDRAULIC GRADIENT then drives a transverse current which crosses the channel at the bed after plunging near the outer bank and assisting bed scour and undermining of the bank. Secondary flows represent an important control of the spatial distribution of erosion and deposition within a river channel, and of channel pattern change.                                 KSR

**sediment, fluvial**  Has been defined as particles derived from rock or biological material that are, or have been, transported by water. It has provided the focus for numerous studies by physical geographers, although a dichotomy exists between studies of contemporary *transport* and studies of the *deposits* associated with contem-

porary and past fluvial activity. The movement of fluvial sediment also has important practical implications for the problems of channel management, river intake and irrigation canal maintenance, reservoir and harbour siltation, debris dumping and floodplain accretion that stem directly from its deposition. Furthermore there is an increasing awareness of the important role of fine sediment in the transport of contaminants such as heavy metals and pesticides by rivers.

Information on the properties of fluvial sediment is of significance in all these contexts, and particle size must rank as the most important since it exerts a major control on entrainment, transport and deposition processes. The size of particulate material transported by rivers ranges between fine clay and colloidal particles of less than 0.5 $\mu$m in diameter to large boulders moved during flood events. Documentary evidence points to boulders weighing up to 7.5 t being transported by the Lynmouth Flood, which occurred in North Devon, UK, in 1953. A useful distinction may be made between material moving as BEDLOAD and that transported as SUSPENDED LOAD. The latter generally involves particles < 0.2 mm in diameter and the diagram provides several examples of typical particle size distributions of suspended sediment from rivers in different areas of the world. Although such particle size analyses commonly relate to the size of the discrete sediment particles, it is important to recognize that many of these will be transported within larger aggregates. In the case of suspended sediment, the

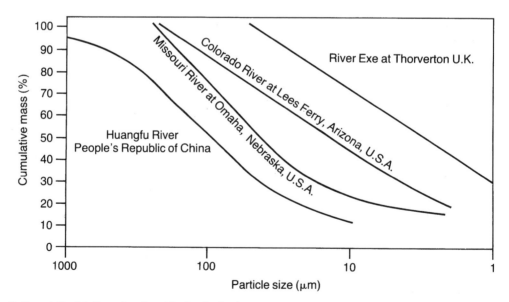

**Sediment, fluvial.** Examples of particle size distribution.

Average chemical composition of inorganic suspended sediment and surficial rocks

| | Concentration (mg g$^{-1}$) | | | | | | | |
|---|---|---|---|---|---|---|---|---|
| | Al | Ca | Fe | K | Mg | Na | Si | Ti |
| Tropical rivers | 114 | 7.5 | 62 | 18 | 9.6 | 5.1 | 264 | 7.3 |
| Temperate and Arctic rivers | 72 | 36 | 45 | 23 | 12 | 8.6 | 293 | 4.9 |
| World rivers | 90 | 25 | 52 | 21 | 11 | 7.1 | 281 | 5.8 |
| Surficial rocks | 70 | 45 | 36 | 24 | 16 | 14 | 275 | 3.8 |

particle size characteristics are largely governed by the nature of the source material. With the larger particles comprising the bedload, the boundary shear stress or force available to initiate movement is the dominant control, although the range of sizes available may also be important. The degree of sorting will vary according to the precise character of the fluvial environment and the transport distances involved. Measurements of particle shape and roundness have also been employed to demonstrate downstream changes in bed material character.

Fluvial sediment transported as bedload consists almost entirely of inorganic material and this will closely resemble the parent rock in terms of mineral composition. The finer sediment transported in suspension may, however, incorporate a considerable proportion of organic material. Its mineralogy may differ considerably from that of the parent rock due to the chemical weathering processes involved in the disintegration of the rock and the selectivity of the detachment and transport processes. This selectivity will result in the suspended sediment showing considerable enrichment in clay-sized particles and inorganic matter, when compared to the source material. Organic matter contents are typically of the order of 10 per cent, although values as high as 40 per cent have been encountered in some rivers. Enrichment in fine material has important implications for sediment-associated transport of contaminants, because clay-sized particles exhibit considerably greater specific surface areas (typical values are $200 \, m^2 \, g^{-1}$ for clay, $40 \, m^2 \, g^{-1}$ for silt and $0.5 \, mm^2 \, g^{-1}$ for sand) and cation exchange capacities.

Information on the average chemical composition of inorganic suspended sediment transported by world rivers and its relation to that of surficial rocks, abstracted from Martin and Meybeck (1979) and Meybeck (1981) is listed above.

These data indicate that in general suspended sediment is enriched in Al, Fe and Ti with respect to parent rock, while Na, Ca and Mg are strongly depleted. This tendency is more marked for tropical rivers than for rivers in temperate and Arctic regions because of the greater efficacy of chemical weathering in tropical areas.

Several workers have attempted to estimate the total annual sediment load transported from the landsurface of the globe to the oceans, although in the absence of data on bedload transport these values relate to suspended sediment. The estimate of Milliman and Meade in 1983 pointed to a mean annual transport to the ocean of $13.5 \times 10^9 \, t \, year^{-1}$. This is equivalent to a sediment yield of approximately $135 \, t \, km^{-2} \, year^{-1}$ from the landsurface of the globe and is 3.6 times greater than the total dissolved load transport to the oceans of $3.72 \times 10^9 \, t \, year^{-1}$ suggested by Meybeck (1979).

Existing estimates of the relative importance of bedload and suspended load in the total transport of fluvial sediment to the oceans are extremely tentative, but they place the bedload contribution at about 10 per cent of the suspended load. This proportion varies markedly for individual rivers. In Arctic streams on Baffin Island measurements suggest that coarse material or bedload constitutes 80–95 per cent of the total transport of fluvial sediment, whereas data from the Volga river in the former USSR indicate that 98–99.7 per cent of the sediment transported is composed of fine material in suspension.

Studies of fluvial sediment deposits have been largely concerned with the coarse fraction which moves relatively slowly through a river system and which may come to rest temporarily or permanently in depositional sinks or stores. These deposits are commonly classified on a genetic basis and they include, for example, point, longitudinal and marginal bars, lag deposits, splays and alluvial fans. Whereas most channel deposits are composed of relatively coarse material (bedload), finer material may be deposited by rivers on floodplains and in lakes or reservoirs, and where fine sediment is transported downslope by unconcentrated surface run-off this may frequently be deposited before reaching the stream channel.

The dichotomy between studies of fluvial sediment transport and of the associated deposits can usefully be bridged by considering the *sediment budget* of a drainage basin. With this concept the transport of sediment out of the basin is seen as the result of the various processes involved in mobilizing sediment within the basin and of the deposition and storage of the sediment within the basin. Only a small proportion of the sediment mobilized may be transported out of the basin and some material may be deposited in temporary storage to be remobilized on a subsequent occasion, for example, during a higher magnitude event. The deposits are therefore treated as an integral part of the conveyance system.          DEW

### Reading and References
Abrahams, A.D. and Marston, R.A. eds. 1993: Drainage basin sediment budgets. *Physical Geography* 14, pp. 221–320. · American Society of Civil Engineers 1975: *Sedimentation engineering*. New York: American Society of Civil Engineers. · Lehre, A.K. 1982: Sediment budget of a small coast range drainage basin in North-Central California. In F.J. Swanson, R.J. Janda, T. Dunne and D.N. Swanson eds, *Sediment budgets and routing in forested drainage basins*. US Forest Service general technical report PNW-141. Pp. 67–77. · Martin, J.M. and Meybeck, M. 1979: Elemental mass-balance of material carried by major world rivers. *Marine chemistry* 7, pp. 173–206. · Meybeck, M. 1979: Concentration des eaux fluviales en éléments majeurs et apports en solution aux océans. *Revue de géologie dynamique et de géographie physique* 21, pp. 215–46. · — 1981: Pathways of major elements from land to ocean through rivers. In *River inputs to ocean systems*. UNEP/UNESCO report. · Milliman, J.D. and Meade, R.H. 1983: World-wide delivery of river sediment to the oceans. *Journal of geology* 91, pp. 1–21. · Richards, K. 1982: *Rivers: form and process in alluvial channels*. London: Methuen. · Statham, I. 1977: *Earth surface sediment transport*. Oxford: Clarendon Press. · Trimble, S.W. 1981: Changes in sediment storage in the Coon Creek Basin, Driftless Area, Wisconsin, 1853 to 1975. *Science* 214, pp. 181–3. · UNESCO 1982: *Sedimentation problems in river basins*. Paris; UNESCO.

## sediment budget
A sediment budget is the difference between sediment input to a given area and sediment output from that area, over a given amount of time. Thus, it can apply to any geomorphological process that involves the transport of sediment, whether in the atmosphere, on land, or in water. It follows directly from the CONTINUITY EQUATION in recognizing that mass cannot be created or destroyed: if the budget is negative, there must have been erosion in the system; if the budget is positive, there must have been deposition in the system. One definition of an EQUILIBRIUM system is one where the sediment budget is zero, implying that the system is just capable of transporting all the sediment applied to it. However, it is exceptionally rare for inputs to be equal to outputs over all time-scales, and this is where the sediment budget can be useful. As a non-zero sediment budget implies that the area delimited by the budget must be eroding or depositing, identification of those time-scales at which the sediment budget is either negative or positive can indicate the time-scales associated with landform change within that system.

This diagnostic property is of immense use to the geomorphologist and may be central to management: Pethick (1996) shows how cliff recession along the Holderness coast in the medium term is regulated by the difference between sediment inputs from cliff recession and sediment outputs driven by longshore drift. Determination of the sediment budget can be either process-based or morphologically based. Process-based approaches are suitable where sediment transport rates are readily measured. For instance, by measuring the concentration of SUSPENDED LOAD and DISCHARGE at two points in a river with no tributary inputs, it is possible to estimate a budget for suspended load. However, the transport of coarser sediment as BEDLOAD in rivers and processes such as landsliding are much more difficult to measure directly. However, it may be possible to estimate a sediment budget by repeat monitoring of landform morphology. This is sometimes referred to as an inverse treatment of the problem. An example is provided by Lane (1997) who repeatedly surveyed the surface of a gravel-bed river in the Swiss Alps to determine sediment budgets, and then used this information to estimate the bedload transport rate upstream required to produce these budgets.          SNL

### References
Pethick, J.S. 1996. Coastal slope development: temporal and spatial periodicity in the Holderness cliff recession. In M.G. Anderson and S.M. Brooks eds, *Advances in hillslope processes*. Chichester: Wiley. · Lane, S.N. 1997: The reconstruction of bed material yield and supply histories in gravel-bed streams. *Catena* 30, pp. 183–96.

## sediment fabric
This refers to the composition and organization of a sedimentary deposit. Identification of sediment fabrics follows from the idea that different generating processes produce markedly different sediment fabrics. For instance glacial, fluvial and aeolian deposits are each typically associated with different fabrics, and thus identification of fabric in the field is central to reconstructing the environment in which the sediment was deposited.          SNL

## sediment yield
The total mass of particulate material reaching the outlet of a drainage basin. Values of sediment yield are commonly

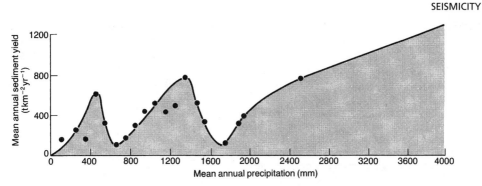

**Sediment yield.** The relationship between mean annual sediment yield and mean annual precipitation.
*Source*: D.E. Walling and A.H.A. Kleo 1979: Sediment yields in areas of low precipitation: a global view. *Proceedings of the Canberra Symposium on the Hydrology of Areas of Low Precipitation.* IAHS Publication 128, pp. 479–93.

evaluated on an annual basis (t year$^{-1}$) and may also be expressed as specific sediment yields or yields per unit area (t km$^{-2}$ year$^{-1}$). The total sediment yield comprises material transported both as SUSPENDED LOAD and BEDLOAD and separate measurements of the two components will generally be necessary. However, where the sediment load of a river is deposited in a lake or reservoir, it may be possible to estimate the total yield directly by monitoring the volume of deposited sediment. The sediment yield from a drainage basin will commonly represent only a small proportion of the gross erosion within the basin. Much of the eroded material will be deposited before reaching the outlet of the basin and the ratio of sediment yield to gross erosion is termed the sediment delivery ratio.

The magnitude of the sediment yield from a drainage basin will reflect control by several factors including climate, topography, lithology, vegetation cover and land use. Maximum values in excess of 20,000 t km$^{-2}$ year$^{-1}$ have been recorded in the severely eroded loess areas of the Middle Yellow River basin in China and in the high rainfall areas of South Island, New Zealand.                                    DEW

**Reading**
Laronne, J.B. and Mosley, M.P. eds 1982: *Erosion and sediment yield.* Benchmark papers in geology 63. Stroudsburg, Pa: Hutchinson Ross. · Walling, D.E. and Webb, B.W. 1983: Patterns of sediment yield. In K.J. Gregory ed., *Background to palaeohydrology.* Chichester: Wiley.

**sedimentary rock** Rock composed of the fragments and particles of older rocks which have been eroded and the debris deposited by wind or water often as distinct strata. Some sedimentary rocks may be of organic origin.

**seeding of clouds** See WEATHER MODIFICATION.

**segregated ice** Ice formed by the migration of pore water to the freezing plane where it forms into discrete lenses, layers or seams ranging in thickness from hairline to greater than 10 m. Segregated ice commonly occurs in alternating layers of ice and soil. Ice structure tends to be parallel to the freezing surface (i.e. usually dominantly horizontal) with air bubbles tending to be elongated and aligned normal to the horizontal layering. When dealing with massive icy bodies, as in the Mackenzie Delta, or within pingos, it is sometimes difficult to differentiate segregated ice from injection ice.                                    HMF

**Reading**
Mackay, J.R. 1972: The world of underground ice. *Annals of the Association of American Geographers* 62, pp. 1–22.

**seiche** The oscillation of a body of water at its natural period. Coastal measurements of sea level often show seiches with amplitudes of a few centimetres and periods of a few minutes due to oscillations of the local harbour, estuary or bay, superimposed on the normal tidal changes.                                    DTP

**seif** Arabic term, meaning *sword*, for a LINEAR DUNE.

**seismicity** The intensity and frequency of earthquakes in an area. Earthquake intensity is measured on the logarithmic Richter scale, the largest shocks having a magnitude of a little over 8.5. For each decrease in unit magnitude the frequency of earthquakes increases by a factor of between 8 and 10. About 25 shocks a year with a magnitude of 7 or more are registered, but the total annual number of earthquakes of all magnitudes exceeds one million. Most seismic activity is located along plate boundaries (see PLATE TECTONICS) and is particularly concentrated in the circum-Pacific belt, which accounts for about 80 per cent of global seismicity.                                    MAS

**Reading**
Bolt, B.A. 1978: *Earthquakes: a primer.* San Francisco: W.H. Freeman. · Gubbins, D. 1990: *Seismology and plate tectonics.* Cambridge: Cambridge University Press. · Wyss, M., ed. 1979: Earthquake prediction and seismicity patterns. *Pure and applied geophysics* 117, pp. 1079–315.

**self-mulching**  A process whereby swelling and shrinking in soils, resulting either from alternate wetting and drying or from freezing and thawing, gives rise to a surface layer, composed of well-aggregated granules or fine blocks, which does not crust.

**selva**  See TROPICAL FOREST.

**semi-arid**  These areas are the most extensive DRYLANDS, covering almost 18 per cent of the world's land surface, including large parts of the western interior USA and Canada, interior Asia, southern Africa, parts of the Sahel, Australia and, in Europe, Spain. They comprise areas where P/PET values fall between 0.2 and 0.5 (see DRYLANDS for methodology). Semi-arid areas have distinctively seasonal rainfall regimes and are DROUGHT prone, but because mean annual rainfall amounts can be up to 800 mm in summer rainfall areas and 500 mm in winter rainfall zones, they can and do support sizeable human populations. This in turn makes their populations and livelihoods drought prone and the environments susceptible to DESERTIFICATION.                              DSGT

**semi-desert**  A semi-arid region.

**sensible heat**  A measure of the heat content of a substance; physically it is termed enthalpy. Climatologically it is most frequently used as a term in the HEAT BUDGET equation where the available NET RADIATION is utilized in terms of a sensible heat flux, a LATENT HEAT flux and a soil heat flux. These fluxes of sensible and latent heat from the earth's surface are produced by the turbulent motions in the lower atmosphere, such as CONVECTION. Sensible heat fluxes are particularly important over hot, dry surfaces, such as deserts, where there is a steep temperature gradient between the surface and the air above.                              PS

**Reading**
Hartmann, D.L. 1994: *Global physical climatology.* San Diego: Academic Press.

**sensible temperature**  Used in the context of thermal comfort, it is the indoor temperature that would produce the same sense of comfort or discomfort to a lightly clothed person, as the actual outdoor weather environment. It is dependent upon windspeed, humidity and the radiation balance as well as the air temperature.

Sensible heat is the same as enthalpy which represents the total heat or total energy content of a substance per unit mass. In the atmosphere a change of sensible heat of a mass of gas is the heat gained or lost by the gas in an exchange at constant pressure. The transport of sensible heat horizontally in the atmosphere is of fundamental importance to the general circulation of the atmosphere as heat is transferred from the tropics towards the poles.                              JET

**sensitivity**  Of a clay soil is the ratio of undisturbed, undrained strength to the remoulded, undrained strength when tested at the same moisture content and (dry) density. A reduction in strength on remoulding is usually seen for most normally consolidated soils (see CONSOLIDATION) but it is rarely greater than 10 per cent and for over-consolidated soils it is usually near zero. However, for certain types of clay (quickclays) the value may be 100 or more. This catastrophic decrease in strength can be caused in the field by an earthquake, for example, and has led to some very large slope failures, in Canada and southern Scandinavia in particular. Loesses may also exhibit a certain amount of sensitivity.                              WBW

**Reading**
Maerz, N.H. and Smalley, I.J. 1985: The nature and properties of very sensitive clays: a descriptive bibliography. *University of Waterloo, Ontario, Bibliography* 12.

**serac**  Pinnacles and cuboid masses of ice associated with rapid glacier flow, as for example found on ice falls and SURGING GLACIERS.

**seral community**  A development or successional plant or animal community. In a community sequence or SERE, all the communities not at the stable terminal (climax) stage are seral. Seral communities tend to be dominated by opportunistic species adapted for rapid dispersal and growth, often small and capable of completing their life cycle comparatively rapidly (*r*-selection). They may also be short-lived communities, of low diversity, small biomass, high net production, with short food chains and poorly developed homeostatic mechanisms. According to E.P. Odum, many of their features are also characteristic of agricultural plant communities which are similarly unstable ecologically. (See also SUCCESSION.)                              JAM

**Reading and Reference**
Drury, W.H. and Nisbet, I.C.T. 1973: Succession. *Journal of the Arnold Arboretum* 54, pp. 331–68. · Odum, E.P. 1969: The strategy of ecosystem development. *Science* 164, pp. 262–70.

**serclimax**  A plant community forming a relatively stable stage in a SERE. It is a long-persisting SERAL COMMUNITY that gives at least an impression of permanence, as particular species or environmental factors may stabilize a community well before the climax stage. (See also ALLELOPATHY; CLIMAX VEGETATION; SUCCESSION).

JAM

**Reading**
Connell, J.H. and Slatyer, R.O. 1977: Mechanisms of succession in natural communities and their role in community stability and organization. *American naturalist* 111, pp. 1119–44. · Moravec, J. 1969: Succession of plant communities and soil development. *Folia geobotanica et phytotaxonomica* 4, pp. 133–64.

**sere**  A sequence of communities, usually plant communities, at a particular site. It is a series of stages that follows from the process of SUCCESSION. Each sere is made up of SERAL COMMUNITIES and may eventually terminate in a stable community. Different seres in a landscape have for long been viewed as at least partially convergent; thus the later stages in a dry environment (xerosere) are supposed to become increasingly similar to the later stages in a wet environment (hydrosere). Many factors, both natural and anthropogenic, may arrest, disturb or deflect such a simple pattern of development.

JAM

**Reading**
Daubenmire, R. 1968: *Plant communities: a textbook of plant synecology.* New York and London: Harper & Row. · Matthews, J. A. 1979: A study of the variability of some successional and climax plant assemblage-types using multiple discriminant analysis. *Journal of ecology* 67, pp. 255–71.

**serir**  A REG. A desert with a surface mantled by sheets of pebbles.

**seston**  'Includes all of the mineral particles and non-living organic matter that is suspended in the flows, whether derived from allochthonous (tributary, etc., external) or autochthonous (living, internal) sources' in rivers (Petts 1985, p. 88).

ASG

**Reference**
Petts, G.E. 1985: *Impounded rivers.* Chichester: Wiley.

## Seventh Approximation for soil classification

A system for the classification of the varieties of soil types recognized in the United States, but applied elsewhere. Principally during the 1950s, the US system of soil classification underwent a series of revisions and the resulting systems were called 'Approximations' to the ideal taxonomy. The seventh and latest approximation was released in 1960. The goal was to find a rigorous system for soil classification, whose categories allowed useful predictions to be made about the suitability of the soils for various uses. Refinement and revision are ongoing, and the system has been outlined in the *Soil Taxonomy* (1988) volume listed below. The classification system recognizes 11 soil orders, together with sub-orders (55), great groups (238), subgroups (>1,200), families (>7,500), and series (>18,500). The names given to the soils in this system are generally based on Greek and Latin roots and are used to convey information on the properties and development of the soil.

The hierarchical system of classification is based upon the identification of a series of diagnostic features in a soil. These include HORIZONS of various kinds, both surface (termed *epipedons*) and subsurface, whose identification is based upon colour, texture, chemical composition and other properties. Examples are the *mollic* epipedon (Latin, *mollis*, 'soft'), a surface layer rich in organic matter (the definition requires > 0.6 per cent organic carbon) and hence soft. Soils displaying this feature are widespread under native grasslands, and are grouped into the *Mollisol* order.

Because of their importance to plant growth, to the degree of oxygenation of the soil, and for the breakdown of organic materials, both the temperature and the wetness of soils are used as criteria for classification. In terms of wetness, soils may range from *aquic*, having features related to saturation and poor aeration, to *xeric*, when long periods of dryness occur. In a similar way, the temperature criterion recognizes soils ranging from *pergelic*, where the mean annual soil temperature is below freezing, to *hyperthermic* if the mean annual soil temperature is above 22 °C. The annual range of temperature is also taken into consideration.

The soil *orders*, the highest level of the taxonomy, are based upon dominant soil forming processes, recognized primarily from the diagnostic horizons. They are set out in the table. Note that all the order names end in -*sol*, from the Latin, *solum*, 'soil'.

In naming suborders and subsequent subdivisions, part of the order name is retained, so indicating immediately the order to which they belong. For example, subdivisions of the *Molli*sols retain the -*oll* component. These include the Aqu*olls* suborder, the Argiaqu*olls* great group and the typic Argiaqu*olls* subgroup.

In general, the sub-orders are groupings of soils belonging to one order that all show some common developmental process, and terms relating to the temperature and wetness regimes described above are thus used in naming suborders.

Soil orders and their characteristics in the Seventh Approximation for soil classification

| Soil order | Primary characteristics |
| --- | --- |
| Entisols | Very little profile development |
| Inceptisols | Embryonic profile differentiation |
| Mollisols | Mollic epipedon |
| Alfisols | Argillic or natric horizons, with moderate to high base saturation |
| Ultisols | Argillic horizon with low base saturation |
| Oxisols | Oxic horizon; highly weathered |
| Vertisols | Rich in fine, swelling clays |
| Aridisols | Dry soils, often with a pale epipedon |
| Spodosols | Spodic horizon (strongly eluviated) |
| Histosols | Peaty soils (> 30 % organic matter) |
| Andisols | Soils on volcanic cinders or ash |

Great groups are identified within a sub-order primarily by the recognition of diagnostic horizons. Great groups are further subdivided into subgroups, of which the primary representatives from the *typic* sub-order.

The soil families are a finer subdivision of the subgroups, and identify soils whose kindred properties result in similar management needs when cultivated. Particular instances of the soil families, often named to record the location where they were first mapped, form the soil *Series*.                                                      DLD

**Reading**
Brady, N.C. and Weil, R.R. 1996: *The nature and properties of soils.* 11th edn. New Jersey: Prentice-Hall. · United States Department of Agriculture, Soil Conservation Service. 1988: *Soil taxonomy.* Florida: Krieger Publishing Co.

**shakehole**   A roughly circular depression in the landscape in which water drains into an underground limestone cave system. The term is used synonymously with DOLINE and swallowhole, but should be restricted in usage to a depression formed by the collapse of underlying limestone strata (Warwick 1976). In practice, shakeholes, etc. are produced by a combination of solution and collapse processes. The term derives from Derbyshire, England, but is now used extensively for any circular depression in limestone regions.                                                      PAB

**Reference**
Warwick, G.T. 1976: Geomorphology and caves. In C.H.D. Cullingford and T.D. Ford eds, *The science of speleology.* London: Academic Press. Pp. 61–126.

**shale**   A compacted sedimentary rock composed of fine-grained particles usually clay-sized. Shales are characteristically fissile in the plane of their bedding.

**shape**   In general, the shape of an object is its external appearance. It is often rather loosely defined, e.g. fan-shaped, cumulus cloud, prismatic peds. A more precise definition is usually required and the ratios of three orthogonal axes which run through the object are conveniently used. How these axes are defined and used depends upon the complexity and size of the object as well as the purpose. A spherical ball has the three orthogonal axis lengths equal, although an object with such three axes equal in length may be much more complex in detail. A golfball is one such example. A Zingg diagram is often used to show relative shapes of sedimentary particles. Here, for each particle, the ratio of intermediate (b) axial length to long axis (a) length is plotted as ordinate against the ratio of short (c) axis to intermediate (b). This procedure is hard to achieve with very small particles. Similar plots of ratio lengths can be used to define landforms for analysis of MORPHOMETRY.

As the external morphology of an object is surprisingly difficult to define, more than one parameter may have to be used, e.g. angularity-roundness and surface texture. In one scheme 'shape' is only one component of a particle's 'FORM'. For very complex forms, techniques such as FOURIER ANALYSIS or FRACTAL DIMENSION may give a better indication of shape rather than axial length ratios.          WBW

**Reading**
Orford, J.D. and Whalley, W.B. 1991: Quantitative grain form analysis. In J.P.M. Syvitski. ed., *Principles, methods, and applications of particle size analysis.* Cambridge: Cambridge University Press. Pp. 88–108.

**shear box**   An apparatus used to measure the shear strength of a soil. In its simplest form it consists of a split box into which the soil to be tested is placed. A normal stress is applied to the top of the sample via weight and, while one half of the box is moved parallel to the base to provide the shear stress, the other half resists the movement, the magnitude of the resistance being measured by a 'proving ring'. In this form the test is strictly called the direct shear test as it differs from simple shear which is an angular, rather than a linear, displacement. At least three tests are performed with different values of normal stress; a plot of the resisting stress corresponding to the normal stress gives values of COHESION and FRICTION which can be used in the MOHR–COULOMB EQUATION. The device can also be used to measure the RESIDUAL STRENGTH of the soil. Different sizes of box are required – the larger the size of grain under test the larger the size of the box; thus 30 cm square boxes may be required for testing gravelly soils. Although mainly a laboratory test, some versions are for *in situ* field determinations. The TRIAXIAL APPARATUS does a similar job but has advantages over the direct shear box.　　WBW

**shear strength**　A measure of the ability of a material to resist shear stress. This is an important parameter in determining the engineering and geomorphic properties of materials. The shear strength of a soil is controlled by components of the MOHR–COULOMB EQUATION. Soils have a maximum strength value (peak strength) usually determined by a triaxial test or SHEAR BOX test. Various types of test conditions may be used to determine the shear strength. These relate to consolidation and drainage conditions. Particularly important are the 'undrained test' where no PORE WATER PRESSURE dissipation is allowed during shearing and the 'drained test' where such drainage is allowed. Generally speaking, the undrained test, which is done with a high strain rate, gives a lower shear strength value.　　WBW

**Reading**
Atkinson, J.H. and Bransby, P.L. 1978: *The mechanics of soils.* London: McGraw-Hill. · Whalley, W.B. 1976: *Properties of materials in geomorphological explanation.* Oxford: Oxford University Press.

**shear stresses**　Two perpendicular loads or stresses (force per unit area) applied parallel (tangential) to the surface of a body. They are themselves perpendicular to the NORMAL STRESS. The diagram illustrates their orientation with respect to a cube of material. Shear stresses produce angular deformation (shear strain) in the body.　　WBW

**Reading**
Statham, I. 1977: *Earth surface sediment transport.* Oxford: Clarendon Press.

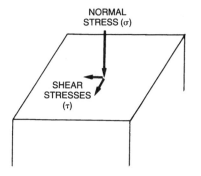

NORMAL
STRESS (σ)

SHEAR
STRESSES
(τ)

**sheet erosion**　The stripping of soil material from a relatively broad area such as a field or hillside to a uniform depth and without the incision of discrete channels as large as rills or

gullies. In reality all hillslope run-off involves some degree of concentration of flow into small proto-channels or preferred flow lines, but if the soil material is resistant to scour or the shearing forces created by the water flow remain below the level required to entrain soil particles on a large scale, then the surface may be lowered without significant incision taking place. The primary eroding force in sheet erosion is probably raindrop splash (see RAINDROP IMPACT EROSION), and for this to be the case, the erosion process must be acting on a surface of low gradient. Steep gradients are associated with faster flow in proto-channels and hence with a greater likelihood that sheet erosion will give way to rill erosion and other forms of channelized run-off. Once flow becomes channelized, flow depth is increased, the surface roughness that retards the flow becomes less effective, and flow speeds increase. Increasing flow speeds are in turn associated with increasing shear stresses at the soil surface, and a greater likelihood that incision will be initiated. It has been suggested (Moss and Green 1983) that on low gradients, the impact of raindrops may result in the creation of widespread local mixing of the flow that tends to restrict the development of flow concentrations, and which may thus help to account for the occurrence of sheet erosion. (See also RILL; GULLY.)　　DLD

**Reading and Reference**
Moss, A.J. and Green, P. 1983: Movement of solids in air and water by raindrop impact. Effects of drop-size and water-depth variations. *Australian journal of soil research* 21, pp. 257–69. · —Walker, P.H. and Hutka, J. 1979: Raindrop-stimulated transportation in shallow water flows: an experimental study. *Sedimentary geology* 22, pp. 165–84.

**sheet flow**　Surface run-off that is not concentrated into rills or other small channels, but which traverses the surface at shallow depth. Sheetflow is one of the forms of flow that define the interrill phase of surface run-off. Such flow can only occur on relatively smooth surfaces having little MICROTOPOGRAPHY. Shallow flows are relatively slow, owing to frictional drag, and thus they generate limited shearing forces at the soil surface that restrict soil entrainment. The flow is also susceptible to the disturbing influence of raindrops, which can generate marked agitation of the moving water, so further slowing its motion and limiting its potential to scour the surface. Nevertheless, INTERRILL FLOW or sheetflow provides the means by which eroded particles are delivered from the unchannelized parts of the landscape into rills. The primary dislodgment of particles is caused by RAIN SPLASH processes, and the drop impacts, by agitating the flow and re-suspending particles that settle out, also assist in the transportation of the eroded particles. These mechanisms are jointly referred to as RIFT (raindrop-induced flow transportation) or rain-flow transportation (Moss 1988), and they account for much of the erosion that takes place in sheet flow.                    DLD

**Reading and Reference**
Gascuel-Odoux, C., Cros-Cayot, S. and Durand, P. 1996: Spatial variations of sheet flow and sediment transport on an agricultural field. *Earth surface processes and landforms* 21, pp. 843–51. · Moss, A.J. 1988: Effects of flow-velocity variation on rain-driven transportation and the role of rain impact in the movement of solids. *Australian journal of soil research* 26, pp. 443–50.

**sheeting**  The formation of joints in a massive rock such that the outer layers of the rock separate in shells or spalls and exfoliate (i.e. peel away from the parent rock mass). The shells are large, with minimum areas of several square metres and thicknesses of tens of centimetres. They commonly form, and may be responsible for, the shape of rounded outcrops and boulders. Successive exfoliation of concentric rock sheets results in the maintenance of rounded forms of some tors and bornhardts.

Sheeting commonly occurs in massive rocks which have once been deeply buried and have subsequently been brought to the ground surface by removal of overburden or overlying ice. The uncovered rock undergoes release of confining stresses and expands outwards towards the ground surface and bounding joints. Extension fractures form perpendicular to the minimum stress direction at the time of failure: this occurs because the original maximum stress direction is controlled by the overburden; as

**Sheeting** developed on rock outcrops in the Namib Desert in Namibia. The rounded form of inselbergs may in part result from the development of dilatation joints associated with this process.

the overburden is eroded the maximum stress converts to being the minimum stress and sheeting develops nearly parallel to the unloading surface. Well-known examples occur in the granites of Yosemite Valley, California and curved forms in sandstones of the Colorado Plateau are described by Bradley (1963).    MJS

**Reading and Reference**
Bradley, W.C. 1963: Large-scale exfoliation in massive sandstones of the Colorado Plateau. *Bulletin of the Geological Society of America*, 74, pp. 519–27.

**shelf**  The continental shelf. The sea floor lying between the coast and the steeper slope down to the deep ocean.

**shell pavements**  Accumulation of shell valves which occur in all places where mixtures of sand and shell are subjected to selective erosional winnowing, leaving a superficial lag deposit of the coarse shell material. Pavements of this type are common in coastal situations where, for example, fine sediment can be moved by tidal currents or by aeolian processes above high-water mark.    ASG

**Reading**
Carter, R. W. G. 1976: Formation, maintenance, and geomorphological significance of an aeolian shell pavement. *Journal of sedimentary petrology* 46, pp. 418–29.

**shield**  A continental area of exposed Precambrian rocks within a CRATON and bordered by a platform area covered by post-Precambrian sedimentary strata. Shields are highly stable low-lying areas which have experienced little deformation or volcanic activity since the Precambrian. Examples include the Canadian Shield of North America and the Baltic Shield in northern Europe.    MAS

**Reading**
Spencer, E.W. 1977: *Introduction to the structure of the earth.* 2nd edn. New York: McGraw-Hill. · Windley, B.F. 1984: *The evolving continents* 2nd edn. London and New York: Wiley.

**shield volcano**  Shield volcanoes are igneous rock constructs, with low gentle slopes rarely exceeding 6°. They are typically composed of predominantly basaltic lava flows. They range in size from diameters of less than 10 km, to Hawaiian edifices which have diameters on the sea floor in excess of 300 km. Viewed from the side, shield volcanoes have the convex profile of a shield lying flat on the ground. Shield volcanoes form from the effusive eruption of fluid (low viscosity) LAVAs that flow away from the summit or are erupted from flank fissures which ensure dispersal of the flows over a wide area. Most shield volcanoes have summit *cal-*

*deras* which are thought to form by collapse, when magma is withdrawn from high level reservoirs to feed flank eruptions. The largest shield volcanoes on earth occur in the Hawaiian islands. The big island of the Hawaiian archipelago is composed of five shield volcanoes, of which Mauna Loa (4169 m asl) and Mauna Kea (4206 m asl) are the largest. Mauna Loa, reaches a height of 10 km above the sea floor and is a very recent feature, having been built during the Quaternary. It is one of the largest topographic features on the surface of the earth. The massive Hawaiian shield volcanoes cause sagging of the Pacific LITHOSPHERE. Shield volcanoes are important topographic features on Mars and Venus as well as on earth. On Mars the shield volcano, Olympus Mons, is the largest known volcano in the solar system with a height of 23 km and a summit caldera complex 80 km in diameter. AD

**Shields parameter** A dimensionless SHEAR STRESS, relating the fluid drag on a particle to the particle's immersed weight; sometimes called Shields beta:

$$\beta = \frac{\tau_0}{(\rho_s - \rho)gD}$$

where $\tau_0$ is bed shear stress, $\rho_s$ and $\rho$ are sediment and fluid density, respectively, $g$ is gravity, and $D$ is grain diameter. The most common application of the Shields parameter is in the estimation of the threshold shear stress for the initiation of sediment movement. This is done using a Shields diagram and Shields curve to define the threshold conditions for different grain sizes. DJS

**Reading**
Middleton, G.V., and Southard, J.B. 1984: *Mechanics of sediment movement*. 2nd edn. Tulsa, Okla.: Society of Economic Paleontologists and Mineralogists.

**shoal** Area of shallow water in a lake or sea. A sand bank which lies just beneath the surface of a lake or sea.

**shore** The area of land immediately adjacent to a body of water.

**shore platforms** Intertidal rock surfaces of low slope angle. The term is preferred to 'wave cut platform' because processes other than mechanical wave action can play a role in their formation, not least weathering and bio-erosion. The form of the platforms depends on the nature of the main processes operative upon them, the nature of the rocks and their structures, on tidal characteristics, on their age, and on their history. ASG

**Reading**
Trenhaile, A.S. 1980: Shore platforms: a neglected coastal feature. *Progress in physical geography*. 4, pp. 1–23.

**sial** A term introduced by Suess to describe that part of the earth's crust with a granitic-type composition dominated by minerals rich in silicon (Si) and aluminium (Al). It is contrasted with the term SIMA. Sial forms at least the upper part of the crust of the continents and has a mean density of about 2700 kg m$^{-3}$ and a silica content of between 65 and 75 per cent. MAS

**sichelwannen** Bow-shaped furrows, with arms generally pointing in the direction of flow of the glacier that created them. They are of the order of 1–2 m long, relatively shallow, and tend to occur in large localized assemblages on glaciated surfaces. (See also P-FORM.) ASG

**Reading**
Allen, J.R.L. 1984: *Sedimentary structures*. Amsterdam: Elsevier. Pp. 264–6.

**sideways looking airborne radar** (SLAR) A microwave remote sensor used to derive images of the earth's surface.

The SLAR senses the terrain to the side of an aircraft's track. It does this by pulsing out long, up to radio, wavelengths of ELECTROMAGNETIC RADIATION and then recording, first, the strength of the pulse return to the aircraft, to detect objects, and, secondly, the time it takes for the pulse to return, to give the range of objects from the aircraft. The name radar is the acronym of these functions of radio detection and ranging. As these pulses are emitted at right angles to the aircraft track, the movement of the aircraft enables pulse lines to be built up to form an image.

Like a MULTISPECTRAL SCANNER or THERMAL INFRARED LINESCANNER, the SLAR possesses collectors, detectors and recorders and, in addition, a transmitter and antenna.

The transmitter produces pulses of microwave energy which are timed by a synchronizer and standardized to a known power by a modulator. For a fraction of a second the transmit/receive switch is switched to transmit, as the transmitter releases a microwave pulse from the antenna. The transmit/receive switch then returns to its original position and the antenna continues to receive pulses that have been backscattered from the earth's surface. These pulses are converted to a form suitable for amplification and further processing by a mixer and local oscillator before being passed to a receiver. The receiver amplifies the signal before passing it to the detector which produces an electronic signal suitable for recording on to photographic film or analogue or digital tape. To improve the spatial resolution of this sensor a larger antenna can be synthesized electronically. Such a SAR, which is termed a synthetic aperture radar (SLAR), is the primary SLAR for environmental research. The SAR is carried by aircraft and several UNMANNED EARTH RESOURCES SATELLITES, most notably ERS-1 and JERS-1.

Four characteristics of SLAR imagery determine its fields of application. They are its relatively high cost, its rapid rate of data acquisition (as it is unhindered by cloud or nightfall) and its sensitivity to both surface roughness and surface moisture content. SLAR was used initially for geological exploration as the likely financial returns were high, the areas to be covered were large and surface roughness and moisture content often varied between the areas of interest.

Today SLAR imagery is used in geomorphology, the mapping of soil moisture and vegetation, the estimation of forest biophysical properties and the location of oil pollution and sea ice. PJC

**Reading**
Trevett, J.W. 1986: *Imaging radar for resources surveys.* London: Chapman & Hall. · Ulaby, F.T., Moore, R.K. and Fung, A.K. 1981: *Microwave remote sensing, active and passive. Volume 1: Fundamentals and radiometry.* Reading, Mass. and London: Addison-Wesley.

**sieve deposits** On alluvial fans are characterized by a very steep lobe front, coarse materials throughout the deposit but especially at the lobe front, and a very high infiltration capacity for the deposit. Transportation of the debris ceases when the flow infiltrates. 'Because water passes through rather than over such deposits, they act as strainers or sieves by permitting water to pass while holding back the coarse material in transport' (Hooke 1967, p. 454). Sieve deposits lack the fine matrix characteristic of debris-flow deposits. Although named in deserts, sieve deposits have also been noted in alpine settings.

The existence of sieve deposits poses problems for the current paradigm of rheology in debris-flow generation. This is because the transportation of large clasts by just water does not fit current models that require the role of fine materials as a transporting agency. The existence of sieve deposits lacking fines has forced some investigators to argue that fines have been washed out of sieve deposits over time. But this argument ignores the overwhelming evidence from studies of STONE PAVEMENT formation that desert sieve deposits are effective dust traps. RID

**Reading and Reference**
Hooke, Roger LeB. 1967. Processes on arid-region alluvial fans. *Journal of geology* 75, pp. 438–60. · Krainer, K. 1988. Sieve deposition on a small modern alluvial fan in the Lechtal Alps (Tyrol, Austria). *Zeitschrift für Geomorphologie* NF 32, pp. 289–98.

**silcrete** A highly siliceous indurated material formed at, or near, the earth's surface through the silicification of bedrock, weathering products or other deposits by low temperature physicochemical processes. Silcrete of Cainozoic age is particularly well developed in areas of inland Australia, in southern Africa and in north-west Europe. It may attain a thickness in excess of 5 m, and through its resistance to weathering and erosion it plays an important role in armouring erosion surfaces. It forms in areas of minimal local relief under both semi-arid and humid climatic regimes. MAS

**Reading**
Langford-Smith, T. ed. 1978: *Silcrete in Australia.* Armidale, NSW: Department of Geography, University of New England. · Summerfield, M.A. 1983: Silcrete. In A.S. Goudie and K. Pye eds, *Chemical sediments and geomorphology*, London: Academic Press. Pp. 59–91.

**sill** A tabular sheet of igneous rock injected along the bedding planes of sedimentary or volcanic formations.

**silt** Fine sediment contained in a soil. More specifically, silt is a SOIL TEXTURE and a PARTICLE SIZE between CLAY and SAND. For the latter, the precise size range regarded as silt varies according to different classification schemes. The lower size limit in all schemes is 2 microns but the upper limit ranges from 20 microns in the Atterberg system, 50 microns in the USDA system, and 63 microns in the British Standards, Soil Service of England and Wales, and MIT schemes.

Silt is readily transported in suspension in wind and water, and is a common component in fluvial and lacustrine deposits, and in DUST and LOESS. DSGT

**siltation** The accumulation of fine sediment (strictly speaking of silt) in a body of water. Applied technically to the settling out of fine particles in water and more generally to the filling or choking of lakes, reservoirs or water courses. JL

**sima** A term introduced by Suess to describe that part of the earth's crust with a basaltic-type composition dominated by minerals rich in silicon (Si) and magnesium (Mg). It is contrasted with the term SIAL. It forms the crust of the ocean basins and the lower portion of the crust of the continents. Sima has a higher mean density than sial ($2800$–$3400\,\text{kg}\,\text{m}^{-3}$) and a lower silica content (less than 55 per cent).     MAS

**simulation** Hypothesis-testing in physical geography often makes use of the vicarious experiment procedure of simulation in which genuine phenomena are represented by scale, analogue or mathematical models. Scale models are small-scale hardware replications such as flumes, wave tanks or kaolin glaciers. Analogue models involve equivalent systems; for example, an electrical potential model of groundwater is feasible because of the mathematical equivalence of the laws governing current flow in a circuit and flow in a porous medium. Computer-based mathematical simulation models (Thornes and Brunsden 1977, pp. 157–71), which may be deterministic or probabilistic, are often used to test alternative hypotheses by comparing model outputs under different assumptions with actual behaviour. Links between system components are represented by mathematical relations, whose constants are adjusted by PARAMETERIZATION procedures until the best fit with reality is achieved. (See also RANDOM-WALK NETWORKS.)     KSR

**Reference**
Thornes, J.B. and Brunsden, D.B. 1977: *Geomorphology and time.* London and New York: Methuen.

**singing sands** When in motion certain dune sands generate clearly audible sounds that have been variously reported as roaring, booming, squeaking, musical and singing (Curzon 1923). A unique combination of granulometric properties appears to be responsible, including a high sorting value, uniform grain size and a high degree of roundness of grains (Van Rooyen and Verster 1983).     ASG

**References**
Curzon, G.N. 1923: The singing sands. In *Tales of travel.* London: Hodder & Stoughton. Ch. 11. · Haff, P.K. 1986: Booming dunes. *American scientist* 74, pp. 376–81. · Van Rooyen, T.H. and Verster, E. 1983: Granulometric properties of the roaring sands in the south-eastern Kalahari. *Journal of arid environments* 6, pp. 215–22.

**sinkhole** A roughly circular depression in the landscape into which water drains and collects. Specifically, it is a depression in limestone terrain, often connecting with an underground cave system through which the water drains. It is used synonymously with SHAKEHOLE and DOLINE and was originally an American term.     PAB

**Reading**
Beck, B.F. and Wilson, W.L. 1987: *Karst hydrology: engineering and environmental applications.* Rotterdam: Balkema.

**sinter** A precipitate of silica or calcium carbonate associated with geysers and hot springs.

**sinuosity** The degree of wandering or winding, applied especially to river channels. It may be defined as the ratio of actual channel distance between identified points compared to the straight or down-valley distance.     JL

**siphon** A vertical or inverted U-shaped portion of a subterranean stream channel in which the water is in hydrostatic equilibrium.

**skerry** A rocky islet shaped primarily by glacial erosion, common along many high latitude coastlines. They show little reworking by marine processes, although they may be submerged at high tides. Skerries often occur as chains or fields of islands, and are widespread along the STRANDFLAT coasts of Norway, Iceland, and Greenland and across the mouths of fjords. Skerry coasts also occur in arctic Canada and Finland.     DJS

**Reading**
Bird, E.C., and Schwartz, M.L. eds 1985: *The world's coastline.* New York: Van Nostrand Reinhold.

**skewness** A statistical measure which is widely used for PARTICLE SIZE analysis. It measures the degree of asymmetry of a statistical distribution as well as whether the distribution has an asymmetrical tail to the left or right.     ASG

**Reading**
Folk, R.L. 1974: *Petrology of sedimentary rocks.* Austin, Texas: Hemphill.

**slab failure** A term usually used of strong rocks (although it can occur in muds and weak rocks) where the TRANSLATIONAL SLIDE is along discontinuities (cracks, joints, etc.) which dip outwards from the face. Failure is largely controlled by the FRICTION between the blocks so that the angle of the discontinuities needs to be high enough to allow most of the frictional strength to be exceeded but less than the cliff slope angle. The final 'trigger' which causes failure may be caused by ice wedging, where the blocks moved are small, or by (cleft) water pressure which changes the EFFECTIVE STRESS conditions. Quantitative assessment of cliff stability can be made with various

ROCK QUALITY INDICES. (See also TOPPLING FAILURE.) WBW

**Reading**
Attwell, P.B. and Farmer, I.W. 1976: *Principles of engineering geology*. London: Chapman & Hall; New York: Wiley. · Selby, M.J. 1993: *Hillslope materials and processes*. 2nd edn. Oxford: Oxford University Press.

**slack** A depression or hollow in an area of sand dunes or mud banks.

**slackwater deposit** A deposit of riverine sediment laid down during a flood at a location sheltered from the main current, and often later employed to estimate the size of the flood. Slackwater deposits may accumulate in the regions of a river bend where the main current follows one bank leaving the other more sheltered, or near the junctions of a trunk stream and lesser tributaries whose flow is held back by the depth of flow in the main channel. These deposits are primarily used to infer the water depth or *stage* reached by major floods, especially floods occurring through the last few thousand years, in order to extend the effective period of observational data on flood magnitudes and their relation to catchment area (Kochel and Baker 1988). Datable materials within or buried by the slackwater sediments are used to determine the age of the flood event that created them. In some areas, the analysis of slackwater deposits laid down by palaeofloods has indicated the former passage of flows considerably larger than have been monitored during modern times, but elsewhere, the magnitudes of inferred palaeofloods fall within the range of modern events. DLD

**Reference**
Kochel, R.C. and Baker, V.R. 1988: Paleoflood analysis using slackwater deposits. In V.R. Baker, R.C. Cochel and P.C. Patton eds, *Flood geomorphology*. New York: Wiley. Ch. 21, pp. 357–76.

**slaking** The disintegration of a loosely consolidated material on the introduction of water or exposure to the atmosphere.

**slickenside** A polished or scratched rock surface produced by the friction during faulting.

**slide** A landslide. The landform produced by mass movement under the influence of gravity.

**slip face** When sand accumulation has built up to a critical angle on the lee side of a SAND DUNE, it fails under the influence of gravity to form a slip (or avalanche) face. Individual grain

fall, grain flow or avalanching dominate the down slope movement of sand on dry slip faces, while slumping can occur if sand is damp. Slip faces form at the angle of repose for sand. DSGT

**slip off slope** The more gently sloping bank of a river on the inside of a meander.

**slope** A word with two applications in physical geography:

1 In a general sense it is used to refer to the angle which any part of the earth's surface makes with a horizontal datum. Synonyms for this usage include: inclination, declivity, and gradient.

2 In geomorphology 'slope' refers to any geometric element of the earth's solid surface whether that element is above or below sea level. Slope elements thus form entire landscapes. In a more restricted use the term is often applied to escarpments and valley sides, and thus excludes floodplains, terrace surfaces and other nearly horizontal elements. To avoid confusion the more explicit word 'hillslope' is in common use.

Hillslopes are regarded as three-dimensional forms produced by weathering and erosion with basal elements which may be either depositional or erosional in origin. The development of hillslopes is consequently the principal result of denudation and the study of such features is a major part of geomorphology. MJS

**Reading**
Carson, M.A. and Kirkby, M.J. 1972: *Hillslope form and process*. Cambridge: Cambridge University Press. · Finlayson, B. and Statham, I. 1980: *Hillslope analysis*. London: Butterworth. · Selby, M.J. 1993: *Hillslope materials and processes*. 2nd edn. Oxford: Oxford University Press. · Young, A. 1972: *Slopes*. Edinburgh: Oliver & Boyd.

**slope replacement** A model of slope evolution formulated by the German geomorphologist, Walther Penck, in which the maximum slope angle decreases through time as a result of replacement from below by gentler slopes, causing the majority of the slope profile to become occupied by a concavity.

**slope wind** See ANABATIC FLOWS; KATABATIC FLOWS.

**smog** A term originally used to describe a combination of smoke and FOG but now used for any visibly polluted air. Dr Harold Antoine Des Voeux first used the word in 1911 to describe a series of pollution episodes in Glasgow, Scotland, during the autumn of 1909. Photochemical smog forms when hydrocarbons, originating from vaporized gasoline and other petroleum products, combine with nitrogen oxide molecules, also emitted by combustion engines, and water vapour in the presence of ultraviolet sunlight. WDS

**Reading**
Lewis, H.R. 1965: *With every breath you take.* New York: Crown.

**snout, glacial** The terminus of a glacier.

**snow** Solid precipitation composed of single ice crystals or aggregates known as snowflakes. Ice crystals are most frequent when temperatures are much below freezing and the moisture content of the air is small. As temperatures increase towards 0 °C, the ice crystals grow and cluster into flakes. Snow is difficult to measure accurately as it tends to block standard rain gauges or be blown out. PS

**snow line** The altitudinal limit on land separating areas in which fallen snow disappears in summer from areas in which snow remains throughout the year. The altitudinal distribution over the globe is the same as for the FIRN LINE but unlike the latter it is not restricted to glaciers. DES

**snow patch** An isolated area of snow which may last throughout the summer and initiate processes associated with NIVATION.

**snowblitz theory** A popular and extreme version of the view that glaciations may start rapidly as a result of positive feedback processes related to the increased albedo of a high latitude continent covered with persistent snow. A few years of excess snow could modify atmospheric circulation patterns and enhance snow accumulation (Lamb and Woodroffe 1970).

Developing the concept for Britain, Calder (1974, p. 118) wrote:

In the snowblitz the ice sheet comes out of the sky and grows, not sideways, but from the bottom upwards. Like airborne troops, invading snowflakes seize whole counties in a single winter. The fact that they have come to stay does not become apparent, though, until the following summer. Then the snow that piled up on the meadows fails to melt completely. Instead it lies through the summer and autumn, reflecting the sunshine. It chills the air and guarantees more snow next winter. Thereafter, as fast as the snow can fall, the ice sheet gradually grows thicker over a huge area.

The theory is far from being established for Arctic Canada, let alone the lush meadows of Britain. Yet evidence of rapid ice sheet build-up as a result of sudden changes in atmospheric and oceanic circulation is emerging. (See also ICE AGE.) DES

**Reading and References**
Calder, N. 1974: *The weather machine and the threat of ice.* London: BBC publications. · Denton, G.H. and Hughes, T.J. 1983: Milankovitch theory of ice ages: hypothesis of ice-sheet linkage between regional insolation and global climate. *Quaternary research* 20, pp. 125–44. · Imbrie, J. and Imbrie, K.P. 1979: *Ice ages.* London: Macmillan. · Ives, J.D., Andrews, J.T. and Barry, R.G. 1975: Growth and decay of the Laurentide ice sheet and comparisons with Fenno-Scandinavia. *Die Naturwissenschalt* 62, pp. 118–25. · Lamb, H.H. and Woodroffe, A. 1970: Atmospheric circulation during the last Ice Age. *Quarternary research* 1, pp. 29–58.

**snowmelt** The part of run-off that is generated by melting of a snowpack on the ground surface. Portions of the snowpack that do not melt may remain on the surface long enough to be compressed by subsequent snowfalls into glacial ice. It may not melt over a period of many years and thus become a semi-permanent snow body, or its mass may be lost through sublimation and deflation.

The quality of the snowpack refers to the potential amount of run-off that may be generated by snowmelt and is measured as the weight of ice divided by the total weight of a unit volume of the snowpack. When the snowpack nears the time for melting, its quality is usually in the 0.90 to 1.00 range (US Army Corps of Engineers 1960). WLG

**Reference**
US Army Corps of Engineers 1960: *Runoff from snowmelt.* US Army Corps of Engineers engineering manual 1110–2–1406.

**soil** The material composed of mineral particles and organic remains that overlies the bedrock and supports the growth of rooted plants.

**soil classification** See SEVENTH APPROXIMATION

**soil erosion** The natural process of removal of top soil by water and wind. It is a process whose rates may be magnified by humans (accelerated erosion). On a global scale the fastest rates occur in zones with highly seasonal precipitation, as in monsoonal, Mediterranean and semi-arid climates (Walling and Kleo 1979). There is a long history of the study of accelerated soil erosion (see e.g. Marsh 1864;

Bennett 1938), and in spite of the introduction of soil conservation measures such as those discussed by Hudson (1971), it continues to be a serious environmental problem (Carter 1977; Pimentel 1976). In the USA soil erosion on agricultural land operates at a rate of about $30 \, t \, ha^{-1} \, year^{-1}$. Water run-off delivers around four billion tonnes of soil to the rivers of the forty-eight contiguous states, and three-quarters of this comes from agricultural land. Another billion tonnes of soil are eroded by the wind, a process which created the Dust Bowl of the 1930s. In addition to the soil erosion caused by deforestation and agriculture, urbanization, fire, war and mining are often significant in accelerating erosion of the soil.                    ASG

**Reading and References**

Bennett, H.H. 1938: *Soil conservation*. New York: McGraw-Hill. · Carter, L.J. 1977: Soil erosion: the problem still persists despite the billions spent on it. *Science* 196, pp. 409–11. · Hudson, N. 1971: *Soil conservation*. London: Batsford. · Marsh, G.P. 1864: *Man and nature*. New York: Scribner. · Morgan, R.P.C. 1986: *Soil erosion and conservation*. Harlow: Longman. · Pimentel, D. 1976: Land degradation: effects on food and energy resources. *Science* 194, pp. 149–55. · Walling, D. and Kleo, A.H.A. 1979: *Sediment yields of rivers in areas of low precipitation: a global view*. International Association of Scientific Hydrology publication 128, pp. 479–93.

**soil moisture deficit**  'Soil moisture deficits are considered to have been set up when evapotranspiration exceeds precipitation and vegetation has to draw on reserves of moisture in the soil to satisfy transpiration requirements' (Grindley 1967).

The evaluation of soil moisture deficit is essential to the estimation of irrigation need since it provides an estimate of the degree to which soil moisture content has dropped below field capacity. Field capacity is the soil moisture condition when excess water has drained out of a saturated or near-saturated soil and it is thought to be the soil moisture condition which will promote maximum plant growth, with transpiration occurring at the potential rate (i.e. transpiration is not limited by moisture availability). Any reduction of soil moisture content below field capacity will create a soil moisture deficit which may be removed by irrigation.

Soil moisture deficits can be estimated using field instruments or evaporation estimation equations combined with evaluation of the soil WATER BALANCE. LYSIMETERS and NEUTRON PROBES allow changes in soil moisture storage to be directly observed so that the deficit may be calculated but other instruments, including evaporation pans and atmometers, as well as the majority of evaporation equations, provide esti-

mates of open water evaporation (Eo) or potential evapotranspiration (PEt), which may be actual evapotranspiration (Et) and a soil water budget for the site. In calculating Et from PEt or Eo, it is necessary to take into account the degree to which a decrease in soil moisture content will reduce the Et rate below the PEt rate. This is a controversial topic which is reviewed by Baier (1968) but the method adopted by the UK Meteorological Office (Grindley 1967) employs the concept of root constants and related drying curves described by Penman (1949). Grindley (1967) explains the way in which the difference between PEt and precipitation in consecutive time periods can be partitioned to calculate Et and soil moisture surplus or soil moisture deficit.                    AMG

**Reading and References**

Baier, W. 1968: Relationship between soil moisture, actual and potential evapotranspiration. In *Soil moisture*. Proceedings of the Hydrology Symposium 6, University of Saskatchewan, 15–16 November 1967. Ottawa: Queen's Printer. Pp. 155–91. · Calder, I.R., Harding, R.J. and Rosier, P.T.W. 1983: An objective assessment of soil moisture deficit models. *Journal of hydrology* 60, pp. 329–55. · Grindley, J. 1967: The estimation of soil moisture deficits. *Meteorological magazine* 96, pp. 97–108. · Penman, H.L. 1949: The dependence of transpiration on weather and soil conditions. *Journal of soil science* 1, pp. 74–89.

**soil profile**  The full sequence through the soil zone from the surface down to the unaltered bedrock.

**soil structure**  The grouping of aggregates within a soil. Individual structures have a variety of forms (see diagram) and may range in size from tiny granules to large blocks. Among the controls of soil structure are the presence of clay, humus and soluble silts. Soil aggregates are sometimes referred to as peds.                    ASG

**soil texture**  The character of the soil imparted by the proportions of sand, silt and clay within a sample. Two examples of soil textural classification schemes are shown in the diagram (at p. 450).                    ASG

**solar constant**  The rate at which solar radiation is received outside the earth's atmosphere on a surface normal to the incident radiation, and at the earth's mean distance from the sun. Its exact value is still a little uncertain but it is nearly $1380 \, W \, m^{-2}$. Despite its name it is probably slightly variable in time.                    BWA

**solfatara**  A small volcanic vent through which acid gases are emitted, usually in areas where violent volcanism has ceased.

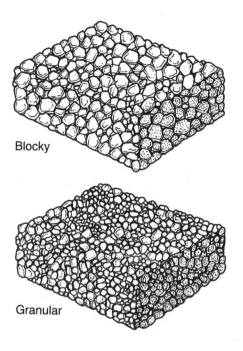

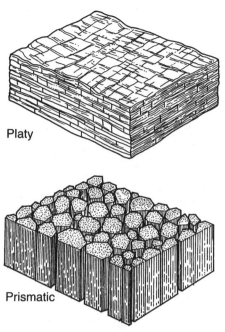

**Soil structure.** Some common forms.

**solifluction** A term first used by J.G. Andersson in 1906 to describe the 'slow flowing from higher to lower ground of waste saturated with water' which he observed in the Falkland Islands. It has subsequently been applied elsewhere to the slow gravitational downslope movement of water-saturated, seasonally thawed materials. In contrast to gelifluction, solifluction does not require permafrost for its occurrence, but modern use of the term does imply the existence of cold climate conditions. It is a form of mass wasting (i.e. viscous flow), faster than soil creep, often in the order of 0.5–5 cm year$^{-1}$. Features produced by solifluction include uniform sheets of locally derived materials, tongue-shaped lobes, and alternating stripes of coarse and fine sediment. When associated with the active layer (i.e. in permafrost regions) the term gelifluction should be used.      HMF

**Reading and Reference**
Andersson, J.G. 1906: Solifluction, a component of subaerial denudation. *Journal of geology* 14, pp. 91–112. · Benedict, J.B. 1970: Downslope soil movement in a Colorado alpine region: rates, processes and climatic significance. *Arctic and alpine research*, 2, pp. 165–226. · Washburn, A.L. 1979: *Geocrylogy: a survey of periglacial processes and environments*. New York: Wiley.

**solonchak** A group of soils which are the result of salinization, and occur where there is an accumulation of soluble salts of sodium, calcium, magnesium and potassium in the upper horizon. The anions found are mostly sulphate and chloride. In contrast to SOLONETZ soils, which are highly alkaline, solonchaks, often called white alkali soils, are only slightly alkaline, their pH seldom rising much above pH 8.
ASG

**solonetz** An intrazonal group of soils which have surface horizons of varying degrees of friability underlain by dark, hard soil characterized by a columnar structure. The hard layer is usually highly alkaline, with the high pH resulting from the adsorbed sodium and the presence of sodium carbonate. The soil colloids, both inorganic and organic, become dispersed and tend to move slowly down the profile, while the frequently observed dark colour of the surface crust is due to dissolved organic matter. These soils, sometimes known as black alkali soils, occur in semi-arid and subhumid areas.
ASG

**solstice** The day of maximum or minimum declination of the sun. Either the longest or shortest day of the year.

**solum** The soil zone above the weathered parent material, in effect the A and B horizons.

**solutes** All natural waters contain organic and inorganic material in solution or solutes. The oceans constitute about 97 per cent of the

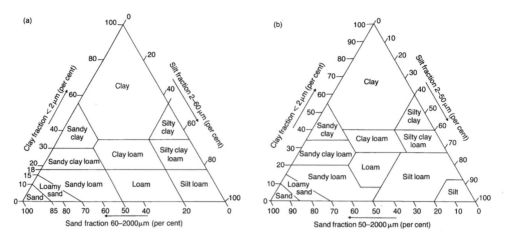

**Soil texture.** Triangular classification based on the limits laid down by: (a) Soil Survey of England and Wales; (b) US Department of Agriculture.

hydrosphere and its average chemical composition is therefore essentially that of seawater, with a total solute concentration of approximately $34,558\,mg\,l^{-1}$. The solute content of the remaining water associated with the terrestrial phase of the hydrological cycle exhibits considerable spatial and temporal variation and is of greater interest to the physical geographer. It has been widely studied as a means of investigating chemical weathering processes and rates of chemical denudation, evaluating nutrient cycling by vegetation communities, and elucidating the processes and pathways involved in the movement of water through the drainage basin system. The major solutes contained in these waters are $Ca^{2+}$, $Mg^{2+}$, $Na^+$, $K^+$, $Cl^-$, $HCO_3^-$, $SO_4^{2-}$, $NO_3^-$ and $SiO_2$.

Precipitation inputs to the landsurface contain significant concentrations of solutes as a result of rain-out and wash-out of atmospheric aerosols. The magnitude and composition of this solute content will vary according to the relative importance of terrestrial and marine aerosols. Highest total solute concentrations are found in coastal areas where marine aerosols dominate and where $Na^+$, $Cl^-$, $Mg^{2+}$ and $K^+$ are the dominant ions. The overall solute content of precipitation declines inland and, at

distances in excess of about 100 km from the coast, terrestrial aerosols predominate as solute sources. Here the major ions are $Ca^{2+}$ and $SO_4^{2-}$. Meybeck (1983) provides estimates of the average solute content of precipitation over coastal and inland zones as shown in table 1.

As it moves through the vegetation canopy and the soil and rock of a drainage basin the solute content of the water will increase and its chemical composition will frequently change. Solutes will be leached and washed from vegetation and solute levels within the soil will be influenced by concentration and precipitation mechanisms, interactions with the soil matrix, release of solutes through chemical weathering, and biotic uptake and release of nutrients.

**Table 1**  Average solute content of precipitation

|  | Concentration (mg $^{-1}$) | | | | | |
|---|---|---|---|---|---|---|
|  | $Ca^{2+}$ | $Mg^{2+}$ | $Na$ | $K$ | $Cl^-$ | $SO_4^{2-}$ |
| Coastal | 0.29 | 0.45 | 3.45 | 0.17 | 6.0 | 1.45 |
| Inland | 0.43 | 0.19 | 0.37 | 0.15 | 0.75 | 1.73 |

*Source:* Meybeck 1983.

**Table 2**  Range of concentrations of solutes in major world rivers

|  | Concentration (mg l $^{-1}$) | | | | | | | |
|---|---|---|---|---|---|---|---|---|
|  | $Ca^{2+}$ | $Mg^{2+}$ | $K$ | $Na$ | $HCO_3^-$ | $Cl^-$ | $SO_4^{2-}$ | $SiO_2$ |
| Minimum | 2.0 | 1.0 | 0.5 | 1.0 | 10 | 1.0 | 1.5 | 2.0 |
| Maximum | 55 | 15 | 4 | 40 | 170 | 45 | 65 | 20 |

*Source:* Meybeck 1983.

Further evolution of the solute content may occur within the groundwater body.

The solute content of streamflow will therefore reflect the characteristics of the upstream drainage basin, including its geology, topography and vegetation cover, and the pathways and RESIDENCE TIME associated with water movement through the basin. Concentrations will vary through time in response to hydrological conditions and will frequently exhibit a DILUTION EFFECT during storm run-off events.

At the global scale, climate and lithology exert a major influence on the solute content of river water. Solute concentrations commonly exhibit an inverse relationship with mean annual runoff and demonstrate marked contrasts between major rock types. Total solute concentrations in rivers draining basins underlain by sedimentary rocks are on average about five times greater than from basins underlain by crystalline rocks and about 2.5 times greater than from basins underlain by volcanic rocks. Typical ranges of concentrations associated with individual solute species in major world rivers are listed in table 2.

Calcium is the dominant cation and $HCO_3^-$ the dominant anion in nearly all major rivers. A greater degree of variation in solute concentrations is to be found when considering data from small streams.

Measurements of the solute input into a drainage basin and the output in streamflow provide a means of establishing a solute budget for the basin. The net solute yield (t year$^{-1}$) (output-input) reflects the production of solutes within the basin, which may in turn be related to the products of chemical weathering, the uptake of atmospheric $CO_2$ by weathering reactions and the mineralization of organic material. On a global basis, approximately 50 per cent of the solutes found in river water represent the products of chemical weathering, but this value will vary markedly between individual catchments. In catchments underlain by resistant crystalline rocks the contribution from chemical weathering may be negligible, whereas in areas of sedimentary rocks this contribution will be dominant.

Interest in the solute content of natural waters has necessitated the development of a wide range of analytical methods, which are documented in a number of laboratory manuals. Many of these methods are now semi-automated and provide a means of dealing with the large numbers of samples produced by automatic samplers or intensive manual sampling programmes. Measurements of specific CONDUCTANCE are widely employed as a simple means of estimating the total solute content of a sample and this parameter may be continuously recorded using simple equipment. Progress has also been made in the development of apparatus for continuous monitoring of individual solute species using specific ion electrodes.                    DEW

### Reading and Reference

American Public Health Association 1971: *Standard methods for the examination of water and wastewater.* New York: American Public Health Association. · Golterman, H.L., Clymo, R.S. and Ohnstad, M.A.M. 1978: *Methods for chemical analysis of fresh waters.* Oxford: Blackwell Scientific. · Hem, J.D. 1970: *Study and interpretation of the chemical characteristics of natural water.* US Geological Survey water supply paper 1473. · Likens, G.E., Bormann, F.H., Pierce, R.S., Eaton, J.S. and Johnson, N.M. 1977: *Biogeochemistry of a forested ecosystem.* New York: Springer-Verlag. · Meybeck, M. 1983: Atmospheric inputs and river transport of dissolved substances. In *Dissolved loads of rivers and surface water quantity/quality relationships.* IAHS publication 141. Pp. 173–92. · Walling, D.E. 1980: Water in the catchment ecosystem. In A.M. Gower ed., *Water quality in catchment ecosystems.* Chichester and New York: Wiley.

**solution, limestone**   The change of limestone from the solid state to the liquid state by combination with water. When water charged with carbon dioxide comes in contact with limestone (either as free $CO_2$ or as $HCO_3^-$) it dissolves the rock (CORROSION). When this occurs beneath the water table in the phreatic zone it dissolves bedding planes or joints to produce characteristically oval cave passages. When it occurs in normal stream situations it produces notches at the side of the stream on a line approximating the water level. Limestone solutional processes can also produce many different microfeatures (KARREN).                    PAB

### Reading

Picknett, R.G., Bray, L.G. and Stenner, R.D. 1976: The chemistry of cave water. In T.D. Ford and C.H.D. Cullingford eds, *The science of speleology.* London: Academic Press. Pp. 213–66. · Trudgill, S.T. ed. 1986: *Solute processes.* Chichester: Wiley.

**sorting**   of a particle size distribution is a measure of the standard deviation of the sample. It relates to the way in which material is differentially removed by particular geomorphic agencies, e.g. wind action tends to leave a well-sorted residual which is represented by a low sorting value, i.e. a predominance in a narrow size range roughly around the mean for a log normal distribution. It can be obtained graphically from phi ($\phi$) percentiles by:

$$So = \frac{\phi90 + \phi80 + \phi70 - \phi30 - \phi20 - \phi10}{5.3}$$

451

A method of calculation using moment measures is also available.                    WBW

**Reading**
Briggs, D. 1977: *Sediments*. London: Butterworth. · Tucker, M.E. 1981: *Sedimentary petrology: an introduction*. Oxford: Blackwell Scientific.

**source area**   The area of a catchment which is physically producing OVERLAND FLOW at any time. This area is changing dynamically during and after storms. Its estimation is central to PARTIAL AREA MODELS, and the term is generally used in the context of overland flow produced by subsurface saturation rather than because the INFILTRATION capacity has been exceeded. Before a storm the previous rainfall establishes the pattern of the SATURATED ZONE within a catchment. Storm rainfall is added to this layer of saturated water, and increases the source area during the storm. After the storm THROUGH-FLOW gradually diminishes the saturated wedge and the source area declines with it. The source area typically consists of river floodplains, together with a narrow strip along the base of concave hillsides and a larger area of converging flow in stream-head hollows. The total area involved varies from 1 to 3 per cent of catchments under dry conditions up to 10–50 per cent during and immediately after major storms, although there are wide differences between catchments.                    MJK

**Reading**
Kirkby, M.J. 1978: *Hillslope hydrology*. Chichester: Wiley.

**source bordering dune**   A sand DUNE that is close to the source of sediment from which material is deflated by aeolian processes, the term is usually used with reference to dunes bordering fluvial systems in drylands but can equally apply to a LUNETTE dune on the margin of a PAN or to COASTAL DUNES.                    DSGT

**Southern Oscillation**   The fluctuation of atmospheric mass (pressure) between the eastern and western hemispheres of the tropical South Pacific Ocean. Atmospheric pressure varies in a see-saw fashion between the two areas, and when one experiences high pressure the other tends to have lower pressure. The oscillation itself occurs on an irregular, interannual basis and is strongly associated with sea surface temperatures.

Under 'normal' conditions, sea level pressure tends to be relatively high in the tropical South Pacific Ocean and comparatively low in the western Pacific and eastern Indian oceans. Similarly, sea surface temperatures are usually much higher in the western Pacific than in the east. Under these conditions the Walker circulation operates in its normal fashion, the net transport of air at the surface is from east to west, and rainfall is much higher in the western Pacific.

At irregular intervals the sea level pressure difference across the Pacific decreases dramatically or even reverses. This weakens the surface easterlies and causes them to retreat eastward, disrupting the Walker circulation. These developments are usually, but not always, accompanied by EL NIÑO conditions, wherein the warm pool of water (and thus rainfall) normally located in the western Pacific migrates eastward.

The state of the Southern Oscillation is expressed in the form of a numerical index. This Southern Oscillation Index (SOI) is defined as the difference between the standardized sea level pressure at Papeete, Tahiti, and Darwin, Australia. Low index values are typically associated with El Niño conditions.                    RSV

**speciation**   See DARWINISM; EVOLUTION.

**species–area curve**   A graph of the relationships between plant or animal numbers and the area of sample plots. In general, the number of species present will increase as area increases within any given community. Eventually, the number of new species found in successively larger plots will become progressively fewer and the species–area curve will flatten and become approximately horizontal. Species–area curves are nevertheless useful guides in the determination of a satisfactory quadrat size for sampling and for comparing the size of the fauna or flora of different islands or various sized landmasses.                    PAS

**Reading**
Hopkins, B. 1957: The concept of minimal area. *Journal of ecology* 45, pp. 441–9. · Kershaw, K.A. 1973: *Quantitative and dynamic plant ecology*. 2nd edn. London: Edward Arnold. · Krebs, C.J. 1978: *Ecology: the experimental analysis of distribution and abundance*. 2nd edn. New York: Harper & Row. · Preston, F.W. 1962: The canonical distribution of commonness and rarity. *Ecology* 43, pp. 185–215, 410–32. · Randall, R.E. 1978: *Theories and techniques in vegetation analysis*. Oxford: Oxford University Press.

**species–energy theory**   Suggests that the present-day species richness of plants and animals for largish regions can be explained in terms of available energy. The hypothesis is that subject to water supply and other factors not being limiting, diversity in terrestrial habitats is to a great extent controlled by the amount of solar energy available, declining with increasing latitude in accordance with the poleward reduction in the receipt of solar radiation (Wright 1983).                                    ASG

**Reference**
Wright, D.H. 1983: Species–energy theory: an extension of the species–area theory. *Oikas* 41, pp. 496–506.

## specific conductance    See CONDUCTANCE, SPECIFIC.

## specific retention    The volume of water which a rock or soil retains against the influence of gravity if it is drained following saturation. The difference between POROSITY and specific retention is the SPECIFIC YIELD.            PWW

**Reading**
Ward, R.C. and Robinson, M. 1990: *Principles of hydrology*. 3rd edn. Maidenhead: McGraw-Hill.

## specific yield    The volume of water that a water-bearing rock or soil releases from storage under the influence of gravity. In an unconfined AQUIFER, it is expressed as the volume per unit surface area of aquifer per unit decline in the WATER TABLE.            PWW

**Reading**
Freeze, R.A. and Cherry, J.A. 1979: *Groundwater*. Englewood Cliffs, NJ: Prentice-Hall.

## spectral analysis    A tool for identifying structure in either time series or space series. It is the direct corollary of time-series analysis. Whilst the latter looks for underlying temporal patterns in a dataset, spectral analysis looks for dominant frequencies. Analysis in the frequency domain assumes that variation in a series can be represented as the sum of a set of periodic components. In the simplest terms, a variable $X$ that varies as a function of $t$ may be represented as:

$$X_t = Z_t + \sum_{i=1}^{k} R_i \cos wi + qi$$

where: $Z_t$ is a stationary random series; $R_i$ is the amplitude of a periodic component of angular frequency $wi$, $k$ is a finite number of (time or space) scales; and $qi$ is a phase shift. The angular frequency is related to the frequency by:

$$f = \frac{w}{2\pi}$$

The most common output of a spectral analysis is the spectral density function, which plots the contribution to the total variance of the series which is accounted for by a given frequency range against frequency. The variance contribu-

tion is normally normalized by the total series variance. Thus, inspection of the spectral density function is a critical means of identifying dominant scales of behaviour within a time or space series.            SNL

**Reading**
Chatfield, C. 1980: *The analysis of time-series: an introduction*. London: Chapman and Hall.

## speleology    The scientific study of caves, their formation and processes. It includes studies of speleogenesis (cave formation processes), cave survey, biology, geology and chemistry. The science used to be conducted by amateurs or semi-professionals but, since 1970, in response to the increasing technicalities of the subject, most of the pertinent research is carried out by full-time scientists and researchers.            PAB

**Reading**
Bögli, A. 1980: *Karst hydrology and physical speleology*. New York: Springer-Verlag. · Ford, T.D. and Cullingford, C.H.D. eds 1976: *The science of speleology*. London: Academic Press.

## speleothem    A general term for depositional features which include stalactites, stalagmites, columns, flowstone, helictites and curtains. Speleothems are commonly calcareous, crystalline deposits but can be made of a number of different materials; silica, gypsum, peat and ice have all been recorded. Calcareous speleothems are formed when rainwater seeps through organic rich soils and absorbs carbon dioxide. On contact with limestone it dissolves some of the rock, but when it reaches a cave roof it comes in contact with air that is not charged with as much carbon dioxide. The air absorbs some of the carbon dioxide, the water becomes less aggressive (indeed, supersaturated) and deposits some calcium carbonate in the cave.            PAB

**Reading**
Warwick, G.T. 1962: Cave formations and deposits. In C.H.D. Cullingford ed. *British caving*. London: Routledge & Kegan Paul. Pp. 83–119.

## sphagnum    PEAT mosses are characteristic peat-formers in ombrotrophic mires, particularly raised mires. In undisturbed raised mires, the characteristic hummock-hollow microtopography is characterized by specific Sphagnum species such as *Sphagnum cuspidatum* in peat hollows and *Sphagnum magellanicum* in peat hummocks.

Sphagnum leaf cells are not of uniform size but consist of narrow chlorophyll cells for assimilation sandwiched between larger hyaline cells. Sphagnum cells take up water through their pores until they are full and release it very slowly; in effect they function like a sponge: the anatomical structure of Sphagnum cells allows water absorption up to 15 to 30 times dry weight. Sphagnum grows as billions of plantlets side-by-side in a raised MIRE forming a mattress of moss growing upward by a few millimetres each year and decaying below the ground surface as a result of lack of light. Other species surviving in the raised mire environment must adapt to this upward growth habit.          ALH

**sphenochasm and sphenopiezm**  The former is the triangular gap of oceanic crust separating two cratonic blocks with fault margins converging to a point, and is interpreted as having originated by the rotation of one of the blocks with respect to the other (e.g. the Bay of Biscay). By contrast, the latter is a wedge of crust caused by the squeezing together of blocks (e.g. the Pyrenees).          ASG

**sphericity**  The degree to which a particle tends toward the shape of a sphere.

**spheroidal weathering**  Exfoliation, onion-weathering. The disintegration of a rock by the peeling of the surface layers which tends to round boulders and cobbles.

**spits**  Spits are generally linear desposits of beach material attached at one end to land and free at the other. There are many different types, including single, recurved, looped, hooked, complex and double spits. They occur where LONGSHORE DRIFT carries material beyond a change in orientation of the coast or at river mouths. They usually have a narrow proximal part and a broader distal end, where recurves are common. Spits may be called bay head, mid bay or bay mouth according to their position in an embayment. They occur mainly on indented coasts where abundant sediment can move alongshore freely. Nearly all spits have formed since sea level stabilized about 4000 years ago, and many are much younger.          CAMK

**Reading**
Schwartz, M.L. ed. 1972: *Spits and bars*. Stroudsburg, Penn: Dowden, Hutchinson and Ross. · Zenkovich, V.P. 1967: *Processes of coastal development*. Edinburgh: Oliver & Boyd.

**spring line**  See SPRINGS.

**spring mounds**  These are formed by the evaporation of mineralized water that reaches the ground surface via artesian springs. The salts that are precipitated on evaporation build up a raised mound, which may be several metres high. Different forms exist, including conical, pinnacles and ridges, which may or may not have a central crater. They are well described from Lake Eyre, Australia and central Tunisia.
          DSGT

**Reference**
Roberts, C.R. and Mitchell, C.W. 1987: Spring mounds of central Tunisia. In L. Frostick and I. Reid eds, *Desert sediments, ancient and modern*. Geological Society of London special publication 35, pp. 321–34.

**springs**  Concentrated point discharges of groundwater outflow, which can occur in terrestrial, intertidal and submarine settings. Springs form one end of a spectrum of ways in which groundwater emergence takes place, with more diffuse seepage areas forming the other. Springs are sometimes called resurgences if they represent the point of emergence of a known surface stream and exsurgences if the headwaters are unknown.

Springs can be classified in a number of ways. For example, cold water springs can be distinguished from geothermal springs, and mineral, hardwater and saline springs can be identified on the basis of water quality. Spring discharge can also be used as a classificatory criterion. Discharge from the majority of springs is variable, with those that flow throughout the year termed perennial whilst those that flow for part of the year are classed as intermittent. Perennial springs can, in turn, be termed as ebbing and flowing if there are alternating high and low pulsating flows on a diurnal or seasonal basis, whilst ARTESIAN springs tend to have more constant flow.

Variations in the amount of flow are usually a direct result of variations in the amount of water in GROUNDWATER storage. This varies due to a combination of climatic conditions which determine the seasonal or annual height of the WATER TABLE, and the characteristics of the AQUIFER. Spring flow from thick, highly porous aquifers tends to be relatively constant because the volume of seasonal storage change is small when compared to the aquifer's total storage volume. Flows from some thin superficial aquifers, for example, scree or glacial gravels, may be highly variable, with discharge only occurring for short periods after rainfall. Springs may occur in groups or clusters, often in lines along the foot of hills and where lithologies of differing permeabilities are juxtaposed. In the case of those draining the same aquifer, the springs at lower altitude are likely to be larger and faster flowing (underflow springs) whilst those at higher elevation are more likely to be intermittent or overflow springs which act as outflow points for the aquifer when the water table is high.

Less commonly, springs may experience a reversal in flow direction and become an inflow point rather than a water outflow. Such springs are called estavelles and are found in KARST terrain. Under normal circumstances, the water table rises upstream of the spring, beneath the slope from which spring water emerges. However, sometimes a low-lying or enclosed basin (a POLJE in karst terrain) becomes so full of water that the water level is temporarily higher than the water table in the adjacent hill. If this is the case, the hydrological gradient is reversed, with subterranean fractures and joints allowing the waters to exit the depression via the estavelle.                                        DJN

**Reading**
Price, M. 1996: *Introducing groundwater.* 2nd edn. London: Chapman & Hall.

**squall line**  A few cumulonimbus storms in a row, organized to produce strong along-line winds and heavy rain. It may be hundreds of kilometres long but only a few wide, so wind speeds increase very rapidly, followed quickly by heavy rain over a large transverse distance, causing widespread damage. Two broad categories of 'tropical' and 'mid-latitude' are not necessarily confined to those regions. Extensive cirrus sheets seen on satellite pictures help diagnosis in regions where data are sparse.                                    JSAG

**Reading**
Ludlam, F.H. 1980: *Clouds and storms.* Englewood Cliffs, NJ: Pennsylvania State University Press.

**stability**  The ability of an ecosystem to maintain or return to its original condition following a natural or human-induced disturbance. This concept of stability has been widely used by scientists in recent years, but many other meanings have been attached to the term 'stability' (Orians 1975). For example, the term has been used in reference to the constancy or PERSISTENCE of species populations or ecosystems. Two major aspects of ecosystem stability in relation to disturbance have received most attention. Even in this context a confusing variety of terms has been used. The first property, which is often labelled 'resistance', is the ability of a system to remain unaffected by disturbances. This property is referred to as 'inertia' by Orians (1975) and Westman (1978) and as 'resilience' by Holling (1973). The second attribute is usually termed 'resilience' and is the ability of the system to recover to its original state following a disturbance. The

more general term 'stability' has also been applied to this property by May (1973) and Holling (1973).

The concepts of resistance and resilience are of considerable interest not only to scientists engaged in basic research but also to environmental planners and managers. Knowledge of the varying ability of ecosystems to resist change, or to recover quickly following disturbance, is of obvious value in planning development projects and assessing potential damage from pollutants.

Resistance and resilience can be measured in a variety of ways. The particular ecosystem characteristics analysed will depend on the nature of the disturbance and the goals of the research, and may range from a focus on individual species populations to overall system properties such as species diversity, primary production and nutrient losses in drainage water. Resistance to a disturbance can be measured by the magnitude of the system response and by the time delay before a response occurs. These parameters provide an assessment of the relative resistance of an ecosystem to different types of stress or alternatively of different ecosystems of the same stress. For example, a study by Vitousek *et al.* (1981) of nitrate losses from disturbed forest plots revealed a major peak in nitrate losses within six months of disturbance in Indiana maple and oak forests, whereas much smaller losses occurred in hemlock and Douglas-fir forests in Oregon. A pine forest in Indiana exhibited an extended delay in response, with substantial losses beginning almost two years later.

Westman (1978) examines a variety of measurement problems associated with resilience and suggests that four aspects of this component can be evaluated. Elasticity refers to the rapidity of the system's return to its original state; amplitude to the zone from which the system can recover; hysteresis, the extent to which the recovery pathway differs from the pattern of disruption which occurred in response to the disturbance; and malleability, the degree to which the new stable state established after disturbance differs from the original steady state. The amplitude aspect of resilience is of particular interest because it deals with a threshold beyond which the system cannot recover to its initial state. It may be possible to suggest that a system is reaching a threshold by studying the rate of change of various characteristics in relation to the range of intensity of a particular stress. Baker (1973) has studied the amplitude response of saltmarsh vegetation to oil pollution. His data suggest that recovery was good when the vegetation was exposed to not more than

455

four oil spillages. Substantial damage and very slow recovery occurred after 8–12 successive oilings, suggesting the presence of a threshold of recovery.

Ecosystem resistance and resilience may not necessarily be closely linked. The initial degree to which a system is altered by disturbance may in some cases be a poor indicator of the ultimate ability of the system to cope with stress. In the Great Lakes many species of fish including herring, walleye and lake trout withstood fishing pressure for many years without any obvious signs of decline, but all these species experienced sudden collapses of populations to near-extinction levels without any advance warning (Holling 1973).

Much research has been devoted to the study of relationships between stability and other ecosystem properties, particularly species diversity (Goodman 1975, Pimm 1984). The notion that more complex systems involving large numbers of interacting species should be more stable than simple systems with few species is intuitively attractive since there should be more alternative pathways for feedback and adjustment to disturbance in the complex system. MacArthur's (1955) hypothesis that stability was a function of the complexity of feeding linkages between organisms in an ecosystem was therefore rapidly accepted and several lines of evidence were used to support the relationship. This evidence, which is reviewed by Elton (1958), involved data from laboratory experiments with one prey/one predator systems, which revealed the occurrence of large population fluctuations followed by rapid extinction. Emphasis was also placed on the vulnerability of simple agricultural systems to pest outbreaks and the contrast between prominent population oscillations in Arctic tundra and the apparent lack of such fluctuations in the complex and species-rich tropical rain forests. A more critical evaluation of the linkage between stability and diversity in recent years has underlined the weakness of this evidence. For example, instability in the laboratory predator–prey system does not provide a valid analogy to the real world since even very simple ecosystems contain many different species. Similarly, the instability of crop monocultures can probably be attributed to the absence of coevolution over long time periods of the species involved.

Empirical studies suggest that there is no simple link between diversity and stability (Goodman 1975). The total species diversity of an ecosystem may not be an adequate measure of complexity, which can take a variety of forms in relation to the trophic level involved and the spatial organization of the system. Watt (1968) has proposed that stability at any herbivore or carnivore trophic level increases with the number of competitor species at that level, decreases with the number of competitor species that feed upon it, and decreases with the proportion of the environment containing useful food. The question of the relationship between diversity and stability is further complicated by evidence that some ecosystems contain a single species high in the food web which can influence the system structure (Paine 1969). The stability of such systems would be highly dependent on the effects of disturbance on the 'keystone' species rather than on the overall species diversity of the system. Extensive examination of the diversity–stability hypothesis has been undertaken, using mathematical models (May 1973). Defining stability as the ability of the system to return to equilibrium after stress, May found that complex model systems are less stable than simple ones. In view of the complexities revealed by recent research and the fact that existing evidence is contradictory, many ecologists now regard the equating of stability with diversity as a tentative hypothesis rather than an axiom. The influence of other ecosystem properties on stability has generally been neglected, although it is likely that the resistance and resilience of systems to human-induced stresses can be affected by the spatial organization of the ecosystem and a variety of linkages between biological and abiotic components (Hill 1975).                    ARH

### Reading and References

Baker, J.M. 1973: Recovery of salt marsh vegetation from successive oil spillages. *Environmental pollution* 4, pp. 223–30. · Elton, C.S. 1958: *The ecology of invasions by animals and plants*. London: Methuen. Ch. 8, pp. 143–53. · Goodman, D. 1975: The theory of diversity–stability relationships in ecology. *Quarterly review of biology* 50, pp. 237–66. · Hill, A.R. 1975: Ecosystem stability in relation to stresses caused by human activities. *Canadian geographer* 19, pp. 206–20. · Holling, C.S. 1973: Resilience and stability of ecological systems. *Annual review of ecology and systematics* 4, pp. 1–23. · MacArthur, R.H. 1955: Fluctuations of animal populations and a measure of community stability. *Ecology* 36, pp. 633–6. · May R.M. 1973: *Stability and complexity in model ecosystems*. Princeton, NJ: Princeton University Press. · Orians, G.H. 1975: Diversity, stability and maturity in natural ecosystems. In W.H. Van Dobben and R.H. Lowe-McConnell eds, *Unifying concepts in ecology*. The Hague: W. Junk. Pp. 139–50. · Paine, R.T. 1969: A note on tropic complexity and community stability. *American naturalist* 103, pp. 91–3. · Pimm S.L. 1984: The complexity and stability of ecosystems. *Nature* 307, pp. 307, 321–6. · Vitousek, P., Reiners, W.A., Melillo, J.M., Grier, C.C. and Gosz, J.R. 1981: Nitrogen cycling and loss following forest perturbation: the components of response. In G.W. Barrett and R. Rosenberg eds, *Stress effects on natural ecosystems*. London and New York: Wiley. Pp. 114–27. · Watt, K.E.F. 1968:

*Ecology and resource management*. New York and London: McGraw-Hill. Ch. 3, pp. 39–50. · Westman, W.E. 1978: Measuring the inertia and resilience of ecosystems. *Bioscience* 28, pp. 705–10.

### stability analysis

The procedure for examining the likelihood of failure of a soil or rock slope. The types of analysis and the way they are carried out depend largely upon the nature of the materials to be investigated but they generally require a knowledge of the COHESION and FRICTION properties of the slope material as well as the jointing characteristics if the material is a rock slope. Slope geometry is also crucial, as is a knowledge of the water availability as pore water pressure (in soils) or cleft water pressure (in rocks). Many stability analyses are two-dimensional but digital computing methods can now make three-dimensional analyses relatively easy. In soil mechanics such analyses are used to determine a factor of safety so that a safe design can be produced, but they can be used in a more geomorphological way to help determine the characteristics once a slope has failed.   WBW

#### Reading
Bell, F.G. 1992: *Engineering properties of soils and rocks*. 3rd edn. Oxford: Butterworth-Heinemann. · Lambe, T.W. and Whitman, R.V. 1981: *Soil mechanics*. New York: Wiley.

### stable equilibrium

A condition of a system in which very limited displacement in any direction is followed by a return to a persistent state or condition. Huggett (1980, figure 1.3) employs the mechanical analogy (from Spanner 1964) of a ball resting in a deep cup: the ball may move from side to side or round and round if the cup is shaken, but will always ultimately return to the bottom of the cup as long as the stable equilibrium condition prevails.   BAK

#### References
Huggett, R. 1980: *Systems analysis in geography*. Oxford: Clarendon Press. · Spanner, D.C. 1964: *Introduction to thermodynamics*. London: Academic Press.

### stack

A free-standing pinnacle of rock, usually in the sea, which represents an outlier of a coastal cliff.

### stadial

A short cold period with smaller ice volumes than the full glacial stages of an ice age. The warmer intervals between them are called interstadials.

### staff gauge

An instrument for determining water depth at a site on a river system. It is often located at a stream gauging site to provide a datum for setting and checking continuous stage recorders. The staff gauge consists of a

**A spectacular stack,** the Old Man of Hoy, developed in Old Red Sandstone in the Orkney Islands. Its height is 140 m.

plate or board with painted or engraved elevation divisions, which is firmly fixed in the water at a river cross-section so that water levels can be read by eye from the divisions on the staff gauge at all flow stages, usually to an accuracy of 0.5 cm. Single staff gauges are usually installed to stand vertically and close to one bank of the river, but in some locations more than one staff gauge will be installed to cover different ranges of stage and occasionally the staff gauge will be inclined to lean against the bank so that flow disturbance is minimized and so that the depth scale may be more easily read. (See also DISCHARGE; HYDROMETRY.)   AMG

### stage

The term used to describe water depth or the elevation of the water surface at a location on a river system. Instruments which continuously monitor water surface elevation at a gauging station site are called stage recorders. The river stage fluctuates through time in response to precipitation events. In a stable and straight river section a relationship can be established between discharge and stage so that discharge can be estimated for any water stage, and this relationship between stage and discharge is known as the discharge rating curve for that site. (See also DISCHARGE.) AMG

**stagnant ice topography** The view that many glacial deposits in formerly glaciated areas are the result of stagnant ice downwasting *in situ* was championed in North America in the 1920s (Cook 1924; Flint 1929), in Scandinavia in the 1940s and 1950s (Mannerfelt 1945; Hoppe 1959) and in Scotland (Sissons 1967). The idea is that the ice, by virtue of shallow surface gradients or by its isolation from the main ice mass, no longer flows actively but melts from the surface downwards, the last remnants being preserved in depressions. The landforms that are formed depend both on the shape of the underlying topography and on the debris characteristics of the glacier. In places, MORAINE landforms of disintegration are dominant and consist of irregular mounds and KETTLES built of varying quantities of basal and supraglacial TILL. The basal till may have reached the glacier surface as a result of COMPRESSING FLOW before being redistributed through the action of slumping and surface sediment flows as the ice melts away. Such conditions also favour considerable meltwater activity. One fine example of such topography with a relief of 30–45 m occurs on the Canadian prairies and covers areas of thousands of square kilometres (Prest 1983).

In hillier parts of the world the role of valleys in influencing the location of the stagnant remnants of ice seems to favour another association of landforms. Once isolated from the ice sheet such stagnant ice masses are plugged by glaciofluvial deposits associated with meltwater streams from both the ice surface and the surrounding hills and their courses are marked by KAME TERRACES, KAMES and associated ESKERS and irregular mounds. Mixed with the glaciofluvial deposits may be irregular mounds of till. Such landscapes abound in upland Scandinavia and Scotland. DES

**Reading and References**
Cook, J.H. 1924: The disappearance of the last glacial ice-sheet from eastern New York. *Bulletin of the New York State Museum* 251, pp. 158–76. · Flint, R.F. 1929: The stagnation and dissipation of the last ice sheet. *Geographical review* 19, pp. 256–89. · Hoppe, G. 1959: Glacial morphology and inland ice recession in north Sweden. *Geografiska annaler* 41, pp. 193–212. · Mannerfelt, C.M.

1945: Några glacialmorfologiska Formelement. *Geografiska annaler* 27, pp. 1–239. (English summary and figure captions.) · Moran, S.R., Clayton, L., Hooke, R. Le B., Fenton, M.M. and Andriashek, L.D. 1981: Glacier-bed landforms of the prairie region of North America. *Journal of glaciology* 25.93, pp. 457–76. · Prest, V.K. 1983: *Canada's heritage of glacial features*. Geological Survey of Canada miscellaneous report 28. · Sissons, J.B. 1967: *The evolution of Scotland's scenery*. Edinburgh: Oliver & Boyd.

**stalagmite, stalactite** See SPELEOTHEM.

**star dune** A type of sand dune that has three or more radial arms extending in various directions from a central high point. Star dunes develop in multidirectional wind regimes, where seasonal changes in the overall sand transporting direction cause sand to accumulate vertically, sometimes through the merging of other dune types. Star dunes are therefore sand accumulating forms – that tend to occur at the depositional centres of sand seas or where airflow is modified by topographic barriers – and can attain heights up to 300–400 m. The morphology of the individual arms of star dunes can vary seasonally in response to sand transport shifts such that they can behave like transverse or linear dunes. Star dunes are sometimes found in chains, and if one transport direction is stronger than the others, slow migration may result. (See DUNE.) DSGT

**Reading**
Lancaster, N. 1989: Star dunes. *Progress in physical geography* 13, pp. 67–91.

**steady flow** The flow of water in open channels is classified according to its temporal and spatial variability, and steady flow occurs when the water depth, discharge and therefore the velocity are temporally constant. Since open channels have a free surface exposed to the atmosphere the flow responds to rainfall and run-off inputs and is naturally temporally unsteady during storm hydrographs. The rate of change of depth and discharge is often sufficiently slow for a steady flow to be assumed during the time interval under consideration, and this forms the basis for the development and application of simplified FLOW EQUATIONS. Temporally steady flow may be spatially uniform or varied, and varied flow may be gradually varied or rapidly varied. KSR

**steady state** The notion that the input, output and properties of a SYSTEM remain constant over time. The concept of steady state is a powerful means of simplification which allows the mathematical modelling of complex natural systems. DES

**steam fog**   Fog formed when cold air passes over warm water. Both heat and water vapour are added to the air which quickly becomes saturated. Any additional water vapour evaporated from the warm water will rapidly condense forming a swirling, steam-like fog. Steam fog is common over heated swimming pools in winter, above lakes in autumn and early winter mornings, over thermal ponds, such as those in Yellowstone National Park, all the year round, and above open water in polar regions, where it is called arctic sea smoke.                     WDS

**Reading**
Ahrens, C.D. 1982: *Meteorology today*. St. Paul, Minn.: West.

**stem flow**   The drainage of intercepted water down the stems of plants. Precipitation may be intercepted by vegetation and subsequently lost through evaporation (INTERCEPTION or interception loss) or it may drip through the vegetation canopy (THROUGHFALL) or it may drain across the leaves and stems of plants or down the branches and trunks of trees as stem flow. Stem flow is a means by which precipitation may reach the ground surface and it may form quite an important route for water, particularly if the structure of the vegetation encourages this form of drainage.                     AMG

**Reading**
Courtney, F.M. 1981: Developments in forest hydrology. *Progress in physical geography* 5, pp. 217–41. · Sopper, W.E. and Lull, H.W. eds 1967: *International symposium on forest hydrology*. Oxford and New York: Pergamon.

**step-pool systems**   Commonly occur in mountain streams with steep gradients and coarse bed materials. They are formed by high-magnitude, low-frequency flood events, and the staircase-like structure tends to be relatively stable for long periods of time during low flows. The steps and pools alternate to produce a characteristic, repetitive sequence, with the steps composed of an accumulation of cobbles and boulders that are transverse to the channel. Finer materials fill the pools. Logs may contribute to the formation of steps.                     ASG

**Reading**
Chin, A. 1989: Step pools in stream channels. *Progress in physical geography* 13, pp. 391–407.

**steppes**   Mid-latitude grasslands with few trees. The Russian equivalent of the North American prairies and Argentinian pampas.

**stick slip**   The jerky motion by which glaciers slide over bedrock. Sudden slip phases of 1–3 cm displacement are interspersed with longer quiescent phases. Each slip phase is highly localized beneath a glacier and probably relates to the failure of a local bond between part of the glacier and the bed.                     DES

**Reading**
Robin, G. de Q. 1976: Is the basal ice of a temperate glacier at the melting point? *Journal of glaciology* 16.74, pp. 183–96.

**stilling well**   A large diameter tube installed at the edge of a river channel, or in a river bank, in order to obtain accurate measurements of river level from a still water surface. The tube is connected to the river by intake pipes, in order to ensure that the water level in the well is identical to that in the channel. The diameter of these pipes should be sufficiently small to damp out turbulence or short-term oscillations, but must be large enough to permit instantaneous response to changes in river level. Provision must be made for flushing or clearing the intake pipes. A FLOAT RECORDER is often associated with a stilling well.                     DEW

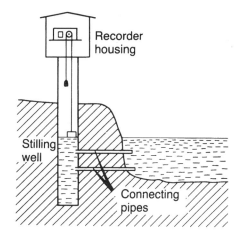

**stillstand**   A period of stability between two phases of tectonic activity in the earth's history. Also a period during which mean sea level is constant.

**stochastic models**   Any mathematical model which represents a stochastic process – a phenomenon whose temporal or spatial sequence is characterized by statistical properties – is a stochastic model. The incorporation in the model of a sequential time or space function of the probability of occurrence is what distinguishes a stochastic model from a purely probabilistic model (such as the probability distribution of floods of different magnitude). The selection of an appropriate model depends on the nature of the process being modelled and the type of data collected to summarize the process numerically (Thornes and Brunsden 1977, pp. 5–7, 70–87). A continuous process is observed continuously through time or space, even if the measured variable only takes a discrete set of values, as in the case of a binary variable denoting presence or absence. For convenience, a continuous process is often represented by data obtained at discrete time or space intervals (usually equal), either as discrete sampled data read at specific points, or as discrete aggregate data which are summed or averaged over a period (e.g. daily rainfall). Another class of stochastic phenomena involves point processes, in which either the frequency of *events* in successive discrete time periods is counted, or the distribution of time *intervals* between events is assessed.

Discrete approximations of continuous time or space series may be modelled using the general linear random model:

$$z_t - \phi_1 z_{t-1} - \ldots - \phi_p z_{t-p} = e_t - \theta_1 e_{t-1}$$
$$\ldots - \theta_q e_{t-q}$$

where the $z$s are values of a variable measured at times $t$, $t-1$, etc., the $e$s are random 'shocks' at these times, and the $\phi_1$ and $\theta_1$ are coefficients. In fitting this mixed autoregressive-moving average model, the objective is to minimize its complexity by reducing the 'order' of dependency defined by the lags $p$ and $q$ (Chatfield 1975; Richards 1979). The stochastic point process (Cox and Lewis 1966) may be described by appropriate probability distributions such as the binomial or Poisson distributions, and modelled as a series using the theory of queues.  KSR

### References

Chatfield, C. 1975: *The analysis of time series: theory and practice.* London: Chapman & Hall. · Cox, D.R. and Lewis, P.A.W. 1966: *The statistical analysis of series of events.* London: Methuen. · Richards, K.S. 1979: *Stochastic processes in one-dimensional series: an introduction*; Catmog 23. Norwich: Geo Abstracts. · Thornes, J.B. and Brunsden, D. 1977: *Geomorphology and time.* London: Methuen.

### stochastic process

A statistical phenomenon in which the evolutionary sequence in time and/or space follows probabilistic laws. 'Stochastic', from the Greek word meaning 'guess', implies a chance process which contrasts with a deterministic phenomenon whose future values can be predicted with certainty if the existing values of the controlling variables are known. In a stochastic process exact prediction is impossible because dependence is partly on past conditions, and partly on random influences. Nevertheless, stochastic models can be used to

represent the process mathematically, and to provide both efficient forecasts and the distribution of forecast errors. Natural phenomena may be inherently stochastic, but often the randomness apparent in their behaviour reflects the scientist's incomplete understanding, and inaccurate measurement (Mann 1970). In practice, many phenomena display the mixed deterministic-stochastic behaviour typified by climatic and hydrological processes (Yevjevich 1972). For example, daily river flows vary seasonally with a fundamentally deterministic cycle related to annual variation of radiation receipt and evaporation loss. Superimposed on this is the random occurrence of sharp increases of flow caused by individual storm inputs, followed by the gradual decrease of discharge in the flood recession curve, caused by water retention in the drainage basin and consequent slow outflow. Thus the stochastic component of the hydrological process involves both random 'shocks' and a system 'memory', which can be modelled by an autoregressive STOCHASTIC MODEL. Note that the hydrologist treats rainfall as a random input, whereas the meteorologist seeks to explain rainfall deterministically.  KSR

### References

Mann, C.J. 1970: Randomness in nature. *Bulletin of the Geological Society of America* 81, pp. 95–104. · Yevjevich, V. 1972: *Stochastic processes in hydrology.* Fort Collins, Col.: Water Resources Publications.

### stock

A large, irregularly shaped intrusion of igneous rock.

### stock resources

NATURAL RESOURCES that have finite availability within the earth system, relative to the time they have taken to form and human lifespans and expectations. In their broadest definition they include 'all minerals and land' (Rees 1990, p. 14) in which three further subdivisions can be recognized. First, there are all mineral elements which are theoretically recoverable but in practice are not because of their geographical distribution or their dispersion amongst other material. Second, there are those resources that are consumed by use, that is the process of using them as a resource destroys the mineral that is being used. This category includes all minerals that are used for fuel, for example coal and oil. Third are minerals that, whilst used as a resource, are theoretically recoverable, even though usage has changed their form. This group includes all metal minerals which can, if disposal and economic factors are favourable, be recycled. Finally, through unsustainable usage, some FLOW RESOURCES can become stock resources.  DSGT

## Reference

Rees, J. 1990: *Natural resources: allocation, economics and policy.* 2nd edn. London: Routledge.

## Stokes' surface

A sedimentary unconformity, recognized by Stokes (1968) where wind scour deflates sediment down to the level of a near-surface water table. Applies principally to deflation from dry lake beds or in coastal settings.

DSGT

### Reading and Reference

Stokes, W.L. 1968. Multiple parallel truncation bedding planes: a feature of wind-deposited sandstone formations. *Journal of sedimentary petrology* 38, pp. 510–15. · Fryberger, S.G., Schenk, C.J. and Krystinik, L.F. 1988. Stokes surfaces and the effects of near surface groundwater table on aeolian deposition. *Sedimentology* 35, pp. 21–41.

## stone line

A horizon of gravel-sized rock fragments within a soil profile or accumulation of relatively fine-grained sediments.

## stone pavement

An area, often planar and level or only gently sloping, across which stone materials resembling paving cover most of the surface. In DESERTS, stone pavements (also termed desert pavements) are composed of abundant stone fragments of pebble size, in which the stones are often embedded within a loamy soil material. The stones, which may carry a desert varnish, are very closely packed so that little soil material is exposed at the surface. In many areas pavements of this kind overlie deep silty soils that contain few or no stones. This seems to require that the stones have been concentrated at the surface following upward migration through the underlying material, probably in response to forces caused by wetting and drying within the soil. In some areas, frost action may be involved (Wainwright *et al.* 1995). Deflation of surface fines by wind and erosion by water may also be involved.

Stone pavements also occur within glaciated limestone landscapes. Large areas of flat-lying bare rock, crossed by solutionally enlarged joints, reflect glacial stripping to the level of a bedding plane, and possibly the subsequent removal of overlying soil materials. Landforms of this kind are referred to as limestone pavements.

DLD

### Reading and Reference

Cooke, R.U. 1970: Stone pavements in deserts. *Annals of the Association of American Geographers* 60, pp. 560–77. · Wainwright, J., Parsons, A.J. and Abrahams, A.D. 1995: A simulation study of the role of raindrop erosion in the formation of desert pavements. *Earth surface processes and landforms* 20, pp. 277–91.

**Stone pavements.** Limestone pavement.

**Stone pavements**. Desert pavement.

**storage** Describes the stores or reservoirs of water included in the hydrological cycle. We can think of surface storage, soil moisture storage and groundwater storage as locations where dynamic reservoirs of water may exist in a drainage basin. In hydrological modelling complex distributions of stored water may be represented as one or more mathematically defined stores or reservoirs, the simplest of which is the linear store. In a linear store water outflow (Q) is directly proportional to water storage (S):

$$S = kQ$$

where $k$ is the storage coefficient.

The storage equation is combined with a continuity equation which states that the difference between inflow (I) to the store and outflow from the store is accommodated by a change in the amount of water in storage:

$$\frac{\partial S}{\partial t} = I - Q$$

AMG

**Reading**
Kirkby, M.J. 1975: Hydrograph modelling strategies. In R. Peel, M. Chisholm and P. Haggett, eds, *Progress in physical and human geography*. London: Heinemann. · — ed. 1978: *Hillslope hydrology*. Chichester: Wiley.

**storm beach** A BEACH with a profile flattened by storm-wave induced erosion. Wave steepness during storm conditions is relatively high, and this condition is considered erosive, or destructive. Beach material is eroded from the foreshore and moved offshore to form a nearshore bar. The storm beach has a low-gradient, concave-upward profile, and its landward margin may be marked with a storm berm. Storm beaches are usually in the dissipative morphodynamic regime, because most wave energy is dissipated through waves breaking in the SURF zone.      DJS

**Reading**
King, C.A.M. 1972: *Beaches and coasts*. 2nd edn. London: Edward Arnold.

**storm run-off** See HYDROGRAPHS; RUN-OFF.

**storm surges** Changes in sea level generated by extreme weather events. They appear on sea-level records as distortions of the regular tidal patterns, and are most severe in regions of extensive shallow water. When maximum surge levels coincide with maximum high-water levels on spring tides, very high total sea levels result. Low-lying coastal areas are then vulnerable to severe flooding. In tropical regions severe surges are occasionally generated by cyclones, hurricanes or typhoons: the actual levels depend on the intensity of the meteorological disturbance, the speed and direction with which it tracks towards the coast, and the simultaneous tidal levels. Areas at risk include the Indian and Bangladesh coasts of the Bay of Bengal, the south-east coast of the USA and the coast of Japan. Satellite and radar tracking of the weather patterns are used to give advanced warning of imminent flood danger. Extra-tropical surges, generated by meteorological disturbances at higher latitudes, usually extend over hundreds of kilometres, whereas the major effects of tropical surges are confined to within a few tens of kilometres of the point where the hurricane meets the coast. Flood-warning systems for extra-tropical surges must take account of the total response of a region to the weather patterns. DTP

**Reading**
Pugh, D.T. 1987: *Tides, surges and mean sea level.* Chichester: Wiley.

**stoss** The direction from which wind, water or ice moves. The windward side of a sand dune.

**strain** A measure of the deformation of a body when a load or STRESS is applied. It is usually expressed as a dimensionless value (ratio or percentage) as, for linear strain, it is the change in length divided by the original length. Similarly, areal and volumetric strains can be defined. In the SHEAR BOX and TRIAXIAL APPARATUS, the tests are normally done at a constant deformation rate and are called 'strain controlled' tests.
WBW

**strain rate** The rate at which a body deforms in response to stress. It is often represented by the symbol $\dot{\varepsilon}$. (See also STRAIN.)

**strandflat** The term was introduced to describe an undulating rocky lowland up to 65 km wide in western Norway (Reusch 1894; Nansen 1922). It is partly submerged, forming an irregular outer belt of skerries, and is backed by a steeply rising coast. Similar features may have been recognized in Iceland, Svalbard, Novaya Zemlya, East and West Greenland, Baffin Island and the Antarctic Peninsula. Nansen's view that the strandflat was cut by freeze–thaw processes adjacent to the shoreline has survived to the present day, but there is an alternative view, namely that it is a marine or subaerial lowland, subsequently modified by glacial action. DES

**Reading and References**
Gjessing, J. 1966: Norway's paleic surface. *Norsk geografisk tidsskrift* 21, pp. 69–132. · Holtedahl, H. 1960: Mountain, fjord, strandflat: geomorphology and general geology of parts of western Norway. In J.A. Dons ed., *Guide to excursions A6 and C3.* International Geological Congress twenty-first session. Oslo: Norden. · Nansen, F. 1922: *The strandflat and isostasy.* Oslo: Videnskapssel Skapets

Skrifter 1. · Reusch, H. 1894: The Norwegian coastplain. *Journal of geology* 2, pp. 347–9.

**strandline** Name sometimes given to an old abandoned shoreline of a lake.

**stratified scree** See GRÈZES LITÉES.

**stratigraphy** The study of the order and arrangement of geological strata. Lithostratigraphy (rock stratigraphy) is concerned with the organization of strata into units based on their lithological characteristics. Biostratigraphy is concerned with the organization of strata into units based on their fossil content. Chronostratigraphy (time stratigraphy) is concerned with the organization of strata into units based on their age relationships. ASG

**Reading**
Bowen, D.Q. 1978: *Quaternary geology.* Oxford: Pergamon.

**stratocumulus** See CLOUDS.

**stratosphere** The layer from heights of about 10–30 km immediately above the TROPOSPHERE, in which temperature is nearly independent of height. Heat transfer is dominated by thermal RADIATION which tends to eliminate temperature differences. Emden's theory of radiative equilibrium suggests a stratospheric temperature some $2^{-1/4}$ that of the troposphere which, at 215 K (0.84 times a tropospheric temperature of 255 K), is about right. JSAG

**Reading**
Goody, R.M. and Walker, J.C.G. 1972: *Atmospheres.* Englewood Cliffs, NJ: Prentice-Hall.

**strato-volcano** A composite volcano. A volcano that emits both molten and solid material and builds up a steep-sided cone.

**stream capture** The process by which one stream erodes more aggressively than a neighbouring stream during the course of drainage basin evolution so that it intersects the channel of that stream and captures its discharge. The removal of discharge from the stream which has been 'beheaded' by capture usually results in its source moving down the original valley leaving an abandoned valley, or 'wind gap', with its floor marked by older fluvial deposits. The beheaded stream may also be left as a misfit stream as it is now too small for the valley.

The capturing stream may be eroding more aggressively for a variety of reasons, which may act singly or in combination. For example, it may have a higher discharge due to differences in precipitation and vegetation cover, as might occur on either side of a mountain range, or climate change may have increased local rainfall. It may have a steeper gradient, possibly due to a change in base level, which will increase its erosive potential, or it may be cutting through a less resistant lithology. Capture takes place as a result of progressive headward erosion by the capturing stream, with headwater channels cutting into the hillside so that valleys and channels extend towards and eventually beyond a drainage divide. Many rivers or streams which have experienced capture as part of their evolution exhibit an abrupt change in channel direction at the point of capture, frequently of the order of 90°, termed the 'elbow of capture'.

River capture has been important in the evolution of many of the world's river systems. For example, the easternmost tributary of the Indus was captured by the Ganges in geologically recent times diverting drainage from a large area of the Himalayas from Pakistan to India. Understanding and reconstructing sequences of river capture may be of economic importance. For example, if an ancient river system contained mineral placer deposits but has been beheaded, it will be necessary to be able to trace the old river course to find areas of abandoned alluvial fill in headwater regions.     DJN

**stream ordering**  See ORDER, STREAM.

**stream power**  The rate of energy expenditure in flowing water. The energy possessed by flowing water is held by virtue of elevation above some BASE LEVEL towards which the water can flow, and the elevation of water above this in turn is derived from the solar-driven atmospheric processes that lift water vapour and deliver precipitation over the land. Along a stream, potential energy is progressively transformed into other forms, notably the kinetic energy of the flowing mass, together with energy dissipated as frictional heat, sound, and in moving sediment particles.

Some workers have employed the total stream power per unit length of channel, 1/2, as a useful measure of the ability of a stream to do landscape work (e.g. Graf 1983). Total power is given by

$$\frac{1}{2} = \rho g Q s$$

where $\rho$ is the density of water, g is the gravitational acceleration, $Q$ is discharge, and $s$ is the energy slope, often approximated as the channel gradient.

As originally presented by R.A. Bagnold, however, stream power was expressed in terms of amount of energy expended per unit area of the bed, a measure that was sought for studies seeking to explain sediment transport. This was written

$$\omega = \rho g Q s / w$$

so that

$$\omega = \rho g d v s$$

in which $d$ is flow depth and $v$ is flow velocity (and $Q = w d v$). The value of $\omega$ ranges from $< 1 \, \mathrm{J\,m^{-2}\,s^{-1}}$ in interrill flow (Parsons et al. 1998) to $> 12,000 \, \mathrm{J\,m^{-2}\,s^{-1}}$ in riverine flood flows (Rajaguru et al. 1995), the latter sufficient to move boulders metres in diameter.

Another widely-used way to express the stream power is in terms of the power per unit weight of water. This is termed *unit stream power* and is written

$$\omega^* = \rho g Q s / w d = \rho g v s$$

Various attempts have been made to employ stream power in sediment transport equations (Yang and Stall 1976; Bagnold 1977), where it forms an intuitively appealing alternative to the many other flow parameters that have been evaluated in attempts to understand sediment motion, which include stream discharge, velocity, and bed shear stress. The stream power approach has also been applied to shallow overland flows (e.g. Moore and Burch 1986).     DLD

**Reading and References**
Bagnold, R.A. 1977: Bed load transport by natural rivers. *Water resources research* 13, pp. 303–12. · Graf, W.L. 1983: Downstream changes in stream power in the Henry Mountains, Utah. *Annals of the Association of American Geographers* 73, pp. 373–87. · Moore, I.D. and Burch, G.J. 1986: Sediment transport capacity of sheet and rill flow: application of unit stream power theory. *Water resources research* 22, pp. 1350–60. · Parsons, A.J., Stromberg, G.L. and Greener, M. 1998: Sediment-transport competence of rain-impacted interrill overland flow. *Earth surface processes and landforms* 23, pp. 365–75. · Rajaguru, S.N., Gupta, A., Kale, V.S., Mishra, S., Ganjoo, R.K., Ely, L.L., Enzel, Y. and Baker, V.R. 1995: Channel form and processes of the flood-dominated Narmada River, India. *Earth surface processes and landforms* 20, pp. 407–21. · Yang, C.T. and Stall, J.B. 1976: Applicability of unit stream power equation. *Journal of the Hydraulics Division, American Society of Civil Engineers* 102(HY5), pp. 559–68.

**streamline**  A line whose tangent at any point in a fluid is parallel to the instantaneous velocity of the fluid at that point. A map of streamlines gives an instantaneous 'snapshot' of the flow. Alternatively one may think of such a map as being one frame in a moving film of the flow. The streamline pattern changes with time. Only in a steady state flow do the streamlines coincide with the trajectories of the fluid particles.     BWA

**strength** See INTACT STRENGTH; MASS STRENGTH.

**strength equilibrium slopes** Are formed on exposed bedrock which has an inclination adjusted to the MASS STRENGTH of the rock. This type of slope is controlled by processes of erosion and by geomorphic resistance operating at a scale of individual joint blocks. A distinction is made between (1) equilibrium slopes, and (2) those which are formed with critical angles for stability dipping out of the slope – these slopes fail by large-scale landsliding along the critical joints and have forms controlled by this process. MJS

Reading
Selby, M.J. 1982: Controls on the stability and inclinations of hillslopes formed on hard rock. *Earth surface processes and landforms* 7, pp. 449–67.

**stress** Is produced by a system of forces in equilibrium tending to produce STRAIN in a body. Stress can be produced in tension or

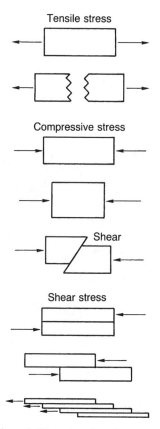

**Stress.** Schematic illustration of tensile, compressive and shear stresses.
*Source*: M.A. Summerfield 1991: *Global geomorphology*. London and New York: Longman Scientific and Technical and Wiley.

compression, by hydrostatic pressure or by shear stress. The units of stress are (force per unit area) newtons per square metre, $N m^{-2}$. In most geomorphological examples, the forces will give values as $kN m^{-2}$ or the equivalent kPa where a pascal (Pa) is equal to a $N m^{-2}$. Tensile stress is an extensional force which tends to stretch or pull material apart. Compressive stress is a force which tends to compress material and thereby change its shape. Shear stress is a force which deforms a mass of material by one part sliding over another along one or more failure plains. WBW

Reading
Whalley, W.B. 1976: *Properties of materials and geomorphological explanation*. Oxford: Oxford University Press.

**stress ecology** Defined by Barrett (1981) as a subdiscipline which attempts to measure and evaluate the impact of natural or foreign perturbations on the structure and function of ecological systems. Perturbations include pesticides, fire, nutrient enrichment and radiation. Stress itself can be defined as a perturbation that is applied to a system by a stressor which is foreign to that system or which may be natural to it but, in the instance concerned, is applied at an excessive level (e.g. phosphorus or water). Stress therefore involves an unfavourable deflection, whereas *subsidy* involves a favourable deflection. ASG

Reference
Barrett, G.W. 1981: Stress ecology: an integrative approach. In G.W. Barrett and R. Rosenberg eds, *Stress effects on natural ecosystems*. Chichester: Wiley.

**striated soil** (also called needle ice, striped ground and striated ground) Consists of a miniature pattern characterized by a distinct alignment of the surface soil particles. The orientation of the stripes does not necessarily coincide with slope gradient. Wind direction and the alignment of the early sun's rays may play a role. NEEDLE ICE is the predominant formative process. ASG

**striation** Scratches etched onto a rock surface by the passage over it of another rock of equal or greater hardness. Striations are characteristic of erosion by glaciers but may also occur beneath snow patches (Jennings 1978) and on coasts affected by sea ice (Laverdière *et al.* 1981; Hansom 1983). Glacial striations are generally up to a few millimetres in width and rarely more than a metre in length. Larger striations grade into grooves. Striations are best displayed on rock surfaces which face up-ice, mainly because pressure melting in these locations forces the rock tools against the bedrock. DES

### Reading and References

Embleton, C. and King, C.A.M. 1975: *Glacial geomorphology*. London: Edward Arnold. · Hansom, J.D. 1983: Ice-formed intertidal boulder pavements in the Sub-Antarctic. *Journal of sedimentary petrology* 53.1, pp. 135–45. · Iverson, N.R. 1991: Morphology of glacial striae: implications for abrasion of glacier beds and fault surfaces. *Bulletin of the Geological Society of America* 103, pp. 1308–16. · Jennings, J.N. 1978: The geomorphic role of stone movement through snow creep, Mount Twynam, Snowy Mountains, Australia. *Geografiska annaler* 60A, pp. 1–8. · Laverdière, C., Guimont, P. and Dionne, J.C. 1981: Marques d'abrasion glacielles en milieu littoral Hudsonien. Québec Subarctique. *Géographie physique et Quaternaire* 35.2, pp. 269–75.

## Strickler equation

In 1923 Strickler analysed data from Swiss gravel-bed rivers lacking bed undulations to develop an equation permitting estimation of the ROUGHNESS coefficient $n$ in the MANNING EQUATION from the measured bed material particle size. It can be shown theoretically that:

$$n = 0.0132 k_s^{1/6}$$

where $k_s$ is a grain roughness height in millimetres. Strickler shows that if $k_s$ is taken as the median grain diameter $D_{50}$, this relationship becomes:

$$n = 0.0151 D_{50}^{1/6}$$

This can be used to estimate the Manning coefficient if grain roughness is the primary source of flow resistance, in wide, flat-bed gravel-floored channels. It is, however, often preferable to define a friction coefficient in terms of a depth: grain size ratio, since the resistance of a grain is partly dependent on the water depth covering it (Richards 1982, p. 66). KSR

### Reading and References

Richards, K.S. 1982: *Rivers: form and process in alluvial channels*. London and New York: Methuen. · Strickler, A. 1923: Beitrage zur Frage der Geschwindigheits formel und der Rauhigkeitszahlen für Strome, Kanole und Geschlossene Leitungen. *Mittelungen des Eidgenössischer Amkes für Wasserwittschaft*. Bern.

## strike

The direction of a horizontal line drawn in the same plane as strata are bedded but at right angles to their dip.

## string bog

An area of water-logged land characterized by ridges of peat separated by water-filled troughs.

## stromatolite

(stromatolith) A term first used in 1908 by E. Kalkowsky to describe some sedimentary structures in the Bunter of North Germany. A currently favoured definition (Walter 1976, p. 1) is that stromatolites are 'organo-sedimentary structures produced by sediment trapping, binding and/or precipitation as a result of the growth and metabolic activity of micro-organisms, principally cyanophytes'. They can develop in marine, marsh and lacustrine environments and, though they form today

A series of cauliflower-like lacustrine stromatolites from Lake Chew Bahir (Stephanie) in southern Ethiopia.

where conditions permit, they reached the acme of their development in the Proterozoic (Hofmann 1973). The largest known forms are mounds several hundreds of metres across and several tens of metres high. Gross morphologies vary in the extreme and range from stratiform crustose forms, through nodular and bulbous mounds and spherical oncoids, to long slender columns, erect to inclined, and with various styles of branching.                                    ASG

### References
Hofmann, H.J. 1973: Stromatolites: characteristics and utility. *Earth science reviews* 9, pp. 339–73. · Kalkowsky, E. 1908: Oolith und Stromatolith im norddeutschen Buntsandstein. *Zeitschrift Deutsche Geologische Gesellschaftz* 60, pp. 68–125. · Walter, M.R. 1976: Stromatolites. *Developments in sedimentology* 20.

### Strouhal number
A dimensionless number relating the frequency of vortex shedding, $f$, from a flow obstruction with a diameter (or other characteristic length), $l$, and the flow speed, $V$: the Strouhal number, $S = fl/V$. For a range of Reynolds numbers from about 103 to 105, $S$ is about 0.2. The Strouhal number has applications for the study of flow over BEDFORMS, for example.                              DJS

### Reading
Schlichting, H. 1979: *Boundary-layer theory.* 7th edn. New York: McGraw-Hill.

### sturzstrom
Very large rock avalanches (with volumes $> 5\,\text{Mm}^3$) initially falling or avalanching from high cliffs may develop low coefficients of internal friction, and therefore travel large horizontal distances (5–30 km) at velocities of $90$–$350\,\text{km}\,\text{h}^{-1}$. They are the most powerful forms of mass-wasting and may be major, but rare, causes of erosion in very high mountain ranges and on large volcanoes. These very large avalanches were named *sturzstroms* by Hsü (1975).                                        MJS

### Reading and Reference
Hsü, K.J. 1975: Catastrophic debris streams (sturzstroms) generated by rockfalls. *Bulletin of the Geological Society of America* 86, pp. 129–40. · Selby, M.J. 1993: *Hillslope materials and processes.* 2nd edn. Oxford: Oxford University Press. Ch. 14.

### subaerial
Occurring or existing at the landsurface.

### suballuvial bench
The lower portion of a rock pediment where it is overlain by alluvial sediments.

### Sub-Atlantic
See BLYTT–SERNANDER MODEL.

### Sub-Boreal
See BLYTT–SERNANDER MODEL.

### subclimax
Any plant community related to and closely preceding the true climax community for an area. Usage is normally in the sense of a stable community resembling the climax but prevented from developing towards it by some disturbance or other arresting factor. If the arresting factor is removed, a subclimax is expected to proceed to the climax stage. The term is also used simply for a long-persisting SERAL COMMUNITY that appears to be climax. Many subclimaxes are the result of the activities of humans and domesticated animals, particularly burning and grazing. (See also CLIMAX VEGETATION; DISCLIMAX; MONOCLIMAX.)      JAM

### Reading
Eyre, S.R. 1966: *Vegetation and soils: a world picture.* London: Edward Arnold. · Oosting, H.J. 1956: *The study of plant communities.* San Francisco: W.H. Freeman.

### subduction zones
These occur where oceanic lithosphere is consumed into the mantle at convergent plate margins. The inclined plane, where the subducting oceanic lithospheric plate slides past the overlying mantle, is seismically active and this inclined plane is called a Benioff zone. Subduction zones involve the destruction of oceanic plate material and this is complementary to the formation of new oceanic plate at constructive plate margins (active ocean ridges); this allows the earth to maintain a constant surface area.

There are two main types of subduction zone: (1) island arcs, where oceanic lithosphere is subducted beneath oceanic lithosphere (for example, the Tonga island arc in the Pacific Ocean), and (2) active continental margins, where oceanic lithosphere is subducted beneath continental lithosphere (for example the Andes along the western coast of South America). Subduction zones are characterized by an oceanic trench and the arcuate plan of the trench and distribution of volcanic islands in an island arc is a function of the geometry of pushing part of the surface of a sphere inwards into its interior. The subducted oceanic lithosphere remains coherent as a distinct relatively rigid slab for depths of up to 700 km (where the deepest earthquakes are recorded) before being consumed into the mantle. The subducted oceanic crust, which forms the uppermost part of the slab, undergoes progressive dehydration as it moves down and this promotes melting in the overlying mantle wedge. These melts segregate as ascending magmas which give rise to arc volcanoes and subduction zones are associated with typically explosive andesitic volcanism. Magmatism at active continental margins can also involve the emplacement of large granitic BATHOLITHS.    AD

**subglacial**  The environment beneath a glacier.

**Reading**
Menzies, J. and Rose, J. eds 1989: Subglacial bedforms – drumlins, rogen moraine and associated subglacial bedforms. *Sedimentary geology* 62, pp. 117–430.

**subglacial bedforms**  A generic term for the range of longitudinal and transverse landforms produced at the base of a GLACIER or ICE SHEET as a result of active ice flow across a sediment base. Whilst individual types of subglacial landforms have often been investigated in isolation, there has been an increasing realization that they all belong to a single family of related landforms, best thought of as a continuum of subglacial bedforms (Rose 1987). This view uses the concept of a single set of processes that may lead to a wide range of landform shapes and characteristics, only some of which conform to our 'classical text book' landform types such as the DRUMLIN. As with aeolian and fluvial BEDFORMS we would expect a combination of distinctive landforms and a range of intermediate morphologies.

Specific landform types that are part of the subglacial bedform system are: FLUTES, drumlins and MEGA-SCALE GLACIAL LINEATIONS, which are all longitudinally streamlined landforms at different scales from metres to tens of kilometres, and are commonly referred to as glacial lineations. ROGEN or RIBBED MORAINE are also subglacial bedforms but are sculptured ridges formed transverse to the ice flow direction. A strong characteristic of all subglacial bedforms is that they occur in discrete and fairly well defined fields or swarms comprising hundreds to thousands of individuals, and that different types such as drumlins and rogen moraine are often found in close association with each other.

An alternative school of thought considers that subglacial bedforms arise from large subglacial flood events, with water rather than ice being the shaping media. (See DRUMLIN for further details.)                                  CDC

**Reading**
Benn, D.I., and Evans, D.J.A. 1998: *Glaciers and glaciation*. London: Arnold. · Rose, J. 1987: Drumlins as part of a glacier bedform continuum. In J. Menzies and J. Rose eds, *Drumlin symposium*. Rotterdam: Balkema. Pp. 103–16.

**subhumid**  One of the five basic humidity provinces recognized in Thornthwaite's climate classification (1948). The classes are defined by calculating a PRECIPITATION efficiency (P/E) index which is the sum of twelve monthly values of the ratio of mean precipitation to mean EVAPORATION. Subhumid falls in the middle of the range so defined, with a P/E value of 32 to 63 and characterizes areas of grassland type vegetation. The savanna regions of Africa which experience an extensive dry season every year fall in this category.                          RR

**Reference**
Thornthwaite, C.W. 1948: An approach to a rational classification of climate. *Geographical review*, 38, pp. 55–94.

**sublimation**  The process of direct deposition of atmospheric water vapour on to an ice surface or evaporation from an ice surface. Cirrus clouds are sometimes formed as a result of a direct phase change from water vapour to ice crystals.                          RR

**sublittoral**  The area of the seas between the intertidal zone and the edge of the continental shelf. Also the deeper parts of a lake in which plants cannot root.

**submarine canyon**  A canyon-like valley form cut into the CONTINENTAL SHELF, continental slope or continental rise. Its resemblance to a subaerial valley extends to the existence of tributary valleys and the presence of knickpoints along its profile. It can extend beyond the continental slope almost as far as the deep ocean floor. Most submarine canyons contain sediments apparently deposited by high-density flows initiated by submarine slides known as turbidity currents. These are probably capable of preventing the canyons from filling with sediment, but the initial formation of most submarine canyons is probably attributable to fluvial erosion prior to subsidence.                          MAS

**Reading**
Barnes, N.E., Bouma, A.H. and Normark, W.R. 1985: *Submarine fans and related turbidity systems*. New York: Springer-Verlag. · Kennett, J.P. 1982: *Marine geology*. Englewood Cliffs, NJ and London: Prentice-Hall. · Shanmugan, G. and Miola, R.J. 1988: Submarine fans: characteristics, models, classification and reservoir potential. *Earth science reviews* 24, pp. 383–428. · Shepard, F.P. and Dill, R.F. 1966: *Submarine canyons and other sea valleys*. Chicago: Rand McNally. · Whitaker, J.D. McD. 1974: Ancient submarine canyons and fan valleys. In R.H. Dott Jr and R.H. Shaver eds, *Modern and ancient geosynclinal sedimentation*. Society of Economic Paleontologists and Mineralogists special publication 19, pp. 106–25.

**subsequent stream**  A stream which follows a course determined by the structure of the local bedrock.

**subsere**  A secondary successional sequence of plant communities. It is a series of community stages, the result of SUCCESSION on incompletely bared surfaces, or beginning with a community not truly climax in status (e.g. a SERAL COMMUNITY, SUBCLIMAX or DISCLIMAX). The essential characteristic of any subsere is its initiation from at least the vestige of a previous community at the site. This may involve residual species, seedlings, existing soil with its seed bank, or a complete community from which some controlling factor has been removed. In many instances, therefore, the subsere may be viewed as a manifestation of the recovery process in damaged ecosystems. (See also PRISERE.)

JAM

**Reading**
Cairns, J. ed. 1980: *The recovery process in damaged ecosystems*. Ann Arbor: Ann Arbor Science. · Fontaine, R.G. Gomez-Pompa, A. and Ludlow, B. 1978: Secondary successions. In *Tropical forest ecosystems: a state of knowledge report prepared by UNESCO/UNEP/FAO*. Paris: UNESCO. Pp. 216–32.

**subsidence** Landsurface sinking resulting from such processes as the withdrawal of groundwater, geothermal fluids, oil and gas; the extraction of coal, salt, sulphur and other solids through mining; the hydrocompaction of sediments; oxidation and shrinkage of organic deposits (notably peats); the development of thermokarst in areas underlain by permafrost; and karstic collapse. ASG

**Reading**
Johnson, A.I. 1991: Land subsidence. *IAHS publication* 200, pp. 1–690.

**succession** The sequential change in both the form and composition of an ecological community over time. The concept was originally developed in North America by, among others, H.C. Cowles (1901), but F.E. Clements, whose monograph entitled *Plant succession: an analysis of the development of vegetation* was published in 1916, is regarded as most influential in the consolidation and widespread adoption of the concept. Clements observed changes in lakeshore dune vegetation with time and developed an explanation and corresponding comprehensive system of terminology to describe the process. Although much of this terminology is rarely employed today, the concept itself has had such a profound influence on ecological thought that it achieved the status of virtual acceptance and, indeed, survived well into the 1970s. Clements defined succession as a sequence ('sere') of plant communities ('seral stages') characterized by increasing complexity and culminating in the CLIMAX VEGETATION. The model recognized various 'sub-climaxes' in which, because of prevailing local conditions such as soil chemistry, succession was delayed, although it would proceed eventually to the true climax.

Human activity was seen to influence the course of succession and could result in a 'deflected sere' with a 'plagioclimax', while subsequent 'secondary succession' might ultimately restore the regional climax vegetation with time.

The essence of the Clementsian succession concept is as follows (after Burrows 1990):

1 succession begins on newly formed substrates (primary) or following disturbance (secondary);
2 sequences occur: pioneer → seral stages → climax;
3 secondary succession culminates in the re-establishment of the climax;
4 frequent disturbance may maintain vegetation in permanent no-maturity (proclimax);
5 climax vegetation is stable and self-perpetuating;
6 plant size, diversity and structural complexity all increase during succession;
7 the process is predominantly driven by AUTOGENIC processes, in other words, it is changes promulgated by the vegetation community itself that facilitate further change, although external ALLOGENIC factors may also be important;
8 species replacement during succession is a result of competition.

A key element of the theory is the emphasis on reaction; the establishment of later colonizers is facilitated by previous occupation of a site by species that modify their habitat to the extent that they bring about their own replacement. Attractively simple though the concept is, and it was certainly widely adopted, there are numerous inherent problems. Most fundamental of these is the idea that succession is progressive, deterministic and, in following a regular sequence and ultimately producing a climax community, predictable. One development of the initial idea mooted that uniform climax communities do not occur across large areas and, alternatively, there is a mosaic of different forms known as the POLYCLIMAX. This extension of the theory fails to explain the really dynamic nature of vegetation in time and space. Although vegetation change per se is undeniable, the ecological critique of Clements' concept has centred on the reality that vegetation communities are fundamentally dynamic entities. It is now clearly recognized that different species assemblages develop through time in any climatic region, forming a complex mosaic pattern depending on age since disturbance, on intrinsic and extrinsic ecological factors and on particular site characteristics. This is not to deny that vegetation change over time may follow a pattern and that this pattern may be repeated under similar conditions. Predictable, directional change over any length of time is, however, rare and not a generally applicable mechanism.

Degradation of communities frequently occurs due, for example, to intrinsic changes in soil characteristics induced by vegetation. In this sense, the succession is not progressive in the way that Clements argued. Finally, cyclic changes in vegetation communities are common and the deterministic uni-directional changes of succession cannot be supported by current observations of nature. There are many varied causes of vegetation change not considered important in the Clementsian view. For example, adult plants may be killed or become senescent and die, revealing gaps of varying sizes at intervals in communities which can then be opportunistically colonized by a different array of species. In such circumstances, the manifestation of change would appear more random than deterministic and predictable. 'Seres' do not exist as vegetation is in a constant state of flux, responding to the dynamics of variations in climate, autogenic and allogenic developments within the community and, of course, human-induced disturbances. In the modern, dynamic view of vegetation communities, there is no requirement, for example, that the tallest species will dominate in mature stages. The preferred term now for phases before maturity is 'transitional' and 'vegetation change' has emerged as a better description of reality than succession and the complex descriptive terminology of Clements.                    MEM

**Reading and References**
Burrows, C.J. 1990: *Processes of vegetation change.* London: Unwin Hyman. · Clements, F.E. 1916: *Plant succession: an analysis of the development of vegetation.* Washington: Carnegie Institute Publications 242. · Cowles, H.C. 1901: The physiographic ecology of Chicago and vicinity. *Botanical gazette* 31, pp. 73–108, 145–82. · Luken, J.O. 1990: *Directing ecological succession.* London: Chapman and Hall.

**suction**   Describes the energy state of soil moisture under unsaturated conditions. Soil moisture suction is often called the matric potential or capillary potential of the soil moisture and it is described as a suction because it is a negative pressure potential resulting from the capillary and adsorptive forces due to the soil matrix holding moisture in the soil at a pressure less than atmospheric pressure. The magnitude of soil moisture suction can be indicated by a negative head of water (or mercury) which can be determined using a tensiometer and which is often expressed as a pF value. (See also CAPILLARY FORCES.)                    AMG

**Suess effect**   The relative change in the $^{14}C/C$ or $^{13}C/C$ ratio of any carbon pool or reservoir caused by the addition of fossil-fuel $CO_2$ to the atmosphere. Fossil fuels are devoid of $^{14}C$ because of the radioactive decay of $^{14}C$ to $^{14}N$ during long underground storage and are depleted in $^{13}C$ because of isotopic fractionation eons ago during photosynthesis by the plants that were the precursors of the fossil fuels. Carbon dioxide produced by the combustion of fossil fuels is thus virtually free of $^{14}C$ and depleted in $^{13}C$. The term 'Suess effect' originally referred to the dilution of the $^{14}C/C$ ratio in atmospheric $CO_2$ by the admixture of fossil-fuel produced $CO_2$, but the definition has been extended to both the $^{14}C$ and $^{13}C$ ratios in any pool or reservoir of the carbon cycle resulting from human disturbances.                    ASG

**Reading**
Keeling, C.D. 1979: The Suess effect: $^{13}$carbon – $^{14}$carbon interrelations. *Environment international* 2.6, pp. 229–300.

**suffosion**   An erosional process occurring in areas where limestone bedrock is overlain by unconsolidated superficial materials. The sediments slump down into widened joints and cavities in the bedrock surface, producing an irregular landsurface. It has been likened to an egg-timer effect.

**sulphation**   The reaction between materials containing calcium carbonate and sulphur dioxide in humid atmospheres. The sulphur dioxide is oxidized to sulphur trioxide in a reaction which can be catalysed, for example, by vanadium oxide produced by internal combustion or by iron oxides on the material surface. Sulphur trioxide then reacts with the calcium carbonate to form gypsum and carbon dioxide.

$$CaCo_{3(s)} + SO_{2(g)} + 1/2O_{2(g)} + 2H_2O_{(g)} \rightarrow$$
$$CaSO_4.2H_2O_{(s)} + CO_{2(g)}$$

This reaction particularly occurs on calcareous stonework in polluted urban environments which are protected from rainwash. A layer of gypsum can form on these surfaces which can incorporate combustion particles to form a *black crust*. Gypsum from these films may be washed into underlying porous stonework by periodic wetting and can contribute to stone disruption by *salt weathering* mechanisms.                    BJS

**summit plane**   See EROSION SURFACE.

**sunspots**   Vortex-like disturbances with large associated magnetic fields that afflict the sun. They are relatively dark regions on the disc of the sun, with an inner 'umbra' of effective radiation temperature about 4500 K, and an outer 'penumbra' of somewhat higher temperature. Their frequency is quasi-periodic, with an average period of around 11 years. Various relationships between sunspot activity and climatic fluctuations have been proposed.

**supercontinent** An assemblage of two or more continents. The present continents are North America, Eurasia, Africa, Australia, Antarctica and South America. Because of PLATE TECTONICS, this arrangement is temporary. In the late Palaeozoic, the four latter continents were joined together (with Arabia and India) as the supercontinent of Gondwana; North America and Eurasia were joined as Laurasia. At least from 260 Ma to 180 Ma (million years) ago, these two were joined as Pangaea (also Pangea), incorporating most of the continental crust on Earth. Momentous though Pangaea's RIFTING apart to form the present continents was, it is unlikely to have been a unique event, since it is now known that plate tectonic processes started in the first half of the 4,600 Ma history of the earth. Evidence exists for earlier supercontinents about 600 Ma ago (Gondwana), 1100 Ma (Rodinia), and before.

Supercontinents are likely to be metastable because of the amount of geothermal heat trapped in the underlying mantle: heat escapes more easily through (thin) oceanic crust than through (thick) continental crust. Build-up of heat leads to thermal uplift, active rifting and fragmentation, creating new ocean basins by sea-floor spreading. The fragmented continents move apart, only to collide with each other on the other side of the earth, where a new supercontinent is formed. A further instability is that as oceanic crust moves away from the mid-ocean ridges where it is generated, it cools and becomes denser. No present oceanic crust is older than 200 Ma, and it may be that the ASTHENOSPHERE cannot support the weight of crust any denser: this may start a new SUBDUCTION zone, turning a PASSIVE MARGIN into an ACTIVE MARGIN and reversing the enlargement of the spreading ocean. The hypothetical supercontinent cycle seems to take about 500 Ma, and there may have been as many as seven cycles in the earth's history. This has many implications for mountain building, glaciation, volcanism, sea level, ocean currents and climatic change.

IE

**Reading**
Condie, K.C. 1997: *Plate tectonics and crustal evolution.* 4th edn. Oxford: Butterworth-Heinemann. · Dalziel, I.W.D. 1995: Earth before Pangea. *Scientific American* 272, pp. 58–73. · Murphy, J.B. and Nance, R.D. 1992:

Mountain belts and the supercontinent cycle. *Scientific American* 266, pp. 84–91.

**supercooling** Denotes the presence of a substance in the liquid phase at a temperature which is below its normal freezing point. It is a common occurrence in the atmosphere where cloud droplets often occur in the supercooled state down to $-20\,°C$ and even down to $-35\,°C$ in rare cases. RR

**superimposed drainage** The pattern of a drainage network which developed on a landscape or bedrock which has since been removed by erosion, the network being preserved on the new landsurface. With antecedence it is one of the main reasons why drainage may appear to be unadjusted to present structures.

**superimposed ice** Formed when water comes into contact with a cold glacier surface and freezes. Under such circumstances it is a process of glacier accumulation and it is common in dry continental climates, such as northern Canada, where 90 per cent of the glacier ice may have formed in this way. Here summer temperatures are high enough to melt the winter snow, but it refreezes when it comes into contact with the glacier surface, which has been chilled by low winter temperatures. DES

**Reading**
Koerner, R.M. 1970: The mass balance of the Devon Island ice cap, NWT, Canada, *Journal of glaciology* 9. 57, pp. 325–36.

**superposition, law of** The law which states that the upper strata in sedimentary sequences postdate those which they overlie.

**supersaturation** A metastable state occurring when a solution contains more of a solute than is necessary to saturate it. The term is usually applied to a solution of limestone and is important in the process of SPELEOTHEM growth in caves.

**supraglacial** See ENGLACIAL.

**surf** Surf is formed on coasts as waves initially break on the shore. It consists of complex, transitional forms of waves and develops in the surfzone on beaches, between the breaker and swash zones. The surfzone is also where longshore currents develop.

**surface detention** The part of precipitation which remains in temporary storage during or immediately after a storm before it moves downslope by OVERLAND FLOW. The majority of surface detention water will form a part of the storm hydrograph but some of the water may infiltrate the soil or may be evaporated after the storm ends. (See also DEPRESSION STORAGE; SURFACE STORAGE.)                                    AMG

**surface run-off**  See RUN-OFF.

**surface storage**  Water stored on the ground surface within a drainage basin. In some circumstances the volume of water stored in this way may be very large so surface storage is of particular importance in drainage basins containing large reservoirs, natural lakes or swamps. Frozen surface water in the form of ice and snow also forms part of surface storage. In relation to the Horton (1933) model of run-off generation, surface storage is considered to be the sum of DEPRESSION STORAGE and SURFACE DETENTION (Chorley 1978).                                    AMG

**References**
Chorley, R.J. 1978: The hillslope hydrological cycle. In M.J. Kirkby ed., *Hillslope hydrology*. Chichester: Wiley. · Horton, R.E. 1933: The role of infiltration in the hydrological cycle. *Transactions of the American Geophysical Union* 14, pp. 446–60.

**surface tension**  The force required per unit length to pull the surface of a liquid apart. It is the result of the change in orientation of molecular bonds as the interface with another substance is approached. A molecule beneath the surface of a liquid is attracted in all directions by surrounding molecules within its 'sphere of molecular attraction' and so the resultant force on it is zero. However, a molecule on the surface of the liquid has a resultant force towards the main body of the liquid because there are more molecules within the part of its 'sphere of molecular attraction' within the liquid than there are in the part which falls within the overlying vapour.                                    AMG

**Reading**
Baver, L.D., Gardner, W.H. and Gardener, W.R. 1991: *Soil physics*. 5th edn. New York: Wiley.

**surge**  See STORM SURGES.

**surging glacier**  A glacier which flows at a velocity of an order of magnitude higher than normal. Whereas normal glaciers flow at 3–300 m year$^{-1}$, surging glaciers may flow at velocities of 4–12 km year$^{-1}$. Some glaciers flow permanently at surging velocities as is the case of the Jakobshavn Isbrae in West Greenland which, nourished by the Greenland ice sheet, flows at 7–12 km year$^{-1}$. Others experience periodic surges whereby a wave of ice moves downglacier at velocities of 4–7 km year$^{-1}$ and may represent a velocity of 10–100 times higher than the presurge velocity. In such cases the jump from normal to surging flow takes place very suddenly. The wave of fast moving ice may plough into formerly stagnant ice near the glacier margin or extend beyond it. The wave is associated with strong COMPRESSING flow in front and extending flow behind and, following its passage, has the effect of lowering the glacier long profile. After a surge, the glacier experiences a quiescent phase, often of several decades, while it builds up to its original profile before surging once more.

One possible explanation of surging behaviour is that there are two modes of basal sliding, one normal and one fast. Budd (1975) has related the discharge of glaciers to glacier velocity and suggested that large outlet glaciers from big ice sheets have sufficient ice supply to maintain the fast mode of sliding permanently. Normal glaciers only have enough ice to maintain normal flow. Periodically surging glaciers occupy an intermediate position and may have enough ice supply for them to be able to cross the threshold from normal to fast flow, but then they cannot provide enough ice to maintain the fast mode of flow and so revert to normal flow. The two modes of flow probably relate to water thicknesses beneath the glacier and the development of cavities (Hutter 1982).                                    DES

**Reading and References**
Many articles on glacier surges are contained in a special theme volume: *Canadian journal of earth sciences* 6.4 (1969), pp. 807–1018. · Budd, W.F. 1975: A first simple model for periodically self-surging glaciers. *Journal of glaciology*. 14, pp. 3–21. · Hutter, K. 1982: Glacier flow. *American scientist* 70 pp. 26–34. · Kamb, B., Raymond, C.F., Harrison, W.D., Engelhardt, H., Echelmeyer, K.A., Humphrey, N., Brugman, M.M., and Pfeffer, T. 1985: Glacier surge mechanism: 1982–83. Surge of variegated glaciers, Alaska, *Science* 227, pp. 469–79. · Sharp, M. 1988: Surging glaciers: behaviour and mechanisms. *Progress in physical geography* 12, pp. 349–70.

**susceptible drylands**  Those DRYLAND areas that are susceptible to reduced productivity through the impact of unsustainable land use practices and other human actions. In the context of DESERTIFICATION the term has been applied within the United Nations Environment Programme (UNEP) (see Middleton and Thomas 1997) to dry-SUBHUMID, SEMI-ARID and arid environments, HYPER-ARID areas; the true deserts, are omitted, since they have extremely low natural productivity. Thus, with only a few exceptions, human actions are unlikely to make productivity even lower than natural levels (and therefore conditions even more 'desert like').

DSGT

**Reference**
Middleton, N.J. and Thomas, D.S.G. 1997: *World atlas of desertification*. 2nd edn. London: UNEP/Edward Arnold.

**suspended load** Sediment transported by a river in suspension. The material is carried within the body of flowing water, with its weight supported by the upward component of fluid turbulence. Particles are commonly less than 0.2 mm in diameter and in many rivers the suspended load will be dominated by silt- and clay-sized particles (i.e. < 0.062 mm diameter). This fine-grained material is frequently referred to as wash load and is supplied to the river by erosion of the catchment slopes. The coarser particles usually represent suspended bed material and are derived from the channel perimeter. Transport rates are supply controlled and the suspended load is therefore a non-capacity load. Measurements of suspended sediment concentration expressed in $mg^{-1}$ or $kg\,m^{-3}$ are required to calculate transport rates. (See also SEDIMENT YIELD.) DEW

**sustained yield** Some resources are, in theory, renewable. They can be infinitely recycled through the biosphere and through human societies, either because they are basically unchanged by their use (e.g. water) or because they are self-regenerating (e.g. plants and animals). Sustained yield is a management concept which aims to regulate the system providing the resources so as to maintain the yield at a desired level into the foreseeable future and perhaps longer. In the case of agroecosystems, for example, it means the maintenance of soil structure and nutrient status, the control of weeds, pests and diseases and the selection of biota to respond to, for example, climatic change. In the case of whaling, it means avoiding the rate of cull that undermines the species' ability to reproduce themselves, an attempt that has so far been notably unsuccessful. Calculation of rates of sustained yield requires a thorough empirical knowledge of the ecosystem in question and possibly sophisticated modelling as well. It is made much more complicated when, as in the case of fish populations, for example, the natural condition seems to be one of considerable cyclic fluctuation. The optimum yield for one year, then, may be markedly different from the next and neither science nor politics may be able to cope with this amplitude of coming and going. IGS

**Reading**
Dasmann, R.F., Milton, J.P. and Freeman, P.H. 1973: *Ecological principles for economic development*. London and New York: Wiley.

**Sverdrup** A unit of discharge used to describe ocean currents, named for oceanographer H.U. Sverdrup. It is abbreviated sv, and represents a flow of $1,000,000\,m^3\,s^{-1}$. The discharge of the Gulf Stream, for example, is 18.5 sv.

**swale** A depression in regions of undulating glacial moraine, a trough between beach ridges produced by erosion, or an area of low ground between dune ridges.

**swallet** A sinkhole. A hole in limestone bedrock which has been produced by solution and through which stream water disappears.

**swallow hole** (or swallet) A feature through which surface water goes underground in a limestone area. They have local names (e.g. AVEN) and a range of forms, from deep shafts (like Gaping Gill in Yorkshire) to less obvious zones in a stream bed where discharge is lost as a result of downward percolation. ASG

**swamp** A type of wetland, dominated by woody plants. By comparison the term 'marsh' tends to be reserved for wetland dominated by graminoids, grasses or herbs. ASG

**Reading**
Gore, A.J.P. ed. 1983: *Mires: swamp, bog, fen and moor*. Amsterdam: Elsevier. 2 vols. · Williams, M. ed. 1990: *Wetlands: a threatened landscape*. Oxford: Blackwell.

**symbiosis** The 'living together' of organisms in close association, which occurs when one or both members of two species come to depend on the presence of the other for survival and reproduction. There can be different degrees of attachment in such a relationship. The weakest, in which one species benefits but the other does not (and neither is harmed) is termed COMMENSALISM. When the bonds between the two are stronger, and positive gains flow in both directions, MUTUALISM may be said to have developed. Lichens, which are composite structures of algae and fungi, provide good examples of this. Another instance may be found in the well-known linkages between the nitrogen-fixing bacterial genus *Rhizobium*, representatives of which live within the root nodules of legumes, including alfalfa, peas, beans and clover; and, as they do so, they obtain from the plants the energy which they need to give rise to the N-fixing reaction:

$$2N_2 + 6H_2O \rightarrow 4NH_3 + 3O_2$$

Some of the ammonia which is produced is in turn made available to the legumes for the synthesis of amino-acids (Delwiche 1970). The specific use of legumes in crop-rotation schemes on deficient soils in Nigeria has given rise to a symbiotic fixation of nitrogen of up to $36.4 \, kg \, ha^{-1} year^{-1}$ (Nye and Greenland 1960). Other plant species associated with different symbiotic bacteria, which together are capable of N-fixation, include cycads, gingkoes, alders, sea buckthorn, bog myrtle, *Ceanothus* shrubs, and the tropical *Casuarina* (Watts 1974). The symbiotic relationships between fungi and roots (*mycorrhiza*) are also important for the expeditious transfer of nutrients between soil and plants in environments as different as tropical rain forest and heather moorland in Europe.

When symbiotically associated species come to rely on each other totally, this may be termed *obligate mutualism*. This often involves *evolutionary symbiosis*, over long periods of time (see COEVOLUTION). Probably the best example relates to the evolution of lobeliads in Hawaii which, in the absence of their normal insect pollinators, emerged along with a group of nectar-eating birds, the honey-creepers (family: Drepanididae) (Amadon 1950). A further extreme form of symbiosis involves PARASITES.

DW

### References
Amadon, D. 1950: The Hawaiian honey-creepers. *Bulletin of the American Museum of Natural History* 95, pp. 1–262. · Delwiche, C.C. 1970: The nitrogen cycle. *Scientific American* 223, pp. 136–47. · Nye, P.H. and Greenland, D.J. 1960: *The soil under shifting cultivation.* Harpendon: Commonwealth Bureau of Soils. · Watts, D. 1974: Biogeochemical cycles and energy flows in environmental systems. In I.R. Manners and M.W. Mikesell eds., *Perspectives on environment.* Washington, DC: Association of American Geographers.

**sympatry** Originating in or occupying the same geographical area. The term sympatric is used to describe species or populations with overlapping geographical ranges which are, therefore, not spatially isolated. Where the ranges are adjacent, but not exactly overlapping, the term used is parapatric. Species isolated are said to be allopatric. When brought together artificially, allopatric species often hybridize quite freely. (See also BIOTIC ISOLATION and ISOLATION, ECOLOGICAL.)

PAS

### Reading
Stebbins, G.L. 1950: *Variation and evolution in plants.* New York and London: Columbia University Press. · — 1977: *Processes of organic evolution.* 3rd edn. Englewood Cliffs, NJ: Prentice-Hall.

**syncline** A trough in folded strata.

**synecology** The ECOLOGY of whole communities (plant, animal or biotic communities) as opposed to individuals or single species. Emphasis is given to the reciprocal relationships between communities and their environments. The unit of study (the community) is at a higher level of organization than the individual organism or the species population but it is a lower level of organization than the ECOSYSTEM or the BIOGEOCOENOSIS. (See also AUTECOLOGY.)    JAM

### Reading
Daubenmire, R. 1968: *Plant communities: a textbook of plant synecology.* New York and London: Harper & Row. · Dice, L.R. 1968: *Natural communities.* Ann Arbor: University of Michigan Press. · Whittaker, R.H. 1970: *Communities and ecosystems.* London and Toronto: Collier-Macmillan.

**synforms**   See ANTIFORMS.

**synoptic climatology** An aspect of climatology concerned with the description of local or regional climates in terms of the properties and motions of the atmosphere rather than with reference to arbitrary time intervals, such as months. There are two stages to a synoptic climatological study: the determination of categories of atmospheric circulation type (often referred to as WEATHER TYPE) and, secondly, the assessment of mean, modal and other statistical parameters of the weather elements in relation to these categories.

Although the first investigations of synoptic patterns were made in the nineteenth century, in many countries longer-standing popular weather lore had already associated cold, heat or precipitation with particular wind directions. Modern synoptic climatology developed during and after the Second World War in response to the needs of military operations to assess likely weather conditions (Barry and Perry 1973). Since then both subjective and objective means of classifying the totality of weather patterns have been widely used.

With the advent of high-speed computers and digitized grid-point data sets of sea-level pressure and geopotential height fields, there has been a transformation in procedures for preparing catalogues of synoptic types since the 1960s (Perry 1983). Two approaches have been widely adopted in developing classifications: (1) a determination of pattern similarity based on correlation methods, (2) the use of one of a range of statistical techniques to extract components of the fields, perhaps combined with a clustering approach to obtain pattern types. New types of data, e.g. satellite imagery (see SATELLITE METEOROLOGY), are now being employed in the classification process and are helping to extend the synoptic approach to the tropics where hitherto it has been little used.

Classification catalogues exist for several areas of the world for the available period of synoptic weather maps (Smithson 1986). In addition to being used in the description and analysis of such persistence and recurrence models as singularities, weather spells and natural seasons, they have found applications in studies of air chemistry, climatic fluctuations and the reconstruction of past climatic states. Synoptic climatology can also serve as an important check on computer-derived numerical climate studies and because the method relates local climate conditions to the atmospheric circulation it provides a realistic basis for much climatological teaching and project work.     AHP

### Reading and References
Barry, R.G. and Perry, A.H. 1973: *Synoptic climatology: methods and applications*. London: Methuen. · Perry, A.H. 1983: Growth points in synoptic climatology. *Progress in physical geography* 7, pp. 90–6. · Smithson, P.A. 1986: Dynamic and synoptic climatology. *Progress in physical geography* 12, pp. 119–29.

## synoptic meteorology
An aspect of meteorology concerned with the description of current weather and the forecasting of future weather using synoptic charts which provide a representation of the weather at a particular time over a large geographical area.

Brandes was the first to develop the idea of synoptic weather mapping by comparing meteorological observations made simultaneously over a wide area. It was not until the invention of the electric telegraph that the rapid preparation of a map of weather observation became possible; the first British daily weather map was sold at the Great Exhibition of 1851. The loss of numerous lives at sea encouraged the issue of regular gale warnings by Admiral Fitzroy who was appointed to the first official meteorological post in the UK. From the 1860s onwards national weather services were established in many countries and by the 1890s the first upper-air soundings gave a better understanding of the vertical structure of weather systems (Bergeron 1981). Just after the First World War the Norwegian meteorologists J. Bjerknes and H. Solberg produced a synoptic model of the mid-latitude frontal cyclone, using observations from a dense network of observing stations. A knowledge of such synoptic models, and a combination of experience, skill and judgement, was the mainstay of the weather forecaster up to the 1950s. The current weather situation was analysed on surface maps by drawing isobars and fronts by hand and distinguishing areas of significant weather, while on upper air charts pressure contours and THICKNESS lines were indicated. Examination of the circulation

patterns and extrapolation of the movement and development of the weather systems was also part of the forecasting process.

The development of weather prediction by numerical methods using high-speed electronic computers has transformed synoptic meteorology in the past forty years, and the role of the meteorologist today is that of monitoring the output of computer-based forecasts, modifying its products in the light of experience of atmospheric behaviour.

New sources of data, including radar echoes of rainfall intensity, and global satellite cloud pictures, have widened the information available to the forecaster.

Regional meteorological centres, such as the one at Bracknell in the UK, collect coded synoptic observations and transmit regional sets to three world centres which in turn produce global data sets for distribution to national meteorological centres. At the Meteorological Office in Bracknell numerical forecasts are produced twice daily for the following 24 hours and the programme centres continue to give a medium-range forecast for the next 72 hours, as well as a six-day prediction. A further forecast model is run to predict the weather in greater detail over the British Isles for the following 36 hours.

AHP

### Reading and Reference
Bergeron, T. 1981: Synoptic meteorology: an historical review. In G.H. Liljequist ed., *Weather and weather maps*. Stuttgart: Birkhauser Verlag. · Hardy, R., Wright, P., Gribbin, J. and Kingston, J. 1982: *The weather book*. London: Michael Joseph.

## systems
Simply 'sets of interrelated parts' (Huggett 1980, p. 1). They are defined as possessing three basic ingredients: 'elements, states, and relations between elements and states'. There can be both concrete and abstract systems; for example, the hot water system of a house or the set of moral values of a society; and the elements may therefore be real things or concepts, each of which is held to possess a variety of properties (or can be said to exist in a variety of states). The overall state of the whole system is then defined by the character of these properties at a given moment. Because the system is defined as a *set* of parts, it follows that there is some boundary which distinguishes that particular set from all other possible sets; and the boundedness of systems is both an important theoretical attribute and a source of immense practical difficulty.

While it is possible to view some systems as completely isolated from all external influences, which in physical examples implies that no movement of energy or mass can occur across their boundaries, most are defined as either *open* or *closed*. An open system exchanges both energy and mass with its surroundings, whereas a closed system is open only to the transfer of energy. These *inputs* of mass and/or energy are termed *forcing functions* and they are generally of considerable importance in the concrete systems of interest to physical geographers. The *through-put* of energy and/or mass creates the linkages or relations between the system elements, which may adjust in the process, either by *negative* feedback mechanisms (*homeostasis*) so that the system state remains unchanged, or by *positive* feedback, so that a net change in the system state results. The outcome of the transfers is termed the system *output*, which may be energy and/or mass and/or a new system state.

Finally, systems are regarded as hierarchical sets: the whole system at any one level being merely a component or element of some higher-order set and its own elements being, in reality, smaller-scale systems in their own right. To give an example: a drainage basin with a single stream channel may be studied as a first-order drainage system; yet its slopes and stream channel may equally well be viewed as individual systems; and the whole basin readily becomes just one element of a larger drainage network.

From this description of systems it should be clear that the concept may be applied to an infinitely wide range of phenomena. This, indeed, was seen as its chief methodological merit by Von Bertalanffy, who introduced the notion in 1950, with the explicit hope that a focus on such a general feature of the physical and mental environment would encourage the sciences, in particular, to adopt a unified methodology. From this came the idea of GENERAL SYSTEM THEORY. The concept of the open system was introduced into geomorphology by Strahler (1950) and its merits were widely advocated by Strahler's pupil, Chorley (see e.g. Chorley 1960, 1967; Chorley and Kennedy 1971; Bennett and Chorley 1978). An extremely persuasive and influential use of the concept has been made by yet another of Strahler's pupils, Schumm (1977), and several other substantial works of a more general nature have also appeared (e.g. Huggett 1980; Sugden 1982).

More influential, however, has been the application of the idea of systems in the field of ecology, where the concept of the ecosystem – as a formal statement – goes back to Tansley (1935), although its basics can readily be traced into the far earlier views of Darwin and Haeckel (Stoddart 1967). Studies such as H.T. Odum's monumental analysis of the Silver Springs ecosystem (1957) remain classic examples of the

possibilities and limitations of what has come to be known as systems analysis. It seems clear that it was in large measure the apparent success with which the systems concept was used as an analytical device by ecologists in the 1950s which spurred its adoption in mainstream physical geography. The growing realization of the problems posed by identification and isolation of ecosystems as objects of analysis has, similarly, been accompanied by a reduction in the emphasis placed upon their investigation in other fields.

The system concept seems to have made four different kinds of appearance in physical geography. First, it is now widely and loosely used as a source of jargon and terminology: the ideas of various forms of equilibrium are particularly persistent and poorly defined borrowings. Secondly, the idea has been adopted as a pedagogic framework into which the results of earlier studies, or the fruits of different concepts, can be slotted for ease of exposition: Chorley and Kennedy (1971) and Sugden (1982) are examples of the category. Thirdly, the concept may be taken as the basis for a substantive investigation of the workings of some portion of reality, in line with ecological studies such as that of Odum (1957): these inquiries are most commonly directed towards drainage basin hydrology, ecology and climatology and few of them have actually been undertaken by geographers (although human geographers have made more explicit use of systems in this particular way); an exception is the study of soils (e.g. Huggett 1975). Finally, the idea of systems has been adopted as a useful framework in which to view questions relating to environmental management and planning: a fairly early, clear and comprehensively explained example is Hamilton *et al.* (1969) and the whole approach has been exhaustively discussed by Bennett and Chorley (1978).

The application of the systems concept to any substantive investigation of the actual or potential workings of portions of the physical environment encounters two principal difficulties. The first is the need to define system boundaries. While arbitrary lines can, of course, be drawn anywhere to define a system of any magnitude, the very nature of the concept is of an interrelated set of elements which is in some real sense functionally or at least morphologically distinguishable from all adjacent sets. In practice, even lakes, islands and river basins – the most clearly definable of the physical units of interest to geographers – do not actually possess sharp and precise boundaries. The resulting difficulties of definition are horribly but understandably akin to those related to the definition of regions: they are equally hard to resolve.

The second problem stems from the need to evaluate and interpret the results of system analyses (see Kennedy 1979). Langton (1972) points out that the only effective way of comparing different systems is with reference to their success in attaining some preferred or most efficient state of operation or morphology. It is clear that we can use this criterion very readily to deal with engineered, planned or 'control' systems, since it is possible to specify the preferred or 'best' outcome. It is very far from clear that it is justifiable to regard natural systems in such a light. What is the 'goal' of an ecosystem? Or a drainage basin? Or the general circulation of the atmosphere? Could this be the reason why substantive studies of the workings of the natural systems of interest to the physical geographer are actually rather thin on the ground?

The concept of the system is, in essence, simple and it can be widely applied. The actual applications in physical geography to date have been most numerous in pedagogic and applied areas and there little fresh insight has been gained into the actual workings of the natural environment. The greatest impact seems to have been made in the wholesale introduction of the terminology of systems analysis.                                          BAK

**Reading and References**
Bennett, R.J. and Chorley, R.J. 1978: *Environmental systems: philosophy, analysis and control*. London: Methuen. · Chorley, R.J. 1960: *Geomorphology and general systems theory*. US Geological Survey professional paper 500–B. · — 1967: Models in geomorphology. In R.J. Chorley and P. Haggett eds, *Models in geography*. London: Methuen. Pp. 59–96. · —and Kennedy, B.A. 1971: *Physical geography: a systems approach*. London: Prentice-Hall International. · Hamilton, H.R. *et al.* 1969: *Systems simulation for regional analysis. An application to river-basin planning.* Cambridge, Mass.: MIT Press. · Huggett, R.J. 1975: Soil landscape systems: a model of soil genesis. *Geoderma* 13, pp. 1–22. · Huggert, R.J. 1980: *Systems analysis in geography*. Oxford: Clarendon Press. · Kennedy, B.A. 1979: A naughty world. *Transactions of the Institute of British Geographers* 4, pp. 550–8. · Langton, J. 1972: Potentialities and problems of a systems approach to the study of change in human geography. *Progress in geography* 4, pp. 125–79. · Odum, H.T. 1957: Trophic structure and productivity of Silver Springs, Florida. *Ecological monographs* 27, pp. 55–112. · Schumm, S.A. 1977: *The fluvial system*. London: Wiley. · Stoddart, D.R. 1967: Organism and ecosystem as geographical models. In R.J. Chorley and P. Haggett eds, *Models in geography*. London: Methuen. Pp. 511–48. · Strahler, A.N. 1950: Equilibrium theory of erosional slopes approached by frequency distribution analysis. *American journal of science* 248, pp. 673–96, 800–14. · Sugden, D.E. 1982: *Arctic and Antarctic: a modern geographical synthesis*. Oxford: Basil Blackwell; Totowa, NJ: Barnes & Noble. · Tansley, A.G. 1935: The use and abuse of vegetational concepts and terms. *Ecology* 16, pp. 284–307. · Von Bertalanffy, L., 1950: The theory of open systems in physics and biology. *Science* 111, pp. 23–9.

**syzygy** One of the two points at which the moon or a planet is aligned with the earth and the sun.

# T

**tafoni** The plural of tafone, which refers to a weathering hollow in a vertical rock face. The term tafoni is frequently used to refer to large ($m^2$) features, while alveoli (honeycomb features) is used for smaller ($cm^2$) weathering hollows that may cover rock surfaces, usually in sandstone or granite, in profusion. Tafoni and alveoli are often seen in coastal environments and DRYLANDS, such that salt weathering has been viewed as a major formative process. However, microorganisms and diurnal thermal contrasts may also contribute to their development, which may exploit lines or zones of weakness in rocks. DSGT

**taiga** The most northerly coniferous forest of cold temperate regions. It does not exist as a zone in the southern hemisphere, and generally refers to open woodland lying to the south of TUNDRA and to the north of the dense BOREAL FOREST. It has also been used more broadly to include the entire area covered by coniferous forest of high latitudes and high mountain slopes. More literally (from the Russian), it is characterized by open, rocky landscapes dominated by conifers but with scattered deciduous trees, such as birch and alder, locally dense along river valleys, with a fairly continuous carpet of lichens and heathy shrubs. The area is often poorly drained and peat-filled. PAF

**Reading**
Larsen, J.A. 1980: *The boreal ecosystem*. New York: Academic Press.

**takyr** A desert soil with a bare, parquet-like surface, broken up by a network of splits into numerous polygonal aggregates. Takyrs are typical landscape elements of the deserts of Central Asia, but comparable claypans are found in the deserts of Australia, Iran and North Africa. They have no higher vegetation, a crusted surface, occur in the lower parts of piedmont plains, and their formation requires a seasonal flooding of the surface by a thin layer of water, carrying suspended clay material and soluble salts (Kovda *et al.* 1979). ASG

**Reference**
Kovda, V.A., Samoilova, E.M., Charley, J.L. and Skujiņš. J.J. 1979: Soil processes in arid lands. In D.W. Goodhall, R.A. Perry and K.M.W. Howes eds, *Arid-land ecosystems: structure, functioning and management*. Vol. I. Cambridge: Cambridge University Press. Pp. 439–70.

**talik** A layer of unfrozen ground below the seasonally frozen surface layer and above or within PERMAFROST.

**talsand** ('valley sand') A widely used concept in north German Quaternary geology. Talsands are large-scale sandy infillings of ice marginal valleys which consist of fluvial beds with a capping of windborne sands. ASG

**Reference**
Schwan, J. 1987: Sedimentologic characteristics of a fluvial to aeolian succession in Weichselian Talsand in the Emsland (FRG). *Sedimentary geology* 52, pp. 273–98.

**talus** A SCREE slope; the term is sometimes also used simply to refer to scree material.

**tank** A man-made pond or lake. A natural depression in bare rock that is filled with water throughout the year.

**tarn** A small mountain lake (northern England). Compare Scottish 'lochan' and Welsh 'llyn'.

**taxon cycle** A concept introduced by P.J. Darlington in 1943 and so named in 1961 by the island biogeographer E.O. Wilson; it attempts to portray the stages through which an invading species-population passes when colonizing an archipelago or particular islands. It examines the way in which the population might adapt to new habitats and undergoes evolutionary diversification in the process. The cycle has been demonstrated for a number of groups of organisms, including ants in Melanesia and birds in the Solomon Islands. Most cycles so far described exhibit three or four stages, usually involving the expansion of a species without geographical divergence, the fragmentation of the distribution with speciation or subspeciation and a contracting stage, in which species are limited to relict areas, leading to extinction or to a secondary expansion. PAS

**Reading**
Pielou, E.C. 1979: *Biogeography*. New York and Chichester: Wiley. · Ricklefs, R.E. and Cox, G.W. 1972: Taxon cycles in the West Indian avifauna. *American naturalist* 106, pp. 195–219.

**taxonomy** The study, description and classification of variation in organisms, including the causes and consequences of such variation. This is a modern wide definition which makes taxonomy synonymous with the now interchangeable term, systematics. Traditionally, taxonomy was often restricted to the narrower activity of the classification and naming of organisms; in this sense, it was only a part of systematics.

The origins of taxonomy lie in human need to classify the discontinuities of variations to be seen in nature. Scientific taxonomy reached its great flowering in the eighteenth century with the still fundamental contributions of the Swedish biologist, Carolus Linnaeus (1707–1778). These included his *Systema Naturae* (1735 and many later editions), in which he classified all the then known animals, plants and minerals.

Today two major approaches to biological classification are recognized. The first is *phenetic* which expresses the relationships between organisms in terms of their similarities in characters, without taking into account how they came to possess these. The second is *phylogenetic*, or evolutionary, in which some aspects of their evolution are taken into account when making the classification, including their CLADISTIC relationship, which refers to the pathways of ancestry. These pathways are usually expressed in the form of a tree-diagram or cladogram. (See also VICARIANCE BIOGEOGRAPHY.) PAS

**Reading**
Jeffrey, C. 1977: *Biological nomenclature*. 2nd edn. London: Edward Arnold. · — 1982: *An introduction to plant taxonomy*. 2nd edn. Cambridge: Cambridge University Press. · Ross, H.H. 1974: *Biological systematics*. Reading, Mass.: Addison-Wesley. · Stace, C.A. 1980: *Plant taxonomy and biosystematics*. London: Edward Arnold.

**tear fault** A fault with a vertical fault plane, the blocks either side of which move horizontally. A transcurrent or strike fault.

**tectonics** The study of the broad structures of the earth's lithosphere and the processes of faulting, folding and warping that form them. Tectonics focuses on structures at the regional scale and above, the investigation of smaller-scale structural features usually being described as structural geology. Tectonics is concerned not only with the determination of the three-dimensional form of geological structures but also their history, origin and relationship to each other. Tectonic landforms are those, such as fault-scarps, produced directly by tectonic mechanisms and those larger features, such as warped erosion surfaces, which owe their general, though not detailed, form to such processes. (See also PLATE TECTONICS.) MAS

**Reading**
Kearey, P. and Vine, F.J. 1990: *Global tectonics*. Oxford: Blackwell Scientific. · Ollier, C. 1981: *Tectonics and landforms*. London: Longman. · Spencer, E.W. 1977: *Introduction to the structure of the earth*. 2nd edn. New York: McGraw-Hill.

**tektites** Small glassy spherules ($< 1$ mm to 0.2 m diameter) that show distinct effects of streamlining during their molten stage. They are found mainly in five 'strewnfields', designated Australasian, Ivory Coast, Czechoslovakian, North American, or Libyan. Tektites within a given strewnfield are related mineralogically, and in terms of age and shape. The most common shape is teardrop, but spheres, discs, and dumbbell shapes are also found. The source of tektites has been debated, but consensus is that the particles represent air-cooled, molten ejecta from meteorite or comet impacts with the earth. Tektites found in Haiti and Mexico have been linked to the CHICXULUB IMPACT. DJS

**Reading**
King, E.A. 1977: The origin of tektites: a brief review. *American scientist* 65, pp. 212–18.

**teleconnections** Fundamentally, these are long-distance linkages between the weather patterns of widely separated regions of the earth. Teleconnections are normally described by statistical relationships depicting the amount of shared variance between two regions or between two atmospheric phenomena. Descriptively, a teleconnection might associate changes in the surface temperatures or in the upper circulation of one region with, for example, the precipitation regime of another part of the earth such that variations in the first create changes in the second. The identification and analysis of teleconnections is currently one of the most important areas of climate research today.

Many of the most studied teleconnections in climate since the late 1970s have been those associated with the South Pacific phenomena known as ENSO. Interest in these teleconnections has been fuelled by the marked global changes linked to the powerful ENSO event of 1982–3, the subsequent long-lasting event in the early 1990s and the high amplitude event in 1997–8. One extreme of ENSO, EL NIÑO, a massive atmospheric–oceanic event characterized by abnormally warm sea surface temperatures in the equatorial Pacific and corresponding changes in the upper air circulation of that region, promotes many corollary changes in weather across the planet. For example, the high magnitude El Niño event of 1997–8 has been linked to abnormally wet winter conditions in the southern United States, to massive floods in western South America, to extensive severe drought in Indonesia, and substantial reduction in the number of Atlantic hurricanes in 1997. All of these climate linkages with ENSO are characterized as teleconnections.

Recently, scientists have also identified teleconnections existing between the other extreme of ENSO, LA NIÑA, abnormally cool waters in the equatorial Pacific, and various regions of the earth. In general, those teleconnections are opposite of those associated with El Niño – dryer conditions in the southern United States and western South America, wetter conditions in Australia and Indonesia and higher numbers of hurricanes in the Atlantic Ocean.

Teleconnections research has focused on two main areas of investigation. The first is the identification of mechanisms that influence and initiate the core phenomena, such as ENSO, while the second is the identification of other weather phenomena that may teleconnect to the weather of distant parts of the world.

With regard to the second area of study, many other weather phenomena are being analysed for their teleconnections to other regions. For example, the so-called PNA (or Pacific North America) pattern is a circulation pattern defined as having a large upper level ridge over the western part of the continent and a trough over the eastern half of the continent. Such a pattern leads to a dichotomy of conditions between the two halves of the continent with generally warmer and dryer conditions in the west and cooler and wetter conditions in the east. Another cyclic regional climate phenomenon thought to have teleconnective linkages to the climate of other parts of the world is the North Atlantic Oscillation (NAO), a periodic circulation existing between Iceland and the Azores.

Long-term climate teleconnections have also been hypothesized. A regional weather circulation known as the 'Greenland above' pattern has been suggested as a possible favourable circulation for initiating an ice age climate regime (Crowley 1984). Development of a large ridge over Greenland forces trough development to the west. Such a circulation is favourable for sustained snow production over eastern Canada. Crowley hypothesizes that if such a pattern were to remain in place for a long period of time, enough ice could accumulate over eastern Canada to force permanent climate change.

The concept of teleconnections is grounded in those associated with chaos theory as applied to climate change by Edward Lorenz. The general principle of climatic chaos theory is that internal instabilities in the climate system can cause complex behaviour elsewhere in the system. Such would appear to be the case with teleconnections. We are now discovering that many internal instabilities, specifically regional weather phenomena like ENSO, can indeed lead to disruptions of the entire climate system. This concept, carried to its most extreme theoretical form (the 'butterfly' principle), suggests that even the small-scale influence of butterfly wings could conceivably produce large-scale disturbances elsewhere on the planet. Such causative linkages between remote locations on the planet can be termed teleconnections and remain a central focus of current scientific investigation.                                        RSC

**temperate ice** Ice which is at the pressure melting point. In temperate glacier ice, water is present throughout and generally amounts to between 0.1 and 2 per cent of the total volume (Lliboutry 1976). *Temperate* ice is contrasted with *cold* ice which is below the pressure melting point. It is common to classify whole glaciers as temperate or cold, but this is misleading since both types of ice are common in most glaciers. For example, a glacier which consists wholly of temperate ice in summer may have a cold surface layer of ice in winter, while it is also likely that cold patches exist at the bottom as a result of pressure changes around obstacles (Robin 1976). Furthermore many 'cold' glaciers have ice at the pressure melting point at depth.                                        DES

**Reading and References**
Lliboutry, L. 1976: Physical processes in temperate glaciers. *Journal of glaciology* 16, pp. 151–8. · Paterson, W.S.B. 1981: *The physics of glaciers*. London and New York: Pergamon. · Robin, G. de Q. 1976: Is the basal ice of a temperate glacier at the pressure melting point? *Journal of glaciology* 16, pp. 183–96.

**temperature** To the general public temperature is a confusing parameter. When used to measure how warm or how cold substances feel it is confusing to have different temperature scales, so that, for instance, the temperature at which pure ice melts can be stated as 0 °Celsius or 32 °Fahrenheit or 273.15 °Kelvin. In science the Kelvin scale is used because it avoids negative temperatures and is a more accurate reflection of the energy possessed by the molecules in a substance. Temperature is really a measure of the molecular kinetic energy of a substance, in other words the average speed at which the molecules are moving in gas, or vibrating in a solid. The Celsius and Fahrenheit scales are used for convenience in that they give easier numbers for people to handle for the range of temperatures normally encountered at the earth's surface.                                        JET

**temperature humidity index**    Originally called the Discomfort Index when it was introduced by E.C. Thom (1959) in the USA, the temperature humidity index (THI) is a simplified form of effective temperature:

$$THI = 0.4(T_{DB} + T_{WB}) + 4.8$$

when the dry-bulb and WET-BULB TEMPERATURES are recorded in °C. Wind-speeds and solar radiation are not taken into consideration and so the index must be used with caution. The THI is widely used in weather forecasts for the general public in the USA during summer. At a THI below 70 there is no discomfort. When the THI reaches 80 everyone feels uncomfortable.                                                       DGT

**Reference**
Thom, E.C. 1959: The discomfort index. *Weatherwise* 12, pp. 57–60.

**temperature inversion**    An increase of temperature with height, the inverse of the normal decrease of temperature with height that occurs in the TROPOSPHERE. When the temperature is getting warmer with height, natural buoyancy is limited and the air is described as being stable. Under these conditions, vertical dispersion of air pollution and visibility are both restricted. Inversions can form in a variety of different ways. (1) Most frequently, inversions form near the surface as a result of radiative cooling during clear nights; more energy is lost from the surface than is gained by counter-radiation from the atmosphere, so most cooling takes place at the surface. They are normally destroyed by surface heating the following morning. (2) When warm air passes over a cold surface an inversion will form. If the air is moist or the surface very cold, persistent fog may form. This often happens when warm air is drawn from an ocean over a snow covered surface. (3) Inversions can form dynamically as a result of ADIABATIC subsidence of air associated with ANTI-CYCLONES. The temperature may decrease initially from the surface but later increases as the effects of the descending warming air are felt. The TRADE WIND inversion of the subtropics is the best example of this type. As it may persist for days or weeks, dispersal of air pollution can be a problem, as in Los Angeles. (4) Radiative heating in the upper atmosphere can produce strong temperature inversions. The absorption of ultra-violet light by ozone in the stratosphere is a good example.                 PS

**Reading**
Barry, R.G. and Chorley, R.J. 1998: *Atmosphere, weather and climate.* 7th edn. London: Routledge. · McGregor, G.R. and Nieuwolt, S. 1998: *Tropical climatology.* Chichester: Wiley.

**tensiometer**    An instrument for estimating the matric or capillary potential of soil moisture. A tensiometer consists of a porous pot at the required point in the soil and connected through watertight tubing to a manometer or pressure gauge. The whole instrument is filled with water before the cup is placed in a carefully augured hole, refilled as near to its former condition as possible. Water may pass through the walls of the pot in response to suction forces in the surrounding soil. As water passes into the soil a partial vacuum builds up inside the instrument and this is monitored as a negative head of water by the vacuum gauge or manometer. Water will move into or out of the pot until the soil moisture suction is balanced by the strength of the partial vacuum or the weight of a negative head of water (or, more usually, mercury). Tensiometers are only suitable for use in comparatively moist soil because of problems in keeping the instrument airtight at high suctions.    AMG

**Reading**
Burt, T.P. 1978: *An automatic fluid-scanning switch tensiometer system.* British Geomorphological Research Group technical bulletin 21. Norwich: Geo Abstracts. · Curtis, L.F. and Trudgill, S. 1974: *The measurement of soil moisture.* British Geomorphological Research Group technical bulletin 13. Norwich: Geo Abstracts.

**tepee**    An overthrust sheet of limestone which appears as an inverted V in a two-dimensional exposure, so named because of its two-dimensional resemblance to the hide dwellings of early American Indians. Tepees are found in tidal areas, around salt lakes, and in CALCRETE, developing as a result of deformation or desiccation and contraction processes related to fluctuations in water levels and in the nature of chemical precipitation.                                        ASG

**Reading**
Kendall, C.G. St C. and Warren, J. 1987: A review of the origin and setting of tepees and their associated fabrics. *Sedimentology* 34, pp. 1007–27. · Warren, J.K. 1983: Tepees, modern (Southern Australia) and ancient (Permian – Texas and New Mexico) – a comparison. *Sedimentary geology* 34, pp. 1–19.

**tephigram**    A meteorological thermodynamic chart used for plotting and analysing dry-bulb and dewpoint temperature from a RADIOSONDE ascent, usually from the surface to the lower STRATOSPHERE. The vertical axis is atmospheric PRESSURE decreasing upwards on a logarithmic scale and TEMPERATURES are plotted with reference to the horizontal axis although the isotherms are skewed to run from bottom left to top right of the diagram. Dry adiabats, saturation adiabats and lines of constant humidity mixing ratio are also printed in the chart. The tephigram is basically an aid to forecasting; for example, fog formation or shower development.                                          RR

**tephra** The solid material ejected from a volcano which includes dust, ash, cinders and volcanic bombs.

**terlough** Depressions, with a sinkhole, which fill with water when the water table rises. The rise may be associated with tidal effects. They are a feature of parts of western Ireland.

**terminal grades** The fine fraction due to a glacial comminution which reflects the mineral components of the debris. Dreimanis and Vagners (1971) suggested that once a glacial TILL has been broken down to mineral-sized fragments it experiences relatively little further breakdown, thus meriting the use of the term terminal grades. This grain-size group comprises one of the bimodal grain-sized products of abrasion, the other being rock fragments. Under some circumstances, however, fine material fragments do break down further to finer material (Haldorsen 1981).                       DES

**References**
Dreimanis, A. and Vagners, U.J. 1971: Bimodal distribution of rock and mineral fragments in basal tills. In R.P. Goldthwait ed. *Till: a symposium.* Ohio State University Press. Pp. 237–50. · Haldorsen, S. 1981: Grain-size distribution of subglacial till and its relation to glacial crushing and abrasion. *Boreas* 10, pp. 91–105.

**terminal moraine** The moraine at the terminus of a glacier. (See also MORAINE.)

**terminal velocity** An object falling through a fluid (wind or water) is retarded by viscous drag forces at its surface. Initially, the force of GRAVITY exceeds the drag force and the object accelerates downwards. As velocity increases, viscous drag forces also increase and acceleration continues until the gravity and drag forces become equal. At this point the object is said to have reached terminal velocity and no further acceleration will take place without an additional external force being applied.           GFSW

**terminations** The boundaries in deep sea core sediments that separate pronounced oxygen isotopic maxima from exceptionally pronounced minima. They are in effect rapid deglaciations, and are conventionally numbered by Roman numerals in order of increasing age. The segments bounded by two terminations are called glacial cycles. Nine terminations have occurred in the past 0.7 million years.    ASG

**Reading**
Kukla, G.J. 1977: Pleistocene land-sea correlations. I. Europe. *Earth science reviews* 13, pp. 307–74.

**termites** Of which there are several thousand species, are insects of the Isoptera order, and about four-fifths of the known species belong to the Termitidae family (Harris 1961). They vary in size according to their species, from the large African Macrotermes, with a length of around 20 mm and a wing span of 90 mm, down to the Middle Eastern Microcerotermes which are only around 6 mm long with a wing span of 12 mm. Major recent taxonomic and ecological surveys include that of Brian (1978), while Lee and Wood (1971) provide a detailed study of the effects of termites on soils, and Goudie (1988) reviews their geomorphological impact.

Termites, though 'fierce, sinister and often repulsive' (Maeterlinck 1927), are remarkable for having been adapted to living in highly organized communities for as long as 150–200 million years, and much of their success is due to their development of elaborate architectural, behavioural, morphological and chemical strategies for colony defence. They occur in great numbers: 2.3 million $ha^{-1}$ in Senegal and 9.1 million $ha^{-1}$ in the Ivory Coast (UNESCO/UNEP/FAO 1979). Maeterlinck (1927) regarded them as 'the most tenacious, the most deeply rooted, the most formidable, of all the occupants and conquerors of this globe'. The vast majority of termite species are found in the tropics.                                  ASG

**Reading and References**
Brian, M.V. ed. 1978: *Production ecology of ants and termites.* Cambridge: Cambridge University Press. · Goudie, A.S. 1988: The geomorphological role of termites and earthworms in the tropics. In H.A. Viles ed., *Biogeomorphology.* Oxford: Blackwell. Pp. 166–92. · Harris, W.V. 1961: *Termites: their recognition and control.* London: Longman. · Lee, K.E. and Wood, T.G. 1971: *Termites and soils.* London and New York: Academic Press. · Maeterlinck, M. 1927: *The life of the white ant.* London: Allen & Unwin. · UNESCO/UNEP/FAO 1979: *Tropical grazing and land ecosystems.* Paris: UNESCO.

**terra rossa** Red soils developed on the iron-oxide-rich residual material on limestone bedrock, particularly in warm temperate regions.

**terracette** Miniature terrace or ridge extending across a slope, usually normal to the direction of maximum slope. Terracettes are rarely more than 0.5 m wide and deep. Their origin is still a matter for debate. Some may be animal tracks, but as others occur in areas where animals are very rare it would seem that some other mechanisms are involved. They are probably a consequence of *soil* mantle instability on steep slopes.                                                    ASG

**terrane** 'A mappable structural entity which has a stratigraphic sequence and an igneous, metamorphic and structural history quite distinct from those of adjacent units. Each terrane is separated from its neighbours by a structural break which may take the form of a normal fault, a reverse fault, a wrench fault, or an overthrust' (Barber 1985, p. 116).

During the 1980s the terrane concept became an important one for many earth sciences (Howell 1989). It was developed most notably in the context of the North American Cordillera system where, it was maintained, substantial portions of the Cordillera were 'exotic' blocks and slithers that had 'docked' onto the North American Craton. Subsequently, the importance of terranes has been recognized with respect to areas as diverse as Highland Scotland and the archipelagos of south-east Asia. Terranologists hold that major oceanic belts often consist of 'collages' of fault-bounded crustal and/or lithospheric fragments, of diverse origins and different sizes. However, as Sengòr and Dewey (1991) point out, reactions to the concept range from enthusiastic applause to abusive rejection. They themselves express some cogent doubts, suggesting, for example (p. 6) that 'terranology not only does not go beyond plate tectonics, it takes a backward step; by confusing primary and secondary collage components, it confuses also their genetic implications.' They continue (p. 17) 'the word terrane is a lump term for a number of older and more informative non-genetic (block and sliver) and genetic (fragment, nappe, strike-slip duplex, microcontinent, island arc, suture, etc.) terms. Because it is less informative, it is less useful than any of these and also because, historically, the term "terrane" has a number of different meanings it is best avoided. Terrane analysis is neither a new way of looking at orogenic belts, nor a particularly helpful one.'                              ASG

**Reading and References**
Barber, A. 1985: A new concept of mountain building. *Geology today* 1, pp. 116–21. · Howell, D.G. 1989: *Tectonics of suspect terranes*. London: Chapman & Hall. · Sengòr, A.M.C. and Dewey, J.F. 1991: Terranology: vice or virtue? In J.F. Dewey, I.G. Gass, G.B. Curry, N.B.W. Harris and A.M.C. Sengòr eds, *Allochthonous terranes*. Cambridge: Cambridge University Press.

**terrestrial magnetism** The natural magnetism of the earth, also referred to as geomagnetism. The shape of the earth's mainly dipole magnetic field suggests that it is related to circular electrical currents flowing approximately normal to the axis of rotation. These electrical currents may be induced by slow convective movements within the partially molten iron-rich core of the earth, with large-scale eddies producing the regional variations in the main field. Secular changes in the field include the continuous movement of magnetic north, variations in the strength of the field and periodic reversals of polarity.                              MAS

**Reading**
Parkinson, W.D. 1983: *Introduction to geomagnetism*. Edinburgh: Scottish Academic Press; Amsterdam: Elsevier. · Smith, P.J. 1981: The earth as a magnet. In D.G. Smith ed., *The Cambridge encyclopaedia of earth sciences*. Cambridge: Cambridge University Press; New York: Crown Publishers. Pp. 109–23.

**territory, animal** An area held and defined by an animal or group of animals. Territories may contain resources such as food or mates, or may be a display or rutting stand, and may or may not be fixed in space and/or time. Notable examples of animals exhibiting marked territorial behaviour are tawny owls, which have a fixed exclusive spatial area; and cats, which have fixed exclusive areas in time. Such territorial behaviour must have certain advantages. The concept of 'economic defendability' suggests that an animal should only defend a territory if there is a net benefit, in terms of propagating its genes, in doing so. Single parameters such as food, mates and predation will act together to determine spacing behaviour which will probably always be a compromise moulded by these selective pressures. The consequences of territorial behaviour may be to limit populations by the establishment of exclusive feeding or breeding rights in an area, the animals excluded then being culled by exposure, starvation and predation. Spacing by territories may also be advantageous to the survival of camouflaged prey.                              KEB

**Reading**
Krebs, C.J. 1985: *Ecology*. 3rd edn. New York and London: Harper & Row. · Krebs, J.R. and Davies, N.B. eds 1984: *Behavioural ecology: an evolutionary approach*. Oxford: Blackwell Scientific.

**Tertiary** Refers to the first part or period of the CAINOZOIC era, comprising the Palaeocene through to the Pliocene.                              ASG

**Tethys Ocean** An enormous seaway initially formed in the Palaeozoic era and attaining its maximum development during the Mesozoic, which extended from what is now the Mediterranean eastwards as far as south-east Asia. Beginning about 75 million years ago extensive volcanism, followed by intense deformation and uplift of sediments accumulated in the Tethys

Ocean throughout the Mesozoic, led to the formation of the present-day Alpine-Himalayan mountain belt. The opening and closure of the Tethys Ocean (see MESSINIAN SALINITY CRISIS) is currently interpreted in terms of CONTINENTAL DRIFT and the operation of PLATE TECTONICS. MAS

**Reading**
Sonnenfield, P. ed. 1981: *Tethys: the ancestral Mediterranean*. Stroudsburg, Pa.: Dowden, Hutchinson & Ross.

**thalassostatic** A term used to describe river terraces which are produced by aggradation during periods of rising or high sea level and by incision at times of low sea level.

**thalweg** The line of maximum depth along a river channel. It may also refer to the line of maximum depth along a river valley or in a lake.

**thaw lake** A shallow, rounded lake occupying a depression resulting from the melting of ground ice (see also THERMOKARST). Thaw lakes are extremely common on the North American and Siberian Arctic lowlands and are ubiquitous wherever there is a flat lowland with silty alluvium and a high ice content. Most lakes are less than 300 m in diameter and less than 3–4 m deep. Following random exposure of ground ice, a water-filled depression soon develops into a roughly circular lake. Eventually vegetation protects the banks from further thawing and in a matter of 2000–3000 years or so the lake is infilled and the cycle of development complete. Many thaw lakes are elongate in shape with a systematic orientation of the long axis at right angles to the prevailing wind (Carson and Hussey 1963). DES

**Reading and References**
Carson, C.E. and Hussey, K.M. 1963: The oriented lakes of Arctic Alaska: a reply. *Journal of geology* 71, pp. 532–3. · French, H.M. 1974: *The periglacial environment*. London: Longman. · Washburn, A.L. 1979: *Geocryology*. London: Edward Arnold.

**thematic mapper (TM)** A sensor which acquires multispectral satellite images of use in a wide range of geographical applications. The name derives from its suitability for the production of land-cover maps divided into themes such as arable, forestry, and built-land, i.e. thematic maps. The images cover a ground area of approximately 185 × 185 km and with a PIXEL size of 30 m. Thematic mapper images have been continuously acquired from the American-owned Landsat satellites since 1982, originally run by NASA and now in commercial ownership. This image type is commonly referred to as Landsat TM.

The amount of reflected and emitted radiation is measured in seven regions of the electromagnetic spectrum: three in the visible portion (blue, 0.45–0.52 $\mu$m; green, 0.52–0.60 $\mu$m; red, 0.63–0.69 $\mu$m), and one in each of the near infrared (0.76–0.90 $\mu$m), mid-infrared (1.55–1.75 $\mu$m) and thermal infrared (10.42–12.50 $\mu$m) and a further in mid-infrared (2.08–2.35 $\mu$m). These wavebands are numbered from 1 to 7 respectively. Whilst the nominal resolution, or pixel size is 30 m, the thermal waveband, band 6, is an exception with a pixel size of 120 m. The broad sampling of the electromagnetic spectrum makes TM a versatile tool and is perhaps the most widely used of satellite images in geographical research. Pixel values are recorded as 8-bit numbers, providing a data range from 0 to 255, which can be calibrated to radiance or reflectance units if required.

The TM sensor is carried aboard the Landsat satellites 4 and 5, which orbit the earth at an altitude of 705 km providing complete earth coverage between 81 °N–81 °S. The minimum revisit period for any point on the ground is 16 days, although in practice the presence of cloud cover may reduce this from months to years. Launch of an enhanced thematic mapper is imminent. CDC

**Reading**
Lillesand, T.M. and Kiefer, R.W. 1994: *Remote sensing and image interpretation*. 3rd edn. New York: Wiley.

**thermal depression** A region of low surface pressure which is generated by strong solar heating on fine days over land in the summer. Thermal depressions can range from the large, seasonal thermal low of the south Asian monsoon to the short-lived lows that sometimes form over England on hot summer days. Their cyclonic inflow tends to be strongest near the surface and to weaken with height since they are shallow features, usually a few kilometres deep.

Fine summer weather in Britain occasionally breaks down in association with an extending thundery thermal depression over France. RR

**thermal efficiency** The term was used by C.W. Thornthwaite (1931) in his first classification of climate. The thermal efficiency index (TEI) indicates the plant growth potential of a location and is calculated by summing the 12 monthly values of (T – 32)/4, where T = mean monthly temperature in °F. The index ranges from zero on the polar limit of the tundra ('frost climate') to over 127 in the tropics. Six temperature provinces were defined. In this classification Thornthwaite made moisture, in the form of precipitation effectiveness (PEI), the primary factor for a T-E Index of over 31 (taiga/cool temperate boundary). In his second classification of climate, Thornthwaite (1948) used potential evapotranspiration as a measure of thermal efficiency. DGT

**Reading and References**
Thornthwaite, C.W. 1931: The climates of North America according to a new classification. *Geographical review* 21, pp. 633–55. · —1933: The climates of the earth. *Geographical review* 23, pp. 433–40. · —1948: An approach toward a rational classification of climate. *Geographical review* 38, pp. 55–94.

**thermal equator** Variously defined, but most commonly the line which circumscribes the earth and connects all points of highest mean annual temperature for their longitude. Sometimes the seasonal or monthly variation of this line is considered. The annual mean position departs by as much as 20° latitude from the equator. In recent years the term thermal equator or heat equator has also been taken as synonymous with the INTERTROPICAL CONVERGENCE ZONE, the belt along which the thermally driven trade winds of the two hemispheres converge and are forced to rise. WDS

**thermal infrared linescanner** An optical remote sensor used to derive thermal images of the earth's surface. Early thermal infrared linescanners have two thermal detectors and record an image onto a photographic film, while more recent thermal infrared line-scanners are often part of a MULTISPECTRAL SCANNER which records data digitally. With both these systems the physical geographer must choose the waveband to be used, the time of day when the image is to be recorded and the method of approximate calibration. The two thermal infrared wavebands used by thermal infrared linescanners are defined by the two transmitting atmospheric 'windows' located between the wavelengths of 3–5 $\mu$m and 8–14 $\mu$m. The choice of which of these wavebands to employ depends upon the application. The peak of radiant emission from the earth's surface occurs in the 8–14 $\mu$m region and this waveband has proved to be the most popular in physical geography.

Thermal infrared linescanners are used most frequently at night when there is no interference from reflected solar radiation. The usual flying time is just before dawn, when the effects of differential solar heating are minimized. Flights are occasionally made during daylight hours, either because AERIAL PHOTOGRAPHY is to be taken or because it is advantageous to have terrain details enhanced by differential solar heating and shadowing. Owing to the variation in emissivity within a scene and the presence of a thermally variable atmosphere between the sensor and the ground, it is not possible to calibrate absolutely the radiant temperature sensed by the detector. Various approximate calibration methods have been employed of which the most accurate involves repeated flights at a range of altitudes. By using the temperature of thermally stable objects of known emissivity as standards, a graph is constructed of temperature versus height. This is extrapolated to the ground surface to give the temperature of enough points to enable calibration of the imagery.

Thermal infrared linescanner data can be presented in image form to be interpreted like an aerial photograph. The interpreter generally uses the images not to map an area but to search for thermal patterns that give a clue to some past, present or future environmental process such as soil movement, frost hollows, water stress in crops, vulcanism or thermal pollution of water. PJC

**Reading**
Cracknell, A.P. and Hayes, L.W.B. 1991: *Introduction to remote sensing*. London: Taylor & Francis. · Lillesand, T.M. and Kiefer, R.W. 1987: *Remote sensing and image interpretation*. 2nd edn. New York: Wiley.

**thermal pollution** The pollution of water by increasing its temperature. Many fauna are affected by temperature so that this environmental impact has some significance. Among the main sources of thermal pollution of stream waters are condenser cooling water released from electricity generating stations, the urban 'heat island effect', reservoir construction, shade removal by deforestation, and changes in the width–depth ratios of channels. The effects of thermal pollution are especially severe at times of low flow. ASG

**Reading**
Langford, T.E.L. 1990: *Ecological effects of thermal discharges*. London: Elsevier. · Pluhowski, E.J. 1970: Urbanization and its effects on the temperature of the streams on Long Island, New York. *United States Geological Survey professional paper* 627-D.

**thermal wind** ($v_t$) The vector difference between the geostrophic wind at two levels in the atmosphere. It is calculated by subtracting the lower level wind ($v_l$) from the upper level wind ($v_u$) and is therefore not a real wind, which would be observed in the intervening layer, but an expression of the shear of the horizontal wind within the layer.

It is termed 'thermal' because its strength and direction are dependent on the thermal pattern of the layer involved. Thus ($v_t$) is aligned parallel to the layer's mean isotherms with cold air to its left and warm air to its right. Its magnitude or length is proportional to the strength of the layer's mean temperature gradient so that thermal winds are strongest in regions where steep horizontal temperature gradients occur in depth, in the polar front zone for example.

Thermal winds are also therefore parallel to THICKNESS contours and can in fact be calculated precisely with reference to thickness variations across a map. RR

**thermistor** A type of semi-conductor resistor which has a high (usually negative) temperature coefficient of resistance. Thus they are often used as temperature sensors or measuring devices. The resistance response to temperature is not linear but can be linearized either in hardware or software in the measuring instrument or subsequent data processing. Despite this disadvantage, it is often preferred to the linear response of the platinum resistance thermometer. WBW

**thermoclasty** See INSOLATION WEATHERING.

**thermocline** A layer of water within a lake or ocean through which the rate of decrease of temperature with depth is much greater than in adjacent layers. It is particularly well developed in tropical oceans where a *permanent* thermocline, up to a few hundred metres in thickness, lies with its upper boundary some 25–150 m below the ocean surface. The temperature gradient may reach $10\,^{\circ}\mathrm{C}\ (100\,\mathrm{m})^{-1}$ in the upper part of the layer. Towards mid-latitudes the permanent thermocline becomes thicker and less intense with its upper surface as much as 600 m deep; it is entirely absent at latitudes greater than about 60°. The relatively stable stratification of the thermocline layer inhibits interchange between the warm waters of the surface mixed layer and the deeper cold waters forming the main body of the ocean (see GLOBAL OCEAN CIRCULATION).

A *seasonal* thermocline readily develops in both lakes and oceans whose surface temperatures undergo a significant annual variation. Lying close to the surface, its depth and intensity will depend on the amount of summer insolation and extent of turbulent mixing by the wind. JEA

**Reading**
Harvey, J.G. 1976: *Atmosphere and ocean*. London: Artemis Press. · Open University Oceanography Course Team 1989: *Seawater: its composition, properties and behaviour*. Oxford: Pergamon; Milton Keynes: Open University Press.

**thermocouple** A temperature measuring device. It employs the phenomenon that when a metal wire is interposed in a wire of a different metal then if one junction is heated relative to the other an electromotive force is produced (Seebeck effect). Generally, one junction is used as the sensor, the other being held at 0 °C. Various metal combinations produce different emfs and are used for different purposes. WBW

**thermodynamic diagram** A chart on which are plotted observations of pressure, temperature and humidity from a given RADIOSONDE ascent. In its simplest form it is a diagram whose vertical axis is pressure and horizontal axis temperature with the former being logarithmic with pressure decreasing upwards and the latter a linear scale with temperature increasing to the right.

Operational thermodynamic diagrams are more complex, however, and are designed so that area on the chart represents energy; the TEPHIGRAM is one example. This is useful when a forester has to assess the likelihood of fog clearance on the basis of the amount of energy required to evaporate it – and whether solar heating at various times of year will be great enough to meet the requirement.

A variety of derived quantities are easily obtained once the temperatures (dry-bulb and dewpoint) have been plotted, for example the POTENTIAL TEMPERATURE and both relative and absolute humidity.

Thermodynamic diagrams are also used in the graphical estimation of the nature and intensity of VERTICAL STABILITY in the atmosphere for the time and place represented. RR

**Reading**
Atkinson, B.W. 1968: *The weather business*. London: Aldus Books.

**thermodynamic equation** Expresses a fundamental relationship in meteorology in which the time rate of change of an air parcel's TEMPERATURE as it moves through the atmosphere is related to both ADIABATIC expansion or compression and to diabatic heating. It is a predictive equation.

In a typical middle latitude disturbance, a cyclone for example, air parcels at middle levels undergo adiabatic temperature changes of about $30\,\mathrm{K\,d}^{-1}$ while the diabatic changes (due to absorption of solar radiation, absorption and emission of infrared radiation, latent heat release, etc.) tend to compensate one another and have a net value of about $1\,\mathrm{K\,d}^{-1}$. RR

**thermograph** A meteorological instrument which is housed in a weather screen and provides a continuous record of air temperature.

The response of a sensor, for example a bi-metallic coil, to fluctuations of air temperature is magnified by a long pen to which it is connected mechanically. The pen traces a line on a daily or weekly strip chart which is fixed to a slowly revolving clockwork-driven drum.      RR

**thermokarst**     Topographical depressions resulting from the thawing of ground ice (Washburn 1979). There are many kinds of thermokarst, including collapsed PINGOS, ground-ice mudslumps, linear and polygonal troughs, THAW LAKES and ALASes. Some thermokarst features result from climatic change, but most relate to minor environmental changes which promote thawing, for example, the shift of a stream channel, natural and human-induced disturbance to tundra vegetation. Thermokarst is sometimes used in a wider sense also to include thermal erosion by flowing water at a temperature above 0 °C (French 1974). As such it would include thermo-erosional niches and overhangs and various features associated with slopewash.      DES

**Reading and References**
French, H.M. 1974: *The periglacial environment*. London: Longman. · Washburn, A.L. 1979: *Geocryology*. London: Edward Arnold.

**thermoluminescence**     The light emitted in addition to incandescence from an insulating crystal when a sample is heated. Thermoluminescent properties of a material accumulate progressively following exposure to continuous radiation. As a result, the thermoluminescence intensity at any given heating temperature is a product of the amount of radiation received. Specific minerals (e.g., quartz, sodium and potassium feldspar) are found to emit at characteristic excitation temperatures, and in spectra of specific wavelengths. The thermoluminescence (TL) signal is usually determined by heating the sample at a constant rate from room temperature up to 500 °C (producing a characteristic 'growth curve'). Independently determining the amount of radiation received means that it is possible to establish the age of the commencement of signal accumulation, providing that a time-zero event is identified (i.e. a time at which all previously accumulated radiation damage has been removed). (See also LUMINESCENCE DATING METHODS.)      SS

**Reading**
Aitken, M.J. 1990: *Science-based dating in archaeology*. London: Longman.

**thermopile**     A radiation detecting device which uses a series of thin wire THERMOCOUPLES

to measure radiation incident on the hot junctions of the thermocouples. The cold junctions are shielded and kept at a uniform, measured temperature. This device differs from a bolometer where a blackened platinum foil is heated by the radiation and the increase in resistance is measured.      WBW

**thickness**     The difference in height above mean sea level of two PRESSURE surfaces above a given point. It is proportional to the mean TEMPERATURE of the layer in question so that the larger values indicate relatively warm air and smaller values relatively cold air. Thickness values are obtained from RADIOSONDE observations and are plotted on a base map which commonly depicts the variations for the 1000–500 mb layer. Contours are drawn, say, every 60 m to produce a thickness analysis (which is also effectively an isotherm analysis of the layer involved) which is of importance in synoptic meteorology.      RR

**Reading**
Atkinson, B.W. 1968: *The weather business*. London: Aldus Books.

**Thiessen polygon**     Defines the horizontal area which is nearer to one rain gauge than to any other in a rain gauge network. The areas of the polygons are used to weight rain-gauge catches when calculating the average areal precipitation. Thiessen polygons are constructed from the perpendicular bisectors of the horizontal projections of straight lines joining adjacent rain gauges. These perpendicular bisectors are extended and joined to leave each rain-gauge site at the centre of a polygon and the area of the polygon is known as its Thiessen weight. The method is widely applied because it is easy to use and makes allowance for the uneven distribution of rain gauges.      AMG

**Reading**
Damant, C., Austin, G.L., Bellon, A. and Broughton, R.S. 1983: Errors in the Thiessen technique for estimating areal rain amounts using weather radar data. *Journal of hydrology* 62, pp. 81–94. · Thiessen, A.H. 1911: Precipitation averages for large areas. *Monthly weather review* 39, pp. 1082–4.

**tholoid**     A volcanic cone situated within a large volcanic crater or caldera.

**threshold, geomorphological**     A threshold of landform stability that is exceeded either by intrinsic change of the landform itself, or by a progressive change of an external variable (Schumm 1979). The concept is closely bound up with the view of a landform as a system or part of a system in which there is normally some sort of balance between the morphology and the processes involved.

An *intrinsic* threshold implies that changes can take place within the system without a change in an external variable. An example is a SURGING GLACIER which may exhibit periodic surges although the input of snow remains identical over many decades. In this case there is a build-up of snow and ice to a critical level which causes a sudden change in the process of basal sliding. The sudden transition to a fast mode of flow lowers the ice surface until the glacier reverts to a slow mode of flow and the cycle starts once more. In this particular case the threshold is peculiar to a particular glacier and probably relates to different modes of sliding related to critical basal conditions (Budd 1975). Other normal glaciers may never quite reach the initial threshold allowing fast flow. Other geomorphological examples of intrinsic thresholds involve river channel patterns and river terrace changes (Schumm 1973, 1979).

An *extrinsic* threshold describes an abrupt change which is triggered by a progressive change in an external variable. Well-known examples are the threshold velocities required to set in motion particles of a given size. With a continuous change in velocity the response of river channel bedforms or of aeolian forms will suddenly change. Another example is the progressive depletion of vegetation which may allow the threshold of gully formation and thereby active soil erosion to be crossed suddenly.

The widespread adoption of a systems approach in geomorphology went hand in hand with the concept of some sort of balance between process and form (Chorley and Kennedy 1971). Systems were seen as hunting for an equilibrium and they were analysed on the assumption that they operated in STEADY STATE. Such an approach made it difficult to study landforms which were evolving rapidly over time. Also it was difficult to predict the response of a landform to change without knowledge of its sensitivity or indeed its proximity to a threshold of change, a vital requirement if geomorphology is to be applied successfully. Threshold analysis offers a powerful means of tackling these limitations. It focuses on one of the mechanisms of change and also opens up the prospect of an effective applied geomorphology which can predict and prevent problems before they occur. Viewed in such a light the study of thresholds is a way of developing the potential of a systems approach to geomorphology.

There is currently much interest in the nature of threshold changes and the possible applications of CATASTROPHE theory which offers a mathematical means of modelling such discontinuities in a system. There are several different types of discontinuity and the nature of a particular threshold may vary according to the direction of change in a system. Thus the initial water thickness needed to initiate surging glacier velocities may be different from the thickness required to maintain or stop them (Hutter 1982). Another example concerning ephemeral streams is provided by Thornes (1982).      DES

**Reading and References**
Budd, W.F. 1975: A first simple model for periodically self-surging glaciers. *Journal of glaciology* 14.70, pp. 3–21. · Chorley, R.J. and Kennedy, B.A. 1971: *Physical geography: a systems approach.* Englewood Cliffs, NJ: Prentice-Hall. · Hutter, K. 1982: Glacier flow. *American scientist* 70, pp. 26–34. · Schumm, S.A. 1973: Geomorphic thresholds and complex response of drainage systems. In M. Morisawa ed., *Fluvial geomorphology.* Binghampton, NY: State University of New York Press. Pp. 299–310. · — 1979: Geomorphic thresholds: the concept and its applications. *Transactions of the Institute of British Geographers* n.s. 4.4, pp. 485–515. · Thornes, J.B. 1982: Structural instability and ephemeral channel behaviour. *Zeitschrift für Geomorphologie*, pp. 233–44. · — 1983: Evolutionary geomorphology. *Geography* 68, pp. 225–35.

**threshold slopes**   Hillslopes with inclinations controlled by the resistance of their soil cover to a dominant degradational process. Such slopes are recognized within areas of consistent rock and soil types and erosional processes by nearly uniform inclinations of characteristic hillslope units. It has been postulated that these units are at the maximum inclination, for temporary stability, permitted by the soil strength and pore water pressures within the soil cover.

Three characteristic maximum hillslope angles have been identified for areas prone to erosion by landsliding processes (Carson 1976): (1) a frictional threshold slope angle where the soil is a dry rock rubble and the angle of inclination equals the angle of plane sliding friction of the rubble; (2) a semifrictional threshold slope angle for cohesionless soils where the water table can rise to the surface and seepage is parallel to it; the slope angle then approximates to half that of the effective angle of plane sliding friction of the soil; and (3) an artesian condition in which water flows out of the soil and the slope angle is less than that of case (2).

The concept of threshold slopes applies primarily to straight slope segments between upper convexities and lower concavities. In areas of rapid uplift and incision of stream channels slope angles may be steeper than threshold angles until landsliding reduces the slope angle to the threshold angle. Threshold slopes will eventually be eliminated by creep, wash and other processes: they are consequently temporary features of a landscape dominated by landsliding.  MJS

**Reading and Reference**
Carson, M.A. 1976: Mass-wasting, slope development and climate. In E. Derbyshire ed., *Geomorphology and climate*. London: Wiley. Pp. 101–30. · —and Kirkby, M.J. 1972: *Hillslope form and process*. Cambridge: Cambridge University Press. · Selby, M.J. 1993: *Hillslope materials and processes*. 2nd edn. Oxford: Oxford University Press.

**throughfall**  The net rainfall below vegetation cover excluding STEM FLOW. Throughfall comprises both precipitation which falls straight through the vegetation canopy and precipitation which has been intercepted by the vegetation but then drips onto the ground.  AMG

**Reading**
Ford, E.D. and Deans, J.D. 1978: The effects of canopy structure on stemflow, throughfall and interception loss in a young Sitka spruce plantation. *Journal of applied ecology* 15, pp. 905–17. · Sopper, W.E. and Lull, H.W. eds 1967: *International symposium on forest hydrology*. Oxford and New York: Pergamon.

**throughflow**  (or subsurface flow) Downslope flow within the soil. Where there are well-defined aquifers downward percolation is commonly rapid enough to prevent appreciable throughflow. Where the bedrock is not highly permeable, lateral flow within the soil is, for many sites, the most effective form of downslope flow because the soil is more permeable and more porous than the bedrock. This is particularly so in open-structured soils like those beneath many mature woodlands. Infiltrated water percolates downward until it meets an impeding horizon, where it is diverted laterally as saturated throughflow. Impedance may be due to general saturation below, or to the reduction in permeability with depth which is a normal feature of soils, with the greatest reduction near the base of the 'A' horizon.  MJK

**Reading**
Kirkby, M.J. ed. 1978: *Hillslope hydrology*. Chichester: Wiley.

**throw of a fault**  The vertical displacement of a fault.

**thrust**  A low-angle reverse fault. Also a fault occurring on the overturned limb of a fold.

**thufur**  A soil hummock found in periglacial environments.

**Reading**
Schunke, E. and Zoltai, S.C. 1988: Earth hummocks (thufur). In M.J. Clark ed., *Advances in periglacial geomorphology*. Chichester: Wiley. Pp. 231–45.

**thunderstorm**  Storm accompanied by lightning and therefore thunder, rarely heard more than about one minute after the flash. Separation of electrical charge probably demands, at some stage, water droplets going down colliding with ice crystals going up and reinforcing an initial electrostatic field. More comprehensive theories are numerous, elaborate and uncertain. The name is often used for any severe convective storm of middle latitudes. Tropical storms are rarely accompanied by thunder.  JSAG

**Reading**
Lane, F.W. 1966: *The elements rage*. Newton Abbot: David & Charles. · Ludlam, F.H. 1980: *Clouds and storms*. University Park, Pa.: Pennsylvania State University Press.

**tidal currents**  The periodic horizontal motions of the sea, generated by the gravitational attraction of the moon and sun. They are linked hydrodynamically with tidal changes of sea level and have similar spring to neap modulations. Places which have a large TIDAL RANGE also have large tidal currents. Large tidal currents may also occur where tidal ranges are small, for example near amphidromes, or through narrow straits which separate two regions having different tidal regimes. Typically, tidal currents on the continental shelf have speeds of $1\,\mathrm{m\,s^{-1}}$.  DTP

**tidal prism**  The total amount of water that flows in or out of a coastal inlet with the rise and fall of the tide, excluding any freshwater discharges. For any given period it is the product of the mean and the high- and low-water surface areas of the bays behind the inlet entrance and the TIDAL RANGE in each segment.  ASG

**tidal range**  The vertical distance between tidal low water and high water. It varies between spring and neap tides over a period of 14 days. Ocean tidal ranges are usually less than a metre, but ranges increase as the tides spread onto the shallower continental shelves. Here typical ranges are 2–5 m, but there are many local variations. In exceptional cases where large spring tides excite local hydrodynamic resonance, as in the Minas Basin of the Bay of Fundy, ranges in excess of 15 m can occur. Tidal amplitude is a half of the tidal range.  DTP

**tides**  The regular movements of the oceans and seas, generated by the gravitational attraction of the moon and sun. They are most easily observed as changes in coastal sea levels, but the associated horizontal currents are equally important for mariners. There are also tidal movements of the atmosphere and of the solid earth which are not apparent to the casual observer.

On average the gravitational attraction between the earth and moon balances the orbital centrifugal force. On the side of the earth nearest to the moon the gravitational force is slightly greater than the centrifugal force, whereas on the opposite side it is weaker. This gives two tidal maxima each day (semi-diurnal tides) as the earth rotates about its axis; however, the times of the maximum lunar tides are later by an average of 52 minutes each day because of the advance of the moon on its orbit. Changes of declination cause daily (diurnal) tides. Longer period tides are generated by varying lunar and solar distances.

Solar tides have average amplitudes which are 46 per cent of the lunar tides, but their maximum values at a site occur at the same times each solar day. Every 14 days, at new and full moon, when the lunar and solar tidal maxima coincide, the combined spring tidal range is large. Between, small neap tide ranges occur when the solar tides tend to cancel the lunar tides.

Tides calculated directly from gravitational theory, the equilibrium tides, are not observed in the ocean because of several additional effects: these include the land boundaries which prevent their westward movement, the deflection of tidal currents caused by the earth's rotation, the tidal movements of the solid earth, and the natural periods of oscillation of the oceans and shelf seas. If the natural period is near to a period in the tidal forcing, large resonant tides are produced. Because the oceans have natural periods close to 12 hours, the observed tides are predominantly semi-diurnal.

From the oceans the tides spread to the adjacent continental shelf regions. Here they are modified by amplification, by local resonances and by reflections at land boundaries. A reflected tidal wave can interfere with an incoming wave to produce zero tidal range at a distance of a quarter wavelength from the reflecting coast; because of the earth's rotation the tides circulate around an amphidromic point of zero tidal amplitude. Tidal energy is eventually dissipated by the frictional drag of the sea bed in shallow water. Schemes to use tidal power have been developed over many centuries. Early examples of tidal mills are found in East Anglia and along the coast of New England. Large schemes have been proposed for the Bristol Channel and the Bay of Fundy. The first scheme to use modern technology is La Rance, near Saint Malo in France, which began operating in 1966.     DTP

**Reading**
Cartwright, D.E. 1977: Oceanic tides. *Reports on Progress of Physics* 40, pp. 665–708. · Pugh, D.T. 1987: *Tides, surges and mean sea level.* Chichester: Wiley.

**till**     A Scottish word, popularly used to describe a coarse, bouldery soil, which was adopted to describe unstratified glacier deposits

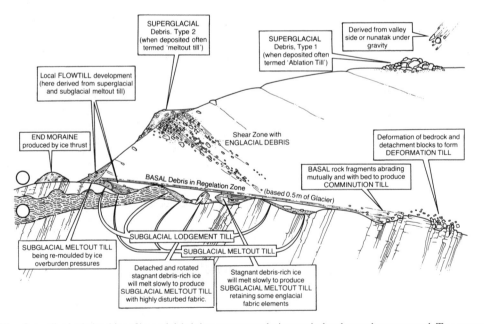

**Till**.   Generalized relationships of ice and debris in a temperate glacier: vertical scale greatly exaggerated. Two cases of end moraine formation shown: (a) glacier piles up ridge as it slides over bedrock; (b) glacier rests on thick saturated till and produces 'squeeze-up' moraines.
*Source*: E. Derbyshire, K.J. Gregory and J.R. Hails 1979: *Geomorphological processes.* London: Butterworth. Figure 5.56.

by A. Geikie (1863). The term is now understood to refer to the sedimentary material deposited directly by the action of a glacier. As such it supplants the former, oversimplistic term boulder clay.

Till covers the landsurface in many former land-bound sectors of mid-latitude ice sheets in the former USSR, northern and north-western Europe, Canada and the northern USA. It has been the focus of much interest among geomorphologists and geologists both because of its importance in the understanding of glacier activity and because of its engineering implications. Till has spawned an enormous literature including several significant symposium volumes (Goldthwait 1971; Legget 1976; *Boreas* 1977).

The table shows a recent classification of till based on the processes of debris release and the position of the debris deposition. Following Lawson (1981) the processes are subdivided into those which are primary and influence the nature of the sediment directly and those which are closely related secondary processes which modify the sediment.

*Meltout* till forms by the direct release of debris from a body of stagnant, debris-rich ice by melting of the interstitial ice. When it occurs subglacially, the debris is let down onto the underlying bed without modification. Thus it retains a fabric (see TILL FABRIC ANALYSIS) inherited from transport in the ice, with preferentially orientated pebbles. The deposit frequently consists of structureless, pebbly, sandy silt, discontinuous laminae, and bands of sediment which may be deposited over large clasts. These latter characteristics are the direct result of the melting of debris-bearing REGELATION ice. Meltout also occurs at the surface, but here secondary processes of flow and slumping may affect the sediment.

In cold, arid environments, such as Victoria Land in Antarctica, the interstitial ice may be lost by sublimation rather than by melting (Shaw 1979). This second primary process produces *sublimation* till, which also retains textural and structural characteristics inherited from glacial transport.

The third primary depositional process is subglacial and produces *lodgement* till. Particles lodge when the frictional resistance with the bed exceeds the drag of the moving ice. Since the tills are deposited beneath the weight of overlying ice, they tend to be characterized by a high degree of compaction, high SHEAR STRENGTH and low porosity. Foliation, slip planes, fine lamination and horizontal joints are also typical. The fabric of the till usually shows a preferred orientation with elongated pebbles parallel to the direction of ice flow. Occasionally, large boulders may have been moulded into minor ROCHE MOUTONNÉE shapes as a result of overriding during lodgement (Sharp 1982).

The secondary processes are intimately associated with till deposition. *Deformation* till has been deformed by glacier movement after its primary deposition. The most common occurrence is beneath glaciers where lodgement occurs. Such a layer of deformation till is massive, relatively poorly consolidated and, when saturated, easily collapses beneath the weight of a person! This is the 'mud' surrounding so many glaciers, and it is commonly up to 70 cm thick. Deformation till and lodgement till are two types of subglacial till which have in the past been called *ground moraine*.

*Sediment flows* and *gravitational slumps* are the result of surface processes which modify supraglacial debris. Sediment flows occur when fine-grained debris is exposed on the surface of a glacier (see COMPRESSING FLOW). They are frequently called *flow tills* (Hartshorn 1958; Boulton 1968). The till becomes saturated and flows down the local ice slope. Lawson (1982) studied such processes on the Matanuska Glacier in Alaska and recognized four main types of sediment flow, depending on the proportion of water they contained and the amount of sorting present. On Matanuska Glacier sediment flows account for 95 per cent of till deposition. Gravitational slumps occur when the surface debris contains insufficient fine material to flow. Instead it is stable on ice slopes up to *c.*35° but further steepening causes it to slump down the ice slope.

A further secondary process is settling through a water column. This is the situation common around the Antarctic where basal debris melts out from the bottom of an ICE SHELF or ICEBERG and falls to the sea floor. Anderson *et al.* (1980) distinguish such a glaciomarine deposit

Classification of till based on the processes of debris release and position in relation to the glacier

| | Primary | Secondary |
|---|---|---|
| Subglacial | Meltout | Deformation |
| | Sublimation | Settling through |
| | Lodgement | standing water |
| Supraglacial | Meltout | Sediment flow |
| | Sublimation | Gravitational |
| | | slumping |
| | | Settling through |
| | | standing water |

*Source:* Lawson 1981.

from normal basal tills on several criteria. The most diagnostic characteristics are the horizontal randomly orientated pebble fabrics of the dropstones and the distinctive marine microfauna. DES

**Reading and References**
Anderson, J.B., Kurtz, D.D., Domack, E.W. and Balshaw, K.M. 1980: Glacial and glacial marine sediments of the Antarctic continental shelf. *Journal of geology* 88.4, pp. 399–414. · Boreas 1977: *A symposium on the genesis of till. Boreas* 6.2. · Boulton, G.S. 1968: Flow tills and related deposits on some Vestspitzbergen glaciers. *Journal of glaciology* 7.51, pp. 391–412. · Geikie, A. 1863: On the phenomena of the glacial drift of Scotland. *Transactions of the Geological Society of Glasgow* 1, pp. 1–190. · Goldthwait, R.P. ed. 1971: *Till, a symposium*. Ohio State University Press. · Hartshorn, J.H. 1958: Flowtill in southeastern Massachusetts. *Bulletin of the Geological Society of America* 69, pp. 477–82. · Lawson, D.E. 1981: Distinguishing characteristics of diamictons at the margin of the Matanuska Glacier, Alaska. *Annals of glaciology* 2, pp. 78–84. · — 1982: Mobilization, movement and deposition of active subaerial sediment flows, Matanuska Glacier, Alaska. *Journal of geology* 90, pp. 279–300. · Legget, R.F. ed. 1976: *Glacial till: an interdisciplinary study*. Royal Society of Canada special publication 12. · Sharp, M.J. 1982: Modification of clasts in lodgement tills by glacial erosion. *Journal of glaciology* 28.100, pp. 475–81. · Shaw, J. 1979: Tills deposited in arid polar environments. *Canadian journal of earth sciences* 14.6, pp. 1239–45.

**till fabric analysis** Measurement of the direction and dip of elongated stones in glacial till (see FABRIC). Elongated stones in lodgement TILL and subglacial meltout till tend to be deposited with their long axes parallel to the direction of ice flow. DES

**Reading**
Andrews, J.T. 1971: Techniques of till fabric analysis. *British Geomorphological Research Group technical bulletin* 6. · Glen, J.W., Donner, J.J. and West, R.G. 1957: On the mechanism by which stones in till become orientated. *American journal of science* 255, pp. 194–205.

**tillite** A consolidated sedimentary rock formed by LITHIFICATION of glacial till, especially pre-Pleistocene till.

**timberline** (or treeline) The upper (altitudinal) or polar (latitudinal) margins of tree growth. Timberlines may be sharply defined or diffuse, responding to increasing climatic constraints through lower temperatures and greater exposure. Generally the trees thin out and become progressively smaller and more stunted, having KRUMMHOLZ characteristics. In the northern hemisphere, the timberline is approximately coincident with the Arctic Circle. On mountains, the timberlines generally decrease in altitude with distance from the equator,

where they are found at around 3300–4000 m, but the height varies greatly with local conditions. PAF

**timebound** A data set covering a known period of time. The term is useful when, for example in climatic data, mean values of a certain parameter are being compared. In a timebound data set, the means for different locations are derived for data covering the same period. In an un-timebound set, the means may be derived for total data sets covering a range of periods, such that the effect of a widespread climatic event (e.g. a 10-year drought) is included in the data for some locations but not all. DSGT

**tolerance** The ability of organisms to withstand environmental conditions. Plants and animals within a particular environment have limits of tolerance beyond which they cannot exist. These tolerance limits reflect a range between minimum and maximum values for essential materials such as heat, light, water and nutrients, which are necessary for growth and reproduction. The relative degree of tolerance is expressed by a series of terms that utilize the prefix 'steno' meaning narrow and 'eury' meaning wide. For example, stenothermal and eurythermal refer to narrow and wide temperature tolerance respectively. ARH

**Reading**
Odum, E.P. 1971: *Fundamentals of ecology*. 3rd edn. Philadelphia and London: W.B. Saunders. Ch. 5, pp. 106–39.

**tombolo** A bar or spit connecting an island to the mainland or to another island.

**topographic dune** A DUNE that accumulates where a sand-carrying wind encounters a hill or other obstacle that causes the sand carrying capacity to be reduced, through flow separation, leading to the deposition of sediment and the accumulation of a dune. In some circumstances a valley may also cause the transport capacity of a wind to fall, leading to aeolian deposition. (See CLIMBING DUNE; ECHO DUNE; FALLING DUNE; LEE DUNE; SAND RAMP.) DSGT

**toposequence** A sequence or grouping of related soils that differ from each other on account of their topographical position. (See also CLINOSEQUENCE and CATENA.)

**toppling failure** A type of slope failure (usually in rocks) characterized by overturning of columns of rock as they fall from a cliff. The mode of failure is common where bedding planes and joints are inclined to the valley side but dip downwards only to a maximum of around 35°. Beyond this value sliding is more-common, in which case SLAB FAILURE results. Triggering of falls may be due to water pressures in the joints or, for small blocks, ice wedging.

WBW

**Reading**
de Freitas, M.H. and Watters, R.J. 1973: Some examples of toppling failure. *Geotechnique* 23, pp. 495–514.

**topset beds** Horizontal sedimentary layers laid down on the surface of inclined beds, as in deltaic environments and aeolian sands.

**tor** 'An exposure of rock *in situ*, upstanding on all sides from the surrounding slopes . . . formed by the differential weathering of a rock bed and the removal of the debris by mass movement' (Pullan 1959, p. 54). This definition is basically the same as that used by Caine (1967), 'residuals of bare bedrock usually crystalline in nature, isolated by freefaces on all sides, the result of differential weathering followed by mass wasting and stripping'. The definition employed by Linton (1955, p. 476) introduced some genetic implications that have not been accepted as desirable: 'a residual mass of bedrock produced below the surface by a phase of profound rock rotting effected by groundwater and guided by joint systems, followed by a phase of mechanical stripping of the incoherent products of chemical action'. Some workers would not recognize a stage of prior deep chemical weathering as being a *sine qua non* for tor development, point-ing to the possible role of physical disintegrative processes or the concurrent operation of weath-ering and stripping processes on rock of variable strength (e.g. Palmer and Radley 1961).

Although Linton's definition has problems, his description of what tors are like is indeed graphic:

They rise as conspicuous and often fantastic features from the long swelling skylines of the moor, and dominate its lonely spaces to an extent that seems out of all proportion to their size. Approach one of them more closely and the shape that seemed large and sinister when sil-houetted against the sunset sky is revealed as a bare rock mass, surmounted and surrounded by blocks and boulders; rarely will the whole thing be more than a score or so feet high. (Linton 1955, p. 470)

Though he was talking about the granite tors of Dartmoor, south-west England, comparable forms occur on a wide range of rock types else-where in Britain.

ASG

**References**
Caine, N. 1967: The tors of Ben Lomond, Tasmania. *Zeitschrift für Geomorphologie* NF 4, pp. 418–29. · Linton, D.L. 1955: The problem of tors. *Geographical journal* 121, pp. 470–87. · Palmer, J. and Radley, J. 1961: Gritstone tors of the English Pennines. *Zeitschrift für Geomorphologie* NF 5, pp. 37–51. · Pullan, R.A. 1959: Notes on perigla-cial phenomena: tors. *Scottish geographical magazine* 75, pp. 51–5.

**toreva blocks** Large masses of relatively unfractured rock that have slipped down a cliff or mountain side, rotating backwards towards the cliff in doing so.

**tornado** A violent rotating storm with winds of $100 \mathrm{m\,s^{-1}}$ circulating round a funnel cloud some 100 m in diameter which includes aerial debris such as doors, bushes and frogs. It is associated with violent cumulonimbus of right-hand parity, identifiable on radar, and is a menace to the mid-western USA. Houses with closed windows may explode, due to sudden imposition of low external pressure. JSAG

**Reading**
Lane, F.W. 1966: *The elements rage*. Newton Abbot: David & Charles.

**torrent** A swift, turbulent flow of water or lava.

**tower karst** Residual limestone hills rising from a flat plain. They are distinguished from KEGELKARST in that the hills have near vertical slopes and are separated from each other by an alluvial plain or swamp. The extremely steep sides of the hills may be caused by marginal solution or fluvial erosion at the edge of the swampy plains. PAB

**Reading**
McDonald, R.C. 1976: Hillslope base depressions in tower karst topography of Belize. *Zeitschrift für Geomor-phologie* NF supplementband 26, pp. 98–103.

**tracers** A general term given to any substance that can be measured, tagged or retrieved in order to infer the operation of an environmental process. A variety of techniques have been developed to investigate the mechanisms, rates and pathways of the transfer of earth surface materials. The essence of a good tracer is that it closely mimics the environmental behaviour of the property of interest but is considerably more efficient to measure or identify. Strategies involve either seeding the phenomenon of inter-est with a suitable tracer substance or using existing distinct physical or geochemical proper-ties to interpret system dynamics. For example, river gravels can be seeded with identifiable

CLASTS which can be traced downstream. At its simplest this might involve painting a number of clasts which are mapped in subsequent surveys. Higher rates of retrieval are likely if magnets are inserted within clasts and can be detected using a search loop. Further information about the conditions under which bedload movement occurs might be obtained if individual clasts can be radio-tagged (Schmidt and Ergenzinger 1992). Process dynamics can also be inferred from pre-existing properties in environmental systems. For example, the radionuclide caesium-137, produced from nuclear weapons testing, is strongly adsorbed near the soil surface, such that its environmental mobility is largely controlled by movement of the soil material. Consequently the measurement of caesium-137 can be used to infer rates of SOIL EROSION and deposition (Ritchie and McHenry, 1990). Other properties of sediments, such as mineralogy, PARTICLE SIZE, organic content, mineral magnetic characteristics and radionuclide inventories, are often strongly related to either geological or source conditions. The properties of sediments transported can therefore be used to 'fingerprint' the relative contribution from different parts of the catchment or different sources (e.g. surface or subsurface soils, contrasting land-use types). DH

**Reading and References**
Foster, I.D.L. ed. 2000: *Tracers in geomorphology*. Chichester: Wiley. · Ritchie, J.C. and McHenry, J.R. 1990: Application of radioactive fallout cesium-137 for measuring soil erosion and sediment accumulation rates and patterns: a review. *Journal of environmental quality* 19, pp. 215–233. · Schmidt, K.-H. and Ergenzinger, P. 1992: Bedload entrainment, travel lengths, rest periods – studied with passive (iron, magnetic) and active (radio) tracer techniques. *Earth surface processes and landforms* 17, pp. 147–65.

**tractive force**   The drag force exerted when a fluid moves over a solid bed. In UNIFORM STEADY FLOW in open channels this force is the effective component of the gravity force acting on the water body in the direction of flow. For a reach of length $L$, cross-section area $A$ and slope $S$, this is $\gamma_f ALS$, where $\gamma_f$ is the unit weight of water. The average value of tractive force per unit of bed area (the mean bed shear stress, $\tau_0$) is:

$$\tau_0 = \gamma_f \, ALS/PL = {}_\gamma RS \approx \gamma_f \, dS$$

if the wetted perimeter is $P$. The simplification follows because the HYDRAULIC RADIUS $R = P/A$, and is approximated by depth $d$ in wide channels. Channel perimeter sediments have a maximum permissible tractive force or shear stress, the critical or threshold shear stress, $\tau_{0c}$. If the flow exerts a stress in excess of this,

entrainment of erosion (traction) occurs at a rate dependent on the excess stress (see DU BOYS EQUATION). It is theoretically possible to design channels to carry clear water with no sediment transport by ensuring that the perimeter sediments are everywhere at or below the threshold state: this is the tractive force theory of channel design (Richards 1982, pp. 281–6). KSR

**Reference**
Richards, K.S. 1982: *Rivers: form and process in alluvial channels*. London and New York: Methuen.

**trade winds**   Winds with an easterly component which blow from the subtropical high pressure areas around 30° of latitude towards the equator. Although only surface winds, they are a major component in the general circulation of the atmosphere as they are the most consistent wind system on earth. Together the north-east and south-east trade winds occupy most of the tropics. BWA

**transfer function**   The transfer function ($S$) is the operator that defines the relationship between the time-series (or space-series) of inputs ($X_t$) to a SYSTEM and the time-series (or space-series) of outputs ($Y_t$) from the system:

$$Y_t = SX_t$$

Thus, $S$ represents the effects of a given system upon the variable $Y$. Transfer functions may be developed through a BLACK BOX approach, in which two variables are related statistically, under the assumption that one variable ($Y$) responds to ($X$). However, transfer functions may have a strong physical basis, and may be derived from experimentation, as parameters within equations derived from first physical principles. SNL

**Reading**
Bennett, R.J. and Chorley, R.J., 1978: *Environmental systems: philosophy, analysis and control*. London: Methuen.

**translational slide**   Occurs where the failure of the soil or rock is along planes of weakness (such as bedding planes or joints) which are approximately parallel to the ground surface. This term is often used in a broad sense and can include a variety of types of mass movement such as mudflows, debris flows, etc. Solid rock movements such as wedge failures can also be translational. WBW

**transpiration**   Plant perspiration, or the loss of water vapour mainly from the cells of the leaves through pores (stomata) but also from the leaf cuticle and through lenticels of the stem. The cooling effect is secondary to the fundamental role of the transpiration, bringing a stream of water, and dissolved mineral nutrients, from the root hairs through the stem vessels (xylem) to the leaves, which is maintained by the vapour pressure gradient of the transpiring cell surfaces. The velocity of the transpiration stream varies from $1-2\,m\,h^{-1}$ in coniferous trees to $60\,m\,h^{-1}$ in some herbs. The main control mechanism on transpiration is the opening and closing of the stomata, induced by osmotic pressure changes consequent upon water balance changes due to high temperatures or other factors.                                          KEB

**Reading**
Etherington, J.R. 1982: *Environmental and plant ecology.* 2nd edn. Chichester: Wiley.

**transverse dune**   A type of sand dune, found in desert and sometimes coastal environments, that forms perpendicular to the dominant sand transporting wind direction, in environments where one general direction of sand transport occurs. Isolated BARCHAN forms or more continuous barchanoid or transverse ridges may occur, depending on the supply of sediment for dune building and the wind energy in the environment. Transverse dunes usually have distinct slip-faces and gentle stoss slopes, and migrate in a down wind direction, the rate of transport depending on wind energy and the volume of sand in the dune, with, generally, large dunes moving more slowly than small ones. Net migration rates of up to $63\,m\,yr^{-1}$, for 3 m high dunes, have been measured, while in a dune field small barchan dunes may migrate on to, or coalesce with, larger, slower forms, even forming compound megabarchans, tens of metres high, in extreme cases. In cases where extreme ($c.180°$) seasonal wind direction changes occur, the direction of migration may reverse, resulting in a reversing dune. See DUNE.                                          DSGT

**Reading**
Thomas, D.S.G. 1997: Sand seas and aeolian bedforms. In D.S.G. Thomas ed., *Arid zone geomorphology.* Chichester: Wiley. Pp. 371–412.

**transverse rib**   A narrow ridge of well-imbricated pebbles and cobbles, which lies transverse to streamflow direction. Transverse ribs normally form a series of regularly spaced ridges, apparently associated with the development of antidune breaking waves.                                          JM

**treeline**   The upper altitudinal limit to which trees can grow, and this depends on such factors as latitude, aspect, exposure and soil type. In arid areas there may be a lower treeline, the position of which is largely controlled by moisture availability, a commodity which tends to become scarcer at lower altitudes.

**triaxial apparatus**   A device used to measure the SHEAR STRENGTH of a soil according to the MOHR–COULOMB EQUATION criterion. Unlike the SHEAR BOX, failure is not along a predetermined line but takes place in a cylinder loaded axially across the ends by a compressive stress. In its simplest form (for cohesive soils only) the curved surfaces of the cylinder are at atmospheric pressure and failure produced by the axial compression gives an 'unconfined' strength. More usually, a surrounding pressure is applied to the sample, tested in a water-filled cell. This surrounding pressure is changed so that three tests give failure at different axial loads. A plot involving the applied load and cell pressure is a Mohr circle construction from which the strength of the soil can be derived. In clays a shear plane can develop in the soil but granular materials tend to bulge and a given STRAIN value is used to indicate failure (often 10 per cent).                                          WBW

**Reading**
Lambe, T.W. and Whitman, R.V. 1981: *Soil mechanics.* New York: Wiley.

**Trombe's curves**   A graph portraying the relationship between the calcium content of saturated solutions at different temperatures and the pH. As pH falls from alkali to acid conditions Trombe suggests that there is a curvilinear increase in the amount of calcium able to be held in solution. These curves are now superseded by more recent work.                                          PAB

**Reading**
Sweeting, M.M. 1972: *Karst landforms.* London: Macmillan.

**trophic levels**   Literally, 'nourishment' or feeding levels within a biological system, which represent stages in the transfer of energy through it: the concept links in with that of the FOOD CHAIN or food web. Thus, in the grazing food chain, all producer organisms (green plants, blue-green and other algae, phytoplankton, etc.) are placed in the first trophic level, which contains the maximum store of energy which is available for use in any given system; the second is comprised of HERBIVORE consumers, the third of CARNIVORE consumers, the fourth of top carnivore consumers, and so on.

The amount of energy present in any trophic level is determined by the constraints set by the first two laws of thermodynamics, the first of

| To next trophic level (k cal/m²/yr) | | | | Respiratory loss (k cal/m²/yr) |
| --- | --- | --- | --- | --- |
| Exported | Retained | | | |
| 8 | 0 | 21 | (top Carnivores) | 13 |
| 46 | 21 | 383 | (Carnivores) | 316 |
| 1555 | 383 | 8428 | (Herbivores) | 6490 |
| 405 | 8428 | 20810 | (Producers) | 11977 |

which (the law of the conservation of energy) may be expressed by:

$$\Delta E = Q + W$$

where $\Delta E$ refers to changes in the internal energy of that level, $Q$ represents the heat given off by it (mainly, the heat associated with RESPIRATION), and $W$ is the work done within it (i.e. the energy retained in cells for growth). The second states that most energy will eventually degrade to heat. The application of the theoretical relationships to real-life situations may be seen by reference to the classic series of experiments conducted at Silver Springs, Florida, by H.T. Odum (1957), a summary of the results of which is given in the diagram. In this small aquatic system the amount of energy (E) fixed, mainly through photosynthesis, in the first trophic level amounted to 20,810 kcal m$^{-2}$ year$^{-1}$; respiration (Q) was equivalent to the loss of 11,977 kcal m$^{-2}$ year$^{-1}$, and a small quotient went into a subsidiary, DECOMPOSER food chain. Some 8428 kcal m$^{-2}$ year$^{-1}$ were stored at this level (W), this being its NET PRIMARY PRODUCTIVITY or, in other words, the quantity available for passage to the second trophic level, at which point a new 'first law' balance comes into operation. The process is then repeated at subsequent trophic levels, causing a further diminution in the amount of available energy in each, until eventually all the energy within the system has been utilized. Normally, the amount of usable energy passed from one trophic level to another is of the order of 10 per cent of net production, but reference to the diagram shows that this can vary widely even in the same system, and it is known to range generally from 5 per cent to 80 per cent (Kormondy 1976). Marine systems are customarily more efficient at transferring energy between trophic levels than are land systems. Both result in a characteristic 'pyramid' of energy flow.

In view of the above the number of trophic levels which any biological system can support is clearly limited. The usual maximum on land is four or five. But in marine environments, where the ratio of plant to animal BIOMASS is much more balanced, and energy transfer efficiencies are greater, there may be up to seven. Unstable and unpredictable communities tend in general to have fewer trophic levels than do stable ones (May 1975); and so also do polluted ones. (See also BIOLOGICAL PRODUCTIVITY; ECOLOGICAL ENERGETICS.)                                            DW

**References**
Kormondy, E.J. 1976: *Concepts of ecology.* 2nd edn. Englewood Cliffs. NJ: Prentice-Hall. · May, R.M. 1975: *Stability and complexity in model ecosystems.* Princeton, NJ: Princeton University Press. · Odum, H.T. 1957: Trophic structure and productivity of Silver Springs, Florida. *Ecological monographs* 27, pp. 55–112.

**tropical cyclones**    A collective term which refers to the intense cyclonic vortices that are observed principally across tropical oceans and exhibit maximum sustained surface winds of 33 m s$^{-1}$ (64 kn). They are known as hurricanes

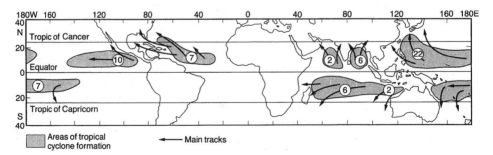

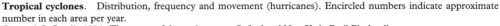

**Tropical cyclones.**  Distribution, frequency and movement (hurricanes). Encircled numbers indicate approximate number in each area per year.
*Source*: A.S. Goudie 1984: *The nature of the environment.* Oxford and New York: Basil Blackwell.

in the Caribbean, typhoons in the north-west Pacific and cyclones in the Indian Ocean. They are formed from pre-existing disturbances most frequently in the 10°–15° latitude band and are known to be favoured by the following conditions:

1 Strong low-level cyclonic relative VORTICITY.
2 A reasonably large CORIOLIS FORCE in order for an organized circulation to develop.
3 A small difference between the disturbance velocity and the vertical profile of the horizontal wind of the large-scale surrounding current (this small 'ventilation' aids the concentration of heating in a vertical column).
4 Sea-surface temperature warmer than 27 °C.
5 An unstable LAPSE RATE from the surface to middle levels.
6 High humidity at mid-tropospheric heights.

Tropical cyclones are characterized by cyclonic inflow (counterclockwise in the northern hemisphere) which is strongest in the lowest 2 km, with inward spiralling cloud bands hundreds of kilometres long and ten kilometres or so wide. These bands converge towards the deep wall cloud which surrounds the eye of the system and is the zone of strongest WINDS (up to $100 \, \mathrm{m \, s^{-1}}$ in extreme cases) and heaviest PRECIPITATION ($50 \, \mathrm{cm \, d^{-1}}$ in vigorous cyclones). The eye is a circular region in which the air subsides, winds decrease and precipitation ceases and it is typically 10–15 km in diameter.

These systems are classically asymmetric in plan view because one flank (right in the northern and left in the southern hemisphere) is characterized by stronger flow as a result of the compounding effect of the large scale steering current when 'added' to the disturbance's wind pattern. The elevated water levels of storm surges, for example, in the Gulf of Mexico and Bay of Bengal, are associated with a cyclone's low central pressure (typically 920–950 mb) and damage is also caused by the high winds and heavy precipitation which commonly occur in tropical cyclones.

The strength of the inflow decreases to a minimum at middle levels (near 6 km) where the ascent is strongest while at the top of the circulation (above 9 km) air spirals out anticyclonically (clockwise in the northern hemisphere) to balance roughly the mass flowing in at low levels. The high level outflow is marked by an extensive shield of cirrus which is clearly visible from weather satellites. In general, trop-

ical cyclones are around 650 km in diameter and thus substantially smaller than middle latitude cyclonic disturbances.

For the period 1958–77 an annual average of 54 tropical cyclones were observed in the northern and 24 in the southern hemispheres. Of this total 33 per cent occurred in the north-west Pacific, 17 per cent in the north-east Pacific, 13 per cent in the Australasian area, 11 per cent in the north-west Atlantic, 10 per cent in the South Indian Ocean and 8 per cent in both the South Indian and Pacific Oceans. The time of maximum frequency coincides with the period of highest sea-surface temperatures: 72 per cent occur from July to October in the northern hemisphere (21 per cent in September) and 68 per cent occur from January to March in the southern hemisphere (25 per cent in January).                                    RR

**Reading**
Pielke, R.A. 1990 *The hurricane*. London: Routledge.

**tropical forest**   Defined literally, an area lying between the lines of the tropics, with trees as the dominant life form. In practice, similar forests extend outside the tropics. There are three broad groups of tropical forest: evergreen, deciduous and mangrove. Evergreen forests are the most widespread, characterized by leaf-exchange mechanisms and little or no bud protection. They occur in lowlands below around 1000 m, in mountain and cloud formations and in water-saturated areas. Deciduous forests range from areas with predominantly evergreen subcanopy trees to dominantly deciduous trees throughout the vertical structure of the forest. Mangrove forests have a range of adapations to saline conditions and waterlogging.       PAF

**Reading**
Golley, F.B., Leith, H. and Werger, M.J.A. eds 1982: *Tropical rain forest ecosystems*. Amsterdam: Elsevier. · Longman, K.A. and Jenik, N. 1987: *Tropical forest and its environment*, 2nd edn. London: Longman. · Mather, A.S. 1990: *Global forest resources*. London: Belhaven Press. · Whitmore, T.C. 1975: *Tropical rainforests of the Far East*. Oxford: Clarendon Press.

**tropical rain forest**   Forest formations of the permanently moist, perhumid, tropics and subtropics. The term was originally coined (as *tropische Regenwald*) by the German botanist A.W.F. Schimper in 1898 (English translation 1903). Schimper in fact identified four major forest types within the tropics, being thorn and savanna forest at the drier end of the climatic continuum and monsoon and rain forest at its wetter end. Rain forests, then, are typically associated with areas that lack seasonality in terms of either temperature or rainfall. The tropical rain

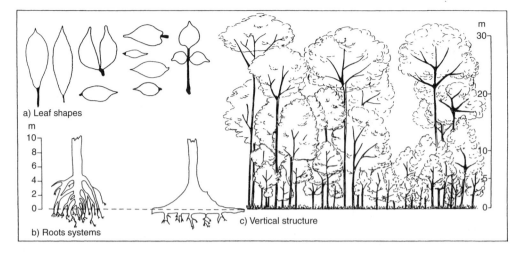

**Tropical rain forest.** Vegetation characteristics in the rain forest, showing: (a) typical leaf shapes with drip-tips; (b) stilt rooted *Uapaca* species (left) and heavy plank buttressing in *Piptadeniastrum africanum* (right); (c) profile diagram of mixed forest in Guyana (only trees above 4.9 m shown).
*Source*: Park 1992, based on originals by Goudie and Pears.

forest climate is consistently warm, mean temperature of the coolest month exceeding 18 °C and moist, with mean annual precipitation usually at least 2000 mm and no month with less than 100 mm. Tropical rain forests occur in all three tropical land areas: the Amazon and Orinoco basins of South America, the Zaire basin of West and Central Africa, and the eastern tropics of Indo-Malesia. Strictly speaking they are confined to areas between the tropics of Cancer and Capricorn, although there are localities immediately beyond the tropics which, due to particular combinations of deeper soil and groundwater, may support closed forest of equivalent diversity and structure.

Tropical rain forest is the most luxuriant of all plant communities; trees are tall, perhaps in excess of 45 m and even up to 80 m (*Araucaria hunsteinii* of the New Guinean rain forest is one of the world's tallest trees at 89 m), and form a closed, evergreen canopy. Conventionally the forest is identified as having three tree layers or strata (see figure), a group of emergents lying over a main stratum at around 20 m down to a layer dominated by smaller, shade-tolerant trees below 15 m. In terms of physiognomy, the large trees are dominated by those with clear, cylindrical boles, often with plank or stilt buttresses. Although there is great diversity in leaf shape and size (see figure), they are most frequently mesophyll in structure; pinnate forms are prominent and so-called drip-tips are also common features. Cauliflory, i.e. flowers and fruits borne directly on stems or boles, is an intriguing feature of a number of species. Large woody climbers, either free-hanging or bole-climbing, are especially abundant in these forests.

Perhaps more than any other biogeographical characteristic, it is the species diversity of tropical rain forests that has attracted most attention. It is estimated that these regions contain more than 100,000 flowering plant species, more than 40 per cent of the global flora. Locally, extraordinary richness is evident, for example, there may be as many as 180 different tree species per hectare and as many as 330 species if the other forms of plant life are included. Animal diversity reaches extraordinary proportions in the tropical rain forest, although precise numbers of species are impossible to calculate because most are insects found in the tree canopy and remain undiscovered. Terry Erwin's work in the rain forest suggests insect diversity at an unfathomable scale, with as many as 163 different species of beetle within the canopy of a single type of tree (Erwin 1983). If this is representative of insect diversity throughout the rain forests, then most estimates of global BIODIVERSITY would have to be increased by an order of magnitude or more. The factors underlying these levels of diversity have been the object of considerable debate. Earlier interpretations were based on the presumed ancient origin and environmental stability of these forests, although it is now realized that environmental fluctuations over time have impacted rain forests. Perhaps, then, the nature of such changes, together with the relatively sedentary nature of many of the constituent forest species, has favoured ALLOPATRIC speciation and resulted in elevated species numbers.

498

The issue as to how such luxuriant and diverse forests are sustained is also of interest and importance. Much has been made of the fact that high levels of BIOLOGICAL PRODUCTIVITY appear to be based on relatively infertile soils, certainly in the case of forests developed on geologically ancient land surfaces or highly leached soils. This has led to suggestions that nutrient cycling in the rain forest must be highly efficient and that the majority of nutrients are maintained within the BIOMASS itself and recirculated rapidly and efficiently by decomposition. More recently, however, it has been realized that there is great variability in the distribution of mineral nutrients in these ecosystems and a substantial proportion of the nutrients may even be supplied from external sources.

Human impact on the tropical rain forests has been very problematic and has, especially since the late 1970s, taken the form of wholesale clearance in many areas, often for subsistence or plantation agriculture. In forests associated with impoverished soils, this means not only the loss of BIODIVERSITY, but also the removal of the physically protective vegetation cover and reduction in total nutrient content. As a consequence, such agriculture is often unsustainable for more than a few seasons and the land may subsequently be abandoned and exposed to erosion. Rates of clearance are subject to much speculation, but it is clear that the effects of development in many remote areas of, for example, Rondoñia State in Amazonia, are threatening biological diversity of arguably the world's most important ecological entity.          MEM

**Reading and References**
Erwin, T. 1983: Beetles and other insects of tropical forest canopies at Manaus, Brazil, sampled by insecticidal fogging. In S.L. Sutton, T.C. Whitmore and A.C. Chadwick eds, *Tropical rain forest ecology and management.* Oxford: Blackwell. pp. 59–75. · Park, C.C. 1992: *Tropical rainforests.* London: Routledge. · Proctor, J. ed. 1989: *Mineral nutrients in tropical forest and savanna ecosystems.* Oxford: Blackwell. · Whitmore, T.C. 1991: *An introduction to tropical rain forests.* Oxford: Clarendon Press.

**tropopause**   The boundary between the TROPOSPHERE and the STRATOSPHERE, usually revealed by a fairly sharp change in the LAPSE RATE of temperature. The change is in the direction of increased atmospheric stability from regions below to regions above the tropopause. The height of the tropopause is about 20 km in the tropics and 10 km in polar regions, the decline with increasing latitude being step-like, with jet-streams occupying the steep rises. The tropopause is frequently difficult to locate, lar-

gely due to its comprising several 'leaves', giving rise to the idea of a multiple tropopause, rather than a single continuous surface.          BWA

**troposphere**   The portion of the atmosphere lying between the earth's surface and the TROPOPAUSE. Due to the varying height of the tropopause, tropospheric depths vary, on average, from 10 km to 20 km, the low values being in polar regions, the high in tropical regions. Within the troposphere the mean values of temperature, water vapour content and pressure decrease with height. Horizontal WIND speeds increase with height and VERTICAL MOTION is substantial. As a result, the troposphere is the part of the atmosphere that contains all the WEATHER we experience from day to day.          BWA

**trottoir**   (from the French word for a pavement or sidewalk) A narrow organic reef constructed by such organisms as *Lithophyllum tortuosum, Vermetidae* and *Serpulidae.* Trottoirs are common in the Mediterranean and in low latitudes and develop in the intertidal zone. ASG

**Reading**
Tzur, Y., and Safriel, U.N. 1979: Vermetid platforms as indicators of coastal movements. *Israel journal of earth sciences* 27, pp. 124–7.

**trough**   In meteorology this term virtually always relates to PRESSURE. Thus a pressure trough is an elongated area of relatively low pressure: the opposite is a ridge, of relatively high pressure. The words trough and ridge clearly derive from the valleys and ridges familiar to us in the solid earth. Troughs occur on scales ranging from mesoscale to continental, the most familiar being those appearing on the synoptic weather map on television. The very largest troughs, found within the planetary, or ROSSBY WAVES are critical to the formation of extra-tropical CYCLONES.          BWA

**truncated spur**   Steepened bluff on the side of a glacial trough in between tributary valleys. It arises through the widening and straightening of a pre-existing sinuous river valley by glacial action. Such features have long been regarded as characteristic of glacial erosion.          DES

**tsunami**   Popularly called tidal waves, are seasurface waves generated by submarine earthquakes and volcanic activity. Physically they propagate as long waves with a speed given by:

$$(water\ depth \times gravitational\ acceleration)^{1/2}$$

In ocean depths they can only be detected by sensitive bottom-pressure measurements. When they reach shallow coastal regions, amplitudes may increase to several metres. Around the Pacific Ocean, which is particularly vulnerable to tsunami, a network of tide gauges is coordinated to give advanced warning of their arrival. Tsunami damage results from flooding, and from wave impacts on coastal structures coupled with the erosion of their foundations.            DTP

**Reading**
Bernard, E.N. 1991: *Tsunami hazard. A practical guide to Tsunami hazard reduction*. Dordrecht: Kluwer.

**tufa**   A freshwater carbonate deposit which, according to Pentecost (1981: p. 365), is 'a soft, porous, calcareous rock formed in springs, waterfalls and lakes in limestone regions'. The term is often used interchangeably with *travertine*, although some authors consider one as being a special case of the other. Pentecost (1981, p. 365), for example, states that 'travertine is identical in composition to calcareous tufa but is a hard non-porous variety used for building'. *Sinter* or *calc-sinter* is another commonly used term but is usually restricted to inorganically precipitated deposits. Tufas can form significant landforms (terraces, barrages, dams etc.) and may contain much palaeoenvironmental information. Organic processes (e.g. precipitation by blue-green algae) probably play a major role in their development.            ASG

**Reading and Reference**
Pentecost, A. 1981: The tufa deposits of the Malham District, North Yorkshire. *Field studies* 5, pp. 365–87. · Viles, H.A. and Goudie, A.S. 1990: Tufas, travertines and allied carbonate deposits. *Progress in physical geography* 14, pp. 19–41.

**tuff**   Consolidated clastic material ejected from volcanoes with a predominance of fragments less than 2 mm in diameter.

**tundra**   Vast, level, treeless and marshy regions, usually with permanently frozen subsoil. Originally derived from northern Eurasia, the term has expanded to include all Arctic and Antarctic areas polewards of the TAIGA, and also to similar alpine environments above the TIMBERLINE on mountains. Drier tundra sites are characterized mainly by herbaceous perennials, with occasional trees and scattered heath plants, grasses, lichens and mosses, while cotton grass, hygrophytic sedges and willows are typical of wet sites. Cryptophytic communities develop in snow and ice.            PAF

**Reading**
Bliss, L.C., Heal, D.W. and Moore, J.T. eds 1981: *Tundra ecosystems: a comparative analysis*. Cambridge and New York: Cambridge University Press. · Ives, J.D. and Barry, R.T. eds 1979: *Arctic and alpine environments*. London and New York: Methuen.

**tunnel valleys**   Form by subglacial stream action. They tend to have flat floors, steep sides and irregular long profiles, and are a feature of northern Germany and Denmark.

**tunnelling, tunnel gully erosion**   A form of erosion, initiated by subsurface water movement, which often causes surface collapse, leading to open gullying. Water movement through soil cracks eluviates material, thereby leading to the development of tunnels which continue to erode as gullies, following tunnel collapse. It is thus related to PIPES.            ASG

**Reading**
Lynn, I.H. and Eyles, G.O. 1984: Distribution and severity of tunnel gully erosion in New Zealand. *New Zealand journal of science* 27, pp. 175–86.

**turbidite**   A sedimentary sequence, fining upwards from coarse sands to clays, deposited by a turbidity current. These submarine flows of sediment are usually triggered on continental slopes by earthquakes, with deposition covering extensive areas of the continental rise and abyssal plains.            PSh

**turbidity current**   A density current. A sinking mass of sediment-laden air or water. Their erosive activity is thought to contribute to the formation of some submarine canyons on the continental shelves. Sediments deposited by turbidity currents are known as turbidites.

**turbulence**   In wind or water, in contrast to LAMINAR FLOW which occurs in parallel layers, turbulence consists of a series of apparently random, quasi-periodic eddies of differing size and velocity which mix the flow over a wide range of scales. Deemed by Bradshaw (1971) the most common, most important and most complicated kind of fluid motion it is significant in all natural fluid flows at scales ranging from the ENTRAINMENT of individual DUST particles in the atmosphere to the large-scale turbulent currents in the oceans.

Turbulence may be viewed as a cascading energy transfer (Clifford and French 1993) whereby energy from the mean flow is extracted by large eddies and dissipated into small (microscale) eddies. The range in eddy sizes allows the local flow direction at a point within TURBULENT FLOW to be different from that of the mean flow direction.            GFSW

**References**
Bradshaw, P. 1971: *An introduction to turbulence and its measurement*. Oxford: Pergamon. · Clifford, N.J. and French, J.R. 1993: Monitoring and modelling turbulent flow: historical and contemporary perspectives. In N.J. Clifford, J.R. French and J. Hardisty eds, *Turbulence:*

*perspectives on flow in sediment transport.* Chichester: Wiley. Pp. 1–34.

## turbulent flow

A fluid flow (air or water) characterized by a mean forward direction but consisting of a series of eddies of various sizes moving in a random manner. The exchange of momentum throughout a turbulent BOUNDARY LAYER is achieved through the mixing motion of gusts and turbulent eddies. Such momentum exchange is far more efficient than the molecular exchange seen in LAMINAR FLOW and this is represented by a steeper velocity gradient, and hence higher SHEAR STRESS toward the surface. Turbulent flow is characterized by a logarithmic increase in velocity away from the surface as a result of bed ROUGHNESS producing a drag at the base of the boundary layer.

Turbulent flow arises when the flow inertia swamps the effects of fluid viscosity and the REYNOLDS NUMBER exceeds 2000. Natural windflow is almost always turbulent because air has a low viscosity and boundary layer depths are high. Streamflow may not be turbulent in thin or very slow flows.

The measurement of turbulent flow requires time-averaged measurements of velocity which smooth out the instantaneous peaks and troughs of the eddies, which are themselves a measure of the turbulence intensity of the flow. A fluid flow with a high turbulence intensity is more efficient at eroding and transporting sediment as maximum peaks in velocity will be much greater than the mean velocity. GFSW

### Reading
Clifford, N.J., French, J.R. and Hardisty, J. 1993. *Turbulence: perspectives on flow in sediment transport.* Chichester: Wiley. · Wiggs, G.F.S. 1997. Sediment mobilisation by the wind. In D.S.G. Thomas ed., *Arid zone geomorphology.* 2nd edn. London: Wiley. pp. 351–72.

## turbulent flow structures

TURBULENT FLOW is composed of a hierarchy of coherent flow structures or EDDIES. A coherent eddy may be defined as an event in a turbulent flow that has a repetitive and correlatable structure to its observed and detectable flow behaviour, such as correlation in space and time of velocity or the intensity of turbulence. Turbulent flow structures exist at a range of scales in all turbulent fluids, from the smallest scales, which are limited by viscous dissipation, to vortices that may scale with the largest dimensions of the flow, for instance flow depth or BOUNDARY LAYER thickness. Turbulent flow structures generated in the basal zones of a fluid include patches of low momentum fluid that are uplifted from near the wall ('ejection' or 'burst' events), packets of high momentum fluid that are sourced from the outer flow and move towards the bed ('sweeps') and flow-parallel longitudinal vortices (streaks). These motions are responsible for generating a large proportion of the turbulent Reynolds stresses within a flow and are critical to sediment transport. Large-scale turbulent flow structures include those that are generated along layers of shear between fluids moving at different velocities or in different directions. These include shear layer instabilities generated around flow separation zones in the leeside of BEDFORMS such as CURRENT RIPPLES or DUNES, the sheltered region of air flow behind forest canopies or islands, and shear generated along the mixing interface between two flows, such as at open-channel junctions and in the Gulf Stream. These largest coherent flow structures may have dimensions ranging from centimetres to kilometres. JLB

### Reading
Ashworth, P.J., Bennett, S.J., Best, J.L. and McLelland, S.J. eds 1996: *Coherent flow structures in open channels.* Chichester: Wiley. · Clifford, N., French, J.R. and Hardisty, J. eds 1993: *Turbulence: perspectives on flow and sediment transport.* Chichester: Wiley.

## typhoon

See TROPICAL CYCLONES.

# U

**ubac** The side of a hill or valley that is most shaded from the sun.

**unconformity** A discontinuity between sedimentary strata which testifies to a temporary interruption in the process of accumulation.

**underfit stream** A stream which is much smaller than expected from the size of its valley. An underfit stream could have occurred as a result of river capture when the beheaded CONSEQUENT STREAM would be smaller than expected. However, Dury (1977) has shown that underfit streams are a widespread occurrence, that they reflect the impact of climatic change, and that the wavelength of VALLEY MEANDERS may be three to ten times greater than the wavelengths of the underfit stream meanders. *A manifestly underfit stream* is an underfit stream which meanders within a more amply meandering valley and an *osage type* of stream has a much smaller pool–riffle spacing than would be expected from the size of the valley meanders.                              KJG

### Reference
Dury, G.H. 1977: Underfit streams: retrospect, perspect and prospect. In K.J. Gregory ed., *River channel changes*. Chichester: Wiley. Pp. 281–93.

**underplating** A potentially highly important tectonic process caused when magma generated over a mantle plume is accreted to the base of the crust. According to the underplating model, the addition of volcanic rock in this way thickens the crust and the resulting isostatic adjustment leads to the formation of a broad *hot-spot* swell up to 2000 km across and with an increase in surface elevation of up to 2000 m. This has substantial implications for river network evolution.                              ASG

### Reading
Summerfield, M.A. 1991: *Global geomorphology*. London and New York: Longman Scientific and Technical and Wiley.

**uniclinal** Pertaining to a formation of rock strata which dip uniformly in one direction.

**uniclinal shifting** The process whereby a stream or river flowing in an asymmetric valley in an area of gently dipping rocks migrates down the dip slope of the valley, cutting back the steeper scarp slope.

**uniform steady flow** Exists when the water depth is equal at every section in a channel reach. Unsteady uniform flow would require the water surface to remain parallel to the channel bed as discharge changes, which is practically impossible. Accordingly, spatially uniform flow is temporally steady. Discharge, and flow depth, width, cross-section area and velocity are all constant from section to section, and the ENERGY GRADE LINE, water surface and bed profile are all parallel. This is rare in natural channels with variable width and POOL AND RIFFLE bedforms.

The CHÉZY EQUATION and MANNING EQUATION define the mean velocity of uniform flow as a function of depth, slope and ROUGHNESS, and are often applied to short natural river reaches where uniform flow can be assumed. If local velocities at every point in the cross-section are constant along a reach, the entire velocity distribution is uniform, the turbulent BOUNDARY LAYER is fully developed, and the logarithmic vertical velocity profile occurs.                              KSR

**Uniformitarianism** A practical tenet held by all modern sciences concerning the way in which we should choose between competing explanations of phenomena. It rests on the principle that the choice should be the simplest explanation which is consistent both with the evidence and with the known or inferred operation of scientific laws. Uniformitarianism is therefore applicable to both historical inference (or 'postdiction') and to prediction of the future outcome of the operations of natural processes (Goodman 1967). It is, in consequence, as Shea (1982, p. 458) has forcibly emphasized, a concept 'with no substantive content – that is, it asserts nothing whatever about nature. Uniformitarianism must be viewed as telling us how to behave as scientists and not as telling nature how it must behave.'

In physical geography Uniformitarianism is usually linked with James Hutton's demonstration (1788) that the simplest explanation for the nature of the earth's surface topography and rock strata was not the invocation of divine intervention at a single moment of creation and then again by the biblical flood, but rather the assumption that processes of erosion, lithification and uplift comparable to those whose operations could be observed or inferred in the modern world, acting over immensely long periods of time, were responsible. Shea (1982) points out that Hutton was not the first to adopt this viewpoint, but as he certainly was the first to provide an extensive working-out of its implications it is reasonable to regard him as the founder of modern earth science. Hutton's views conflicted sharply with those of other natural philosophers, notably Werner and Kirwan, who came to be termed Catastrophists (see Chorley *et al.* 1964). These latter produced interpretations, often incredibly complex, of rocks and relief which derived from an implicit belief that the biblical chronology was sacrosanct and that God had intervened directly to control the mechanisms of earth sculpture.

One of the major sources of confusion which has come to enshroud discussions of Uniformitariansim in earth science in general, and in geomorphology in particular, has derived from a change in the interpretation of 'Catastrophism'. Increasingly, the term has been taken to imply a belief that large, sudden and (to human eyes) 'catastrophic' events have more significance in earth history than the slow and virtually continuous operation of 'normal' processes (see e.g. the extraordinary influential paper by Wolman and Miller 1960). This change in meaning has left Uniformitarianism apparently standing for a view in which the *simplest* explanation is equated with that which requires the slowest and/or most *uniform* rate of operation of processes: this fallacy – termed 'gradualism' (see Hooykaas 1963) – is a complete misinterpretation of the Uniformitarian tenet. Moreover, any careful reading of Hutton, or Playfair (1802), or any edition of Lyell's *Principles of geology* (first published 1830–3), make it abundantly clear that all those early Uniformitarians ascribe a very important role to what would now be termed 'high magnitude, low frequency' events. This tendency becomes particularly marked in the later editions of Lyell's *Principles* (e.g. the 9th, 1853) as his congenital reluctance to believe that 'normal' fluvial processes are *actually* responsible for substantial earth sculpture, leads him to an increasing emphasis on the role of sudden, large and locally 'catastrophic' occurrences. It cannot be emphasized too forcibly that Uniformitarianism does not, as a principle, require any presupposition about the rates of operation of processes, other than those limits apparently fixed by the laws of physics and chemistry.

Nor does the concept involve – as another fallacy proposes – the belief that only processes which can actually be observed in operation may be properly invoked as explanations. In consequence, it equally does not assume that the nature and rates of operation of processes have remained unchanged over time. It is, for example, both apparent and entirely consistent with the Uniformitarian tenet, that the nature and rates of processes on the earth must have been very different from today either before the evolution of land plants, or at the height of one of the Pleistocene glacial advances.

Probably the most instructive example of the application and misapplication (or misconstruction) of the Uniformitarian principle, and of the conflict which can be generated, is the case of J.H. Bretz (1923) and his interpretation of the channelled scabland of the north-west USA as the product of an almost unimaginably large flood. Baker (1981) has provided a fascinating (and sobering) collection and discussion of the major papers in the channelled scabland debate, which should be required reading for all geomorphologists.

The essence of this controversy concerns the most probable origin of the huge complex of deep channels cut through loess and basalt in the Columbia Plateau region of eastern Washington State (including the site of the Grand Coulee Dam). Using the evidence from painstaking field studies, Bretz in a series of papers from 1923 onwards argued that the simplest interpretation of the data called for a great flood or 'debacle', which cut channels over a $40,000\,km^2$ area. The source of the water for this 'Spokane flood', Bretz found in the site of glacial Lake Missoula: it was suggested that the ice damming the lake had been suddenly breached, releasing some $500\,mile^3$ of water into the scabland tract. Bretz's explanation was entirely consonant with Uniformitariansim. He observed scabland features which, while huge, were clearly products of running water and all that was required to explain them was, therefore, a way of providing a very large flow of water in a very short period. Glacial lake dam bursts are well-documented occurrences.

The Spokane flood theory was attacked – and very viciously attacked – on the grounds both that the proposed explanation smacked far too much of the Diluvial Catastrophists' interpretations of *all* valleys in terms of the mighty waters of Noah's flood and, in addition, because no flood as large as the one hypothesized had ever been observed. Both lines of argument depend upon fallacious interpretations of the Uniformitarian principle. Their proponents, attempting to provide 'permissible' explanations for the channelled scablands without Bretz's flood, tied themselves in increasingly complicated knots.

Ironically, it was ultimately rather small-scale and geomorphologically undramatic evidence to which the Uniformitarian principle was correctly applied, which led to the vindication of Bretz's earlier hypothesis. The vital evidence consisted of the discovery of giant current ripple marks both within the Grand Coulee area and on the floor of the former Lake Missoula. The hydraulic and hydrodynamic relationships between depth and velocity of water movement and the dimensions of bedforms such as current ripples are, in fact, so well established that the simplest and therefore Uniformitarian explanation for the giant examples was a water body with all the characteristics of Bretz's Spokane flood.

While the details of Bretz's explanation of the channelled scabland have been modified by later studies (there were, it seems, several different dam bursts and floods), in essence his 1923 views have been accepted. Indeed, his 'outrageous' mechanism has since been used to explain the apparently similar 'channelled scablands' seen on the planet Mars (see Baker 1981), as it is a very sound application of the Uniformitarian tenet to assume that direct analogy may be the simplest explanation of apparently directly analogous forms.

Nevertheless, it must be emphasized again that Uniformitarianism is a guiding tenet of science and *not* a rule of nature. As theories about the operation of nature change, so it is possible – and, indeed, inevitable – that one 'Uniformitarian' explanation will come to replace another. BAK

**Reading and References**

Baker, V.R. ed. 1981: *Catastrophic flooding*. Stroudsburg, Pa.: Dowden, Hutchinson & Ross. · Bretz, J.H. 1923: The channelled scablands of the Columbia Plateau. *Journal of geology* 31, pp. 617–49. · Chorley, R.J., Dunn, A.J. and Beckinsale, R.P. 1964: *The history of the study of landforms*. Vol. I. London: Methuen. · Goodman, N. 1967: Uniformity and simplicity. In C.C. Albritton ed., *Uniformity and simplicity*. Geological Society of America special paper 89, pp. 93–9. · Hooykaas, R. 1963: *Natural law and divine miracle: the principle of uniformity in geology, biology and theology*. Leiden: E.J. Brill. · Hutton, J. 1788: Theory of the earth; or an investigation of the laws observable in the composition, dissolution and restoration of land upon the globe. *Transactions of the Royal Society of Edinburgh* I, part II, pp. 209–304. · Lyell, C. 1830–3: *Principles of geology*. 3 vols. London: John Murray. · — 1853: *Principles of geology*. 9th edn. London: John Murray. · Playfair, J. 1802: *Illustrations of the Huttonian theory of the earth*. London: Cadell & Davies. · Shea, J. 1982: Twelve fallacies of Uniformitarianism. *Geology* 10, pp. 455–60. · Wolman, M.G. and Miller, J.P. 1960: Magnitude and frequency of forces in geomorphic processes. *Journal of geology* 68, pp. 54–74.

**unit hydrograph** A characteristic or generalized hydrograph for a particular drainage basin.

A unit hydrograph of duration $t$ is defined as the hydrograph of direct run-off resulting from a unit depth of effective rainfall generated uniformly in space and time over the basin in unit time. The unit depth was originally one inch but is now usually one centimetre and $t$ is chosen arbitrarily according to the size of the basin and to the response time to major events and can be 1, 6 or 16 hours, for example. The technique was developed by L.K. Sherman in 1932 and it has been used to predict hydrographs for the engineering design of reservoirs, flood detention structures and urban stormwater drainage. Since many streams are still ungauged, the discharge records from all stations in an area can be analysed and synthetic unit hydrographs developed from the relations between unit hydrograph parameters and drainage basin characteristics. The drainage basin characteristics of the basin above an ungauged site can then be used to obtain the synthetic unit hydrograph for that site. To compare drainage basins of different areas dimensionless unit hydrographs can be constructed with the discharge ordinate expressed as the ratio to the peak discharge and the time ordinate expressed as the ratio to the lag time. The instantaneous unit hydrograph is a mathematical abstraction produced when the duration of the effective precipitation becomes infinitesimally small and this is used in the investigation of rainfall–run-off dynamics. Unit hydrograph theory depends upon a number of assumptions including the HORTON OVERLAND FLOW MODEL and, with the advent of the PARTIAL AREA MODEL of run-off formation, it has been necessary to revise the use and analysis of the unit hydrograph. KJG

**Reading and Reference**

Dunne, T. and Leopold, L.B. 1978: *Water in environmental planning*. San Francisco: W.H. Freeman. Pp. 329–50. · Shaw, E.M. 1983: *Hydrology in practice*. Wokingham: Van Nostrand Reinhold. Pp. 326–44. · Sherman, L.K. 1932: Stream flow from rainfall by the unit graph method. *Engineering news record* 108, pp. 501–5.

**unit response graph** The theoretical quickflow hydrograph produced by 1 inch of effective rainfall and derived from the actual quickflow hydrography by assuming a linear extension such as is carried out in the derivation of simple unit hydrographs (Walling 1971). The derivation of a unit response graph is similar to the derivation of a unit hydrograph but flow separation is based upon the method proposed by Hibbert and Cunningham (1967) (see HYDROGRAPHS) and each unit response graph will vary in shape in relation to the contributing area generating storm run-off.

Classic unit hydrograph theory assumes that the whole catchment contributes to storm run-off and so the form of the hydrograph will reflect the characteristics of the whole catchment and any variation in hydrograph form will result entirely from variations in the time distribution of effective rainfall. Hydrograph separation can be achieved using any consistent technique although Linsley *et al.* (1982) stress the importance of using a separation technique which ensures that the time base of direct or storm run-off remains relatively constant from storm to storm. An hour unit hydrograph will be storm run-off response to a unit of effective rainfall (usually 1 cm or 1 in) falling with even intensity over the entire catchment during a period of $n$ hours. Because of the assumed linear relationship between effective rainfall and storm run-off, it is possible to derive standard hydrographs for different rainfall intensities and for different rainfall durations by applying simple transformations to a unit hydrograph or to a derivative of a unit hydrograph such as an *S*-curve or an instantaneous unit hydrograph. The unit hydrograph concept is explained in detail by Linsley *et al.* (1982).

In the case of the unit response graph, the magnitude of the ordinates of the quickflow hydrograph are adjusted so that there is a volume of run-off equivalent to a unit of effective rainfall over the catchment, but unit response graphs to storms of the same duration would not be expected to have the same form because of the influence of the size and shape of the contributing area of the speed with which water can drain from the catchment.          AMG

**References**
Hibbert, A.R. and Cunningham, G.B. 1967: Streamflow data processing opportunities and application. In W.E. Sopper and H.W. Lull eds, *International symposium on forest hydrology.* Oxford and New York: Pergamon. Pp. 725–36. · Linsley, R.K., Kohler, M.A. and Paulhus, J.L.H. 1982: *Hydrology for engineers.* 3rd edn. New York: McGraw-Hill. · Walling, D.E. 1971: Streamflow from instrumented catchments in south-east Devon. In K.J. Gregory and W.I.D. Ravenhill eds, *Exeter essays in geography.* Exeter: University of Exeter. Pp. 55–81.

**unloading**   The stripping of rock or ice from a landscape and the resulting effects the release of pressure has on the exhumed landsurface.

**unmanned earth resources satellites**   Satellites carrying a range of REMOTE SENSING devices for the production of images of the earth's surface. There are five groups of unmanned earth resources satellites. Groups one and two both record radiation in visible and near visible wavelengths. Group one comprises the Landsat series, which were the first generation of earth resources satellites, and group two comprises the second generation of earth resources satellites and includes SPOT. Group three carries sensors that record thermal wavelengths and includes HCMM; group four carries sensors that record microwave wavelengths including Seasat, ERS-1 and Radarsat; and group five comprises the 'polar platform' satellites that will be providing physical geographers with a major source of environmental data until well into the next century.

### The Landsat series

After the success of MANNED EARTH RESOURCES SATELLITES the National Aeronautics and Space Administration (NASA) of the USA and the US Department of Interior developed an experimental earth resources satellite series to evaluate the utility of images collected from an unmanned satellite. The first satellite in the series carried two types of sensor, a four waveband multispectral scanning system (MSS) and three return beam vidicom (RBV) television cameras. When launched in July 1972 it was called ERTS, the Earth Resources Technology Satellite, a name that it held until January 1975 when it was renamed Landsat. The main advantages offered by the imagery collected from this satellite were: ready availability, low cost, repetitive multispectral coverage and minimal image distortion. Landsat 1 had a high and fast orbit at an altitude of 900 km and a speed of 6.5 km s$^{-1}$. The orbit was circular, flying within 9° of the north and south poles and sun-synchronous as it kept pace with the sun's westward progress as the earth rotated. To obtain repeat coverage of an area the orbits were moved westwards each day and this enabled an image to be taken of each area of the earth's surface every 18 days. Landsat 1 lasted for almost six years until January 1978 and for part of its life shared the heavens with Landsat 2 which was launched in 1975. Landsats 3, 4 and 5 were launched in 1978, 1982 and 1984. Landsats 6 and 7 were launched in 1993 and 1997. For Landsats 4 and 5 the satellite body was changed to increase stability and payload capability and the orbit altitude was lowered to 705 km, thus giving a faster repeat cycle of 16 days and a changed orbit spacing.

The Landsat satellites all carry, or have carried, two sensors: a multispectral scanning system (MSS) and either a thematic mapper (TM) or RBV television cameras. The MSS records four images of a scene, each covering a ground area of 185 × 185 km at a nominal spatial resolution of 79 m. These four images cover green, red, near infrared and infrared wavebands and were identified by the channels they occupied in the satellite's telemetry system, which were 4, 5, 6 and 7 respectively. The MSS sensor has undergone very little change since the launch of the first Landsat. The three important changes are, first, the addition of an extra waveband, known as band 8, to the MSS of Landsat 3. This recorded thermal infrared images but as it failed shortly after launch few images have been used. Secondly, to compensate for the lower orbit altitude of Landsats 4 and 5, the spatial resolution of the MSS images was decreased by 3 m to 82 m and the field of view was increased by 3.41° to 14.93°. Thirdly, the numbering of the MSS wavebands was modified from Landsat 4 onwards. Landsat MSS data were initially used to obtain a synoptic view of a large area of the earth's surface for visual interpretation. Today, owing to the availability of DIGITAL IMAGE PROCESSING, digital Landsat MSS data are frequently used for classifying land cover, estimating characteristics of the earth's surface and for monitoring change.

The thematic mapper (TM) carried by Landsats 4 and 5 records image areas of 185 × 185 km in seven wavebands with a spatial resolution of around 30 m in six of them. The TM is an important sensor for physical geographers and is used extensively throughout the subject.

Return beam vidicom (RBV) television cameras were carried on Landsats 1, 2 and 3. On Landsats 1 and 2, three cameras were used, each filtered into a different waveband, camera 1 into green, camera 2 into red and camera 3 into near infrared. Unfortunately, the RBV on board Landsat 1 returned only 1690 images before it was turned off in August 1972. The RBV on Landsat 2 returned even fewer images and so for Landsats 1 and 2 the MSS images were their primary product. Landsat 3 carried two RBVs and these were both filtered to a broad red to near infrared waveband. Their design was similar to the RBVs carried by Landsats 1 and 2 except for their focal length, which was increased to give a nominal ground resolution of around 30 m and an image area of 98 × 98 km.

### Second generation of earth resources satellites

Like the Landsat satellites, these will carry optical sensors but they will be linear array multispectral scanners. The satellites so far named include the French satellite Système Probatoire d'Observation de la Terre (SPOT). SPOT 1 and SPOT 2 were launched in 1986 and 1990 and SPOT 3 and 4 were launched in 1993 and 1996. The satellites are operated by the Centre National d'Etudes Spatiales and have a near-polar, sun-synchronous, 832 km high orbit,

which will provide repeat ground coverage every 26 days. It carries two high-resolution visible (HRV) sensors, two tape recorders and telemetry equipment to transmit data to earth. The HRV records an area of 60 × 60 km with a spatial resolution of 10 m in one panchromatic waveband or 20 m in green, red and near infared wavebands. These images can be recorded obliquely to give a stereoscopic view of the terrain, avoid patchy cloud and decrease the revisit time.

Japan has launched two earth resources satellites. The first, in 1990, was the Marine Observation Satellite (MOS-1) and the second, in 1992, was the Japanese Earth Resources Satellite (JERS-1). MOS-1 has an orbit and altitude similar to the Landsat series of satellites and carries three sensors, the most important of which is a linear array multispectral scanner named the Multispectral, Electronic, Self-scanning Radiometer (MESSR). This senses in four wavebands from green to near infrared, has a spatial resolution of 50 m and an image area of 200 × 200 km. JERS-1 carries a stereoscopic linear array sensor and also a SIDEWAYS LOOKING AIRBORNE RADAR (type SAR).

The Indian Remote Sensing (IRS) satellites carry Linear Imaging Self-scanning (LISS) multispectral scanners with design specifications similar to those of the sensors on board Landsat. IRS-la and IRS-lb were launched in 1988 and 1991 with further IRS launches in 1993, 1994 and 1996.

### Satellites carrying thermal sensors

Landsats 3, 4 and 5 and the Heat Capacity Mapping Mission (HCMM) carried thermal sensors with a low spatial resolution. The HCMM satellite was launched in April 1978 and lasted until September 1980. Its orbit was near-circular at an altitude of 620 km. The satellite contained a scanning radiometer which recorded in a visible and near infrared waveband and a thermal infrared waveband. The orbits of the satellite were arranged to ensure that images were obtained of each scene during times of maximum and minimum surface temperature for the determination of thermal inertia. The multispectral scanner had a very wide scan angle of 60° resulting in an image width of 720 km. The spatial resolution decreased from around 0.6 km at the centre of the image to around 1 km at the edge of the image. The data from the HCMM sensors were intended for geological mapping but they have also been used for microclimatology, pollution monitoring and hydrology.

### Satellites carrying microwave sensors

The first unmanned earth resources satellite to carry a sideways looking airborne radar (type

SAR) was Seasat. This was an experimental satellite designed by NASA to establish the utility of microwave sensors for remote sensing of the oceans. Images of the land were also obtained giving physical geographers a synoptic view of the earth in microwave wavelengths. Seasat had a circular non-sun-synchronous orbit at an altitude of 800 km, sensing the earth's surface from 72° N to 72° S, orbiting the earth 14 times a day and passing over the same area every 152 days. The satellite carried two sensors of potential interest to physical geographers: a multispectral scanner and a sideways looking airborne radar (type SAR). The multispectral scanner recorded in two wavebands, visible at a spatial resolution of 2 km and thermal infrared at a spatial resolution of 4 km. The sideways looking airborne radar (type SAR) produced images of 100 km wide swaths with a nominal spatial resolution of 25 m. These data have been used for many applications such as the monitoring of sea state and the mapping of vegetation, sea ice and urban form.

The first remote sensing satellite to be launched by the European Space Agency (ESA) was the Earth Resources Satellite (ERS-1) in 1991. It was launched into a sun-synchronous orbit at an altitude of around 700 km with a great cycle of three days. It carries several sensors, the most important as far as physical geographers are concerned being the sideways looking airborne radar (type SAR). The Japanese Earth Resources Satellite (JERS-1), launched in 1992, carries a sideways looking airborne radar (type SAR). This sensor records at longer wavelengths than that on board ERS-1 and can be used to sense beneath vegetation canopies.

Canada launched the satellite Radarsat in 1995. It had a sideways looking airborne radar (type SAR). The major application of data collected from the satellite was for mapping ice, especially in the offshore oil drilling areas of northern Canada.

### Satellites carrying a suite of environmental sensors

A series of major earth resources satellites is being developed. These 'polar platforms' will carry suites of sensors that will enable the simultaneous estimation of fundamental environmental processes on land, in the oceans and in the atmosphere. The first two of these satellites were launched in 1998. The US Earth Observing System (EOS) satellite AM-1 had four sensors and the ESA Polar Orbit Earth Observing Mission (POEM) satellite EN-VISAT seventeen sensors.                                   PJC

**Reading**
Barrett, E.C. and Curtis, L.F. 1992: *Introduction to environmental remote sensing*. 2nd edn. London and New York: Chapman & Hall. · CEOS 1992: *The relevance of satellite missions to the study of the global environment*. Committee on Earth Observation Satellites. London: British National Space Centre. · Cracknell, A.P. and Hayes, C.W.B. 1991: *Introduction to remote sensing*. London: Taylor & Francis. · Curran, P.J. 1985: *Principles of remote sensing*. Harlow: Longman Scientific and Technical. · Mather, P.M. ed. 1992: *TERRA-1 understanding the terrestrial environment*. London: Taylor & Francis.

**unstable channels** A river or tidal channel that is shifting through erosion and deposition. Some writers restrict the term to those that are shifting rapidly, changing their pattern or adjusting to changed conditions, because many channels quite normally shift their courses without being in a state of imbalance or disequilibrium with environmental controls. By contrast, engineering design commonly aims to achieve stable channels or canals that remain fixed in position and which will not require costly maintenance works. (See also CHANNELS.)        JL

**unstable equilibrium** If we had two spheres, one larger than the other, and we placed the smaller one upon the larger one in such a way that, upon letting it go, it remained where we put it, then, in the absence of any disturbing force, the two spheres would be in equilibrium. But we are all aware that it would be extraordinarily difficult to achieve the above result and that, even should we succeed, the merest hint of a breath would disturb the equilibrium, sending the smaller sphere increasingly further away from its original position in equilibrium. Thus we had initially a state of unstable equilibrium. In many natural systems, and particularly in fluids, this type of equilibrium may exist. For example, very warm parcels of air may remain near the ground until some small disturbance triggers their release.        BWA

**unsteady flow** Occurs in an open channel (e.g. a river or canal) when the depth and discharge of water at different points along a reach change through time because of the passage of a flood wave or surge along the channel. Simplified FLOW EQUATIONS cannot be applied to such translatory wave processes. Analysis of changes of flow conditions at a section, or of the shape of the flood wave as it travels downstream, therefore require the application of wave theory or FLOOD ROUTING methods.        KSR

**upper westerlies** These refer to the global circulation pattern, above the boundary layer, in the mid-latitudes (30° to 60° of latitude), of both the northern and southern hemispheres. The boundary layer extends upward at least 1 kilometre above the surface. In this layer winds are significantly affected by surface friction, which decreases with height above the surface. Depending on the topography, the 'upper air' roughly begins between 850 and 500 millibars (mb) and extends upwards to the tropopause. The upper westerlies form a river of air blowing generally from west to east, which separates cold polar air masses from warmer subtropical air masses. Incursions of polar air moving equatorward (troughs) and warm air moving poleward (ridges) create a wavelike pattern called Rossby waves (also called longwaves or stationary waves). Above the boundary layer, the winds are approximately geostrophic where the pressure gradient force is perfectly counterbalanced by the CORIOLIS effect. Resultant winds consequently blow parallel to the isobars. The upper level flow tends to steer surface low-pressure systems around the ridges and troughs. Surface storms are also generated by upper level disturbances (shortwaves) superposed on the Rossby waves. NJS

**upwelling** The vertical movement of deeper water towards the sea surface. It occurs where a divergence of surface currents must be compensated for by vertical flow, e.g. wind-driven offshore currents may be balanced by coastal upwelling. This deeper water is often rich in nutrients, which allow a high productivity of phytoplankton near the surface. As a result, many of the world's most important fisheries are in areas of upwelling. These include the seas off north-west Africa, Oregon and Peru. Every five years or so the Peruvian upwelling is inhibited when the tropical Pacific Ocean responds to a relaxation of the trade winds. The phenomenon, known as the EL NIÑO effect, has serious consequences for the Peruvian fishing industry. DTP

**ural (type) glacier** A small glacier developed in the lee of prevailing winds of a mountain or plateau. Snow is deposited in a 'rotor' in the lee (as a large snowdrift). A glacier formed in this way may continue to exist even though the mountain and its ural glacier may be below the regional snowline. Named after the Ural mountains but other examples exist, e.g. in northern Iceland and in the Colorado Rockies (USA). WBW

**uranium series dating** Uranium-series dating methods are numerical dating techniques which exploit the properties of the RADIOISOTOPIC decay of uranium. They are applicable to a variety of materials over the time range from $10^4$ to $10^6$ years and represent a valuable component of GEOCHRONOLOGY investigations. They may be used to date a range of inorganic (e.g., secondary carbonate and salt) and biogenic (e.g., enamel, bone, shell) deposits. The principle of the method is based on the observation that uranium in surface sediments is typically bound in silicate and oxide minerals. Over time (a few million years) it will approach secular equilibrium with its decay chain protégé products. Weathering of uranium-bearing minerals produces water-soluble complexes that may become separated from their less soluble protégé products. When this mobile uranium precipitates (inorganically or biogenically) as a trace constituent of surficial minerals it develops a new succession of protégé isotopes. There are two uranium decay series, each commencing with a different parent uranium ISOTOPE; $^{238}U$ and $^{235}U$. Each series undergoes a series of transitions to a stable isotope of lead. Methods employing $^{230}Th$ (the protégé isotope of $^{234}U$ from within the $^{238}U$ decay series) are most frequently used due to its convenient HALF-LIFE ($T_{1/2}$) of 75.4 ka, and corresponding age range of approximately 350 ka. By measuring the activity ratio of protégé isotopes to parents (e.g., $^{230}Th/^{234}U$) and knowing the half-lives of the daughter products it is possible to determine the time which has elapsed since precipitation. The method assumes that a sample has behaved as a closed system since the time of formation/deposition (i.e., there has been neither loss or gain of any isotopes except by radioactive decay), and that at the time of formation, the activity ratios of the isotopes being used to determine the age were either zero or some determinable level.

The conventional means of activity measurement is by alpha spectroscopy (the particles emitted during the decay of key isotopes being measured). More recently, ratios of uranium isotopes have been directly counted by thermal ionization mass spectrometric techniques (TIMS). This improves precision of analyses, increases the age range of the method, reduces minimum sample size requirements, and makes possible the measurement of entirely new types of precipitated terrestrial deposits. SS

**Reading**
Taylor, R.E. and Aitken, M.J. 1997: *Chronometric dating in archaeology: advances in archaeological and museum science* Vol. 2. New York: Plenum Press.

**urban ecology** A branch of ecology dealing with the environment and its organisms specifically in the context of urbanized areas. Although once thought of as ecological deserts, it is now recognized that heavily built-up areas support a variety of plant and animal species, some well-adapted to the environmental constraints. The impacts of human activities on the urban environment, together with the identification of the conservation potential of open (or 'green') spaces in urban areas provide the focus of urban ecology. In recent years, urban planners have adopted a still broader definition in recognizing the potential of ecological principles in promoting the sustainable management of urban ecosystems. This involves, for example, revising land use practices, restoring degraded habitats and promoting recycling.          MEM

**Reading**
Breuste, J., Feldmann, H. and Ohlmann, O. eds 1998: *Urban ecology*. Berlin: Springer-Verlag.

**urban hydrology**  The study of the hydrological cycle and of the water balance within urban areas. Extensive impervious areas mean that surface storage is reduced, infiltration is not possible and evapotranspiration is much less than in rural areas. Impervious areas increase the amount of surface run-off and this is accentuated by the stormwater drainage system which collects water from roads, roofs and other impervious surfaces. Modern stormwater drainage systems are installed separately from foul water drainage systems but in the past a single system was often employed in urban areas. Stream discharge from urban areas tends to have higher peak flows and lower base flows than discharge from rural areas and the FLOOD FREQUENCIES along rivers draining urban areas will be significantly changed from the time when the urban area did not exist. Increased flooding has often been observed within and downstream from urban areas as urbanization has occurred and the larger and more frequent floods may have led to increased river channel erosion. Problems of increased frequency and extent of flooding have often been mitigated by engineering works. The area of Moscow, Russia is shown (Lvovich and Chernogaeva 1977) to have a decrease of evapotranspiration of 62 per cent, of groundwater run-off of 50 per cent, and an increase in total run-off of 155 per cent. In urban hydrology it is not simply a modification of the rural hydrological cycle but there can also be a series of other components supplying or reducing water. Urban areas also generate a characteristic water quality with water temperatures often higher than those of rural areas, with higher solute concentrations reflecting additional sources including pollutants, and suspended sediment concentrations particularly high during building activity but much lower when the urban area is established and the sources of suspended sediments are no longer

exposed. (See also LAND USE; HYDROLOGICAL CYCLE.)          KJG

**Reading and References**
Hollis, G.E. 1979: *Man's impact on the hydrological cycle in the United Kingdom*. Norwich: Geo Books. · Kilber, D.R. ed. 1982: *Urban stormwater hydrology*. Water resources monograph 7. Washington DC: American Geophysical Union. · Lvovich, M.I. and Chernogaeva, G.M. 1977: The water balance of Moscow. *Effects of urbanization and industrialization on the hydrological regime and on water quality*. International Association of Hydrological Sciences publication 123. Pp. 48–51. · Walling, D.E. 1981: Hydrological processes. In K.J. Gregory and D.E. Walling eds, *Man and environmental processes*. London: Butterworth. Pp. 57–81.

**urban meteorology**  The study of atmospheric phenomena attributable to the development of human settlements. It encompasses work on the process involved (physical, chemical and biological), the resulting climate effects, and the application of this knowledge to the planning and operation of urban areas. It is one of the clearest examples of man's role in climate modification.

Urban development disrupts the climatic properties of the surface and the atmosphere. These, in turn, alter the exchanges and budgets of heat, mass and momentum which underlie the climate of any site. Every land clearance, drainage, paving and building project leads to the creation of a new microclimate in its vicinity, and the collection of these diverse, human-affected microclimates is what constitutes the urban climate in the air layer below roof level (henceforward called the urban canopy layer, or UCL, Oke 1987). These very localized effects tend to be merged by turbulence above roof level where they form the urban boundary layer (UBL) which appears like a giant urban plume over and downwind of the city.

A city exerts both roughness and thermal influences on winds. When synoptic winds are strong the greater roughness produces greater turbulence (by 10–20 per cent), increased frictional drag, slower winds (by about 25 per cent), cyclonic turning, and a general tendency towards uplift. In the downwind rural area the near surface flow recovers its original characteristics but urban effects are detectable in the elevated UBL plume for tens of kilometres. The drag may even retard the passage of weather fronts. In windy conditions, flow in the UCL is extremely variable. While some areas are sheltered, others may be experiencing strong across-street vortices, gustiness or jets (especially near tall buildings). When synoptic winds are light or absent, thermal effects associated with the heat island (see below) become evident. The city may generate its own thermal circulation, analogous to SEA/LAND BREEZE, with 'country' breezes converging on the city centre, rising and diverging aloft to form a counter flow. Urban thermal effects can also lead to acceleration near the surface both as a result of the heat pressure field and because thermal turbulence helps transport momentum downwards.

Considering the major changes in the physical environment wrought by urban development, the changes in the energy (heat) budget are surprisingly small. For example, despite all the radiant fluxes being altered (by pollution or changed surface properties) the net radiation in cities is usually within 5 per cent of that of their rural surroundings. It is true that the city's energy budget is supplemented by heat released by combustion, but though this heat source may be important to climate in some locations, in most places it is minor (Kalma and Newcombe 1976). Usually more important is the fact that the city channels more energy into sensible rather than LATENT HEAT. This is because of the removal of many sources of water for EVAPO-TRANSPIRATION. As a result more heat is used to warm the air and ground (including buildings etc.). The relative warmth of the city is called its 'heat island' (Landsberg 1981; Oke 1982). On an annual basis the canopy layer of a large city (10 million inhabitants) is typically 1–3 °C warmer than its surrounding countryside. This may seem small, but the magnitude of the heat island varies diurnally (largest near midnight, the smallest in the afternoon) and in response to weather (largest with calm and no cloud). The difference is also related to city size (measured by its population or better by the geometry of its central street canyons, Oke 1982). On the most favourable nights in a large city differences of 10 °C and more have been recorded. Spatial variation of temperature within the UCL bears a strong relation to land use and building density and there is a sharp gradient at the urban/rural boundary. The city's warmth extends down into the underlying ground and upward into the UBL above. At night the heat island maintains a weak mixed layer above the city (tens to hundreds of metres deep) when rural areas are stable.

The exchange of moisture between the surface and the air is altered by changes in the availability of water and energy, and in the perturbed airflow. Normally values of atmospheric moisture in the daytime UCL are lower than in the country (on account of less evapotranspiration and greater mixing), but the reverse holds at night (because of decreased dewfall and the release of water vapour from combustion). The effects seen in the UCL are also evident in the UBL plume. An exception is provided by high latitude cities in winter where evaporation from frozen surfaces is very small so that humidity is largely governed by vapour from combustion, with the result that the city is more humid by both day and night. At temperatures below −30 °C ice fog is a common, and unpleasant, fact of urban life. Above freezing, urban effects on fogs are complex: extra warmth may decrease their frequency but extra condensation nuclei may increase their density and severity. AEROSOL is also responsible for a general increase in daytime haze in the subcloud layer of the UBL, and a deterioration of visibility (Braham 1977).

Urban modification of PRECIPITATION is a subject that has received considerable research study, especially through Project METROMEX in St Louis (Changnon 1981). There seems to be a consensus view that cities enhance precipitation in their downwind areas. These effects seem to be most marked in relation to summer convective rainfall, especially heavy rain, and severe weather (thunder- and hail-storms) rather than frontal precipitation. Annual increases of up to 10 per cent are commonly reported, but the exact role of urban versus non-urban influences is often hard to determine. There is also difficulty in isolating the most important causes. It is possible that the microphysics of urban clouds is altered (e.g. cloud droplet sizes and numbers) and/or that cloud dynamics are changed by the UBL (e.g. strength of uplift, height of mixed layer) leading to more favourable precipitation conditions. The latter changes seem most important in St Louis but much more work is needed.

The field, which began in the early nineteenth century with descriptive studies, is now engaged in the study of meteorological processes and attempts to devise models which link processes and effects. Its most significant deficiencies are in having little knowledge of tropical urban climates and its failure to develop applied science aspects (Page 1970).                    TRO

**Reading and References**

Braham, R.R. Jr 1977: Overview of urban climate. *Proceedings of the Conference on Metropolitan Physical Environment: USDA Forest Service general technical report NE-25.* Upper Darby, PA. Pp. 1–17. · Changnon, S.A. Jr ed. 1981: METROMEX; a review and summary. *Meteorological monographs.* 18.40. Boston: American Meteorological Society. · Kalma, J.D. and Newcombe, K.J. 1976: Energy use in two large cities: a comparison of Hong Kong and Sydney, Australia. *Environmental studies* 9, pp. 53–64. · Landsberg, H.E. 1981: *The urban climate.* New York: Academic Press. · Oke, T.R. 1987: *Boundary layer climates.* 2nd edn. London: Methuen. · — 1982: The energetic basis of the urban heat island. *Quarterly journal of the Royal Meteorological Society* 108, pp. 1–24. · Page, J.K. 1970: *The fundamental problems of building climatology considered from the point of view of decision-making by the architect and urban designer.* WMO technical note 109. Geneva: World Meteorological Organization.

**urstromtäler** An anastomosing pattern of meltwater channels in northern Germany. Individual channels may be hundreds of kilometres long and often more than 100 m deep with irregular long profiles. Some channel patterns are buried by later glacial deposits (Ehlers 1981). Although there is still discussion about their origin, it seems that they were cut primarily by subglacial meltwater erosion.                 DES

**Reference**

Ehlers, J. 1981: Some aspects of glacial erosion and deposition in northern Germany. *Annals of ecology* 2, pp. 143–6.

**uvala** A depression or large hollow in limestone areas produced when several sinkholes coalesce.

# V

**vadose** That zone of the groundwater system in which the available pore spaces are not fully saturated, i.e., located between the ground surface and the WATER TABLE. The zone below the water table is correspondingly termed the *phreatic* zone. In the vadose zone, water-filled pores in the soil may be tension-saturated, that is, the water is not under positive hydrostatic pressure, but is tightly held to the walls of the pore spaces. Below the water table, all pressures are positive.                                            DLD

**Reading**
Dingman, S.L. 1994: *Physical hydrology.* New Jersey: Prentice-Hall.

**valley bulges** (valley-bottom bulges) Consist of strata that have bulged up in the base of a valley as a result of erosive processes. They are widespread in the sedimentary rock terrains of the English Midlands, where limestones, sandstones and clays occur in close juxtaposition. The mechanism of formation invoked for those of the Stroud area in Gloucestershire (Ackermann and Cave 1977) is that during the Pleistocene severe erosion and valley incision occurred at a time when permafrost conditions pertained. At the end of the cold period the rocks thawed out and the susceptible clays, silts and sands, highly charged with water, became plastic, and under the weight of the more competent limestones above were extruded through the weakest points of the recently developed valley floors. Cambering of strata would occur on the valley sides.          ASG

**Reference**
Ackermann, K.J. and Cave, R. 1977: Superficial deposits and structures, including landslip, in the Stroud District, Gloucestershire. *Proceedings of the Geologists' Association of London* 78, pp. 567–86.

**valley meanders** Meanders which are usually cut in bedrock and which usually have a greater wavelength than that of the contemporary river pattern. Dury (1977) has shown that valley meanders were produced during periods of higher run-off and higher peak discharges before stream shrinkage which led to contemporary underfit streams. Some writers have suggested that the valley meanders may not indicate stream shrinkage but are rather related to rare high magnitude events, to the contrast between bedrock and fluvial deposits, and to the pattern of stream migration (see Dury 1977 and papers cited therein).                                            KJG

**Reading and Reference**
Dury, G.H. 1976: Discharge prediction, present and former from channel dimensions. *Journal of hydrology* 30, pp. 219–45. · — 1977: Underfit streams: retrospect, perspect and prospect. In K.J. Gregory ed., *River channel changes.* Chichester: Wiley. Pp. 281–93.

**valley wind** The up-valley flow which develops during the day, especially in north–south orientated valleys during anti-cyclonic weather in summer. The flow is induced by strong heating of the valley air, making it much warmer than the air at the same elevation over the adjacent plain. Valley winds are usually as reliable as the mountain, or down-valley, winds which develop at night.                                            WDS

**Reading**
Atkinson, B.W. 1981: *Meso-scale atmospheric circulations.* London: Academic Press.

**valloni** The drowned river valleys of a Dalmatian-type coastline.

**Van't Hoff's rule** The rule which states that when a system is in thermodynamic equilibrium a lowering of temperature will promote an exothermic reaction and a raising of the temperature an endothermic one.

**vapour pressure** The pressure exerted by the molecules of a given vapour. In meteorology, the vapour in question is usually water vapour and the pressure is the partial pressure, i.e. the contribution by water vapour to the total pressure of the atmosphere. It may be calculated by using a humidity slide rule or tables in conjunction with values of dry-bulb and WET-BULB TEMPERATURE and is expressed in millibars. Water vapour's concentration is highly variable in the atmosphere, so vapour pressure changes substantially in time and space with the highest values (15–20 mb) being found in the humid tropics and the lowest (1–2 mb) across wintertime high latitude continents.                                            RR

**varves** Traditionally defined as being sedimentary beds or lamina deposited in a body of still water within the course of one year. The term has normally been applied to thin layers, usually deposited by meltwater streams in a body of water in front of a glacier. A glacial varve normally includes a lower 'summer' layer consisting of relatively coarse-grained sand or silt, produced by rapid ice melt in the warmer months, which grades upwards into a thinner 'winter' layer composed of finer material deposited from suspension in quiet water while the streams feeding the lake are frozen. However, it is becoming increasingly clear that varves may be deposited in a wide range of environments, both lacustrine and marine, and an alternative term, rhythmite, is now widely used. ASG

**Reading**
O'Sullivan, P.E. 1983: Annually laminated sediments and the study of Quaternary environmental changes – a review. *Quaternary science review* 1, pp. 245–313. · Schlüchter, Ch. 1979: *Moraines and varves: origins, genesis, classification.* Rotterdam: Balkema.

**vasques**  Wide (up to several decimetres), shallow pools with flat bottoms, which form a network consisting of a tiered, terrace-like series of steps on limestone coastal platforms, especially in aeolianite. The pools are separated from each other by winding, narrow, lobed ridges, 10–200 mm in height, and running continuously for tens of metres. They develop between high and low tide levels, especially in intertropical and Mediterranean climatic regions.                                                   ASG

**Reading**
Battistini, R. 1983: La morphogénèse des plateformes de corrosion littorale dans les grès calcaires (plateforme supériere et plateforme à vasques) et le problème des vasques, d'après des observations faites à Madagascar. *Revue de géomorphologie dynamique* 30, pp. 81–94.

**vauclusian spring**  See SPRINGS.

**vector**  Beyond its general use as a word for a line with a fixed length and direction but no fixed location, vector is now used to describe one of the two main types of GEOGRAPHIC INFORMATION SYSTEM (GIS) for storing and analysing spatial data. All the objects of interest are represented as either points, lines or areas, each of which can have attributes recorded. In the case of streams: for example, each section of stream would be stored as a line, which might have information on pH and water temperature recorded. Vector GIS are particularly good at handling information about discrete objects (see RASTER) and are used in areas such as local government and the utilities.                                       SMW

**veering**  See WIND.

**velocity area method**  A widely used method of measuring the discharge of a river, in which a series of verticals is used to subdivide the cross-section into a number of segments, the discharge of each segment is determined as the product of average *velocity* and cross-sectional *area*, and the total discharge is calculated as the sum of the values for the individual segments. Verticals are spaced at intervals of no greater than 1/15th of the width. Measurements of mean velocity in the vertical are obtained by using a rotating current meter and these are assumed to be representative of the average velocity in the adjacent segment. (See also DISCHARGE.)                                       DEW

**Reading**
British Standards Institute 1964: *Methods of measuring liquid flow in open channels. Part 3: Velocity area methods.* BS 3680. London: British Standards Institution.

**velocity profile and measurement**    Any fluid (e.g. water or air) moving over or adjacent to a fixed boundary will illustrate a vertical variation in velocity, increasing from zero or near zero at the boundary, to higher values in the main body of the flow. Understanding how velocity varies for a given flow over a boundary is crucial to estimating erosion and deposition of sediment. Most natural flows have a high REYNOLDS NUMBER (greater than 2000) and can therefore be classified as turbulent. The zone of influence of the boundary upon the flow is called the BOUNDARY LAYER. If the flow is of limited depth, and the boundary layer extends throughout this depth, the flow is said to be fully developed. Close to the bed, there is a thin zone where LAMINAR FLOW occurs, called the laminar sub-layer. Above this, there is a transitional buffer layer and then the fully turbulent zone. Field measurements have suggested that in this zone velocity increases regularly with elevation above the bed (see BOUNDARY LAYER). This is supported by basic theory (Prandtl, 1952) and is commonly known as the law of the wall:

$$\nu = \frac{1}{\kappa}\sqrt{\frac{\tau_o}{\rho_w}}\ln\frac{y}{y_o}$$

where: $\nu$ = velocity at elevation $y, \kappa$ = Von Karman's constant (which is often assumed to be 0.4); $\tau_o$ = shear stress exerted by the flow upon the bed; $\rho_w$ = density of water; and $y_o$ = the roughness height. Applying this equation to a measured velocity profile allows estimation of bed SHEAR STRESS, although there is debate over how far up into the flow this equation holds. If it is assumed that it holds throughout the flow, integrating the equation shows that the depth-averaged velocity is approximately $0.38h$ above the bed, where $h$ is water depth. This can be used in the VELOCITY AREA METHOD for

DISCHARGE estimation in rivers where a CURRENT METER is available that can measure point velocities.                                                                    SNL

**Reading and References**
Carson, M.A., 1971: *The mechanics of erosion*. London: Pion. · Herschy, R.W. (ed.) 1978: *Hydrometry, principles and practices*. Chichester: Wiley. · Prandtl, L., 1952: *Essentials of fluid dynamics: with applications to hydraulics, aeronautics, meteorology and other subjects*. London: Blackie.

**ventifact**  A stone that has been shaped by the wind, especially in arid and polar areas. Abrasion is achieved by sand, dust or snow, and the stones become shaped (often into three-sided DREIKANTER), and have surface textures that may be polished, pitted or fluted. They have some utility for estimating past and present wind directions.                                          ASG

**Reading**
Whitney, M.I. and Dietrich, R.V. 1973: Ventifact sculpture by windblown dust. *Bulletin of the Geological Society of America* 84, pp. 2561–81.

**vertical motion**  (in the atmosphere). The vertical component of air velocity. Persistent net horizontal flow out of a region, called horizontal DIVERGENCE, would result in the depletion of air in the region, so that its density would decrease; but observed density changes are small, so that vertical motion must result. Such persistent outflow leads to downward motion above a flat surface (where the vertical motion is necessarily zero). In the upper troposphere persistent horizontal divergence usually leads to upward motion because the great stability of the stratosphere above is unfavourable for vertical motions. Persistent horizontal convergence leads to vertical motions of the opposite sign.

Vertical motion in the troposphere is normally upwards in CYCLONES and downwards in ANTICYCLONES. Magnitudes of vertical motions in such systems are typically only a few centimetres per second, but, because they persist, large vertical displacements of air are involved. The vertical motion in frontal zones is greater than the average values within depressions, being typically a few tens of centimetres per second.

Vertical motions within cumulus clouds are of the order of a metre per second, but in cumulonimbus cloud they may be as much as several tens of metres per second.                                 KJW

**Reading**
Battan, L.J. 1974: *Fundamentals of meteorology*. Englewood Cliffs, NJ: Prentice-Hall.

**vertical stability/instability**  In the atmosphere, terms that characterize the response of a parcel of air to its vertical displacement. A return of the parcel to its original position indicates vertical stability, while continued vertical movement of the parcel indicates vertical instability. When initially possessing a temperature that is equal to that of the surrounding environment, a parcel of rising air that cools at a faster (slower) rate than does the surrounding environment will be colder (warmer) than the environment, and stable (unstable). The rate at which an air parcel cools is directly related to the amount of water vapour that the parcel contains.                                             AWE

**vertisol**  One of the eleven orders of the US system of soil taxonomy. Vertisols are soils rich in clays ($> 30\%$), especially 2:1 lattice clays belonging to the smectite group, which shrink and swell upon wetting and drying. Consequently, vertisols are soils that frequently display deep cracking, especially in environments where seasonal moisture conditions are variable. Materials from the upper parts of the soil profile may mix downward into the soil by falling into these cracks. Distinctive polished surfaces (SLICKENSIDES) may be produced in vertisols by the rubbing of adjacent blocks of material.

                                                                        DLD

**Reading**
Brady, N.C. and Weil, R.R. 1996: *The nature and properties of soils*. 11th edn. New Jersey: Prentice-Hall.

**vesicular**  Possessing numerous large pores and internal voids.

**vicariance biogeography**  In general terms, the study of groups of plants or animals which are descended from a common ancestor but which are now spatially isolated from each other in disjunct or endemic distributions; more specifically, a recently developed school of biogeography which traces its ancestry through the writings of de Candolle, Croizat and Hennig, focusing on allopatric differentiation (see SYMPATRY) and endemism, linking the study of CLADISTICS and TAXONOMY with that of biogeography.

At the simplest level, many vicariants, or vicarious taxa, form species-pairs, such as the American *Maianthemum canadense* and the Eurasian *Maianthemum bioflium* or *Drosera uniflora* in South America and *Drosera arcturi* in New Zealand. But others have a more complex grouping and represent multiple vicariism, a phenomenon particularly common on isolated oceanic islands. A fine example of such multiple vicariism is provided by the 30 or so species of the palm genus, *Pritchardia*, on the Hawaiian Islands, where the geographical isolation of the taxa is a result of both island separation and the internal topography of the islands themselves.

On one island, for example, there are nine different species, each of which grows in a separate valley.

This classic concept of vicariance involves allopatric species which have descended from a common ancestral population and which attained some degree of spatial isolation. Vicariance is essentially a passive condition when contrasted with ADAPTIVE RADIATION, a process involving active selection by different environmental conditions. This traditional form of vicariance is described as horizontal or geographical vicariance. The differences in the isolated pairs of groups will have arisen through the lack of gene exchange, chance differences in genetic composition and, possibly, genetic drift.

Three other forms of vicarious distribution have also been recognized. The first of these is altitudinal vicariance in which the related taxa form lowland/highland pairs. The second is habitat or ecological vicariance where the species-pairs occupy different environmental niches. This is well exemplified by the sea arrow-grass (*Triglochin maritima*) and the marsh arrow-grass (*T. palustris*), which inhabit respectively saltwater and freshwater marshes. Paradoxically, the term has also been used to describe taxa unrelated from the evolutionary standpoint, but which appear to be adapted to the same ecological niche in different locations. Finally, phenological vicariants have been recognized in which the separation is seasonal, with the related taxa flowering or breeding at different times in the year. Such taxa may actually be sympatric.

The proponents of the new school of vicariance biogeography seek a far more comprehensive philosophy and methodology and see themselves in direct and crusading opposition to what they term 'dispersalist' biogeography in which, they argue, the traditional explanations for allopatric species depend on the long-range dispersal of taxa from centres of origin. They claim that this 'misguided' approach is based on the views and writings of Charles Darwin, whose *On the origin of species* (1859) and ideas have been dismissed as 'piffle' by one of the main protagonists of the new school, Leon Croizat. Unfortunately, the whole debate has been marred by totally unnecessary and often gratuitous *ad hominem* abuse, so much so that it has frequently proved very difficult to grasp the thrust of the arguments involved. The contrived opposition of vicariance versus dispersal also seems to many biogeographers somewhat forced, as these are by no means the only processes involved in the development of plant and animal areas.

Two particular developments underpin the growth of the new school, which is most fully documented in the pages of the journal, *Systematic zoology*. The first is traced through the phylogenetic systematics, now normally referred to as cladistics, of the entomologist, Willi Hennig (1966). Cladistics is simply a method of classification, giving rise to graphs (cladograms) of relative affinity. They make no *a priori* assumptions about the nature of the relationships involved. On replacing the taxa represented by cladograms with the localities they inhabit, cladograms of affinities of areas or biological area-cladograms result. These are crucial to the new approach, for, as Nelson (in Nelson and Rosen 1981) is at pains to stress, vicariance biogeography begins by asking the question: 'Is there a cladogram of areas of endemism?'

Secondly, the new biogeography implicitly accepts that the motor of vicariance is continental movement, brought about through sea-floor spreading and plate tectonics, and it is this geological or geophysical model of the earth which has given the concept of vicariance biogeography such a boost since the geologists' crucial synthesis of plate tectonics in the 1960s. Nevertheless, most vicariance biogeographers insist that the sequence of distributional events must be derived from the taxa themselves *before* the geological evidence is introduced and that the biological evidence must stand whether or not it matches the 'fashionable' geology.

Thus, the new biogeography attempts above all to link recent developments in systematics with biogeography. It is first and foremost concerned with the coincidence of pattern, a point clearly emphasized by the essential character of Croizat's original methods in what are seen as founder publications, particularly *Panbiogeography* (1958) and *Space, time, form: the biological synthesis* (1964). His idea was to map distributions of whole varieties of taxa and then to establish concordances of pattern in order to identify ancestral biotas. This then necessitated explanations for the fragmentation of previously more extensive distributions. The phenomenon of fragmentation was vicariance. If a given type of distribution – 'an individual track' – recurs in group after group of organisms, the region delineated by the coincident distributions – 'the generalized track' – becomes statistically and thus geographically significant.

The method is well exemplified by the study of Parenti (1981) on cyprinodontiform fishes. Cladograms of areas are derived from cladograms of taxa and then from the coincidence of areal patterns for different taxa 'a pattern of earth history is suggested'. It should be especially noted that this pattern is independent of geological hypotheses although, it is admitted, a geological model may be found to fit the pattern. Above all, the absence of such a model is not seen as invalidating the biogeography.

Many biogeographers, of course, regret this rejection of independent sources of information and are far from happy with the newest manifestations of the underlying cladism (e.g. Ridley 1983). Moreover, they regard the whole method of vicariance biogeography as inductive, mechanical, unrelated to process, lacking in any true sense of geography beyond that of simple geographical coordinates, and over-encumbered with a new and less than euphonious jargon. Undoubtedly, the case has not been helped by the stridency of some of its proponents. Nevertheless, vicariance is an important feature of plant and animal distributions and, as with dispersal, biogeography would be considerably the poorer if it were not given its due prominence.                                      PAS

**Reading and References**
Croizat, L. 1958: *Panbiogeography, or an introductory synthesis of zoogeography, phytogeography, and geology.* Caracas: privately published. · —1964: *Space, time, form: the biological synthesis.* Caracas: privately published. · Hennig, W. 1966: *Phylogenetic systematics.* Urbana: University of Illinois Press. · Nelson, G. 1978: From Candolle to Croizat: comments on the history of biogeography. *Journal of the history of biology* 11, pp. 269–305. · —and Platnick, N.I. 1981: *Systematics and biogeography: cladistics and vicariance.* New York: Columbia University Press. · —and Rosen, D.E. eds 1981: *Vicariance biogeography: a critique.* New York: Columbia University Press. · Parenti, L.R. 1981: A phylogenetic and biogeographic analysis of cyprinodontiform fishes (Teleostei, Atherinomorpha). *Bulletin of the American Museum of Natural History* 168, pp. 335–557. · Patterson, C. 1981: Biogeography: in search of principles. *Nature* 291, pp. 612–13. · Ridley, M. 1983: Can classification do without evolution? *New scientist* 1 December, pp. 647–51. · Stoddart, D.R. 1977, 1978, 1981, 1983: Biogeography. *Progress in physical geography.* 1, pp. 537–43; 2, pp. 514–28; 5, pp. 575–90; 7, pp. 256–64. · Stott, P.A. 1984: History of biogeography. In J.A. Taylor ed., *Themes in biogeography.* London: Croom Helm. Pp. 1–24.

**vigil network**   See REPRESENTATIVE AND EXPERIMENTAL BASINS.

**virgation**   The formation of trails of ice crystals falling from a cloud.

**viscosity**   The property of fluids by virtue of which they resist flow. Newton's law of viscous (non-turbulent) flow is given by:

$$F = \eta\, A\, dv/dx$$

where $F$ is the tangential force between two parallel layers of liquid of area $A$, a distance $dx$ apart, moving with relative velocity $dv$, and $\eta$ is the coefficient of viscosity, dynamic viscosity or just viscosity. It is measured in $N\ s\,m^{-2}$. It

should be carefully distinguished from the kinematic viscosity $v$ which is $\eta$ divided by the fluid density $\rho$; the units here are in $m^2\,s^{-1}$.         WBW

**Reading**
Whalley, W.B. 1976: *Properties of materials and geomorphological explanation.* Oxford: Oxford University Press.

**vlei**   A term, Afrikaans in origin, used in southern Africa, especially South Africa, to describe valley-bottom areas such as flood-plains, valley grasslands and other low-lying wetlands. Vlei is used to describe permanently, seasonally or sporadically flooded land, most frequently in direct association with river valleys, although the presence of a river channel per se is not a precondition, in which case use of the term is equivalent to that of DAMBO more commonly used further north in Africa.         MEM

**void ratio**   The ratio of the volume of interstitial voids in a portion of sedimentary rock to the volume of that portion.

**volcano**   An opening or vent through which magma, molten rock, ash or volatiles erupt on to the earth's surface, or the land form produced by the erupted material. Volcanoes tend to be conical in shape but can have a variety of forms, depending on the nature of the erupted material (particularly its viscosity), the character of recent eruptive activity and the extent of post-erupted modification by erosion (see illustration on p. 518). Most volcanoes are concentrated at convergent and divergent plate boundaries (see PLATE TECTONICS) but others, located in the interior of plates, are associated with HOT SPOTS.         MAS

**Reading**
Francis, P.W. 1976: *Volcanoes.* Harmondsworth: Penguin. Ollier, C.D. 1988: *Volcanoes.* Oxford: Blackwell. · Williams H. and McBirney, A.R. 1979: *Volcanology.* San Francisco: Freeman, Cooper.

**volkerwanderungen**   The time of the migrations of the Huns and other barbarians who invaded Europe contributing to the decline of the Roman Empire. The stimulation of these migrations has been under debate, however the assertion holds that climate change was the underlying factor. Many scientists believe that the drying up of pastures used by the nomads in the heart of the Eurasian continent unsettled the barbarian tribes and started the peoples migrating westwards into Europe (approximately 140 BC to AD 451). In Europe and the Near East, cooler and wetter conditions existed from about 4500 years BP until around 2500 BP. Then, during the dawn of the Roman Empire (around 100 BC) and extending until around AD 400, a climate

Types of volcanic eruptions

| Type | Characteristics |
| --- | --- |
| Icelandic | Fissure eruptions, releasing free-flowing (fluidal) basaltic magma; quiet, gas-poor; great volumes of lava issued, flowing as sheets over large areas to build up plateaux (Colombia). |
| Hawaiian | Fissure, caldera and pit crater eruptions; mobile lavas, with some gas; quiet to moderately active eruptions; occasional rapid emission of gas-charged lava produces fire fountains; only minor amounts of ash; builds up lava domes. |
| Strombolian | Stratocones (summit craters); moderate, rhythmic to nearly continuous explosions, resulting from spasmodic gas escape; clots of lava ejected, producing bombs and scoria; periodic more intense activity with outpourings of lava; light-coloured clouds (mostly steam) reach upward only to moderate heights. |
| Vulcanian | Stratocones (central vents); associated lavas more viscous; lavas crust over in vent between eruptions, allowing gas build-up below surface; eruptions increase in violence over longer periods of quiet until lava crust is broken up, clearing vent, ejecting bombs, pumice and ash; lava flows from top of flank after main explosive eruption; dark ash-laden clouds, convoluted, cauliflower-shaped, rise to moderate heights more or less vertically, depositing tephra along flanks of volcano. (Note: ultravulcanian eruption has similar characteristics but results when other types, e.g. Hawaiian, become phreatic and produce large steam clouds, carrying fragmental matter.) |
| Vesuvian | More paroxysmal than Strombolian or Vulcanian types; extremely violent expulsion of gas-charged magma from stratocone vent; eruption occurs after long interval of quiescence or mild activity; vent tends to be emptied to considerable depth; lava ejects in explosive spray (glow above vent), with repeated clouds (cauliflower) that reach great heights and deposit tephra. |
| Plinian | More violent form of Vesuvian eruption; last major phase is uprush of gas that carries cloud rapidly upward in vertical column for miles; narrow at base but expands outward at upper elevations; cloud generally low in tephra. |
| Peléean | Results from high-viscosity lavas; delayed explosiveness; conduit of stratovolcano usually blocked by dome or plug; gas (some lava) escapes from lateral (flank) openings or by destruction or uplift of plug; gas, ash and blocks move downslope in one or more blasts as nuées ardentes or glowing avalanches, producing directed deposits. |
| Katmaian | Variant of a Peléean eruption characterized by massive outpourings of fluidized ash flows; accompanied by widespread explosive tephra; ignimbrites are common end products, also hot springs and fumaroles. |

*Source*: Short, N.M. 1986: Volcanic landforms. In N.M. Short and R.W. Blair eds, *Geomorphology from space – a global overview of regional landforms*. Washington, DC: NASA. Table 3–1, p. 187.

change occurred whereby conditions became drier and warmer. The migrations at this time were from east to west, which suggested trouble in the centre of the Eurasian continent. Winters in that region are always cold and harsh so the change that provided the impetus for the migrations was determined to be drought. Climatic research has shown that the fifth century AD was abnormally arid across Europe in the region west of the Ural Mountains with annual precipitation levels averaging 50–100 mm below normal. Further and more direct evidence of the dryness was the lowering of the Caspian Sea level over the period from 500 BC to AD 500, with the trend accelerated after AD 200.                                                   LN

**Von Karman constant**  A universal constant that relates the mixing length in turbulent flow to the distance from a solid boundary. The mix-

ing length is a measure of the size of turbulent eddies, or in other words of the average distance travelled by small volumes of water following random paths that are superimposed on the main forward flow direction of a stream. If the mixing length is 1 and distance from the boundary $y$, then the relation is

$$1 = \kappa l y$$

where $\kappa l y$ is von Karman's constant. This is taken to have a value of 0.4 in clear water, but its value is thought to be lowered in the presence of high concentrations of suspended sediment particles, because these impede the free movement of the turbulent eddies.                      DLD

**Reading**
Richards, K. 1982: *Rivers*. London: Methuen. · Wang, S. 1981: Variation of Karman constant in sediment-laden flow. *Proceedings of the American Society of Civil Engineers, Journal of the Hydraulics Division* 107, pp. 407–17.

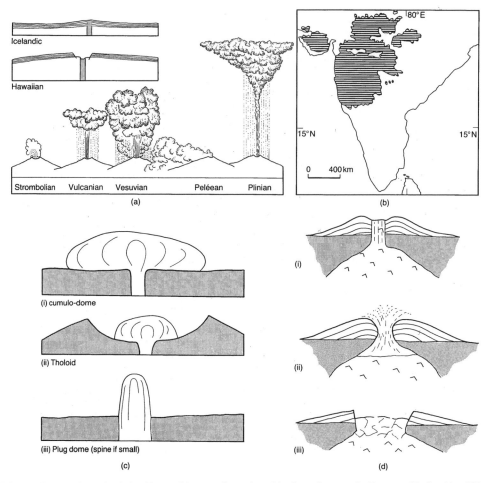

**Volcano.** Some major volcanic landforms: (a) types of eruption; (b) a lava plateau – the Deccan of India; (c) acid lava (viscous) extrusion forms; (d) the stages in the formation of a caldera by collapse: (i ) initial volcano; (ii) explosion; (iii) collapse.
*Source*: A.S. Goudie 1984: *The nature of the environment*. Oxford and New York: Basil Blackwell.

**vorticity** A microscopic measure of rotation in a fluid. It is a vector quantity defined as the curl of the velocity and has dimensions of $(\text{time})^{-1}$. In cartesian coordinates the $x$, $y$ and $z$ components of vorticity ($\zeta$) are given by:

$$\zeta_x = \frac{\partial w}{\partial y} - \frac{\partial v}{\partial z}; \zeta_y = \frac{\partial u}{\partial z} - \frac{\partial w}{\partial x}$$

$$\zeta_z = \frac{\partial v}{\partial x} - \frac{\partial u}{\partial y};$$

where $u$, $v$ and $w$ are the $x$, $y$ and $z$ components of velocity. The absolute vorticity is given by the curl of the absolute velocity and relative vorticity is given by the curl of the velocity relative to the earth. In the northern hemisphere relative vorticity in a cyclonic sense is positive and in an anti-cyclonic sense is negative. Fluid in solid rotation with angular velocity $\omega$ has vorticity $2\omega$.

In large-scale motions in the atmosphere the vertical components of absolute and relative vorticity are of chief importance and these terms are often used without the explicit modifier 'vertical component of'. The difference between the vertical component of absolute and relative vorticity is given by the vertical component of vorticity of the earth due to its rotation, being $2\Omega \sin \phi$, where $\Omega$ is the angular velocity of the earth and $\phi$ is latitude.

In adiabatic, frictionless motion a quantity called potential vorticity is conserved, this being expressed by:

$$(\text{absolute vorticity}) \times \frac{\partial \theta}{\partial p} = \text{constant}$$

where $\theta$ is potential temperature and $p$ is pressure. KJW

**Reading**
Atkinson, B.W. ed. 1981: *Dynamic meteorology: an introductory selection*. London and New York: Methuen. · Gill, A.E. 1982: *Atmosphere–ocean dynamics*. London: Academic Press. · Houghton, J. 1986: *The physics of atmospheres*. Cambridge: Cambridge University Press.

**vugh** A void or cavity within a rock which can be lined with mineral precipitates.

**vulcanism** The movement of magma or molten rock and associated volatiles onto or towards the earth's surface. Extrusions of material onto the earth's surface can occur through a vent known as a VOLCANO or through linear openings in the crust, called fissures. Intrusion of magma or molten material into the upper part of the crust can give rise to large masses of igneous rock, such as BATHOLITHS, which may cause domal uplift of the overlying strata. The term vulcanism is also used as a synonym of volcanism and in this sense refers only to extrusion of material onto the earth's surface. MAS

**Reading**
Williams, H., and McBirney, A.R. 1979: *Volcanology*. San Francisco: Freeman, Cooper.

519

# W

wadi   An Arabic word, sometimes also spelt *oued*, generally used as a term for ephemeral river channels in desert areas. Wadis may flow only occasionally, and then sometimes discontinuously, along their courses.

A desert river system (wadi) developed in Sahra el-Arabiya, Egypt. In spite of low annual rainfall totals, fluvial activity is a potent force in many desert areas.

Walker circulation   A large atmospheric circulation cell oriented east–west along the equator in the tropical Pacific. This circulation pattern is one of the two main features of atmospheric motion in the Pacific. The other major feature is the HADLEY circulation, which is more pronounced and consists of a cell oriented in a north–south direction.

Under 'normal' conditions, the Walker circulation involves rising air (low pressure) over the warm maritime continent of the western tropical Pacific, sinking air (high pressure) over the cold ocean currents of the eastern Pacific, westerly winds aloft between these centres of vertical motion, and easterly winds at the surface. The

surface easterlies of the Walker circulation result in the upwelling of cold ocean water along the South American coast, as well as higher sea surface temperatures and higher mean sea levels in the western Pacific. Rainfall is also higher in the western part of the basin.

The Walker circulation weakens during EL NIÑO events. In an El Niño, the pressure gradient between the eastern and western Pacific decreases, and the easterly trade winds relax. The weakened easterlies allow the warm surface waters of the western Pacific to migrate eastward. As the warm water migrates, the ascending branch of the Walker cell (and thus the zone of maximum precipitation) moves into the normally arid central and eastern Pacific. In theory, the weakening of the Walker circulation can both cause and be caused by changes in sea surface temperatures. In reality, the precise 'trigger' of an El Niño event is not completely understood.

The term 'Walker circulation' usually refers to the east–west wind regime over the tropical Pacific Ocean. However, similar zonal wind systems exist in both the tropical Atlantic and Indian Oceans. Technically the term 'Walker cell' applies to these circulation patterns as well.
RSV

Wallace's line   A zoogeographical boundary, originally put forward in 1858 by Alfred Russel Wallace, which runs through the middle of the Malay archipelago and which, Wallace argued, marked the meeting of the Asian and Australian faunas. The original Wallace line ran between the islands of Bali and Lombok and then between Borneo and Celebes (Sulawesi). In 1910 Wallace moved his line to the east, so that it lay between Celebes and the Moluccas. Many alternative lines have been suggested and, in 1928, the whole island area with its complex distributions was designated by Dickerson a separate zoogeographical region called Wallacea.

Wallace's realms   A division of the world into six zoogeographical regions, defined by their distinctive faunas and proposed by Alfred Russel Wallace in his classic of zoogeography, *The geographical distribution of animals* (1876); also called the Sclater–Wallace system of zoogeographical regions. The names of the six regions were adapted from the continents or were of a classical form, and are normally given as: the Pal(a)earctic, the Nearctic, the Neotropical, the Ethiopian, the Oriental and the Australian (see FAUNAL REALMS for a full description and illustration). Wallace based his system on the earlier work of Philip Lutley Sclater (1858), who had likewise recognized six regions based on his study of the distribution of birds. Wallace chose mammals to verify Sclater's six divisions and, in doing so, he drew on his wealth of personal exploration, which ranged from his visits to the Amazon and Rio Negro (1848–52) to south-east Asia (1854–62). Though frequently modified and expanded, the Sclater–Wallace regions remain the basis of our understanding of the great 'realms of life'. (See also FLORISTIC REALMS; WALLACE'S LINE.) PAS

**Reading**
George, W. 1962: *Animal geography*. London: Heinemann. · — 1964: *Biologist philosopher: a study of the life and writings of Alfred Russel Wallace*. London: Abelard-Schuman. · Sclater, P.L. 1858: On the general distribution of the members of the class *Aves*. *Journal of the Linnean Society of London* 2, pp. 130–45. · Smith, C.H. 1983: A system of world faunal regions. 1. Logical and statistic derivation of the regions. *Journal of biogeography* 10, pp. 455–66. · Udvardy, M.D.F. 1969: *Dynamic zoogeography*. New York: Van Nostrand Reinhold.

**Walther's law** A key to the interpretation of vertical stratigraphic sequences and environmental reconstructions. Only laterally contiguous FACIES (sedimentary subenvironments) can form vertically contiguous facies. Walther's law assumes that facies migrate laterally with local changes in depth, and thereby transpose horizontal associations into vertical associations. This relationship is useful for reconstructing the lateral associations from vertical sequences obtained from cores or exposures. DJS

**Reading**
Middleton, G.V. 1973: Johannes Walther's law of the correlation of facies. *Geological Society of America bulletin* 84, pp. 979–88.

**waning slopes** Depositional hillslope units formed at the base of a talus (scree) slope as weathering and rain wash fine-grained particles from the talus and deposit them as a concave unit which may become progressively flatter (Wood 1942). L.C. King (1957) adopted the basic definitions of slope units, proposed by Wood, and the waning slope is identified as a rock-cut pediment with a veneer of sediment produced by surface wash. MJS

**References**
King, L.C. 1957: The uniformitarian nature of hillslopes. *Transactions of the Edinburgh Geological Society* 17, pp. 81–102. · Wood, A. 1942: The development of hillside slopes. *Proceedings of the Geologists' Association* 53, pp. 128–40.

**warm front** A frontal zone in the atmosphere where, from its direction of movement, cool air is being replaced by rising warm air. As the front approaches, cirrus clouds are gradually replaced by cirrostratus, altostratus and finally nimbostratus clouds. Precipitation usually occurs within a wide belt up to about 400 km ahead of the surface front. As the surface front passes, temperatures and dewpoint increase, winds veer and pressure stops falling. Most fronts exhibit considerable differences from this standard model outlined, partly depending upon the rate of uplift within the warm air. Well-developed warm fronts are relatively rare in the southern hemisphere extra-tropical cyclone belt where sources of warm air for the warm sector are limited. Kinematically and dynamically there is no fundamental difference between cold and warm fronts, despite the views of the Norwegian School of Meteorology which made a sharp distinction between their roles in the precipitation process. PS

**warm ice** See TEMPERATE ICE.

**warm occlusion** See OCCLUSION.

**warm sector** The area of warm air lying between the warm and cold fronts of an EXTRA-TROPICAL CYCLONE. It eventually disappears from the surface as the cyclone evolves to become occluded. Temperatures in the warm sector are noticeably higher than in the preceding and following air streams. Cloud and precipitation are very variable. With a strong ridge of high pressure there may be clear skies but more frequently the skies are overcast. Heavy rain is likely over upland areas if the warm sector is potentially or conditionally unstable. PS

**warping** The bending and deformation of extensive areas of the earth's crust without the formation of folds or faults.

**washboard moraine** Morainic ridges a few metres high lying transverse to the direction of former ice flow. Also called *cross-valley*, and *De Geer*. There are almost as many theories about transverse moraines as there are researchers, but one common characteristic is that they are often associated with the presence or former presence of lakes. They may be a subaqueous example of a push moraine (perhaps annual). Others may relate to subglacial thrusting as proposed for rogen moraines. (See also MORAINE.) DES

**Reading**
Embleton, C. and King, C.A.M. 1975: *Glacial geomorphology*. London: Edward Arnold.

**water balance** The water balance or water budget of an area over a period of time represents the way in which precipitation during the time period is partitioned between the processes of evapotranspiration and run-off, taking account of changes in water storage. The water balance summarizes the changes in the components of the HYDROLOGICAL CYCLE during a particular time period and may be expressed for a drainage basin as:

$$P = Q + Et \pm \Delta SS \pm \Delta SMS \pm \Delta AZS \\ \pm \Delta GS \pm DT$$

where all the variables are expressed as a depth of water over the catchment area for the time period studied. $P$ is precipitation, $Q$ is run-off, $Et$ is evapotranspiration losses, $\Delta SS$ is change in surface storage, $\Delta SMS$ is change in soil moisture storage, $\Delta AZS$ is change in aeration zone storage, $\Delta GS$ is change in groundwater storage, and $DT$ is deep transfer of water across the watershed.

The water balance can be evaluated by direct field measurements, or a climatic water balance can be calculated by making simplifying assumptions about the role and operation of different stores. A water balance can be calculated for any size of area from small LYSIMETERS up to whole continents and it can be used to isolate the hydrological effects of man's activities, but water balance evaluation is most widely applied at the drainage basin scale. A water balance can also be calculated for any time period, but it is simplest to evaluate for a time period where storage is approximately the same at the start and end of the period. In the UK the *water year* which runs from 1 October to 30 September is often used for water balance calculations because this starts and ends at the period of minimum storage in most years. If the storage can be assumed to be the same at the start and end of the required period the water balance equation simplifies so that precipitation is only partitioned between evapotranspiration and run-off.

Estimation of the water balance of individual basins is often used to check or to estimate the magnitude of a water balance component that is difficult to measure. Evapotranspiration loss has been frequently studied in this way and formed the basis of the Institute of Hydrology's original Plynlimon catchment study, where losses from an afforested catchment and a grassland catchment were estimated using a water balance approach and were compared with potential evapotranspiration rates calculated from observations from automatic weather stations. Water balance studies are also useful in identifying the impact of catchment modifications on hydrological processes. This is well illustrated by Clarke and Newson's analysis (1978) of the impact of the 1975–6 drought in the Institute of Hydrology's research catchments and the implications for the water balance of drainage basins under different vegetation cover.

Calculation of the water balance from climatic data was first attempted by Thornthwaite (1948) in his classification of world climates. The evaluation of a climatic water balance requires observations or estimates of precipitation and potential evapotranspiration (see POTENTIAL EVAPORATION) and a means of correcting potential evapotranspiration to an actual evapotranspiration rate by taking account of the degree to which evapotranspiration is limited by availability of soil moisture. Baier (1968) reviewed various proposals for the relationship between the ratio of actual to potential evapotranspiration and percentage soil moisture availability. The UK Meteorological Office (Grindley 1967) assumes a mix of riparian, short-rooted and long-rooted vegetation. The riparian vegetation is assumed continually to transpire at the potential rate and the short- and long-rooted vegetation are assumed to transpire at the potential rate up to soil moisture deficits of 75 mm and 200 mm respectively, and then at a gradually reducing rate. Thus, the water budget which partitions precipitation between evapotranspiration, soil moisture storage and rainfall excess can be calculated from an area and for consecutive time periods, most frequently producing a monthly water budget.

The evaluation of the water balance provides an essential stage in the estimation of the water resources of an area and so mapping of water balance components, notably precipitation, evapotranspiration and run-off have received a great deal of attention. Gurnell (1981) reviews attempts at evapotranspiration mapping and many of these maps fall within a context of water balance mapping, ranging from maps of the water balance of administrative areas within countries, to national maps and continental maps (Doornkamp *et al.* 1980), to maps for the whole world (Baumgartner and Reichel 1975; UNESCO 1978a and b). The table summarizes the water balance of the continents according to Baumgartner and Reichel (1975). Mapping the water balance for such large areas requires the combined use of measurements of precipitation, run-off and climatic variables which allow the estimation of evapotranspiration; the evaluation of climatic water balances and the use of regional relationships between altitude and each of the water balance variables.

Water balance components in millimetres for world continents

| Continent | Precipitation | Evapotranspiration | Run-off |
|---|---|---|---|
| Europe | 657 | 375 | 282 |
| Asia | 696 | 420 | 276 |
| Africa | 696 | 582 | 114 |
| Australia (with islands) | 803 | 534 | 269 |
| Australia (without islands) | 447 | 420 | 27 |
| North America | 645 | 403 | 242 |
| South America | 1564 | 946 | 618 |
| Antarctica | 169 | 28 | 141 |

Source: Baumgartner and Reichel 1975.

In this way isoline maps of the water balance components over large areas may be produced.

Water balance information is a useful tool in water resources planning because it can represent both long-term average and extreme conditions at a site. Extreme values of soil moisture deficit and water surplus may give an indication of the drought and flood producing potential of an area and temporal trends in the water balance may reflect climatic change or the influence of catchment modification. The water balance equation and the concept of the hydrological cycle are the foundations of hydrological studies.                                                 AMG

**References**
Baier, W. 1968: Relationship between soil moisture, actual and potential evapotranspiration. In *Soil moisture*. Proceedings of the Hydrology Symposium 6. University of Saskatchewan, 15–16 November 1967. Ottawa: Queen's Printer. · Baumgartner, A. and Reichel, E. 1975: *The world water balance: mean annual global, continental and maritime precipitation, evaporation and runoff*. New York: Elsevier. · Clark, R.T. and Newson, M.D. 1978: Some detailed water balance studies of research catchments. *Proceedings of the Royal Society* series A, 363, pp. 21–42. · Doornkamp, J.C., Gregory, K.J. and Burn, A.S. eds 1980: *Atlas of drought in Britain 1975–1976*. London: Institute of British Geographers. · Grindley, J. 1967: The estimation of soil moisture deficits. *Meteorological magazine* 96, pp. 97–108. · Gurnell, A.M. 1981: Mapping potential evapotranspiration: the smooth interpolation of isolines with a low density station network. *Applied geography* 1, pp. 167–83. · Thornthwaite, C.W. 1948: An approach toward a rational classification of climate. *Geographical review* 38, pp. 55–94. · UNESCO 1978a: *World water balance and water resources of the earth*. Paris: UNESCO. ——1978b: *Atlas of the world water balance*. Paris. UNESCO.

**water mass**   A body of water having approximately uniform characteristics typically acquired in a source region in contact with the atmosphere. Oceanic water masses are usually identified from their values of temperature and salinity which are conservative properties, although non-conservative properties – for example, the concentrations of dissolved oxygen or of nutrients such as phosphate and silica – can be useful in tracing their movements. Upper water masses are found above the THERMOCLINE where their formation is controlled by the pattern of surface currents (see CURRENTS, OCEAN). For example, the subtropical GYRES enclose central waters of different types all of which have relatively high temperatures and salinities. The source regions of deeper water masses are predominantly at high latitudes where the thermocline is weak or absent. Most of the Atlantic Ocean basin is occupied by North Atlantic deep water; south of about 30 °N Antarctic bottom water is found close to the sea bed and Antarctic intermediate water at depths in the range 500–1500 m. These deep water masses of the Atlantic are characterized by values of temperature $(T)$ and salinity $(S)$ that vary within very narrow limits; such water masses may be designated *water types*. Other water masses, including the central waters mentioned above, have a characteristic range of values of $T$ and $S$ (see table on p. 524). Investigations of the stratification of the oceans and the identification, movement and mixing of water masses is greatly facilitated by the plotting of data values on a graph of $T$ as ordinate against $S$ as abscissa (a $T$–$S$ diagram) on which lines of constant density (isopycnals) are convex upwards. (See pp. 525–6 for diagrams.)                      JEA

**Reading**
Harvey, J.G. 1976: *Atmosphere and ocean*. London: Artemis Press. · Mamayev, O.I. 1975: *Temperature–salinity analysis of world ocean waters*. Amsterdam: Elsevier. · Open University Oceanography Course Team 1989: *Seawater: its composition, properties and behaviour*. Oxford: Pergamon; Milton Keynes: Open University Press. · Sverdrup, H.U., Johnson, M.W. and Fleming, R.H. 1942: *The oceans*. Englewood Cliffs, NJ: Prentice-Hall. · Tolmazin, D. 1985: *Elements of dynamic oceanography*. Boston and London: Allen & Unwin.

**water table**   The surface defined by the level of free standing water in fissures and pores at the top of the saturated zone. It is an equilibrium

Major water masses of the world oceans and their T–S characteristics

| Location | Atlantic Ocean | | | Indian Ocean | | | Pacific Ocean | | |
|---|---|---|---|---|---|---|---|---|---|
| | Name | T(°C) | S(%) | Name | T(°C) | S(%) | Name | T(°C) | S(%) |
| Central waters | North Atlantic | 20.0 | 36.5 | Bay of Bengal | 25.0 | 33.8 | Western North Pacific | 20.0 | 34.8 |
| | South Atlantic | 18.0 | 35.9 | Equatorial | 25.0 | 35.3 | Eastern North Pacific | 20.0 | 35.2 |
| | | | | South Indian | 16.0 | 35.6 | Equatorial Western South Pacific | 20.0 | 35.7 |
| Intermediate waters | Atlantic subarctic | 2.0 | 34.9 | – | | | Pacific subarctic | 5–9 | 33.5– 33.8 |
| | Mediterranean intermediate | 11.9 | 36.5 | Red Sea intermediate | 23.0 | 40.0 | North Pacific intermediate | 4–10 | 34.0– 34.5 |
| | Antarctic intermediate | 2.2 | 33.8 | Timor Sea intermediate | 12.0 | 34.6 | South Pacific intermediate | 9–12 | 33.9 |
| | | | | Antarctic intermediate | 5.2 | 34.7 | Antarctic intermediate | 5.0 | 34.1 |
| Deep and bottom waters | North Atlantic deep and bottom | 2.5 | 34.9 | Antarctic deep and bottom | 0.6 | 34.7 | Antarctic deep and bottom | 1.3 | 34.7 |
| | Antarctic deep | 4.0 | 35.0 | | | | | | |
| | Antarctic bottom | −0.4 | 34.66 | | | | | | |

*Source*: Tolmazin 1985, table 7.1. After Mamayev 1975; Sverdrup *et al*. 1942.

surface at which fluid pressure in the voids is equal to atmospheric pressure. The equivalent term in continental European literature is the PIEZOMETRIC SURFACE. A potential source of confusion arises because the latter is sometimes used in English literature to describe the water-level elevations in wells tapping a confined artesian AQUIFER. The term *potentiometric surface* is preferred for this (Freeze and Cherry 1979).

PWW

**Reference**
Freeze, R.A. and Cherry, J.A. 1979: *Groundwater*. Englewood Cliffs, NJ: Prentice-Hall.

**water transfer**  The movement of water away from natural water courses, to satisfy human needs. This is a common feature in DRYLANDS, to provide water for agricultural irrigation. It is also carried out to meet the demands of urban areas. The term water transfer is increasingly used however to refer to the movement of water from one drainage basin or catchment, where there is a perceived surplus related to human needs, to another where there is a deficit. To achieve such transfers it is necessary not only to divert and store water, but to allow its passage across, or through, the divides or watersheds that separate catchments. Water transfers are a feature of the drier western USA, especially in order to satisfy the demands of heavily populated areas in, for example, California.  DSGT

**waterfall**  A stream that falls from a height. Waterfalls are often the sites of greatest concentration of energy dissipation along the course of a stream and have generally been regarded as

forming where a soft rock is eroded from beneath a harder rock (the caprock model). However, in reality, waterfalls have more diverse forms than this simple model would suggest. It is likely that it is applicable only in areas with gently dipping strata. In addition to such structural control, waterfalls depend for their development on such factors as glacial overdeepening, tectonic changes and base-level change.  ASG

**Reading**
Young, R.W. 1985: Waterfalls: form and process. *Zeitschrift für Geomorphologie*, supplementband 55, pp. 81–95.

**watershed**  The boundary which delimits a drainage basin as the basic hydrological unit. On large-scale topographical maps it is often drawn as a line according to the contour information, but this surface watershed may not correspond with the subsurface boundary of the basin delimited according to the WATER TABLE as the PHREATIC DIVIDE. Sometimes, especially in North America, watershed is used synonymously with catchment or drainage basin area.  KJG

**waterspout**  A vortex disturbance that forms in the atmosphere over a body of water when the atmosphere is unstable and the air stagnant. Waterspouts are favoured by conditions of high temperature and humidity, and still air, and can cause damage to and even the foundering of large boats.  ASG

**Reading**
Gordon, A.H. 1951: Waterspouts. *Weather* 6, pp. 364–71.

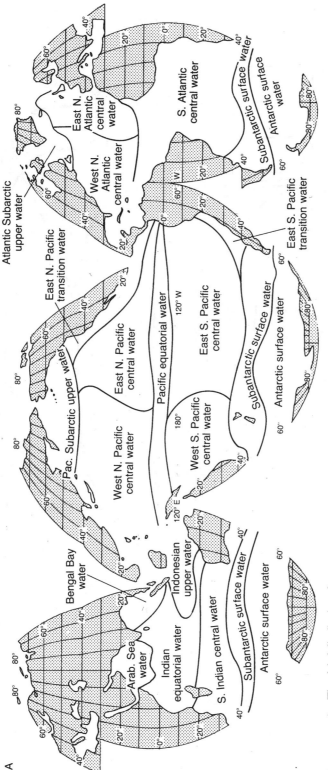

**Water mass: 1.**   The global distribution of upper water masses.

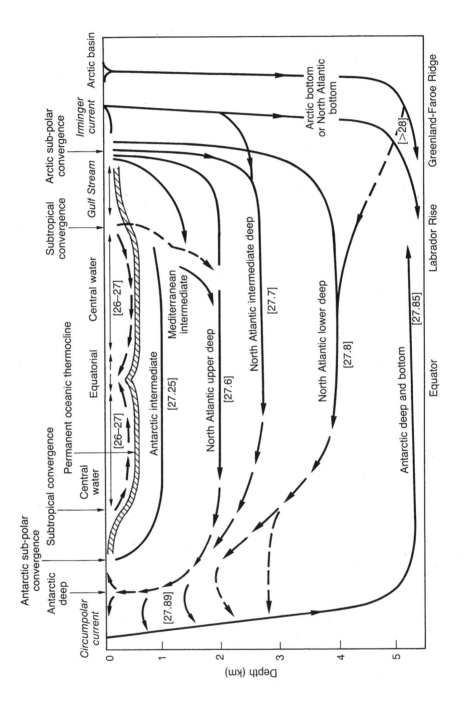

**Water mass: 2.** Schematic representation of the meridional circulation in the Atlantic Ocean. The numerical values in square brackets are mean density values ($\sigma_t$).
*Source:* Tolmazin 1985.

**watten**  Tidal marshland between the mainland and an offshore sand bar.

**wave cut platform**  See SHORE PLATFORMS.

**waves**  Regular oscillations in the water surface of large water bodies. They are produced by the pushing effect of wind on the water surface. Energy is transferred from the wind to the water and eventually dissipated when the waves hit a shoreline. Waves are of great geomorphological importance because of the role they play in coastal sedimentation and erosion processes. In deep water, wave form can be predicted by a variety of equations, and movement of individual water particles takes the form of circular orbits within the wave. On approaching shallow water, wave forms become more complex and there is a general forward trend to the movement of water particles.                        HV

**Reading**
Carter, R.W.G. 1988: *Coastal environments*. London: Academic Press.

**waxing slopes**  Hillslope units forming crests of convex profile at the intersection of the cliff and the hilltop or plateau surface. It is assumed by Wood (1942) that these slope units will increase in significance in the landscape as cliffs (free faces) and talus (scree) slopes are eliminated, and that they will lengthen to become major parts of the total hillslope as a result of weathering and soil creep.               MJS

**Reference**
Wood, A. 1942: The development of hillside slopes. *Proceedings of the Geologists' Association* 53, pp. 128–40.

**weather**  The overall state of the atmosphere on a time-scale of minutes to months, with particular emphasis on those atmospheric phenomena that affect human activity. Thus sunshine, temperature, rainfall, wind, cloud contribute to weather whereas air density does not. In contrast to weather, CLIMATE is concerned with the long-term behaviour of the atmosphere.     BWA

**weather forecasting**  The science of predicting the future state of the atmosphere from very short periods of less than one hour up to seven or even ten days ahead. In recent years the science has been revolutionized by the application of modern technology, e.g. by satellites and radar, and by the development of advanced mathematical models which can be run on the most powerful of modern electronic computers.

In many of the more advanced forecasting centres operations are carried out as an automatic routine with many thousands of observations flowing in continuously from around the globe. These come from manned surface and upper air stations, automatic data buoys in remote areas, flight level reports from commercial aircraft, and polar orbiting and geostationary satellites. The data are accepted automatically into the computing system and are first assimilated from their real geographically scattered pattern into a regular global network of so-called gridpoints at typically fifteen levels from the surface to the STRATOSPHERE. These points number around 36,000 at a given level or approximately 500,000 in total. Given that seven variables are assimilated at each point (including TEMPERATURE, WIND and HUMIDITY), it is apparent that about $3.5 \times 10^7$ numbers are stored to define the state of the atmosphere at the given observation time.

At every point a set of predictive equations is solved in order to calculate the values of temperature, wind, humidity, etc. for a sequence of 'time-steps' to produce a forecast for, say, 12, 24, 48 and 72 hours ahead. The longer term or medium range (up to seven days) forecasts are based on observations taken at the main synoptic hours of 00 and 12 GMT. The most powerful computers in existence are used operationally and are currently capable of carrying out $5 \times 10^7$ instructions per second; a ten-day global weather forecast involves around $5 \times 10^{11}$ operations executed over a four-hour period.

Modern weather forecasting by this means provides a vast range of products for appraisal by forecasters who are still required to interpret and to communicate the information to the user. Predicted fields include the traditional mean sea-level pressure charts and upper air charts along with more modern fields of 'large-scale' and 'convergence scale' precipitation totals on a regular grid and forecast TEPHIGRAMS. Very short-term prediction of precipitation (nowcasting) in the period nought to six hours ahead is underway in Britain, based on the linear extrapolation of radar network observations of precipitation patterns.

Weather forecasting services are provided for many users, for example offshore oil developments (sea and swell, deck level winds), civil aviation (wind, temperature and clear air turbulence along the flight), ship routeing (optimum course for a given journey) and agriculture (e.g. crop spraying operations and weather-related disease outbreak).                        RR

**Reading**
Aktinson, B.W. 1968: *The weather business*. London: Aldus Books. · Browning, K.A. ed. 1982. *Nowcasting*. London: Academic Press. · Gadd, A.J. 1981: Numerical modelling of the atmosphere. In B.W. Atkinson ed.,

*Dynamical meteorology; an introductory selection.* London and New York: Methuen. Pp. 194–204. · Wickham, P.G. 1970: *The practice of weather forecasting.* London: HMSO.

**weather modification** This refers to both intentional and inadvertent alteration of natural atmospheric processes by human intervention. Most commonly, intentional weather modification research involves cloud seeding to enhance precipitation, suppress hail formation, lessen tropical storm intensity, disperse supercooled fog, and increase understanding of cloud physics and dynamics. Additionally, increasing evidence is linking anthropogenic activities to unintentional weather modification. For example, urban development has replaced natural surfaces with high thermal mass materials, such as asphalt and concrete, causing minimum temperatures to increase in the heat island effect. Industrial and automobile pollutant gases act as condensation nuclei, increasing rainfall downwind of urban and industrial centres. The weekly build-up of pollutant gases has also been linked to changes in the temporal patterns of precipitation along the Atlantic coast of the United States. Excessive grazing, DEFORESTATION and DESERTIFICATION decrease moisture retaining surface vegetation, leading to higher temperatures and drier conditions. A recent study has linked anti-radar chaff releases by the military to the suppression of lightning in intense convective storms in the south-west United States. NJS

**weather type** The categorization of the principal modes or patterns of atmospheric circulation, on a local, regional or hemisphere scale produces a series of classes known as weather types. Each type tends to be associated with a particular type of weather, and weather typing represents the first stage in a SYNOPTIC CLIMATOLOGY study. Objective methods of typing pressure patterns are superseding subjective methods, as computer resources and suitable gridded data sets become more available (Barry 1980; Perry 1983). Among the longest daily weather type catalogues are those of the UK (Lamb 1972) and for central Europe. AHP

**References**
Barry, R.G. 1980: Recent advances in synoptic and dynamic climatology. *Progress in physical geography* 4, pp. 88–96. · Lamb, H.H. 1972: *British Isles weather types and a register of the daily sequences of circulation patterns 1861–1971.* Meteorological Office Geophysical memoir 118. London: HMSO. · Perry, A.H. 1983: Growth points in synoptic climatology. *Progress in physical geography* 7, pp. 90–6.

**weathering** One of the most important of geomorphological and pedological processes, occurring when the rocks and sediments in the top

Classification of weathering processes

*Processes of disintegration (physical or mechanical weathering)*

Crystallization processes
  salt weathering (by crystallization, hydration and thermal expansion)
  frost weathering
Temperature change processes
  insolation weathering (heating and cooling)
  fire
  expansion of dirt in cracks
Wetting and drying (especially of shales)
Pressure and release by erosion of overburden
Organic processes (e.g. root wedging)

*Processes of decomposition (chemical weathering)*

Hydration and hydrolysis
Oxidation and reduction
Solution and carbonation
Chelation
Biological chemical changes

metres of the earth's crust are exposed to physical, chemical and biological conditions much different from those prevailing at the time the rocks were formed. Two main types are recognized (see table). Mechanical weathering involves the breakdown or disintegration of rock without any substantial degree of chemical change taking place in the minerals which make up the rock mass, while chemical weathering involves the decomposition or decay of such minerals. In most parts of the world both types of weathering may operate together though in differing proportions, and one may accelerate the other. For example, the mechanical disintegration of a rock will greatly increase the surface area that is then exposed to chemical attack. The rate at which weathering occurs will depend both on climatic conditions (e.g. the frequency of frost, the amount of water available for chemical reactions, etc.) and rock character (e.g. porosity, the density of jointing, and the susceptibility of the minerals). Consequences of weathering are duricrust formation, the development of karst, and the provision of material for removal by erosive processes. ASG

**Reading**
Goudie, A.S. 1981: Weathering. In A.S. Goudie ed., *Geomorphological techniques.* London: Allen & Unwin. Pp. 139–55. · Ollier, C.D. 1969: *Weathering.* Edinburgh: Oliver & Boyd. · Yatsu, E. 1988: *The nature of weathering.* Tokyo: Sozosha.

**weathering front** The limit of the zone of weathering of bedrock beneath the landsurface.

**weathering index** A quantitative or semi-quantitative indicator of the degree of weathering.

Indices of both chemical and physical weathering have been proposed. Chemical indices of the alteration of rocks, $W_I$, may be expressed as the ratio of unweathered to weathered minerals in a volume of material, or as the ratio of the chemically more mobile to the chemically less mobile species, in the general form:

$$W_I = \frac{\text{Proportion of chemical reactants}}{\text{Proportion of residual products}}$$

or more specifically in a form such as:

$$W_I = \frac{\text{feldspar} + \text{mica} + \text{calcite}}{\text{clay minerals} + \text{quartz}}$$

Physical indices of weathering have been expressed in a number of ways such as: capacity of unweathered material to take up water compared with the capacity of weathered material; the softening effect of water on material; the degree of swelling and slaking which results from water absorption; changes in strength or hardness of minerals; and velocity of ultrasound (seismic) waves through materials.                                                                    MJS

**Reading**
Selby, M.J. 1993: *Hillslope materials and processes*. 2nd edn. Oxford: Oxford University Press. Ch. 8.

## weathering potential index
This is a measure of the ease with which rocks decay chemically. It is the ratio of the mole percentage of alkaline and alkaline earth metals (Ca, Na, Mg; K) less combined water to the mole percentage total metals (including Fe and Ti) less water. It provides a means of assessing the Goldich weathering stability series in numerical form. However, the computed value of quartz (zero) is not in agreement with its generally stable nature.                                                              WBW

**Reference**
Wahlstrom, E.E. 1948: Pre-Fountain and recent weathering on Flagstaff Mountain near Boulder, Colorado. *Geological Society of America bulletin* 59, pp. 1173–89.

## weathering profile
The complete cross-section of the chemically altered zone above fresh bedrock; essentially developed *in situ* by weathering processes. Combined with the near-surface, transported layer also termed REGOLITH. It commonly extends below the conventional *soil profile* with its A, B, C horizons contained within 1–5 m of the ground surface, and may reach depths of 100 m or more. At the base of the profile is found the *weathering front*, where fundamental physico-chemical changes cause the rock to become fissile, and disaggregate, transforming up-profile to a mixture of sand and clay usually called *saprolite*. In susceptible rock affected by intense weathering, as below well drained sites within the forested tropics, almost complete transformation to saprolite can occur within a narrow zone of $m^{-1}$, and the weathering front can appear as a *basal weathering surface*. In other situations a zone of gradual transition from saprolite to bedrock may occur over $m^1$. Such transitional materials with a low clay content ($< 2$–7%), are called *saprock* or *grus* (German).

Chemical alteration generally increases towards the surface and the content of fresh rock declines. Metal cation mobility during weathering allows DURICRUSTS to form under stable conditions. These crusts are dominated by one or more metallic oxide: aluminium (alucrete or bauxite), iron (ferricrete or laterite), silica (silcrete) and calcium (calcrete or caliche) are the most common. A six-fold division of the weathering profile is often used (esp. in engineering) (see table and figure).                        MFT

Scale of weathering grades of rock mass

| Term | Description | Grade |
|---|---|---|
| Fresh | No visible sign of rock material weathering; perhaps a slight discolouration on major discontinuity surfaces. | I |
| Slightly weathered | Discolouration indicates weathering of rock material and discontinuity surfaces. All the rock material may be discoloured by weathering. | II |
| Moderately weathered | Less than half of the rock material is decomposed or disintegrated to a soil. Fresh or discoloured rock is present either as a continuous framework or as corestones. | III |
| Highly weathered | More than half of the rock material is decomposed or disintegrated to a soil. Fresh or discoloured rock is present either as a discontinuous framework or as corestones. | IV |
| Completely weathered | All rock material is decomposed and or disintegrated to soil. The original mass structure is still largely intact. | V |
| Residual soil | All rock material is converted to soil. The mass structure and material fabric are destroyed. There is a large change in volume, but the soil has not been significantly transported. | VI |

| | | |
|---|---|---|
| Humus/topsoil | | |
| VI Residual soil | All rock material converted to soil; mass structure and material fabric destroyed. Significant change in volume. | |
| V Completely weathered | All rock material decomposed and/or distintegrated to soil. Original mass structure still largely intact. | |
| IV Highly weathered | More than 50% of rock material decomposed and/or disintegrated to soil. Fresh/discoloured rock present as discontinuous framework or corestones. | |
| III Moderately weathered | Less than 50% of rock material decomposed and/or disintegrated to soil. Fresh/discoloured rock present as continuous framework or corestones. | |
| II Slightly weathered | Discolouration indicates weathering of rock material and discontinuity surfaces. All rock material may be discoloured by weathering and may be weaker than in its fresh condition. | |
| IB Faintly weathered | Discolouration on major discontinuity surfaces | |
| IA Fresh | No visible sign of rock material weathering | |

A. Idealized weathering profiles without corestones (left) and with corestones (right)    B. Example of a complex profile with corestones

Rock decomposed to soil
Weathered / disintegrated rock
Rock discoloured by weathering
Fresh rock

**Reading**
Fookes, P.G. 1997: *Tropical residual soils*. Geological Society Professional Handbooks. London: The Geological Society. · Ollier, C.D. 1984: *Weathering* 2nd edn. London: Longman. · Gerrard, A.J. 1988: *Rocks and landforms*. London: Unwin Hyman. · Thomas, M.F. 1994: *Geomorphology in the tropics*. Chichester: Wiley.

**weathering rinds**    Oxidation phenomena which stain the parent rock red-yellow when exposed to air or near-surface groundwater for some time. They may extend for more than a millimetre into the rock, whereas DESERT VARNISH is generally much thinner. The thickness of weathering rinds may have some utility for relative dating of surface outcrops (Anderson and Anderson 1981).    ASG

**Reference**
Anderson, L.W. and Anderson, D.S. 1981: Weathering rinds on quartz arenite clasts as a relative age indicator and the glacial chronology of Mount Timpanogos, Wasatch Range, Utah. *Arctic and alpine research* 13, pp. 25–31.

**weathering-limited slopes**    Hillslopes on which the potential rate at which weathered soil and debris can be removed by erosional processes exceeds the rate at which the material can be produced by weathering. The rate of ground loss is consequently controlled by the rate of weathering, and the form of the hillslope unit is controlled by the relative resistance to weathering of the rock masses on which the slope is formed.    MJS

**Reading**
Young, A. 1972: *Slopes*. Edinburgh: Oliver & Boyd.

**wedge failure**    The removal of blocks of rock (or, more unusually, clay) where two or three fractures, often joints, intersect at a high angle in a cliff, and dip down towards the valley. The size of blocks from this kind of fall is rarely greater than a few $10 \, m^3$. (See also SLAB FAILURE; TOPPLING FAILURE.)    WBW

**weeds**    See ALIENS.

**weir**    A structure built across a river or stream channel in order to measure the flow. It may be built of concrete, metal or wood and possesses two major features. First, it ponds back the flow to create a pool with flow velocities. Secondly, it incorporates a crest or notch over or through which the water flows freely from the upstream pool. A unique and stable relationship exists between the depth or head of water above the weir crest and water discharge. Values of discharge may be obtained from measurements of water stage by using published formulae or calibration curves established using other methods of flow measurement. (See also DISCHARGE.)    DEW

Reading
Ackers, P., White, W.R., Perkins, J.A. and Harrison, A.J.M. 1978: *Weirs and flumes for flow measurement.* Chichester: Wiley.

**wells** Vertical dug or bored shafts penetrating to the saturated zone for the purpose of exploiting GROUNDWATER. In some old literature and on old maps, wells may refer to SPRINGS used as water supplies. PWW

**westerlies** Belts of winds, generally southwesterly in the northern hemisphere and north-westerly in the southern hemisphere, with average position between about latitudes 35° and 60°. These belts move poleward in the winter and equatorward in summer and are part of the GENERAL CIRCULATION OF THE ATMOSPHERE, being the low-level branch of the FERREL CELL. KJW

Reading
Lutgens, F.K. and Tarbuck, R.E.J. 1982: *The atmosphere.* 2nd edn. Englewood Cliffs, NJ: Prentice-Hall.

**wet-bulb temperature** The temperature at which a sample of air will become saturated by evaporating pure water into it at constant pressure. It is measured by a thermometer whose bulb is covered with muslin which is kept constantly wet with distilled water. The lower the humidity is, the stronger the evaporation from the muslin and the stronger the cooling of the thermometer bulb. Wet-bulb temperature is an indicator of absolute humidity and its difference from the dry-bulb is a measure of relative humidity. RR

**wetland** Defined by Maltby (1986) as 'a collective term for ecosystems whose formation has been dominated by water, and whose processes and characteristics are largely controlled by water. A wetland is a place that has been wet enough for a long time to develop specially adapted vegetation and other organisms.' They include areas of marsh, fen, peatland or water, whether natural or artificial, permanent or temporary, with water that is static or flowing, brackish or salt, including marine water the depth of which at low tide does not exceed 6 m. ASG

Reference
Maltby, E. 1986: *Waterlogged wealth. Why waste the world's wet places?* London: Earthscan. · Williams, M. ed. 1990: *Wetlands: a threatened landscape.* Oxford: Blackwell.

**wetted perimeter** The perimeter of a river channel which is covered by water at a specific stage of flow. Such a wetted perimeter ($p$) is used together with the cross-sectional area ($a$) of the water at the same stage to give the HYDRAULIC RADIUS as $a/p$. KJG

**wetting front** The subsurface limit to which a soil, particularly in a desert region, becomes saturated by infiltrating rainwater.

**whirlwind** Loosely used term to describe rotating winds of scales up to that of a tornado. They are usually manifestations of intense convection over very small areas. The winds are frequently strong enough to raise surface materials (causing dust devils) and can become extremely hazardous when associated with large fires, such as may result from heavy bombing, or as occur in semi-arid bush areas. JSAG

**width–depth ratio** A simple measure of the shape of a river channel cross-section usually obtained as $w/d$ where $w$ is the top width of the cross-section to be characterized and $d$ is derived as an average value by dividing the cross-sectional area ($a$) by $w$ in the form:

$$d = \frac{a}{w}$$

The ratio is conventionally evaluated for the bankfull stage, and Riley (1976) suggested two alternative measures: namely bed width, and the exponent $x$ in the equation:

$$w_1 = cd^x$$

where $w_1$ is width at stage $d_t$. KJG

Reference
Riley, S.J. 1976: Alternative measures of river channel shape and their significance. *Journal of hydrology* (NZ) 15, pp. 9–16.

**wilderness** A wild, uncultivated area. Ideally, wilderness areas should never have been subject to human activity which has resulted in manipulation, either deliberate or unconscious, of the ecology of the area (Simmons 1981, p. 62). ASG

Reference
Simmons, I. 1981: *The ecology of natural resources.* 2nd edn. Oxford: Blackwell.

**Wilson cycle** One of the hypothesized long-term tectonic cycles related to the mechanisms of plate tectonics, and named in honour of J. Tuzo Wilson. Modern ocean basins are no more than 200–300 million years old, and many will close, as a result of plate movement, within a similar period. Plate tectonic processes though are thought to have been active over > 2 billion years, allowing sufficient time for many cycles of opening and closing of ocean basins. When oceans close and continents are sutured together, the larger mass of continental crust insulates the mantle beneath, which warms slowly until convection is triggered. Continental fragmentation then begins anew, with rifting, sea floor spreading, and ultimately the next closing of the oceans occurring in a sequential or cyclic manner. This sequence of events, spanning 500 Ma or so, is the Wilson cycle.   DLD

**Reading**
Kearey, P. and Vine, F.J. 1990: *Global tectonics*. Oxford: Blackwell.

**wind**   Air in motion relative to the surface of the earth. It is important to note that this definition covers more than the layman's view of simple horizontal airflow. First, the motion may also have a substantial vertical component, and indeed it is this component that is a primary cause of all the WEATHER we experience on this planet. Secondly, it is vital to appreciate that the motion is *relative* to the surface of the earth. Thus, in 'calm' conditions, when air is stationary relative to the surface, it is still moving through space at the same velocity as the surface of the earth. This characteristic has important repercussions on the direction of airflows and upon the momentum budget of the atmosphere.

Wind is specified in terms of speed and direction. Speeds are reported in knots or metres per second, the reading being taken over a few minutes to reduce the effects of very short duration gusts, and direction (which is the direction from which the air flows) is reported to the nearest $10°$ of the compass. If wind direction varies with time a clockwise shift is called veering and an anticlockwise one backing. At the surface of the earth winds are measured by ANEMOMETERS, with the actual sensor preferably being at a height of 10 m. At higher levels winds are measured by pilot balloons and RADIOSONDES. In both cases the measurements are made of only the horizontal component of the wind. Hence we have no routine direct measurements of vertical winds, largely because they are notoriously difficult to measure at all, let alone accurately. This is so because of the smallness of vertical windspeeds: the speed of vertical winds is usually of the order of $cm\,s^{-1}$, whereas that of horizontal winds is of the order of $m\,s^{-1}$. This lack of information on vertical motion is most frustrating both for the theoretician and for the practical meteorologist, such as a forecaster. It is precisely the phenomenon that we wish to measure that, up to now, has proved unmeasurable.

Although observations of winds have been, and continue to be, vital to our knowledge of the atmosphere, it is true to say that deeper understanding has emerged from theoretical analyses of airflow. These analyses all rest on Newton's Second Law, which states that if a

force is applied to a mass, it will accelerate that mass in the direction of the force. This law can readily be applied to a particle of air in the atmosphere. The value in doing this lies in the acceleration part of the law. Acceleration is the change of velocity with time.

Hence, if we can derive an acceleration of air at a given time from the application of Newton's law, we can then derive an air velocity at a future time. This has the double advantage not only of giving information about the velocity of the airflow, which has been shown to be fundamental to weather and climate, but also of giving its value at a future time, that is a forecast. The acceleration of the air largely results from the application of three forces: the pressure gradient force, the CORIOLIS FORCE and the FRICTION force. Hence the equation of motion describing the flow in a horizontal direction – say $x$ – is:

$$\frac{\partial u}{\partial t} = -\frac{1}{\rho}\frac{\partial p}{\partial x} + 2\Omega v \sin\phi$$

*Acceleration   Pressure   Coriolis force*
*gradient force*

$$+\frac{\partial}{\partial z}\left(\frac{K\partial u}{\partial z}\right)$$

*Frictional force*

where $u$ is the air velocity in $x$ direction, $t$ is time, $\rho$ is air density, $p$ is air pressure, $\Omega$ is the angular velocity of the earth's rotation, $v$ is the air velocity along a direction perpendicular to $x$, $\phi$ is latitude, $z$ is the vertical coordinate, and $K$ is the EDDY DIFFUSIVITY of momentum.

In large-scale meteorology it is customary to make the $x$-direction east–west, the $y$-direction north–south and the $z$ direction vertical. The equation is thus complemented by similar ones in the $y$ and $z$ directions, but for present purposes it suffices.

The fact that four terms appear in the equation does not mean that all four are important to all airflows at all times. This was appreciated by Jeffreys in the 1920s and by scrutiny of each term he was able to classify winds dynamically. In addition Jeffreys showed that certain types of wind were associated with circulations of certain sizes. This classic paper deserves to be more widely known among non-meteorologists. Jeffreys argues that the pressure gradient force is an important term, being the force responsible for most air motion relative to the earth's surface. If, of the remaining three terms, the Coriolis force is much larger than both the acceleration and frictional terms, then, in the equation, the pressure gradient term is balanced by the Coriolis term and the resultant wind is known as the geostrophic wind, a term suggested by Sir N. Shaw. Such a wind blows parallel to the isobars, with low pressure to the left and high pressure to the right in the northern hemisphere and the converse in the southern hemisphere. If, secondly, the Coriolis and frictional terms are small in comparison with the acceleration term, then the latter balances the pressure gradient term to give Eulerian winds. If, thirdly, the frictional terms exceed the Coriolis and accelerational terms, then friction must balance the pressure gradient. The wind blows along the pressure gradient and the friction, assumed to act opposite to the flow, is sufficient to prevent the velocity of a particular mass of air from increasing steadily throughout its journey. Jeffreys termed such a wind antitriptic.

Crude as was Jeffreys' classification it provided a firm, rational foundation for an understanding of air motion *per se* and the organization of air flow into recognizably different configurations. Jeffreys himself noted that there would be many cases in which the number of terms in the equations of motion comparable with the pressure gradient term is two or three. Such a case is the wind in the lowest kilometre or so of the atmosphere, where the pressure gradient, Coriolis and frictional forces are in balance. In this boundary layer the windspeed is zero at the ground, but increases in value (and direction veers) with height until the geostrophic condition is reached at the top of the friction layer. A further case occurs when the isobars are curved as opposed to straight. This means that a centripetal force enters the equation and the resultant (friction free) wind is known as the GRADIENT WIND, a flow resulting from a balance of pressure gradient, Coriolis and centripetal forces. Clearly one can further complicate each of these basic types by allowing the relative magnitude of any term to change.

Observations have confirmed the basic validity of Jeffreys' results. The general atmospheric circulation comprises entities ranging in characteristic horizontal dimension from thousands of kilometres (the ROSSBY WAVES) to a few kilometres or even a few hundreds of metres (locally induced flows such as cold air drainage). For the Coriolis term to have a major influence upon airflows, their characteristic horizontal size must be greater than 70 km (at the poles) and 400 km (10° from the equator). Thus, cyclones, anti-cyclones and Rossby waves are manifestations of geostrophic and gradient flow. Systems smaller than the critical size (mesoscale and smaller) are largely unaffected by the Coriolis force (the sea breeze is the notable exception). They fall into Jeffreys' 'antitriptic' category if friction is an important force: sea breezes, mountain breezes exemplify this type. If, however, friction may be ignored but the accelera-

tional terms may not, then these small systems fall into Jeffreys' Eulerian category. Severe local storms, such as thunderstorms and tornadoes, exemplify this type of airflow.

As noted earlier, this fundamental approach to the classification of winds also allows the possibility of calculating future values from equations such as the one above. We now know that certain combinations of forces produce certain types of circulation which, in turn, have certain values for the accelerational term $\partial u/\partial t$. Consequently, if we know (from observations or, in a theoretical treatment, simply by specification) initial distributions of air velocity, and pressure, the estimated acceleration allows us to forecast a future distribution of air velocity. From this a forecaster can infer the weather that is likely to ensue. If the equations are suitably modified and the calculations pushed far into the future on a global scale it is possible to 'create' the whole general atmospheric circulation – or global climate. This is now being done by several research groups throughout the world, with most promising results. (See also GENERAL CIRCULATION MODELLING.)     BWA

**Reading**
Atkinson, B.W. 1981: *Dynamical meteorology – an introductory selection*. London and New York: Methuen. Esp. ch. 1. · —1986: *Meso-scale atmospheric circulations*. New York and London: Academic Press. · —1982: Atmospheric processes. In J.M. Gray and R. Lee eds. *Fresh perspectives in geography*. Special publication 3. Department of Geography. Queen Mary College, University of London. · Barry, R.G. 1992: *Mountain weather and climate*, 2nd edn. London and New York: Routledge. · Barry, R.G. and Chorley R.J. 1993: *Atmosphere, weather and climate*. 6th edn. London and New York: Routledge. · Defant, F. 1951: Local winds. In T.F. Malone ed. *Compendium of meteorology*. American Meteorological Society. pp. 655–72. · Jeffreys, H. 1922: On the dynamics of winds. *Quarterly journal of the Royal Meteorological Society* 48, pp. 29–46. · Meteorological Office 1978: *A course in elementary meteorology*. 2nd edn. London: HMSO. · Yoshino, M.M. ed 1976: *Local wind bora*. Tokyo: University of Tokyo Press.

**wind chill** An index of the degree of atmospheric cooling experienced by a person. The amount of heat loss or gain can be measured either by the actual heat exchange in watts per square metre of skin exposed or as an equivalent temperature, i.e. the temperature in still air that would correspond to the cooling (or heating) generated by the particular combination of temperature and windspeed. Other indices have been devised to take account of clothing type and thickness.     PS

**Reading**
Dixon, J.C. and Prior, M.J. 1987: Wind chill indices – a review. *Meteorological magazine* 116, pp. 1–17.

**wind rose** Illustrates graphically the climatic characteristics of *wind* direction, and also frequently windspeed, at a particular location. Radii are drawn outwards from a small circle, proportional to the frequency of wind from each direction. The percentage frequency of calms is usually indicated inside the circle. Windspeed can be depicted by varying the thickness of the radiating lines. There are many variations of this basic method. A wind rose can be adapted to show the relationship between wind direction and other meteorological variables, such as precipitation or the incidence of thunderstorms. DGT

**Reading**
Monkhouse, F.J. and Wilkinson, H.R. 1971: *Maps and diagrams*. London and New York: Methuen. pp. 240–3.

**wind shadow** The region downwind of an obstacle, shielded from the wind. Hedges are less of an obstacle than walls and hence reduce the strength of lee eddies while still providing shelter. The effect persists a distance downwind about thirty times the height of the obstacle in neutrally stratified conditions, much further for strong inversions. It may lead to complete stagnation of air in city basins, as in Los Angeles, Rome and London. JSAG

**Reading**
Monteith, J.L. 1973: *Principles of environmental physics*. London: Arnold.

**World Heritage Site** Site considered officially to be of great significance to global cultural or natural heritage and designated as such under the Convention Concerning the Protection of the World Cultural and Natural Heritage (more commonly known as 'The World Heritage Convention'). The convention was adopted by the General Conference of UNESCO in 1972, although it came into force only in 1975. To date, more than 150 countries are signatories to the Convention, making it one of the most universal international legal instruments for the protection of the world's cultural and natural heritage. The Convention operates through a World Heritage Committee that establishes, keeps up-to-date, and publishes a World Heritage List of cultural and natural properties, known as World Heritage Sites. There are three international non-governmental or inter-governmental organizations named in the Convention to advise the Committee in its deliberations. These bodies are, the IUCN, the World Conservation Union, (for natural sites), and the International Council on Monuments and Sites (ICOMOS) together with the International Centre for the Study of the Preservation and Restoration of Cultural Property (ICCROM) (for cultural sites). Nominations for such sites are submitted by the member states and must be considered to be of universal cultural or natural value in order to be accorded this status. The Convention is legally binding on signatory countries that must help identify, protect, conserve, and transmit to future generations World Heritage properties. When a site is nominated, experts conduct a careful investigation into its merits. The World Heritage Fund helps give technical cooperation, and emergency assistance in the case of properties severely damaged by specific disasters or threatened with imminent destruction.

For the purposes of the Convention, the following are considered as 'cultural heritage': monuments: architectural works, works of monumental sculpture and painting, elements or structures of an archaeological nature, inscriptions, cave dwellings and combinations of features, which are of outstanding universal value from the point of view of history, art or science; groups of buildings: groups of separate or connected buildings which, because of their architecture, their homogeneity or their place in the landscape, are of outstanding universal value from the point of view of history, art or science; sites: works of humans or the combined works of nature and humans, and areas including archaeological sites, which are of outstanding universal value from the historical, aesthetic, ethnological or anthropological point of view.

For the purposes of the Convention, the following are considered as 'natural heritage': natural features consisting of physical and biological formations or groups of such formations, which are of outstanding universal value from the aesthetic or scientific point of view; geological and physiographical formations and precisely delineated areas which constitute the habitat of threatened species of animals and plants of outstanding universal value from the point of view of science or conservation; natural sites or precisely delineated natural areas of outstanding universal value from the point of view of science, conservation or natural beauty.

Articles Four and Five of the Convention deal with the responsibilities of the signatory states. Each must recognize that the duty of ensuring the identification, protection, conservation, presentation and transmission to future generations of the cultural and natural heritage situated on its territory, belongs primarily to that state and that it is charged with doing all it can to this end. To ensure that effective and active measures are taken for the protection, conservation and presentation of the cultural and natural heritage situated on its territory, each signatory must endeavour, as far as possible:

1 to adopt a general policy which aims to give the cultural and natural heritage a function in the life of the community and to integrate the protection of that heritage into comprehensive planning programmes;

2 to set up within its territories, where such services do not exist, one or more services for the protection, conservation and presentation of the cultural and natural heritage with appropriate human resources;

3 to develop scientific and technical studies and research and to work out operating methods to counteract the dangers that threaten cultural or natural heritage;

4 to take the appropriate legal, scientific, technical, administrative and financial measures necessary for the identification, protection, conservation, presentation and rehabilitation of this heritage; and

5 to foster the establishment or development of national or regional centres for training in the protection, conservation and presentation of the cultural and natural heritage and to encourage scientific research in this field.

There are now more than 500 World Heritage Sites, including Australia's Great Barrier Reef, India's Taj Mahal, the Grand Canyon of the United States, South America's Iguaçu Falls and the Giant's Causeway in Britain. There are approximately three times as many cultural sites as those identified for their natural significance. The Convention has its headquarters in Paris under the auspices of UNESCO.          MEM

**Reference**

UNESCO 1975: *Convention concerning the protection of the world cultural and natural heritage*. Paris: UNESCO.

# X

**xerophyte** A plant adapted to living in conditions of aridity such that it is able to survive and reproduce in the long-term absence or scarcity of water. A range of specific metabolic and morphological adaptations are associated with xerophytes and include drought avoidance and drought-tolerance strategies. Adaptations include the presence of deep roots (so-called tap-roots, e.g. *Welwitchschia mirabilis* of the Namib Desert), the ability to store water (e.g. numerous cactus species and succulents), the ability to imbibe water from condensed atmospheric moisture (as in some highly specialized Chilean desert plants) and the ability, as in EPHEMERAL plants, to complete the life-cycle in the short period of moisture availability following rain. MEM

# Y

**yardangs** Streamlined wind-erosion forms with their long axes parallel to the wind. They are often described as resembling an inverted ship's hull, although in many cases they are flat-topped. Length to width ratios are commonly 3:1 or greater. Height varies from a few metres to 200 m and length from several metres to several kilometres. The windward face of the yardang is typically blunt-ended, steep and high, whereas the leeward end declines in elevation and tapers to a point. However, yardangs may take on many forms. Yardangs form parallel to one another and are separated by either U-shaped or flat-bottomed troughs.

Yardangs form in a broad range of geological materials including sandstones, limestones, claystones, granites, gneisses, schists and lacustrine sediments. Variations in form result from differences in material. Yardangs develop by abrasion, deflation, or a combination of these processes, and are further modified by fluvial erosion and mass movement. The ultimate form varies according to lithology, structure and regional climatic history.

Yardangs are uncommon, as they require conditions of great aridity, unidirectional or seasonally reversing winds, and in some cases, a favourable material and some assistance from weathering to form. In Africa, yardangs occur in the Arabian Peninsula, Egypt, Libya, Chad, Niger, and the Namib Desert. Asia has several yardang fields, located in the Taklimakan Desert in the Tarim Basin, the Qaidam depression of Central Asia, and the Lut Desert, Iran. Yardangs are found along the coastal desert of Peru in South America and as minor groups in North America. Large yardang fields also occur on Mars. JL

**Reading**
Breed, C.S., McCauley, J.F., Whitney, M.I., Tchakerian, V.P. and J.E. Laity. 1997: Wind erosion in drylands. In D.S.G. Thomas ed., *Arid zone geomorphology: process, form and change in drylands*. 2nd edn. Chichester: Wiley. pp. 437–64. · Laity, J.E. 1994: Landforms of aeolian erosion. In A.D. Abrahams and A.J. Parsons eds, *Geomorphology of desert environments*. London: Chapman & Hall. pp. 506–35. · Ward, A.W. and Greeley, R. 1984: Evolution of the yardangs at Rogers Lake, California. *Bulletin of the Geological Society of America* 95, pp. 829–37.

**yazoo** A tributary stream that runs parallel to the main river for some distance.

**Younger Dryas** Between around 11,000 and 10,000 BP there appears to have been a cold phase of climatic deterioration characterized by major ice advance in north-west Europe. This phase has been called the Younger Dryas (termed the Loch Lomond stadial in the British Isles) named after the Alpine plant Mountain Avens (*Dryas octopetula*), the distribution of which was more southerly in north-west Europe. It has traditionally been referred to as a north-west European climatic anomaly; however, indications of this episode of cooler climatic conditions are found eastwards into Russia, Spain and Portugal, North Africa and North America. This has been interpreted as having resulted from precipitation variations due to latitudinal changes in the positions of the polar atmospheric and oceanic fronts in the north Atlantic. Outside the glaciated areas in the Younger Dryas time, periglaciation, with discontinuous permafrost, was both extensive and highly effective. AP

**Reading**
Dawson, A.G. 1992: *Ice age earth: late Quaternary geology and climate*. London: Routledge.

# Z

**zeuge** A tabular mass of rock perched on a pinnacle of softer rock as a result of erosion, usually by wind, of the underlying materials. Although common in textbooks describing desert landscapes, zeuges are in reality rather rare.

**zibar** A low dome-shaped sand dune that is usually formed of coarse sand and has no slip face.

**zonal circulation** Any flow along latitude circles. The term is used in a more specific sense to indicate a high index circulation in mid-latitudes with a strong westerly wind between 35 and 55° and little meridional air mass exchange.

At the surface, pressure systems have a dominantly east–west orientation and in middle latitudes unsettled weather with alternating troughs of low pressure and ridges of high pressure move in quick succession giving changeable conditions on any specified area.                          AHP

**zonal soil** A soil which occurs over a wide area and owes its characteristics to climatic rather than local topographical or geological factors.

**zonation** In biogeography and ecology, one of the most important patterns of floral and faunal distribution. On a world scale, characteristic groups of organisms occupy different zones of the terrestrial and aquatic surface, which are primarily determined by climate. Each zone forms an idealized latitudinal band around the earth, although this is frequently modified by local factors (BIOMES). There are also approximately concentric zones around MOUNTAINS, consisting of lateral bands of plants and animals typical of the changing environment from base to summit. The term has also been extended to include any ecological unit with spatial dimensions, e.g. the tension or boundary zone between different, competing biota.        PAF

**Reading**
Furley, P.A. and Newey, W.W. 1983: *Geography of the biosphere*. London: Butterworth.

**zoogeography** The science of the distribution of animals, linking the subject matter of the discipline of zoology with the viewpoint of geog-
raphy. Since the work of Hesse in 1924 it is now common to draw a distinction between ecological animal geography and historical animal geography, the first focusing on the relationships between animals and the environment, the second on the definition and interpretation of the spatial distribution of animals over the surface of the earth. This latter subject is sometimes known as chorology or faunistics.

The origins of zoogeography are clearly traceable, along with the first scientific biology, to Aristotle, but it was the great scientific expeditions of the nineteenth century that really laid the foundations of the subject as we know it today. The first important organizer of zoogeographical knowledge was the German biologist and philosopher Ernst Haeckel, who introduced the term ecology into the subject (1866). Early synthesists of historical zoogeography were P.L. Sclater (1858), T.H. Huxley (1868) and J.A. Allen (1871), but the modern 'Father of zoogeography' is usually regarded as Alfred Russel Wallace, who published his now classic work, *The geographical distribution of animals*, in 1876 (see WALLACE'S LINE; WALLACE'S REALMS). Another pioneer, this time in the ecological approach, was K. Möbius (1877), who developed the concept of biotic communities in his paper on the oyster.

Early interpretations of animal distributions were based on the account of Noah's ark given in Genesis. This view was first challenged by Augustinus *c*.AD 400, when he postulated separate creations, after the biblical flood, to account for the unique faunas of islands (see ISLAND BIOGEOGRAPHY). With the discovery of America a further impetus was given to this view of separate creations, e.g. Paracelsus, although many others talked of land-bridges linking the continents. With the evolutionary thinking of Darwin, Wallace, Huxley, Lamarck and many others and with the publication of *On the origin of species by means of natural selection* (1859), the whole framework of thought changed and zoogeography grew in stature as a scientific discipline. The earliest writers in this new mould (e.g. Wallace) still believed in the permanency of the continents, a view soon to be challenged by Wegener's theory of continental drift, developed in the early part of the twentieth century. The essential vindication of this theory in the 1960s with the models of plate tectonics and sea-floor spreading has, in contrast, provided a completely new insight into the causes of animal distributions and has led to a lively and often bitter debate on the relative merits of dispersal and vicariance, the latter term meaning the separate development of faunas from common ancestors on the separating continents. This development has been accompanied by the rise of a new school of zoogeography, called VICARIANCE BIOGEOGRAPHY which, in turn, is linked with the development of phylogenetic systematics and CLADISTICS in systematics (see TAXONOMY). PAS

### Reading
George, W. 1962: *Animal geography*. London: Heinemann. · Illies, J. 1974: *Introduction to zoogeography*. London: Macmillan. · Nelson, G. and Rosen, D.E. 1981: *Vicariance biogeography: a critique*. New York: Columbia University Press. · Udvardy, M.D.F. 1969: *Dynamic zoogeography*. New York: Van Nostrand Reinhold. · See also the journal *Systematic zoology*, 1970 onwards.

**zooplankton**  Small or microscopic organisms found in marine and fresh waters comprising protozoa and small crustaceans, jellyfish, worms, and molluscs, together with the eggs and larvae of the many animal species inhabiting marine and fresh waters. The most important protozoan groups in the zooplankton are dinoflagellates and foraminifera. These planktonic organisms may be BENTHIC or PELAGIC in habit. Upon death, the insoluble remains, typically rich in calcium carbonate, may accumulate as sediments or oozes on the ocean floor. MEM

# Index

**Notes**
1 Page numbers in **bold** indicate definitions. Those in *italics* indicate illustrations, maps and tables.
2 Specific *places* and *species* have been omitted as these are too numerous to include. Names of *people* are also omitted, unless they are important historically and mentioned often.
3 Acronyms and initials are arranged alphabetically.